# Accident Prevention Manual for Industrial Operations

## Administration and Programs

# Occupational Safety and Health Series

The National Safety Council's OCCUPATIONAL SAFETY AND HEALTH SERIES is composed of four volumes written to help readers establish and maintain safety and health programs. The latest information on establishing priorities, collecting and analyzing data to help identify problems, and developing methods and procedures to reduce or eliminate illness and accidents, thus mitigating injury and minimizing economic loss resulting from accidents, is contained in all volumes in the series:

*ACCIDENT PREVENTION MANUAL FOR INDUSTRIAL OPERATIONS (2-Volume Set)*
*Administration and Programs*
*Engineering and Technology*
*FUNDAMENTALS OF INDUSTRIAL HYGIENE*
*INTRODUCTION TO OCCUPATIONAL HEALTH AND SAFETY*

Other hardcover books published by the Council include:
*MOTOR FLEET SAFETY MANUAL*
*SUPERVISORS GUIDE TO HUMAN RELATIONS*
*SUPERVISORS SAFETY MANUAL*

# Accident Prevention Manual for Industrial Operations

## Administration and Programs

Ninth Edition

Printed in the United States of America.
93 92 91 90 5 4 3 2

Library of Congress Cataloging in Publication Data
National Safety Council.
Accident Prevention Manual for Industrial Operations: Administration and Programs
International Standard Book Number 0-87912-135-1
Library of Congress Catalog Card Number: 86-63578
7.5M290 Product Number 121.42

# Preface

The ninth edition of the *ACCIDENT PREVENTION MANUAL FOR INDUSTRIAL OPERATIONS* is published in two volumes. The *Administration and Programs* volume encompasses management techniques, governmental regulations, and programs for safety and health professionals. The *Engineering and Technology* volume covers more technical information vital to the safety and health professional. The National Safety Council's *FUNDAMENTALS OF INDUSTRIAL HYGIENE,* third edition, and *INTRODUCTION TO OCCUPATIONAL HEALTH AND SAFETY* are additional volumes needed to complete readers' safety and health libraries. Contact the National Safety Council for more information.

The ninth edition of the *ACCIDENT PREVENTION MANUAL FOR INDUSTRIAL OPERATIONS* is the cumulation of facts and ideas that have become part of the safety movement and should be used to organize and transmit information of value to safety and health professionals—indeed, anyone—committed to preventing accidents and preserving well-being. Covering a broad spectrum of subjects, this *Manual* pinpoints problem areas and directs the reader to the appropriate sources of help.

As used in the *ACCIDENT PREVENTION MANUAL,* the term "accident" means that occurrence in a sequence of events that usually produces unintended injury, illness, death, and/or property damage. Prevention of such occurrences should be the responsibility of employees of every level. A second term, "safety and health professional," is used to mean all those interested in or affected by occupational safety and health.

### New material

The ninth edition has expanded and updated material in every chapter. References in each chapter have also been revised to reflect current sources. Some specific changes are listed below.

Chapter 1—expanded history of the safety movement, including developments in health. Chapter 2—completely updated, with discussions of three new topics: medical access, the right-to-know, and environmental impact. Chapter 3—new material on four topics: explanation and illustration of the assignment of risk assessment code (RAC) or single risk number, a rating method for estimating the relative severity of hazard exposures, hazard control, and productivity improvement; expansion of discussions of off-the-job safety and purchasing. Chapter 4—emphasizes protection of the individual rather than mere compliance with governmental requirements; expands hazard analysis to include "reducing exposure to injury"; more clearly explains Permissible Exposure Levels (PEL) and Threshold Limit Values (TLV); discusses concentration measurement values as an evaluation of all the measurements taken as a whole. Chapter 5—a totally new chapter discussing how the occupational safety and health information systems can be used more efficiently and effectively and the computer programs available; includes a new section on what the safety professional should know about the special hazards of loss control and security in a computer room. Sections in the eighth edition Chapter 5 have been distributed to other chapters in the ninth edition: the sections on machine design and reducing exposure to injury are now in the completely revised Chapter 10, the job safety analysis discussion has been added to Chapter 9, and the section on purchasing is now in Chapter 3. Chapter 6—completely rewritten to reflect OSHA's recordkeeping requirements as of 1987. Chapter 7—new sections on off-the-job injury costs, identifying causal factors and selecting corrective actions, and a new quick reference Guide. On-the-job disabling injury costs are more clearly described and three sample computations have been added. This section complements the program portion of on-the-job injuries that has been enlarged in Chapter 3. Chapter 8—expanded sections on injuries and diseases covered by workers' compensation and medical benefits; includes the latest interpretation of the "exclusive remedy" provision and the protection against termination for workers filing worker compensation claims. Chapter 9—the discussion on job safety analysis has been expanded and moved from the former Chapter 5. Chapter 10—completely rewritten to incorporate latest research on ergonomics; new sections on identifying ergonomic-related problems, overexertion low back injuries, upper extremity cumulative trauma disorders, and establishing an ergonomics program. Chapter 11—new material on feedback and behavioral management; discussions of motivation and emotion have been expanded. Chapter 23—a new section on legal and social restrictions on hiring; expansion of driver performance measurements. Chapter 24—completely updated and verified by the sources listed.

### Contributors

The *ACCIDENT PREVENTION MANUAL FOR INDUSTRIAL OPERATIONS* is unique—the compilation of the experience and expertise of contributors from all major occupations and industries. Each of the reviewers and contributors is a practicing expert. To assure uniformity and accuracy, the final version of the text was reviewed by William J. Larson, PE, CSP. The National Safety Council and the editors wish to express their appreciation and gratitude to Mr. Larson and each of the contributors who devoted many hours to updating and checking the accuracy of this publication. These contributors to the *Administration and Programs* volume include: A. G. Baker, Eva Barnard, RN, David Brigham, Alan Carpenter, Min K. Chung, Campbell Dewey, Robert Elam, Nigel Ellis, Raymond C. Ellis, Jr., Robert Firenze, Charles R. Goerth, Esq., Gary Hahn, Dr. Harold Holmes, Robert D. Jordan, David W. Klonicke, Thomas R. Krause, Gary E. Lovested, Robert Meyer, William M. Montante, James M. Palmer, John Polhemus, Peter Rickert, Charles Simpson, John Szwarc, Paul Tamburelli, Elliot Tanz, Larry Volin, Harry Von Heubon, and Adrienne Whyte.

The National Safety Council also wishes to thank members of its staff who contributed significantly to this *Manual:* George Benjamin, MD, Thomas Danko, Barbra Jean Dembski, Alan Hoskin, Larry Huey, Joseph Kelbus, Maureen Kerwin, Joseph Koeberl, Ronald Koziol, Joseph Lasek, John Laumer, Robert J. Marecek, Russell E. Marhefka, Robert O'Brien, Carl Piepho, Barbara Plog, Douglas Poncelow, Cynthia Pondel, and Philip E. Schmidt. The National Safety Council expresses special gratitude to Frank E. McElroy, PE, CSP, for his valuable contributions to this edition.

# Contents

# 1

# Occupational Safety: History and Growth



The mission of the National Safety Council is to educate and influence society to adopt safety and health policies, practices, and procedures that prevent and mitigate human and economic losses arising from accidental causes and adverse occupational and environmental health exposures. (Approved by the Board of Directors, October 18, 1983.)

THE GOAL OF THE NATIONAL SAFETY COUNCIL is to work for the well-being of each person on a 24-hour-a-day basis in all environs, both on and off the job. In this context, safety and health can no longer be considered as separate entities. It is virtually impossible to separate occupational illness from occupational injury.

In an imperfect world there will always be risks, but the National Safety Council will continue to strive to reduce the number and severity of those risks—no matter the cause—as much as possible. As the Council works to free persons from those risks that result in accidental death or injury, it seeks ways to provide everyone with a safe and healthy environment.

In this, the ninth edition of the *Accident Prevention Manual for Industrial Operations,* the National Safety Council presents a compilation of facts and ideas that are a part of the safety movement's general heritage. For special information regarding occupational health and industrial hygiene, see two other National Safety Council books: *Introduction to Occupational Health and Safety* and the *Fundamentals of Industrial Hygiene,* respectively.

## PHILOSOPHY OF ACCIDENT PREVENTION

In medieval days, the master craftsman tried to instruct apprentices and journeymen to work skillfully and safely, because he knew the value of high quality and uninterrupted production. However, it took the Industrial Revolution, which began in England during the 18th century, to create the conditions which led to the development of accident prevention as a specialized field.

The industrial safety philosophy developed because the tremendous forces of production which were released resulted in numerous injuries and deaths. Without a deterrent to stop this waste of personnel and resources, the number of accidents and injuries that would otherwise have occurred would have boggled the imagination.

One way to enlighten management to accept responsibility for preventing accidents was to pass workers' compensation laws. This "new" line of thinking held the employer responsible for a share of the economic loss suffered by the employee who was involved in an accident.

It was a rather short step from this to the realization that a large proportion of accidents could be prevented and that the same industrial brain power that could develop ways to produce vast quantities of goods also could be used to develop ways to prevent accidents. Industry soon discovered that efficient production and safety were related. From this beginning grew the safety movement as it is known today.

The progress in reducing the number of accidents and injuries in the relatively short period of time since this movement began has exceeded the highest expectations of the early safety pioneers. The accidental death rate per 100,000 persons in the United States has decreased 59 percent during the last 75 years.

Experience has shown that there is virtually no hazard that cannot be overcome by practical safety measures. To further that belief, the National Safety Council continues its concerted efforts to prevent accidents and occupational illnesses.

In summary, here are five reasons to work hard to prevent accidents and occupational illnesses:

1. Needless destruction of life and health is a moral evil.
2. Failure to take necessary precautions against predictable accidents and occupational illnesses involves moral responsibility for those accidents and occupational illnesses.
3. Accidents and occupational illnesses severely limit efficiency and productivity.
4. Accidents and occupational illnesses produce far-reaching social harm.
5. The safety movement has demonstrated that its techniques are effective in reducing accident rates and promoting efficiency.

## THE BEGINNINGS OF SAFETY AND HEALTH AWARENESS

The written history of health and safety began about the time of the building of the Egyptian pyramids. The *Ebers Papyrus* and the *Edwin Smith Papyrus,* both found in 1862 and dating from about 3000 B.C., were, respectively, collections of household and medical recipes to cope with various traumatic events like crocodile bites, burns, and the removal of foreign objects (splinters); and a "textbook of surgery," which discussed a variety of injuries and treatments involving splints, dressings, and ointments.

About 2000 B.C., Hammurabi, a Babylonian ruler, revised the old laws of the land and produced a Code of some 280 paragraphs. It covered bodily injury and physicians' fees and probably was the first document that included a beginning of what today is known as workers' compensation laws. Two of the clauses that would be of most interest to safety and health professionals are:

"§199. If [a man] has caused the loss of the eye of a gentleman's servant or has shattered the limb of a gentleman's servant, he shall pay half his price."

"§206. If a man has struck a man in a quarrel, and has caused him a wound, that man shall swear 'I do not strike him knowingly,' and shall [be responsible for] the doctor."

Ramses III in about 1500 B.C. hired physicians to care for mine and quarry workers as well as those engaged in the construction of public works such as canals and large temples. His decision was far more to retain a healthy work force than to be loved by his subjects.

Hippocrates, usually called the father of medicine, about 400 B.C., described tetanus. About 200 B.C., the effects of lead poisoning were described by the Greek poet and physician Nicander.

Various Roman writers from 100 B.C. through the second century A.D. described the plague of Athens, the ill effects of their environment on mine workers, and the unhealthful effects of using lead for water piping and containers for blending wine.

As early as the first century A.D., Pliny the Younger mentioned lead poisoning as a disease present among mine slaves. Pliny the Elder wrote about the use of ox bladders as primitive respirators used by workers producing vermillion to keep the mercury fumes out of their breathing zone.

### The Middle Ages

Although workers must have suffered from the ill effects of working with pigments, grinding of metalware, and the silvering of mirrors, their ordinary living conditions so shortened their lives that tuberculosis and the various plagues took their toll before occupational diseases resulted in death.

In the seventh century in ancient Lombardy, King Rothari codified existing laws in 388 chapters, which was probably the origin of the basic principles of compensation for injury.

The edict applied to personal injuries received in brawls, fights, and feuds, and payments for disability and death were established.

In the eleventh century, King Canute, King of Denmark, Norway, and England, stated the principles of compensation for particular injuries. The importance of the loss of a thumb was recognized—the compensation for its loss was twice that for the loss of the second digit and two-and-one-half times that given for the loss of the third digit.

In 1473 Ulrich Ellenbog, an Austrian physician, wrote a tract directed toward goldsmiths and other handlers of metal, warning against the burning of coal in confined spaces and the inhalation of vapor arising from the heating of metals such as lead, antimony, silver, and mercury. This is considered to be the first writing devoted exclusively to industrial metal poisoning.

Six years after the death in 1555 of George Agricola, a Saxon physician, his book, *De Re Metallica,* was published. It emphasized the need to ventilate mines and illustrated various devices that would force air below ground. Other illustrations depicted personal protective devices—gloves, leggings, and masks. The work was of such prominence that it would endure for centuries.

In 1567 Philippus Aurelous, a.k.a. Theophrastus Bombastus von Hohenheim, who later called himself Paracelsus, had a treatise published, *On the Miners' Sickness and Other Miners' Diseases,* in which he distinguished between acute and chronic poisoning. This was the first monograph dealing with the diseases of a specific occupational group. Paracelsus grew up in Switzerland, studied medicine in Italy, and practiced medicine as an itinerant teacher, visiting mines and workshops.

In the early part of the 18th century, Bernardino Ramazzini published the classic *Discourse on the Diseases of Workers,* which still applies today. Ramazzini pointed out that in addition to the standard questions asked of a patient, one more should be added: "What is your occupation?"

Ramazzini, dubbed the father of occupational medicine, summarized the two causes which he believed were responsible for the occupational diseases of workers of his day: "The first and most potent is the harmful character of the materials that they handle, for these emit noxious vapors and very fine particles inimical to human beings and induce particular diseases; the second cause I ascribe to certain violent and irregular motions and unnatural postures of the body, by reason of which the natural structure of the vital machine is so impaired that serious diseases gradually develop therefrom."

### Mass production appears

Until the 1700s, production methods were labor-intensive, with work being done by hand in cottages.

Three developments were to change this way of life: In England in 1764, the spinning jenny was developed and in 1784, the power loom was perfected. In America, Eli Whitney added

**Figure 1-1.** Back in "the good old days," getting the children to school was a greater problem than teaching them to be safe at school. Here, barefoot children toiled in the mills.

his invention, the cotton gins, in 1792. These and other innovations ushered in what would later be called the Industrial Revolution. What began in Britain in the 18th century and spread to the Continent and the United States transformed the life of Western man, the nature of society, and the relationship between people.

Specifically, the innovations encountered in the processes and organization of production changes included:

- The substitution of inanimate for animal sources of power, particularly steam power through the combustion of coal.
- The substitution of machines for human skills and strength.
- The invention of new methods for transforming raw materials, particularly in the making of iron and steel, and industrial chemicals.
- The organization of work in large units, such as factories or forges or mills, making possible the direct supervision of the process and an efficient division of labor.

Paralleling these production changes were the altered technologies employed in agriculture and transportation.

Initially, this was termed the "factory system," but later, when it reached a larger and more complex scale, was designated the Industrial Revolution by A. Toynbee (Toynbee, 1884), whose nephew, Arnold J. Toynbee, is described as "the first economic historian to think of, and to set out to describe, the Industrial Revolution as a single great historical event, in which all the details come together to make an intelligent and significant picture."

Because these changes in production methods with their concomitant need for masses of workers brought with them hazards never before encountered, the history of occupational safety and health was greatly affected. The increasing need for hazard control was recognized.

### The Industrial Revolution comes to America

The effects of the Industrial Revolution were first felt in the United States about a century after it started in Great Britain. Before the 19th century, most families in the United States lived and worked on farms. Some industries had developed, namely printing, shipbuilding, quarrying, cabinetmaking, bookbinding, clockmaking, and the production of paper, chocolate, and cottonseed oil. However, it was the textile industry that saw the beginning of the factory system in America, especially in New England where hundreds of spinning mills shot up (see Figure 1-1). As the Industrial Revolution continued its unbounded growth, the toll on workers began to show. (Felton, 1986.)

## HISTORY OF U.S. SAFETY AND HEALTH MOVEMENT

During the last half of the 19th Century, American factories were expanding their product lines and producing at heretofore unimagined rates. While the factories were far superior in terms of production to the preceding small handicraft shops, they were often inferior in terms of human values, health, and safety.

In terms of human values alone, the 1900 census showed 1,750,178 working children between 10 and 15 years inclusive—25,000 were employed in mines and quarries; 12,000 in making chewing tobacco and cigars; 5,000 in sawmills; 5,000 at or near steam-driven planers and lathes; 7,000 in laundries; 2,000 in bakeries; and 138,000 as servants and waiters in hotels and restaurants. (See Figure 1-1.)

These deficiencies were probably inevitable. The tools of mass production had to be invented and applied before anyone could begin to imagine the problems they would create, and the problems had to be known before corrective measures could be considered, tested, and proved. Deaths and injuries were accepted as being part of "industrial progress."

While this change in the work environment was taking place, the thinking of the public, management, and the law was still reflecting the past, when the worker was an independent craftsman or a member of the family-owned shop. Common law provided the employer with a defense that gave the injured worker little chance for compensation. The three doctrines of common law that favored the employer were:

*Fellow servant rule*—Employer was not liable for injury to employee that resulted from negligence of a fellow employee.

*Contributory negligence*—Employer was not liable if the employee was injured due to his own negligence.

*Assumption of risk*—Employer was not liable because the employee took the job with full knowledge of the risks and hazards involved.

In large industrial centers, the ugly results of industrial accidents and poor occupational health conditions became more and more obvious. Voices of protest were raised. Though there were employers who denied the existence of the problem, wiser management people began to do something about it.

As early as 1867, Massachusetts had begun to use factory inspectors, and ten years later that state had a law requiring the safeguarding of hazardous machinery. During 1877, Massachusetts also passed the Employer's Liability Law that made employers liable for damages when a worker was injured. However, court decisions based on common law often let the employer escape liability.

From 1898 on, there were additional efforts to make the employer financially liable for accidents. In his Presidential message of 1908, Theodore Roosevelt stated: "The number of accidents which result in the death or crippling of wage earners . . . is simply appalling. In a very few years it runs up a total far in excess of the aggregate of the dead in any major war."

His message was echoed when his social legislation passed that year in Congress. That first workers' compensation law covered only federal employees and set a precedent for state laws to follow.

In 1911, the first effective workers' compensation act was passed in Wisconsin and declared constitutional by the Wisconsin Supreme Court within a few months. New Jersey and Washington also passed laws that year.

While a bill for workers' compensation (the Wainwright Law) had been passed in New York in 1910, it was declared unconstitutional by the New York Court of Appeals on the grounds that the law violated both the federal and New York State Constitutions, "because it took property from the employer and gave it to his employee without due process of law." After the legislature adopted an amendment to the state constitution and it was approved in 1913 at the general election, a compulsory Workmen's Compensation Act finally became effective in mid-1914.

Coincident with the declaration of unconstitutionality of the early act, a fire occurred in a clothing factory in New York City that took the lives of 146 employees. This disaster, called the Triangle Fire, unified the demand for factory legislation and gave force to reform because the fire took place on the same day (March 25, 1911) that the Court of Appeals decision was publicized.

Other such laws were, at first, declared invalid because of conflict with the due process of law provisions of the 14th Amendment. After the U.S. Supreme Court in 1916 declared it to be constitutional in *New York Central Railroad Co.* v. *White,* 243 U.S. 188, many states passed compulsory laws on workers' compensation.

In the late 1800s and early 1900s, the railroads conquered the West while extracting a heavy toll in human life. It was said that a man was killed for each mile of track laid. By 1907, annual railroad employee deaths had reached 4,353.

Progress was made on the technical side of the problem. The railroads adopted the air brake and the automatic coupler well before the turn of the century. They also worked on guarding and fire prevention.

Next came the recognition that guarding was not the total solution and that people's actions were important factors in creating accident situations.

Insurance companies began relating the cost of premiums for workers' compensation insurance to the cost of accidents. Management began to understand the close relationship between successful production and safe production.

During the first decade of the 20th Century, two giant industries, railroads and steel, began the first large-scale organized safety programs. From this period comes one of the great and historic documents of safety. In 1906, Judge Elbert Gary, president of the United States Steel Corporation, wrote:

"The United States Steel Corporation expects its subsidiary companies to make every effort practicable to prevent injury to its employees. Expenditures necessary for such purposes will be authorized. Nothing which will add to the protection of the workmen should be neglected."

The Association of Iron and Steel Electrical Engineers, organized soon after this announcement, devoted considerable attention to safety problems.

### Birth of the National Safety Council

Nineteen hundred twelve was to be an historic year for accident prevention. The previous year, a request came from the Association of Iron and Steel Electrical Engineers (which had been formed in 1907) to call a general industrial safety conference on a national scale. The result was the First Cooperative Safety Congress, which met in 1912 in Milwaukee. This gathering called for

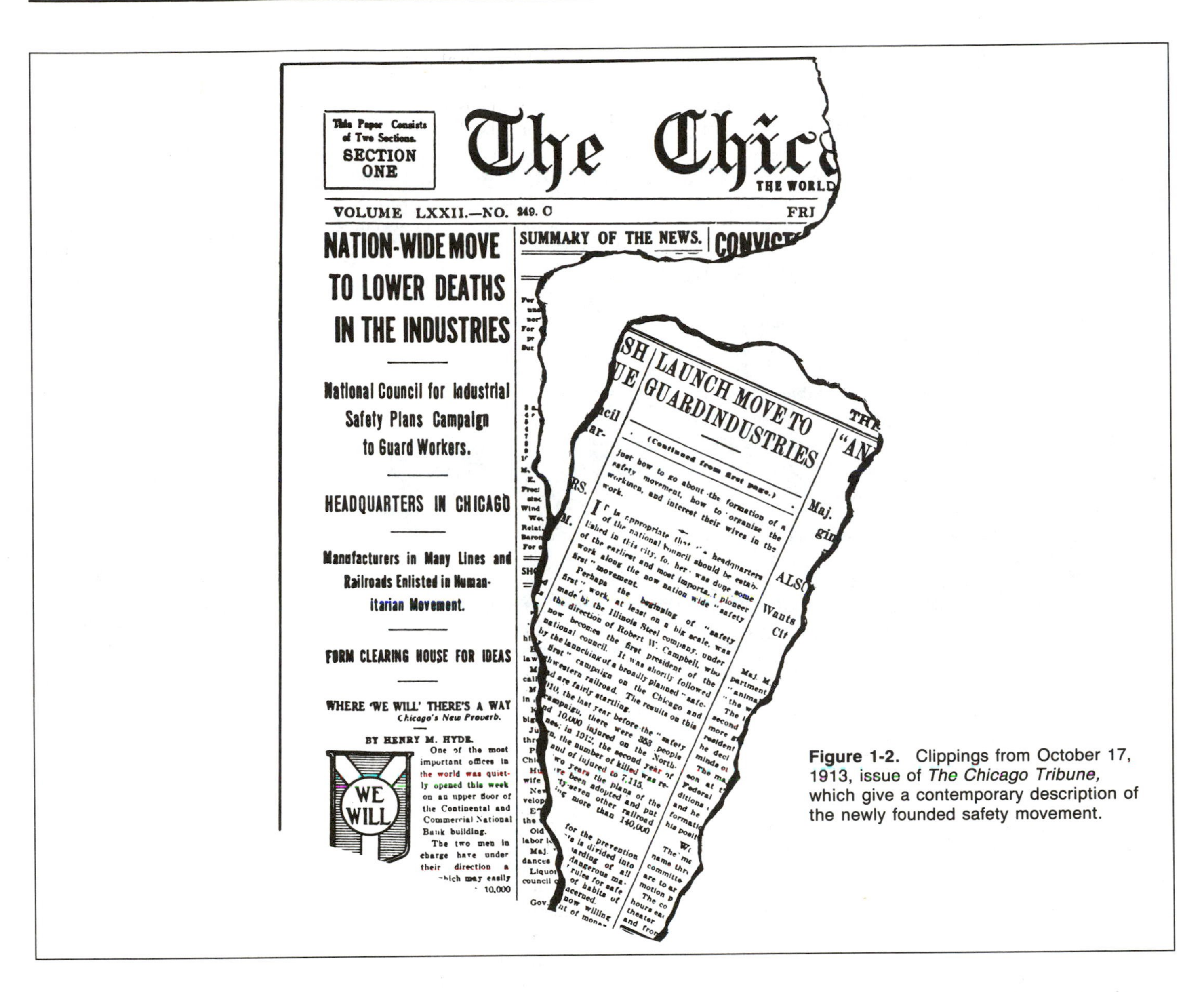

This Paper Consists of Two Sections. SECTION ONE

The Chic[ago]

THE WORLD

VOLUME LXXII.—NO. 249. C

NATION-WIDE MOVE TO LOWER DEATHS IN THE INDUSTRIES

National Council for Industrial Safety Plans Campaign to Guard Workers.

HEADQUARTERS IN CHICAGO

Manufacturers in Many Lines and Railroads Enlisted in Humanitarian Movement.

FORM CLEARING HOUSE FOR IDEAS

WHERE 'WE WILL' THERE'S A WAY
Chicago's New Proverb.

BY HENRY M. HYDE.

WE WILL

One of the most important offices in the world was quietly opened this week on an upper floor of the Continental and Commercial National Bank building.

The two men in charge have under their direction a ...

SUMMARY OF THE NEWS.

LAUNCH MOVE TO GUARD INDUSTRIES

(Continued from first page.)

Just how to go about the formation of a safety movement, how to organize the workmen, and interest their wives in the work.

It is appropriate that the headquarters of the national council should be established in this city, for here was done some of the earliest and most important pioneer work along the now nation wide "safety first" movement.

Perhaps the beginning of "safety first" work, at least on a big scale, was made by the Illinois Steel company, under the direction of Robert W. Campbell, who now becomes the first president of the national council. It was shortly followed by the launching of a broadly planned "safety first" campaign on the Chicago and Northwestern railroad. The results on this road are fairly startling.

**Figure 1-2.** Clippings from October 17, 1913, issue of *The Chicago Tribune*, which give a contemporary description of the newly founded safety movement.

another meeting in New York the following year, and at that meeting the National Council for Industrial Safety was organized. Shortly afterward, the organization's name was changed to the National Safety Council, and its program was broadened to include all aspects of accident prevention and, today, occupational health. Yet it must be remembered that the Council was the creation of industry and that its activities have always been heavily concentrated on industrial safety. (See Figure 1-2.)

The group that met in Milwaukee and New York was composed of a few safety "professionals," some management leaders, public officials, and insurance specialists. Their one point in common was a desire to attack a problem which most people thought to be either unimportant or could not be solved. Because these people were determined, the safety movement as we know it today was designed and built.

Actually the underlying objective when the National Safety Council was formed in 1913 was standardization. The primary purpose of the Council to provide an avenue of communication, an exchange of views, and solutions to common problems in accident prevention was an expression of the need for standards.

In 1918 the Council conducted a national survey of state, federal, and municipal regulations together with a study of insurance recommendations, technical association recommendations, and the practices of industry. The survey depicted utter chaos in industrial safety. The need for unified methods and practices was clear.

Realizing its own limitations, the Council consulted the National Bureau of Standards, which agreed to call a conference to discuss the establishment of procedures for standardizing safety methods and practices. Meeting in Washington, D.C., in 1919, the attendees expressed the feeling that uniformity had become, not only extremely desirable, but almost imperative.

The conference voted to formulate safety standards under the auspices and procedures of the American Engineering Standards Committee (AESC), which had been formed in 1918 by five engineering societies and three governmental departments.

### American Standards Association beginnings

In 1920, the National Safety Code Program was brought into the AESC. This caused the first reorganization of the Committee and was the beginning of what later was to become the

**Figure 1-3.** One of the first safety committees was formed with mill employees at Kimberly-Clark Co., Neenah Paper Co. The plaque in foreground is dated 1915.

American Standards Association (ASA). A national code committee was organized to suggest the initial safety code projects. This later became the Safety Codes Correlating Committee, the first of the ASA group of 18 standards boards. Bringing manufacturing companies and trade associations into AESC membership also initiated a broader program of engineering standards. As a result, an enlarged national standardization program was launched.

In 1928, recognizing that the extending activities called for a more formal type of organization, the member groups reorganized the AESC as the American Standards Association, now known as the American National Standards Institute (ANSI).

ASA continued to be an important partner in the safety movement. This group handled the "things" of safety (see Figure 1-7) while the National Safety Council has worked with the "people" portion of accident and occupational illness prevention.

### Accident prevention discoveries

As industry developed some experience in safety, it discovered that engineering could prevent accidents, that employees could be reached through education, and that safety rules could be established and enforced. Thus the "Three E's of Safety"—Engineering, Education, and Enforcement—were developed.

Among the breakthroughs made during the 1900-1980 era were the identification and the efforts to control certain occupational health diseases such as mercury and lead poisoning. Asbestos acting in company with cigarette smoke was found to be a carcinogen causing lung cancer, or without smoke, mesothelioma. Chromium compounds and beryllium also were studied.

There were other discoveries, too. Safety departments had often argued that savings in compensation costs and medical expenses would many times repay safety expenditures. Thoughtful business leaders soon learned that these savings were only a fraction of the financial benefits to be derived from accident prevention work. Newer, more effective techniques have been discovered and are described elsewhere in this volume. See especially Chapters 3 through 7, and 10.

### Acceleration of the drive for safety and health

Industrial safety received wide acceptance in the years between the two world wars. Conservation of manpower during World War II intensified the safety growth, and the federal government encouraged safety activities by its contractors. As industry expanded to meet the needs of the war effort, additional safety personnel were hastily trained to try to keep pace. The acceptance of safety as part of the industrial picture did not diminish with the end of the war. By then, the importance of safety to quality production was well established, and the small handful of dedicated people in 1912 had grown to tens of thousands.

In 1948, for example, Admiral Ben Moreell, then president of Jones and Laughlin Steel Corporation, wrote:

"Although safe and healthful working conditions can be justified on a cold dollars-and-cents basis, I prefer to justify them on the basic principle that it is the right thing to do. In discussing safety in industrial operations, I have often heard it stated

**Figure 1-4.** World War I era poster published by National Safety Council. Note the "Universal Safety" emblem.

that the cost of adequate health and safety measures would be prohibitive and that 'we can't afford it.'

"My answer to that is quite simple and quite direct. It is this: 'If we can't afford safety, we can't afford to be in business.' "

A discussion of current federal safety legislation follows later in this chapter under Safety and the Law, and also in Chapter 2, Governmental Regulation and Compliance.

A by-product of organized safety activity has increased interest in safety engineering in colleges and universities. Many schools offer degrees and advanced courses in this subject and are contributing to a higher standard of knowledge among professionals in the field.

The World War II labor shortage dramatically brought home to management the magnitude and seriousness of the problem of off-the-job accidents to industrial employees. The wartime theme of the National Safety Council, "Save Manpower for Warpower," focused attention on efficient and safe production.

Today, an increasing number of employers are including off-the-job safety in their overall safety programs. Companies realize their operating costs and production schedules are affected almost as much when employees are injured away from work as when they are injured on the job. Off-the-job safety is an extension of a company's on-the-job safety program and is intended to educate the employee to follow the safe practices he uses on the job in his outside activites. Companies have found that on-the-job and off-the-job programs complement each other.

From the earliest days of industrial safety, it has been difficult to make a clear separation between health and accident hazards. Is dermatitis an accident or a disease? What about hernias, hearing loss, and heart trouble? Inevitably, safety professionals have become interested in many health problems that are on the borderline between diseases and accidents. In 1939, the American Industrial Hygiene Association was established to promote the recognition, evaluation, and control of environmental stresses arising in or from the workplace.

## EVALUATION OF ACCOMPLISHMENTS

Since the factors are complex, no simple rating scale can indicate all the answers to the question, "What has the safety movement accomplished?" In the absence of such a rating scale, an attempt to answer the question must be made by assembling several kinds of data.

First, the question must be asked, "Has the safety movement, in fact, done anything to prevent accidents?" To that question can be answered a clear "Yes!"

If the annual accidental death rate per 100,000 of population which held in 1912 had continued, over 2,600,000 more accidental deaths would have occurred. Since 1912, the death rate for persons of normal working age—25 to 64 years—declined fifty

**Figure 1-5.** Back in 1910, this was a modern medical facility. Note sterilizer on work table—it looks more like a large coffee maker. (Courtesy Norton Company, Worcester, Mass.)

nine percent while the rate of all ages of the entire population declined fifty two percent. Medical progress accounts for some of this gain, but the larger part is certainly the product of organized safety work.

Since World War II, the number of work-related deaths per 100,000 population, standardized to the age distribution of the population in 1940, has decreased steadily (Figure 1-8). This indicates that the risk of on-the-job death has declined for the population as a whole. Part of the progress made in lowering the overall death rate, however, can be attributed to the rapid growth in recent years of the service-producing sector of the economy with its lower death rate and the decline of some relatively risky parts of the goods-producing economic sector. In 1945, 43 percent of the non-agricultural workforce was in production industries (mining, construction, and manufacturing). By 1985 the proportion had declined to 26 percent.

A clearer picture emerges by looking at the trends on a more detailed level. Table 1-A shows the significant improvement in the death rates of the major industry groups. Death rates in five of the seven private sector groups have been reduced by 50 percent or more over the last 40 years. This clearly indicates, by one criterion, the effectiveness of the organized safety movement.

**Table 1-A.** Work Deaths per 100,000 Workers

| *Industry Group* | *1945* | *1985* | *Percent Change* |
|---|---|---|---|
| Agriculture, forestry, and fishing | 53 | 49 | − 8% |
| Mining and quarrying | 187 | 50 | −73% |
| Construction | 126 | 37 | −71% |
| Manufacturing | 19 | 6 | −68% |
| Transportation and public utilities | 52 | 29 | −44% |
| Wholesale and retail trade | 10 | 5 | −50% |
| Services | 20 | 6 | −70% |

Source: National Safety Council, *Accident Facts*, 1946 and 1986 editions.

Long-term trends in non-fatal occupational injury rates cannot be examined because of a break in continuity of the historical statistical series. Until the early 1970s, injury rates were based on the voluntary *American National Standards Method of Recording and Measuring Work Injury Experience,* ANSI Z16.1. With the passage of the Occupational Safety and Health

**Figure 1-6.** Atlantic Refining Company's fire brigade practices scaling a storage tank. Compare this World War I Era equipment with the modern equipment shown in Chapter 18, "Planning for Emergencies."

Act of 1970, it became mandatory for most private sector employers in the USA to keep occupational injury and illness records in accordance with OSHA recordkeeping requirements. Most employers dropped the voluntary standard rather than keep two sets of records.

A clear trend has not yet emerged in the occupational injury and illness incidence rates published by the Bureau of Labor Statistics since 1972. Business cycles and changes in the distribution of the labor force among industries can mask any short-term changes in rates due to more effective or more intensive safety efforts.

### The dollar values

It has been estimated that the annual cost of occupational accidents in the United States exceeds $37 billion. If the 1912 accident rates had continued unchanged and if there had been no organized safety movement, this annual cost would have easily been two or three times as great, even in constant dollars.

Against such dollar savings, the relatively small expenditures for safety throughout America provide a striking contrast. Each dollar spent for safety by American industry is probably returning a clear profit of several hundred percent.

### Industry and nonwork accidents

Directly and indirectly, industry is bearing a substantial part of the cost of nonwork accidents and their prevention. While the National Safety Council is the creation of industry and largely supported by it, the Council as well as state and local safety organizations play a major role in the fight against such accidents. Industry supports a large part of the job of informing the general public on these problems through the press, radio, and television.

The effectiveness of the nonwork accident prevention campaign is shown by the fact that, from the time records were first kept in 1921, both home and public accident death rates have generally declined. (See Figure 1-8.)

If industry has been a large contributor to this successful work, it has also been a heavy beneficiary from it. Disruption of the labor force, worry and hardship among employees, loss of purchasing power by consumers, and heavy tax burdens for the support of hospitals and relief agencies all result, in part, from nonwork accidents.

## SAFETY'S RESOURCES

*Statistics* measure what has been accomplished and *resources* describe the tools, methods, and knowledge that have been developed for use by safety and health professionals to meet future accident and occupational illness problems.

### Know-how

This *Accident Prevention Manual for Industrial Operations,* for example, is an accumulation of facts and experiences that are a part of the safety and health movement. Its purpose is to present key points of specific, as well as general, knowledge to safety-interested people, from students, to those just recently involved in this type work, to advanced and experienced practitioners.

Today, an individual using this Manual can find better answers to a wider range of industrial safety problems than were available to the wisest and best-trained professional safety practitioner several decades ago. Yet, even this Manual can't contain all of the knowledge available to fight the never-ending war against accidents and occupational illnesses.

Other material may be found in numerous pamphlets, books, and periodicals published by safety and health organizations, government agencies, and insurance companies, and in the studies and directives of individual industrial concerns. The literature of various trades and professions is likewise rich in safety information. A list of handbooks is presented in Chapter 19, Safety Engineering Tables, of the *Engineering and Technology* volume. At the end of this chapter is a list of the general safety books that are used as sources of questions for the examination for Certified Safety Professional.

The National Safety Council offers a series of training courses, at both the beginning and advanced levels, for professionals. The Council also offers extensive consulting services.

Finally through conferences, technical seminars, newsletters, and other publications professional safety engineers, executives, supervisors, and rank-and-file employees exchange safety information regularly. The annual Congress and Exhibition, held in alternating years in Chicago and in cities across our nation, is an excellent means of professional development. (See Figure 1-9.)

### The heritage of cooperation

The safety movement would be far less effective if its members had hoarded and concealed their discoveries from their colleagues in competitive companies.

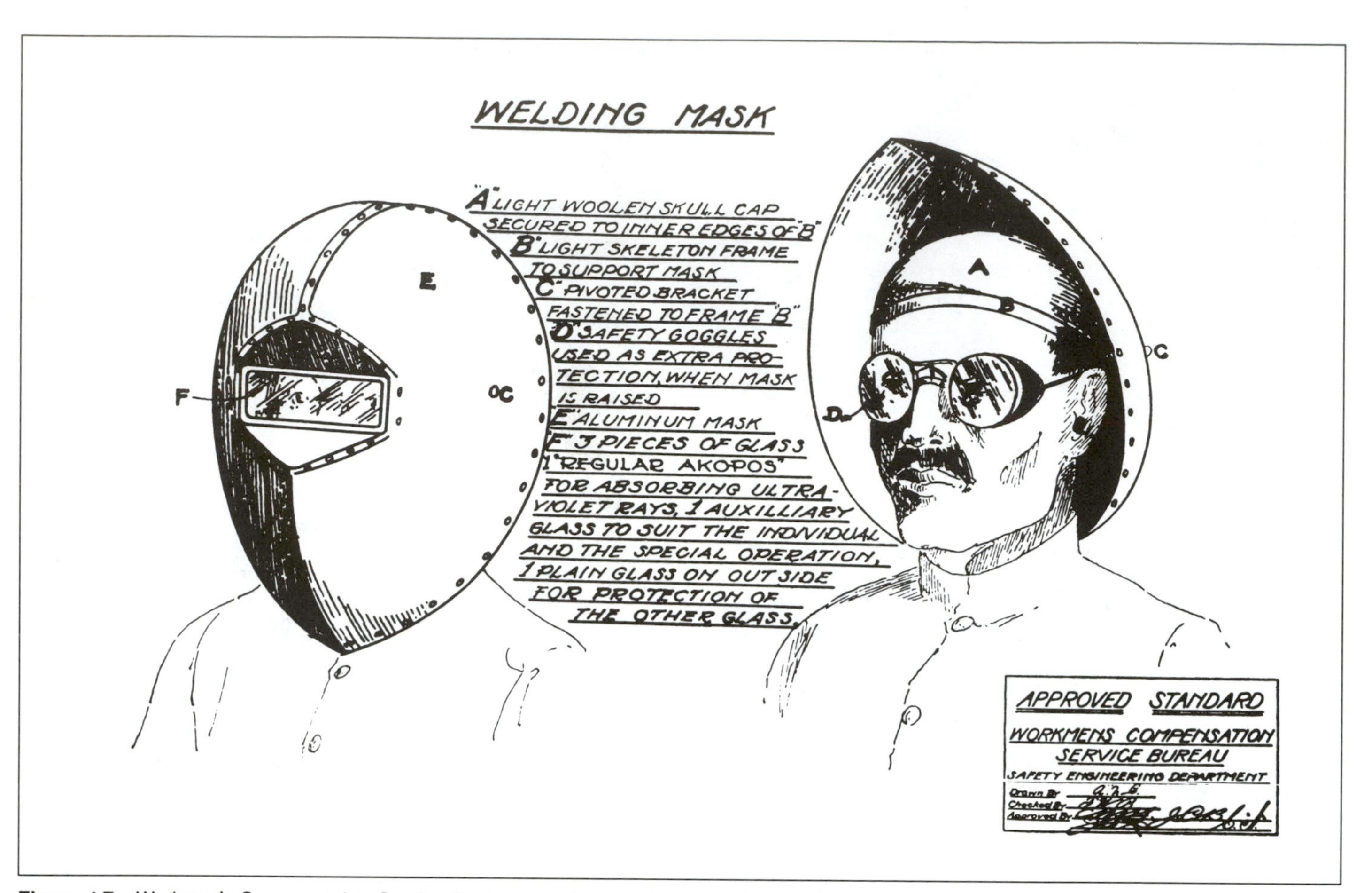


**Figure 1-7.** Workmen's Compensation Service Bureau gave "approval only of the principal" of protective equipment and clothing. Its "Universal Safety Standard," dated 1913, did push for double protection even back then. The National Safety Council publishes no standards; these are left to the American National Standards Institute, which was co-founded with the National Safety Council to handle standards, while the Council worked with people.

It was teamwork that created the safety activities of the Association of Iron and Steel Electrical Engineers. It was broadened teamwork that called for and attended the first Milwaukee Conference which led to the formation of the National Safety Council and other safety organizations.

Effective accident prevention requires cooperation. Through the Council and other safety organizations, safety professionals met to exchange ideas, develop safety publications, and stimulate one another in friendly competition.

The tradition that there should be "no secrets in safety," no denial of help even to a competitor when it involves saving life, is one of the great strengths in the safety movement.

### Good will

In its early days, safety did not have a high priority with management. But today, no small part of the safety professional's capital is the prestige and good will built up for safety proposals and expenditures over the years. Where the pioneers had to battle every step of the budgetary way, safety professionals today have a far more receptive hearing from management.

### Professionalism

Dedicated safety and health professionals continue to be accident and occupational illness prevention's most valuable asset. Their ranks have grown to the point where, in the mid-eighties, membership in the American Society of Safety Engineers (ASSE) is approaching 21,000. This organization, dedicated to both their interests and their professional development, has approximately 120 chapters in the U.S. and Canada. Individual membership is worldwide.

In addition to the ASSE members, there are many other qualified safety professionals who, together with thousands of specialists and technicians, carry out a limited scope of activities within the field. Numerous others devote less than 50 percent of their time to safety functions.

In 1968, the ASSE was instrumental in forming the Board of Certified Safety Professionals (BCSP). Its purpose is to provide a means of giving professional status to qualified safety people by certification after meeting strict educational and experience requirements and passing an examination. Similarly in the hygiene field, the professional certification of industrial hygienists (CIH) was sponsored by the American Industrial Hygiene Association. Both the ASSE and the AIHA are described in Chapter 24, Sources of Help.

### Advancement of knowledge

The tremendous increase in scientific knowledge and technological advancement since the close of World War II has added to the complexities of safety work.

The approach has oscillated between one that emphasizes environmental control or engineering, and one that emphasizes human factors. From this, several imporatnt trends in safety work and the safety professional's development have emerged. All are discussed in subsequent chapters of this volume.

- First, increasing emphasis is developing toward analyzing the loss potential of the activity with which the safety professional is concerned. Such analysis will require greater ability (1) to predict where and how loss- and injury-producing events will occur and (2) to find the means of preventing such events.
- Second, factual, unbiased, and objective information about loss-producing problems and accident causation is increasingly being developed so that those who have ultimate decision-making responsibilities can make sound decisions.
- Third, there is increasing use of the safety and health professional's knowledge and assistance in developing safe products. Applying the principles of accident causation and control to the product is becoming more important because of the increase in product liability cases, the sudden emphasis in law of the entire field of negligent design, and the obvious impact a safer product would have on the overall safety of the environment.

To identify and evaluate the magnitude of the safety problem, safety professionals must be concerned with all facets of the problem—personal and environmental, transient and permanent—in order to determine the causes of accidents or the existence of loss-producing conditions, practices, or materials. From the collected and analyzed information they propose alternate solutions, together with recommendations based upon their specialized knowledge and experience, to those who have decision-making responsibilities.

Therefore, application of this knowledge—whether to industry, transportation, the home, or in recreation—makes it imperative that those in this field be trained to utilize scientific principles and methods to achieve adequate results. Of prime importance are the knowledge, skill, and ability to integrate machines, equipment, and environments with people and their capabilities.

Safety and health professionals in performing these functions draw upon specialized knowledge in both the physical and social sciences. They apply the principles of measurements and analysis to evaluate safety performance and are required to have fundamental knowledge of statistics, mathematics, physics, chemistry, and engineering.

They need training in the field of behavior, motivation, and communications and knowledge of management principles as well as the theory of business and government organization. Specialized knowledge must include a thorough understanding of the causative factors contributing to accident and occupational illness occurrence as well as methods and procedures designed to control such events.

Safety professionals also need diversified education and training, if they are to meet future challenges. The population explosion, the problems of urban areas and future transportation systems, the weakening of the family, the decline of respect for authority, and the cloudy economy, as well as the increasing complexities of everyday life, will create many problems and stretch safety professionals' creativity to its maximum, if they are to successfully provide the knowledge and leadership to conserve life, health, and property.

Training for the safety and health professional of the future can no longer be the "on-the-job," one-on-one type, only. It must include specialized undergraduate education that leads to a bachelor's or higher degree.

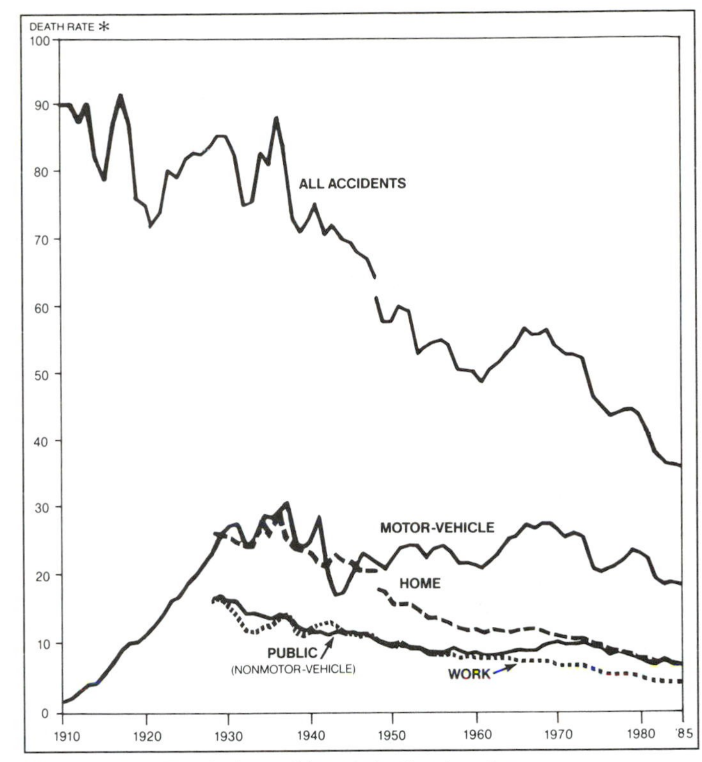


**Figure 1-8.** Trends in accidental death rates drop, both on and off the job. Shown here are deaths per 100,000 population, adjusted to the 1940 age distribution. The break in 1948 shows the estimated effect of classification changes.

Training courses, such as those offered by the National Safety Council, will continue to educate a large number of individuals who began performing safety functions and must receive initial training or advanced training in certain specialized areas.

Approximately 200 four-year colleges and universities offer courses in safety and health, and several dozen offer a bachelor's or higher degree in safety. Two-year, community colleges are offering associate degrees or certificates for courses designed for the safety technician or part-time administrator. Governmental agencies and ASSE are also conducting such professional development programs. (See Chapter 9, Safety Training.)

### Summary of achievements

The safety movement has helped save tens of thousands of lives. It is saving industry and its employees billions of dollars a year.

It faces the future with numerous resources for eliminating accidents and occupational illnesses—resources in know-how, teamwork, good will, and education programs that produce trained and dedicated safety workers.

It has, therefore, done much to meet the double challenge presented to it: to deal with accidents and occupational illnesses now and to build soundly for the long-range attack upon these problems in the future.

## SAFETY TODAY

To answer the question, "What has the safety movement accom-

**Figure 1-9.** Main entrance to the Council's annual Congress and Exposition. Photo shows 1986 crowd at morning admission time. Inset shows the Council exhibit; for additional photo, see Chapter 9, Safety Training. At the Congress, ideas are exchanged by speakers, seminars, workshop sessions, and through informal conversations; at the Exposition, registrants find out what goods and services are available commercially.

plished?" we look at continued growth in safety awareness and accident reduction. To answer the question, "How does the safety movement stand today?" we must look at what is wrong, as well as what is right, with the present situation. The answer can be found in an appraisal of how the safety movement stands in relation to how it ought to stand. The first point to be considered is simple and grim:

- Accidents still bleed this country of more than 90 thousand lives a year, cause about 9 million disabling injuries, and account for a total financial loss of more than $100 billion.
- Work accidents destroy more than 11,000 lives a year, about half these deaths occurring in what is normally considered industry. Work accidents injure about 2 million persons annually and cost more than $37 billion.

In recent years, the ratio of off-the-job deaths to on-the-job deaths was about 3 to 1 and more than half of the injuries suffered by employees occurred off the job.

In terms of time loss, all injuries to workers, both on and off the job, caused a loss of about 100 million man-days of work.

Within the industrial community, there are very large variations in accident rates from industry to industry and from company to company.

The wholesale and retail trade, services, and finance, insurance and real estate industries all have occupational injury and illness incidence rates that are below the private sector average. Rates are above average in the construction, agriculture, manufacturing, transportation and public utilities, and mining industries.

Injury incidence rates by size of establishment are lowest for the very small establishments (1 to 19 employees), and rise steadily until they reach a maximum in establishments of 100 to 249 employees. The rates then decline steadily as establishment size continues to increase.

### Small establishments

It has been stated that small businesses, those with 100 to 250 employees, have proportionately more work injuries than large corporations or very small, 1 to 19 employee, companies. However, since many small companies do not accurately record and report their experience, it is difficult to establish any valid data comparing work injury experience. It is safe to say that companies, large or small, that ignore safety and health efforts will, in the main, have more than their share of accidents and occupational illnesses.

The seriousness of the small-enterprise problem is widely recognized and the National Safety Council has devoted much effort to meeting it. One way has been through the establishment of liaison between the National Safety Council and the trade associations representing many small conpanies.

Certain aspects of the small-company problem can be stated with assurance:

1. The small establishment may not need or cannot employ specialized safety and health personnel to deal with the accident and occupational health problem.

2. The number of accidents or the financial position of many small companies makes it difficult to convince them that spending the money for proper equipment, layout, guarding, and other elements is important.

3. Managers of small operations deal with a host of problems and seldom have the expertise or time for the proper study of accidents and occupational illnesses and their causes.

4. In small units, statistical measures of performance are unreliable, so it is difficult to produce clear-cut evidence of the cost of accidents versus the effectiveness of accident prevention work. In other words, a small operation may have, by luck, a good or bad accident record over a few years, whether or not its safety program is sound.

These are obstacles to progress—real and serious ones. They are not, of course, excuses for failure to try to prevent accidents and occupational illnesses. The trade association approach offers the best hope for improvement of this group.

### Labor-management cooperation

From its inception, one of the prime goals of organized labor has been the safety and health of its members. Many of today's international unions were organized originally to deal with extremely hazardous situations in the workplace, and have a sincere desire to work together with management on methods to prevent occupational injury and illness.

In 1949, the National Safety Council issued a policy statement declaring the common interest of labor and management in accident prevention. Even before this date, representatives of leading labor organizations served as members of the Council's governing boards. In 1955, a Council Labor Department and a Labor Conference (now known as Labor Division) were formally established.

The Labor Division continues to serve as a vital link between industry management and the nation's labor unions. By distributing Council safety and health materials at trade and industry shows, the Division is an invaluable resource for union members.

Labor Division representatives review products, training materials, and policy statements. The division shares information with labor leaders and more than 550 volunteers from international and local labor unions. The combined efforts of these groups reach the nearly 20 million members of organized labor with information to help them improve the quality of their lives, both on and off the job.

Labor has also been instrumental in promoting legislation such as the Occupational Safety and Health Act of 1970 (OSHAct), described in the next chapter.

Some unions have done extensive safety work and published printed matter and released films of benefit to the safety movement.

Through the auspices of the National Safety Council, the Labor Divison and the Industrial Division, along with other affected divisions, frequently have cooperated in preparing Council positions on matters pertaining to standards action, oversight testimony, publicity releases, and other areas which bear on occupational safety and health. As a result of these joint committees, Council positions are being recognized as more representative of all elements of society, allowing them to have even greater impact on administrative agencies and legislative bodies.

Through a program begun in January 1978 with a symposium of leaders from government, industry, and organized labor, the National Safety Council has begun an extensive program to determine causal factors of injuries. Realizing that most past data have only given us the numbers of types of injuries, the focus of this program is to change investigatory and reporting methods, as well as to provide an information exchange bank of the factors that have actually caused injury or occupational illness.

### Statistics, standards, and research

Statistical data on industrial accidents have been compiled by the National Safety Council for more than 50 years. Analyses computed annually and published in industry rate pamphlets and the Council's *Accident Facts* are of utmost importance in evaluating leading accident causes. For example, a recent study of the effects of raising the drinking age to 21 in 10 states by the Council's Statistics Department revealed a statistically significant reduction in the fatal accident involvement rate of drivers affected by the legislation.

Staff members also provide continual support for other Council departments, including: a summary and analysis of data on work-related motor vehicle injuries, compiled for a committee of the board of directors; an estimate of the effects of lower gasoline prices on travel and motor vehicle-related deaths for the Highway Traffic Safety Division; tabulations and a summary of data from a survey of truck drivers for the Motor Transportation Division; and tabulations of data from a survey of meat industry safety management attitudes and practices for the Industrial Division. There are more than 300 full-time technical/professional employees at the Council.

Some industries through their trade associations have recorded accident rates for almost 60 years. In most instances, even the divisions of an industry can establish their positions with regard to number and types of accidents and can determine their experience in comparison with national averages.

There are a large number of ANSI and other standards relating to safety. Continuing research has been necessary over the years to keep these standards in line with current industrial development and the development of new products and materials.

Special research projects, such as those making studies of walking surfaces and safety belts, can be and have been financed by private sources and coordinated by the National Safety Council. Recent Council research surveys have been conducted on the training programs for young and/or new drivers and on student reaction to the *Defensive Driver Handbook*. The results of these projects are given as a summary of findings.

### Safety and the law

Early legal action in industrial safety took the form of laws to regulate and investigate. The next phase was largely concerned with workers' compensation payments.

The following years have seen a gradual growth by federal, state, and local governments in regulating industry on safety matters. The Walsh-Healey Act, which deals with companies having supply contracts with the federal government, is an example of such regulation.

In certain industries—notably mining and transportation—U.S. federal government regulation and inspection have been extensive. The Construction Safety Act, which was passed in 1969, deals with the particular problems of that industry.

In 1970, the Williams-Steiger Occupational Safety and Health Act was passed and, for the first time, the United States had a *national* safety law. Every business, with one or more employees, which is affected by interstate commerce is covered by the law. Safety in this country took on a new direction and meaning as a result of the Occupational Safety and Health Act.

Concern with health and safety of workers has become a major priority for management. It goes beyond the obvious benefits of less lost time, reduced costs for worker compensation insurance, and lower medical and administrative expense resulting from disability, death, and impaired productivity. Failure to comply with the federal law's requirements brings citations, which (at the least) result in administrative costs, but could also lead to serious monetary penalties. Gross disregard of the law has also led to criminal sanctions against employers and even against managers individually. The criminal action has not only come from federal and state job safety and health agencies. Local prosecutors have been successful convicting individual managers for murder and aggravated assault as a result of death and injury to workers.

Managements must deal with serious emerging issues in worker health and safety law. They include ways to deal with the special problem of the employee who is at risk in the work environment because of physical condition, language problems, or particular susceptibility to injury or disease. Another issue is complying with the burgeoning paperwork arising from the recordkeeping requirements, the Medical Access Standard, and the Hazard Communication Standard. More details are in Chapter 2, Governmental Regulation and Compliance.

Today industry accepts almost without exception the idea of financial responsibility for work injuries. Not all of industry, however, is convinced of the effectiveness of government regulation of safety procedures.

A recent development in some states has been the establishment of laws making health and accident insurance compulsory to cover employee disabilities from diseases or accidents which originate off the job.

This compulsory insurance might be considered either a drastic extension of the principle of workers' compensation or an extension of social security legislation. It differs from workers' compensation in that it puts a financial burden upon management for diseases and accidents which are products of conditions beyond its control.

Whatever the theory, the result of these laws is to give the employer a direct financial stake in dealing with the off-the-job accident problem. See Chapter 3, Hazard Control Program Organization.

### Safety and occupational health

Even though medical and safety cooperation in accident prevention began during the earliest days of the safety movement, interest in safety on the part of the medical profession and, conversely, interest in employee and public health on the part of the safety professional is increasing.

Part of this interest in workers' health results from concern with occupational disease, noise, radiations, and other problems beyond the former concepts of occupational accident prevention. Part of the interest in protecting the health of citizens of nearby communities comes from such highly publicized events as the 3-Mile Island accident, the Chernobyl, Ukraine, meltdown, and the Bhopal, India, disaster.

**Work environment.** Over time, health and safety professionals became aware of the relationship between physical illness and the job, as workers in certain industries exhibited a higher than normal incidence of such problems as dermatitis, musculoskeletal problems, pulmonary disease, mental illness, and cancer.

The health and safety of today's workers results from a concerted effort by a safety, industrial hygiene, and occupational health team working with a management dedicated to the idea that an organization's primary asset is healthy and safe workers.

For more details, refer to the Council's *Fundamentals of Industrial Hygiene,* part of the Occupational Safety and Health Series, and the *Introduction to Occupational Health and Safety* manual.

**Community environments.** Increasingly in America, the public has demanded a larger role in the management of community environmental risks. Both public and private risk managers realize that providing avenues for public participation is a necessary part of their decision-making process. The problem is *how* both to assure public involvement and improve the quality of decisions.

To help fill the gap in credible risk communication on environmental health and safety issues the National Safety Council established the Environmental Health and Safety Institute. This special-purpose organization is led by a Board of Governors and operates mostly through philanthropic fundings from concerned corporations, foundations, labor unions, and individuals. It will develop accurate and objective information on environmental and public health risks, improve public knowledge about these risks, and disseminate this information to the public.

**Handicapped workers.** The utilization by progressive companies of workers with disabilities and federal and state "equal opportunity" laws has modified the practice of pre-employment examinations to screen out unfit or undesirable prospects. Medical personnel perform a preplacement examination to determine what physical or mental restrictions are appropriate to the prospective employee but do not determine fitness for a specific job. The job description must specify realistic physical and mental requirements to which the medical restrictions can be matched by the personnel department. If possible, modifications in the job must be made to accommodate the handicapped. (See the discussion in Chapter 20, Workers with Disabilities.)

### Psychology and "accident proneness"

Safety professionals who are thoughtfully looking for ways to improve their work encounter a great deal of challenging information in modern psychological writing—and also a great deal of careless and misleading generalizations.

Concern about the so-called "accident prone" individual in industry is as old as the safety movement. Statistical information suggests existence of such individuals, though clear and sharp data demonstrating this point are remarkably hard to come by. Too many alleged "proofs" turn out to be statistically deceptive, or based on inadequate samples, or the result of highly subjective diagnoses.

The elusiveness of statistical support for the existence of accident-prone individuals suggests to some thoughtful safety professionals that accident proneness may be a passing phase in the individual rather than a permanent characteristic, or at most, a problem encountered in an insignificant minority.

Realistically, objective analysis might disclose some supervisory deficiency or procedural weakness which may aggravate the hazard of certain operations or performance of individuals or groups of workers.

The same observation applies to psychological tests used as screening devices for new employees. Spectacular claims have been made from time to time for the effectiveness of such tests in predicting accident proneness, but none has established itself to the general satisfaction of the safety profession.

In the past, the work of psychologists like Dunbar and the Menningers aroused great interest among safety professionals. However, contributions to the practical day-to-day fight against accidents tend to come from the disciplines of engineering and behavioral psychology, such as human factors engineering, system safety, and risk management or assessment.

Refer to the discussion in Chapters 9 and 11.

### Summary

The present situation in the field of industrial safety is one of progress and improvement, largely through the continued application of techniques and knowledge slowly and painfully acquired through the years.

There appears to be no limit to the progress possible through the application of the universally accepted safety techniques of education, engineering, and enforcement.

Yet large and serious problems remain unsolved. A number of industries still have high accident rates. There are still far too many instances where management and labor are not working together or have different goals for the safety program.

The resources of the safety movement are great and strong—an impressive body of knowledge, a corps of able professional safety people, a high level of prestige, and strong organizations for cooperation and exchange of information.

## CURRENT PROBLEMS

Some problems of the safety movement are directly related to traditional strengths and weaknesses. Some of these problems are social and political in nature. Still others are essentially organizational.

### Technology and public interest

There is no reason for the safety professional to view the public's interest in product safety, a better environment, and general technological trends with alarm. Emphasis upon automation and more refined instrumentation will probably continue. New specific problems will arise, but they will be of a type that well-established methods of safety engineering are competent to solve.

The use of new materials and techniques—particularly radioactive materials and lasers—is likely to present more serious difficulties to the safety professional. However, even here, there is considerable experience.

Safety professionals need to stay tuned into the rapid developments in communications and computerization as they impact on their safety field. (See Chapter 5.)

### Political problems

On the political side remains the timeworn problem of industry-government relations. The key issue here is how much government should regulate and which aspects of American life it should regulate.

### Organizational problems

On the national scale, a wide variety of organizations are attacking specific aspects of occupational safety and health problems. The National Safety Council is, of course, the giant in the field—a strong, constructive, and nonpolitical, noncommercial giant. It has repeatedly sought and often achieved cooperative division of labor between itself and other organizations in the safety and health field.

One of the guiding principles of the Council has been that there was work enough and credit enough for all.

It remains to be seen whether the best organizational forms have been found for participation by all businesses in safety and health work. Safety and health professionals should be ready to consider new ideas and work to solve new problems.

### A look to the future

The greatest reasons for intensifying the safety effort are humane and moral. Our neighbors, our friends, our family—their worth cannot be measured in dollars or coded into computer records.

During the coming decade, we will see the American population grow, but more slowly than in the past. By 1990 there will be about 240 million living Americans. The average age of these persons will increase due to longevity thanks, in part, to health care advances. The "Baby Boom" generation will be among those aging adults and the generation crowding the workforce.

The shift from extractive and manufacturing to service field areas of employment will continue. This fact will hopefully contribute to a decline in occupational death rates since the number of people employed in high-risk industry will be lower.

The workforce continues to undergo major changes as minorities and women continue their upward mobility. Safety implications from the increasing employment of women are complicated and a number of difficult choices will still have to be made.

As social conditions, including a high divorce rate, one-parent families, and two-working-parent households continue to mushroom, the effect on the traditional American family is felt. Today there is reduced respect for either parental or social authority, a factor seen and felt in the workplace.

A survey of National Safety Council Board members ranked the major safety and health issues facing the nation. Drinking drivers came in first, followed by occupant restraints, a national uniform minimum age of 21 for drinking, a national 55 mph speed limit, and the transportation of hazardous materials.

These are issues of major importance and will need the expertise and guidance of all of us in the safety and health field. The future is, as it always has been, most uncertain. By working together all those in the safety and health community can take some of the uncertainty out of it by helping to make our workplaces and our off-the-job environs safer, healthier places to be.

**Opportunities for the safety professional.** The 1980s are an exciting time for the safety professions. Safety management, safety engineering, industrial hygiene, occupational medicine, and the new field of holistic medicine are discovering new and compelling reasons for drawing together in ever closer cooperation and opening new areas of employment for the safety and health professional.

Other expanding employment opportunities for safety professionals lie in the safety departments of international (and some local) labor unions, on the staffs of a number of trade associations, and, of course, in government service. Safety consulting has expanded rapidly, both in the form of individuals offering their talents on a contract basis and in the form of consulting service offered by nonprofit associations and by a few industrial concerns.

The availability of advanced education for safety professionals registered a large percentage increase in the '80s, but it started from a small base. The numerical growth remained inadequate to meet the expanding need. The result was that there was a serious shortage of highly qualified safety professionals. At the same time, the population bulge in the prime years of life created a surplus of the less-well-trained people in management. This had two effects on safety positions in companies and in government. One was an increase in the competition for senior safety positions. The other was an increase in the demand for advanced safety training to a degree beyond the immediate capacity of the technical colleges. One of the greatest growth opportunities for highly qualified safety professionals lies, in fact, in the staffing of college courses in safety.

These are the challenges of tomorrow—improved performance on the job and profession, coupled with the necessary education to compete and work effectively in the safety and health field.

## GENERAL SAFETY BOOKS

In addition to this manual, the following are currently available books on the basics of occupational safety and health that are used to obtain the questions on the management aspects section of the examination for Certified Safety Professional qualification.


Browning, Robert L. *The Loss Rate Concept in Safety Engineering.* New York, N.Y.: Marcel Dekker, 1980.

DeReamer, Russell. *Modern Safety and Health Technology.* New York, N.Y.: John Wiley & Sons, 1981.

Ferry, Ted S. *Modern Accident Investigation and Analysis.* New York, N.Y.: John Wiley & Sons, 1981.

Firenze, Robert J. *The Process of Hazard Control.* Dubuque, Iowa: Kendall-Hunt, 1978.

Gilmore, Charles L. *Accident Prevention and Loss Control.* New York, N.Y.: The American Management Association, 1970.

Grimaldi, John V. *Safety Management,* 4th ed. Homewood, Ill.: Richard D. Irwin, Inc., 1984.

Heinrich, H. W. *Industrial Accident Prevention,* 5th ed. New York, N.Y.: McGraw-Hill Book Co., 1980.

Petersen, Daniel C. *Analyzing Safety Performance.* New York, N.Y.: Garland STPM Press, 1980.

——. *Safety by Objectives.* River Dale, N.J.: Aloray Publishers, 1978.

——. *Techniques of Safety Management.* New York, N.Y.: McGraw-Hill Book Co., 1978.

Tarrents, William E. *The Measurement of Safety Performance.* New York, N.Y.: Garland STPM Press, 1980.


## REFERENCES


American Engineering Council. *Safety and Production.* New York, N.Y.: Harper & Brothers Publishers, 1928.

American National Standards Institute, 1430 Broadway, New York, N.Y. 10018.

Andrews, E. W. "The Pioneers of 1912." *National Safety News, 66:24-25,* 64-65 (July 1952).

Beyer, David Stewart. *Industrial Accident Prevention,* 3rd ed. Boston and New York: Houghton Mifflin Co., 1928.

Campbell, R. W. "The National Safety Movement." *Proceedings of the Second Safety Congress of the National Council for Industrial Safety,* pp. 188-192, 1913.

DeBlois, Lewis A. *Industrial Safety Organization for Executive and Engineer.* New York, N.Y.: McGraw-Hill Book Company, 1926.

Eastman, Crystal. *Work Accidents and the Law.* New York, Charities Publication Committee, 1910. Reprint. New York: Arno Press, 1969.

Felton, Jean S. "History of Occupational Health and Safety," Chapter 3 of *Introduction to Occupational Health and Safety,* ed. by Joseph LaDou. Chicago, Ill.: National Safety Council, 1986.

Heinrich, H. W. *Industrial Accident Prevention,* 1st ed. New York, N.Y.: McGraw-Hill Book Company, 1931.

Holbrook, Steward H. *Let Them Live.* New York, N.Y.: The Macmillan Co., 1939.

Menninger, K. A. *Man Against Himself.* New York, N.Y.: Harcourt, Brace and World, Inc., 1956.

Meyer, Robert L. Series commemorating the Diamond Anniversary of the Council. *Safety and Health,* (January through October 1987).

Mock, H. E. *Industrial Medicine and Surgery.* Philadelphia, Pa.: W. B. Saunders, 1920.

National Safety Council, 444 North Michigan Ave., Chicago, Ill. 60611.

*Accident Facts.* Issued annually.

"Golden Anniversary Issue." *National Safety News,* 87:5 (May 1963)

*Proceedings of the First Co-Operative Safety Congress,* 1912.

*Proceedings of the National Safety Congress.* Issued annually from 1914-1925.

*Proceedings of the Second Safety Congress of the National Council for Industrial Safety,* 1913.

*Safety and Health.* Issued monthly.

Ramazzini, B. *Diseases of Workers,* trans. from the Latin text *De Morbis Artificum* of 1713, by W. C. Wright. New York, N.Y.: Hafner Publishing, 1964.

Schaefer, Vernon G. *Safety Supervision.* New York, N.Y.: McGraw-Hill Book Company, 1941.

Schulzinger, Morris S. "Accident Syndrome—A Clinical Approach." *Archives of Industrial Health,* 11:66-71, 1955. Chicago, Ill., American Medical Assn.

Schwedtman, Ferd., and James A. Emery. *Accident Prevention and Relief.* New York, N.Y.: National Association of Manufacturers in the United States of America, 1911.

Toynbee, A. *The Industrial Revolution.* (First published in 1884.) Boston, Mass.: Beacon Press, 1956.

# 2

# Governmental Regulation and Compliance



THE TWO MAJOR PIECES of federal legislation having an impact on occupational safety and health passed by the Congress in the last two decades are the Occupational Safety and Health Act of 1970 and the Federal Mine Safety and Health Act of 1977. Although other federal legislation affecting occupational safety and health to a lesser degree remains on the books, this chapter will focus on the Occupational Safety and Health Act and the Federal Mine Safety and Health Act.

Note that to save space and to be legally correct, the citations of the various references to laws, regulations, and court decisions are in legal terminology. This is all explained in the last section of this chapter.

## PART I
## The Occupational Safety and Health Act

A new national policy was established on December 29, 1970, when President Richard M. Nixon signed into law the Occupational Safety and Health Act of 1970 (Public Law No. 91-596 found at 29 U.S.C. §§651-678). The Congress of the United States declared that the purpose of this piece of legislation is "to assure so far as possible every working man and woman in the Nation safe and healthful working conditions and to preserve our human resources."

The OSHAct took effect April 28, 1971. It was coauthored by Senator Harrison A. Williams (Dem.-N.J.) and the late Congressman William Steiger (Rep.-Wis.) and hence is sometimes designated as the Williams-Steiger Act. The Act is regarded by many as landmark legislation since it goes beyond the present workplace and considers the working environment of the future as related to health hazards.

The information provided in Part I of this chapter focuses on federal OSHA programs. State OSHA programs may differ from the federal program in certain areas. However, unless specifically instructed to the contrary, the recommendations in this chapter can be followed whether the jurisdiction rests at the federal or state level.

### LEGISLATIVE HISTORY

Historically, the enactment of safety and health laws has been left to the states. Prior to the 1960s only a few federal laws (such as the Walsh-Healey Public Contracts Act and the Longshoremen's and Harbor Workers' Compensation Act) directed any attention to occupational safety and health. The decade of the sixties, however, saw significant congressional action in this arena. A number of pieces of legislation passed by the Congress during the sixties, including the Service Contract Act of 1965, the National Foundation on Arts and Humanities Act, the Federal Metal and Nonmetallic Mine Safety Act, the Federal Coal Mine Health and Safety Act, and the Contract Workers and Safety Standards Act (Construction Safety Act), directed attention to occupational safety and health.

Each of these federal laws was applicable to a limited number of employers. These laws were directed at those who had obtained federal contracts or they zeroed-in on a specific industry. Even collectively, all the federal safety legislation passed prior to 1970 was not applicable to the majority of employers or employees. Until 1970, congressional action related to occupational safety and health was, at best, sporadic, covering only specific sets of employers and employees with little attempt for an omnibus coverage that is a part of the OSHAct.

Proponents of more significant federal presence in occupational safety and health, mostly represented by organized labor, based their position primarily on the following:

1. With few exceptions, the states failed to meet their obligation in regard to occupational safety and health. Only a few of the states had safety and health legislation that was considered reasonable or adequate. Many states legislated safety and health only in specific industries. In general, states had inadequate safety and health standards, inadequate enforcement procedures, inadequate staff with respect to quality and quantity, and inadequate budgets.
2. In the late 1960s, approximately 14,300 employees were being killed annually on or in connection with their job and more than 2.2 million employees suffered a disabling injury each year as a result of work-related accidents. The injury/death toll was considered by most to be much too high and therefore not acceptable.
3. The nation's work injury rates in most industries were increasing throughout the decade of the sixties. Since the trend was in the wrong direction, proponents of federal presence felt that federal legislation would assist in reversing this trend.

The act evolved amid a stormy atmosphere in both houses of Congress. Highly controversial issues were involved. Such issues were responsible for sharply drawn lines between political parties and between the business community and organized labor. After three years of political hassle, numerous compromises were made; this ultimately enabled the passage of the OSHAct of 1970 by both houses of Congress.

## ADMINISTRATION

Administration and enforcement of the OSHAct are vested primarily with the Secretary of Labor and the Occupational Safety and Health Review Commission (discussed later). With respect to the enforcement function, the Secretary of Labor performs the investigation and prosecution aspects of the enforcement process and the Review Commission performs the adjudication portion of the enforcement process.

Research and related functions and certain educational functions are vested in the Secretary of Health, Education, and Welfare (now known as the Secretary of Health and Human Services) and are, for the most part, carried out by the National Institute for Occupational Safety and Health established within the Department of Health and Human Services (DHHS). Compiling injury and illness statistical data is handled by the Bureau of Labor Statistics, U.S. Department of Labor.

To assist the Secretary of Labor, the Act authorizes the appointment of an Assistant Secretary of Labor for Occupational Safety and Health. This position is filled by Presidential appointment with the advice and consent of the Senate. The Assistant Secretary is the chief of the Occupational Safety and Health Administration (OSHA) established within the Department of Labor (DOL). The Assistant Secretary acts on behalf of the Secretary of Labor. For the purpose of this chapter, OSHA is also synonymous with the term Secretary or Assistant Secretary of Labor.

The primary functions of the four major governmental units assigned to carry out the provisions of the Act are described in this section on administration.

### Occupational Safety and Health Administration

The Occupational Safety and Health Administration (OSHA) came into existence officially on April 28, 1971, the date the Williams-Steiger Occupational Safety and Health Act became effective. This agency was created by the Department of Labor to discharge the Department's responsibilities assigned by the Act.

**Major areas of authority.** The Act grants OSHA the authority, among other things, (1) to promulgate, modify, and revoke safety and health standards; (2) to conduct inspections and investigations and to issue citations, including proposed penalties; (3) to require employers to keep records of safety and health data; (4) to petition the courts to restrain imminent danger situations; and (5) to approve or reject state plans for programs under the Act.

The Act also authorizes OSHA (1) to provide training and education to employers and employees; (2) to consult with employers, employees, and organizations regarding prevention of injuries and illnesses; (3) to grant funds to the states for identification of program needs and plan development, experiments, demonstrations, administration and operation of programs; and (4) to develop and maintain a statistics program for occupational safety and health.

**Major duties delegated.** In establishing the Occupational Safety and Health Administration, the Secretary of Labor delegated to the Assistant Secretary for Occupational Safety and Health the authority and responsibility for safety and health programs and activities of the Department of Labor, including responsibilities derived from:

1. Occupational Safety and Health Act of 1970
2. Walsh-Healey Public Contracts Act of 1936, as amended
3. Service Contract Act of 1965
4. Public Law 91-54 of 1969 (construction safety amendments)
5. Public Law 85-742 of 1958 (maritime safety amendments)
6. National Foundation on the Arts and Humanities Act of 1965
7. Longshoremen's and Harbor Workers' Compensation Act (Title 33, Chapter 18, §§901, 904, U.S. Code; Act of March 4, 1927, Chapter 509, 44 Stat. 1424)
8. Federal safety program under Title 5 U.S. Code §7902

Similarly, the Commissioner of the Bureau of Labor Statistics was delegated the authority and given the responsibility for developing and maintaining an effective program for collection, compilation, and analysis of occupational safety and health statistics, providing grants to the states to assist in developing and administering programs in such statistics, and coordinating functions with the Assistant Secretary for Occupational Safety and Health.

The Solicitor of Labor is assigned responsibility for providing legal advice and assistance to the Secretary and all officers of the Department in the administration of statutes and Executive Orders relating to occupational safety and health. In enforcing the Act's requirements, the Solicitor of Labor also has the responsibility for representing the Secretary in litigation before the Occupational Safety and Health Review Commission, and, subject to the control and direction of the Attorney General, before the federal courts.

To assist in carrying out its responsibilities, OSHA has established ten regional offices in the cities of Boston, New York, Philadelphia, Atlanta, Chicago, Dallas, Kansas City, Denver, San Francisco, and Seattle. (See Directory of Federal Agencies at the end of this chapter.) The primary mission of the regional office chief, known as the Regional Administrator, is to supervise, coordinate, evaluate, and execute all programs of OSHA in the region. Assisting the Regional Administrator are Assistant Regional Administrators for (1) training, education, consultation, and federal agency programs, (2) technical support, and (3) state and federal operations.

Area offices have been established within each region, each headed by an Area Director. The mission of the Area Director is to carry out the compliance program of OSHA within designated geographic areas. The area office staff carries out its activities under the general supervision of the Area Director with guidance of the Regional Administrator, using policy instructions received from the national headquarters. The real action for implementing the enforcement portion of the OSHAct is carried out by the area offices in those states that do not have an approved state plan. The area office monitors state activities in those states that have a state plan. (See Federal–State Relationships, pages 32 and 33.)

## Occupational Safety and Health Review Commission

The Occupational Safety and Health Review Commission (OSHRC) is a quasi-judicial board of three members appointed by the President and confirmed by the Senate. The Commission is an independent agency of the Executive Branch of the U.S. Government and is not a part of the Department of Labor. The principal function of the Commission is to adjudicate cases resulting from an enforcement action initiated against an employer by OSHA when any such action is contested by the employer or by his employees or their representatives.

The Commission's actions are limited to contested cases. In such cases, OSHA notifies the Commission of the contested cases and the Commission hears all appeals on actions taken by OSHA concerning citations, proposed penalties, and abatement periods, and determines the appropriateness of such actions. When necessary, the Commission may conduct its own investigation and may affirm, modify, or vacate OSHA's findings.

There are two levels of adjudication within the Commission: (1) the administrative law judge, and (2) the three-member Commission. All cases not resolved on informal proceedings are heard and decided by one of the Commission's administrative law judges. The judge's decision can be changed by a majority vote of the Commission if one of the members, within 30 days of the judge's decision, directs that the judge's decision be reviewed by the Commission members. The Commission is the final administrative authority to rule on a particular case, but its findings and orders can be subject to further review by the courts. (For further information, see Contested Cases later in this chapter.)

The headquarters of the OSHRC is located at 1825 K Street, NW, Washington, D.C. 20006.

## National Institute for Occupational Safety and Health

The National Institute for Occupational Safety and Health (NIOSH) was established within the HEW (currently known as the Department of Health and Human Services) under the provisions of the OSHAct. Administratively, NIOSH is located in DHHS's Center for Disease Control in Atlanta, Ga. NIOSH is the principal federal agency engaged in research, education, and training related to occupational safety and health.

The primary functions of NIOSH are to (1) develop and establish recommended occupational safety and health standards, (2) conduct research experiments and demonstrations related to occupational safety and health, and (3) conduct education programs to provide an adequate supply of qualified personnel to carry out the purposes of the OSHAct.

**Research and related functions.** Under the OSHAct, NIOSH has the responsibility for conducting research for new occupational safety and health standards. NIOSH develops criteria for the establishment of such standards. Such criteria are transmitted to OSHA which has the responsibility for the final setting, promulgation, and enforcement of the standards.

The OSHAct also requires NIOSH to publish an annual listing of all known toxic substances and the concentrations at which such toxicity is known to occur. While the entry of a substance on the list does not mean that it is to be avoided, it does mean that the listed substance has a documented potential of being hazardous if misused and, therefore, care must be exercised to control the substance. Conversely, the absence of a substance from the list does not necessarily mean that a substance is nontoxic. Some hazardous substances may not qualify to be listed because the dose that causes the toxic effect is not known.

**Education and training.** NIOSH also has the responsibility to conduct (1) education and training programs which are aimed at providing an adequate supply of qualified personnel to carry out the purpose of the Act and (2) informational programs on the importance and proper use of adequate safety and health equipment. The long-term approach to an adequate supply of training personnel in occupational safety and health is found in the colleges and universities and other institutions in the private sector. NIOSH encourages such institutions, by contracts and grants, to expand their curricula in occupational medicine, occupational health nursing, industrial hygiene, and occupational safety engineering.

**Employer and employee services.** Of principal interest to individual employers and employees are the technical services offered by NIOSH. The five main services that are provided upon request to NIOSH's Division of Technical Services, Cincinnati, Ohio 45226, are:

1. Hazard evaluation—Provides on-site evaluations of potentially toxic substances used or found on the job.
2. Technical information—Provides technical information concerning health or safety conditions at workplaces, such as the possible hazards of working with specific solvents, and when to use protective equipment.
3. Accident prevention—Provides technical assistance for controlling on-the-job injuries including the evaluation of special problems and recommendations for corrective action.
4. Industrial hygiene—Provides technical assistance in the areas of engineering and industrial hygiene, including the evaluation of special health-related problems in the workplace and recommendations for control measures.
5. Medical service—Provides assistance in solving occupational

medical and nursing problems in the workplace including the assessment of existing medically related needs and the development of recommended means for meeting such needs.

NIOSH approves coal mine dust personal sampler units, gas detector tube units, and respiratory protective devices including self-contained breathing apparatus, gas masks, supplied-air respirators, chemical-cartridge respirators, and dust, fume, and mist respirators. NIOSH also has a certification program for sound level meters.

NIOSH representatives, although not authorized to enforce the OSHAct, are authorized to make inspections and to question employers and employees in order to carry out those duties assigned to the DHHS under the Act.

It has both warrant and subpoena power, if necessary, to obtain the information needed for its investigations. It may gain access to employee records, but only with the consent of employees or if methods are used to maintain the employee's right of privacy in respect to the information in the records.

### Bureau of Labor Statistics

The responsibility for conducting statistical surveys and establishing methods used to acquire injury and illness data is placed in the Bureau of Labor Statistics (BLS). Questions regarding recordkeeping requirements and reporting procedures can be directed to any of the OSHA regional or area offices or the BLS regional offices. (See the directory at the end of this chapter.)

### Advisory committees

The Act established a 12-member National Advisory Committee on Occupational Safety and Health (NACOSH) to advise, consult with, and make recommendations to the Secretaries of Labor and Health and Human Services (HHS) with respect to the administration of the Act. Eight members are designated by the Secretary of Labor and four by the Secretary of HHS. Members include representatives from management, labor, occupational safety and health professions, and the public.

The Act also authorizes the appointment of 15-member advisory committees to assist OSHA in the development of standards. The Standards Advisory Committees on Construction Safety and Health and on Cutaneous Hazards are the two currently in place.

## MAJOR PROVISIONS OF THE OSHAct

### Coverage

Except for specific exclusions, the Act is applicable to every employer who has one or more employees and who is engaged in a business affecting commerce. The law applies to all 50 states, the District of Columbia, Puerto Rico, and all U.S. possessions.

Specifically *excluded* from coverage are all federal, state, and local government employees. There are, however, special provisions in the Act for federal employees and potential coverage for state and local government employees. The Act requires each federal agency head to establish and maintain an occupational safety and health program consistent with the standards promulgated by the Secretary of Labor. Executive Orders setting requirements for federal programs have been issued, the last (up to the time this chapter was written) being Executive Order 12196 issued February 26, 1980. OSHA regulations implementing the Executive Order are found in 29 C.F.R. Part 1960 (1983).

Employees of states and political subdivisions of the states are excluded from the federal OSHAct. However, states with approved state plans are required to provide coverage for these public employees. Public employees in states that do not have approved plans are not covered by the OSHAct in any manner.

However, two states—Connecticut and New York—have state plans covering only public employees. The District of Columbia and the states of Mississippi, New Hampshire, New Jersey, Rhode Island, and Wisconsin have laws specifically providing job health and safety protection to public employees.

The OSHAct is also not applicable to those operations where a federal agency (and state agencies acting under the Atomic Energy Act of 1954), other than the Department of Labor, has statutory authority to prescribe or enforce standards or regulations affecting occupational safety or health and is performing that function. An example of this exclusion is specific issues covered by Department of Transportation regulations in the railroad industry.

Also excluded from the OSHAct are operators and miners covered by the Federal Mine Safety and Health Act of 1977 (see Part II of this chapter) which is applicable to mines of all types; coal and noncoal, surface and underground.

Curiously, inspection restrictions can vary from year to year, depending on the whims of Congress at the time it considers appropriations for the Department of Labor. In its yearly appropriation bills since 1977, Congress has placed restrictions on OSHA enforcement. For example, items exempt from inspection in fiscal year (FY) 1987 include:

- Farmers with ten or fewer employees on the day of inspection and the 12 months preceding the day of inspection
- Any work activity in any recreational, hunting, fishing or shooting area
- Employers with ten or fewer employees in industries with three-digit Standard Industrial Classification injury/illness rates of less than seven per 100 employees

The exemption is inapplicable to situations involving, among others, employee complaints, imminent dangers, health hazards, accidents resulting in a fatality or hospitalization involving five or more employees, or discrimination complaints.

OSHA clarified its interpretation of coverage with respect to certain employees by issuing a regulation. (This policy regarding "Coverage of Employees Under the Williams-Steiger Occupational Safety and Health Act of 1970" is contained in *Code of Federal Regulations* (C.F.R.), Title 29, Chapter XVII, Part 1975.) OSHA has also said that churches and religious organizations, with respect to their religious activities, are not regarded as employers. Likewise, persons who in their own residences employ others to perform domestic household tasks are not regarded as employers. Further, any person engaged in agriculture who is a member of the immediate family of the farmer is not regarded as an employee and hence is not covered by the Act.

### Employer and employee duties

Each employer covered by the Act:

1. Has the general duty to furnish each of his employees employment and places of employment which are free from recognized hazards that are causing or likely to cause death or serious physical harm (this is commonly known as the "general duty clause")

2. Has the specific duty of complying with safety and health standards promulgated under the Act

Each employee, in turn, has the duty to comply with the safety and health standards and all rules, regulations, and orders that are applicable to his own actions and conduct on the job.

For employers, the general duty provision is used only where there are no specific standards applicable to a particular hazard involved. A hazard is "recognized" if it is a condition that is generally recognized as a hazard in the particular industry in which it occurs and is detectable (1) by means of the human senses, or (2) there are accepted tests known in the industry to determine its existence which should make its presence known to the employer. An example of a "recognized hazard" in the latter category is excessive concentration of a toxic substance in the work area atmosphere, even though such concentration could only be detected through use of measuring devices.

During the course of an inspection a compliance safety and health officer is concerned primarily with determining whether the employer is complying with the promulgated safety and health standards. However, he will also direct attention to determining whether the employer is complying with the general duty clause.

The law provides for sanctions against the employer in the form of citations and civil and criminal penalties if the employer fails to comply with his two duties. However, there is no provision for government sanctions against an employee for failure to comply with the employee's duty. While some may view the latter as unjust, significantly, it was not one of the controversial issues in the formative stages of the Act.

Both management and organized labor have long agreed that safety and health on the job is a management responsibility. The business community generally did not want the law structured to provide for government sanctions against an erring employee because there are measures which management can invoke against an employee who obstructs the employer's efforts to provide a safe workplace.

While the law expressly places upon each employee the obligation to comply with the standards, final responsibility for compliance with the requirements of the Act remains with the employer. Employers thus should take all necessary action to assure employee compliance with the promulgated standards and establish within their safety system a means whereby they become aware of situations where employees are not complying with applicable standards.

The duty of an employer to protect employees against health hazards in addition to safety hazards is gaining emphasis. The growing awareness of the hazard of chemical exposure has caused OSHA to focus on the employer's obligation to monitor the work environment, provide periodic medical examinations, and make available a range of protective measures to guard against hazards of the work environment.

Emerging as an issue is the question of how far an employer must go to protect employees who are at increased risk of occupational injury or illness. The increased risk may come from lack of experience, language problems, special sensitivity to chemicals, or being a woman of child-bearing age. Medical screening is one method employers have begun to use to identify present and prospective employees who are at increased risk. Denial of employment, or assignment or reassignment, based on the results of the screening or awareness of at-risk status has become controversial. Can such actions intended to protect the employee become discriminatory or invade privacy? The early 1980s saw the use of the job-discrimination provisions of the Civil Rights Act to protect women workers in general, and pregnant women in particular, against termination and detrimental reassignment because of hazardous working conditions.

### Employer rights

An employer has the right to:

1. Seek advice and off-site consultation as needed by writing, calling, or visiting the nearest OSHA office
2. Request and receive proper identification of the OSHA compliance safety and health officer prior to inspection
3. Be advised by the compliance safety and health officer (CSHO) of the reason for an inspection
4. Have an opening and closing conference with the CSHO
5. File a Notice of Contest with the OSHA area director within 15 working days of receipt of a notice of citation and proposed penalty
6. Apply to OSHA for a temporary variance from a standard if unable to comply because of the unavailability of materials, equipment, or personnel needed to make necessary changes within the required time
7. Take an active role in developing safety and health standards through participation in OSHA Standards Advisory Committees, through nationally recognized standards-setting organizations, and through evidence and views presented in writing or at hearings
8. Avail himself, if a small business employer, of long-term loans through the Small Business Administration to help bring the establishment into compliance, either before or after an OSHA inspection
9. Be assured of the confidentiality of any trade secrets observed by an OSHA compliance officer.

## On-Site Consultation

Congress has authorized, and OSHA now provides through a state agency or private contractors, free on-site consultation service for employers in every state. These consultants help employers identify hazardous conditions and determine corrective measures.

The service is available upon employer request. Priority is given to businesses with fewer than 150 employees, which are generally less able to afford private sector consultation. Emphasis is placed on highly hazardous jobs.

The consultative visit consists of an opening conference, a walk-through of the company's facility, a closing conference and a written summary of findings. During the walk-through, the employer is told which OSHA standards are applicable and what they mean. The employer is told of any apparent violations of those standards and, where possible, is given suggestions on how to reduce or eliminate the hazard.

Because employers, not employees, are subject to legal sanctions of the OSHA standards, the employer controls the extent of participation by employees or their representatives in the visit. However, the consultant must be allowed to confer with individual employees during the walk-through in order to identify and judge the nature and extent of hazards.

In the early 1980s, OSHA introduced voluntary compliance programs as a way of cutting down the amount of inspecting it had to do. Employers with good safety and health records and formal safety programs are exempt from inspection unless there are employee complaints or serious accidents. These voluntary programs carry names such as "Star" and "Try."

### Employee rights

Although the employee has the legal duty to comply with all the standards and regulations issued under the OSHAct, there are many employee rights that are also incorporated in the Act. Since these rights may affect labor relations as well as labor negotiations, employers as well as employees should be aware of the employee rights contained in the Act. Employee rights fall into three main areas and are related to (1) standards, (2) access to information, and (3) enforcement.

**With respect to standards:**

1. Employees may request OSHA to begin proceedings for adoption of a new standard or to amend or revoke an existing one.
2. Employees may submit written data or comments on proposed standards and may appear as an interested party at any hearing held by OSHA.
3. Employees may file written objections to a proposed federal standard and/or appeal the final decision of OSHA.
4. Employees must be informed when an employer applies for a variance of a promulgated standard.
5. Employees must be afforded the opportunity to participate in a variance hearing as an interested party and have the right to appeal OSHA's final decision.

**With respect to access to information:**

1. Employees have the right to information from the employer regarding employee protections and obligations under the Act and to review appropriate OSHA standards, rules, regulations, and requirements that the employer should have available at the workplace.
2. Affected employees have a right to information from the employer regarding the toxic effects, conditions of exposure, and precautions for safe use of all hazardous materials in the establishment by means of labeling or other forms of warning where such information is prescribed by a standard.
3. If employees are exposed to harmful materials in excess of levels set by the standards, the affected employees must be so informed by the employer and the employer must also inform the employees thus exposed what corrective action is being taken.
4. If a compliance safety and health officer determines that an alleged imminent danger exists, he must inform the affected employees of the danger and that he is recommending that relief be sought by court action if the imminence of such danger is not eliminated.
5. Upon request, employees must be given access to records of their history of exposure to toxic materials or harmful physical agents that are required to be monitored or measured and recorded.
6. If a standard requires monitoring or measuring hazardous materials or harmful physical agents, employees must be given the opportunity to observe such monitoring or measuring.
7. Employees have the right of access to (1) the list of toxic materials published by NIOSH, (2) criteria developed by NIOSH describing the effects of toxic materials or harmful physical agents, and (3) industrywide studies conducted by NIOSH regarding the effects of chronic, low-level exposure to hazardous materials.
8. On written request to NIOSH, employees have the right to obtain the determination of whether or not a substance found or used in the establishment is harmful.
9. Upon request, the employees should be allowed to review the Log and Summary of Occupational Injuries (OSHA No. 200) at a reasonable time and in a reasonable manner.

**With respect to enforcement:**

1. Employees have the right to confer in private with the compliance safety and health officer and to respond to questions from the CSHO, in connection with an inspection of an establishment.
2. An authorized employee representative must be given an opportunity to accompany the compliance officer during an inspection for the purpose of aiding such inspection. (This is commonly known as the "walk-around" provision.) Also, an authorized employee has the right to participate in the opening and closing conferences during the inspection.
3. An employee has the right to make a written request to OSHA for a special inspection if the employee believes a violation of a standard threatens physical harm, and the employee has the right to request OSHA to keep his identity confidential.
4. If an employee believes any violation of the Act exists, he has the right to notify OSHA or a compliance officer in writing of the alleged violation, either before or during an inspection of the establishment.
5. If a request is made for a special inspection and it is denied by OSHA, the employee must be notified in writing by OSHA, together with the reasons, that the complaint was not valid. The employee has the right to object to such a decision and may request a hearing by OSHA.
6. If a written complaint concerning an alleged violation is submitted to OSHA and the compliance officer responding to the complaint fails to cite the employer for the alleged violation, OSHA must furnish the employee or his authorized representative a written statement setting forth the reasons for its final disposition.
7. If OSHA cites an employer for a violation, employees have the right to review a copy of the citation which must be posted by the employer at or near the place where the violation occurred.
8. Employees have the right to appear as an interested party or to be called as a witness in a contested enforcement matter before the Occupational Safety and Health Review Commission.
9. If OSHA arbitrarily or capriciously fails to seek relief to counteract an imminent danger and an employee is injured as a result, that employee has the right to bring action against OSHA for relief as may be appropriate.
10. An employee has the right to file a complaint to OSHA within 30 days if he believes he has been discriminated against because he asserted his rights under the Act.
11. An employee has the right to contest the abatement period fixed in the citation issued to his employer by notifying the OSHA Area Director that issued the citation within 15 working days of the issuance of the citation.

### The OSHA poster

The OSHA poster (OSHA 2203, see Figure 2-1) must be prominently displayed in a conspicuous place in the workplace where notices to employees are customarily posted. The poster informs employees of their rights and responsibilities under the Act.

### Occupational safety and health standards

The Act authorizes OSHA to promulgte, modify, or revoke occupational safety and health standards. The rules of procedure for promulgating, modifying or revoking standards are spelled out in the *Code of Federal Regulations* (C.F.R.), Title 29, Chapter XVII, Part 1911. The current requirements are available at all OSHA area and regional offices.

OSHA is responsible for promulgating legally enforceable standards which may require conditions, or the adoption or use of practices, means, methods, or processes that are reasonably necessary and appropriate to protect employees on the job. It is the employers' responsibility to become familiar with the standards applicable to their establishments and to make sure that employees have and use personal protective equipment required for safety. In additon, employers are responsible for complying with the Act's general duty clause.

In order to get the initial set of standards in place without undue delay, the Act authorized OSHA to promulgate any existing federal standard or any national consensus standard without regard to the usual rulemaking procedures prior to April 28, 1973. The initial set of standards, Part 1910, appeared in the *Federal Register* of May 29, 1971. Subscriptions to the *Federal Register* are obtained through the Government Printing Office, Washington, D.C. 20402.

Standards contained in Part 1910 are applicable to general industry. Those contained in Part 1926 are applicable to construction. Standards applicable to ship repairing, shipbuilding, shipbreaking, and long-shoring are contained in Parts 1915 through 1918 respectively. Because standards cannot remain static due to use of new equipment, methods, and materials, all are subject to updating via modification.

OSHA standards incorporate by reference other standards adopted by standards-producing organizations. Standards incorporated by reference in OSHA standards in whole or in part include, but are not limited to, standards adopted by the following standards-producing organizations:

American Conference of Governmental Industrial Hygienists
American National Standards Institute
American Petroleum Institute
American Society of Agricultural Engineers
American Society of Mechanical Engineers
American Society for Testing and Materials
American Welding Society
Compressed Gas Association
Crane Manufacturers Association of America, Inc.
Institute of Makers of Explosives
National Electrical Manufacturers Association
National Fire Protection Association
National Plant Food Institute
National Institute for Occupational Safety and Health
Society of Automotive Engineers
The Fertilizer Institute
Underwriters Laboratories Inc.
U.S. Department of Commerce
U.S. Public Health Service

OSHA has the authority to promulgate emergency temporary standards where it is found that employees are exposed to grave danger. Emergency temporary standards can take effect immediately upon publication in the *Federal Register.* Such standards will remain in effect until superseded by a standard promulgated under the procedures prescribed by the Act. The law requires OSHA to promulgate a permanent standard no later than six months after publication of the emergency temporary standard.

Any person adversely affected by any standard issued by OSHA has the right to challenge its validity by petitioning the U.S. Court of Appeals within 60 days after its promulgation.

**Input from the private sector.** Occupational safety and health standards promulgated by OSHA will never cover every conceivable hazardous condition that could exist in any workplace. Nevertheless, new standards and modification of existing standards are of significant interest to employers and employees alike. Industry organizations as well as individuals and employee organizations should express their views by responding to (1) OSHA's advance notice of proposed rulemaking, which usually calls for information upon which to base proposed standards, and (2) OSHA's proposed standards, since it is within the private sector that most of the expertise and the technical competence lies. To do less means that industry and employees are willing to let the standards development process rest in the hands of OSHA.

Sources used by OSHA for the revision of existing occupational safety and health standards or the development of new standards are standards advisory committees appointed by the Secretary of Labor and NIOSH criteria documents.

In order to promulgate, revise, or modify a standard, OSHA must first publish in the *Federal Register* a notice of any proposed rule that will adopt, modify, or revoke any standard and invite interested persons to submit their views on the proposed rule. The notice will include the terms of the new standard and will provide an interval of at least 30 days from the date of publication and usually 60 days or more for interested persons to respond. Interested persons may file objections to the rule and are entitled to a hearing on their objections if they request that a hearing be held. However, objections must specify the parts of the proposed rule to which they object and the grounds for such objection. If a hearing is requested, OSHA must hold one. Based on (1) the need for control of an exposure to an occupational injury or illness, and (2) the reasonableness, effectiveness and feasibility of the control measures required, OSHA may issue a rule promulgating an additonal standard or modify or revoke an existing standard.

### Recordkeeping requirements

Most employers covered by the Act are required to maintain in each establishment records of recordable occupational injuries and illnesses. Regulations pertaining to recording and reporting injuries and illnesses are codified in Title 29, C.F.R., Chapter XVII, Part 1904. Such records consist of:

- A log and summary of occupational injuries and illnesses, OSHA Form 200
- A supplementary record of each occupational injury or illness, OSHA Form 101
- An annual summary of occupational injuries and illnesses—OSHA Form 200 to be used in preparing the summary. The

annual summary must be posted by February 1 of the following year and remain posted until March 1. (See Figure 2-1.)

For details concerning recording and reporting occupational injuries and illnesses, see Chapter 6, Accident Records and Incidence Rates.

OSHA Forms 200 and 101 are available at all Bureau of Labor Statistics (BLS) regional offices and OSHA area and regional offices. If your state has an OSHA approved plan, be sure to check for any additional recordkeeping requirements.

In an effort to relieve small businesses from recordkeeping requirements, OSHA has ruled that an employer who had no more than ten employees at any time during the calendar year immediately preceding the current calendar year, need not comply with the recordkeeping requirements. However, if an employer, regardless of size, has been notified in writing by the Bureau of Labor Statistics that he has been selected to participate in the statistical survey of occupational injuries and illnesses, then he will be required to maintain the log and summary and to make reports for the period of time specified in the notice. Further, no employer is relieved of his obligation to report to the nearest OSHA area office any fatalities or multiple hospitalization accidents.

In 1987, employers who engaged in retail trade, finance, insurance, real estate and services were exempted from most of the recordkeeping requirements.

### Reporting requirements

Within 48 hours after the occurrence of an accident which is fatal to one or more employees or which results in the hospitalization of five or more employees, the employer must report the accident either orally or in writing to the nearest area director of OSHA. In states with approved state plans, the report must be made to the state agency which has the enforcement responsibilities for occupational safety and health. If an oral report is made, it shall always be followed with a confirming letter written the same day. The report must relate the circumstances of the accident, the number of fatalities, and the extent of any injuries.

### Variances from standards

There will be some occasions when, for various reasons, standards cannot be met. In other cases, the protection already afforded by an employer to employees is equal to or superior to the protection that would be granted if the standard were followed strictly to the letter. The Act provides an avenue of relief from these situations by empowering OSHA to grant variances from the standards, providing the granting of such variances would not degrade the purpose of the Act. The detailed "Rules of Practice for Variances, Limitations, Variations, Tolerances, and Exemptions" are codified in Title 29, C.F.R., Chapter XVII, Part 1905, 11(b).

There are two types of variances—temporary and permanent. An employer may apply for an order granting a temporary variance provided he establishes that (1) he cannot comply with the applicable standard because of unavailability of personnel or equipment or time to construct or alter facilities; (2) he is taking all available steps to protect his employees against exposure covered by the standard; and (3) his program will effect compliance with the standard as soon as possible.

Employer applications for an order for a temporary variance must contain at least the following:

1. Name and address of the applicant
2. Address(es) of the place(s) of employment involved
3. Identification of the standard from which the applicant seeks a variance
4. A representation by the applicant that he is unable to comply with the standard and a detailed statement of reasons therefor
5. A statement of the steps the applicant has taken and will take, with dates, to protect employees against the hazard covered by the standard
6. A statement of when the applicant expects to be able to comply with the standard and what steps he has taken, with dates, to come into compliance with the standard
7. A certification that he has informed his employees of the application. A description of how employees have been informed is to be included in the certification. Information to employees must also inform them of their right to petition for a hearing

An employer may also apply for a permanent variance from a standard. A variance order can be granted if OSHA determines that an employer has demonstrated by a preponderance of evidence that he will provide a place of employment as safe and healthful as that which would prevail if he complied with the standard.

Employer applications for an order for a permanent variance must contain at least the following:

1. Name and address of the applicant
2. Address(es) of the place(s) of employment involved
3. A description of the countermeasures used or proposed to be used by the applicant
4. A statement showing how such countermeasures would provide a place of employment which is as safe and healthful as that required by the standard for which the variance is sought
5. Certification that he has informed his employees of the application
6. Any request for a hearing
7. A description of how employees were informed of the application and of their right to petition for a hearing

An employer may request an interim order permitting either kind of variance until his formal application can be acted upon. Again, the request for an interim order must contain statements of fact or arguments why such interim order should be granted. If a request for an interim order is denied, the applicant will be notified promptly and informed of the reasons for the decision. If the order is granted, all concerned parties will be informed and the terms of the order will be published in the *Federal Register*. In such cases, the employer must inform the affected employees regarding the interim order in the same manner used to inform them of the variance application.

Upon filing an application for a variance, OSHA will publish a notice of such filing in the *Federal Register* and invite written data, views, and arguments regarding the application. Those affected by the petition may request a hearing. After review of all the facts, including those presented during the hearing, OSHA publishes its decision regarding the application in the *Federal Register*.

Beginning in the early 1980s, OSHA authorized its regional administrators to make "interpretations" of standards that had the effect of becoming variances for individual employers. Such interpretations had no effect on other employers and reflected

**Figure 2-1.** OSHA poster (OSHA 2203), "Safety and Health Protection on the Job," must be posted conspicuously at every plant, job site, or other establishment. At left is the annual Log and Summary of Occupational Injuries and Illnesses, OSHA Form 200, that must be posted by February 1 of the following year and remain in place until March 1.

specific conditions at a particular workplace. OSHA has also been issuing "clarifications" of standards for employers asking for deviation from standards. While granting less than 10 percent of the requests for variances that employers submitted since enforcement began, about eight times as many clarifications have been issued. (See Figure 2-1.)

### Workplace inspection

Prior to the U.S. Supreme Court's decision on the controversial Barlow case—Marshall v. Barlow's Inc., 436 U.S. 307 (1978)—the Department of Labor's Compliance Safety and Health Officers (CSHO) could enter, at any reasonable time and without delay, any establishment covered by the OSHAct to inspect the premises and all its facilities. (See Title 29, C.F.R., Chapter XVII, Part 1903.) However, the Barlow decision requires the OSHA compliance officer to present a search warrant if the employer demands that he do so. OSHA's entitlement to a warrant does not depend on demonstrating probable cause to believe that conditions on the premises violate the OSHA regulations, but merely that reasonable legislative or administrative standards for conducting an inspection are satisfied. As a general rule, it is not advisable for the employer to refuse entry to a CSHO without a search warrant, since such action normally would result in only delaying an inspection. In the large majority of such cases, OSHA will seek and obtain a search warrant, which may take anywhere from 48 hours to a month, or even more. Since the Barlow decision, OSHA has had to obtain a warrant in about 2.5 percent of the inspections.

The OSHAct authorizes an employer representative as well as an authorized employee representative to accompany the CSHO during the official inspection of the premises and all its facilities. Employee representatives also have the right to participate in both the opening and closing conferences.

Usually the authorized employee representative is the union steward or the chairman of the employee safety committee. Occasionally there may be no authorized employee representative, especially in those establishments that are nonunion shops. In the absence of an employee representative, the CSHO will confer with employees whom he picks at random.

An employer should not refuse to compensate employees for the time spent participating in an inspection tour and for related activities such as participating in the opening and closing conferences. A 1977 amendment to OSHA regulations provided that a denial of pay for the time spent assisting compliance personnel amounts to discrimination, but this may be revoked.

**Inspection priorities.** OSHA has established priorities for assignment of manpower and resources. The priorities are as follows:

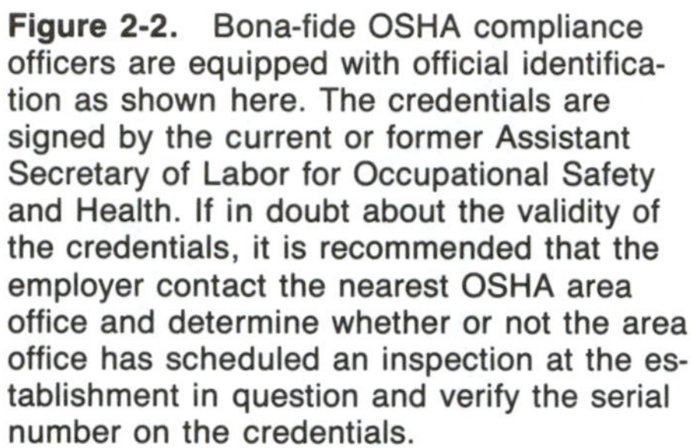

**Figure 2-2.** Bona-fide OSHA compliance officers are equipped with official identification as shown here. The credentials are signed by the current or former Assistant Secretary of Labor for Occupational Safety and Health. If in doubt about the validity of the credentials, it is recommended that the employer contact the nearest OSHA area office and determine whether or not the area office has scheduled an inspection at the establishment in question and verify the serial number on the credentials.

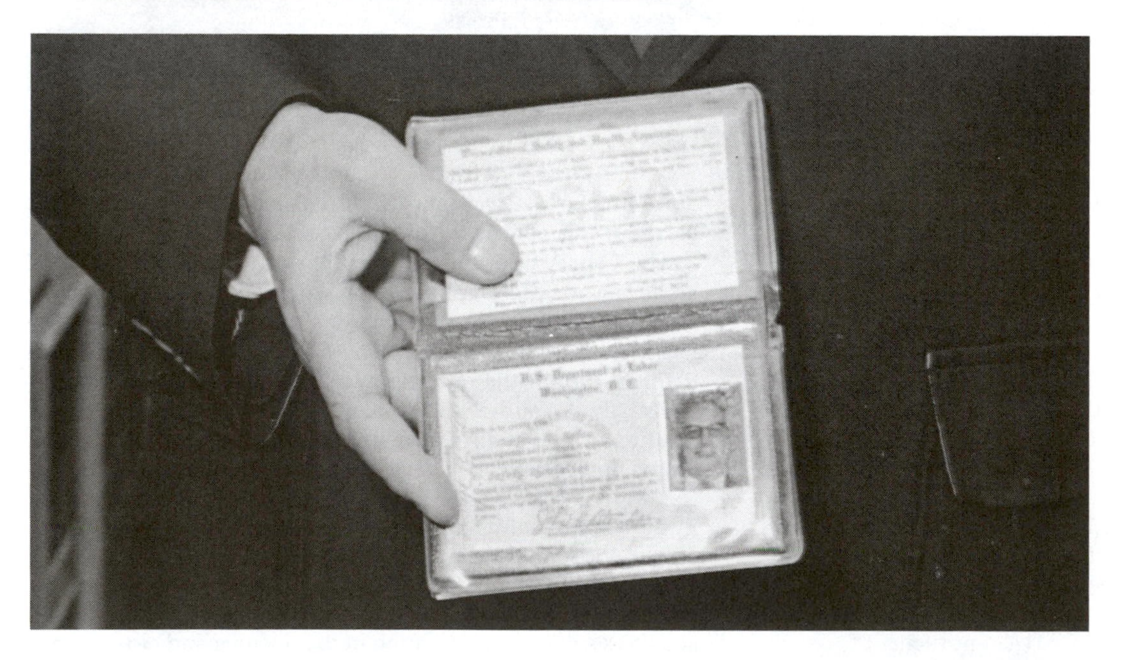

1. *Investigation of imminent dangers.* Allegations of an imminent danger situation will ordinarily trigger an inspection within 24 hours of notification.
2. *Catastrophic and fatal.* Accidents will be investigated if they include any one of the following:
   One or more fatality
   Five or more employees hospitalized for more than 24 hours
   Significant publicity
   Issuance of specific instructions for investigations in connection with a national office special program.
3. *Investigations of employee complaints.* Highest priority is

given those complaints that allege an imminent danger situation. Complaints alleging a "serious" situation are given high priority. If time and resources allow, the CSHO will normally inspect the entire workplace rather than limit the inspection to the condition alleged in the complaint.
4. *Programmed high-hazard inspections.* Industries are selected for inspection based on the death, injury, and illness incidence rates, employee exposure to toxic substances, etc.
5. *Reinspections.* Establishments cited for alleged serious violations normally are reinspected to determine whether the hazards have been abated.

**General inspection procedures.** The primary responsibility of the CSHO, who is under the supervision of the OSHA area director, is to conduct an effective inspection to determine if employers and employees are in compliance with the requirements of the standards, rules, and regulations promulgated under the OSHAct. OSHA inspections are almost always conducted without prior notice.

To enter an establishment, the CSHO will present his credentials to a guard, receptionist, or other person acting in such a capacity. Employers should always insist on seeing and checking the CSHO's credentials carefully before allowing the individual to enter their establishment for the purpose of an inspection (Figure 2-2). Anyone who tries to collect a penalty or promotes the sale of a product or service is not a CSHO.

The CSHO will usually ask to meet an appropriate employer representative. It is recommended that employers furnish written instructions to the security, receptionist, and other affected personnel regarding the right of entry, treatment, who should be notified, and to whom and where the CSHO should be directed so that undue delay can be avoided.

**Opening conference.** The CSHO will conduct a joint opening conference with employer and employee representatives. Where it is not practical to hold a joint conference, separate conferences are to be held for employer representatives. If there is no employee representative, then a joint conference is not necessary. In those instances where separate conferences are held, a written summary of each conference is to be made and the summary made available on request to employer and employee representatives.

Since the CSHO will want to talk with the safety personnel, such personnel should participate in the opening conference. The employer representative who accompanies the CSHO during the inspection should also participate in the opening conference.

At the opening conference, the CSHO will:

1. Inform the employer that the purpose of his visit is to make an investigation to ascertain if the establishment, procedures, operations, and equipment are in compliance with the requirements of the OSHAct.
2. Give the employer copies of the Act, standards, regulations, and promotional materials, as necessary.
3. Outline in general terms:
   The scope of the inspection
   The record he wants to review
   His obligation to confer with employees
   The physical inspection of the workplace
   The closing conference
4. If applicable, furnish a copy of the complaints(s).
5. Answer questions those in attendance might have.

In the opening conference, the employer representative should determine which areas of the establishment the CSHO wishes to inspect. If the inspection includes areas of the plant in which trade secrets are maintained, the employer representative should orally request *confidential treatment* of all information obtained from such areas and follow up with a trade secret letter to the CSHO requesting him to keep information identified in the letter strictly confidential by not discussing any of it or providing copies to any person not authorized by law to receive the information without prior written consent of the employer.

During the course of the opening conference, the CSHO may request to review company records. The CSHO is authorized to review only the records required to be maintained by the OSHAct, regulations and standards. In general, these records include the "Log and Summary of Occupational Injuries and Illnesses" (OSHA Form 200) and the "Supplemental Record of Occupational Injuries and Illnesses" (OSHA Form 101). Such records should be made readily available to the officer.

The CSHO may want to obtain information regarding the safety and health program that the employer now has in operation so that he can evaluate such a program. Naturally, a comprehensive safety and health program that shows evidence of effective performance in accident prevention will be impressive to all concerned.

The CSHO will also ascertain from the employer whether employees of another employer (for example, a contracting employer for maintenance or remodeling) are working in or on the establishment. If so, the CSHO will afford the authorized representative of those employees a reasonable opportunity to accompany him during the inspection of the workplaces where they are working.

During the conference, the CSHO will explain the employee representative's rights and ask for the authorized employee representative. Generally the employee representative will be an employee of the establishment inspected. However, if, in the judgment of the CSHO, good cause has been shown that accompaniment of a third party (such as an industrial hygienist or safety consultant) who is not an employee of the employer (but is, indeed, an authorized employee representative) is reasonably necessary to conduct an effective and thorough inspection, such a third party may accompany the CSHO during the inspection. The final decision will rest with the CSHO.

The employer is not permitted to designate the employee representative. Employee representatives may change as the inspection process moves from department to department. The CSHO may deny the right of accompaniment to any person whose conduct interferes with a full and orderly inspection. If there is no authorized employee representative, the CSHO will consult with a reasonable number of employees concerning matters of safety and health in the workplace during the course of the inspection.

**Inspection of facilities.** The CSHO will normally take the time necessary to inspect all of the operations in the establishment. The inspections have as their primary objective the enforcement of the occupational safety and health standards as well as the enforcement of other promulgated regulations such as the posting of the OSHA poster (Figure 2-1).

The CSHO will have the necessary instruments for checking certain items, such as noise levels, certain air contaminants and toxic substances, grounding, and the like. During the course of inspection, the CSHO will note any apparent violation of the standards and will normally record any apparent violation, including its location, and any comments that he has regarding the violation. He will do the same for any apparent violation of the general-duty clause. The notes will serve as the basis of information for the area director when issuing citations or proposed penalties. For these reasons, the employer representative should ascertain any apparent violations from the CSHO during the actual inspection of the facilities. The employer representative should make notes identical to the CSHO's during the actual inspection so that he will have precisely the same information that the CSHO has.

It should be noted that the CSHO is only required to record apparent violations and is not required to present a solution or method of correcting, minimizing, or eliminating the violation. OSHA, however, will respond to requests for technical information concerned with complying with given standards. In such cases, the employer is urged to contact the regional or area office.

If, during the course of an inspection, the CSHO receives a complaint from an employee regarding a condition which is alleged to be in violation of an applicable standard, the CSHO, even though the complaint is brought to him via an informal process, will normally inspect for the alleged violation.

In the course of his normal inspection, the CSHO may make some preliminary judgments with respect to environmental conditions affecting occupational health. In such cases, he will generally use direct-reading instruments. Should this occur, and if proper instrumentation is available, it would be prudent for the employer to have qualified personnel at the establishment make duplicate tests in the same area at the same time under the same conditions. In addition, the employer representative should again take careful notes on the CSHO's methods as well as the results. If the inspection indicates a need for further investigation by an industrial hygienist, the CSHO will notify the Area Director who may assign a qualified industrial hygienist to investigate further. If a laboratory analysis is required, samples are sent to OSHA's laboratory in Salt Lake City and the results will be reported back to the Area Director.

**Closing conference.** Upon completion of the inspection, the CSHO will hold a joint closing conference with employee representatives and representatives of the employer. If a joint conference is not possible, a separate conference will be held. Again, the employer's safety personnel should be present at the closing conference. It is at this time that the CSHO will advise the employer and employee representatives of all conditions and practices which may constitute a safety or health violation. He should also indicate the applicable section or sections of the standards that may have been violated.

The CSHO will normally advise that citations may be issued for alleged violations and that penalties may be proposed for each violation. Administratively, the authority for issuing citations and proposed penalties rests with the Area Director or his representative.

The employer will also be informed that the citations will fix a reasonable time for abatement of the violations alleged. The CSHO will attempt to obtain from the employer an estimate of the time he feels would be required to abate the alleged violation, and then he will take such estimate into consideration when recommending a time for abatement. Although the employer is not required to do so, it might be advantageous to give the officer copies of any correspondence or orders concerning equipment to achieve compliance since it may help to establish a reasonable abatement period and may reduce the proposed penalty by demonstrating good faith.

The CSHO should also explain the appeal procedures with respect to any citation or any notice of a proposed penalty.

**Informal post-inspection conferences.** Issues raised by inspections, citations, proposed penalties, or notice of intent to contest may be discussed at the request of an affected employer, employee, or employee representative at an informal conference held by the Assistant Regional Director. Whenever an informal conference is requested by either the employer or employee representatives, both parties shall be afforded the opportunity to participate fully.

**Follow-up inspections.** Follow-up inspections will always be made for those situations involving imminent danger or where citations have been issued for serious, repeated or willful violations. Follow-up inspections for all other cases will be conducted at the discretion of the Area Director.

The follow-up inspection is intended to be limited to verifying compliance of those conditions which were alleged to be in violation. The follow-up inspection is conducted with all of the usual formality of the original inspection, including the opening and closing conferences, and the walk-around rights of the employer and employee representative.

## Violations

In addition to the general-duty clause, the occupational safety and health standards promulgated under the OSHAct are used as a basis for determining alleged violations. There are four types of violations: imminent danger, serious, nonserious, and de minimis (very minor).

**Imminent danger.** The OSHAct defines imminent danger as "any condition or practice in any place of employment which is such that a danger exists which could reasonably be expected to cause death or serious physical harm immediately or before the imminence of such danger can be eliminated through the enforcement procedures otherwise provided by this Act." Therefore, for conditions or practices to constitute an imminent danger situation, it must be determined that there is a reasonable certainty that immediately or within a short period of time such conditions or practices could result in death or serious physical harm.

"Serious physical harm" includes the permanent loss or reduction in efficiency of a part of the body, or inhibition of a part of the body's internal system such that life is shortened or physical or mental efficiency is reduced.

Normally a health hazard would not constitute an imminent danger except in *extreme* situations, such as the presence of potentially lethal concentrations of airborne toxic substances that are an immediate threat to the life or health of employees.

If, during the course of inspection, the CSHO deems that the existing set of conditions appears to constitute an imminent danger situation, he will immediately advise the employer or his representative that such a danger exists and will attempt to have the danger corrected immediately through voluntary compliance. Further, if any employees appear to be in imminent danger, they will be informed of the danger and the employer will be requested to remove them from the area of imminent danger.

An employer will be deemed to have abated the imminent danger if he eliminates the imminence of the danger by (1) removing employees from the danger area and assuring the CSHO that employees will not return until the hazardous condition has been eliminated or (2) eliminating the conditions or practices which constitute the imminent danger. Normally abatement is achieved in these two ways. When the employer voluntarily eliminates the danger, no imminent danger procedure is instituted and no Notice of Imminent Danger is issued. However, citations and proposed penalties are, nonetheless, issued.

If the employer refuses to voluntarily abate the alleged imminent danger, the CSHO will inform the affected employees of the danger involved and will inform the employer as well as the affected employees that he will recommend to the Area Director a civil action (in the form of a court order) for appropriate relief (e.g., to shut down the operation). In such cases the CSHO will personally post the imminent danger notice at or near the area in which the exposed employees are working. The federal CSHO has no authority to order the closing down of an operation or to direct employees to leave the area of imminent danger or the workplace.

In such cases, the Area Director normally will request the Regional Solicitor to obtain an injunction permanently restraining the employer's practices. In order to protect the employees until the hearing on the injunction, a temporary restraining order, issued without notice to the employer and effective for up to five days, may be issued. The Act vests jurisdiction in the U.S. district courts to restrain any condition or work practice in imminent danger situations.

**Serious violation.** A serious violation is one where a substantial *probability* of death or serious physical harm could result, and that the employer knew, or should have known, of the hazard.

OSHA's *Field Operations Manual* (Chapter VIII) sets forth four steps for the CSHO to follow to determine whether a violation is serious or other than serious.

1. Determine the type of accident or health hazards that the standard is designed to prevent. (Example: a guard designed to prevent an operator's hand from coming in contact with the circular blade of a table saw.)
2. Determine what type of injury could be reasonably expected to result from the hazard. (In the example stated above, it could be reasonably predicted that contact between the operator's hand and the saw blade could result in the amputation of a finger or fingers, laceration of fingers or of an entire hand, or amputation of a hand.)
3. Determine whether the injury noted in Step 2 is likely to cause death or serious physical injury which involves the loss of use of part of the body or substantial reduction of efficiency of a part of the body. (In the example stated, a deep laceration of the fingers or hand could result in substantial reduction in the efficiency of that limb.)
4. Determine that the employer knew of the violative condition which means that the employer had actual knowledge or could have known of the hazardous conditions if he had exercised reasonable diligence.

It is obvious that the CSHO must make an evaluation that death or serious physical harm could result from a condition which is an alleged violation.

Note that the emphasis in deciding whether or not a condition represents a serious violation is based on the *seriousness* or *severity* of a potential injury that could arise out of the potential accident, rather than on the *probability* that the accident will occur as a result of the violation. In many cases, the decision in determining whether a violation is serious or not will require professional judgment.

The *Industrial Hygiene Technical Manual* is used as a guide for handling health inspections and citations. (See the chapter on Governmental Regulations, *Fundamentals of Industrial Hygiene,* published by the Council.)

**Other than serious violations.** An other than serious violation is one that has a direct relationship to job safety and health, but probably would not cause death or serious physical harm. For example, a violation of housekeeping standards that might result in a tripping accident would be classified as a nonserious violation since the probable consequence of such a condition would be strains or contusions which are not classified as serious physical harm.

**De minimis violations.** De minimis violations are those that have no immediate or direct relationship to safety or health. "De minimis" is short for the legal maxim, *De minimis non curat lex,* "The law does not concern itself with trifles."

**Special types.** There are two other special types of violations: a willful violation, and a repeated violation.

- A willful violation exists where evidence shows that:

1. The employer committed an intentional and knowing violation of the Act and knows that such action constitutes a violation.
2. Even though the employer was not consciously violating the Act, he was aware that a hazardous condition existed and made no reasonable effort to eliminate the condition.

- A repeated violation is where a second citation is issued for a violation of a given standard or the same condition which violates the general-duty clause. A repeated violation differs from a failure to abate in that repeated violations exist where the employer has abated an earlier violation, and upon later inspection, is found to have violated the same standard.

The first criminal actions under OSHA against managers personally occurred in the early 1980s. (In the mid-1970s, a corporate employer was fined because of the death of an employee.) A construction subcontractor spent two months in jail, received probation, and paid a fine after assaulting an inspector. A corporate safety director received a three-month jail term and a $10,000 fine for disabling a piece of equipment to prevent it being tested for safety hazards.

Criticisms of OSHA's laxness in exercising the right to seek criminal sanctions for violations of the law led in the mid-1980s to state and local prosecutors bringing indictments against employers and their managers under the criminal laws for murder and aggravated assault. In one case, the president, plant manager, and a foreman of a plant in Chicago where a worker died from exposure to cyanide were convicted of murder and sentenced to 40 years in prison plus fines. Their cases were on appeal at the time of preparation of this edition.

## Citations

When an investigation or inspection reveals a condition that is alleged to be in violation of the standards or general-duty clause, the employer may be issued a written citation which will describe the specific nature of the alleged violation, the standard allegedly violated and will fix a time for abatement. *Each citation,* or copy thereof, *must be prominently posted by the employer at or near the place where the alleged violation occurred.* All citations will be issued by the Area Director or his designee and will be sent to the employer by certified mail.

A "Citation for Serious Violation" will be prepared to cover those violations which fall into the "serious category." This type of violation *must* be assessed a monetary penalty.

A citation is used for other than serious violations which *may* or *may not* carry a monetary penalty. A citation may be issued to the employer for employee actions which violate the safety and health standards.

A notice, in lieu of a citation, is issued for de minimis violations which have no direct relationship to safety and health. Unlike the citation, the employer is not required to post this notice.

If an inspection has been initiated as a result of an employee complaint, the employee or authorized employee representative may request an informal review of any decision not to issue a citation.

Employees may not contest citations, amendments to citations, penalties, or lack of penalties. They may contest the time for abatement of a hazardous condition specified in a citation. They also may contest an employer's Petition for Modification of Abatement (PMA), which requests an extension of the abatement period. Employees must contest the PMA within 10 working days of its posting or within 10 working days after an authorized employee representative has received a copy.

Within 15 working days of the employer's receipt of the citation, an employee may submit a written objection to OSHA. The OSHA area director forwards the objection to the Occupational Safety and Health Review Commission, which operates independently of OSHA.

Employees may request an informal conference with OSHA to discuss any issues raised by an inspection, citation, notice of proposed penalty, or employer's notice of intention to contest.

**Petition for modification of abatement.** Upon receiving a citation, the employer must correct the cited hazard by the prescribed date. However, factors beyond the employer's reasonable control may prevent the completion of corrections by that date. In such a situation, the employer who has made a good faith effort to comply may file for a Petition for Modification of Abatement date.

The written petition should specify all steps taken to achieve compliance, the additional time needed to achieve complete compliance, the reasons such additional time is needed, all temporary steps being taken to safeguard employees against the cited hazard during the intervening period, that a copy of the PMA was posted in a conspicuous place or near each place where a violation occurred, and that the employee representative (if there is one) received a copy of the petition.

### Penalties

In proposing civil penalties for citations, a distinction is made between serious violations and all other violations. There is no requirement that a penalty be proposed when a violation is not a serious one, but a penalty *must* be proposed for a serious violation. In either case the maximum penalty that may be proposed is $1,000. In case of willful or repeated violations, a civil penalty of up to $10,000 may be proposed. Criminal penalties may be imposed on any employer who, among other things, willfully violates a standard and that violation causes death to any employee. There are no penalties for de minimis violations.

Penalties may be proposed for an alleged violation even though the employer immediately abates or initiates steps to abate the alleged violation. However, actions to abate should be favorably considered when determining the amount of adjustment applied for "good faith."

The information that follows describes the system that OSHA uses to arrive at its proposed penalties. Since it is unlikely that most employers comply with all of the promulgated standards, the employer can use a similar strategy to establish his own priorities for voluntary compliance with the standards.

**Other than serious violations.** For other than serious violations, the penalty may range from 0 to $1,000 for each violation. An "unadjusted penalty" is based on the gravity of the alleged violation. Three factors are used to determine gravity and all factors require professional judgment.

1. The *severity* of injury or illness most likely to result. The severity factors are rated as follows:
   "A" —For conditions in which the injury would require first aid treatment or less, such as minor cuts, bruises or splinters
   "B" —For conditions in which the injury would require treatment by a doctor, such as sutures or setting broken bones in a finger
   "C" —For conditions in which the injury would require hospitalization for 24 hours or more.
2. The *probability* or likelihood that an injury or illness would result from the alleged violation. Consideration is given to the extent that such a condition has already resulted in injury or illness and the number of employees exposed to the substandard condition. The probability is rated as follows:
   "A" —If the likelihood is low
   "B" —If the likelihood is moderate
   "C" —If the likelihood is high
3. The *extent* to which the standard is violated. Here there are two factors involved—(1) standards pertaining to the workplace and (2) standards pertaining to employee procedures. The rating scheme for standards pertaining to the workplace is as follows:
   "A" —If any isolated violations are observed; that is, no more than 15 percent of the units covered by the standards are in violation.
   "B" —If from 15 to 50 percent of the affected units are in violation.
   "C" —If over 50 percent of such units are in violation.

With respect to standards pertaining to employee procedures, the rating system is as follows:
   "A" —If the violation occurs occasionally
   "B" —If the violation occurs frequently
   "C" —If the violation occurs regularly

It is obvious from these measurement schemes that an "A" rating is the least severe and that the "C" rating is the most severe. These ratings are averaged to determine the final rating. (An "X" rating is used when the employer clearly demonstrates a blatant disregard for the violation in question.) Each final letter rating has been assigned the following dollar ranges for the purpose of establishing an "unadjusted penalty."

"A" = none
"B" = $100 to $200
"C" = $201 to $500
"D" = $501 to $1,000

**Penalty reductions:** The "unadjusted penalty" may then be adjusted downward up to 50 percent depending on the employer's "good faith," size of business, and history of previous violations. A reduction of up to 20 percent may be given for "good faith." Evidence of good faith includes awareness of the OSHAct and any overt indications of the employer's desire to comply with the Act. A reduction of up to 10 percent may be given for business size measured in terms of the number of employees employed by the employer. A reduction of up to 20 percent may be given for a favorable history regarding previous violations. Normally, such history is based on the employer's past experience under the OSHAct. However, in certain cases, the employer's past history under other federal or applicable state safety and health statutes may be considered. The penalty adjustment factors are applied to the "unadjusted penalty" to determine the "adjusted penalty."

The "adjusted penalty" is further reduced by 50 percent (the abatement credit) if the employer corrects the violation within the abatement period specified in the citation. This reduction is made at the time the proposed penalty is calculated to determine the proposed penalty to be assessed for each violation.

**Serious violations.** The law requires that any employer who has received a citation for a serious violation must be assessed a proposed civil penalty of up to $1,000 for each such violation. Due to the severity of a serious violation, the amount of the proposed penalty for each cited serious violation is usually calculated from a base of $1,000 (the unadjusted penalty), which is the maximum penalty allowed. The unadjusted penalty may then be adjusted downward by up to 50 percent, depending on the employer's good faith, size of business, and history of violations, just as in the case of other than serious violations. However, the additional 50 percent "abatement credit" applicable to other than serious violations is *not* applicable to serious violations.

**Imminent danger.** Penalties may be proposed in cases of imminent danger even though the employer immediately eliminates the imminence of such danger or initiates steps to abate such danger. If the danger is abated, the situation can be reduced in gravity to the "serious" category, and some to the "nonserious" category—all dependent on what was done to remove the imminence and how much of the hazard is removed.

**Proposed penalties.** Once the proposed penalties have been calculated, each is listed in the "Citation and Notification of Penalty" (OSHA-2 Form), which is used to officially inform the employer of violations found during the inspection and penalties proposed for those violations. This is sent to the employer by certified mail. An information copy is sent to an employee representative or the employee organization.

**Notice of failure to correct.** The Act provides that any employer who fails to correct an uncontested violation within the abatement period may be assessed a proposed penalty of up to $1,000 for each day that the violation continues after the expiration of the abatement period. This penalty provision can be applied when a followup inspection discloses that the employer has not abated a violation for which a citation has been issued and the citation and proposed penalty have become final.

**Time for payment of penalties.** When a citation and/or proposed penalty is uncontested, the payment is due after the lapse of 15 working days following receipt of the citation. When a citation and/or penalty are contested, the payment (if any) is not due until the final order of the Occupational Safety and Health Review Commission or the appropriate Circuit Court of Appeals is issued.

States can adopt their own penalty structure, even exceeding that set up under OSHA. For example, California in 1983 increased the penalties for serious violations from $1,000 to $2,000 a day; willful or repeated serious violations, from $10,000 to $20,000; and failure to correct cited violations from $1,000 to $2,000 a day.

## CONTESTED CASES

An employer has the right to contest an OSHA action if he feels that such action is not justified. The employer may contest a citation, a proposed penalty, a notice of failure to correct a violation, the time allotted for abatement of an alleged violation, or any combination of these. The regulations concerning the rules of procedure for contested cases adopted by the Occupational Safety and Health Review Commission are codified in Title 29, C.F.R., Chapter XX, Part 2200.

An employee or authorized employee representative may contest only the time allotted for an abatement of an alleged violation.

Prior to going through the formality of initiating a contest, employers should request an informal hearing with the Area Director or the Assistant Regional Director. Many times such informal sessions will resolve the questions and the issues; therefore, the formal contested case proceedings can be avoided.

If the informal conference fails to resolve the dispute between OSHA and the employer and the latter elects to contest the case, it must be remembered that affected employees or the authorized employee representative are automatically deemed to be parties to the proceeding. In contesting an OSHA action, the employer must comply with the following which are applicable to the specific case:

1. Notify the Area Office which initiated action that he is contesting. *This must be done within 15 working days from receipt of OSHA's notice of proposed penalty;* it must be sent by certified mail. If the employer does not contest within 15 working days after receipt of the notice of proposed penalty, the citation and proposed assessment of penalties are deemed to be a final order of the Occupational Safety and Health Review Commission and are not subject to review by any court or agency and the alleged violation must be corrected within the abatement period specified in the citation.
2. If any of the employees working on the site of the alleged violation are union members, a copy of the notice of contest must be served upon their union.
3. If employees who work on the site are not represented by a union, a copy of the notice of contest must either be posted at a place where the employees will see it or be served upon them personally.

4. The notice of contest must also contain a listing of the names and addresses of those parties who have been personally served a notice and, if such notice is posted, the addresses of the place the notice was posted.
5. If the employees at the site of the alleged violation are not represented by a union and have not been personally served with a copy of the notice to contest, posted copies most specifically advise the unrepresented employees that they may be prohibited from asserting their status as parties to the case if they fail to properly identify themselves to the Commission or the Hearing Examiner prior to the commencement of the hearing or at the beginning of the hearing.
6. There is no specific form for the notice of contest. However, such notice should clearly identify what is being contested—the citation, the proposed penalty, the notice of failure to correct a violation, or the time allowed for abatement—for each alleged violation or combination of alleged violations.

If the employer contests an alleged violation in good faith, and not solely for delay or variance of penalties, the abatement period does not begin until the entry of the final order by the Review Commission.

When a notice of contest is received by an Area Director from an employer or from an employee or an authorized employee representative, he will file with the Review Commission the notice of contest and all contested citations, notice of proposed penalties, or notice of failure to abate.

Upon receipt of the notice of contest from the Area Director, the Commission will assign the case a docket number. Ultimately, an Adminstrative Law Judge (ALJ) will be assigned to the case and will conduct a hearing at a location reasonably convenient to those concerned. OSHA presents its case and is subject to cross examination by other parties. The party contesting then presents his case and is also subject to a cross examination by other parties. Affected employees or an authorized employee representative may participate in the hearings. The decision by the ALJ will be based *only* on what is in the record. Therefore, if statements are unchallenged, the statements will be assumed to be fact.

Upon completion of the hearings, the ALJ will submit the record and his report to the Review Commission. If no Commissioner orders a review of an ALJ's recommendation, such recommendation will stand as the Review Commission's decision. If any Commissioner orders a review of the case, the Commission itself must render a decision to affirm, modify, or vacate the judge's recommendation. The Commission's orders become final 15 days after issuance, unless stayed by a court order.

Any person adversely affected or aggrieved by an order of the Commission may obtain a review of such order in the U.S. Court of Appeals if sought within 60 days of the order's issuance.

## SMALL BUSINESS LOANS

The Act enables economic assistance for small businesses. It amends the Small Business Act to provide for financial assistance to small firms for changes that will be necessary to comply with the standards promulgated under the OSHAct or standards promulgated by a state under a state plan. Before approving any such financial assistance, the Small Business Administration (SBA) must first determine that the small firm is likely to suffer substantial economic injury without such assistance.

An employer can make an application for a loan under one of two procedures: (1) before he has been inspected in order to come into compliance, or (2) after he has been inspected to correct alleged violations.

When an employer has not been inspected and requests a loan to bring his establishment into compliance before it is inspected, he must submit to the SBA:

A statement of the conditions to be corrected
A reference to the OSHA standards that require correction
A statement of his financial condition that necessitates applying for a loan

The employer should submit this information to the nearest SBA field office along with any background material. The SBA will then refer the application to the appropriate OSHA Regional Office, Office of Technical Support. The OSHA Regional Office will review the application and advise SBA whether the employer is required to correct the described conditions in order to come into compliance and whether his proposed use of funds will accomplish the needed corrections. Direct contact with the applicant will be initiated by OSHA only after clearance with the SBA.

If the employer is making an application after an inspection to correct alleged violations, the procedure is the same as before inspection, except that the applicant also must furnish SBA a copy of the OSHA citation(s). SBA then refers the application to the OSHA Area Office that conducted the inspection. That office will notify SBA whether the proposed use of loan funds will adequately correct cited violations.

Forms for loan applications may be obtained from any SBA field office. In some instances, private lending institutions will be able to provide the form for SBA/bank participation loans.

## FEDERAL-STATE RELATIONSHIPS

The OSHAct encourages the states to assume the fullest responsibility for the administration and enforcement of their own occupational safety and health laws. Any state may assume responsibility for the development and enforcement of occupational safety and health standards relating to any occupational safety and health issue covered by a standard promulgated under the OSHAct. However, in order to assume this responsibility, such state must submit a state plan to OSHA for approval. If such a plan satisfies designated conditions and criteria, OSHA must approve the plan. The regulations pertaining to state plans for the development and enforcement of state standards are codified in *CFR*, Title 29, Chapter XVII, Part 1902.

The following states and possessions have approved plans as of early 1987: Alaska, Arizona, Hawaii, Iowa, Kentucky, Maryland, Minnesota, Tennessee, Utah, Virgin Islands, and Wyoming.

The basic criterion for approval of state plans is that the plan must be "at least as effective as" the federal program. There was no congressional intent to require the state programs to be a "mirror image" of the federal program. Congress believed rules for developing state plans should be flexible to allow consideration of local problems, conditions, and resources.

The Act provides for funding the implementation of the state program, up to half the costs.

A state plan must include any occupational safety and health "issue" (industrial, occupational, or hazard group) for which a corresponding federal standard has been promulgated. A state plan cannot be less stringent, but it may include subjects not covered in the federal standards. However, state plans that do not include those issues covered by the federal program, in effect, surrender such issues to OSHA. For example, a state plan may cover all industry except construction. If such is the case, the state surrenders its jurisdiction for safety and health programs in construction operations to OSHA and it is then OSHA's obligation to enforce the federal standards for those operations not covered by the state plan.

Following approval of a state plan, OSHA will continue to exercise its enforcement authority until it determines on the basis of actual operations that the state plan is indeed being satisfactorily carried out. If the implementation of the state plan is satisfactory during the first three years after the plan's approval, then the federal standards and federal enforcement of such standards under the OSHAct can become inapplicable with respect to issues covered under the plan. This means that for the interim period of dual jurisdiction, employers must comply with the state standards as well as the federal standards.

While the state agencies administering the state plan are vitally concerned with its success, this is not always the case with the members of the state legislature. The state legislature must appropriate not only an adequate budget, but in many cases must pass legislation that will ultimately enable the state agency to carry out all the functions incorporated in the state plan. Should the state agency responsible fail to fully implement the state plan and the state's performance falls short of the mark of being "at least as effective as" the federal program, OSHA has the right and the obligation to withdraw its approval of the state plan and once again reassume full jurisdiction in that state.

## MEDICAL ACCESS AND RIGHT-TO-KNOW

The 1980s saw two important standards come into effect: final rules for access to exposure and medical records, and hazard communication. Both are designed to provide information to employees about the hazardous conditions they are, or have been, exposed to, and to make available to them their own medical and exposure records and documented information about chemicals in use in the workplace.

**Medical access standard.** Employers must maintain records on the exposure employees have had to dangerous substances during their working time. These records, together with an employee's medical records, must be made available to the employee, or a designated representative, upon request. The standard embodies certain restrictions on the extent of the information which must be provided and the procedure for providing it to designated representatives of employees.

**Hazard Communication Standard.** Commonly called "right-to-know" standard, it imposes upon employers engaged in manufacturing an obligation to identify the hazardous chemicals used in the workplace, and to make available to employees information about the chemicals, including their dangers and ways of protecting against them. Part of the requirement is an ongoing education program on hazardous chemicals. It is controversial—some critics think it does not go far enough, others, that it goes too far. In early 1987, the standard was still being litigated.

The principal issues are the limitation of its coverage to the manufacturing sector, the limited number of chemicals it applies to, and the protection it gives to trade secrets.

Employer groups have generally supported the standard as a way to limit the variation in state and local right-to-know legislation that proliferated in the early 1980s. Over 20 states and local communities enacted statutes and ordinances, which in some instances do not go as far as the federal standard, but in other cases go much further. The right-to-know law of one state, New Jersey, has been subjected to court scrutiny. Preliminary rulings, which are being appealed in early 1987, support the state's right to legislate in areas not covered by the federal standard. The indication is that employers will increasingly face a requirement to comply with both federal and state/local laws, where the two do not conflict.

## WHAT DOES IT ALL MEAN?

Congressional action in the form of the OSHAct is only a limited step in achieving the full purpose of the Act. Getting it to work with reasonable efficiency is the second and more difficult task. Achieving the *purpose* will depend on the willingness and cooperation of all concerned—employees and organized labor as well as business and industry.

There is no doubt that the Act has given new visibility to the whole realm of occupational safety and health. Since there are many employee rights incorporated in the OSHAct, it has given the employees a significant part of the action related to occupational safety and health matters. It has moved the laggards from "little or no safety" to "some safety," but not to "optimum safety." It has raised occupational safety and health to a higher priority in business management. It has given new status and responsibility to the occupational safety and health professional. Management is now relying more heavily on the safety profession for advice. And, it has given a new status to nationally recognized consensus standards-producing organizations.

New impetus is given to the field of occupational health, a much more difficult discipline with which to work when compared to occupational safety. Much more needs to be done to determine what kind of exposures are indeed hazardous to humans and under what conditions. Further, much more needs to be done to determine what countermeasures are not only adequate, but also reasonable and feasible to eliminate or minimize exposures to occupational health hazards. There exists a great need for much more research and data in occupational health to achieve optimum occupational safety and health programming.

The OSHAct has encouraged greater training for professionals in occupational safety and health. New curricula and university programs leading to various degrees in safety and health have been inaugurated by several universities and more are yet to come.

The OSHAct added new impetus to the product safety discipline. Until the passage of the Consumer Product Safety Act, the OSHAct was the most significant piece of legislation affecting product safety ever passed by the Congress. Designers and manufacturers of equipment now used by industry have a moral (but not legal) obligation to design, deliver, and install such equipment in accordance with the applicable standards.

The OSHAct, as well as the Occupational Safety and Health Administration, is not without limitations. Mere compliance with the requirements of the Act will not achieve optimum safety and health in terms of cost, benefits, and human values. All concerned must recognize that occupational safety and health cannot be handed to the employer or to the employee by legislative enactment or administrative decree. At best, state or federal occupational safety and health standards can cover only those things that are enforceable—namely control over physical conditions and environment.

As a matter of hard reality, enforcement standards simply do not adequately relate to the human in the human-machine-environment system. Important elements of a complete safety program, such as (1) establishment of work procedures to limit risk, (2) supervisory training, (3) job instruction training for employees, (4) job safety analysis, and (5) human factors engineering, to name a few, have not, for the most part, been included in the standards promulgated under the OSHAct—nor do the standards relate to employee attitudes, morale, or teamwork.

For the most part, the occupational safety and health standards promulgated under the OSHAct are minimal criteria and represent a floor rather than a goal to achieve. Thus, to rely on mere compliance with the occupational safety and health standards is to invite disaster since the residual risk after compliance remains unacceptable. Effective accident prevention and control of occupational health hazards must go beyond the OSHAct.

A violation of a standard is only symptomatic of something wrong with the management safety system. Only complete occupational safety and health programming as described elsewhere in this Manual will achieve a level of risk that is acceptable to employers as well as employees. The *real objective* and the purpose of the OSHAct is better occupational safety and health performance and not more compliance with a promulgated set of standards.

## ENVIRONMENTAL IMPACT

Managers who are responsible for occupational health and safety are increasingly affected by developments in environmental law. In some cases, responsibility for compliance with environmental laws and regulations rests on the manager responsible for occupational health and safety, or on a member of the same department. It is also not possible now to clearly draw a line between health and safety in the plant and the health and safety of the surrounding community. Further, environmental laws and regulations contain provisions which relate to the protection of people coming into contact with dangerous waste products and other hazardous substances, whether workers or members of the public.

The two pieces of legislation of greatest significance to those responsible for occupational health and safety are the Resource Conservation and Recovery Act (RCRA), and the Comprehensive Environmental Response, Compensation and Liability Act (CERCLA), commonly known as Superfund. The 1986 amendments to Superfund, which set up procedures for reporting and controlling those hazardous substances with an environmental impact, are particularly important. These environmental laws impose requirements for the use of protective equipment and procedures, and extensive reporting to federal, state and local government agencies. They also affect the design and operation of many manufacturing processes and waste disposal systems.

As a result, occupational health and safety and environmental-law compliance have become inextricably interrelated—and will continue to be so.

# PART II
# The Federal Mine Safety and Health Act

On November 9, 1977, President Carter signed into law the Federal Mine Safety and Health Act of 1977, Public Law 95-164. The Act became effective March 9, 1978.

The Federal Mine Safety and Health Act of 1977 (subsequently referred to as the Mine Act) is intended to ensure, so far as possible, safe and healthful working conditions for miners. It is applicable to operators of all types of mines, both coal and noncoal and both surface and underground. The Mine Act states that mine operators are responsible for the prevention of conditions or practices unsafe and unhealthful in mines, which endanger the safety and health of miners.

Mine operators are required to comply with the safety and health standards promulgated and enforced by the Mine Safety and Health Administration (MSHA), an agency within the Department of Labor. Like OSHA, MSHA may issue citations and propose penalties for violations. Unlike the OSHAct, miners (employees) are subject to government sanctions for violations of standards relating to smoking. Similarly, employers and other supervisory personnel may be personally liable for civil penalties or may be prosecuted criminally.

## LEGISLATIVE HISTORY

Historically, the Bureau of Mines within the Department of the Interior administered the mine safety and health laws. Before the Congress passed the Mine Act, mine operators were governed by two separate laws, the Federal Coal Mine Health and Safety Act of 1969 and the Federal Metal and Nonmetallic Mine Safety Act of 1966.

Because the Bureau of Mines was also charged with promoting mine production, critics charged that this responsibility produced an inherent conflict of interest with respect to enforcement of safety and health laws. The establishment of the Mine Enforcement Safety Administration (MESA) in 1973 within the Interior Department failed to reduce the criticism. Congress looked for alternative solutions, including the transfer of mine safety and health to the OSHAct. Congress finally settled on a solution by adopting the Federal Mine Safety and Health Act of 1977 which repealed the Federal Coal Mine Health and Safety Act of 1969 and the Federal Metal and Nonmetallic Mine Safety Act of 1966.

## ADMINISTRATION

The administration and enforcement of the Federal Mine Safety and Health Act are vested primarily with the Secretary of Labor and the Federal Mine Safety and Health Review Commission. The agency that administers the investigation and prosecution aspects of the enforcement process is the Mine Safety and Health Administration. The Federal Mine Safety and Health Review Commission, an independent agency created by the Act, reviews contested MSHA enforcement actions.

The Mine Act separates *health* research and *safety* research. Miner *health* research and standards development is the responsibility of the National Institute for Occupational Safety and Health (NIOSH), located in the Department of Health and

Human Services, in cooperation with MSHA. The Department of the Interior is responsible for mine *safety* research in cooperation with MSHA, and the Department of Labor is responsible for mine inspector training.

### Mine Safety and Health Administration

The Mine Safety and Health Administration, located within the Department of Labor, administers and enforces the Mine Act. MSHA is headed by an Assistant Secretary of Labor for Mine Safety and Health and is appointed by the President with the advice and consent of the Senate. The Assistant Secretary acts on behalf of the Secretary of Labor. For the purposes of this chapter, MSHA is also synonymous with the term Secretary or Assistant Secretary of Labor.

MSHA is authorized to adopt procedural rules and regulations to carry out the provisions of the Act. The agency also has the responsibility and authority to perform the following:

Promulgate, revoke or modify safety and health standards
Conduct mine safety and health inspections
Issue citations and propose penalties for violations
Issue orders for miners to be withdrawn from all or part of the mine
Grant variances
Seek judicial enforcement of its orders

Assisting the Assistant Secretary in carrying out the provisions of the Act is, among others, (1) an Administrator for Coal Mine Safety and Health and (2) an Administrator for Metal and Nonmetal Mine Safety and Health. Each administrator is responsible for a Division of Safety and a Division of Health.

### Federal Mine Safety and Health Review Commission

The five-member Federal Mine Safety and Health Review Commission serves as the administrative adjudication body. The Review Commission is completely independent from the Department of Labor. The commission has the authority to assess all civil penalties provided in the act. It reviews contested citations, notices of proposed penalties, withdrawal orders, and employee discrimination complaints. The Commission is appointed by the President for six-year terms with the advice and consent of the Senate. The first commissioners took office for staggered terms of two, four and six years.

The Commission appoints Administrative Law Judges (ALJs) to conduct hearings on behalf of the Commission. The decision of an ALJ becomes a final decision of the Commission 40 days after its issuance unless the Commission directs a review.

### National Institute for Occupational Safety and Health

The functions carried out under the Mine Act by the National Institute for Occupational Safety and Health (NIOSH), Department of Health and Human Services (DHHS), include the following:

Miner health research
Recommending standards to MSHA for adoption
Conducting health hazard evaluations at a mine upon request

To carry out its responsibilities, NIOSH is given authority to enter workplaces for the purpose of gathering information for research and for making health hazard evaluations. NIOSH also has the authority to provide medical examinations for miners at government expense for research purposes, to develop recordkeeping regulations relating to toxic exposure, and to require mine operators to make additional reports from time to time.

NIOSH also has the responsibility for reviewing toxic materials or harmful physical agents which are used or found in mines and to determine whether such substances are potentially toxic at the concentrations found. Further, NIOSH must thereafter review the toxicity of new substances brought to its attention. NIOSH is also required to submit criteria documents on toxic substances to assist MSHA in setting its standards.

### Department of the Interior shares duties

The Mine Act assigns the responsibility for mine safety research to the Department of the Interior and the training of mine inspectors, operators, and miners to the Department of Labor. Mine safety and health inspectors and technical support personnel of MSHA are trained by the DOL's National Mine Health and Safety Academy located in Beckley, West Virginia. The DOL is also authorized to conduct education and training programs for operators and miners in safety and health matters.

## MAJOR PROVISIONS OF THE MINE ACT

### Coverage

The Mine Act covers all mines that affect commerce. The Act defines "mines" as all underground or surface areas from which minerals are extracted and all surface facilities used in preparing or processing the minerals. Structures, equipment, and facilities including roads, dams, impoundments, and tailing ponds used in connection with mining and milling activities are also included.

OSHA and MSHA established an interagency agreement which, among other things, delineates certain areas of authority and provides for coordination between OSHA and MSHA in all areas of mutual interest. (This was published at 44 F.R. 22827, April 17, 1979.) In case of jurisdictional disputes between OSHA/MSHA, the Secretary of Labor is authorized to assign enforcement responsibilities to one of the agencies.

### Advisory committees

The Act requires the Secretary of the Interior to appoint an Advisory Committee on Mine Safety Research.

The Secretary of HHS is required to appoint an Advisory Committee on Mine Health Research.

The Secretary of Labor or the Secretary of HHS may appoint other advisory committees as deemed appropriate to advise in carrying out the provisions of the Act.

### Miners' rights

The Act affords miners (employees) a number of rights, including the following:

- Miners may request in writing an inspection if they believe a violation of a standard or an imminent danger situation exists in the mine. Similarly, written notification of alleged violations or imminent danger situations may be given to an inspector before or during an inspection.
- An authorized representative of miners must be given the opportunity to accompany the inspector during the inspection process. Also, miners have the right to participate in post-inspection conferences held by the mine inspector at the mine.
- At least one representative of the miners who accompanies the inspector during the inspection must be paid at his regular rate of pay for the time spent accompanying the inspector.

- Miners are entitled to observe monitoring and examine monitoring records when the standards require monitoring exposure to toxic materials or harmful physical agents.
- Miners, including former miners, must be provided access to records relating to their own exposures.
- Operators must notify miners who are exposed to toxic substances in concentrations which exceed prescribed limits of exposure. Further, those miners must be informed of the corrective action being taken.
- Miners given new work assignments for medical reasons because of exposure to hazardous substances must be paid at their regular rate if the related standard so provides.
- Miners who are not working because of a withdrawal order are entitled to be compensated subject to certain limits.
- Miners or their authorized representatives may contest the issuance, modification, or termination of any order issued by MSHA or the time period set for abatement.
- Miners adversely affected or aggrieved by an order of the Review Commission may obtain judicial review.
- Miners may file a complaint with the Review Commission concerning compensation for not working arising out of a withdrawal order or for acts of employee discrimination.
- Miners, through their authorized representative, may petition for a variance from mine safety standards.
- MSHA is required to send to the miners' authorized representative copies of proposed safety or health standards. In addition, the mine operator must provide a copy of such standards on its office bulletin board.
- To keep miners informed, mine operators are required to post copies of orders, citations, notices, and decisions issued by MSHA or the Review Commission.
- Miners are entitled to receive training for their specific jobs and must be given refresher training annually and are entitled to normal compensation while being trained. When a miner leaves the operator's employ, he is entitled to copies of his training certificates.
- Operators may not discriminate against miners or representatives of miners.
- Miners suffering from black lung disease are entitled to extensive black lung benefits.

### Duties

Mine operators are required to comply with the safety and health standards and other rules promulgated under the Act and are subject to sanctions for failing to comply. Similarly, every miner is required to comply with the safety and health standards promulgated under the Act. However, no sanctions are imposed against miners except for willful violation of safety standards relating to smoking or to the carrying of smoking materials, matches, or lighters.

### Miner training

Mine operators are required to have a safety and health training program approved by MSHA which provides the following:

- At least 40 hours of instruction for new underground miners. The training must include the statutory rights of miners and their representatives under the Act, use of the self-rescue device and use of respiratory devices, hazard recognition, escapeways, walk-around training, emergency procedures, basic ventilation, basic roof control, electrical hazards, first aid, and the health and safety aspects of the task assignment.
- Twenty-four hours of instruction for new surface miners. The training must include all of the items for underground miners just listed, except escapeways, basic ventilation, and basic roof control, none of which are essential to surface mining.
- At least 8 hours of refresher training for all miners on an annual basis.

The Mine Act requires that the training must be conducted during normal working hours and that the miners must be paid at their normal rate during the training period. Regulations concerning training and retraining of miners are codified at Title 30, C.F.R., Part 48. (See Chapter 9.)

## MINE SAFETY AND HEALTH STANDARDS

The Mine Act authorizes MSHA to promulgate, modify, or revoke mine safety and health standards.

To get the initial set of standards in place without delay under the Mine Act, the safety and health standards under the Coal Mine Health and Safety Act of 1969 were adopted under the Mine Act. These standards are codified in Title 30 C.F.R., Parts 70, 71, 74, 75, 77, and 90.

Similarly, the Mine Act adopted the mandatory standards that prevailed under the Metal and Nonmetallic Mine Safety Act of 1966. Later, many of the advisory standards were adopted as mandatory standards under the Mine Act. All of the noncoal standards are codified in Title 30 C.F.R., Parts 55, 56, and 57. (A list of all of the standards promulgated under authority of the Mine Act are listed in References, at the end of this chapter.)

If MSHA should determine that a standard is needed, it may propose a standard or seek assistance from an advisory committee. The proposed standard must be published in the *Federal Register* and a time period of at least 30 days must be established for public comment. MSHA may hold public hearings if objections are made to a proposed standard. Upon adoption by MSHA, the standard must be published in the *Federal Register.* The new standard becomes effective upon publication or at a date specified.

### Judicial review

Any person adversely affected by any standard issued by MSHA has the right to challenge the validity of the standard by petitioning in the U.S. Court of Appeals within 60 days after promulgation of the standard. The filing of such a petition does not stay enforcement of the standard, but the Court may order a stay before conducting a hearing on the petition. Objections that were not raised during rulemaking will not be considered by the Court, unless good cause is shown for the failure to have raised an objection.

### Input from the private sector

Mine safety and health standards, promulgated by MSHA, can never cover every conceivable hazardous condition that could exist in mines. Nevertheless, new standards, modification or revocation of existing standards are of significant interest to mine operators and miners alike. Mining operator organizations, miner organizations as well as individuals should express their views in the rulemaking process by responding to MSHA's proposed standards since it is within the private sector that most of the expertise and the technical competence lies. To do less means that mining operators and miners are willing to let the standards development process be done by MSHA.

### Emergency temporary standards

MSHA has the authority to publish emergency temporary standards if it deems that immediate action must be taken to protect miners "exposed to grave danger" from toxic substances or physically harmful agents. The emergency temporary standard is effective immediately upon publication in the *Federal Register* and remains in effect until superseded by a permanent standard promulgated under the normal rulemaking procedures. MSHA is required to promulgate a permanent standard within nine months after publication of an emergency temporary standard.

### Variances

Upon petition by an operator or a representative of miners, MSHA may modify the application of any mandatory *safety* standard. The Act does not allow for variances of *health* standards.

A petition may be granted if MSHA deems that an alternative method of compliance will achieve the same measure of protection for miners as the standard would provide or that the standard in question will result in less safety to miners.

A variance petition should be filed with the Assistant Secretary of Labor for Mine Safety and Health. If the mining operator submits a petition, a copy must be served on the miners' representative. Similarly, if the miners' representative petitions for a variance, a copy must be served on the mine operator. The petition must include the name and address of the petitioner and the mailing address and identification, and name or number of the affected mine. It must also identify the standard, describe the desired modification, and state the basis for the request.

MSHA will publish a notice of the petition in the *Federal Register.* The notice will contain information contained in the petition. Interested parties have 30 days to comment. MSHA then will conduct an investigation on the merits of the petition and the appropriate Administrator issues a proposed decision. The proposed decision becomes final 30 days after service unless a hearing request is filed within that time.

### Accident, injury, illness reporting

For the purpose of reporting accidents, injuries and illnesses under the Mine Act, the term "accident" includes:

- A fatality at a mine
- An injury which has the potential to cause death
- Entrapment for more than 30 minutes
- An unplanned ignition or explosion of gas or dust
- An unplanned fire not extinguished within 30 minutes of its discovery
- An unplanned ignition or explosion of a blasting agent or an explosive
- Roof fall in active workings where roof bolts are in use or a roof fall that impairs ventilation or impedes passage
- Coal or rock outbursts that disrupt mining activities for more than one hour
- Conditions requiring emergency action or evacuation
- Damage to hoisting equipment in a shaft or slope that endangers an individual or interferes with use of equipment for more than 30 minutes
- An event at the mine that causes a fatality or bodily injury to an individual not at the mine at the time of occurrence

"Occupational injury" means an injury that results in death, loss of consciousness, administration of medical treatment, temporary assignment to other duties, transfer to another job, or inability to perform all duties on any day after the injury.

"Occupational illness" is an illness or disease that may have resulted from work at a mine or for which a compensation award is made.

All mine operators are required to *immediately* report accidents (as defined earlier) to the nearest MSHA district or subdistrict office. Similarly, operators must investigate and submit to MSHA, upon request, an investigation report on accidents and occupational injuries. The investigation report must include:

- The date and hour of occurrence
- The date the investigation began
- The names of the individuals participating in the investigation
- A description of the site
- An explanation of the accident or injury
- The name, occupation, and experience of any miner involved
- If appropriate, a sketch, including dimensions
- A description of actions taken to prevent a similar occurrence
- Identification of the accident report submitted.

All mine operators must submit to MSHA within 10 days of occurrence a report of each accident, occupational injury or illness on Form No. 7000-1. A separate form is to be prepared for each miner affected.

Accident investigation reports as well as the injury/illness reports filed by means of Form 7000-1 must be retained for five years at the mine office closest to the mine in which the accident/injury/illness occurs.

### Inspection and investigation procedures

Inspections of a mine are conducted by MSHA to determine if an imminent danger exists in the mine and if the mine operator is complying with the safety and health standards and with citations, orders, or decisions issued. Mine inspectors from MSHA or representatives of NIOSH have the right to enter any mine to make an inspection or an investigation. Although a representative of NIOSH has the authority to enter mines, he has no enforcement authority. The Mine Act's provision for warrantless inspection has been held valid.

As in OSHA, the advance notice of inspection for the purpose of ascertaining compliance is prohibited. However, NIOSH may give advance notice of inspections for research and other purposes.

**Frequency.** All underground mines are to be inspected by MSHA in their entirety at least four times a year. Surface mines are to be inspected at least two times a year. Spot inspections must be conducted by MSHA based on the number of cubic feet of methane or other explosive gases liberated during a 24 hour period. The Act authorizes MSHA to develop guidelines for additional inspections based on other criteria.

**Miner complaints.** A miner's authorized representative, or a miner if there is no authorized representative, may request in writing an immediate inspection by MSHA, if there is reasonable grounds to believe that a violation of a standard or an imminent danger situation exists. MSHA will normally conduct a special inspection soon after receiving the complaint of an alleged violation or imminent danger situation. If MSHA determines that a violation does not exist, it must notify the complainant in writing.

Similarly, before or during an inspection, the miners' authorized representative, or a miner if there is no authorized representative, may notify the inspector in writing of any alleged violation or imminent danger situation believed to exist in the mine.

**Health hazard evaluations.** Upon written request of an operator or authorized representative of miners, NIOSH is authorized to enter a mine to determine whether any toxic substance, physical agent or equipment found or used in the mine is potentially hazardous. A copy of the evaluation will be submitted to both the operator and the miners' representative.

**The inspection procedure.** A MSHA inspector will normally begin the inspection at the mine office. He will inform the mine operator why he is there and will state what records he wants to examine. His review of the records will likely focus on the preshift or on-shift examination record. Such records help the inspector determine where he should concentrate his efforts during the inspection.

An operator's representative and a representative authorized by the miners must be given the opportunity to accompany the MSHA inspector during the inspection. Similarly, each must be given the opportunity to participate in the post-inspection conference. One miner representative (who is an employee of the operator) must be paid his regular wage for the time spent accompanying the inspector.

If the inspector observes a condition that he believes is a violation of the standards, a citation must be issued. If, in the opinion of the mine inspector, an imminent danger condition exists, then the inspector must issue a withdrawal order.

After completing the inspection, the inspector will hold a close-out conference with the representatives of the mine operator and the miners to discuss his findings. Occasionally, in the interests of all concerned, a separate closing conference may be held with the mine operator and another with the miners' representative.

### Withdrawal orders

MSHA has the authority under specified conditions, to issue orders to an operator requiring the operator to withdraw the miners from all or part of a mine. Miners idled by such an order are entitled to receive compensation at their regular rate of pay for specified periods of time. All miners in the affected area must be withdrawn except those necessary to eliminate the hazard, public officials whose duty requires their presence in the area, representatives of the miners qualified to make mine examinations, and consultants.

If an imminent danger is found to exist during an inspection, MSHA is required to order the withdrawal of all persons from the affected area except those referred to in Section 104 of the Act until the danger no longer exists. The order must describe the conditions or practices which cause and constitute the imminent danger and the area affected. The withdrawal order does not preclude issuance of a citation and proposed penalty.

There are other situations in which MSHA may issue a withdrawal order such as the following:

- If, during a follow-up inspection, MSHA finds that a mine operator has failed to abate a violation for which a citation has been issued and there is no valid reason to extend the abatement period, MSHA must issue a withdrawal order until the violation is abated.
- If a mine operator fails to abate a respirable dust violation for which a citation has been issued and the abatement period has expired, MSHA must either extend the abatement period or issue a withdrawal order.
- If two violations constituting "unwarrantable failures" to comply with the standards are found during the same inspection, or if the second unwarrantable violation is found within 90 days of the first, a withdrawal order must be issued. An unwarrantable failure violation (second within 90 days) is a situation of such a nature that the operator knew or should have known that a violation existed and failed to take corrective action.
- A miner may be ordered to be withdrawn from a mine if he has not received the safety training required by the Act. Miners withdrawn for this reason are protected by the Act from discharge or loss of pay.

Except for withdrawal orders issued for respirable dust violations and imminent danger situations, an operator or a miner may file a written request for a temporary stay of the order with the Review Commission. Also, temporary relief may be requested from any modification or termination of a withdrawal order. The Review Commission may grant a stay of a withdrawal order if the granting of such relief would not adversely affect the safety and health of miners.

Both operators and miners, or their representatives may contest an imminent danger withdrawal order by filing with the Review Commission an application for review of the order within 30 days of its receipt, or within 30 days of any modification or termination of such an order if the modification or termination is being contested.

### Citations

If a MSHA inspector or his supervisors believe that the mine operator is in violation of any standard, rule, order, or regulation promulgated under the Mine Act, a citation must be issued to the operator with "reasonable promptness." A citation may be issued at the site of the alleged violation. In any case, a citation for all alleged violations will be issued before the inspector leaves the mine property, unless mitigating circumstances exist.

Citations must be in writing, describe the nature of the violation, and include a reference to the provision of the Mine Act, standard, rule, regulation, or order allegedly violated. The citation, based on the inspector's opinion, will establish a reasonable time period for abatement.

The Act requires that the citations be posted on the mine's bulletin board. Copies are sent to the miners' representative, to the state agency charged with administering mine safety and health laws, and to those designated by the operator as having responsibility for safety and health in the mine.

Within 10 days after an operator receives a citation or an order for an alleged violation, the operator and/or representative of the miner has the right to request a health and safety conference with MSHA management. The conferencing process is designed to allow parties an opportunity to present evidence surrounding an alleged incident that may have influenced the action taken by the inspector. The conferencing officer considers information from these sources and has authority to modify, vacate, or leave as issued the citation.

This procedure enables mine operators and/or representatives of miners to resolve some issues prior to the penalty stage.

### Penalties

A civil penalty of not more than $10,000 must be assessed for each violation of the Act. Penalties up to $1,000 per day may be assessed for each day the operator fails to correct a violation for which a citation has been issued. A fine of $25,000 and/or one year imprisonment may be assessed for a willful violation of a standard. A penalty of up to $250 per occurrence may be assessed miners having willfully violated a standard relating to smoking or the carrying of smoking materials, matches or lighters.

After the alleged violation has been abated and after any health and safety conference has been conducted, MSHA will issue the proposed penalty. Non-significant and substantial (commonly termed "Non-S&S") violations that are abated in a timely fashion usually result in a $20 penalty. In determining the amount of the penalty for all violations, MSHA considers six criteria:

1. The operator's history of previous violations
2. The size of the operator's business
3. Whether or not the operator is negligent
4. The effect of the operator's ability to continue in business
5. The gravity of the violation
6. The demonstrated good faith in attempting to achieve rapid compliance after notification of the alleged violation.

Most of the proposed penalty assessments are assigned a range of penalty points based on each of the above criteria. The total points are then converted into a dollar penalty. In addition, violations which are quite grave, or involve negligence, are usually given a special penalty assessment.

## CONTESTED CASES

Operators and miners (or miners' representatives) have 30 calendar days from the receipt to contest a citation, a withdrawal order, or a proposed penalty. The notice of contest must be filed with the Assistant Secretary of Labor for Mine Safety and Health at the MSHA headquarters located at 4015 Wilson Boulevard, Arlington, Va. 22203, by registered or certified mail. A copy of the notice of contest must be sent to the representative of the miners.

If a mine operator fails to notify MSHA within the 30 day period and no notice is filed by any miner or miners' representative, then the citation and/or the proposed penalty is deemed a final order of the Federal Mine Safety and Health Review Commission and is not subject to review by any court or agency. However, it should be understood that the citation and the penalty have separate 30 day periods within which they may be contested. For instance, if a mine operator fails to contest the citation, all is not lost. When the proposed penalty is received at some later date, the mining operator has 30 days within which to contest that penalty. In addition, if the penalty is contested, the citation may be reopened for negotiation at the same time.

A citation may be contested before the operator receives a notice of proposed penalty, even if the alleged violation has been abated. A notice of contest consists of a statement of what is being contested and the relief sought. A copy of the order or citation being contested must accompany the notice of contest.

Upon receiving the notice of contest, MSHA immediately notifies the Review Commission and a docket number and an Administrative Law Judge (ALJ) are assigned to the case. The Review Commission will provide an opportunity for a hearing via the ALJ. The ALJ may hold an informal conference with all parties involved to clarify and settle the issues. If the issues are not resolved, then a formal hearing conducted by the ALJ will take place.

Mine operators, miners or representatives of miners, and applicants for employment may be parties to the Review Commission proceedings. Miners or their representatives may become parties by filing a written notice with the Executive Director of the Review Commission prior to the hearing.

The Review Commission's ALJ's are authorized, among other things, to administer oaths, issue subpoenas, receive evidence, take depositions, conduct hearings, hold settlement conferences, and render decisions. The decision will include findings of facts, conclusions of law, and an order. A copy of the decision will be issued to each of the parties involved and to each of the Commissioners. Any person aggrieved by the decision of the ALJ may, within 30 days of issuance of an order or decision, file a petition for a discretionary review by the Review Commission.

The Review Commission on its own motion and with the affirmative vote of two members may direct review of an ALJ's decision within 30 days of issuance only if the decision may be contrary to law or to Commission policy or if a novel question of policy has been presented.

Any person adversely affected by a decision of the Review Commission, including MSHA, may appeal to the U.S. Court of Appeals within 30 days of issuance. The court may affirm, modify or set aside the Commission's decision in whole or in part.

# Directory of Federal Agencies

## THE OCCUPATIONAL SAFETY AND HEALTH ADMINISTRATION

### National headquarters

*OSHA*

Occupational Safety and Health Administration, U.S. Department of Labor, Department of Labor Building, 200 Constitution Avenue, NW., Washington, D.C. 20210; (202)523-8148.

*NIOSH*

National Institute for Occupational Safety and Health, U.S. Department of Health and Human Services, 1600 Clifton Road NE., Atlanta, Ga. 30333; (404)329-3061.

*BLS*

Bureau of Labor Statistics, U.S. Department of Labor, 200 Constitution Avenue, NW., Washington, D.C. 20210; (202)523-1092.

*OSHRC*

Occupational Safety and Health Review Commission, 1825 K Street, NW., Washington, D.C. 20006; (202)634-7970.

### OSHA regional offices

*Region I* (Connecticut, Maine, Massachusetts, New Hampshire, Rhode Island, Vermont)
16–18 North Street, 1 Dock Square,
Boston, Mass. 02109; (617)223-6710.

*Region II* (New York, New Jersey, Puerto Rico, Virgin Islands, Canal Zone)
1515 Broadway (1 Astor Plaza),
New York, N.Y. 10036; (212)944-3432.

*Region III* (Delaware, District of Columbia, Maryland, Pennsylvania, Virginia, West Virginia)
Gateway Building, 3535 Market Street,
Philadelphia, Pa. 19104; (215)596-1201.

*Region IV* (Alabama, Florida, Georgia, Kentucky, Mississippi, North Carolina, South Carolina, Tennessee)
1375 Peachtree Street, NE.,
Atlanta, Ga. 30367; (404)881-3573.

*Region V* (Illinois, Indiana, Michigan, Minnesota, Ohio, Wisconsin)
J.C. Kluczynski Federal Building,
230 South Dearborn Street,
Chicago, Ill. 60604; (312)353-2220.

*Region VI* (Arkansas, Louisiana, New Mexico, Oklahoma, Texas)
555 Griffin Square Building,
Dallas, Texas 75202; (214)767-4731.

*Region VII* (Iowa, Kansas, Missouri, Nebraska)
Old Federal Office Building,
911 Walnut Street, Kansas City, Mo. 64106; (816)374-5861.

*Region VIII* (Colorado, Montana, North Dakota, South Dakota, Utah, Wyoming)
Federal Building, 1961 Stout Street,
Denver, Colo. 80294; (303)837-3061.

*Region IX* (Arizona, California, Hawaii, Nevada, Guam, American Samoa, Trust Territory of the Pacific Islands)
Federal Building, 450 Golden Gate Avenue,
San Francisco, Calif. 94102; (415)556-7260.

*Region X* (Alaska, Idaho, Oregon, Washington)
Federal Office Building, 909 First Avenue,
Seattle, Wash. 98174; (206)442-5930.

### THE MINE SAFETY AND HEALTH ADMINISTRATION

Mine Safety and Health Administration, U.S. Department of Labor, 4015 Wilson Boulevard, Arlington, Va. 22203; (202) 235-1284

National Institute for Occupational Safety and Health, U.S. Department of Health and Human Services, 1600 Clifton Blvd. N.E., Atlanta, Ga. 30333; (404) 329-3061

National Mine Health and Safety Academy, P.O. Box 1166, Beckley, W.Va. 25801; (304) 255-0451

Federal Mine Safety and Health Review Commission, Skyline 2, 10th Floor, 5203 Leesburg Pk., Falls Church, Va. 22041; (703) 756-6200

## Compilations of Regulations and Laws

The safety specialist and the industrial hygienist should be familiar with three U.S. Government publications:

The *Federal Register* (F.R.)
The *Code of Federal Regulations* (C.F.R.)
The *United States Code* (U.S.C.).

The first two are published by the Office of the Federal Register, National Archives and Records Service, General Services Administration. All three publications are available from the Superintendent of Documents, U.S. Government Printing Office, Washington, D.C. 20402.

### THE 'FEDERAL REGISTER'

The *Federal Register,* published daily Monday through Friday, provides a system for making publicly available regulations and legal notices issued by all federal agencies. In general, an agency will issue a regulation as a proposal in F.R., followed by a comment period, then will finally promulgate or finally adopt the regulation in F.R. Reference to material published in F.R. is usually in the format A F.R. B, whereby A is the volume number, F.R. indicates *Federal Register,* and B is the page number. For example 43 F.R. 58946, indicates volume 43, page 58946.

### THE 'CODE OF FEDERAL REGULATIONS'

The *Code of Federal Regulations,* published annually in paperback volumes, is a compilation of the general and permanent rules and regulations that have been previously released in F.R.

The C.F.R. is divided into 50 different titles, representing broad subject areas of federal regulations, for example, Title 29—"Labor;" Title 40—"Protection of Environment;" Title 49—"Transportation," etc. Each title is divided into chapters (usually bearing the name of the issuing agency), and then further divided into parts and subparts covering specific regulatory areas. Reference is usually in the format 40 C.F.R. 250.XX, meaning Title 40 C.F.R. Part 250 (Hazardous Waste Guidelines and Regulations), or 49 C.F.R. 172.XX, (Hazardous Materials Table and Hazardous Materials Communications Regulations). The "XX" refers to the number of the specific regulatory paragraph.

The *Code of Federal Regulations* is kept up to date by the individual issues of the *Federal Register.* These two publications must be used together to determine the latest version of any given rule or regulation.

### THE 'UNITED STATES CODE' OFFICIAL EDITION

Whereas the two previously described publications contain rules and regulations authorized by a law, the *U.S. Code* describes the actual law. The *U.S. Code* is the current official compilation by subject of the "public, general and permanent laws of the United States in force. . . . No new law is enacted and no law is repealed. It is prima facie the law. It is presumed to be the law. The presumption is rebuttable by production of prior unrepealed Acts of Congress at variance with the Code." (Preface to the *United States Code.)* In other words, the Code is an arrangement by subject of the federal legislation of a public and permanent nature, from 1789 to date in force today; it does not include repealed and expired acts.

**Editorial selection of statutes included.** Inclusion of legislation in the Code is under the supervision of a committee of the

House of Representatives. Code sections included in one edition may be omitted from the next, or may be changed from one title of the Code to another by editorial fiat. To avoid possible confusion, the date or supplement number of the Code edition cited should be given.

**Editorial notes in the Code** are printed with it but not technically a part of it. This material, set out in fine print beneath the respective Code sections to which applicable, has the purpose of aiding in the understanding of those sections.

**Text of Code sections.** Sections are copied from the original enactment, but often with changes of form, not of substance. Introductory words of the act, such as "Provided," or "That," at the beginning of the original statute section may be omitted; and the Code title and section numbers are substituted in the body of the text for the official title and sections of the act from which derived, when these are mentioned in the act. There are some other changes that are permitted. Statutory authority is cited in parentheses at the end of each Code section or group of sections. Such authority may be Congressional acts or joint resolutions, Presidential executive orders, or reorganization plans. The Code is well indexed.

**Tables of contents to Code titles.** Each of the fifty titles into which the Code is divided is preceded by a table of contents, consisting of a table of chapters in that title, by number and caption, and at the beginning of each chapter there is a similar expanded table for that chapter.

## REFERENCES


Bureau of National Affairs, Inc., 1231 25th Street, NW., Washington, D.C. 20037. *Occupational Safety and Health Reporter.*

Commerce Clearing House, Inc., 4025 West Peterson Avenue, Chicago, Ill. 60646. *Employment Safety and Health Guide.*

La Dou, Joseph, Ed. *Introduction to Occupational Health and Safety.* Chicago, Ill.: National Safety Council, 1986.

National Institute for Occupational Safety and Health, 5600 Fisher Lane, Rockville, Md. 20857.
"The Advisor" (newsletter).
"Occupational Safety and Health Directory."

National Safety Council, 444 North Michigan Avenue, Chicago, Ill. 60611.
*Fundamentals of Industrial Hygiene*
*Safety and Health* (magazine)
"OSHA Up-to-Date" (newsletter).

Price, Miles O., and Harry Bitner. *Effective Legal Research,* 4th Ed. Boston, Mass.: Little, Brown, and Co., 1979.

Rothstein, Mark. *Occupational Health and Safety Law,* 2nd Ed., West Publishing Co., Minneapolis, Minn., 1983 (with yearly updates).

Superintendent of Documents, U.S. Government Printing Office, Washington, D.C. 20402.
*Annual List of Toxic Substances.*
*Directory of Federal Agencies.*
*Federal Register.*
*Field Operations Manual.*
*Industrial Hygiene Technical Manual.*
Occupational Safety and Health Act of 1970 (P.L. 91-596).
*Occupational Safety and Health Regulations,* Title 29, *Code of Federal Regulations* (C.F.R.)

Part 11—Department of Labor, National Environmental Policy Act (NEPA) Compliance Procedures.
Part 1901—Procedures for State Agreements.
Part 1902—State Plans for the Development and Enforcement of State Standards.
Part 1903—Inspections, Citations and Proposed Penalties.
Part 1904—Recording and Reporting Occupational Injures and Illnesses.
Part 1905—Rules of Practice for Variances, Limitations, Variations, Tolerances and Exemptions.
Part 1906—Administration Witnesses and Documents in Private Litigation.
Part 1907—Accreditation of Testing Laboratories
Part 1908—On-Site Consultation Agreements.
Part 1910—Occupational Safety and Health Standards.
Part 1911—Rules of Procedure for Promulgating, Modifying, or Revoking Occupational Safety or Health Standards.
Part 1912—Advisory Committees on Standards.
Part 1912a—National Advisory Committee on Occupational Safety and Health.
Part 1913—OSHA Access to Employee Medical Records.
Part 1915—Occupational Safety and Health Standards for Shipyard Employment.
Part 1917—Marine Terminals.
Part 1918—Safety and Health Regulations for Longshoring.
Part 1919—Gear Certification
Part 1920—Procedure for Variations under the Longshoremen's and Harbor Workers' Compensation Act.
Part 1921—Rules of Practice in Enforcement Proceedings under Section 41 of the Longshoremen's and Harbor Workers' Compensation Act.
Part 1922—Investigational Hearings under Section 41 of the Longshoremen's and Harbor Workers' Compensation Act.
Part 1924—Safety Standards Applicable to Workshops and Rehabilitation Facilities Assisted by Grants.
Part 1925—Safety and Health Standards for Federal Service Contracts.
Part 1926—Safety and Health Regulations for Construction.
Part 1928—Occupational Safety and Health Standards for Agriculture.
Part 1949—Office of Training and Education, Occupational Safety and Health.
Part 1950—Development and Planning Grants for Occupational Safety and Health.
Part 1951—Procedures for 23(g) Grants to State Agencies.
Part 1952—Approved State Plans for Enforcement of State Standards.
Part 1953—Changes to State Plans for the Development and Enforcement of State Standards.
Part 1954—Procedures for the Evaluation and Monitoring of Approved State Plans.
Part 1955—Procedures for Withdrawal of State Plan Approval.
Part 1956—Safety and Health Provisions for Public Employees in Non-approved Plan States.
Part 1960—Safety and Health Provisions for Federal Employees.

Part 1975—Coverage of Employees under the Williams-Steiger Occupational Safety and Health Act of 1970.
Part 1977—Discrimination Against Employees Exercising Rights Under the Williams-Steiger Occupational Safety and Health Act of 1970.
Part 1990—Identification, Classification and Regulation of Potential Occupational Carcinogens.
Part 2200—Review Commission Rules of Procedure.
Part 2201—Regulations Implementing the Freedom of Information Act.
Part 2202—Standards of Ethics and Conduct of Occupational Safety and Health Review Commission Employees.
Part 2203—Regulations Implementing the Government in the Sunshine Act.
Part 2204—Implementation of the Equal Access to Justice Act.
Part 2205—Enforcement of Nondiscrimination on the Basis of Handicap.

Federal Mine Safety and Health Act of 1977 (P.L. 95-164).

*Mine Safety and Health Regulations and Standards.*
29 CFR Part 2700—Federal Mine Safety and Health Review Commission, Rules of Procedure.
30 CFR Part 11—Certification of Vinyl Chloride Respiratory Protective Devices.
Part 40—Representatives of Miners at Mines.
Part 41—Notification of Legal Identity of Mine Operators.
Part 43—Procedures for Processing Hazardous Condition Complaints.
Part 44—Procedures for Processing Petitions for Modification of Safety Standards.
Part 45—Independent Contractors.
Part 46—State Grants for Advancement of Health and Safety in Coal and Other Mines.
Part 47—National Mine Health and Safety Academy.
Part 48—Training and Retraining of Miners.
Part 49—Mine Rescue Teams.
Part 50—Notification, Investigation, Reports and Records of Accidents, Injuries, Illnesses, Employment, and Coal Production in Mines.
Part 56—Health and Safety Standards: Metal and Nonmetallic Mines.
Part 57—Health and Safety Standards: Metal and Nonmetallic Underground Mines.
Part 70—Mandatory Health Standards: Underground Coal Mines.
Part 71—Mandatory Health Standards: Surface Work Areas of Underground Coal Mines and Surface Coal Mines.
Part 74—Coal Mine Dust Personal Sampler Units.
Part 75—Mandatory Safety Standards: Underground Coal Mines.
Part 77—Mandatory Safety Standards: Surface Coal Mines and Surface Work Areas of Underground Coal Mines.
Part 90—Procedure for Transfer of Miners with Evidence of Pneumoconiosis.
Part 100—Civil Penalties for Violation of the Federal Mine Safety and Health Act of 1977.
42 CFR Part 37—Specifications for Medical Examinations of Underground Coal Miners.
Part 85—Requests for Health Hazard Evaluations.
Part 85a—NIOSH Policy on Workplace Investigations.

# 3

# Hazard Control Program Organization



To BE EFFECTIVE, a hazard control program must be planned and be logical; it can't just grow by itself. Program objectives and safety policies need to be established. Responsibility for the hazard control program needs to be determined. Specific steps to identify and control hazards need to be performed; these will be discussed later in this chapter. But first, those who will design and participate in the hazard control program need to understand the nature of hazards, their effects on the work process, and the basic causes of accidents and ways they can be controlled.

## ACCIDENTS AND HAZARD CONTROL

### Definition of hazards

Hazards are a major cause of accidents. A workable definition of hazard is any existing or potential condition in the workplace which, by itself or by interacting with other variables, can result in deaths, injuries, property damage, and other losses (Firenze, 1978).

This definition carries with it two significant points.

- First, a condition does not have to exist at the moment to be classified as a hazard. When the total hazard situation is being evaluated, potentially hazardous conditions must be considered.
- Secondly, hazards may result not only from independent failure of workplace components but also from one workplace component acting upon or influencing another. For instance, if gasoline or another highly flammable substance comes in contact with sulfuric acid, the reaction created by the two substances produces both toxic vapors and sufficient heat for combustion.

Hazards are generally grouped in two broad categories: those dealing with safety and injuries and those dealing with health and illnesses. Hazards that involve only property damage must also be considered.

### Effects of hazards on the work process

In a well-balanced operation, workers, equipment, and materials interact within the work environment to produce a product or perform a service. When operations go smoothly and time is used efficiently and effectively, production is at its highest.

What happens when an accident interrupts an operation? Does it carry a price tag? An accident increases the time needed to complete the job, reduces the efficiency and effectiveness of the operation, and raises production costs. If the accident results in injury, materials waste, equipment damage, or other property loss, there is a futher increase in operational and hidden costs and a decrease in effectiveness. These cost factors are discussed in Chapter 7.

### Hazard control and productivity improvement

The process of identifying and eliminating or controlling hazards in the workplace is one way of optimizing a company's human, financial, technological, and physical resources. Optimizing a company's resources results in higher productivity. For purposes here, productivity is defined as producing more output with a given level of input resources.

Hazard control, like productivity improvement, is a strategic process. To be effective, it must be integrated into the day-to-day activities and management systems of the organization and must become institutionalized—an operating norm and a strategic part of the company's culture.

There are other similarities between efforts aimed at hazard control and productivity improvement. To achieve both objectives, a company must intelligently manage its financial and human resources and use the most appropriate technology. It must illustrate innovative, enlightened, and efficient use of its plant, equipment, materials, and work force, and have a trained, educated, and skilled work force.

**The production process** causes an interaction of a company's human and physical resources to accomplish a given task. Time is an important element in the production process. An accident, even if it does not result in injury, property damage, or wasted material, interrupts the process.

An interruption, in turn, increases the time needed to complete the task. It may also reduce the efficiency and effectiveness of the overall operation and increase production costs. Sometimes, a succession of interruptions, or one long one, will also prevent the production schedule or desired product quality from being met. This can lead to failure to attract new business.

**Production accomplishment and control.** Control, by definition, keeps the system on course and prevents unwanted changes from occurring. However, it also implies allowing for variations within the system, providing they remain within controlled limits.

Any production system has built-in control limits, both upper and lower. These limits provide the direction and also the acceptable leeway for the system's operation.

There are many aspects of control including control over the quality of products and services, personnel, capital, energy, materials, and the plant environment. Each of these factors interacts with the other factors to produce the desired effect.

**Determining accident factors.** In order to set realistic goals for its process, the company should first determine the major factors likely to cause loss of control. It should determine where these factors are, how important they are, and what their potential effects are. Control measures can then be instituted. This will help to reduce the uncalculated risk and potential losses. Factors responsible for losses from accidents may be identified by either inspection or detailed hazard analyses. The control measures that follow may be some type of process innovation or a personnel, machine safeguarding, personal protective equipment, training, or administrative change. In addition to the control measures, monitoring systems should be used for continuously assessing the effectiveness of these hazard-reducing controls.

## Controlling hazards: a team effort

Traditionally, management personnel have solely relied on their safety and production people to locate, evaluate, and prescribe methods of controlling hazardous situations. However, the more that is learned about hazard and loss control, the more evident it becomes that the job is too large for any individual or small group to do alone. Accident and hazard reduction requires a team effort by employees and management.

Here is how several departments and employee teams can work together.

- The engineering departments can design facilities to be free of uncontrolled hazards and provide technical hazard identification and analysis services to other departments. Their designs must comply with federal, state or provincial, and local laws and standards.
- Manufacturing departments can reduce hazards through efforts such as effective tool design, changes in processes, job hazard analysis and control, and coordinating and scheduling production.
- Quality control can test and inspect all materials and finished products. It can conduct studies to determine whether alternate design, material, and methods of manufacture could improve quality and safety of the product and safety of the employees making the product.
- Purchasing departments can make sure that materials and equipment entering the workplace meet established safety and health standards, and that adequate protective devices are an integral part of equipment. They should disseminate information received from suppliers to line management and workers about safety and health hazards associated with workplace substances and materials.
- Maintenance can perform construction and installation work in conformance with good engineering practices and with acceptable safety and health criteria. This department also can provide planned preventive maintenance on electrical systems, machinery, and other equipment to prevent abnormal deterioration, loss of service, or safety and health hazards.
- Industrial relations often administers programs directly related to health and safety.

Input also can come from the joint safety and health committee (discussed later in this chapter) and from quality circles and safety circles (see Chapter 12, Maintaining Interest in Safety).

## Hazard control and management

To coordinate these departmental efforts, a program of hazard control is necessary as part of the management process (Windsor, 1979). Such a program provides hazard control with management tools such as programs, procedures, audits, and evaluations. Sometimes hazard control program teams neglect the basics in their rush to be competitive and innovative, to deal with complex employee relations issues and government involvement, and to address the technical aspects of the programs. A program of hazard control assures the old standbys also will be addressed. These basics include sound operating and design procedures, operator training, inspection and test programs, and communicating essential information about hazards and their control.

A hazard control program coordinates responsibility among departments. For example, if one department makes a product and another distributes it, they share responsibility for hazard control. The producer knows the nature of the process, its apparent and suspected hazards, and how to control the hazards. The producer and the distributor are responsible for making sure this information does not end with the production department but is available to the purchaser or the next unit in the manufacturing process.

Coordination is also important when manufacturing responsibility is transferred from one department to another (as when a pilot program becomes a full-blown manufacturing unit). When a process is phased out, coordination is also necessary to assure that experienced personnel with knowledge of the hazards are retained throughout the phase-out and that appropriate hazard control activities are carried out until the process is terminated.

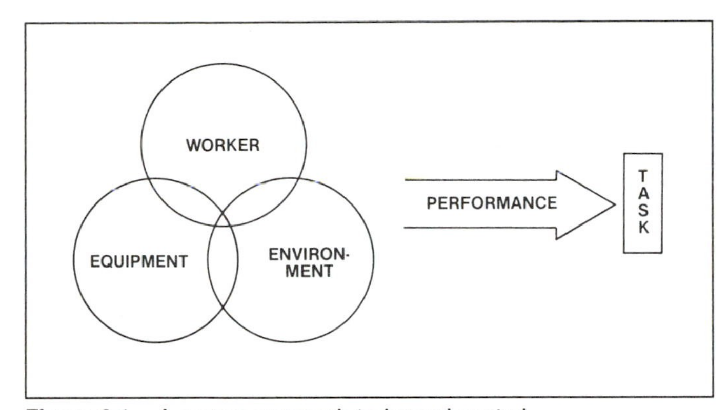


**Figure 3-1.** A system approach to hazard control recognizes the interaction between worker, equipment, and environment in the performance of work.

### Worker-equipment-environment system

Those involved in establishing effective hazard control programs must understand the interrelationships between the worker-equipment-environment system. Chapter 10, Ergonomics: Human Factors Engineering, will examine the system in greater detail. The present chapter explains the elements of the system shown in Figure 3-1. As Figure 3-2 illustrates, an accident can intervene between the system and the task to be accomplished.

**Worker.** In any worker-equipment-environment system, the worker peforms three basic functions:

Sensing
Information processing
Controlling.

- As a *sensor,* the worker serves to monitor or gather information.
- As an *information processor,* the worker uses the information collected to make a decision about the relevance or appropriateness of various courses of action.
- The third function, *control,* flows from the first two. Once information is collected and processed, the worker keeps the situation within acceptable limits or takes the necessary action to bring the system back into an acceptable or safe range.

Evaluating an accident in the light of these three functions can pinpoint the causes.

Did the error occur while the worker was gathering information as a sensor? Was the worker able to gather information accurately, for example, in adequate illumination without glare?

Did the error occur as a result of faulty information processing and decision making?

Did the error occur because an appropriate control option was not available or because the worker took inappropriate action?

In order for the system to move toward its production objectives, the employee must perform work effectively and avoid taking unnecessary risks. To do this, workers must be made aware of the following (Firenze, 1978):

1. The necessary requirements of the task and the steps needed to accomplish it
2. Personal knowledge, skill, and limitations and how they relate to the task
3. What will be gained if the task is attempted and the worker succeeds
4. What will result if the task is attempted and the worker fails
5. What will be lost if the worker does not attempt to accomplish the task at all.

**Equipment** is the second component in the system. All equipment must be properly designed, maintained, and used. Hazard control can be affected by the shape of tools, their size and thickness; the weight of equipment; operator comfort; and the strength required to use or operate tools, equipment, and machinery. These variables influence the interaction between worker and equipment. Other equipment variables important in hazard recognition include speed of operation and mechanical hazards (Plog, 1988).

**Environment.** Special consideration must be given to environmental factors that might detract from the comfort, health, and safety of the worker. Emphasis should be placed on factors such as:

1. Layout; the worker should have sufficient room while performing his assigned task
2. Maintenance and housekeeping
3. Adequate illumination; poorly lit areas increase eyestrain and also the chance of making an accident-causing mistake
4. Temperature, humidity, noise, vibration, and ventilation of toxic materials.

Interpersonal relationship is another system factor that plays an important role in operational effectiveness. The task performed by one worker is related to tasks performed by others. Special consideration must be given to coordinating information, materials, and human effort (Hannaford, 1976).

## ACCIDENT CAUSES AND THEIR CONTROL

Accidents are caused. Close examination of each accident situation shows that it can be attributed, directly or indirectly, to one or more of the following (Firenze, 1978):

1. Oversight or omissions or malfunction of the management system as related to the three following items. (Refer to discussion later in this chapter.)
2. Situational work factors, for example, facilities, tools, equipment, and materials
3. Human factor; either the worker or another person
4. Environmental factors, such as noise, vibration, temperature extremes, illumination.

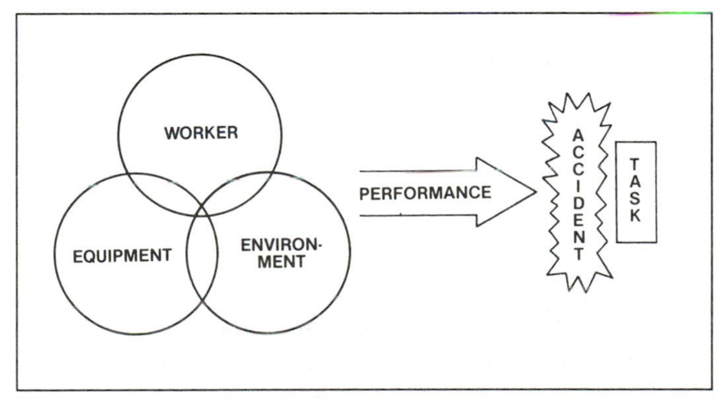


**Figure 3-2.** An accident causes the work system to break down. It intervenes between the worker, equipment, and environment and the task to be performed.

If an adequate management system properly interfaces with the worker, the equipment, and the environment, then the likelihood of accidents occurring in the workplace is greatly reduced.

### Human factors

The human factor is the person who, by commission (actions) or by omission (failure to act), causes an accident. Both workers and management can cause accidents by commission, for example, when a worker sharpens a wood gouge on a grinder without resting the tool on the grinder's rest. The worker can contribute to the cause of an accident by an act of omission when failing to wipe an oil spot from the floor. An unsafe act is generally described as a human action departing from prescribed hazard controls or job procedures or practices; or an action causing a person unnecessary exposure to a hazard.

An unsafe act often is a deviation from the standard job procedures, such as:

1. Using equipment without authority
2. Operating equipment at an unsafe speed or other improper way
3. Removing safety devices, such as guards, or rendering them inoperative
4. Using defective tools.

Unsafe acts can be a deviation from a standard or written job procedure, safety rules or regulations, instructions, or job safety analyses. Why the deviation is the real issue. Some causes and their countermeasures are given in the following examples of management and employee unsafe acts and omissions. When implementing a hazard control program, emphasis should be placed on the countermeasure.

1. There was no known standard for safe job procedure. Countermeasure: Perform a job safety analysis (JSA) and develop a good procedure through job instruction training (JIT).
2. The employee did not know the standard job procedure. Countermeasure: Train in the correct procedure.
3. Employee knew, but did not follow, the standard job procedure. Countermeasure: Consider an employee performance evaluation. Test the validity of the procedure and motivation.
4. Employee knew and followed the procedure. Countermeasure: Develop a safer job procedure.
5. Procedure encouraged risk-taking incentive, such as incentive piecework. Countermeasure: Change unsafe job design or procedure.
6. Employee changed approved job procedure or safety equipment. Countermeasure: Change method so safety measures cannot be bypassed.
7. Employee did not follow correct procedure because of work pressure or supervisor's influence. Countermeasure: Counsel employee and supervisor; consider change in work procedures or job requirements.
8. Individual characteristics, which may involve a disability, made the employee unable or unwilling to follow the correct procedure. Countermeasure: Counsel employee; consider change in work procedures or job requirements. Also consider training.

When a worker is directly involved in an accident, actions can be automatically judged unsafe. A great many accidents actually are the result of someone deviating from the standard job procedures, doing something that the worker is not supposed to do, or failing to do something that should be done. In other situations, however, the worker becomes the target for criticism when, although directly involved in the accident, other factors forced this involvement (Firenze, 1981). The following example illustrates this point.

Suppose a newly hired worker, after receiving what was thought to be sufficient instruction on the use of a table saw guard, is required to make a particular cut that cannot be made with the guard in proper position. In this case the required task causes the worker to remove the guard temporarily so the cut can be made. While removing the guard, the worker's hand slips off the wrench and is cut on the saw blade. Obviously the worker was instrumental in the accident situation, and consequently many people would view the act as unsafe. A closer analysis reveals, however, that the primary cause factor cannot be placed solely on the worker's shoulders.

In this instance, a failure in the management system contributed to the accident. First, the purchasing department should have had better knowledge of the guard's capabilities, limitations, and compatibility with process requirements. Second, the new worker should have been instructed in the use of the guard, as well as how to maintain and remove it when necessary. Most important, a contingency plan should have provided protection if and when the saw would have to be used without adequate safeguarding.

Differentiating between worker error and supervisory error is an important first step in hazard control. Human error is reduced when (1) supervisors and workers know the correct methods and procedures to accomplish given tasks; (2) workers demonstrate a skill proficiency before using the particular piece of equipment; (3) higher management and supervisors consider the relationship between worker performance and physical characteristics and fitness; (4) the entire organization gives high and continuous regard to potentially dangerous situations and the corrective action necessary to avoid accidents; and (5) supervisors provide proper direction, training, and surveillance. The supervisor must be aware of the worker's level of skill with each piece of equipment and process and adjust the supervision of each worker accordingly. The supervisor shapes worker attitudes and actions when it is known that nothing less than safe work practices and as safe a workplace as possible will be accepted.

### Situational factors

Situational factors are another major cause of accidents. These factors are those operations, tools, equipment, facilities, and materials that contribute to accident situations. Examples are unguarded, poorly maintained, and defective equipment; ungrounded equipment that can cause shock; equipment without adequate warning signals; poorly arranged equipment, buildings, and layouts that create congestion hazards; and equipment located in positions that unnecessarily can expose more people to a potential hazard.

Some causes of situational problems are:

1. Defects in design—for example, a lightweight metal, unvented container for use with flammable materials or no guard on a power press;
2. Poor, substandard construction—for example, a ladder built with defective lumber or with a variation in the space between its rungs;
3. Improper storage of hazardous materials—for example, oxygen and acetylene cylinders stored in an unstable manner and ready to topple over with the slightest impact;

4. Inadequate planning, layout, and design—for example, a welding station located near combustible materials or placed where many workers without eye protection are exposed to the intense light of the welding arc.

An example of a situational problem occurred in a light industrial manufacturing plant where maintenance workers too often found themselves replacing a bearing on an expensive machine. Something had to be done to save downtime, labor, and the cost of the bearing. The industrial engineering and maintenance departments jointly devised the solution: a system that fed oil to the bearing at set intervals, keeping it well-lubricated. It was no longer necessary to frequently replace the bearing.

But the solution created new hazards. When oil was fed to the bearing, it dripped onto the floor in the aisle adjacent to the machine. Workers could have slipped on the oil spot and sustained serious injuries. Fork lifts were driven over the oil. With oil on the rubber wheels, the driver might not have been able to stop the vehicle.

Had the maintenance and industrial engineering organizations been thinking of accident causes, they could have avoided situational hazards by correcting their design. As an interim action, they might have collected the oil by placing a pan under the motor where the bearing was housed. This would buy the time needed to install a tube to return the oil to the system, thus saving the oil as well as eliminating the hazard.

### Environmental factors

The third factor in accident causation is the environmental one, the way in which the workplace directly or indirectly causes or contributes to accident situations. Environmental factors fall into three broad categories: physical, chemical, and biological.

**Physical category.** Noise, vibration, radiation, illumination, and temperature extremes are examples of factors having the capacity to influence or cause accidents and illnesses. If operations on a machine lathe, for example, produce high noise levels that can prevent the worker from hearing other sounds, his communications with others may be impaired. Thus, workers may be unable to warn one another of a hazard in time to avoid an accident.

**Chemical category.** Classified under this category are toxic fumes, vapors, mists, smokes, and dusts. In addition to causing illnesses, these elements often impair a worker's skill, reactions, judgment, or concentration. For example, a worker who has been exposed to the narcotic effect of some solvent vapors may experience an alteration of judgment and move too close to the cutting blade of a milling machine.

**Biological category.** Biological factors are those capable of making a person ill from contact with bacteria, viruses, and other micro-organisms or from contact with fungi or parasites. For example, boils and inflammations caused by staphylococci and streptococci, and grain itch caused by parasites.

### Sources of situational and environmental hazards

Situational and environmental hazards enter the workplace from many sources: purchasing agents; those responsible for tool, equipment, and machinery placement and for providing adequate machine guards; and those responsible for maintaining shop equipment, machinery, and tools.

Employee contributions to situational and environmentally caused hazards include disregarding safety rules and regulations by (1) making safety devices inoperative, (2) using equipment and tools incorrectly, (3) using defective tools rather than taking the time to secure serviceable ones, (4) failing to use exhaust fans when required, and (5) using toxic substances in unventilated areas or without proper protection.

Purchasing agents can be instrumental in causing situational and environmental hazards if they give little consideration to hazards. Purchasing agents may acquire tools, equipment, and machinery without adequate guards and other safety devices, especially if such items can be obtained at a bargain. Sometimes toxic and hazardous materials are purchased when less toxic and hazardous materials could be substituted. Sometimes purchasing agents fail to acquire the necessary warning and control information from the vendor and disseminate this data to those in charge of the particular process. However, in many companies, the agent's purchasing is controlled by engineering, safety, and government standards and regulations.

Those involved in layout, design, and placement of equipment and machinery also must consider adequate safeguarding and safety devices or equipment; otherwise, they contribute to hazardous situations in the workplace. Examples are:

1. Placing equipment and machinery with reciprocating parts where workers can be crushed between the equipment and substantial objects
2. Installing electrical control switches on machinery where the operator will be exposed to the hazards of cutting tools or blades in order to start and stop the equipment
3. Installing equipment and machinery guards that interfere with work operations
4. Locating high hazard work stations where they expose workers unnecessarily. For example, placing a welding station in the middle of a floor area instead of locating it in a corner or along a wall where better control over the welding arc light is possible.

Those responsible for maintenance, both management and employees among others, sometimes cause hazards in the workplace. Examples are:

1. Improperly identifying high and low pressure steamlines, compressed air and sanitary lines;
2. Not detecting or replacing worn or damaged machine and equipment parts, such as abrasive wheels on power grinders;
3. Failing to adjust and lubricate equipment and machinery on a scheduled basis;
4. Failing to inspect and replace worn hoisting and lifting equipment;
5. Failing to replace worn and frayed belts on equipment;
6. Over-oiling motor bearings, resulting in oil being thrown onto the insulation of electrical wiring and onto the floor, and possible damage to the bearings;
7. Failing to replace guards;
8. Failing to tag and lock out unsafe equipment.

More details are included in this chapter under the sideheading, Responsibility for the hazard control program.

### Need for a balanced approach

Prior to the development of the concept of hazard control, accidents were viewed as chance occurrences or acts of God, a view still held by some today. A variation of this point of view is that accidents are an inherent consequence of production. Such approaches accept accidents as inevitable and, therefore, yield

**LIST OF TYPICAL INCIDENTS**

An incident is any observable human activity sufficiently complete in itself to permit references and predictions to be made about the persons performing the act.

1. Adjusting and gaging (calipering) work while the machine is in operation.
2. Cleaning a machine or removing a part while the machine is in motion.
3. Using air hose to remove metal chips from table or work (a brush or other tools should be used for this purpose, except on recessed jigs).
4. Using compressed air to blow dust or dirt off clothing or out of hair.
5. Using excessive pressure on air hose.
6. Operating machine tools (turning machines, knurling and grinding machines, drill presses, milling machines, boring machines) without proper eye protection (including side shields).
7. Not wearing safety glasses in a designated eye-hazard area.
8. Failing to use protective clothing or equipment (face shield, face mask, ear plugs, safety hat, cup goggles).
9. Failing to wear proper gloves or other hand protection when handling rough or sharp-edged material.
10. Wearing gloves, ties, rings, long sleeves, or loose clothing around machine tools.
11. Wearing gloves while grinding, polishing, or buffing.
12. Handling hot objects with unprotected hands.
13. No work rest or poorly adjusted work rest on grinder (1/8 in. maximum clearance).
14. Grinding without the glass eye-shield in place.
15. Making safety devices inoperative (removing guards, tampering with adjustment of guard, beating or cheating the guard, failing to report defects).
16. Using an ungrounded or uninsulated portable electric hand tool.
17. Improperly designed safety guard, for example, a wide opening on a barrier guard which will allow the fingers to reach the cutting edge.

**Figure 3-3.**

no information about causation and prevention. Control strategies are limited to mitigating the consequences of the occurrence.

In the early days of hazard control, accident prevention activities focused on the human element. Findings indicated that a small proportion of workers accounted for a significant percentage of accidents. From these findings came the accident proneness theory of causation. Control strategies were devised to reduce human error through training, education, motivation, communication, and other forms of behavior modification. During World War II, industrial psychology was aimed at matching employees to particular jobs, and personnel screening and selection were seen as ways to prevent accidents. The weakness of accident proneness and other behavior models is this: while useful for understanding human behavior, they do not consider the interaction between the worker and the other parts of the system. (See the discussion in Chapters 9 and 11.)

The 1950s and 1960s saw the emphasis change to engineering and control programs aimed at machines and equipment. With the implementation of the OSHAct in 1970, emphasis was placed on preventing accidents through control of the work environment and the elements of the workplace. Specification standards and compliance rules and regulations were spelled out.

### Management oversight and omission

The emphasis of many organizations over the last few decades has been to take into account system defects, which result from management oversight or omission, or malfunction of the management system. A balanced approach to hazard control looks at each component of the system and includes such weaknesses as inadequate training and education, improper assignment of responsibility, unsuitable equipment, or badly budgeted funds. Because managers are the people responsible for the design of systems, system defects can occur because of management errors.

### Examining accident causation

There are two basic approaches to examining how accidents are caused, after-the-fact and before-the-fact.

**After-the-fact.** This approach relies on examining accidents after they have occurred in order to determine cause and develop corrective measures. Evaluation of past performance uses information derived from accident and inspection reports and insurance audits. This approach too often is only used after a serious accident has resulted in injury or damage, or system ineffectiveness. Furthermore, accident frequency and severity rates do not answer the crucial questions *what, why,* and *when.*

**Before-the-fact.** This method relies on inspecting and systematically identifying and evaluating the nature of undesired events in a system. One such method is the critical incident technique.

*Critical incident technique.* This technique can identify the cause of an accident before the loss occurs. To obtain a representative sample of workers exposed to hazards, persons are selected from various departments of the plant. An interviewer questions a number of persons who have performed particular jobs within certain environments. They are asked to describe only existing hazards and unsafe conditions they are aware of. These are called incidents. Figure 3-3 lists some incidents that might be described. Incidents are then classified into hazard categories, and problem areas are identified.

The critical incident technique measures safety performance and identifies practices or conditions that need to be corrected.

Inquiry can be made into the management systems that should have prevented the occurrence of unsafe acts or the existence of unsafe conditions. The technique can lead to improvements in hazard control program management.

The procedure needs to be repeated because the worker-equipment-environment system is not static. Repeating the technique with a new sample of workers can reveal new problem areas and measure the effectiveness of the accident prevention program.

*Safety sampling,* also called behavior or activity sampling, is another technique that uses the expertise of those within the organization to inspect, identify, and evaluate hazards (Pollina, 1962). This method relies on personnel—usually management or safety staff members—who are familiar with operations and well trained in recognizing unsafe practices. While making rounds of the plant or establishment, they record on a safety sampling sheet both the number and type of safety defects they observe. A code number can be used to designate specific unsafe conditions. For example, hands in dies, failure to wear eye protection and protective clothing, failure to lock out source of power while working on machinery, crossing over belt conveyors, working under suspended loads, improper use of tools, transporting unbanded steel.

Observations must be made at different times of the day, on a planned or random basis in the actual work setting, and throughout the various parts of the plant. In a short time the observations can easily be converted to a simple report showing what specific unsafe conditions exist in what areas and what supervisors and foremen need help in enforcing good work practices. The information is unbiased and therefore irrefutable. What has been recorded is what has been observed.

## PRINCIPLES OF HAZARD CONTROL

Hazard control is the function directed toward recognizing, evaluating, and eliminating, or at least reducing, the destructive effects of hazards emanating from human errors and from the situational and environmental aspects of the workplace (Firenze, 1978). Its primary function is to locate, assess, and set effective preventive and corrective measures for those elements detrimental to operational efficiency and effectiveness.

The process exists on three levels:

1. National—laws, regulations, exposure limits, codes, standards of governmental, industrial, and trade bodies;
2. Organizational—management of hazard control program, safety committees;
3. Component—worker-equipment-environment.

Hazard control can be thought of as "looking for failures." In the first place, there are fewer failures than successes. Second, it is easier to agree on what constitutes failure than on what is success. Failure is the inability of a system or a part of a system to perform as required under specified conditions for a specific length of time. The causes of failures often can be determined by answering a series of questions. What can fail? How can it fail? How frequently can it fail? What are the effects of failure? What is the importance of the effects? The manner in which a system, or portion of a system, can exhibit failure is commonly known as the *mode of failure.*

The opposite of failure is not necessarily total success—that error-free performance which is an ideal state, not a reality—but the *minimum acceptable* success. That is the point where processes are accomplished with a tolerable number of losses and interruptions, keeping efficiency and effectiveness of the operation within acceptable limits of control.

Management builds into each of its production systems lower and upper limits of control. Each of these interfacing subsystems—maintenance, quality control, production control, personnel, purchasing, to name a few—is designed to move the system within acceptable limits toward its objective. This concept of keeping operations within acceptable limits gives substance and credibility to the process of hazard control. In addition to familiarizing management with the full consequences of system failures, hazard control can pinpoint hazards *before* failures occur. The anticipatory character of hazard control increases productivity.

## PROCESSES OF HAZARD CONTROL

An effective hazard control program has six steps or processes (Firenze, 1978):

1. Hazard identification and evaluation
2. Ranking hazards by risk
3. Management decision making
4. Establishing preventive and corrective measures
5. Monitoring
6. Evaluating program effectiveness.

Let's discuss each in turn.

### Hazard identification and evaluation

The first step in a comprehensive hazard control program is to identify and evaluate workplace hazards. These hazards are associated with machinery, equipment, tools, operations, and the physical plant.

There are many ways to acquire information about workplace hazards. A good place to begin is with those who are familiar with plant operations and the hazards associated with them. See Chapter 24, Sources of Help, for a description of many organizations that can be of help. The critical incident technique (described on this and the previous page) is useful for obtaining information from workers and supervisors. Insurance company loss control representatives know those hazards most likely to cause damage, injuries, and fatalities. In addition to the National Safety Council (NSC), associations such as the American Society of Safety Engineers (ASSE), American Industrial Hygiene Association (AIHA), and the American Conference of Governmental Industrial Hygienists (ACGIH) have information about safety and health experience. Manufacturers of industrial equipment, tools, and machinery offer information about the hazards associated with their products, as do suppliers of materials and substances. Labor representatives and business agents can offer a perspective on hazards overlooked by others. Safety and health personnel in organizations doing similar work can be of inestimable value.

A second place to look would be old inspection reports, either internal (by a safety and health committee or company management and specialists) or external (by local, state or provincial, or federal enforcement agencies). OSHA can supply information describing violations uncovered in similar operations and outlining compliance regulations. See Chapter 2, Governmental

Regulation and Compliance, and Chapter 24, Sources of Help, for descriptions of state agencies and private concerns that give on-site inspection and consultation services under OSHA and NIOSH.

Hazard information also can be obtained from accident reports. Information explaining how a particular injury, illness, or fatality occurred often will reveal hazards requiring control. Close review of accident reports filed in the past three to five years also will identify the individuals and specific operations involved, the department or section where the accident occurred, the extent of supervision, and possibly the injured person's deficiencies in knowledge and skill.

OSHA incident rates also are useful. Although they are historical and reflect what has happened, not the current status of safety performance, they provide, from a large sample, data that reflects what actually has occurred in the workplace.

Other sources which can be valuable are described in other chapters, in the data sheets of the National Safety Council, in the specifications for particular equipment and machines published by the American National Standards Institute (ANSI), Underwriters Laboratories Inc. (UL), American Society for Testing and Materials (ASTM), and the National Fire Protection Association (NFPA). Information about work activities, facilities, and equipment is distributed by the National Institute for Occupational Safety and Health (NIOSH).

Hazard analysis is another avenue to acquiring meaningful hazard information and a thorough knowledge of the demands of a particular task. Analysis probes operational and management systems to uncover hazards that (1) may have been overlooked in the layout of the plant or the building and in the design of machinery, equipment, and processes; (2) may have developed after production started; or (3) may exist because original procedures and tasks were modified.

The greatest benefit of hazard analysis is that it forces those conducting the analysis to view each operation as part of a system. In doing this, each step in the operation is assessed while consideration is paid to the relationship between steps and the interaction between workers and equipment, materials, the environment, and other workers. Other benefits of hazard analysis include (1) identifying hazardous conditions and potential accidents; (2) providing information with which effective control measures can be established; (3) determining the level of knowledge and skill as well as the physical requirements workers need to execute specific shop tasks; and (4) discovering and eliminating unsafe procedures, techniques, motions, positions, and actions.

The topic of hazard analysis—its underlying philosophy, the basic steps to be taken, and its ultimate use as a safety, health, and decision-making tool—will be treated in Chapter 4.

### Ranking hazards by risk (consequence and probability)

The second step in the process of hazard control is to rank hazards by risk. Such ranking takes into account both the consequence (the severity) and the probability (the frequency). This second process is necessary so hazards can be addressed according to the principle of "worst first." Ranking provides a consistent guide for corrective action, specifying which hazardous conditions warrant immediate action, which have secondary priority, and which can be addressed in the future.

The classification scheme outlined in Figure 3-4 is suggested for rating hazards by consequence.

**Figure 3-4.** Relative consequences of various hazard categories

| *Hazard Consequence Category* | *Explanation* |
|---|---|
| I. | **Catastrophic**—may cause death or loss of a facility. |
| II. | **Critical**—may cause severe injury, severe occupational illness, or major property damage. |
| III. | **Marginal**—may cause minor injury or minor occupational illness resulting in lost workday(s), or minor property damage. |
| IV. | **Negligible**—probably would not affect personnel safety or health and thus, less than a lost workday, but nevertheless is in violation of specific criteria. |

Once hazards have been ranked according to their potential destructive consequences, the next step is to estimate the probability of the hazard resulting in an accident situation. Quantitative data for ranking hazard probability are desirable, but almost certainly they will not be available for each potential hazard being assessed. Whatever quantitative data exist should be part of the risk-rating formula used to estimate probability. Qualitative data—estimates based on experience—are a necessary supplement to quantitative data. Figure 3-5 shows how probability estimates should be made.

**Figure 3-5.** Qualitative probability estimate for use in decision making

*Hazard probability category (qualitative estimate)*

A. Likely to occur immediately or within a short period of time when exposed to the hazard
B. Probably will occur in time
C. Possible to occur in time
D. Unlikely to occur

After estimating both consequence and probability, the next and final step is to estimate worker exposure to the hazard. The exposure classification scheme in Figure 3-6a is suggested for rating exposure.

**Figure 3-6a.** Exposure category

Exposure—the number of persons regularly exposed to the hazard. Here the team must evaluate how many people would ordinarily be exposed to the hazard. The following scheme is used to estimate exposure:

1. Greater than 50 different persons regularly exposed to the hazard
2. From 10 to 49 different persons regularly exposed to the hazard
3. From 5 to 9 different persons regularly exposed to the hazard
4. Less than 5 different persons regularly exposed to the hazard.

### Risk assessment

When the hazards have been ranked according to all three criteria—consequence, probability, exposure—the next step is to assign a single risk number or risk assessment code (RAC). Figure 3-6b illustrates the RAC numbers and their designations.

**Figure 3-6b.** Risk assessment code (RAC)

| *RAC No.* | *Title* |
|---|---|
| 1 | Critical |
| 2 | Serious |
| 3 | Moderate |
| 4 | Minor |
| 5 | Negligible |

**RECORD OF OCCUPATIONAL SAFETY AND HEALTH DEFICIENCIES**

Location Shipping

Pete Varga
Inspector

| Deficiency No. | Date Recorded | Description of Hazardous Condition | Specific Location | Identification of Acceptable Standard | Hazard Rating | | Corrective Action | Estimated Cost of Correction | Date Deficiency Corrected | Resources Used for Correction |
|---|---|---|---|---|---|---|---|---|---|---|
| | | | | | Conse-quence | Proba-bility | | | | |
| S - 1 | 12/11/8- | Ungrounded Tools and Equipment | Throughout Shop | OSHA; Subpart S National Electrical Code, Article 250; 4S | I | A | Provide receptacles with the 3-prong outlet.<br><br>Test each to make certain it is grounded.<br><br>Make sure that all tools (other than double-insulated) have a grounding plug. | $5,000 | 1/9/8- | $4,900 |

**Figure 3-7.** One approach to recording and displaying hazard information for decision making. (Reprinted with permission from RJF Associates, Inc.)

## Management decision making

The third step involves providing management with full and accurate information, including all possible alternatives, so it can make intelligent, informed decisions concerning hazard control. Such alternatives will include recommendations for training and education, for better methods and procedures, equipment repair or replacement, environmental controls, and—in rare cases where modification is not enough—recommendations for redesign. Information must be presented to management in a way that clearly states the actions required to improve conditions. The person who reports hazard information must do so in a manner that promotes, rather than hinders, action.

After management's decision-makers receive hazard reports, they normally have three alternatives:

1. They can choose to take no action.
2. They can modify the workplace or its components.
3. They can redesign the workplace or its components.

When management chooses to take no positive steps to correct hazards uncovered in the workplace, it usually is for one of three reasons:

1. It feels that it cannot take the required action. Immediate constraints—be they financial, crucial production schedules, or limitations of personnel—loom larger than the risks involved in taking no action.
2. It is presented with limited alternatives. For example, it may receive only the best and most costly solutions with no less-than-totally-successful alternatives to choose from.
3. It does not agree that a hazard exists. However, the situation can require additional consultation and study to resolve any problem.

When management chooses to modify the system, it does so with the idea its operation is generally acceptable but, with the reported deficiencies corrected, performance will be improved. Examples of modification alternatives are the acquisition of machine guards, personal protective equipment, or ground-fault circuit interrupters to prevent electrical shock; a change in training or education; a change in preventive maintenance; isolating hazardous materials and processes; replacing hazardous materials and processes with non- or at least less-hazardous ones; and purchasing new tools.

Although redesign is not a popular alternative, it sometimes is necessary. When redesign is selected, management must be aware of certain problems. Redesign usually involves substantial cash outlay and inconvenience. For example, assume that the air quality in a plant is found to be below acceptable standards. The only way to correct this situation is to completely redesign and install the plant's general ventilation system. The cost and inconvenience are obvious.

Another problem is the probability the new design will contain hazards of its own. For this reason, whenever redesign is offered as an alternative, those making the recommendation must establish and execute a plan to detect problems in both design and early stages so the hazards can be eliminated or reduced.

One way to expedite decision making regarding actions for hazard control is to present findings in such a manner that management can clearly understand the nature of the hazards, their location, their importance, the necessary corrective action, and the estimated cost.

Figure 3-7 shows a record of occupational and safety health deficiencies. It illustrates one approach for recording and displaying hazard information for decision making. It indicates the hazard ranking, the specific location and nature of the hazard, and what costs are likely to be incurred. It also specifically states the recommended corrective action. At a glance, it shows if the corrective action has been taken and the final cost.

## Establishing preventive and corrective measures

After hazards have been identified and evaluated and information for informed decisions has been provided, the next step involves the actual installation of control measures.

Controls are of two kinds: administrative (through personnel, management, monitoring, limiting worker exposure, measuring performance, training and education, housekeeping and maintenance, purchasing) and engineering (isolation of source, lockout procedures, design, process or procedural changes, monitoring and warning equipment, chemical or material substitution).

Before control installation takes place, it is essential that those involved in safety and health activities understand how hazards are controlled. Figure 3-8 illustrates the three major areas where hazardous conditions can be either eliminated or controlled.

- The first and perhaps best control alternative is to attack a hazard at its source. One method is to substitute a less harmful agent for the one causing the problem. For example, if a certain solvent is highly toxic and flammable, the first step is to determine whether the hazardous substance can be exchanged for one that is nontoxic, nonflammable, and still capable of doing the job. If a nonhazardous substance meeting these criteria is not available, then a less toxic, less flammable substance can be substituted and additional safeguards employed.
- The second alternative is to control the hazard along its path. This is done by erecting a barricade between the hazard and the worker. Examples of engineering controls are (1) machine guards, which prevent a worker's hands from making contact with the table saw blade; (2) protective curtains, which prevent eye contact with welding arc flashes; and (3) a local exhaust system, which removes toxic vapors from the breathing zone of the workers.
- The third alternative is to direct control efforts at the receiver, the worker. Removing the worker from exposure to the hazard can be accomplished by (1) employing automated or remote control options (for example, automatic feeding devices on planers, shapers); (2) providing a system of worker rotation or rescheduling some operations to times when there are few workers in the plant; or (3) providing personal protective equipment when all options have been exhausted and it is determined that the hazard does not lend itself to correction through substitution or engineering redesign.

Protective equipment may be selected for use in two instances: when there is no immediately feasible way to control the hazard by more effective means, and when it is employed as a temporary measure, while more effective solutions are being installed. There are, however, major shortcomings associated with the use of personal protective equipment:

1. Nothing has been done to eliminate or reduce the hazard.
2. If the protective equipment (such as gloves or an eye shield) fails for any reason, the worker is exposed to the full destructive effects of the hazard.
3. The protective equipment may be cumbersome and interfere with the worker's ability to perform a task, thus compounding the problem.

Chapter 17, Personal Protective Equipment, will treat these subjects more fully.

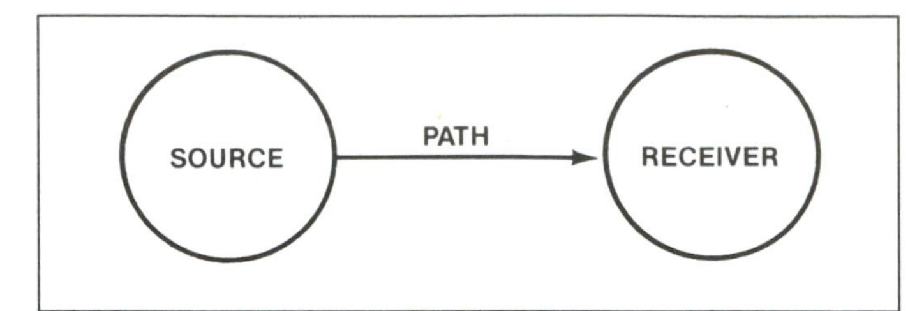


**Figure 3-8.** Three major areas where hazards can be controlled—the contaminant source, the path it travels, and the employee's work pattern and use of personal protective equipment.

### Monitoring

The fifth step in the process of hazard control deals with monitoring activities to locate new hazards and assess the effectiveness of existing controls. Monitoring includes inspection, industrial hygiene testing, and medical surveillance. These subjects will be dealt with in Chapter 4, Acquiring Hazard Information.

Monitoring is necessary (1) to provide assurance that hazard controls are working properly; (2) to make sure that modifications have not so altered the workplace that hazard controls can no longer function adequately; and (3) to discover new or previously undetected hazards.

### Evaluating program effectiveness

The final process in hazard control is to evaluate the effectiveness of the safety and health program. Evaluation involves answering the following questions. How much is being spent to locate and control hazards in the plant? What benefits are being received, for example, reduction of injuries, workers' compensation cases, and damage losses? What impact are the benefits having on improving operational efficiency and effectiveness?

Evaluation examines the program to see if it has accomplished its objectives (effectiveness evaluation) and whether they have been achieved in accordance with the program plan (administrative evaluation, including such factors as schedule and budget).

Evaluation must be adapted to (1) the time, money, and kinds of equipment and personnel available for the evaluation; (2) the number and quality of data sources; (3) the particular operation; and (4) the needs of the evaluators.

Among the criteria management can use to determine the effectiveness of its safety and health program effort are the number and severity of injuries to workers compared with work hours; the cost of medical care; material damage costs; facility damage costs; equipment and tool damage or replacement costs; and the number of days lost from accidents.

An indicator of the effectiveness of a hazard control program is the experience rating given a company by the insurance carrier responsible for paying workers' compensation. Experience rating is a comparison of the actual losses of an individual (company) risk with the losses that would be expected from a risk of such size and classification. Experience rating determines whether the individual risk is better or worse than the average and to what extent the premium should be modified to reflect this variation. Experience modification is determined in accordance with the Experience Rating Plan (ERP) formula, which has been approved by the insurance commissioners in most states. Loss frequency is penalized more heavily than loss severity because it is assumed that the insured can control the small loss more easily than the less frequent, severe loss.

## ORGANIZING AN OCCUPATIONAL SAFETY AND HEALTH PROGRAM

The purposes of a hazard control program organization are to assist management in developing and operating a program designed to protect workers, prevent and control accidents, and increase effectiveness of operations. Figure 3-9 illustrates the major organizational components of a hazard and loss control program.

### Establishing program objectives

Critical to the design and organization of a safety and health program is the establishment of objectives and policy to guide the program's development. If the organization has a joint safety and health committee, it could be the logical body to set objectives. It is assumed that those making recommendations to man-

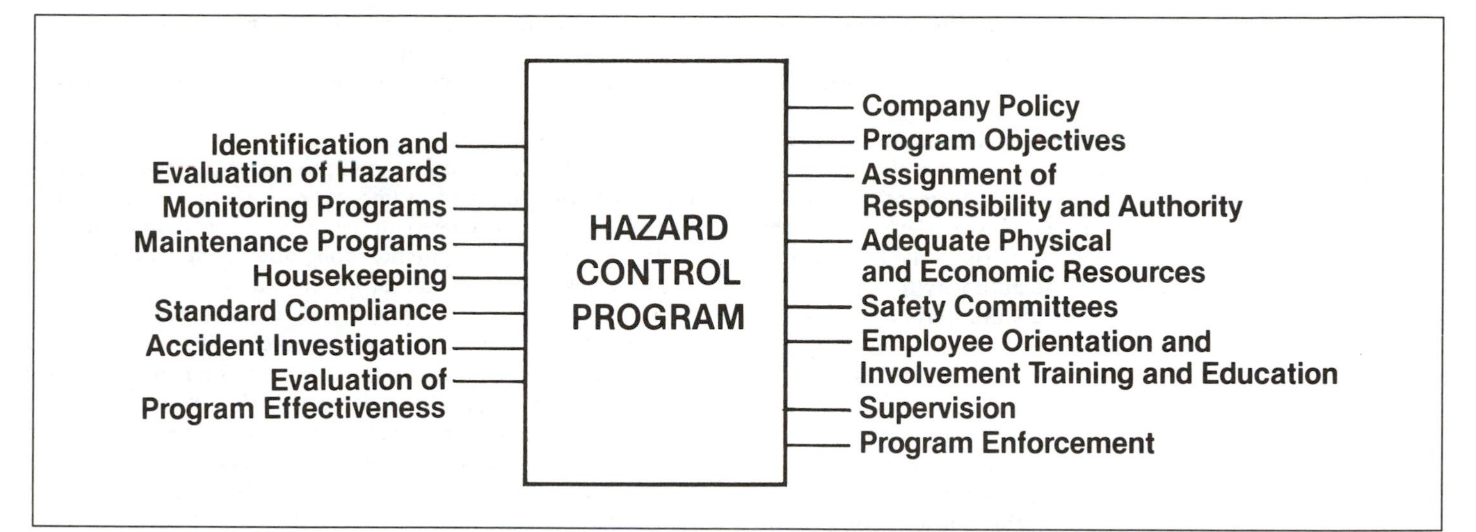


**Figure 3-9.** Major components of a hazard control program. (Reprinted with permission from RJF Associates, Inc.)

agement would be employee representatives, supervisors, middle management, and safety and health professionals (safety directors, managers, supervisors, and administrators; industrial hygiene technicians and professionals; and fire protection engineers).

Among the program objectives should be:

1. Gaining and maintaining support for the program at all levels of the organization
2. Motivating, educating, and training the program team to recognize and correct or report hazards located in the workplace
3. Engineering hazard control into the design of machines, tools, and facilities
4. Providing a program of inspection and maintenance for machinery, equipment, tools, and facilities
5. Incorporating hazard control into training and educational techniques and methods
6. Complying with established safety and health standards.

## Establishing organizational policy

Once the objectives have been formulated, the second step is for management to adopt a formal policy. A written policy statement, signed by the chief administrator of the organization, should be made available to all personnel. It should state the purpose behind the hazard control program and require the active participation of all those involved in the program's operation. The policy statement also should reflect:

1. The importance that management places on the health and well-being of employees
2. Management's commitment to occupational safety and health
3. The emphasis the company places on efficient operations, with a minimum of accidents and losses
4. The intention of integrating hazard control into all operations, including compliance with applicable standards
5. The necessity for active leadership, direct participation, and enthusiastic support of the entire organization.

The policy of the American Telephone and Telegraph Company is probably the shortest statement, but it drives home the point that:

NO JOB IS SO IMPORTANT AND NO SERVICE IS SO URGENT—THAT WE CANNOT TAKE TIME TO PERFORM OUR WORK SAFELY.

The National Safety Council publishes a 12-page Industrial Data Sheet, Management Safety Policies, No. 585, that covers this subject more fully.

After a safety policy has been established, it should be publicized so that each employee becomes familiar with it, particularly how it applies directly to him or her. Meetings, letters, pamphlets, bulletin boards: these are ways to publicize the statement. It should also be posted in management offices to serve as a constant reminder of management's commitment and responsibility.

## Responsibility for the hazard control program

Responsibility for the hazard control program can be established at the following levels: board of directors, chief executive officers, managers, and administrators; department heads, supervisors, foremen, and employee representatives; purchasing agents; housekeeping and maintenance personnel; employees; safety personnel; staff medical personnel; and safety and health committees.

**Management and administration.** Before any hazard control program gets underway, it must receive full support and commitment from top management and administration. The president, board members, directors, and other management personnel have the responsibility for the hazard control program by furnishing the motivation to get the program started and to oversee its operations. Responsibility involves the continuing obligation to monitor the effectiveness of the ongoing safety and health program. Management must initiate discussions with personnel during preplanning meetings and periodically review the performance of its hazard control program. Discussions should cover program progress, specific needs, and a review of company procedures and alternatives for handling emergencies in the event an accident occurs.

Specifically, responsibility at this level consists of setting objectives and policy and supporting safety personnel in their requests for necessary information, facilities, tools, and equipment to conduct an effective hazard control program and establish a safe and healthy work environment. Management must realize that it is not fulfilling its organization's potential efficiency and effectiveness until it brings its operations at least into compliance with mandatory and voluntary federal and state or provincial safety and health standards.

Management and administration must delegate the necessary prerogatives at all organizational levels to assure the success of the hazard control program. Although management cannot delegate to others its responsibility for employee safety and health, it can assign responsibility for certain parts of the hazard control program. Responsibility and the authority to act must be delegated together. While authority always starts with those in the highest administrative levels, it eventually must be delegated to other responsible people at lower levels in order to achieve desired results. If safety professionals, safety committees, department heads, supervisors, foremen, and employee representatives are to conduct a vigorous and thorough hazard control program, and if they are to accept and assert the authority delegated to them when circumstances warrant it, they must be fully confident that they have administrative support.

Management must understand that, although it can assert authority, it may find resistance to this authority unless it has enlisted support from the earliest stages of the program. If supervisors, employees, and their representatives are not aware of the reasons for and the benefits of a thoroughgoing hazard control program, they may resist any changes in their methods of operation and instruction and may do as little as possible to assist the overall program effort.

Management must insist that safety and health information is an integral part of training, methods, materials, and operations. It must guarantee a system where hazard control is considered an important part of equipment purchase and process design, operation, preventive maintenance, and layout and design. It must make sure that effective fire prevention and protection controls exist. Management also has the responsibility to be certain that subcontractors, at the time of negotiating the contract, are fully informed of applicable standards. Management must see that subcontractors comply fully with company and other applicable safety regulations.

Administrators are required to safeguard employees' health by seeing to it that the work environment is adequately controlled. They must be aware that those operations producing airborne fumes, mists, smokes, vapors, dusts, noise, and vibration have the capacity to cause impaired health or discomfort among their workers. Administrators must be aware that occupational illnesses beginning in the workplace can take their toll later, even after a worker retires. To protect the future of its employees, management must maintain a constant industrial hygiene monitoring system.

Management must provide meaningful criteria to measure the success of the hazard control program and to provide information upon which to base future decisions. It must decide the hazard control program goals for reduced accidents, injuries, illnesses, and their associated losses.

There are many concrete ways management can show evidence of its commitment to safety: attending safety meetings, reviewing and acting upon accident reports, reviewing safety records through conferences with department heads and joint employee-management committees, and by setting a good example.

**Department heads, supervisors, foremen, and employee representatives** are in strategic positions within the organization. Their leadership and influence should assure that safety and health standards are enforced and upheld in each individual area and that standards and enforcement are uniform throughout the workplace.

What are some responsibilities of department heads? They make certain that materials, equipment, and machines slated for distribution to the areas under their jurisdiction are hazard-free or that adequate control measures have been provided. They make certain that equipment, tools, and machinery are being used as designed and are properly maintained. They keep abreast of accident and injury trends occurring in their areas and take proper corrective action to reverse these trends. They investigate all accidents occurring within their jurisdiction. They should see to it that all hazard control rules, regulations, and procedures are enforced in their departments. They require that a hazard analysis be conducted for certain operations, particularly those that they regard as dangerous, either from past accident history or from their perception of accident potential. They require that hazard recognition and control information be included in instruction, training, and demonstration sessions for both supervisors and employees. They actively participate in and support the safety and health committee and follow up on its recommendations.

Supervisors, foremen, and employee representatives carry great influence. With their support, top management can be assured of an effective safety and health program. Supervisors have a moral and professional responsibility to safeguard, educate, and train those who have been placed under their direction. Thus, they are generally responsible for creating a safe and healthy work setting and for integrating hazard recognition and control into all aspects of work activities. By their careful monitoring, they can prevent accidents.

For all practical purposes supervisors, foremen, and employee representatives are the eyes and ears of the workplace control system. On a day-to-day basis, they must be aware of what is happening in their respective areas, who is doing it, how various tasks are being performed, and under what conditions. As they monitor their areas, they must prevent accidents from occurring. If, despite their controls, they see danger, they must be prepared to intervene in the operation and take immediate corrective action.

What are the chief responsibilities of safety and health supervisors? They train and educate workers in methods and techniques which are free from hazards. They should be certain that employees understand the properties and hazards of the materials stored, handled, or used by them. They make sure that employees observe necessary precautions, including proper guards and safe work practices. They furnish employees with the proper personal protective equipment, instruct them in its use, and make certain it is worn.

Supervisors demonstrate an active interest in and comply with hazard control policy and safety and health regulations. They actively participate in and support the safety and health committees. They supervise and evaluate worker performance, with consideration given to safe behavior and work methods. They should first try to convince employees of the need for safe performance, and then, unpleasant though it may be, they should administer appropriate corrective action when health and safety rules are violated. Correcting employees requires tact and good judgment. Enforcement should be viewed as education rather than discipline. However, if a supervisor feels that a worker is deliberately disobeying rules or endangering his life and the lives of others, then prompt and firm action is called for. Laxity in the enforcement of safety rules undercuts the entire safety and health program and allows accidents to happen.

Supervisors monitor their area on a daily basis for human, situational, and environmental factors capable of causing accidents. They should make sure that meticulous housekeeping practices are developed and used at all times. They should correct hazards detected in their monitoring or report such hazards to the persons who can take corrective action. They should investigate all accidents occurring within their areas to determine causes.

Foremen and employee representatives share much of the responsibility for safety and health with upper management. Their job is to inspect, detect, and correct. What are the specific responsibilities of foremen and representatives? They should encourage fellow workers to comply with the organization's safety and health regulations. They detect safety violations and hazardous machinery, tools, equipment, and other implements. They take corrective action when possible and report to the supervisor the hazard and the corrective action taken or still required. They participate in accident investigations. They represent the interest of the workers on the safety and health committee.

Practical training aids for employee representatives, foremen, and supervisors do exist. The Council's *Supervisors Safety Manual* and accompanying "Supervisors Development Program," also available through the National Safety Council, have been widely accepted by industry.

**Housekeeping and preventive maintenance** can be regarded as two sides of the same coin. No hazard control program can succeed if housekeeping and maintenance are not seen as integral parts.

*Good housekeeping* reduces accidents, improves morale, and increases efficiency and effectiveness. Most people appreciate a clean and orderly workplace where they can accomplish their tasks without interference and interruption.

An industrial organization, by its very nature, contains tools that must be kept clean. In its operation, it uses flammable substances and materials requiring special storage and removal. It generates dust, scrap metal filings and chips, waste liquids, and scrap lumber that must be disposed of.

Housekeeping is a continuous process involving both workers and custodial personnel. A good housekeeping program incorporates the housekeeping function into all processes, operations, and tasks performed in the workplace. The ultimate goal is for each worker to see housekeeping as an integral part of performance, not as a supplement to the job to be done.

When the workplace is clean and orderly and housekeeping becomes a standard part of operations, less time and effort will be spent keeping it clean, making needless repairs, and replacing equipment, fixtures, and the like. When the worker can concentrate on required tasks without excess scrap material, tools, and equipment interfering with work, operations will be more efficient and the product of higher quality. Time will be used for work, not searching for tools, materials, or parts. When a plant is clean and orderly as well as safe, employee morale is heightened.

When everything has an assigned place, there is less chance that materials and tools will be taken from the plant or misplaced. In a few moments before quitting time, the supervisor can determine what is missing. Different colors of paint can be applied to tools to identify the department to which they belong. Tool racks or holders should be painted a contrasting color as a reminder to workers to return the tools to their proper places. When stored on a rack, the space directly behind each tool should be painted or outlined in color to call attention to a missing tool.

Money is saved and efficiency is increased when workers treat materials with the care they deserve by: minimizing spillage and scrap, saving pieces of material for use in future projects, and returning even small parts to their storage area. When aisle and floor space is uncluttered, movement within the plant is safer, and machinery and equipment can be easily cleaned and maintained. When the plant has adequate work space and when oil, grease, water, and dust are removed from floors and machinery, workers are less likely to slip, trip, fall, or inadvertently come into contact with dangerous parts of machinery.

When a workplace is kept free from accumulations of combustible materials that can burn upon ignition or, in the case of certain material relationships, spontaneously ignite without an external source of ignition, the chances of fires are minimized. Furthermore, an orderly plant permits easy exit by keeping exits and aisles leading to exits free from obstructions. A neat and orderly workplace also makes it easier to locate and obtain fire emergency and extinguishing equipment.

*Preventive maintenance* is the orderly, uniform, continuous, and scheduled action to prevent breakdown and prolong the useful life of equipment and buildings. Preventive maintenance is a shared responsibility. Workers caring for tools and equipment accomplish specific maintenance tasks. Other maintenance duties, such as oiling, tightening guards, adjusting tool rests, and replacing wheels, are routinely performed by workers.

Advantages to be gained from preventive maintenance include safer working conditions, decreased downtime of equipment because of breakdown, and increased life of the equipment.

Satisfactory production depends on having buildings, equipment, machinery, portable tools, safety devices, and the like in operating condition and maintained in a manner that production activities will not be interrupted while repairs are being made or equipment replaced.

Preventive maintenance prolongs the life of the equipment by ensuring its proper use. When tools are kept dressed or sharpened and in satisfactory condition, the right tool will be used for the job. When safe and properly maintained tools are issued, workers have an added incentive to give the tools better care. When repairs are made so equipment is not inoperative for long periods, workers do not need to improvise by using a piece of equipment for a purpose for which it was not intended. Sound and efficient maintenance management anticipates machine and equipment deterioration and sets up overhaul procedures designed to correct defects as soon as they develop. Such a repair and overhaul system obviously requires close integration of maintenance with inspection.

Preventive maintenance has four main components:

1. Scheduling and performing periodic maintenance functions
2. Keeping records of service and repairs
3. Repairing and replacing equipment and equipment parts
4. Providing spare parts control.

Maintenance schedules can be set up on either a time or use basis, whichever comes first. Factors to be considered include:

- The age of the machine
- The number of hours per day the machine is used
- Past experience
- The manufacturer's recommendations.

The manufacturer's specifications provide standards that need to be maintained for safe and economical use of the machine. These specifications give maintenance personnel definite guidelines to follow. Examples of scheduled activities include lubricating each piece of equipment; replacing belts, pulleys, fans, and other parts; and checking and adjusting brakes.

Two types of records should be kept. The first is a maintenance service schedule for each piece of equipment. This schedule indicates the date the equipment was purchased or placed in operation, its cost (if known), where it is used, each part to be serviced, the kind of service required, the frequency of service, and the person assigned to the servicing. Each piece of equipment also requires a repair record, which includes an itemized list of parts replaced or repaired and the name of the person who did the work.

In addition to scheduled adjustments and replacements, maintenance personnel must repair malfunctioning or broken equipment in accordance with the manufacturer's specifications. Sometimes equipment must be sent back to the manufacturer or his service representative for repair. Maintenance personnel should be aware of their limitations and recognize that their experience and expertise are not sufficient for all repairs. Those assigned repair responsibilities require special safety training. Many of the jobs to be performed include testing or working on equipment with guards and safety devices removed. Therefore, a statement of necessary precautions should accompany the repair directive.

Maintenance personnel (along with others) have a responsibility to tag and lock out defective equipment, or tag if the equipment, like a ladder, cannot be locked out.

Another element of the total preventive maintenance program is the survey of spare parts requirements. In order to keep needed repair parts on hand, it is necessary to schedule review and reordering of stocked spare parts. If maintenance personnel keep purchasing agents informed of their anticipated stock needs, downtime while waiting for parts to arrive will be prevented.

The difference between a mediocre maintenance program and a superior one is the former is aimed at maintaining facilities, the latter at improving them. If conditions are good, a mediocre program will keep them that way but will not make them better; if conditions are not good, a mediocre program will not improve them. Preventive maintenance, on the other hand, is a program of mutual support that creates safe conditions, eliminates costly delays and breakdowns, and prolongs equipment life.

**Employees.** Employees make the safety and health program succeed. Well-trained and educated employees are the greatest deterrent to damage, injuries, and health problems in the plant or establishment.

What are the specific ways a hazard control program can be rooted in employee involvement and concern? Employees can observe safety and health rules and regulations and work according to standard procedures and practices. They can recognize and report to the foreman or supervisor hazardous conditions or unsafe work practices in the plant. They can develop and practice good habits of hygiene and housekeeping. They can use protective and safety equipment, tools, and machinery properly. They can report all injuries or hazardous exposure as soon as possible. Employees can help develop safe work procedures and make suggestions for improving work procedures.

What shapes an employee's attitude toward safety? From the day an employee goes to work, whether or not the firm has a formal training program, the employee starts to form attitudes. Substantial if subtle influence is exerted by the attitudes seen in management, supervisors, and fellow workers. If they regard safety as integral to an effective operation, a mark of skill and good sense, if they participate actively and cooperatively in the safety program, if employees are recognized for having good safety records, then the employee will regard safety as something important, not as window-dressing or a gimmick to which lip service is periodically paid.

Introduction to the safety program should come on the employee's first day on the job. A three-pronged approach is suggested (Kane, 1979):

1. Coverage of general company policy and rules; discussion of various benefit programs, for example, hospitalization, pension plan, holidays, sick leave. The personnel department usually is responsible for giving this information immediately after the new employee is added to the payroll.
2. Discussion of general safety rules and the safety program. This part of the program should be the responsibility of a safety professional. The company's safety handbook should be given to the new employee, and the company's safety policy statement should be explained. The reasons behind the general safety rules should be explored with the employee, who is more likely to follow rules when they are understood.
3. Explanation of specific safety rules that are applicable in the new employee's department. At this point the supervisor's role overlaps with that of the safety professional. The supervisor can show how some hazards have been eliminated while others, which could not be designed out of the operation, are guarded against. This discussion provides an opportunity to talk about safe work practices and emergency procedures as well as to show how personal protective equipment can further reduce the effects of the hazard. The employee's responsibilities for safety and health also must be stressed. These responsibilities include reporting all accidents and incidents, checking equipment and tools before use, and operating equipment only with proper authorization and prior instruction.

Figure 3-10 lists the company safety rules developed by the Construction Advancement Foundation SAFE Committee and distributed to construction employees in Indiana.

**Purchasing agents.** Those responsible for purchasing items for organizations are in a key position to reduce hazards associated with operations. The purchasing department has much latitude in selecting machinery, tools, equipment, and materials used in the organization. In maintaining standards of quality, efficiency, and price, the purchasing department must make certain that safety has received adequate attention in designing, manufacturing, and shipping items.

Depending upon the company organization, other departments—such as safety, engineering, quality control, maintenance, industrial hygiene, and medical—should indicate to the purchasing department what equipment and materials meet with their approval. The purchasing department is responsible

**COMPANY SAFETY RULES**

*All Employees Will Abide By The Following Rules:*

1. Report unsafe conditions to your immediate supervisor.
2. Promptly report all injuries to your immediate supervisor.
3. Wear hard hats on the jobsite at all times.
4. Use eye and face protection where there is danger from flying objects or particles, such as when grinding, chipping, burning and welding, etc.
5. Dress properly. Wear appropriate work clothes, gloves, and shoes or boots. Loose clothing and jewelry should not be worn.
6. Never operate any machine unless all guards and safety devices are in place and in proper operating condition.
7. Keep all tools in safe working condition. Never use defective tools or equipment. Report any defective tools or equipment to immediate supervisor promptly.
8. Properly care for and be responsible for all personal protective equipment.
9. Be alert and keep out from under overhead loads.
10. Do not operate machinery if you are not authorized to do so.
11. Do not leave materials in aisles, walkways, stairways, roads or other points of egress.
12. Practice good housekeeping at all times.
13. Do not stand or sit on sides of moving equipment.
14. The use of, or being under the influence of, intoxicating beverages or illegal drugs while on the job is prohibited.
15. All posted safety rules must be obeyed and must not be removed except by management's authorization.
16. Comply at all times with all known federal, state and local safety laws as well as employer regulations and policies.
17. Horseplay causes accidents and will not be tolerated.

Violations of any of these rules may be cause for immediate disciplinary action.

**Figure 3-10.** Reprinted with permission from the Construction Advancement Foundation SAFE Committee.

for soliciting such guidance and direction. (Purchasing responsibilities are covered in depth later in this chapter.)

First, the purchasing agent must make certain all items comply with federal and state or provincial regulations and with local ordinances. A statement to this effect must be part of the purchase order. Purchasing agents also will be guided by (1) codes and standards of the American National Standards Institute (ANSI), Canadian Standards Association (CSA), and other standards and specifications groups; (2) products approved or listed by such agencies as Underwriters Laboratories Inc. and the National Fire Protection Association; and (3) recommendations by such agencies as the National Safety Council, insurance carriers or associations, the Factory Mutual System in its *Factory Mutual Handbook of Industrial Loss Prevention* and *Loss Prevention Data,* and trade or industrial organizations. (See Chapter 24, Sources of Help.)

Second, the purchasing agent must make certain tools, equipment, materials, and machinery are purchased with adequate regard for safety. This requirement applies even to ordinary items such as boxes, cleaning rags, paint, and common hand tools. It is essential to compare safety features among the various brands when purchasing personal protective equipment and larger items, especially machines. Sometimes the cost of an adequately guarded machine seems out of proportion to that of an unguarded machine to which makeshift guards can be added. But experience has repeatedly proven that the best time to eliminate or minimize a hazard is in the design stage. Safeguards that are integral parts of a machine are the most efficient and durable.

Third, the purchasing agent must be cost-conscious, realizing that every accident has both direct and indirect costs. The agent should understand that the organization cannot afford bargains which later result in accident losses and occupational disease.

### Professionals in hazard control

The roles of various professionals in the hazard control program are described in National Safety Council's *Fundamentals of Industrial Hygiene,* 3rd ed. The following discussion, therefore, describes briefly the responsibilities of the safety and health professional, the industrial hygienist, and staff medical personnel.

**Safety and health professionals (hazard control specialists).** To assure the continuity of the safety program, management usually places program administration in the hands of a safety director or manager of safety and health. The number of full-time safety professionals is increasing as the nature of their duties is better understood.

To effectively administer a safety program requires considerable training and many years of experience. A safety program has many facets: occupational health, product safety, machine design, plant layout, security, damage control, fire prevention. Safety as a profession combines engineering, management, preventive medicine, industrial hygiene, and organizational psychology. It requires knowledge of system safety analysis, job safety analysis, job instruction training, human factors engineering, biomechanics, and product safety. The professional must have thorough knowledge of the organization's equipment, facilities, manufacturing process, and workers' compensation, and must be able to communicate and work with all types of people. Other desirable traits are the ability to see both management and employee viewpoints, and being a good trainer. A list of tasks performed by occupational safety and health profes-

## THE SCOPE OF THE PROFESSIONAL SAFETY POSITION

A safety and health professional brings together those elements of the various disciplines necessary to identify and evaluate the magnitude of the safety problem. He is concerned with all facets of the problem, personal and environmental, transient and permanent, to determine the causes of accidents or the existence of loss producing conditions, practices or materials.

Based upon the information he has collected and analyzed, he proposes alternate solutions, together with recommendations based upon his specialized knowlege and experience, to those who have ultimate decision-making responsibilities.

The functions of the position are described as they may be applied in principle to the safety professional in any activity.

The safety professional in performing these functions will draw upon specialized knowledge in both the physical and social sciences. He will apply the principles of measurement and analysis to evaluate safety performance. He will be required to have fundamental knowledge of statistics, mathematics, physics, chemistry, as well as the fundamentals of the engineering disciplines.

He will utilize knowledge in the fields of behavior, motivation, and communications. Knowledge of management principles as well as the theory of business and government organization will also be required. His specialized knowledge must include a thorough understanding of the causative factors contributing to accident occurrence as well as methods and procedures designed to control such events.

The safety professional of the future will need a unique and diversified type of education and training if he is to meet the challenges of the future. The population explosion, the problems of urban areas, future transportation systems, as well as the increasing complexities of man's every day life will create many problems and extend the safety professional's creativity to its maximum if he is to successfully provide the knowledge and leadership to conserve life, health, and property.

### Functions of the Professional Safety Position

The major functions of the safety professional are contained within four basic areas. However, application of all or some of the functions listed below will depend upon the nature and scope of the existing accident problems, and the type of activity with which he is concerned.

The major areas are:

A. Identification and appraisal of accident and loss producing conditions and practices and evaluation of the severity of the accident problem.
B. Development of accident prevention and loss control methods, procedures, and programs.
C. Communication of accident and loss control information to those directly involved.
D. Measurement and evaluation of the effectiveness of the accident and loss control system and the modifications needed to achieve optimum results.

### A. Identification and Appraisal of Accident and Loss Producing Conditions and Practices and Evaluation of the Severity of the Accident Problem

These functions involve:

1. The development of methods of identifying hazards and evaluating the loss producing potential of a given system, operation or process by:
   a. Advanced detailed studies of hazards of planned and proposed facilities, operations and products.
   b. Hazard analysis of existing facilities, operations and products.
2. The preparation and interpretation of analyses of the total economic loss resulting from the accident and losses under consideration.
3. The review of the entire system in detail to define likely modes of failure, including human error and their effects on the safety of the system.
   a. The identification of errors involving incomplete decision making, faulty judgment, administrative miscalculation and poor practices.
   b. The designation of potential weaknesses found in existing policies, directives, objectives, or practices.
4. The review of reports of injuries, property damage, occupational diseases or public liability accidents and the compilation, analysis, and interpretation of relevant causative factor information.
   a. The establishment of a classification system that will make it possible to identify significant causative factors and determine needs.
   b. The establishment of a system to ensure the completeness and validity of the reported information.
   c. The conduct of thorough investigation of those accidents where specialized knowledge and skill are required.
5. The provision of advice and counsel concerning compliance with applicable laws, codes, regulations, and standards.
6. The conduct of research studies of technical safety problems.
7. The determination of the need of surveys and appraisals by related specialists such as medical, health physicists, industrial hygienists, fire protection engineers, and psychologists to identify conditions affecting the health and safety of individuals.
8. The systematic study of the various elements of the environment to assure that tasks and exposures of the individual are within his psychological and physiological limitations and capacities.

### B. Development of Accident Prevention and Loss Control Methods, Procedures, and Programs

In carrying out this function, the safety professional:

1. Uses specialized knowledge of accident causation and control to prescribe an integrated accident and loss control system designed to:
   a. Eliminate causative factors associated with the accident problem, preferably before an accident occurs.

Figure 3-11. (Continued on next page.)

b. Where it is not possible to eliminate the hazard, devise mechanisms to reduce the degree of hazard.
c. Reduce the severity of the results of an accident by prescribing specialized equipment designed to reduce the severity of an injury should an accident occur.

2. Establishes methods to demonstrate the relationship of safety performance to the primary function of the entire operation or any of its components.
3. Develops policies, codes, safety standards, and procedures that become part of the operational policies of the organization.
4. Incorporates essential safety and health requirements in all purchasing and contracting specifications.
5. As a professional safety consultant for personnel engaged in planning, design, development, and installation of various parts of the system, advises and consults on the necessary modification to ensure consideration of all potential hazards.
6. Coordinates the results of job analysis to assist in proper selection and placement of personnel, whose capabilities and/or limitations are suited to the operation involved.
7. Consults concerning product safety, including the intended and potential uses of the product as well as its material and construction, through the establishment of general requirements for the application of safety principles throughout planning, design, development, fabrication and test of various products, to achieve maximum product safety.
8. Systematically reviews technological developments and equipment to keep up to date on the devices and techniques designed to eliminate or minimize hazards, and determine whether these developments and techniques have any applications to the activities with which he is concerned.

**C. Communication of Accident and Loss Control Information to Those Directly Involved**

In carrying out this function the safety professional:

1. Compiles, analyzes, and interprets accident statistical data and prepares reports designed to communicate this information to the appropriate personnel.
2. Communicates recommended controls, procedures, or programs designed to eliminate or minimize hazard potential, to the appropriate person or persons.
3. Through appropriate communication media, persuades those who have ultimate decision-making responsibilities to adopt and utilize those controls which the preponderance of evidence indicates are best suited to achieve the desired results.
4. Directs or assists in the development of specialized education and training materials and in the conduct of specialized training programs for those who have operational responsibility.
5. Provides advice and counsel on the type and channels of communications to insure the timely and efficient transmission of useable accident prevention information to those concerned.

**D. Measurement and Evaluation of the Effectiveness of the Accident and Loss Control System and the Needed Modifications to Achieve Optimum Results**

1. Establishes measurement techniques such as cost statistics, work sampling or other appropriate means, for obtaining periodic and systematic evaluation of the effectiveness of the control system.
2. Develops methods that will evaluate the costs of the control system in terms of the effectiveness of each part of the system and its contribution to accident and loss reduction.
3. Provides feedback information concerning the effectiveness of the control measures to those with ultimate responsibility, with the recommended adjustments or changes as indicated by the analyses.

**Figure 3-11.** (Concluded.)

sionals is included in the Board of Certified Safety Professionals, Curricula Development and Examination Study Guidelines, Technical Report No. 1. (See References.)

The passage of the Occupational Safety and Health Act of 1970 requires that certain safety standards be met and maintained. Generally, organizations with moderate or high hazards and/or employing 500 or more persons need a full-time safety professional. The nature of the operation may indicate the need for a full-time professional, regardless of the number of people employed.

The growing number of safety professionals is reflected in the growing membership of the American Society of Safety Engineers (ASSE), which is about 21,000. This organization has approximately one hundred chapters in the U.S. and Canada.

In 1968 the ASSE was instrumental in forming a new organization, the Board of Certified Safety Professionals (BCSP). Its purpose is to provide professional status by certification to qualified safety people who meet strict educational and experience requirements and pass an examination. Several thousand safety professionals have been certified since the BCSP was formed. (Details of both organizations are given in Chapter 24, Sources of Help.)

The safety professional—whether called a safety engineer, safety director, hazard control specialist, loss control manager, a safety and health professional, or some other title—normally functions as a specialist on the management level. The hazard control program should enjoy the same position as other established activities of the organization, such as sales, production, engineering, or research. Its budget reflects top management's commitment to the safety and health of its employees and includes the safety professional's salary, the salary of staff to help him, travel allowance, cost of safety equipment, cost of training and continuing education, and other related items. Safety professionals must define needs of their program and, according to priorities, make short- and long-range budget projections. With such projections in hand, they are able to pre-

**Figure 3-12.** Each year, purchasing agents and safety and health professionals glean product and service information at the National Safety Congress and Exhibition.

sent their needs to those with fiscal responsibility and stand a better chance of acquiring what they need to make their programs function.

Figure 3-11 is the ASSE's description of the duties of the safety and health professional, who does not need to do all of them all the time, however.

In general, the safety professional advises and guides management, supervisors, foremen, employees, and such departments as purchasing, engineering, and personnel on all matters pertaining to safety. Formulating, administering, monitoring, evaluating, and improving the accident prevention program are other responsibilities of the safety and health professional.

The safety professional usually investigates serious accidents personally or through the staff, reviews supervisors' accident reports, checks corrective actions taken to eliminate accident causes, makes necessary reports to management, and maintains the accident report system, including files. The safety and health professional recommends safety provisions in (1) plans and specifications for new building construction, (2) repair or remodeling of existing structures, (3) safety equipment, (4) designs of new equipment, and (5) processes, operations, and materials. The safety professional provides (or cooperates with the training supervisor to provide) safety training and education for employees. The safety staff also take continuing education courses and attend professional meetings like the National Safety Congress.

Further, the safety professional makes certain that federal, state or provincial, and local laws, ordinances, orders, and regulations relating to safety and health are complied with and that standards, whether mandatory or recommended, are met; accepts responsibility for preparing the necessary reports (for example, OSHA Forms 100 and 102) for management. On jobs involving subcontractors, the safety professional informs each subcontractor of relevant hazard control responsibilities; this is often spelled out in the contract. (See the discussion in the *Engineering and Technology* volume, Chapter 2, Construction and Maintenance of Plant Facilities.)

The safety professional supervises disaster control, fire prevention, and firefighting activities when they are not responsibilities of other departments and stimulates and maintains employee interest in and commitment to safety.

It is not uncommon to find the safety professional has been given the authority to order immediate changes on fast-moving and rapidly changing operations or in cases where delayed action could endanger lives (as in construction, demolition, or emergency work, fumigation, and some phases of manufacture of explosives, chemicals, or dangerous substances). Where the safety professional exercises this authority, it is done with discretion by accepting accountability to management for errors in judgment but realizing that errors on the side of caution are more easily justified than unnecessary risk-taking.

**Industrial hygienists.** The industrial hygienist is trained to recognize, evaluate, and control health hazards—particularly chemical, physical, and biological agents—that exist in the workplace and have injurious effects on workers. Specialists work in fields such as toxicology, epidemiology, chemistry, ergonomics, acoustics, ventilation engineering, and statistics.

The American Board of Industrial Hygiene (ABIH) examines and certifies persons in industrial hygiene. Many certified industrial hygienists are also certified safety professionals. Physicians, nurses, and safety professionals can move part or all of the way into industrial hygiene functions.

What are the specific responsibilities of the industrial hygienist in the hazard control program? The industrial hygienist: (1) recognizes and identifies those chemical, physical, and biological agents that can adversely affect the physical and mental health and well-being of the worker; (2) measures and documents levels of environmental exposure to specific hazardous agents; (3) evaluates the significance of exposures and their relationship to occupationally and environmentally induced diseases; (4) establishes appropriate controls and monitors their effectiveness; and (5) recommends to management how to correct unhealthy (or potentially unhealthy) conditions.

**(Staff) medical personnel.** Occupational health services vary greatly from one organization to another, depending on number of employees, nature of the operation, and the commitment of the employer. One plant may include a full-time physician, nurses, and technicians, with treatment rooms and a dispensary. Another may only have the required first aid kit and a person trained to adequately render first aid and cardiopulmonary resuscitation (CPR).

The small plant has special need for persons thoroughly trained in first aid and CPR to take care of the employees on all shifts. Sometimes small companies and plants in the same locality share the services of a qualified physician on either a part- or full-time basis.

How much responsibility a part- or full-time nurse assumes generally depends on the availability of a licensed physician. It is not unusual for an occupational nurse to be solely and completely responsible for the occupational safety and health program with medical direction from a physician who rarely visits the workplace.

The American Association of Occupational Health Nurses in 1977 defined occupational health nursing as "the application of nursing principles in conserving the health of workers in all occupations. It involves prevention, recognition, and treatment of illness and injury, and requires special skills and knowledge in the fields of health, education, and counseling, environmental health, rehabilitation, and human relations." In addition to being responsible for health care, the occupational nurse may on occasion monitor the workplace, perform industrial hygiene sampling, act as a consultant on sanitary standards, and be responsible for health education.

Chapter 19, Occupational Health Services, deals specifically with the duties of the industrial physician, nurse, and first aid attendants. Regardless of size, health programs share common goals: to maintain the health of the work force, to prevent or control diseases and accidents, and to prevent or at least reduce disability and resulting lost time.

Staff medical personnel are a vital part of the total health and safety program. Their most obvious contribution is to provide emergency medical care for employees who are injured or become ill on the job. An outgrowth of this responsibility is to provide follow-up treatment of employees suffering from occupational disease or injuries. For many, periodic examinations are required by OSHA regulations. A further outgrowth is medical personnel's promotion of health education programs for employees and their families. Assisting with problems of alcohol and drug abuse; keeping immunizations up to date; promoting mental health, weight control, and regular exercise: these are only some ways that medical personnel can encourage the health and well-being of employees.

Medical personnel foster a healthful environment by occasional tours of the workplace. Such tours also familiarize medical personnel with the materials and processes used and procedures performed by the employees under their care.

Another way that medical personnel are important to the health and safety organization is in the placement of employees. Proper placement matches the worker to the demands of the job. Often a preplacement examination gives the medical staff an opportunity to recommend where a prospective worker is to be assigned. This placement should take into consideration the physical capacity and mental ability of the employee so no one is subjected to unnecessary safety and health risks.

## PURCHASING

The safety department should have excellent liaison not only with the engineering department but also with the purchasing department, as discussed in the previous section.

It should be the duty of the safety department to devise and put in writing the safety standards that will guide the purchasing department. These standards should be set up so the hazards involved in a particular kind of equipment or material are eliminated, for example, by the substitution of a safe material for a dangerous one, or safeguarded for the protection of the worker, the machine, and the product.

The purchasing agent is not closely concerned with educational and enforcement activities, but is vitally concerned with many phases of the engineering activities. It is the agent's duty to select and purchase the various items of machinery, tools, equipment, and materials used in the organization. It is also the agent's responsibility—in part or to a considerable degree—to see that in design, manufacture, and particulars of shipment of all these items, safety and health have received adequate attention.

In one example, a lead hazard occurred in the unloading of litharge (PbO) shipped in 10-gallon paint pails with covers. These pails arrived, either in trucks or in boxcars, with a film of litharge on the outside. When they were moved, a lead concentration was produced in the air at 30 to 40 times the permissible limit.

Several solutions were considered and tried, but the answer was to eliminate the hazard by having the purchasing department specify a rubber gasket under the pail lid as a part of the purchasing requirements. Thus the leakage, which created a serious health hazard, was easily controlled.

### Specifications

The engineering department, with the help of the safety department, should specify the necessary safeguarding to be built into a machine before it is purchased.

Persons responsible for purchasing in an industrial plant are necessarily cost conscious. Consequently, the safety professional must become aware of the accident losses to the company in terms of specific machines, materials, and processes. For instance, to recommend the expenditure of several thousand dollars for a superior grade of tool, the evidence must justify the investment.

Because of highly competitive marketing, manufacturers of machine tools and processing equipment often list safety devices as accessories. It is important that the safety professional be familiar with OSHA-required auxiliary equipment and be able to justify its inclusion in the original order.

In some large organizations, the safety professional is charged with checking all plans and specifications for machinery and other equipment. In many organizations, particularly where certain items, such as goggles or safety shoes, are to be reordered from time to time, standard lists have been prepared through the cooperation of various operating officials, and purchases are selected only from among the types and companies shown on these approved lists. In still other establishments, the responsibility for design, quality, safety, and other features rests with the employees who are to use the articles. In such cases, the purchasing agent is responsible only for price, date of delivery, and similar details.

In many companies where purchases are made in enormous quantities and at a great investment of money, important duties are placed in three coordinate departments: (1) the engineering department where plans and specifications are prepared for all machinery and equipment to be purchased; (2) the safety department where these plans and specifications are carefully checked for safety, and final inspections of articles purchased are carried out; and (3) the purchasing department, which still has latitude in making selections as well as in determining standards of quality, efficiency, and price.

Still another variable must be mentioned. Many companies have both a full-time purchasing agent and a full-time safety professional. However, in many other companies, especially smaller ones, these important duties are assumed by executives who devote part of their time to other activities. Nevertheless, the measures that should be taken to prevent accidents in the small plant are substantially the same as those in the large plant. The part the purchasing agent can take in the safety program is similar; the interest will be the same and the activities will vary only by degree. Opportunity will lie in accepting as fully as possible all the suggestions that are presented here and in cooperating as closely as possible with others for the safety of all the workers in the company.

**Specification of shipping methods.** When materials are ordered, it may be desirable to specify they be shipped in a particular manner. If safe and efficient shipping methods are worked out and then specified in the orders, the suppliers will be better able to deliver materials on time, in good condition, and in a shape or form that can be easily and safely handled by employees. Labeling of hazardous materials should be specified. Use DOT-authorized shipping labels.

Materials Safety Data Sheets must be supplied with all chemicals or products that contain chemicals.

## Codes and standards

In purchasing, the safety professional will find a need for thorough knowledge of the accident history of the plant, the costs involved in accidents, and the probable benefits of the changes suggested.

To fulfill this function in cooperation with the purchasing department, it is important that the safety professional be familiar with codes and standards. When a specific item of equipment is recommended, the safety professional should be able to state that it is a type approved by authoritative bodies and that it meets OSHA requirements.

It is usually necessary for the safety professional to consult with everyone concerned before setting up company standards for the guidance of the purchasing department.

There are many guides and standards that can be used as models. Accordingly, the safety professional (and all others concerned with setting company standards) should be familiar with the following:

1. Codes and standards approved by the American National Standards Institute and other standards and specifications groups—see Chapter 24, Sources of Help.
2. Codes and standards adopted or set by federal, state, and local governmental agencies, such as the Occupational Safety and Health Administration, the Bureau of Mines, and the National Bureau of Standards.
3. Codes, standards, and lists of approved or tested devices published by agencies such as the National Institute of Occupational Safety and Health, the Mining Safety and Health Administration, Underwriters Laboratories Inc., and fire protection organizations. For fire protection, the standards and codes of the National Fire Protection Association should be followed. (See Chapter 24.)
4. Safe practice recommendations of such agencies as the National Safety Council, insurance carriers or their associations, and trade and industrial organizations (see Figure 3-12).

## Purchasing-safety liaison

With a background of knowledge gathered from materials listed above, the safety professional should be well prepared to advise the purchasing department when required to do so.

**What purchasing can expect.** The purchasing agent can reasonably expect that the safety professional will:

1. Give specific information about process and machine hazards that can be eliminated by change in design or by installing manufacturer-designed guarding.
2. Supply similar information about other equipment, tools, and materials along with facts about injuries caused.
3. Give specific information about health and fire hazards in the workplaces.
4. Provide information on federal and state safety requirements.
5. Supply on request additional special information on accident experience with machines, equipment, or materials when such articles are about to be reordered.
6. Request assistance in the investigation of accidents that may have been caused by faulty equipment or material.

**What the safety department can expect.** Where there is effective liaison between the safety and purchasing departments, the safety professional can expect that the purchasing agent will:

1. Become familiar with the departmental and plant process hazards, especially in relation to machinery, equipment, and materials.
2. Ask the safety department for information on hazards and accident costs, for federal and state safety requirements,

for lists of approved devices and appliances before making purchases.
3. Become acquainted with the specific location and departmental use of machinery or equipment about to be ordered.
4. Participate in accident investigations where injuries may have been caused through the failure of machinery, equipment, or materials.

### Safety considerations

In the purchase of many articles, there is no need to consider safety. Some items, however, have a more important bearing upon safety than may be suspected.

Utmost caution should be observed when purchasing personal protective equipment, such as eye protection, respirators, and masks; equipment to move suspended loads, such as ropes and chains; equipment to move and store materials; and miscellaneous substances and fluids for cleaning and other purposes that might constitute or aggravate a fire or health hazard. Adequate labeling that identifies contents and calls attention to hazards should be specified.

Investigation, however, may show that unsuspected hazards also lie in the purchase of very ordinary items, such as common hand tools, reflectors, tool racks, cleaning rags, paint for shop walls and machinery, and even filing cabinets. Among the factors to be considered by the purchasing agent are, for example, maximum load strength; long life without deterioration; reduction of sharp, rough, or pointed characteristics; less frequent need for adjustment; ease of maintenance; reduction of fatigue-causing characteristics; and minimal hazard to the workers' health.

Here are a few examples of hazards created by purchased items that were thought to be safe. Goggles supplied to one group of workers were found to have imperfections in the lenses that caused eyestrain and headache, which led to fatigue and accidents. The toes of a laborer were crushed because his safety shoe had an inferior metal cap and collapsed under a weight that would have been easily supported by a well-made shoe. In another plant, workers were supplied with wooden carrying boxes, when a proper type of metal box could have eliminated the hazard of splinters and perhaps an infected hand.

It is the purchase of larger items, however, and especially in the buying of machines, that the more spectacular examples of purchasing for safety are found. Today, machines of many types are manufactured and can be purchased with adequate safeguards in place as integral parts of the machines. The enclosed motor drive is an outstanding example of engineering machine construction for safety.

When an order for equipment is about to be placed, the purchasing agent, if possible, should not consider any machine which has been only partly guarded by the manufacturer and which needs to be fitted with makeshift safeguards. The agent should be in frequent consultation with the safety department before making any purchase where safety is a factor. The agent also should be particularly careful to see that every purchased machine complies fully with the safety regulations of the state in which it is to be operated.

### Price considerations

When considering plant purchases, there is a constant struggle in the mind of the purchasing agent to reconcile quality, work efficiency, and safety with the price of an item.

Sometimes it seems that the cost of an adequately safeguarded machine is out of proportion to the cost of an unguarded machine, including the estimated cost of adding home-made safeguards. But the experience of many industrial plants has proved again and again that the best time to safeguard a machine or process is in the design stage. Safeguards planned and built as integral parts of a machine are the most efficient and durable.

The purchasing agent, through accident information supplied by the safety department, is familiar with the costs of specific accidents in the plant, especially those in which an unsafe condition caused by a purchased item has been found to contribute to an accident.

The purchasing agent should be able to defend the decision to buy a slightly higher priced component, if that is necessary. Cooperation between the safety, engineering, manufacturing, and purchasing departments is absolutely necessary if accidents are to be eliminated.

These are arguments that must appeal to all executives who are responsible for the success of the industrial organization. The executive who is already sold on safety is agreeable to expenditures reasonably justified in the interest of accident prevention. If an executive does not have this attitude, means of increasing interest and of broadening understanding of the accident situation must be sought. In this undertaking, the purchasing agent can undoubtedly count upon the active cooperation of the safety professional.

In some instances, the purchase of machinery or equipment involves important engineering details. For such purchases, the company undoubtedly will have a system whereby definite specifications, perhaps including drawings, will first be prepared by engineers. These plans and specifications will then be carefully checked by the safety professional before bids and estimates of cost are solicited.

The purchasing agent will have the plans and specifications at hand when asking for prices; and will want to keep in close touch with the safety professional throughout the negotiations in order to use the latter's knowledge and experience in accident prevention.

After the purchase order has been made out and before it is signed, one other most important detail should not be overlooked. This is a statement, in language that cannot possibly be misinterpreted, that the articles ordered must comply fully with the applicable federal and state safety laws and regulations of the locality in which they are to be used.

This statement must be made a part of the purchase order.

## SAFETY AND HEALTH COMMITTEES

Safety and health committees can be invaluable to the hazard control program by providing the active participation and cooperation of many key people in the organization. They also can be unproductive and ineffective. The difference between success and failure lies with the original purpose of the committee, its staffing and structure, and the support it receives while carrying out its responsibilities.

A safety and health committee is a group that aids and advises both management and employees on matters of safety and health pertaining to plant or company operations. In addition, it performs essential monitoring, educational, investigative, and evaluative tasks.

Committees may represent various constituencies or levels within the organization. Another division is based on function: management or workplace committees. The joint safety and health committee (discussed here) is responsible for:

- Actively participating in safety and health instruction programs and evaluating the effectiveness of these programs
- Regularly inspecting the facility to detect unsafe conditions and practices and hazardous materials and environmental factors
- Planning improvements to existing safety and health rules, procedures, and regulations
- Recommending suitable hazard elimination or reduction measures
- Periodically reviewing and updating existing work practices and hazard controls
- Assessing the implications of changes in work tasks, operations, and processes
- Field-testing personal protective equipment and making recommendations for its use or alteration based on the findings
- Monitoring and evaluating the effectiveness of safety and health recommendations and improvements
- Compiling and distributing safety and health and hazard communications to the employees
- Immediately investigating any workplace accident
- Studying and analyzing accident and injury data.

The OSHAct in Section 2(b)(13) clearly contemplates the possibility of joint safety and health initiatives as a supplementary approach to more effectively accomplishing OSHA's objectives. Joint committees have considerable potential for reducing injuries and illnesses, thus leaving OSHA free to target enforcement according to the worst-first principle.

The joint committee concept stresses cooperation and a commitment to safety as a shared responsibility. Employees can become actively involved in and make positive contributions to the company's safety and health program. Their ideas can be translated into actions. The committee serves as a forum for discussing changes in regulations, programs, or processes and potential new hazards. Employees can communicate problems to management openly and face to face. Information and suggestions can flow both ways. The knowledge and experience of many persons combine to accomplish the objectives of creating a safe workplace and reducing accidents. With many minds addressing a problem simultaneously, with so much thought power concentrating on an issue, effective solutions are more easily produced in the give-and-take atmosphere. Because they facilitate communication and cooperation, joint committees usually result in higher morale as well.

Even though a joint committee represents both employees and management, analyses and recommendations of the committee—whether they pertain to policy or practice—should be reviewed and confirmed when they relate to specialized areas by those with expertise (for example, electrical safety, exposure levels).

Labor-management cooperation was discussed in Chapter 1, Occupational Safety: History and Growth. Committee organization and operation are covered in the Council publication "You Are the Safety and Health Committee."

## OFF-THE-JOB SAFETY PROGRAMS

There is a certain amount of confusion as to what off-the-job (OTJ) safety really includes. Essentially, off-the-job safety involves employees, and is a term used by employers to designate that part of their safety program directed to the employees when they are not at work.

The principle aim of off-the-job safety is to get an employee to follow the same safe practices while pursuing outside activities as used on the job. Experience indicates, however, that many individuals tend to leave their safety training at the workplace when they go home. Therefore, off-the-job safety should not be a separate program, but rather an extension of a company's on-the-job safety program.

One of the basic reasons for a company to become involved in off-the-job safety is employee-power. While companies now have a legal responsibility to prevent injuries on the job, they have a moral responsibility to try to prevent injuries away from the job. All injuries are a waste of a valuable resource—people. Injuries and fatalities happen to people who call on customers, make a product, service equipment, keep the books, and do many other jobs involved in running a business.

The other reason for an off-the-job safety program is cost. Operating costs and production schedules are affected as much when employees are injured away from work as when they are injured on the job. (These costs are discussed in detail in Chapter 7, Accident Investigation, Analysis, and Costs.)

### What is off-the-job safety?

Off-the-job safety is a logical extension of the occupational safety program. Accident/illness prevention at work makes good business sense (it saves dollars), while fulfilling an organization's moral and legal responsibilities. An effective off-the-job safety program meets these same needs: reduction of costly employee absences due to accidents, injuries, or deaths; and commitment to employee well-being.

In today's society, the traditional family has changed and continues to change. The high cost of living, changes in educational attainment, and societal expectations often result in more than one wage earner per family. Therefore, if both adults are employed outside the home, accidents, injuries, and deaths to members of an employee's immediate family may result in an employee absence.

The objective of off-the-job safety is to prevent accidents, injuries, illnesses, and deaths of employees and their families. Preventing off-the-job incidents that could result in injury/illness can be accomplished by using methods proven successful for increasing safety awareness at work.

### Promoting OTJ safety

There is nothing special about techniques for promoting off-the-job safety. The same principles and techniques used to put across safety on the job are employed. From a safety standpoint, operating power tools at home is the same as operating the same equipment at work; driving the family car is the same as driving a company vehicle.

A company *does* have to depend more on education and persuasion to get its message across, because once an employee leaves the office, plant, or job site, the assumption of risk is thought to be a personal right, and supervision is no longer a mitigating factor. An employee must realize that accidents do not always happen to other people.

As with any other program—whether it be attendance, quality control, or waste reduction—management support and guidance is essential. Once management has been

**Figure 3-13.** During this off-the-job safety activity, hunting instructors show the recommended method of crossing a fence with a weapon. One person holds the weapons while his companion goes over or through the fence. The weapons are passed across the fence, muzzle up, then the other person crosses.

shown the seriousness of the problem (through experience and cost records), there should be little difficulty in obtaining support.

### Program benefits

Many benefits are derived by developing or revitalizing an off-the-job safety program. These include:

- Fewer off-the-job accidents, injuries, and deaths;
- Fewer employee absences;
- Reduced operating costs;
- Safety awareness at the home is brought back to the employee's workplace;
- Improved work efficiency and performance;
- Enhanced employee and employer relationships;
- Participation of employees and their families in community safety actions;
- Positive, viable demonstration of organization's commitment to employee-family well-being and social issues affecting the employee's family.

As with any viable occupational safety program, management involvement and participation in off-the-job safety must be vocal, visible, and continuous—from management through all supervisory levels.

### Off-the-job safety policy

Communicate management commitment to employees and their families. For example, a written policy statement concerning off-the-job safety or reference to off-the-job safety should be in the occupational safety policy statement signed by the top organizational official.

### Getting started

Various methods and sources of information on off-the-job accidents are available to an organization. For example, records may be kept of the causes of employee absences (accidents); health and accident insurance claims may record the accident cause on the form for payment; the National Safety Council's annual publication *Accident Facts* can be used to pinpoint the leading causes of accidental deaths and injuries nationally (these data can be used to ascertain the most prevalent types of accidents for a particular organization).

It is important to tailor the off-the-job safety program to the particular needs of an organization. The location of the plant site, its environment, and the special interests of employee groups such as skiing (water and snow), boating, mountain-climbing, spelunking (cave exploring), hunting, camping, or flying are factors that can help in selection of topics and activities.

### Select program details

A good topic breakdown provides a solid structural framework around which off-the-job safety program content can be developed. For example, content may be based on seasonal hazards; on home, traffic, and public accidents; on health risk assessment, such as exercise and fitness, stress management, alcohol and drugs, community right-to-know; or according to accident types and causes.

Timely, interesting, and practical topics will attract the attention of employees, create and maintain enthusiasm, and encourage active participation to develop patterns of safe behavior.

A seasonal emphasis outline might include:

- Spring—good housekeeping, lawn mowers, garden tools, do-it-yourself activities, pruning/planting trees, plowing, bicycles;
- Summer—sunburn, swimming, camping, boating, hiking, field sports, fishing, poison ivy, insects, vacation hazards;
- Fall—hunting (see Figure 3-13), home power tools, back-to-school hazards, home-heating equipment, yard cleanup, repair and storage of tools;
- Winter—winter sports, Christmas holiday safety, severe weather exposure, overexertion, winter driving.

Topics can tie into programs of national scope or interest. These national programs provide radio, TV, newspaper, and other forms of publicity that help promote your program content. *National Child Passenger Safety Awareness Week* (February), *National Safe Boating Week* in June, *National Fire Prevention Week* in October, and *National Drunk and Drugged Driver Awareness Week* in December are examples.

A program on Home Fire Safety might include showing a film on the proper use of fire extinguishers, or general fire prevention or safety as well as handout literature on fire safety topics. Combined with this program could be a company discount for employee purchase of smoke alarms, fire extinguishers, and fire escape ladders for home, workshop, and auto use.

### Employee, family, and community involvement

The assistance of a special off-the-job safety group may be valuable. Membership in this group can include employees, family members of employees, local civic and school groups, or community persons with an interest or role in safety in general. Organizations that use such a committee often find that its members contribute immeasurably to the success of the program by providing a perspective that represents their background and understanding of off-the-job safety. Another benefit of the group is that members share serious safety convictions with peers, friends, neighbors, and their families.

Every community has special interest groups and organizations already concerned with various phases of off-the-job safety that will lend their resources and personnel to assist in company activities. Among such groups are the local safety council; Chamber of Commerce including the Jaycees; service clubs; Red Cross chapters; local papers; radio and TV stations; health, police, and fire departments; parent-teacher associations; rescue squads; emergency service groups.

The local women's clubs can be an especially strong source of support for off-the-job safety. The clubs may adopt home, traffic, or recreation safety as a project for the year and, in so doing, secure the participation of many homemakers and family members who are not exposed to a formal safety program.

A representative from safety and health disciplines, such as the American Society of Safety Engineers, can present an off-the-job safety program to an organization's members or employees. Assistance can be obtained from state safety organizations; medical, visiting nurse, and other associations; poison control centers; state police; insurance companies; and public health groups.

A few of these groups maintain accident records, some participate in special programs, some publish bulletins on health and safety subjects, and some conduct courses in subjects related to off-the-job safety. Safe-behavior booklets and posters on seasonal activities are available from the National Safety Council.

An organization can cooperate with the municipal recreation department in promoting swimming classes or courses in boating safety. The American National Red Cross can be enlisted to conduct a series of first aid courses. Local police can be asked to supervise an auto inspection clinic. Insurance companies can often provide the necessary equipment for testing physical qualifications and driver skills.

Organization personnel can, in turn, provide leadership and support for community activities. For instance, employees can help form local safety councils, assist schools and churches by making safety inspections upon request, or act as volunteer members of local fire departments.

Several small companies in a community may consider the possibility of pooling their resources and talents in a joint off-the-job safety program, as is sometimes done with disaster and rescue programs. Advance publicity assures employee support by communicating objectives, plans, and activities. An informal letter from top management sent to employees' homes personalizes the organization's concern for the safety of not only the employees but their families as well.

The initial meeting for employees can be followed by other meetings that include the members of his family. Picnics and other outings, featuring various types of entertainment and safety exhibits, are held by many organizations as a means of reaching the families.

An organization's newsletter, magazine, or paper can carry the word to employees. Company bulletin boards can reinforce these messages.

The community-at-large can be kept informed of plans and progress through spot announcements and stories on home, traffic, and recreational safety carried by the local radio and TV stations and the local newspapers. Preparation of such publicity can be financed by several companies together or by the local safety council.

Meetings of local organizations, such as church groups, women's and service clubs, PTA, and similar groups, can also be used as a means of informing the community-at-large about off-the-job safety programs and activities.

### Programs that worked

**Contest.** Offering some incentive, as minimal as a savings bond or as generous as a college scholarship, can usually elicit good participation from employees' youngsters. The resulting posters, essays, calendars, or slogans can then become vehicles for promoting safety.

The two most widely used contest ideas involve essays (with a limit on the number of words) or posters (with restrictions on size and materials), such as, "What My Dad's/Mother's Safety Means to Me," "Vacation Safety," "Community Safety," and "How Dad/Mother Can Be Safe at Work". Competitors should be divided into age groups—ages running from about 5-7 years old in the youngest, up to 16-18 years for the oldest. Each entrant should receive a token of participation, like a key chain, certificate, embroidered patch, or even a model of the organization's product. To assure impartiality, judges may be from outside the community or the organization.

**Traffic safety.** Offering the National Safety Council's Defensive Driving Program, at the organization's expense and possibly with organization facilities and instructors, to all driving members of the employees' families is one of the most positive home, off-the-job, or community safety efforts an organization can make. It can improve the driving habits of those who drive, and can also promote car pooling as a safe answer to energy, traffic, and pollution problems.

Auto safety checks tie in well with vacation and holiday programs and work best on weekends. Have plenty of qualified inspectors available so participants will not be discouraged by long lines.

**Company picnic.** Safety picnics should involve the whole family and can be as expensive as a completely catered affair or as simple as a family picnic. A safety theme can be included in drawings for door prizes, activities for all ages, and presentations of awards to or recognition of employees for safety achievements.

**Family night.** Quite different from picnics, family-night gatherings are built around the presentation of some discussion or audiovisual presentation on safety, accompanied by refreshments. Sometimes the theme shifts from general safety to a program to make the family aware of and gain support for the safety efforts made at work. Sometimes an "open house" tour of the plant facilities can be combined with the family-night get-together.

Family first aid programs, including cardiopulmonary resuscitation training, can be held at business locations after business hours, or on weekends to involve the family. First aid kits available at a company discount purchase price could be offered for family use in the home, auto, or workshop.

**Youth activities.** Sponsorship of activities like softball and football teams and bicycle rallies usually is not aimed directly at promoting safety, but a poor safety record among young activity participants *can* hurt the sponsor's image. Therefore, financial sponsorship of such activities should not be the end of an organization's participation. Employee leaders, who are knowledgeable in the activity and trained in safety and first aid (at the organization's expense), should be available to participants. The result will be strong, positive community sentiment. Involvement of groups, such as the Boy and Girl Scouts, 4-H Clubs, Campfire, Inc., FFA, and FHA, will expand off-the-job safety efforts.

**Recreational programs.** In any organization, many employees and their families are sports enthusiasts. Their pastimes may include hunting, boating, camping, swimming, fishing, skiing, to name a few. At the season-openings of these activities, organizations may sponsor clinics featuring registered/competent instructors to check equipment and to provide instruction on improving skills. Sources of help include the local gun clubs, powerboat squadrons, the Coast Guard Auxiliary, National Recreation and Park Association, the President's Council on Physical Fitness and Sports, the National Red Cross, YMCA, police and fire departments, and health organizations. All of these groups can provide ideas for safety activities.

**Vacation-holiday program.** Some organizations close down for regular summer vacation; others offer year 'round vacation periods and three-day weekends. A good time to offer safe driving tips and safety literature to employees is at vacation times.

Other promotional methods include:

- Publicity distributed in-house that reinforces safety, both on- and off-the-job (pamphlets, press releases, posters, and billboards are ideal vehicles for in-house promotions). The National Safety Council has a great selection of this material, aimed at a variety of off-the-job safety topics. Employee-generated, original posters are also effective in personalizing safety efforts.
- Nearly every organization has some sort of in-house journal or newsletter that is either given to employees at work or sent to their homes. No other publication enjoys wider readership within an organization, and it is most appropriate to safety education and safety program promotion, because both employees and their families see it. Articles can be written on all aspects of a safety program, and the employee and family can be solicited for safety-related story ideas as well.

### Summary

There are three benefits a company can realize from expanding its safety program to include off-the-job safety. The first is a reduction in lost production time and operating costs from both on- and off-the-job injuries. Second, companies have found that efforts in off-the-job safety have produced an increased interest by employees in their on-the-job safety program. The third benefit, often overlooked, is that of better public relations.

The aim of safety education, namely, changing the employee's behavior, is especially true of off-the-job safety. No asset is more important to a company than its employees. They should be protected not only during working hours, but also be given every incentive to also be safe *off* the job.

## REFERENCES

Blankenship, L.M. Nonoccupational Disabling Injury Cost Study, K/DSA-457. Martin Marietta Energy Systems, Inc., October 1981.

Board of Certified Safety Professionals of the Americas, 208 Burwash, Savoy, Ill. 61874. "Curricula Development and Examination Study Guidelines," Tech. Report No. 1.

Construction Advancement Foundation, Hammond, Ind. *Safety Manual.*

Factory Mutual Engineering Corp., Norwood, Mass. 02062. *Loss Prevention Data.*

Firenze, Robert J. *Guide to Occupational Safety and Health Management.* Dubuque, Ia., Kendall/Hunt Publishing Co., 1973.

———. *The Process of Hazard Control.* Dubuque, Ia., Kendall/Hunt Publishing Co., 1978.

———. *Safety and Health in Industrial/Vocational Education.* Cincinnati, Ohio, National Institute for Occupational Safety and Health, 1981.

Hannaford, Earle S. *Supervisors Guide to Human Relations,* 2nd ed. Chicago, Ill., National Safety Council, 1976.

Johnson, William G. *MORT Safety Assurance Systems.* New York, N.Y., Marcel Dekker, Inc., 1980.

Kane, Alex. "Safety Begins the First Day on the Job," *National Safety News,* January 1979, p. 53.

Manuele, Fred A. "How Effective Is Your Hazard Control Program?" *National Safety News,* February 1980, pp. 53-58.

Maryland, State of, Department of Labor and Industry, Safety Engineering and Education Division. "Safety Program Safety Committee Manual," 1967.

National Association of Suggestion Systems, 230 N. Michigan Ave., Chicago, Ill. 60611.
  *Performance Magazine* (6 times a year).
  "Suggestion Newsletter" (6 times a year).

National Safety Council, 444 N. Michigan Ave., Chicago, Ill. 60611.
  *Accident Facts* (published annually).
  Industrial Data Sheets
    *Management Safety Policies,* No. 585.
    *Off-the-Job Safety,* No. 601.
    *Safety Committees,* No. 631.
  *Supervisors Safety Manual.*
  "You Are the Safety and Health Committee."

Peters, George A. "Systematic Safety," *National Safety News,* September 1975, pp. 83-90.

Plog, Barbara, ed. *Fundamentals of Industrial Hygiene,* 3rd ed. Chicago, Ill., National Safety Council, 1988.

Pollina, Vincent. "Safety Sampling," *Journal of the American Society of Safety Engineers,* August 1962, pp. 19-22.

Tainter, Sarah A., and Monro, Kate M. *The Secretary's Handbook.* New York, The Macmillan Co.

U.S. Department of Human Resources, National Institute for Occupational Safety and Health, Division of Technical Services, Cincinnati, Ohio 45226. *Self-Evaluation of Occupational Safety and Health Programs,* Publication 78-187, 1978.

U.S. Department of Labor, Bureau of Labor Standards. *Safety Organization,* Bulletin 285, 1967.

U.S. Department of Labor, Occupational Safety and Health Administration. *Organizing a Safety Committee,* OSHA 2231, June 1975.

U.S. Department of Transportation, Office of Hazardous Materials, Washington, D.C. 20590. "Newly Authorized Hazardous Materials Warning Labels." (Latest edition.) (Based on Title 49, *Code of Federal Regulations,* sections 173.402, −403, and −404; import or export shipments are covered in Title 14, CFR, section 103.13.)

Windsor, Donald G. "Process Hazards Management," a speech given before the Chemical Section, National Safety Congress, October 17, 1979.

# Acquiring Hazard Information



BEFORE HAZARDS CAN BE CONTROLLED, they must be discovered. Monitoring is an effective means of acquiring hazard information. *Monitoring* can be defined as a set of observation and data collection methods used to detect and measure deviations from plans and procedures in current operations (Johnson, 1980).

Through monitoring it can be ascertained (1) that controls are functioning as intended; (2) that workplace modifications have not altered conditions so that controls no longer function effectively; and (3) that new problems have not crept into the workplace since the most recent controls were introduced (Firenze, 1978).

Monitoring can involve four functions: hazard analysis, inspection, measurement, and accident investigation. Including all four functions means that monitoring is performed *before* the operation begins, *during* the life cycle of the operation, and *after* the system has broken down. A systems approach to hazard control will use each of these methods; this chapter tells how.

## HAZARD ANALYSIS

Data from hazard analysis can be thought of as being the baseline for future monitoring activities. Before the workplace can be inspected to assure that environmental and physical factors fall within safe ranges, the hazards inherent in the system must be discovered. Hazard analysis has proven itself an excellent tool to identify and evaluate hazards.

It is not a new idea to analyze a problem or situation to extract data for decision making. Good workers and their supervisors are always—though sometimes unconsciously—making assessments that guide their actions. Written analyses carry the process one step further. They provide the means to document hazard information.

### Philosophy behind hazard analysis

Written analyses can form the basis for more thorough inspections. They can communicate data about hazards and risk potential to those in command positions. They can educate those in the line and staff organizations who must know the consequences of hazards existing within their operations and the purpose and logic behind established control measures. Management can require formal, written analysis for each critical operation. By doing so, it not only gathers information for immediate use, but it also reaps benefits over the long run. Once important hazard data are committed to paper, they become part of the technical information base of the organization. These documents show the employer's concern for locating hazards and establishing corrective measures before an accident can happen.

Traditionally, when systems have been analyzed to determine failures that detracted from their effectiveness, they were analyzed during their operational phase. Hazard analytical techniques applied during this phase of a system's life cycle returned substantial dividends by reducing both accident and overall system losses.

Recently a shift has taken place. Hazard control specialists no longer concentrate solely on operations. Instead, they are looking at the conceptual and design stages of the systems for which they are responsible. They are using analytical methods and techniques before the process or product is built to identify and judge the nature and effects of hazards associated with their systems. This widened assessment, in many instances, has significantly altered the direction of hazard control efforts. When potential failures can be located prior to the production or on-stream process stage of a system's life cycle, specialists can cut costs and avoid damage, injuries, and death. Systems engineering was initially concerned with increasing effectiveness, not profits. Properly applied, however, it can point out profitable solutions to many of management's most perplexing operational problems.

### What is hazard analysis?

*Hazard analysis* is an orderly process used to acquire specific information (hazard and failure data) pertinent to a given system (Firenze, 1978). A popular adage holds that "most things work out right for the *wrong* reasons." By providing data for informed management decisions, hazard analysis helps things work out right for the *right* reasons. The method forces those conducting the analysis to ask the right questions, and it also helps to answer those questions. By locating those hazards that are the most probable and/or have the severest consequences, hazard analyses produce information essential in establishing effective control measures. Analytic techniques assist the investigator in deciding what facts to seek, determining probable causes and contributing factors, and arranging results so they are orderly and clear.

What are some uses for hazard analysis?

1. It can uncover hazards that have been overlooked in the original design, mock-up, or setup of a particular process, operation, or task.
2. It can locate hazards that developed after a particular process, operation, or task was instituted.
3. It can determine the essential factors in and requirements for specific job processes, operations, and tasks. It can indicate what qualifications are prerequisites to safe and productive work performance.
4. It can indicate the need for modifying processes, operations, and tasks.
5. It can identify situational hazards in facilities, equipment, tools, materials, and operational events (for example, unsafe conditions).
6. It can identify human factors responsible for accident situations (for example, deviations from standard procedures).
7. It can identify exposure factors that contribute to injury and illness (such as contact with hazardous substances, materials, or physical agents).
8. It can identify physical factors that contribute to accident situations (noise, vibration, insuffícent illumination, to name a few).
9. It can determine appropriate monitoring methods and maintenance standards needed for safety.

### Formal methods of hazard analysis

Formal hazard analytical methods can be divided into two broad categories: inductive and deductive.

**Inductive method.** The inductive analytical method uses observable data to predict what can happen. It postulates how the component parts of a system will contribute to the success or failure of the system as a whole. Inductive analysis considers a system's operation from the standpoint of its components, their failure in a particular operating condition, and the effect of that failure on the system.

The inductive method forms the basis for such analyses as failure mode and effect analysis (FMEA) and operations hazard analysis (OHA).

In failure mode and effect analysis, the failure or malfunction of each component is considered, including the mode of failure. The effects of the hazard(s) that led to the failure are traced through the system, and the ultimate effect on the task performance is evaluated. However, because only one failure is considered at a time, some possibilities may be overlooked. Figure 4-1 illustrates the FMEA format used at Aerojet Nuclear Co., Idaho Falls, Idaho. Figure 4-2 illustrates the OHA format used for industrial operations.

Once the inductive analysis is completed and the critical failures requiring further investigation are detected, then the fault tree analysis will facilitate an inspection (see Deductive method).

Chapter 5, Removing the Hazard from the Job, discusses job safety analysis (JSA), an analysis using the inductive method. That chapter explains the basic steps to be taken and the uses of a job safety analysis.

**Deductive method.** If inductive analysis tells us *what* can happen, deductive analysis tells us *how*. It postulates failure of the entire system and then identifies how the components could contribute to the failure.

Deductive methods use a combined-events analysis, often in the form of trees. The positive tree calls for stating the requirements for success; see Figure 4-3. Positive trees are less commonly used than fault trees because they can easily become a list of "you shoulds" and sound moralizing.

Fault trees are reverse images of positive trees and show ways troubles can occur. An undesired event is selected. All the possible happenings that can contribute to the event are diagrammed in the form of a tree. The branches of the tree are continued until independent events are reached. Probabilities are determined for the independent events.

The fault tree requires rigorous, thorough analysis; all known sources of failure must be listed. The fault tree is a graphic model of the various parallel and sequential combinations of system component faults that can result in a single, selected system fault. Figure 4-3 illustrates three types of analytical trees.

Analytical trees have three advantages:

1. They accomplish rigorous, thorough analysis without wordiness. Using known data, the analyst can identify the single and multiple causes capable of inducing the undesired event.
2. They make the analytical process visible, allowing for the rapid transfer of hazard data from person to person, group to group, with few possibilities for miscommunication during the transfer.
3. They can be used as investigative tools. By reasoning backwards from the accident (the undesired event), the investigator is able to reconstruct the system and pinpoint those elements responsible for the undesired event.

**Cost-effectiveness.** The cost-effectiveness method can be used as part of either the inductive or deductive approach. The cost of system changes made to increase safety is compared with the decreased costs of fewer serious failures or with the increased efficiency of the system. Cost-effectiveness frequently is used to decide among several systems, each capable of performing the same task.

**Choosing which method to use.** To decide what hazard ana-

## FMEA Form

| COMPONENT | FAILURE OR ERROR MODE | EFFECTS ON | | SEVERITY INDEX | FAILURE FREQUENCY INDEX | CRITICALITY | DETECTION METHODS | COMPENSATING PROVISIONS AND REMARKS |
|---|---|---|---|---|---|---|---|---|
| | | OTHER COMPONENTS | WHOLE SYSTEM | | | | | |
| | | | | | | | | |
| | | | | | | | | |

*For Reliability:*
- Design Characteristic
- Failure Mode
- Failure Probability
- Effect on System
- Essentiality Code
- Control to Minimize:
  - Frequency
  - Effect

*For Priority Problem Lists:*
- Energy Sources:
  - Kinds
  - Amounts
- Potential Targets
- Barriers, Controls
- Residual Risk
- Failure Mode
- Failure Mechanism
- Consequence Potential
- Frequency, Consequence Matrix Class
- Action-Decision Classes
  - Authority Level
  - Type and Date Action Due

**Figure 4-1.** Failure mode and effect analysis form used by the Aerojet Nuclear Company (Johnson, 1980).

**OPERATIONS HAZARD ANALYSIS**

| Process | Operational Step | Task | Source of Potential Hazard | Triggering Event | Potential Effect on Equip., Material Environment | Personal Injury, Property Damage | RAC | Procedural Requirements | Safety & P.P.E. |
|---|---|---|---|---|---|---|---|---|---|
| Turning steel stock between centers on machine lathe. | Rough turning steel stock. | Select cutting tool and place in tool holder. | Improper tool used for rough cutting operation. | Starting lathe. | Tool jams in stock.<br><br>Stock comes off centers.<br><br>Uneven cut.<br><br>Wasted stock. | Operator is hit in face with flying chips of steel. | 2 | A right-cut tool or roundhouse tool should be held in a straight tool holder.<br><br>Lathe located to minimize exposure to other work stations. | Select proper tool for job.<br><br>Operator to wear face protection while operating lathe. |
| | | Place tool holder in the tool post and adjust cutting tool to proper location. | Tool holder extending too far from tool post. | Starting lathe. | Same as above.<br><br>Breaking cutting tool. | Operator is hit in face with flying chips of steel from stock and broken cutting tool. | 2 | Tool post should be at end of T-slot. Face of tool must be on center and turned slightly away from headstock. | Operator to wear face protection while operating lathe. |

**Figure 4-2.** This Operations Hazard Analysis (OHA) form is used for industrial operations. (Printed with permission from Indiana Labor & Management Council, Inc.)

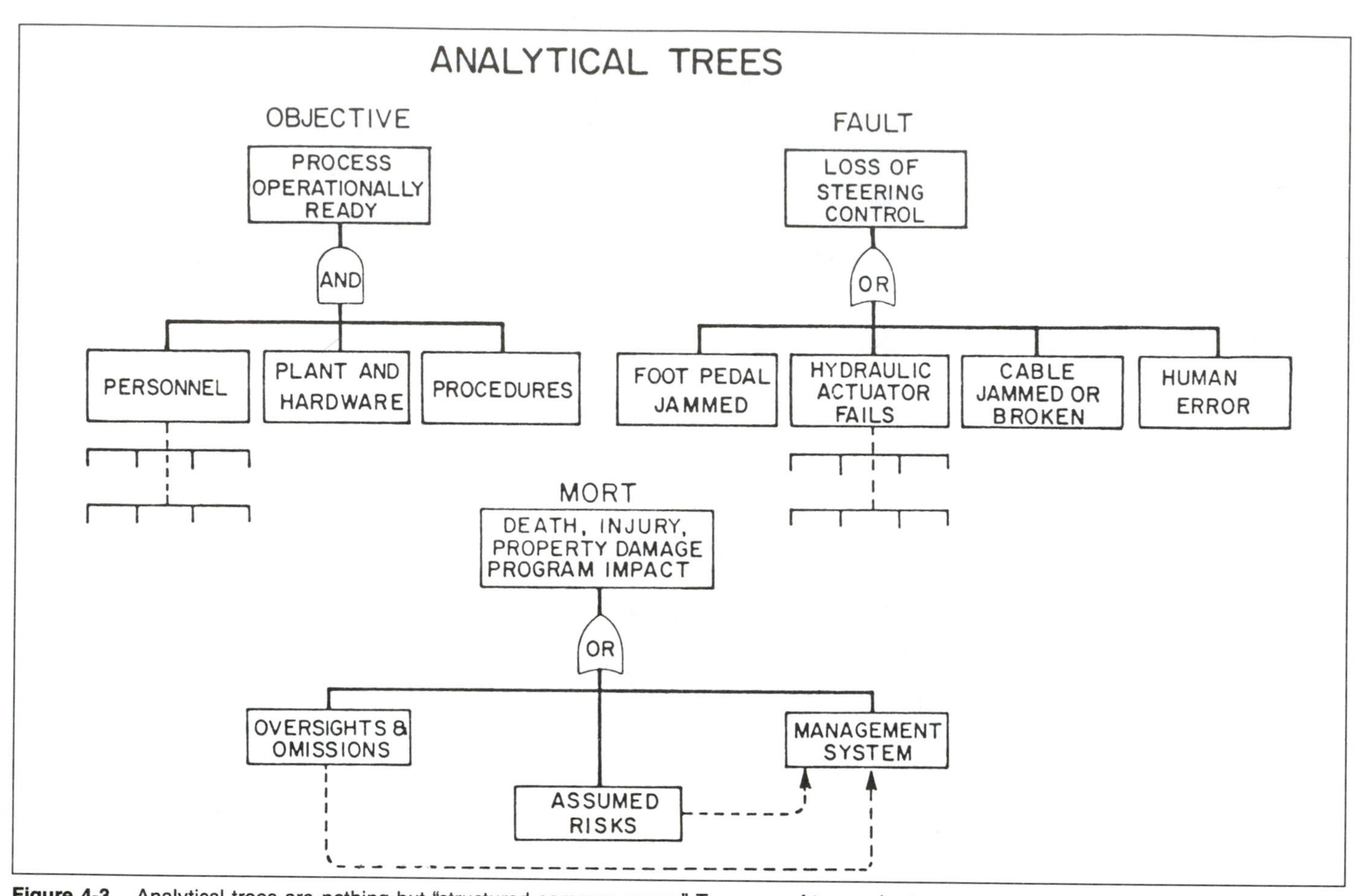


**Figure 4-3.** Analytical trees are nothing but "structured common sense." Trees are of two major types—the objective or positive trees, which emphasize how a job should properly be done, and the fault trees, which chart those things that can go wrong and produce a specific failure. A fault tree structured for one job can be generalized to cover a wide variety of jobs. The MORT diagram describes the ideal safety program in an orderly, logical manner; it is based on three branches: (1) a branch dealing with specific oversights and omissions at the work site, (2) a branch that deals with the management system that establishes policies and makes the entire system go, and (3) an assumed-risk branch visually recognizing that no activity is completely risk-free, and that risk-management functions must exist in any well-managed organization (Nertney, 1977). (Printed with permission from *Professional Safety,* February 1977.)

lytical approach is best for a given situation, the hazard control specialist will want to answer five questions:

1. What is the quantity and quality of information desired?
2. What information already is available?
3. What is the cost of setting up and conducting analyses?
4. How much time is available before decisions must be made and action taken?
5. How many people are available to assist in the hazard analysis, and what are their qualifications?

Conducting a hazard analysis can be expensive. Before a hazard analysis technique is chosen, it is important to determine what information is needed and how important it is.

It is beyond the scope of this Manual to go into detail regarding other applications of system safety. (See Johnson, 1980.) A few other areas in which large-project system safety methods and ideas can, with a little adaptation, be scaled down to apply to industrial jobs are indicated in Figure 4-4 (Nertney, 1977).

### Who should participate in hazard analysis?

A hazard analysis, to be fully effective and reliable, should represent as many different viewpoints as possible. Each person familiar with a process or operation has acquired insights concerning problems, faults, and situations that can cause accidents. These insights need to be recorded along with those of the initiator of the hazard analysis, who usually will be the safety professional. Input from workers and employee representatives can be extremely valuable at this stage.

### What processes, operations, and tasks need to be analyzed?

Many processes, operations, and tasks in any establishment or plant are good candidates for hazard analysis because they have the potential to cause accidents. Eventually, hazard analyses should be completed for all jobs, but the most potentially threatening should have immediate attention. In determining which processes, operations, and tasks receive priority, those making the decisions should take the following into consideration:

1. **Frequency of accidents.** An operation or task that has a history of repeated accidents associated with its performance is a good candidate for analysis, especially if different employees have the same kind of accident while performing the same operation or task.

| System Safety Method or Concept | Scaled-Down Version | Comments |
|---|---|---|
| Hazard Analysis and Risk Projection | 1. Think like an insurance agent—try to reduce hazards to consequences.<br>2. Use insurance companies. "How much would you charge to insure us?" | Safety people too often try to communicate with line managers in terms of "hazards" rather than reducing the hazards to risks and consequences. |
| Biomechanics | Looking at the job from the worker's point of view in terms of physical characteristics. | For example, does the vehicle cab "fit" small men and women? Can the largest person, the smallest person, the person with the shortest reach, etc. operate the equipment effectively—or have the designers given us another "average-man" design? Can the job be done on the hottest and coldest days? Are the people going to be completely exhausted half way through the job, etc.? |
| Human Factors | 1. Simple step-by-step walk-through of the work<br>2. Misuse analysis<br>3. Select equipment that is easy to operate *and* matches existing equipment in operational logic. | Most of the fancy human factors analyses have shown that a simple walk-through will indicate many very obvious booby traps for the workers in terms of job steps that are hard to do or things that invite the people to do it wrong or to misuse equipment. |
| Job Information Systems<br><br>Formal Change Analysis Methods | 1. Do job-site surveillance and monitoring in an orderly manner.<br>2. Pay attention to what the *workers* say.<br>3. Be sensitive to *any changes* in people, hardware, or plans. | "Following one's nose" can too easily lead to oversights and failure to "pick up" on changes. Studies have shown that in a series of six accidents, workers had reported all major contributing factors in questionnaire-type studies performed prior to the accident. Another analysis indicated two typewritten pages of changes which contributed directly to five major accidents. |
| Formal Work-Flow Charting for Safety Reasons | 1. Formal work-flow charting to get the job done properly.<br>2. Job Safety Analysis | 1. Experience is indicating that the benefits of formal use of work and project flow-charting go far beyond safety benefits. This is particularly true in avoiding foul-ups between different working groups.<br>2. Use of step-by-step job safety analysis at worker and first-line supervisor level has been adopted by a number of large business-managed tax-paying industrial firms as a cost-effective, efficient method for producing on-the-job safety. |

**Figure 4-4.** Areas in which system safety and ideas can, with a little adaptation, be applied to industrial jobs (Nertney, 1977). (Printed with permission from *Professional Safety,* February 1977.)

2. **Potential for injury.** Some processes and operations can have a low accident frequency but a high potential for major injury (for example, tasks on a grinder conducted without a tool rest or tongue guard).
3. **Severity of injury.** A particular process, operation, or task can have a history of serious injuries and be a worthy candidate for analysis, even if the frequency of such injuries is low.
4. **New or altered equipment, processes, and operations.** As a general rule, whenever a new process, operation, or task is created or an old one altered (because of machinery, equipment, or other changes), a hazard analysis should be conducted. For maximum benefit the hazard analysis should be done while the process or operation is in the planning stages. No equipment should be put into regular operation until it has been checked for hazards, its operation studied, any necessary additional safeguards installed, and safety instructions or procedures developed. Adhering to such a procedure ensures that employees can be trained in hazard-controlled safe operations, and serious injuries and exposures will be prevented.
5. **Excessive material waste or damage to equipment.** Processes or operations producing excessive material waste or damage to tools and equipment are candidates for hazard analysis. The same problems causing the waste or damage can be the ones that, given the right situation, could cause injuries.

## INSPECTION

Hazard analysis is a process that can take place during the planning, design, and operational phases of the system. *Inspection* can be defined as that monitoring function conducted in an organization to locate and report existing and potential hazards having the capacity to cause accidents in the workplace (Firenze, 1978). Inspection works because it is an essential part of hazard control. It is a vital managerial tool, not a gimmick.

### Philosophy behind inspection

Inspection can be viewed negatively or positively:

- Fault-finding, with the emphasis on criticism
- Fact-finding, with the emphasis on locating potential hazards that can adversely affect safety and health.

The second viewpoint makes the most sense. To be effective, this viewpoint depends on three things: (1) yardsticks adequate for measuring a particular situation, (2) comparison of what is with what ought to be, and (3) corrective steps being taken to achieve desired performance (Firenze, 1978). Failure to analyze inspection reports for *causes* of defects ultimately means the failure of the monitoring function. Corrective action may fix the specific item but fail to fix the system.

What are the purposes of inspection? Its primary purpose is

to detect potential hazards so they can be corrected before an accident occurs. Inspection can determine those conditions which need to be corrected or improved to bring operations up to acceptable standards, both from safety and operational standpoints. Secondary purposes are to improve operations and thus increase efficiency, effectiveness, and profitability.

While management ultimately has the responsibility for inspecting the workplace, authority for carrying out the actual inspecting process extends throughout the organization. Obviously supervisors, foremen, and employees fulfill an inspection function, but so do departments as diverse as engineering, purchasing, quality control, personnel, maintenance, and health care.

## Types of inspection

Inspection can be classified as one of two types—either continuous or at intervals.

**Continuous, ongoing inspection** is conducted by employees, foremen, supervisors, and maintenance personnel as part of their job responsibilities. Continuous inspection involves noting an apparently or potentially hazardous condition or unsafe act and either correcting it immediately or making a report to initiate corrective action. Continuous inspection of personal protective equipment is especially important.

Supervisors continuously make sure that tools, machines, and equipment are properly maintained and safe to use and that safety precautions are being observed. Toolroom employees regularly inspect all hand tools to make sure that they are in safe condition. Foremen are often responsible for continuously monitoring the workplace and seeing that equipment is safe and that employees are observing safe practices. When foremen or supervisors inspect machines at the beginning of a shift to see if they are ready to operate, a safety inspection must be part of the operation.

Continuous inspection is one ultimate goal of a good safety and health program. It means that each individual is vigilant, alert to any condition having accident potential, and willing to initiate corrective action.

Continuous inspection is sometimes called *informal* because it does not conform to a set schedule, plan, or checklist. Critics argue that continuous inspection is erratic and superficial, that it does not get into out-of-the-way places, and that it misses things. The truth is that both kinds of inspections are necessary. They complement one another.

The supervisor's greatest advantage in continuous inspection—familiarity with the employees, equipment, machines, and environment—can also be a disadvantage. Just as an old newspaper left on a table in time becomes part of the decor, a hazard can become so familiar that it is no longer noticed. The supervisor's blind spot is particularly likely to occur with housekeeping and unsafe acts. Poor housekeeping conditions may not be noticed because the change is gradual and the effect is cumulative. A similar phenomenon may occur with unsafe acts such as employees smoking in prohibited areas or failing to wash thoroughly before taking a break for eating or smoking.

No matter how conscientious the supervisors are, they cannot be objective. Inspections of their areas reflect personal and vested interests, knowledge and understanding of the production problems involved in the area, and concern for the employees. A planned periodic inspection of their areas by another supervisor can be used to audit their efforts. Furthermore, the supervisors who inspect another area may return to their own sections with renewed vision. Having looked at the trees day in and day out, they need occasionally to take the long view and see the forest.

Though this section will be devoted primarily to discussing planned inspections, continuous inspections should be regarded as a cooperative, not a competitive, activity.

**Planned inspection at intervals** is what most people think of as "real" safety and health inspection. It is deliberate, thorough, and systematic by design. In many cases specific items or conditions are examined. An established procedure is followed and a checklist usually is used. Planned inspection is of three types: periodic, intermittent, and general.

*Periodic inspection* includes those inspections scheduled to be made at regular intervals. Such inspections can be of the entire plant, a specific area, a specific operation, or a specific type of equipment. They can take place weekly, monthly, semiannually, annually, or at other suitable intervals. Items such as safety guard mountings, scaffolds, elevator wire ropes (cables), two-hand controls, and fire extinguishers, and other items relied on for safety, require frequent inspection. The greater the accident severity potential, the more often the item should be inspected.

Periodic inspections can be of several different types:

1. Inspections by the safety professional, industrial hygienist, and joint safety and health committees.
2. Inspections for preventing accidents and damage or breakdowns (checking mechanical functioning, lubricating, and the like) performed by electricians, mechanics, and maintenance personnel. Sometimes these persons are asked to serve as roving inspectors.
3. Inspections by specially trained certified or licensed inspectors, often from outside the organization (for example, inspection of boilers, elevators, unfired pressure vessels, cranes, power presses, fire extinguishing equipment).
4. Inspections done by outside investigators to determine compliance with government regulations.

The advantage of periodic inspection is that it covers a specific area and allows detection of unsafe conditions in time to provide effective countermeasures. The staff or safety committee periodically inspecting a certain area is familiar with operations and procedures and therefore quick to recognize deviations. A disadvantage of periodic inspection is that deviations from accepted practices are rarely discovered because employees are usually aware of the presence of inspectors.

*Intermittent inspections* are those made at irregular intervals. Sometimes the need for an inspection is indicated by accident tabulations and analysis. If a particular department or location shows an unusual number of accidents or if certain types of injuries occur with greater frequency, then an inspection is called for. When construction or remodeling is going on within or around a facility, an unscheduled inspection may be needed to find and correct unsafe conditions before an accident occurs. The same is true when new equipment is installed or when new processes are instituted or old ones modified.

Another form of intermittent inspection is that made by the industrial hygienist when a health hazard is suspected or present in the environment. This monitoring of the workplace is covered in detail in Measurement and Testing, later in this chapter. It usually involves:

Sampling the air for the presence of toxic vapors, fumes, gases, and particulates
Testing materials for toxic properties
Testing ventilation and exhaust systems for proper operation.

*A general inspection* is a planned inspection of places not receiving periodic inspection. A general inspection covers those areas where no one ever visits and where no one ever gets hurt. It includes parking lots, sidewalks, fencing, and similar outlying regions.

Many out-of-the-way hazards are located overhead, where they are difficult to spot. Overhead inspections frequently disclose the need for repairs to skylights, windows, cranes, roofs, and other installations affecting the safety of both the employees and the physical plant. Overhead devices can require adjustment, cleaning, oiling, and repairing.

Inspections of overhead areas are necessary to make certain all reasonable safeguards are provided and safe practices observed. Inspectors must determine that persons performing overhead jobs are provided with suitable staging, safety belts, and lifelines. They must apply this safety directive to themselves during the inspection. They must look for loose tools, bolts, pipelines, shafting, pieces of lumber, windows, electrical fixtures, and other objects that can fall from building structures, cranes, roofs, and similar overhead locations.

Safety conditions change after dark, when the illumination is artificial. Therefore, when an organization has more than one shift, it is important inspections also be performed at night to make sure adequate illumination is provided and the lighting system is maintained in a satisfactory condition. The safety professional should make this inspection, aided by a photometer and camera, where necessary.

Even when there are no regular night shifts, some employees—maintenance personnel, firefighters, and night watchmen—are required to work after dark. The safety professional occasionally needs to check on night work conditions.

General inspections are usually required before reopening a plant after a long shutdown.

## PLANNING FOR INSPECTION

A safety and health inspection program requires:

1. Sound knowledge of the plant
2. Knowledge of relevant standards, regulations, and codes
3. Systematic inspection steps
4. A method of reporting, evaluating, and using the data.

An effective program begins with analysis and planning. If inspections are casual, shallow, and slipshod, the results will reflect the method. Before instituting an inspection program, the answers to five questions are necessary:

1. What items need to be inspected?
2. What aspects of each item need to be examined?
3. What conditions need to be inspected?
4. How often must items be inspected?
5. Who will conduct the inspection?

### The hazard control inspection inventory

To determine what factors affect the inspection, a hazard control inspection inventory can be conducted. Such an inventory is the foundation upon which a program of planned inspection is based. It resembles a planned preventive maintenance system and yields many of the same benefits.

The entire facility—yards, buildings, equipment, machinery, vehicles—should be divided into areas of responsibility. Once these areas have been determined, they should be listed in an orderly fashion. A color-coded map of the facility or floor plan might be developed. It often is desirable to divide large areas or departments into smaller areas that can be assigned to each first-line supervisor and/or the hazard control department's inspector.

### What items need to be inspected?

Once specific areas of responsibility have been determined, an inventory should be made of those items that can become unsafe or cause accidents. These would include:*

1. Environmental factors (illumination, dusts, gases, sprays, vapors, fumes, noise).
2. Hazardous supplies and materials (explosives, flammables, acids, caustics, toxic materials or by-products).
3. Production and related equipment (mills, shapers, presses, borers, lathes).
4. Power source equipment (steam and gas engines, electrical motors).
5. Electrical equipment (switches, fuses, breakers, outlets, cables, extension and fixture cords, grounds, connectors, connections).
6. Hand tools (wrenches, screwdrivers, hammers, power tools).
7. Personal protective equipment (hard hats, safety glasses, safety shoes, respirators).
8. Personal service and first aid facilities (drinking fountains, wash basins, soap dispensers, safety showers, eyewash fountains, first aid supplies, stretchers).
9. Fire protection and extinguishing equipment (alarms, water tanks, sprinklers, standpipes, extinguishers, hydrants, hoses).
10. Walkways and roadways (ramps, docks, sidewalks, walkways, aisles, vehicle ways).
11. Elevators, electric stairways, and manlifts (controls, wire ropes, safety devices).
12. Working surfaces (ladders, scaffolds, catwalks, platforms, sling chairs).
13. Material handling equipment (cranes, dollies, conveyors, hoists, forklifts, chains, ropes, slings).
14. Transportation equipment (automobiles, railroad cars, trucks, front-end loaders, helicopters, motorized carts and buggies).
15. Warning and signaling devices (sirens, crossing and blinker lights, klaxons, warning signs).
16. Containers (scrap bins, disposal receptacles, carboys, barrels, drums, gas cylinders, solvent cans).
17. Storage facilities and areas both indoor and outdoor (bins, racks, lockers, cabinets, shelves, tanks, closets).
18. Structural openings (windows, doors, stairways, sumps, shafts, pits, floor openings).
19. Buildings and structures (floors, roofs, walls, fencing).
20. Miscellaneous—any items that do not fit in preceding categories.

---

*Adapted from *Facility Inspection* (see References).

**LIST OF POSSIBLE PROBLEMS TO BE INSPECTED**

| | | | | |
|---|---|---|---|---|
| Acids | Closets | Forklifts | Piping | Shapers |
| Aisles | Connectors | Fumes | Pits | Shelves |
| Alarms | Containers | Gas cylinders | Platforms | Sirens |
| Atmosphere | Controls | Gas engines | Power tools | Slings |
| Automobiles | Conveyors | Gases | Presses | Solvents |
| Barrels | Cranes | Hand tools | Racks | Sprays |
| Bins | Crossing lights | Hard hats | Railroad cars | Sprinkler systems |
| Blinker lights | Cutters | Hoists | Ramps | Stairs |
| Boilers | Docks | Horns and signals | Raw materials | Steam engines |
| Borers | Doors | Hoses | Respirators | Sumps |
| Buggies | Dusts | Hydrants | Roads | Switches |
| Buildings | Electric motors | Ladders | Roofs | Tanks |
| Cabinets | Elevators | Lathes | Ropes | Trucks |
| Cables | Explosives | Lights | Safety devices | Vats |
| Carboys | Extinguishers | Mills | Safety shoes | Walkways |
| Catwalks | Eye protection | Mists | Scaffolds | Walls |
| Caustics | Flammables | Motorized carts | Shafts | Warning devices |
| Chemicals | Floors | | | |

**Figure 4-5.** This list of possible problems found in the workplace is adapted from an OSHA publication. (Adapted from *Principles and Practices of Occupational Safety and Health, Student Manual,* Booklet Three, U.S. Department of Labor, OSHA 2215.)

There are many sources of information about items to be inspected. Maintenance employees know what problems can cause damage or shutdowns. The workers in the area are qualified to point to causes of injury, illness, damage, delays, or bottlenecks. Medical personnel in the organization can list problems causing job-related illnesses and injuries. Manufacturers' manuals often specify maintenance schedules and procedures and safe work methods.

Gathering the information about standards, regulations, and codes is a necessary first step in determining what items need to be inspected. A good resource for checking U.S., international, and non-U.S. national standards is the *Index and Directory of Industry Standards,* 3 vols., Information Handling Services, Englewood, CO, latest edition. The *Index* covers 60,000 documents from 362 standards-gathering bodies. Other resources include the national consensus standards published by the American National Standards Institute (ANSI); publications of the National Fire Protection Association (NFPA), Underwriters Laboratories Inc. (UL), the National Bureau of Standards, and the National Safety Council; and the OSHAct and regulations, published by the U.S. Government Printing Office. In countries other than the United States, contact local regulatory agencies.

Building codes, building inspection books, guides to building and plant maintenance: these also will be useful references. Publications of the National Fire Protection Association will help make certain that fire hazards are effectively being controlled. Sometimes insurance company surveys contain checklists to determine the condition of buildings. Research and reference material is contained in subsequent chapters of this Manual. Other publications of the National Safety Council, such as *Accident Facts,* may prove useful.

State or provincial and federal governments also publish accident statistics. The Walsh-Healey Public Contracts Act (41 CFR 50) gives safety and health standards for federal supply contracts. In one of its publications OSHA gave examples of possible problems found in the work area. These are cited in Figure 4-5. The *Federal Register* and the *Code of Federal Regulations,* Title 29, parts 1900-1950, give OSHA regulations. Figure 4-6 shows some special subjects addressed by the subparts of 29 CFR 1910 (General Industry) and the subparts of 29 CFR 1926 (Construction). It is important to remember, however, that usually federal and state or provincial laws, codes, and regulations set up minimum requirements only. To comply with company policy and secure maximum safety, it frequently is necessary to exceed these requirements. OSHA publications indicate not only what standards are required but also what violations are most frequent. These same sources are helpful in the next step—determining critical factors to be inspected.

### What aspects of each item need to be examined?

Particular attention should be paid to the parts most likely to cause the greatest problems when they become unsafe. These parts are most likely to develop unsafe or unhealthy conditions because of stress, wear, impact, vibration, heat, corrosion, chemical reaction, and misuse. Safety devices, guards, controls, work or wearpoint components, electrical and mechanical components, and fire hazards would become unsafe first. For a particular machine, critical parts would include the point of operation, moving parts, and accessories (flywheels, gears, shafts, pulleys, key ways, belts, couplings, sprockets, chains, controls, lighting, brakes, exhaust systems). Also to be checked would be feeding, oiling, adjusting, maintenance, grounding, how attached, work space, and location.

The most critical parts of an item are not always obvious. When the security of a heavy load depends on a cotter pin being in place, then that pin is a critical part. *Poor Richard's Almanac* is relevant to today's hazard control efforts: "A little neglect may breed great mischief. . ."

### What conditions need to be inspected?

The unsafe conditions for each part to be inspected should be described specifically and clearly. A checklist question that reads "Is. . . safe?" is meaningless because it does not define what makes an item unsafe. The unsafe conditions for each item to be inspected must not only be listed but also described. Usually, conditions to look for can be indicated by such words as *jagged, exposed, broken, frayed, leaking, rusted, corroded, missing, vibrating, loose,* or *slipping.* Sometimes exact figures are needed; for example, the maximum pressure in a boiler or the percent

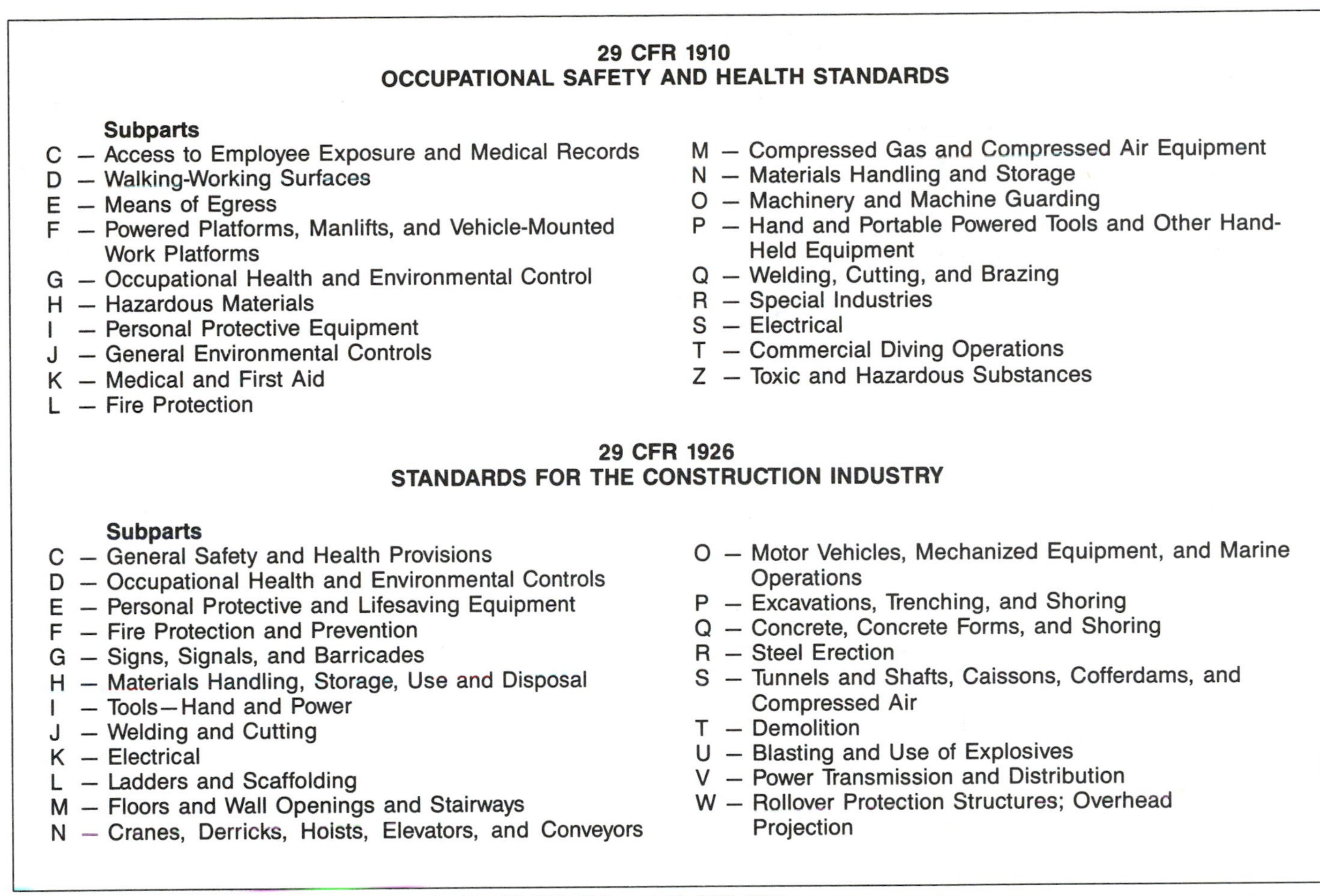

**29 CFR 1910**
**OCCUPATIONAL SAFETY AND HEALTH STANDARDS**

**Subparts**

C — Access to Employee Exposure and Medical Records
D — Walking-Working Surfaces
E — Means of Egress
F — Powered Platforms, Manlifts, and Vehicle-Mounted Work Platforms
G — Occupational Health and Environmental Control
H — Hazardous Materials
I — Personal Protective Equipment
J — General Environmental Controls
K — Medical and First Aid
L — Fire Protection
M — Compressed Gas and Compressed Air Equipment
N — Materials Handling and Storage
O — Machinery and Machine Guarding
P — Hand and Portable Powered Tools and Other Hand-Held Equipment
Q — Welding, Cutting, and Brazing
R — Special Industries
S — Electrical
T — Commercial Diving Operations
Z — Toxic and Hazardous Substances

**29 CFR 1926**
**STANDARDS FOR THE CONSTRUCTION INDUSTRY**

**Subparts**

C — General Safety and Health Provisions
D — Occupational Health and Environmental Controls
E — Personal Protective and Lifesaving Equipment
F — Fire Protection and Prevention
G — Signs, Signals, and Barricades
H — Materials Handling, Storage, Use and Disposal
I — Tools—Hand and Power
J — Welding and Cutting
K — Electrical
L — Ladders and Scaffolding
M — Floors and Wall Openings and Stairways
N — Cranes, Derricks, Hoists, Elevators, and Conveyors
O — Motor Vehicles, Mechanized Equipment, and Marine Operations
P — Excavations, Trenching, and Shoring
Q — Concrete, Concrete Forms, and Shoring
R — Steel Erection
S — Tunnels and Shafts, Caissons, Cofferdams, and Compressed Air
T — Demolition
U — Blasting and Use of Explosives
V — Power Transmission and Distribution
W — Rollover Protection Structures; Overhead Projection

**Figure 4-6.** Lists of inspection subjects required by 29 CFR 1910 and 29 CFR 1926.

spread of a sling hook.

Many different types of monitoring checklists are available, and they can vary in length from thousands of items to only a few. Each type has its place. Generally, the longer checklists refer to OSHA standards. These are useful in determining which standards or regulations apply to individual situations. Once the applicable standards are identified, a checklist can be tailored to organizational needs and uses and computerized for action and follow-up.

The Center for Disease Control, U.S. Department of Health, has devised a suggested checklist for the safety evaluation of shop and laboratory areas. The worksheet is referenced to the OSHA "General Industry Standards."

Checklists serve as reminders of what to look for and as records of what has been covered. They give inspections direction. They allow on-the-spot recording of all findings and comments before they are forgotten. In case an inspection is interrupted, checklists provide a record of what has and what has not been inspected. Without checklists, inspectors can miss things they should see or be unsure, after inspecting an area, that they have covered everything.

Good checklists also help in follow-up. Of course, merely running through a checklist does little to locate or correct problems. Simply checking off the items on the list is not conducting a safety inspection. The checklist must be used as an aid to the inspection process, not as an end in itself. A hazard observed during inspection must be recorded, even though it is not part of the checklist.

The following nine pages comprise a portfolio of various companies' inspection checklists. The first nine are used with computer follow-up on inspection results, actions to be taken, and corrections made. Refer to the following portfolio of checklists.

The amount of detail included in the checklist will vary, depending upon the inspector's knowledge of the relevant standards and the nature of the inspection. An experienced inspector with thorough knowledge of the standards will need only sufficient clues to remind him or her of the items to be inspected. Checklists for infrequent inspection generally will be more detailed than daily or weekly ones.

Checklists should have columns to indicate either compliance or action-date. Space should be provided to cite the specific violation, a way to correct it, and a recommendation that the condition receive more or less frequent attention. Whatever the format of the checklist, space should be provided for the inspector's signature and the inspection date.

Checklists can be prepared by the safety and health committee, by the safety director, or by a subcommittee that includes engineers, supervisors, employees, and maintenance personnel. The safety and health professional and the department supervisor should monitor checklist development and make sure all applicable standards are covered. In their final form the checklists should conform to the inspection route.

*(Text continues on page 87.)*

Mechanical Inspection

Acer. No. 26023-61
Ord. No. 759223
Dept. No. 862

(30)

Koch Dry Ovens and Paint Booth Make Up Air Blowers

Inspec's. Name: ______

Inspec. Date: ______

To Be Inspected- Jan-Mar-May-Jul-Sep-Nov

Inspect For: Security, Condition, Operation, Vibration, Belt Tension, Safety

Dept. 862

| Mach. No. | Bearings | Belts and Pulleys | Belt Guards | Lube & Lube Lines | O.K. | Inspector's Comments |
|---|---|---|---|---|---|---|
| | Sub Assembly Paint System on Mezzanine - North to South | | | | | |
| 7227 | | | | | | |
| 7226 | | | | | | |
| 7186 | | | | | | |
| 7187 | | | | | | |
| 7188 | | | | | | |
| 7225 | | | | | | |
| 7189 | | | | | | |
| | Paint Booth Make Up Air Bldg. - "V" Roof - East Side | | | | | |
| 6790 | | | | | | |
| 6791 | | | | | | |
| 7219 | | | | | | |
| 7221 | | | | | | |
| 7220 | | | | | | |
| 7218 | | | | | | |
| | Work in Process Paint System on Mezzanine - North to South | | | | | |
| 6760 | | | | | | |
| 6758 | | | | | | |
| 6795 | | | | | | |
| 6796 | | | | | | |
| | | | | | | |
| | | | | | | |
| | | | | | | |

| Mach. No. | Bearings | Belts and Pulleys | Belt Guards | Lube & Lube Lines | O.K. | Inspector's Comments |
|---|---|---|---|---|---|---|
| Combine Finish Paint Syst. - Dry Ovens and Paint Booth Make Up Air Units | | | | | | |
| Bldg. - "V1" South to North | | | | | | |
| 7251 | | | | | | |
| 7252 | | | | | | |
| 7259 | | | | | | |
| 7260 | | | | | | |
| 7261 | | | | | | |
| 7262 | | | | | | |
| 7297 | | | | | | |
| 7298 | | | | | | |
| 7300 | | | | | | |
| Touch Up Paint Dry Oven (3-Ovens) Bldg. - "V2" North East Corner | | | | | | |
| 6390 | | | | | | |
| | | | | | | |
| | | | | | | |
| | | | | | | |
| | | | | | | |
| | | | | | | |
| | | | | | | |
| | | | | | | |
| | | | | | | |

MECHANICAL INSPECTION (2)

EXHAUST BLOWERS & AIR CONDI. HEAT EXCHANGER & DOOR SEAL BLOWERS

3-Month Inspection
Feb-May-Aug-Nov

Inspector's Name ____________________ Inspection Date ____________

ACCT. No. 26023-79
ORD. No. 759223
DEPT. No. -862

| Blower Machine Number | Check for Excessive Vibration | Chk. Belts and Pulleys for Ten.& Align.& Condi. | All Safety Shields are in Place and Secure |
|---|---|---|---|
| "V" Bldg. Roof E. Side | | | |
| 7426-Air Condi.Heat Exchanger | | | |
| 7429-Air Condi.Heat Exchanger | | | |
| Sub.Assemb.Paint & Wash Ovens Mezzanine - North to South | | | |
| 7228 | | | |
| 7178 | | | |
| 7179 | | | |
| 7181 | | | |
| 7180 | | | |
| 7183 | | | |
| 7182 | | | |
| 7185 | | | |
| 7224 | | | |
| 7223 | | | |
| 7187 | | | |
| 7190 | | | |
| Flow Coat Booth | | | |
| 7192 | | | |
| 7193 | | | |
| 7194 | | | |
| 7195 | | | |
| 7196 | | | |
| 7197 | | | |
| Spray Paint Booth | | | |
| 7205 | | | |
| 7206 | | | |
| 7208 | | | |
| 7207 | | | |
| 7212 | | | |
| 7213 | | | |
| 7215 | | | |
| 7214 | | | |

| Blower Machine Number | Check for Excessive Vibration | Chk.Belts & Pulleys for Ten.& Align.& Condi. | All Safety Shields are in Place and Secure |
|---|---|---|---|
| W.I.P. Paint Dip | | | |
| 5765 | | | |
| 6767 | | | |
| 6766 | | | |
| 6764 | | | |
| W.I.P. Paint Spray | | | |
| 6784 | | | |
| 6785 | | | |
| 6786 | | | |
| 6787 | | | |
| 6763 | | | |
| W.I.P. Wash Dry Off | | | |
| 6761 | | | |
| 6760 | | | |
| 6759 | | | |
| 6757 | | | |
| W.I.P. Paint Dry | | | |
| 6793 | | | |
| 6794 | | | |
| 6798 | | | |
| 6797 | | | |
| W.I.P. Wash Booth | | | |
| 6755 | | | |
| 6754 | | | |
| 6753 | | | |
| Comb. Finish Paint | | | |
| 7241 | | | |
| 7242 | | | |
| 7243 | | | |
| 7249 | | | |
| Wash Dry Off | | | |
| 7250 | | | |
| 7253 | | | |
| | | | |

| Blower Machine Number | Check for Excessive Vibration | Same as Other | Same |
|---|---|---|---|
| Wash Cool Down | | | |
| 7255 | | | |
| 7256 | | | |
| 7257 | | | |
| Combine Paint | | | |
| 7268 | | | |
| 7269 | | | |
| 7270 | | | |
| 7271 | | | |
| 7272 | | | |
| 7263 | | | |
| 7264 | | | |
| 7265 | | | |
| 7266 | | | |
| 7267 | | | |
| Combine Flash Off | | | |
| 7292 | | | |
| 7293 | | | |
| Combine Paint Dry | | | |
| 7295 | | | |
| 7296 | | | |
| 7299 | | | |
| 7301 | | | |
| Combine Cool Down | | | |
| 7303 | | | |
| 7304 | | | |
| 7305 | | | |
| 7310 | | | |
| 7312 | | | |
| 7306 | | | |
| 7311 | | | |
| Spray & Dry - N.E. Corner-"V2" | | | |
| | | | |
| | | | |
| | | | |

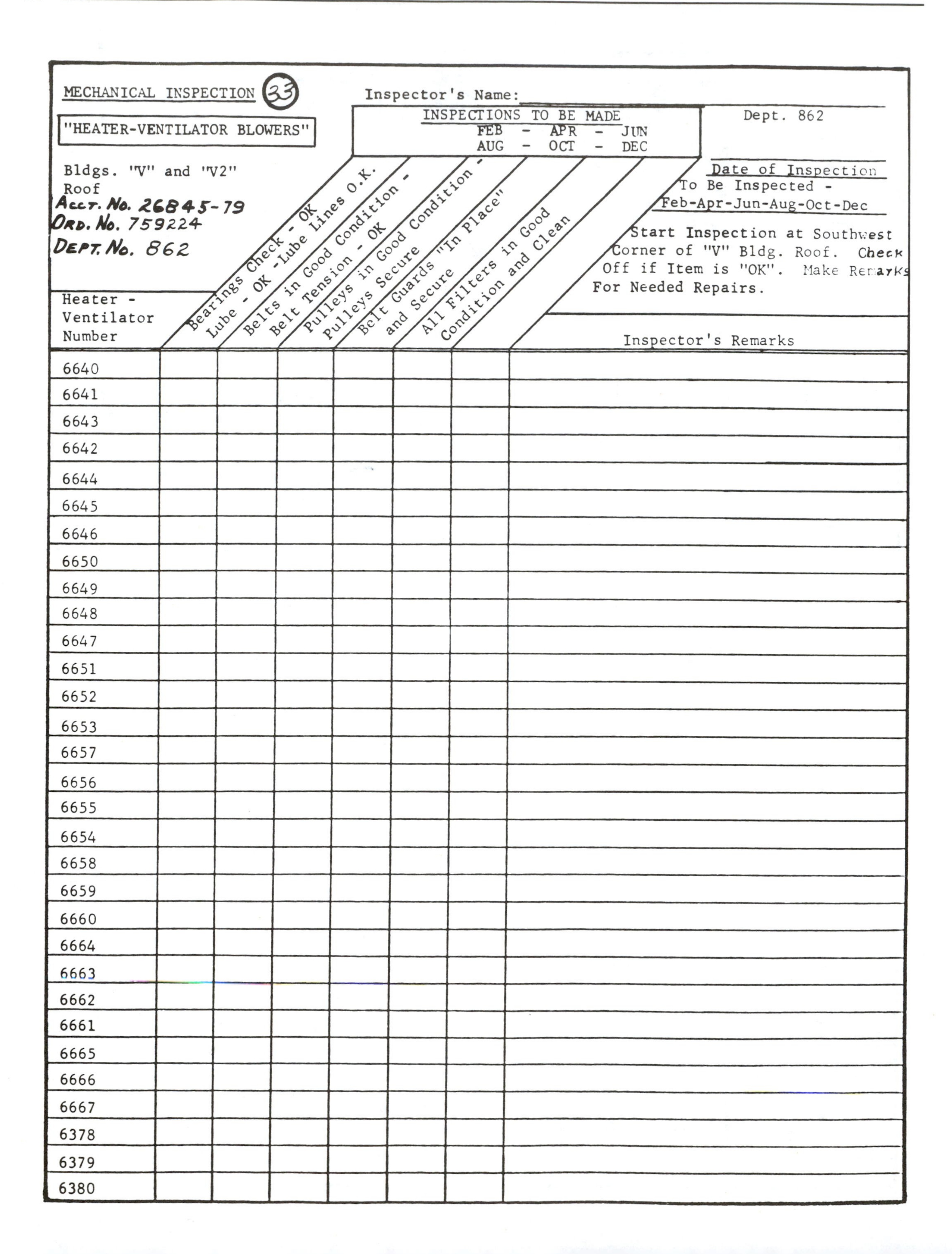

MECHANICAL INSPECTION (33)

"HEATER-VENTILATOR BLOWERS"

Bldgs. "V" and "V2"
Roof
Acct. No. 26845-79
Ord. No. 759224
Dept. No. 862

Inspector's Name:

INSPECTIONS TO BE MADE
FEB - APR - JUN
AUG - OCT - DEC

Dept. 862

Date of Inspection
To Be Inspected -
Feb-Apr-Jun-Aug-Oct-Dec

Start Inspection at Southwest Corner of "V" Bldg. Roof. Check Off if Item is "OK". Make Remarks For Needed Repairs.

| Heater - Ventilator Number | Bearings Check - OK Lube - OK - Lube Lines O.K. | Belts in Good Condition - Belt Tension - OK | Pulleys in Good Condition - Pulleys Secure | Belt Guards "In Place" and Secure | All Filters in Good Condition and Clean | | | Inspector's Remarks |
|---|---|---|---|---|---|---|---|---|
| 6640 | | | | | | | | |
| 6641 | | | | | | | | |
| 6643 | | | | | | | | |
| 6642 | | | | | | | | |
| 6644 | | | | | | | | |
| 6645 | | | | | | | | |
| 6646 | | | | | | | | |
| 6650 | | | | | | | | |
| 6649 | | | | | | | | |
| 6648 | | | | | | | | |
| 6647 | | | | | | | | |
| 6651 | | | | | | | | |
| 6652 | | | | | | | | |
| 6653 | | | | | | | | |
| 6657 | | | | | | | | |
| 6656 | | | | | | | | |
| 6655 | | | | | | | | |
| 6654 | | | | | | | | |
| 6658 | | | | | | | | |
| 6659 | | | | | | | | |
| 6660 | | | | | | | | |
| 6664 | | | | | | | | |
| 6663 | | | | | | | | |
| 6662 | | | | | | | | |
| 6661 | | | | | | | | |
| 6665 | | | | | | | | |
| 6666 | | | | | | | | |
| 6667 | | | | | | | | |
| 6378 | | | | | | | | |
| 6379 | | | | | | | | |
| 6380 | | | | | | | | |

(39)

AIR COMPRESSOR INSPECTION
TIRE ROOM DEPT. 945

MECHANICAL - (862)

ACCT. NO. 26021-67
SHOP ORD. 759226
DEPT. 945

DATE: ____________
INSPECTOR: ____________
SUPERVISOR: ____________

IF ITEM IS O.K. USE [✓] MARK. IF WORK IS NEEDED USE [W] AND EXPLAIN. IF INSPECTOR COMPLETES REPAIR USE [R] & EXPLAIN

| ITEM NO. | PART TO BE INSPECTED | | 1stWK | 2ndWk | 3rdWk | 4thWk | 5thWk |
|---|---|---|---|---|---|---|---|
| 1. | CHECK CRANKCASE OIL LEVEL (ANDEROL-500 OIL) | 7694 | | | | | |
| | | 7840 | | | | | |

| ITEM NO. | PART TO BE INSPECTED | CHECK OFF 7694 | CHECK OFF 7840 |
|---|---|---|---|
| 2. | CHANGE CRANKCASE OIL (ANDEROL-500 OIL) (JAN-MAR-MAY-JUL-SEP-NOV) | | |
| 3. | CHECK CONDITION OF COMPRESSOR AND AIR RECEIVER | | |
| 4. | | | |

| ITEM NO. | PART TO BE INSPECTED | | 1stWk | 2ndWk | 3rdWk | 4thWk | 5thWk |
|---|---|---|---|---|---|---|---|
| 5. | CHECK CONDITION OF AIR RECEIVER CONDENSATE TRAP. DRAIN AS NECESSARY | 7694 | | | | | |
| | | 7840 | | | | | |

| ITEM NO. | PART TO BE INSPECTED | CHECK OFF 7694 | CHECK OFF 7840 |
|---|---|---|---|
| 6. | CLEAN OR REPLACE INTAKE AIR FILTER (JAN-JUL) | | |
| 7. | DISASSEMBLE COMPRESSOR VALVES. CLEAN OR REPLACE ALL PARTS AS NECESSARY - PER INSTRUCTIONS ON PAGE 25 OF I.R. INSTRUCTION FORM NO. 1050-H (JAN-JUL) | | |
| 8. | CHK. CONDITION OF DRIVE BELTS. REPLACE AS NECESSARY. CHECK TENSION. | | |
| 9. | CHK. CONDITION, SECURITY, SOUND-ELEC. MOTOR (LUBE-JAN G-1 GREASE) | | |
| 10. | CLEAN COMPRESSOR & RECEIVER WITH AIR JET (JAN-JUL) (EXTERNAL SURFACE OF CYLINDERS & INTERCOOLER TUBES) | | |

TORQUE CHECK ALL CAP SCREWS

| ITEM NO. | LOCATION OF CAP SCREWS (JAN-JUL) | QUANT. SCREW | TORQUE | CHECK OFF 7694 | CHECK OFF 7840 |
|---|---|---|---|---|---|
| 11. | CONSTANT SPEED UNLOADERS | | | | |
| 12. | AIR HEADS | | | | |
| 13. | CYLINDER BOLTS | | | | |
| 14. | SHAFT END COVER | | | | |
| 15. | CRANKCASE COVER PAN | | | | |
| 16. | DISCHARGE AIR MANIFOLD | | | | |
| 17. | | | | | |
| 18. | | | | | |
| 19. | | | | | |
| 20. | | | | | |

| ITEM NO. | POST ITEM NUMBER AND LIST MAINTENANCE TO BE DONE |
|---|---|
| | |
| | |
| | |
| | |
| | |
| | |
| | |

(44.) GRAIN TANK TRACK
DROP SECTION
"V" BUILDING

Inspect For: Security-Safety-Operation-Condition

INSPECTION DEPT. - 862

ACCT. NO. - 90059-67
ORD. NO. - 759225
DEPT. NO. - 862

If Item is O.K. use ✓ mark. If work is needed use W and explain. If inspector completes repair, use R and explain.

DATE

INSPECTOR'S SIGNATURE

SUPERVISOR'S SIGNATURE

Inspector is to perform lubrications listed in each box at the time indicated.

| ITEM NO. | ITEM FOR INSPECTION | CHECK OFF LOCATION COL. F-19 9190 W. Line | COL. C-19 9191 E. Line | INSPECTOR'S COMMENTS |
|---|---|---|---|---|
| 1. | Drop Section Structural Members and Bolted & Welded Joints | | | |
| 2. | (4) Trolley End Trucks and (2) Drop Section Main Track | | | |
| 3. | Power and Free Drop Track Section with Hardware - Stops - Stop Actuators - Guides - Upper Stops - Carrier Release | | | |
| 4 | (4) Hoisting Cables and (2-Each) Connecting Points and Hardware | | | |
| 5. | Hoist Shaft - (2) Cable Drums - (4) Pillow Block Bearings. Grease Lube: Jan-Apr-July-Oct Use "G-1" Grease | | | |
| 6. | Hoist Shaft Worm Gear Reducer - (3) Drive Couplings - Hydraulic Motor Check Oil level - MONTHLY - Use "O-52" Change Oil in OCTOBER | | | |
| 7. | Hoist Shaft Cam Switch - Drive Sprockets & Chain Hand Oil - MONTHLY - "O-31" | | | |
| 8. | (4) Outboard - Hoist Cable Pulleys Grease - MONTHLY "G-14" | | | |
| 9. | Slack Cable Limit Switch | | | |
| 10. | Hydraulic Drive Unit Check Oil Level-MONTHLY-Change Oil - OCTOBER "O-30" | | | |
| 11. | Horizontal Drive - (2) Drive Wheels-(6) Bearings - Drive Gear Box - Drive Motor. Grease Bearings-January&July "G-1" Chk.Gear Box Oil Level Monthly Change Oil - OCTOBER "O-48" Oil | | | |
| 12. | Electric Control Station with Strain Chain | | | |
| 13. | Power Cable Festoon | | | |
| | | | | |
| | | | | |
| | | | | |
| | | | | |

# PITTSBURGH ROLLS CORPORATION

## CRANE INSPECTION REPORT

Crane No.................................Type.................................... Capacity..........................................

### RUNWAY AND CONDUCTORS

Track Alignment......................Spread........................... Fastenings ..........................................
Line Conductors..................................................Conductor Supports...............................................

### TRUCKS AND MAIN COLLECTORS

Truck Wheels, Flat Spots?..........................Flanges...........................End Play...................................
Axle Bearings.............................................Lubrication ..........................................................
Truck Drive Bearings........................Lubrication................................ Gears ...................................
Gear Screws.......................Pinion..............................Key................. Collectors ..............................

### GIRDERS AND DRIVE

Drive Shaft....................Couplings................Bearings..................... Lubrication ..............................
Foot Brake Shaft..........Couplings................Bearings..................... Lubrication ..............................
Bridge Brake Case.....................Adjustment.......................... Lubrication ...................................
Walkway........................................... Railing ........................ Ladder .................................................
Bridge Motor Support......................Shaft Extension...................... Couplings ...................................
Bridge Drive Gear Case...................................Gears............. Lubrication ...................................

### MOTORS

| Location | Armature | Commutator | Brushes | Brush Holders | Bearings | Lubrication |
|---|---|---|---|---|---|---|
| Bridge........ | | | | | | |
| Hoist........ | | | | | | |
| Aux. Hoist...... | | | | | | |
| Trolley ........ | | | | | | |

### CONTROLLERS

| | Brushes | Brush Holders | Contacts | Wiring | Springs | Resistance |
|---|---|---|---|---|---|---|
| Bridge........ | | | | | | |
| Hoist........ | | | | | | |
| Aux. Hoist ...... | | | | | | |
| Trolley ........ | | | | | | |

Trolley Wheels..................Trolley Wheel Bearings............................ Lubrication ..............................
Trolley Gear Case...............Case Support...............Gears............... Lubrication ..............................
Hoist Gear Case.............Main Gear Train.............Comp. Train...........Lubrication...................................
Mech. Brake...............Drift.......................Elec. Brake............... Adjustment .......... Lubrication.....................
Drum.........................Cable or Chain.........................Cable Pin......................Limit Switch..........................
Limit Switch Adjustment.......................Hook.............. Sheaves ..............Lubrication...............................
Trolley Conductors...................Trolley Collectors..........................Cage Roof....................Door....................
Windows..................Foot Brake Treadle..........................Cont. Levers.......................Load Test.....................
Bell or Signal.....................................................................................................................................
Inspected by.................................................................... Date..........................................................

KEY WORDS—G=Good; F=Fair; W=Worn; A=Need Attention;
C=Need Cleaning; T=Too Tight

BUTLER MANUFACTURING COMPANY
WEEKLY INSPECTION OF FIRE PROTECTIVE EQUIPMENT
Kansas City Plant

Instructions: Fill out this blank while making inspection. Do not report a valve open unless you personally have inspected and tested it. Every valve controlling sprinklers or water supplies to sprinklers should be listed. When the blank is filled out, it should be sent to the Safety Department.

| SPRINKLER VALVES | | | | | | |
|---|---|---|---|---|---|---|
| Valve No. | AREA CONTROLLED | Location | Open | Shut | Sealed | Pressure |
| 1 | Entire West System | Bldg. 57 | | | | |
| 2 | Bldg. 43-43B | Bldg. 57 | | | | |
| 3 | Valves 4-5-6-7-8-9-10-11 | Bldg. 43 | | | | PIV |
| 4 | East End Bldg. 2-2B | Bldg. 2 | | | | |
| 5 | Center 2-2B, Paint Line | Bldg. 2 | | | | |
| 6 | West End Bldg. 2 | Bldg. 2 | | | | |
| 7 | Valves 8-9-10-11 | Bldg. 2 | | | | PIV |
| 8 | Valves 9-10-11 | Bldg. 62 | | | | |
| 9 | Bldg. 3 | Bldg. 62 | | | | |
| 10 | Bldg. 62-62B-6A-5C-5 | Bldg. 62 | | | | |
| 11 | Bldg. 4-4A-5-5A-5C | Bldg. 62 | | | | |
| 12 | Paint Booth P34KC | Bldg. 1 | | | | |
| 13** | Oven | Bldg. 1 | | | | |
| 14 | Paint Booth P57KC | Bldg. 1 | | | | |
| 15 | Paint Booth P58KC | Bldg. 1 | | | | |
| 16 | Paint Booth P59KC | Bldg. 1 | | | | |
| 17 | Paint Booth P56KC | Bldg. 1 | | | | |
| 18 | Locker Room Offices | Bldg. 67 | | | | |
| 19 | Laboratory Paint Room | Bldg. 58A | | | | |
| 20 | Paint Shop Bldg. 23 | Bldg. 23 | | | | |
| 21 | Bldg. 65 | Bldg. 65 | | | | |
| 22 | Paint Booth P49KC | Bldg. 63 | | | | |
| 23 | Paint Booth P710G | Bldg. 63 | | | | |
| 24 | Bldg. 17 | Bldg. 17 | | | | |
| 25 | Bldgs. 53-63 | Bldg. 17 | | | | |
| 26* | Bldg. 57 | Bldg. 57 | | | | |
| 27 | Bldg. 17A Offices | Bldg. 15 | | | | |
| 28 | Bldg. 17A Balcony | Bldg. 15 | | | | |
| 29 | Entire East System | 12th St. | | | | PIV |
| 30* | Bldg. 52 | Driveway | | | | PIV |
| 31* | Bldg. 46 | Driveway | | | | PIV |
| 32 | Valves 21-22-23-24-25 | Driveway | | | | PIV |
| | | | | | | |
| | | | | | | |
| | | | | | | |
| | | | | | | |
| | | | | | | |
| | | | | | | |
| | | | | | | |

* Controls Dry System
** Manually Operated, always shut

GENERAL CONDITIONS

HYDRANTS: In good condition? ________

Clear? ________ Remarks ________

________________

AUTOMATIC SPRINKLERS: Any heads missing? ________ Disconnected? ________

Obstructed by high-piled stock? ______

Any rooms not sufficiently heated to prevent freezing? ________ How many extra heads available? ________

SPRINKLER ALARMS: Tested? ________ In good condition? ________ Do not test hydraulic alarms when temperatures are below freezing.

EXTINGUISHERS, SMALL HOSE: In good condition? ________

FIRE DOORS: All inspected? ________

In good order? ________

HOUSEKEEPING: Good throughout? ________

Combustible waste removed before night? ________

REMARKS on other matters relating to fire hazard: ________

________________

________________

Date ________________ Signed ________________

*Courtesy Butler Manufacturing Company.*

# STATIONARY SCAFFOLD SAFETY CHECK LIST

PROJECT: ______________________________

ADDRESS: ______________________________

CONTRACTOR: ______________________________

DATE OF INSPECTION: ______________ INSPECTOR: ______________

| | Yes | No | Action/Comments |
|---|---|---|---|
| 1. Are scaffold components and planking in safe condition for use and is plank graded for scaffold use? | | | |
| 2. Is the frame spacing and sill size capable of carrying intended loadings? | | | |
| 3. Have competent persons been in charge of erection? | | | |
| 4. Are sills properly placed and adequate size? | | | |
| 5. Have screw jacks been used to level and plumb scaffold instead of unstable objects such as concrete blocks, loose bricks, etc.? | | | |
| 6. Are base plates and/or screw jacks in firm contact with sills and frame? | | | |
| 7. Is scaffold level and plumb? | | | |
| 8. Are all scaffold legs braced with braces properly attached? | | | |
| 9. Is guard railing in place on all open sides and ends above 10' (4' in height if less than 45") | | | |
| 10. Has proper access been provided? | | | |
| 11. Has overhead protection or wire screening been provided where necessary? | | | |
| 12. Has scaffold been tied to structure at least every 30' in length and 26' in height? | | | |
| 13. Have free standing towers been guyed or tied every 26' in height? | | | |
| 14. Have brackets and accessories been properly placed: Brackets? | | | |
| Putlogs? | | | |
| Tube and Clamp? | | | |
| All nuts and bolts tightened? | | | |
| 15. Is scaffold free of makeshift devices or ladders to increase height? | | | |
| 16. Are working level platforms fully planked between guard rails? | | | |
| 17. Does plank have minimum 12" overlap and extend 6" beyond supports? | | | |
| 18. Are toeboards installed properly? | | | |
| 19. Have hazardous conditions been provided for: Power lines? | | | |
| Wind loading? | | | |
| Possible washout of footings? | | | |
| Uplift and overturning moments due to placement of brackets, putlogs or other causes? | | | |
| 20. HAVE PERSONNEL BEEN INSTRUCTED IN THE SAFE USE OF THE EQUIPMENT? | | | |

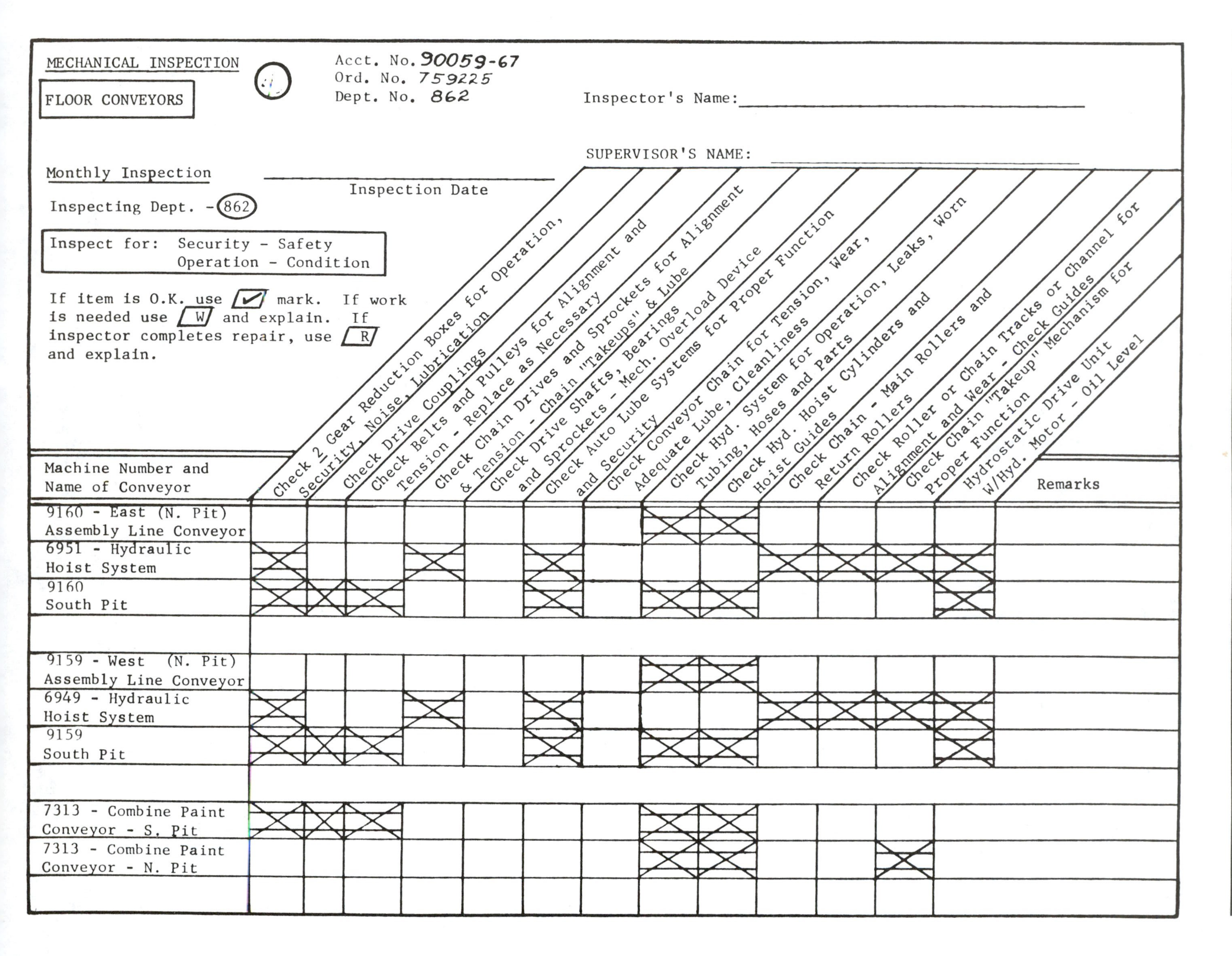

MECHANICAL INSPECTION

FLOOR CONVEYORS

Acct. No. 90059-67
Ord. No. 759225
Dept. No. 862

Inspector's Name: ____________________

SUPERVISOR'S NAME: ____________________

Monthly Inspection

____________________
Inspection Date

Inspecting Dept. - (862)

Inspect for: Security - Safety
Operation - Condition

If item is O.K. use [✓] mark. If work is needed use [W] and explain. If inspector completes repair, use [R] and explain.

| Machine Number and Name of Conveyor | Check 2 Gear Reduction Boxes for Operation, Security, Noise, Lubrication | Check Drive Couplings | Check Belts and Pulleys for Alignment and Tension - Replace as Necessary | Check Chain Drives and Sprockets for Alignment & Tension - Chain "Takeups" & Lube | Check Drive Shafts, Bearings and Sprockets - Mech. Overload Device | Check Auto Lube Systems for Proper Function and Security | Check Conveyor Chain for Tension, Wear, Adequate Lube, Cleanliness | Check Hyd. System for Operation, Leaks, Worn Tubing, Hoses and Parts | Check Hyd. Hoist Cylinders and Hoist Guides | Check Chain - Main Rollers and Return Rollers | Check Roller or Chain Tracks or Channel for Alignment and Wear - Check Guides | Check Chain "Takeup" Mechanism for Proper Function | Hydrostatic Drive Unit W/Hyd. Motor - Oil Level | Remarks |
|---|---|---|---|---|---|---|---|---|---|---|---|---|---|---|
| 9160 - East (N. Pit) Assembly Line Conveyor | | | | | | | | X | X | | | | | |
| 6951 - Hydraulic Hoist System | X | | | X | | X | | | | X | X | X | X | |
| 9160 South Pit | X | X | X | | | X | | X | X | | | | X | |
| | | | | | | | | | | | | | | |
| 9159 - West (N. Pit) Assembly Line Conveyor | | | | | | | | X | X | | | | | |
| 6949 - Hydraulic Hoist System | X | | | X | | X | | | | X | X | X | X | |
| 9159 South Pit | X | X | X | | | X | | X | X | | | | X | |
| | | | | | | | | | | | | | | |
| 7313 - Combine Paint Conveyor - S. Pit | X | X | X | | | | | X | X | | | | | |
| 7313 - Combine Paint Conveyor - N. Pit | | | | | | | | X | X | | | X | | |
| | | | | | | | | | | | | | | |

| DEPARTMENT | UNIT | SUPERVISOR RESPONSIBLE | APPROVED BY | DATE | PAGE NO. |
|---|---|---|---|---|---|
| *Maintenance* | *Workshop* | *J. P. Smith* | *Ralph T. Welles* | *4/16/72* | *1* |

| 1. PROBLEMS | 2. CRITICAL FACTORS | 3. CONDITIONS TO OBSERVE | 4. FREQUENCY | 5. RESPONSIBILITY |
|---|---|---|---|---|
| *1. Overhead hoist* | *Cables, chains, hooks, pulleys* | *Frayed or deformed cables, worn or broken hooks and chains, damaged pulleys* | *Daily—before each shift* | *Operators* |
| *2. Hydraulic pump* | *High pressure hose* | *Leaks; broken or loose fittings* | *Daily* | *Shift leader* |
| *3. Power generator* | *High voltage lines* | *Frayed or broken insulation* | *Weekly* | *Foreman* |
| *4. Fire extinguishers* | *Contents, location, charge* | *Correct type, fully charged, properly located, corrosion, leaks* | *Monthly* | *Area safety inspector* |
| *5. General housekeeping* | *Passageways, aisles, floors, grounds* | *Free of obstructions, clearly marked, free of refuse* | *Daily* | *Shift leader foreman* |

**Figure 4-7.** Responsibility for each inspection should be assigned in a hazard control inspection inventory. (Printed from *Principles and Practices of Occupational Safety and Health, Student Manual,* Booklet Three, U.S. Department of Labor OSHA 2215.)

Choosing the inspection route means inspecting an area completely and thoroughly while avoiding:

Time-consuming backtracking and repetitions

Long walks between items

Unnecessary interruptions of the production process

Distracting employees.

Often a closed-loop inspection will give good results. Sometimes it is valuable to follow the path of the material being processed.

### How often must items be inspected?

The frequency of inspection is determined by four factors.

1. *What is the loss severity potential of the problem?* Ask yourself, "If the item or critical part should fail, what would happen? What injury, damage, or work interruption would result?" The greater the loss severity potential, the more often the item should be inspected. Because a frayed wire rope on an overhead crane block has the potential to cause a much greater loss than a defective wheel on a wheelbarrow, the rope needs to be inspected more frequently than the wheel.
2. *What is the potential for injury to employees?* If the item or critical part should fail, how many employees would be endangered and how frequently? The greater the probability for injury to employees, the more often the item should be inspected. For example, a stairway continually used by many people needs to be inspected more frequently than one that is seldom used.
3. *How quickly can the item or part become unsafe?* The answer to this question depends on the nature of the part and the conditions to which it is subjected. Equipment and tools that get heavy use can become damaged, defective, or worn more quickly than those rarely used. An item located in a particular spot can be exposed to greater damage than an identical item in a different location. The shorter the time in which it can become unsafe, the more frequently the item should be inspected.
4. *What is the past history of failures? What were the results of these failures?* Maintenance and production records and accident investigation reports can provide valuable information about how frequently items have failed and the results in terms of injuries, damage, delays, and shutdowns. The more frequently it failed in the past and the greater the consequences, the more often the item needs to be inspected.

OSHA regulations require inspections at specific intervals; for example, manlifts, not more than 30 days; limit switches, weekly. As a specific example, OSHA regulations dealing with cranes specify daily visual inspection of some aspects, monthly signed reports, semiannual inspection of standby cranes, and periodic inspection of other parts. According to OSHA, the intervals depend on "the nature of the critical components of the crane and the degree of their exposure to wear, deterioration, or malfunction."

The preceding portfolio of checklists contains two inspection reports for cranes. These forms would meet the OSHA requirement for a monthly signed report. Note that the critical parts are listed and the specific conditions named.

Frequency of inspections should be described in specific terms: for example, before every use, when serviced daily, monthly, quarterly, yearly.

### Who will conduct the inspection?

Answering the four previous questions—the items to be inspected, the aspects of each item to be inspected, the conditions to be inspected, and the frequency of inspections—will help define the persons qualified to do the inspecting. No individual or group should have exclusive responsibility for all inspections. Some things will have to be inspected by more than one person. For example, while an area supervisor may inspect an overhead crane weekly and maintenance personnel inspect it monthly, the operator of the crane will inspect it before each use. When grinding wheels are received, they are inspected by the stock room attendant, but they must be inspected again by the operator before they are used.

As part of the hazard control inspection inventory, responsibility for each inspection should be assigned. Figure 4-7 shows how the inventory can designate the proper person by title: area supervisor, operator, foreman, maintenance foreman, and so forth.

A suggested guide for planned inspections is as follows:

*Daily*—area supervisor and maintenance personnel; they also can request suggestions from employees in their various work stations.

*Weekly*—department heads.

*Monthly*—supervisors, department heads, the safety department, and safety and health committees.

The safety department also may be actively involved in monthly, quarterly, semiannual, and annual inspections.

Five qualifications of a good inspector are:

1. Knowledge of the organization's accident experience
2. Familiarity with accident potentials and with the standards that apply to his or her area
3. Ability to make intelligent decisions for corrective action
4. Diplomacy in handling personnel and situations
5. Knowledge of the organization's operations—its workflow, systems, and products.

**Safety professionals.** Clearly the safety professional spearheads the inspection activity. During both individual and group inspections, he or she can perform an educational function. By using on-the-spot examples and firsthand contact, supervisors, foremen, and stewards can be taught the fundamentals of hazard identification. Safety and health committees also can be taught what to look for when making inspections. The organization's fire protection representative or industrial hygienist will work with the hazard control specialist in inspections.

The number of safety professionals depends on the size of the company and the nature of its operation. Large companies with well-organized accident prevention programs usually employ a full-time staff. Sometimes large companies also have designated employees who spend part of their time on inspections.

In organizations where toxic and corrosive substances are present, the industrial hygienist will be part of the inspection team (also see the following section on Measurement and Testing). When an organization uses chemicals, the chief chemist will need to closely cooperate with the safety professional and fire protection representative in establishing inspection criteria. If the organization has no industrial hygienist, the safety professional must possess special training about the hazardous properties of substances, unstable properties of chemicals, and methods of control. An inspection conducted without this knowledge is only perfunctory.

**Company or plant management.** Safety inspections should be considered part of the duties of company or plant management. By participating in inspections, management evidences its commitment to assuring a safe working environment. But the psychological effect of inspection by senior executives goes beyond merely showing an interest in safety. When employees know that management is coming to inspect their area, things get straightened up in a hurry. Conditions which previously seemed "good enough" quickly are found unsatisfactory and corrective action is taken.

**First-line supervisor or foreman.** Because supervisors and foremen spend practically all their time in the shop or plant, they are continually monitoring the workplace. At least once a day, supervisors need to check their areas to see that (1) employees are complying with safety regulations, (2) guards and warning signs are in place, (3) tools and machinery are in a safe condition, (4) aisles and passageways are clear and proper clearances maintained, and (5) material in process is properly stacked or stored. Although such a spot check does not take the place of more detailed inspections, it emphasizes the supervisor's commitment to maintaining safety in the area. A supervisor also should conduct regular formal inspections to make certain all hazards have been detected and safeguards are in use. Such inspections can be performed weekly as an individual and monthly as part of a safety and health committee.

**Mechanical engineer and maintenance superintendent.** Either as individuals or as members of a committee, the mechanical engineer and the maintenance superintendent also need to conduct regular formal inspections. Necessary work orders for guards or for correcting faulty equipment can be written up on the spot.

**Employees.** As mentioned previously, employee participation in continuous inspection is one goal of an effective hazard control program. Before beginning the work day, the employee should inspect the workplace and any tools, equipment, and machinery that will be used. Any defects the employee is not authorized to correct should immediately be reported to the supervisor.

**Maintenance personnel.** Maintenance employees can be of great help in locating and correcting hazards. As they work, they can conduct informal inspections and report hazards to the supervisor, who in turn should encourage the mechanics to offer suggestions.

**Joint safety and health committees.** Joint safety and health committees (discussed in the previous chapter) conduct inspections as part of their function. They give equal consideration to accident, fire, and health exposures. By periodically visiting areas, members notice changed conditions more readily than someone who is there every day. Another advantage provided by the committee is the various backgrounds, experience, and knowledge represented.

If the committee is large, the territory should be divided among teams of manageable size. Large groups going through the plant are unwieldy and distracting. See the Council's *You Are the Safety and Health Committee.*

**Other inspection teams.** If there is no safety and health committee, a planned, formal inspection is still necessary. An inspection team must be assigned, a team including the hazard control specialist, production manager, supervisor, employee representative, fire prevention specialist, and industrial hygienist. The important point is that inspections should be under the direction of a responsible executive who will provide the authority necessary to assure effectiveness.

**Outside inspectors** sometimes are needed to perform inspections. For example, insurance company safety engineers and local, state or provincial, and federal inspectors may perform inspections.

**Contractors' inspection services.** For some technical systems, notably sprinkler systems, contracting companies furnish inspection services. Companies that do not have qualified safety professionals and a well-established maintenance program can avail themselves of such services.

An example will show how such services operate. A sprinkler contractor arranges with the customer for periodic inspection and tests of sprinkler equipment. Frequency of inspection is negotiated between the contractor and the client. In some cases, the inspection will include other items, such as fire extinguishers, hoses, or fire doors. The contractor furnishes a comprehensive written report. The client can request that the contractor send copies of the report to the insurer.

The basic contract does not include maintenance work or materials required for alterations, repairs, or replacement. However, if the report indicates any maintenance needs, the client can have the contractor perform the work.

Although contract service does not relieve management of its primary reponsibility for inspection and maintenance, it does provide excellent inspection for small companies, buildings with mixed tenants, and companies with systems too complex for inspection by its own maintenance staff.

## CONDUCTING INSPECTIONS

### Preparing to inspect

Inspections should be scheduled at a time allowing maximum opportunity to view operations and work practices with minimum interruption. The inspection route should be planned in advance.

Before making an inspection, the inspector or inspection team should review all accidents having occurred in the area. At this brief meeting, team members should discuss where they are going and what they will be looking for. During the inspection, it will be necessary to "huddle" before going into noisy areas in order to avoid arm waving, shouting, and other unsatisfactory methods of communication.

In addition to the regular checklist and accident reports, inspectors should have copies of the previous inspection report for that particular area. Reviewing this report makes it possible to check whether earlier recommendations have been followed and hazards corrected.

Those making inspections should wear the protective equipment required in the areas they enter: safety glasses and shoes, hard hats, acid-proof goggles, protective gloves, respirators, gas masks, and so forth. If inspectors do not have and cannot get special protective equipment, they should not go into the area. They must be careful to "practice what they preach."

Inspectors also should be aware of any special hazards they encounter. For example, because welding crews and other maintenance crews move from place to place, they may be encountered anywhere in the plant. Inspectors should know what precautions are necessary where these crews are working.

### Relationship of inspector and supervisor

Before inspecting a particular department or area, the inspector should contact the department head, supervisor, or other person in charge. This person may have information which is important for the inspection, particularly when conditions are temporarily altered because of construction, maintenance, equipment downtime, employee absence, and so forth.

If no rules prohibit it, the person in charge may want to accompany the inspector. Tactfully, the inspector can agree, but should make it clear no tour guide is needed. The inspector must preserve independence and the opportunity to make uninfluenced observations.

If the supervisor of the area does not accompany the inspector, the supervisor should be consulted before the inspector leaves the area. The inspector should discuss each recommendation with the supervisor. Usually an agreement can be reached as to the relative importance of a recommendation. Obviously an inspector should not pick numerous trivial items merely to make the report look good. On the other hand, the inspector does not have the authority to pass up any condition that might result in an accident.

Even minor items that the supervisor can quickly correct should be reported; though it can be noted, on the written report, that the supervisor promises to correct the particular condition. This keeps the record clear and serves as a reminder to check the condition during the next inspection.

An inspector cannot fail to report hazards because a supervisor interprets such reporting as criticism. If a supervisor becomes defensive or resentful, the inspector can only repeat what the supervisor knows: that the purpose of the inspection is fact-finding, not fault-finding. By retaining objectivity and refusing to let the issue of safety degenerate into the issue of pesonality, the inspector keeps matters on the proper professional footing and maintains an attitude that is firm, friendly, and fair.

Sometimes a supervisor will request the inspector's assistance in recommending new equipment, reassignment of space, or transfer of certain jobs from one department to another. When these suggestions deal with safety, the inspector will want to include them in notes and consider whether to make them part of the report. However the inspector must be careful not to promise either a supervisor or an employee more than actually can be delivered. For example, if a member of the safety and health committee promises that a machine will be replaced by one with an automatic feed and then learns that current budget funds for equipment replacement have been designated, unwittingly the credibility of the entire committee has been undermined.

### Relationship of inspector and employee

Unless company policy or departmental rules prohibit conversation with employees, the inspector can ask questions about operations, being careful, however, not to usurp the responsibility of the supervisor. If, for example, a member of a safety and health committee sees an employee who seems to be working unsafely, it is better to ask the supervisor than the employee about the supposed infraction. The committee member may not fully understand the operation and may be incorrect in the assumption. In another case, the employee may be committing an unsafe act sanctioned by those in authority and could become defensive about carrying out orders. It is the supervisor's job to require compliance with regulations; it is the inspector's job to do the inspecting and reporting. If, however, the situation appears to present an immediate danger, the employee should be notified.

Chapter 3 differentiated between deviations from accepted practices and workplace-induced human error. The inspection team needs to look for both. The inspector is not concerned with identifying the person who is responsible for the unsafe behavior (fault-finding). The goal is to identify the behavior (fact-finding) and see that it is corrected.

Unsafe behaviors will vary from one area to another. Among common items that might be noted are the following:

1. Using machinery or tools without authority
2. Operating at unsafe speeds or in other violation of safe work practice
3. Removing guards or other safety devices or rendering them ineffective
4. Using defective tools or equipment or using tools or equipment in unsafe ways
5. Using hands or body instead of tools or push sticks
6. Overloading, crowding, or failing to balance materials or handling materials in other unsafe ways, including improper lifting
7. Repairing or adjusting equipment that is in motion, under pressure, or electrically charged
8. Failing to use or maintain (or using improperly) personal protective equipment or safety devices
9. Creating unsafe, unsanitary, or unhealthy conditions by improper personal hygiene, using compressed air for cleaning clothes, poor housekeeping, or smoking in unauthorized areas
10. Standing or working under suspended loads, scaffolds, shafts, or open hatches.

Because the inspector's purpose is to locate unsafe acts, not pinpoint blame, the report should not specify any names. When the report states, "An employee in this area was observed. . ." the supervisor has been advised of the need to enforce safe work practices. The inspector should not be seen as a policeman handing out tickets or, worse, as a snoop from "outside." Information derived from inspections should not be used for punitive measures.

Sometimes it is necessary to closely observe workers at work in order to understand a task. The inspector should explain to the worker the need to observe the task and ask permission to watch. When an employee understands that no one is trying to catch an error but rather that he or she has been chosen to demonstrate a task because of exceptional skills, he or she probably will agree to the observation.

### Recording hazards

Inspectors should locate and describe each hazard found during inspection. A clear description of the hazard should be written down and questions and details recorded for later use. It is important to determine which hazards present the most serious consequences and are most likely to occur. The hazard-ranking scheme described in Chapter 3 will simplify the job of classifying hazards.

Properly classifying hazards places them in the right perspective. A significant benefit is that potential consequences and the probability of such consequences occurring are described without the need for long narrative description. Management should be able to understand and evaluate the problems, assign priorities, and quickly reach decisions.

Unsafe conditions or deviations from accepted practices must be described in detail. Machines and operations must be identified by their correct names. Locations must be accurately named or numbered. Specific hazards must be described. Instead of noting "poor housekeeping," for example, the report should give the details: "Empty pallets left in aisles, slippery spots on the floor from oil leaks, a ladder lying across empty boxes, scrap piled on the floor around machines." Instead of noting "guard missing," the report should read, "Guard missing on shear blade of No. 3 machine, SW corner of Bldg. D."

Some plan should be adopted to note intermediate or permanent corrective measures. For example, if intermediate safety measures have been taken, the item could be circled. When permanent measures are taken, the item can be crossed out or marked with an X. Such a system identifies those items requiring further corrective action (see Figure 4-8).

If the inspection is being performed by a committee, one member can be given the task of keeping notes. Without such notes it is almost impossible to write a satisfactory inspection report.

### Condemning equipment

When a piece of equipment presents an imminent danger, the inspector immediately should notify the supervisor and see to it that the machine or equipment is shut down, tagged, or locked out to prevent its further use. Figure 4-9 shows both sides of a danger tag that can be used to prevent further use of equipment or materials that have become unsafe through wear, abuse, or defects.

When danger tags are used, those persons authorized to condemn equipment must sign them. Only the inspector who places the tag should be permitted to remove it and only when satisfied that the hazardous condition has been corrected.

Lockouts may also be necessary. Before equipment is worked on, the main switch or power source must be locked out. For more on lockouts, see Chapters 8 and 15, *Engineering and Technology* volume of this Manual.

No equipment or materials should be placed out of service without notifying the person in authority in the department affected.

### Writing the inspection report

Every inspection must be followed by a clearly written report. Without a complete and accurate report, the inspection would be little more than an interesting sightseeing tour. Inspection reports are usually of three types:

1. *Emergency*—made without delay when a critical or catastrophic hazard is probable. Using the classification system described in Chapter 3, this category would include any items marked IA or IIA.
2. *Periodic*—covers those unsatisfactory nonemergency conditions observed during the planned periodic inspection. This report should be made within 24 hours of the inspection. Periodic reports can be initial, follow-up, final, or a combination of all three.
3. *Summary*—lists all items of previous periodic reports for a given time.

The written report should include the name of the department or area inspected (giving the boundaries or location if needed), date and time of inspection, names and titles of those performing the inspection, date of the report, and the names of those to whom the report was made.

One way to make the report is to begin by copying items carried over from the last report because permanent corrective measures had not been taken. Each item is numbered consecutively. The item number can be followed by the hazard classification (IB, IIIC, and so on). Carryover items can be marked with

**INSPECTION REPORT**

Area Inspected: Building D — Date and Time of Inspection: 11/19/30—11:00 a.m.

Inspector and Title: Ron Baker, Hazard Control Specialist — Date of Report: 11/20/80

Names of Those to Whom Report Is Sent: Bob Firenze (Executive Director); Loren Hall (Department Head); file

No. of Items Carried Over from Previous Report: 3 — No. of Items Added to This Report: 4 — Total No. of Items on This Report: 7

| *Item (asterisk indicates old item)* | *Hazard Classification* | | *Hazard Description* | *Specific Location* | *Supervisor* | *Corrective Action Recommended* | *Corrective Action Taken* |
|---|---|---|---|---|---|---|---|
| | *Consequence* | *Probability* | | | | | |
| *1 | II | B | Guard missing on shear blade #2 machine. Work order issued to engineering for new guard 10/16/80. Wooden barrier guard in temporary use 10/23/79. Guard still missing. | S.W. corner, bay #1 | Jay Rillo | Contact engineering to replace guard | Engineering says they will have guard by 11/24. |
| *2 | IV | C | Window cracked. Work order issued for replacement 10/30/80. | South wall, bay #3 | Joe Whitestone | Have maintenance replace window | Maintenance to replace all broken windows starting next week. |
| 3 | II | B | Oil and trash still accumulated under main motor. Was to be cleaned by 10/30/80. | Pump room | Tony Silva | Clean area; have supervisor talk to men | Cleaned out 11/21. Silva told men to keep area clear. |
| 4 | III | B | Mirror at pedestrian walk out of line | North end of machine shop | Tom Schroeder | Post temporary warning sign; call maintenance for adjustment | Sign posted 11/21—Butler has scheduled adjustment for 12/1 |
| 5 | II | A | Three workers at cleaning tank not wearing eye protection | Electric shop | Hank Beine | Have supervisor give more training and education | Discussed with Beine—he held meeting on 11/25 |
| 6 | I | A | Cable on jib crane badly frayed | Bay #3 | Joe Whitestone | IMMEDIATE ACTION REQUIRED | Tagged crane out of Service Cable to be replaced 11/21 |
| 7 | II | B | Guard rail damaged on stairway to second floor | Bay #1 | Jay Rillo | Issue work order to carpenter shop to make replacement | Work issue order 11/21 |

**Figure 4-8.** Inspection report form simplifies procedures and emphasizes carryovers, new items, and responsibilities. Column at right is for noting corrective action later taken. (Printed with permission from RJF Associates, Inc. Bloomington, Ind.)

an asterisk. The narrative should include the date the hazard was first detected. Each hazard should be described and its location given. After the hazard is listed, the recommended corrective action should be specified and a definite abatement date be established. There should follow a space for noting corrective action taken later. Figure 4-8 is a sample of an inspection report made by the organization's hazard control specialist after a weekly inspection. A report should show what is right as well as wrong.

When the report is that of a committee, it is well to have it reviewed by each member of the inspection team for accuracy, clarity, and thoroughness.

Generally inspection reports are directed to the head of the department or area where the inspection was made. Copies are also directed to executive management and the manager to whom the department head reports.

### Follow-up for corrective action

When the inspection report is written and disseminated, the inspection process starts to return benefits. The information acquired and the recommendations made are without value unless corrective action is taken. Information and recommen-

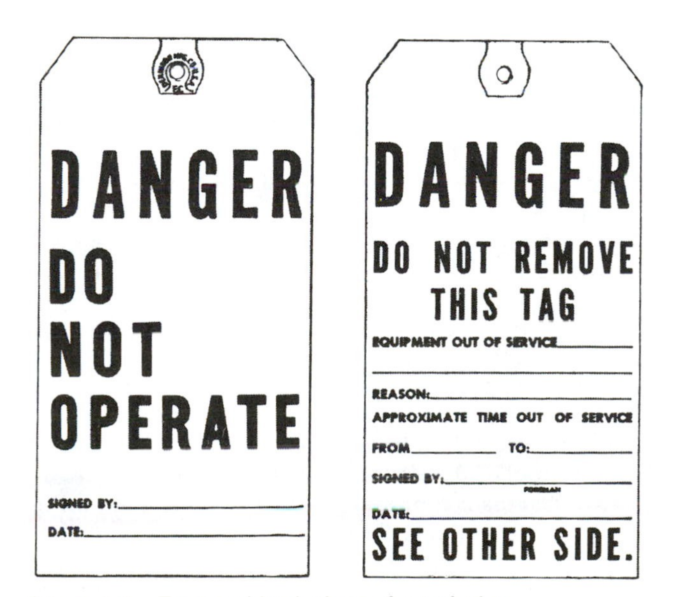


**Figure 4-9.** Front and back views of a typical tag used when equipment is taken out of service because it has become unsafe.

dations provide the basis for establishing priorities and implementing programs that will reduce accidents, improve conditions, raise morale, and increase the efficiency and effectiveness of the operation.

Recommendations can be listed in the order in which the hazards were discovered or grouped according to the individual responsible for their compliance. Recommendations are then sent to the proper official for approval. Where possible, a definite time limit for compliance should be set for each recommendation, and followed up.

Often the safety professional is authorized to make recommendations directly to the affected foreman, supervisor, or department if such recommendations do not require major capital outlay. One company has simplified the process of making individual safety recommendations by devising special forms (see Figure 4-10). These forms—which have a carbon attached so the safety department can keep its records intact—provide a convenient follow-up file.

Some organizations require that inspection reports be reviewed by the safety and health committee, particularly when recommendations apply to education and training and directly affect employees.

In making recommendations, inspectors should be guided by four rules (*Facility Inspection,* 1973):

- Correct the cause whenever possible. Do not merely correct the result, leaving the problem intact. In other words, be sure you are curing the disease, not just the symptom. If you do not have the authority to correct the real cause, bring it to the attention of the person who does.
- Immediately correct everything possible. If the inspector has been granted the authority and opportunity to take direct corrective action, take it. Delays risk accidents.
- Report conditions beyond your authority and suggest solutions. Inform management of the condition, the potential consequences of hazards found, and solutions for correction. Even when nothing seems to come of a recommendation, it can pay unexpected dividends. A company safety and health committee made a detailed proposal about guarding a particularly hazardous location, only to be told that the engineers had planned to move operations to another location. However, instead of feeling that it had wasted its time, the committee pointed out that the organization had serious communication problems, with the right hand not knowing what the left was doing. The committee recommended that effective management techniques be applied to the hazard control program.
- Take intermediate action as needed. When permanent correction takes time, don't just ignore the hazard. Take any temporary measures you can, such as roping off the area, tagging out equipment or machines, or posting warning signs. These measures may not be ideal, but they are preferable to doing nothing.

Some of the general categories into which recommendations might fall are setting up a better process, relocating a process, redesigning a tool or fixture, changing the operator's work pattern, providing personal protective equipment, and improving personnel training methods. Recommendations can also call for improvements in the preventive maintenance system and in housekeeping. Cleaning up a lot of dirt may be considered the janitor's job, but *preventing* its accumulation is part of the hazard control program.

Management must realize that employees are keenly interested in the attention it pays to correcting faulty conditions and hazardous procedures. Recommendations approved by management should become part of the organization's program. At regular intervals, supervisors should report progress in complying with the recommendations to the safety department, the company safety and health committee, or the person designated by management to receive such information. Inspectors should periodically check to see what corrective action progress is being made. Unsafe conditions that are not corrected indicate a breakdown in management communications and program application.

Sometimes management will have to decide among several courses of action. Often these decisions will be based on cost effectiveness. For example, it may be effective and practical, from the standpoint of cost, to substitute a less toxic material that works as well as the highly toxic substance presently in use. On the other hand, replacing a costly but hazardous machine may have to wait until funds can be designated. In this case, the immediate alternative may be to install machine guards.

## MEASUREMENT AND TESTING

Two special sorts of inspection are those made by the industrial hygienist and the medical staff. Testing for exposures to health hazards requires special equipment not always available to the hazard control specialist. In such cases assistance frequently can be secured from the state industrial hygiene division or the provincial department of labor and health. Another source of help can be the industrial hygienists employed by consulting firms and by insurance companies. Conducting physical examinations of employees exposed to occupational health hazards can require apparatus not available to the organization's medical staff.

The following discussion is a summary of recognition, evaluation, and control of health hazards; details are in *Fundamentals of Industrial Hygiene* (see References). It should help in understanding the role of the safety professional.

### Kinds of measurement and testing

Occupational health surveillance monitors chemical, physical, biological, and ergonomic hazards. Four monitoring systems are used: personal, environmental, biological, and medical.

**Personal monitoring.** One example of personal monitoring is measuring the airborne concentrations of contaminants. The measurement device is placed as closely as possible to the site at which the contaminant enters the human body. When the contaminant is noise, the device is placed close to the ear. When a toxic substance is inhaled, the device is placed in the breathing zone.

**Environmental monitoring.** Environmental monitoring measures contaminant concentrations in the workroom. The measurement device is placed in the general area adjacent to the worker's usual work station or where it can sample the general room air.

**Biological monitoring.** Biological monitoring measures changes in composition of body fluid, tissues, or expired air to detect the level of contaminant absorption. For example, blood or urine can be tested to determine excessive lead absorp-

**SAFETY RECOMMENDATION** No. 1053

Date Issued__________19__

Date Ret'd.__________19__

To__________

PLEASE HAVE THE FOLLOWING UNSAFE CONDITION OR DEVIATION FROM STANDARD PROCEDURE CORRECTED:

__________

__________

__________

__________

__________

Please sign and return to Safety Department within ten days, indicating below what disposition was made of this recommendation.

__________
SAFETY DEPARTMENT

RECOMMENDATION FOLLOWED ( ) WORK COMPLETED__________
DATE

RECOMMENDATION REJECTED ( ) FOR FOLLOWING REASON:__________

__________

__________

Copy of this recommendation is on file in the Safety Department. The Safety Department is instructed to send a detailed list of all unanswered recommendations more than ten days old to the General Superintendent the first of each period.

__________
(SIGNED) DEPARTMENT SUPERINTENDENT

**Figure 4-10.** Special form is padded, numbered in pairs, and carboned in order to save office work and permit follow-up until recommended work is complete or procedure correction is made. (Printed with permission from Tennessee Eastman Corporation.)

tion. The phenol in urine sometimes is measured to determine excessive benzene absorption.

**Medical monitoring.** When medical personnel examine workers to see their physiological and psychological response to a contaminant, the process is termed medical monitoring. Medical monitoring can include health and work histories, physical examinations, X rays, blood and urine tests, pulmonary function tests, and vision and hearing tests. The aim of such monitoring is to find evidence of exposure early enough to identify the person especially susceptible and to identify damage before it is irreversible.

Biological and medical monitoring provide information after the exposure already has occurred. However, such programs also encompass arrangement to treat an identified health problem and to take corrective action to prevent further damage.

To understand how industrial hygienists measure for health hazards, it is necessary to define some basic terms, to distinguish between acute and chronic effects, and to see how safe exposure levels are established.

### Measuring for toxicity

**Toxicity.** The toxicity of a material is not identical with its potential for being a health hazard. Toxicity is the capacity of a material to produce injury or harm. Hazard is the possibility that exposure to a material will cause injury or illness when a specific quantity is used under certain conditions.

The key elements to be considered when evaluating a health hazard are:
1. The amount of material to which the employee is exposed
2. The total time of the exposure
3. The toxicity of the substance
4. Individual susceptibility.

Not all toxic materials are hazardous. The majority of toxic chemicals are safe when packaged in their original shipping containers or contained within a closed system. As long as toxic materials are adequately controlled, they can be safely used. For example, many solvents, if not properly used, will cause irritation to eyes, mouth, and throat. Some also are intoxicating and can cause blistering of the skin and other forms of dermatitis. Prolonged exposure can cause more serious illness. But if the solvents are used in a well-ventilated area and the person is provided with protective equipment that prevents the substance from coming in contact with skin, then they can be used without being a hazard.

The toxic action of a substance can be divided into acute and chronic effects:

*Acute effects.* These usually involve short-term, high concentrations that can cause irritation, illness, or death. They can be the result of sudden and severe exposure, during which the substance is rapidly absorbed. Acute effects can be related to an accident, which disrupts ordinary processes and controls. For example, sudden exposure to a high concentration of zinc oxide fumes in the welding shop can cause acute poisoning.

*Chronic effects.* These usually involve continued exposure to a toxic substance over a long time period. When the chemical is absorbed more rapidly than the body can eliminate it, accumulation in the body begins. If the level of contaminant is relatively low, the effects, even if they are serious and irreversible, may go unnoticed for long periods of time. Cancer, for example, may not show up until years after the exposure occurred. Breathing even low concentrations of carbon monoxide for long periods of time can cause damage to the heart muscles and blood vessels.

**Inhalation hazards.** Inhalation of harmful materials may irritate the upper respiratory tract and lung tissue, or the terminal passages of the lungs and the air sacs, depending upon the solubility of the material.

Inhalation of biologically inert gases may dilute the atmospheric oxygen below the normal blood saturation value and disturb cellular processes.

Other gases and vapors may prevent the blood from carrying oxygen to the tissues or interfere with its transfer from the blood to the tissue, producing chemical asphyxia.

Inhaled contaminants that adversely affect the lungs fall into three general categories. Each is discussed later in greater detail.

- Aerosols (particulates) which, when deposited in the lungs, may produce either rapid local tissue damage, some slower tissue reactions, eventual disease, or only physical plugging.
- Toxic vapors and gases that produce adverse reaction in the tissue of the lungs themselves.
- Some toxic aerosols or gases that do not affect the lung tissue locally but (1) are passed from the lungs into the blood stream, where they are carried to other body organs, or (2) have adverse effects on the oxygen-carrying capacity of the blood cells themselves.

An example of the first type (aerosols) is asbestos fiber, which causes fibrotic growth in alveolar tissue, plugging the ducts or limiting the effective area of the alveolar lining. Other harmful aerosols are fungi, such as those found in sugar cane residues, which produce bagassosis.

An example of the second type (toxic gases) is hydrogen fluoride, a gas that directly affects lung tissue. It is a primary irritant of mucous membranes, even causing chemical burns. Inhalation of this gas will cause pulmonary edema, a condition where the lungs fill with fluids in reaction to a burning of the tissue, and direct interference with the gas-transfer function of the alveolar lining.

An example of the third type is carbon monoxide, a toxic gas passed into the blood stream without essentially harming the lung. The carbon monoxide passes through the alveolar walls into the blood, where it ties up the hemoglobin so it cannot accept oxygen, thus causing oxygen starvation of the body. Cyanide gas has another effect—it prevents enzymic utilization of molecular oxygen by cells.

Sometimes several types of lung hazards occur simultaneously. In mining operations, for example, explosives release oxides of nitrogen. These impair the bronchial clearance mechanism, so that coal dust (of the particle sizes associated with the explosions) is not efficiently cleansed from the lungs.

If a compound is very soluble—such as ammonia, formaldehyde, sulfuric acid, or hydrochloric acid—it is rapidly absorbed in the upper respiratory tract and, during the initial phases of exposure, it does not penetrate deeply into the lungs. Consequently, the nose and throat become very irritated and a person is driven out of the exposure area.

*Influence of solubility.* Compounds that are insoluble in body fluids cause considerably less throat irritation than the soluble ones, but may penetrate deeply into the lungs. Thus, a very serious hazard can be present and not be immediately recognized because of a lack of warning that the local irritation would otherwise provide. Examples of such compounds (gases) are nitrogen dioxide and ozone. The immediate danger from these compounds in high concentrations is acute lung irritation or, possibly later, chemical pneumonia.

There are numerous chemical compounds that do not follow the general solubility rule. Such compounds are not very soluble in water and yet are irritating to the eyes and respiratory tract. They also can cause lung damage, even death in many situations.

**Threshold Limit Values.** Individual susceptibility to respiratory toxins is difficult to assess. Nevertheless, certain recommended limits have been established. A Threshold Limit Value (TLV) refers to airborne concentrations of substances and represents an exposure level under which most people can work, day after day, without adverse effect. Because of wide variations in individual susceptibility, however, an occasional exposure of an individual at or even below the threshold limit may not prevent discomfort, aggravation of a preexisting condition, or occupational illness.

The term TLV refers specifically to limits published by the American Conference of Governmental Industrial Hygienists (ACGIH). These TLV limits are reviewed and updated each year. The National Safety Council's *Fundamentals of Industrial*

*Hygiene* explains this subject in detail. A brief overview follows. There are three categories of Threshold Limit Values:

1. *Time-Weighted Average* (TLV-TWA) is the time-weighted average concentration for a normal eight-hour day or 40-hour week. Nearly all persons can be exposed day after day to airborne concentrations at these limits without adverse effect.
2. *Short-Term Exposure Limit* (TLV-STEL) is the concentration to which persons can be exposed for a period of up to 15 minutes continuously without suffering:
   a. Irritation
   b. Chronic or irreversible tissue change
   c. Narcoses of sufficient degree to reduce reaction time, impair self-rescue, or materially reduce work efficiency, provided the daily TLV-TWA is not exceeded. No more than four 15-minute exposure periods per day are permitted with at least 60 minutes between exposure periods.

   A STEL is not a separate, independent exposure limit, rather it supplements the time-weighted average (TWA) limit where there are acute effects from a substance whose toxic effects are primarily of a chronic nature. STELs are recommended only where toxic effects have been reported from high short-term exposures in either humans or animals.
3. *Ceiling* (TLV-C) is the concentration that should not be exceeded even for an instant.
4. *"Skin" notation.* Nearly one-fourth of the substances in the TLV list are followed by the designation "skin." This refers to potential exposure through skin absorption. This designation is intended to suggest appropriate measures for the prevention of cutaneous absorption.

**Permissible exposure levels.** The first compilation of health and safety standards from the U.S. Department of Labor's Occupational Safety and Health Administration appeared in 1970. Because it was derived from then-existing standards, it adopted many of the TLVs established in 1968 by the American Conference of Governmental Industrial Hygienists. Thus Threshold Limit Values—a registered trademark of the ACGIH—became, by federal standards, permissible exposure limits (PELs). These PELs represent the legal maximum level of contaminants in the workplace air.

The General Industry OSHA Standards lists about 400 substances for which exposure limits have been established. These are included in subpart Z, "Toxic and Hazardous Substances," Sections 1910.1000 through 1910.1500.

Most of these exposure limits are tabulated in Section 1910.1000. Tables Z-1 and Z-3 in that regulation were originally part of the 1968 TLV list of the ACGIH. These limits already had been adopted by the U.S. Department of Labor under provisions of the Walsh-Healey Act before passage of the OSHAct of 1970. Table Z-2 contains limits developed by the American National Standards Institute (ANSI). Sections 6(a) and 4(b) of the OSHAct gave OSHA authority to promulgate these previously established standards without the hearings and waiting periods required in Section 6(b). This authority ended in April 1973, two years after the effective date of the act.

Sections 1910.1001 through 1910.1500 give more detailed standards regarding individual substances. These standards have been developed in conformance with Section 6(b) of the OSHAct. They were contested by the affected parties in many cases in the U.S. Courts of Appeals. Some of the substances included in this group are asbestos, vinyl chloride, inorganic arsenic, acrylonitrile, cotton dust, and coke oven emissions.

Section 1910.1000 is reproduced in full as Appendix A-2 in the Council's *Fundamentals of Industrial Hygiene,* 3rd ed. Refer to this book for details on air sampling and industrial toxicology, which are briefly reviewed in this chapter.

**Action level** is that point at which employers must initiate certain provisions: employee exposure measurement, employee training, and medical surveillance. The action level is defined as one-half of the permissible exposure.

Why is the action level set well below the PEL? Simply stated, setting the action level at one-half the permissible exposure protects employees from overexposure. It provides optimum employee protection with the minimum burden to the employer. Where employee exposure measurements indicate that no employee is exposed to airborne concentrations of a substance in excess of the action level, employers in effect are exempted from initiating certain provisions.

The action level recognizes that air samples can only estimate the true time-weighted averages. Both employer and employee can be confident that, if the measured exposure level falls *below* the action level, then there is a very high probability the actual exposure level is below the permissible exposure level. Statistical probability suggests that exposure below the action level will not harm the employee.

### When to measure?

The measurements done by the industrial hygienist can be divided into three phases:

1. Problem definition phase
2. Problem analysis phase
3. Solution phase.

**Problem definition phase.** Frequently measurement is done to see whether there is a problem. Some OSHA regulations require measurement at certain specified intervals or any time there is a change in production, process, or control measures. Measurement often establishes that there is no excessive exposure. Such monitoring of the workplace assures a safe environment.

Monitoring, then, frequently determines that employers are in compliance with OSHAct requirements, state or provincial regulations, commonly accepted standards, Threshold Limit Values, permissible exposure levels, and action levels. Newer health standards published by OSHA usually state:

> Each employer who has a place of employment in which [toxic substance name] is released into the workplace air shall determine if there is any possibility that any employee may be exposed to airborne concentrations of [toxic substance name] above the permissible level. The initial determination shall be made each time there is a change in production, process, or control measures that may result in an increase in airborne concentrations of [toxic substance name].

When any hazardous substances are released into the workplace air, the employer must take the first step in the employee exposure monitoring program. For OSHA-regulated substances, there must be an actual exposure determination to see if any employee has been exposed to concentrations in excess of the recommended levels. This determination should be made even if there is any chance that any employee has been exposed at a level above recommended exposure levels.

But where does sampling begin? Should the sample be taken at the worker's breathing zone? Out in the general air? At the machine or process that is putting out the toxic substance? OSHA requires sampling in the worker's breathing zone; however, sampling at all three sites provides a clearer picture of the situation.

Should the sample be taken for two seconds, two hours, or a whole day? There are two major types of samples:

*The grab sample,* taken over so short a period of time that the atmospheric concentration is assumed to be constant throughout the sample. This usually will cover only part of an industrial cycle. A series of grab samples can be taken in an attempt to define the total exposure.

*The long-term sample,* taken over a sufficiently long period of time so that the variations in exposure cycles are averaged. Usually one sample or a series of samples is taken to represent the employees' eight-hour average exposure. OSHA regulations usually require this type of sampling.

An adequate number of tests should be taken to define the time-weighted average exposure, in order to relate this to recommended or regulatory exposure levels. But samples also must be taken to characterize the peak emissions during various portions of the process cycle.

If employee measurements indicate exposure at or above the action level, then OSHA requires all employees exposed at or above the action level must be identified and their exposure measured. The population at risk is thus identified.

When exposure measurements are at or above the action level but not above PEL or just below the TLV-TWA, the employer needs some statistically reliable means to be certain that exposures exceeding these values are not occurring. Periodic sampling should be done. Medical examinations are necessary to determine if any especially susceptible individuals are exhibiting effects at these exposures.

If employees are exposed above the PEL or TLV, then a more intensive monitoring program is necessary. Medical examinations are required to measure the health effects on those employees exposed in excess of the PEL. Noninhalation exposures—such as skin absorption—also may occur. Therefore, accurate exposure evaluation may require breath, blood, and urine sampling.

The problem definition phase is one of orderly progression. At each step of the process employers can ascertain whether they need to proceed to the next higher step.

**Problem analysis phase.** Once the problem has been defined in the first phase of the measurement process, the causes of the problem must be determined. Opportunities for improvement must be identified. The objectives of the solutions must be set. Alternative solutions need to be determined.

The following eight methods suggest some ways that exposure hazards can be controlled:

1. Substitution of a less harmful material for one that is dangerous to health
2. Change or alteration of a process to minimize worker contact
3. Isolation or enclosure of a process or work operation to reduce the number of persons exposed
4. Wet methods to reduce generation of dust in operations
5. Local exhaust at the point of generation or local dispersion of contaminants
6. Personal protective devices (see Chapter 17 in this volume)
7. Good housekeeping, including cleanliness of the workplace, waste disposal, adequate washing, toilet and eating facilities, healthful drinking water, and control of insects and rodents
8. Training and education. The OSHA hazard communication standard requires training for employees exposed to hazardous chemicals.

**Solution phase.** Once the problem has been analyzed and a number of solutions proposed, the most effective, timely, and practical solution needs to be selected—the solution providing optimum benefits with minimal risks. The details of the solution should be carefully worked out. In effect, a blueprint needs to be developed, describing what needs to be done, how it is to be done, by whom, and in what sequence the actions are to take place.

Once controls are installed, they must be periodically checked to see that they are properly functioning. Follow-up monitoring and inspection are necessary to determine that the solution to a given hazardous exposure is controlling it within the specified limits. In other words, the monitoring function should be regarded as circular, not horizontal; if measurement at the solution phase reveals that controls are inadequate, the industrial hygienist must return to the first phase, that of defining the problem.

### Who will do the measuring?

Not every organization requires or can afford the services of a full-time industrial hygienist. Independent consultants can be hired to accomplish two major objectives:

1. Identify and evaluate potential health risks and accident hazards to workers in the occupational environment
2. Design effective controls to protect the safety and health of workers.

Because any person can legally offer services as an industrial hygiene consultant, it is important that the consultant who is hired be trained, experienced, and competent. A competent industrial hygiene consultant must have detailed knowledge of proper sampling equipment and analytic procedures.

Good sources of information and assistance regarding consultants are the American Industrial Hygiene Association and the American Society of Safety Engineers, the professional associations related to occupational safety and health; see the descriptive listing in Chapter 24, Sources of Help. Regional offices of NIOSH usually have lists of consultants in their area. Many insurance companies have loss prevention programs that employ industrial hygienists. The National Safety Council chapters having offices in major cities can offer assistance. The Council offers a full range of consulting services in safety and occupational health management (see Chapter 24, Sources of Help). For a state-by-state listing of governmental consulting service offices, see Chapter 24 under U.S. Government Agencies.

## ACCIDENT INVESTIGATION

A fourth function of monitoring in the total hazard control system is accident investigation, the subject of Chapter 7. The following discussion demonstrates how accident investigation fits into the systems approach to hazard control.

### Why accidents are investigated

When viewed as an integral part of the total occupational safety and health program, accident investigation is especially important as a means to determine cause, uncover indirect accident causes, prevent similar accidents from occurring, document facts, provide information on costs, and promote safety.

**Determine cause.** Accident investigation determines cause. At what points did the hazard control system break down? Were rules and regulations violated? Did defective machinery or factors in the work environment contribute to the accident? Poor machinery layout, for example, or the very design of a job process, operation, or task can contribute to an undesirable situation. Chapter 3 outlined the three primary sources of accidents: human, situational, and environmental factors. The accident investigation concentrates on gathering all information about the factors leading to the accident.

**Uncover indirect accident causes.** Thorough accident investigation is very likely to uncover problems that indirectly contributed to the accident. Such information benefits accident-reduction efforts. For example, a worker slips on spilled oil and is injured. The oil spill is the direct cause of the accident, but a thorough investigation might reveal other factors: poor housekeeping, failure to follow maintenance schedule, inadequate supervision, faulty equipment (such as a lathe leaking oil).

**Prevent similar accidents.** Accident investigation identifies what action can be taken and what improvements made to prevent similar accidents from occurring in the future.

**Document facts.** Accident investigation documents the facts involved in an accident for use in instances of compensation and litigation. The report produced at the conclusion of an investigation becomes the permanent record of facts involved in the accident. Management can breathe more easily when it knows that an accident situation can be reconstructed months or years after the occurrence because the details of the accident have been recorded properly, accurately, and thoroughly.

**Provide information on costs.** Accident investigation provides information on both direct and indirect costs of accidents. Chapter 7 gives details for estimating accident costs.

**Promote safety.** Accident investigation reaps psychological as well as material benefits. The investigation projects the organization's interests in safety and health. It indicates management's sense of accountability for accident prevention, its commitment to a safe work environment. An investigation in which both labor and management participate promotes cooperation between constituencies too often seen as adversaries.

Despite what many people believe, accident investigation is a fact-finding, not a fault-finding, process. When attempting to determine the cause of an accident, the novice is tempted to conclude that the person involved in the accident was at fault. But if human error is chosen (and it is not the real cause), the hazard which caused the accident will go unobserved and uncontrolled. Furthermore, the person falsely blamed for causing the accident will respond to the unjustified corrective action with resentment and alienation. Further cooperation will be discouraged, and respect for the organization's safety and health program will be undermined. The intent of accident investigation is to pinpoint causes of error and defects so similar accidents can be prevented.

Conducting an accident investigation is not simple. It can be very difficult to look beyond the incident at hand to uncover causal factors, determine the true loss potential of the occurrence, and develop practical recommendations to prevent recurrence.

A major weakness of many accident investigations is the failure to establish and consider *all* factors—human, situational, and environmental—that contributed to the accident. Reasons for this failure are several:

- Inexperienced or uninformed investigator
- Reluctance of the investigator to accept responsibility
- Narrow interpretation of environmental factors
- Erroneous emphasis on a single cause
- Judging the effect of the accident to be the cause
- Arriving at conclusions before all factors are considered
- Poor interviewing techniques
- Delay in investigating accidents.

The trained investigator must be ready to acknowledge as contributing causes any and all factors that may have, in any way, contributed to the accident. What at first may appear to be a simple, uninvolved accident can, in fact, have numerous contributing factors, which become more complex as analyses are completed.

Immediate, on-the-scene accident investigation provides the most accurate and useful information.

### When to investigate accidents

The longer the delay in examining the accident scene, interviewing the injured and witnesses, the greater the possibility of obtaining erroneous or incomplete information. The accident scene changes, memories get fuzzy, and people talk to each other. Whether consciously or not, witnesses may alter their initial impressions to agree with someone else's observation or interpretation. Prompt accident investigation also expresses concern for the safety and well-being of employees.

As a general rule, all accidents, no matter how minor, are candidates for thorough investigation. Many accidents occurring in an organization are considered minor because their consequences are not serious. Such accidents—or incidents, as some people prefer to call them—are taken for granted and often do not receive the attention they demand. Management, safety and health committees, supervisors, and employees must be aware that serious accidents arise from the same hazards as minor incidents. Usually sheer luck determines whether a hazardous situation results in a minor incident or a serious accident.

### Who should conduct the investigation?

Chapter 7 discusses the question of who is to make the investigation: the supervisor or foreman, the safety and health professional, a special investigative committee, or a company safety and health committee. As a supplement to that discussion, the following section will outline the roles played by physicians and management in accident investigation. It also covers the responsibility of the safety professional in preventing further accidents from occurring during the investigation itself.

**Physician.** A physician's assistance is particularly important when human factors have been designated as primary or contributing causes of an accident. The physician can assess the nature and degree of injury and assist in determining the source and nature of the forces that inflicted the injury. The physician also can: (1) determine what special biomedical studies, if any, are needed; (2) establish whether the injured person was physically and mentally fit at the time of the accident and whether the screening, selection, and preplacement process is adequate; (3) help judge the adequacy of safety and health protection procedures and equipment; (4) help evaluate the effectiveness of the plans, procedures, equipment, training, and response of rescue, first aid, and emergency medical care personnel. The physician also can evaluate the effectiveness of measures aimed at early detection of medical conditions, mental changes, or emotional stress.

**Management.** Management and department heads should help investigate accidents resulting in lost work days or major property damage. When management actively participates in accident investigation, it can evaluate the hazard control system and determine whether outside assistance is desired or required to upgrade existing structures and procedures. Management also must review accident reports in order to make informed decisions. When accident investigation reveals the need or desirability of specific corrective action, management has the responsibility to determine whether the recommended action has indeed been implemented.

**Safety during the investigation.** In many cases the accident scene is a dangerous place. Electrical equipment may be damaged. Structural members may be weakened by fire or explosion. Radioactive or toxic materials may have been released.

The safety and health professional must be particularly alert to the hazards encountered by the investigating team. Specialized training enables the safety and health professional to provide proper protective equipment and to explain to other investigators hazards they may encounter and emergency procedures they should follow.

See details in Chapter 16, Planning for Emergencies.

### What to look for

During the accident investigation many questions must be answered. Because of the infinite number of accident-producing situations, contributing factors, and causes, it is impossible to list all questions to apply to all investigations. The following questions are generally applicable, however, and will be considered in most accident investigations (Firenze, 1978).

1. What was the injured person doing at the time of the accident? Performing his assigned task? Maintenance? Assisting another worker?
2. Was the injured employee working on an unauthorized task? Was the employee qualified to perform the task and familiar with the process, equipment, and machinery?
3. What were other workers doing at the time of the accident?
4. Was the proper equipment being used for the task at hand (screwdriver instead of can opener to open a paint can, file instead of grinder to remove burr on a bolt after it was cut)?
5. Was the injured person following approved procedures?
6. Is the process, operation, or task new to the area?
7. Was the injured person being supervised? What was the proximity and adequacy of supervision?
8. Did the injured employee receive hazard recognition training prior to the accident?
9. What was the location of the accident? What was the physical condition of the area when the accident occurred?
10. What immediate or temporary actions could have prevented the accident or minimized its effect?
11. What long-term or permanent action could have prevented the accident or minimized its effect?
12. Had corrective action been recommended in the past but not adopted?

During the course of the investigation, the above questions should be answered to the satisfaction of the investigators. Other questions that come to mind as the investigation continues should be recorded.

### Conducting interviews

Interviewing accident or injury victims and witnesses can be a very difficult assignment if it is not properly handled. The individual being interviewed often is fearful and reluctant to provide the interviewer with accurate facts about the accident. The accident victim may be hesitant to talk for any number of reasons. A witness may not want to provide information that might place personal blame on friends, fellow workers, or the supervisor. To obtain the necessary facts during an interview, the interviewer must first eliminate or reduce fear and anxiety by developing rapport with the individual being interviewed. It is essential that the interviewer clear the air, create a feeling of trust, and establish lines of communication before beginning the actual interview.

Once good rapport has been developed, the following five-step method should be followed during the actual interview.

1. Discuss the purpose of the investigation and the interview (fact-finding, not fault-finding).
2. Have the individual relate his or her version of the complete accident with minimal interruptions. If the individual being interviewed is the one who was injured, ask what was being done, where and how it was being done, and what happened. If practical, have the injured person or eyewitness explain the sequence of events that occurred at the time of the accident. Being at the scene of the accident makes it easier to relate facts that might otherwise be difficult to explain.
3. Ask questions to clarify or fill in any gaps.
4. The interviewer should then repeat the facts of the accident to the injured person or eyewitness. Through this review process, there will be ample opportunity to correct any misunderstanding that may have occurred and clarify, if necessary, any of the details of the accident.
5. Discuss methods of preventing recurrence. Ask the individual for suggestions aimed at eliminating or reducing the impact of the hazards which caused the accident to happen. By asking the individual for ideas and discussing them, the interviewer will show sincerity and place emphasis on the fact-finding purpose of the investigation, as it was explained at the beginning of the interview.

In some cases, contractual agreements may call for an employee representative to be present during any management interview, if the employee so requests.

### Implementing corrective action

Chapters 6 and 7 outline specific ways to record and classify data: how to identify key facts about each injury and the accident that produced it, how to record facts in a form that facilitates analysis and reveals patterns and trends, how to estimate accident costs, how to comply with OSHAct recordkeeping requirements.

An accident in any organization is of significant interest. Employees ask questions that reflect their concern. Is there any potential danger to those in the immediate vicinity? What caused the accident? How many people were injured? How badly?

Those who investigate accidents must be truthful in replying to questions. They should not try to cover up. On the other hand, they must be certain they are authorized to release information, and they must be sure of their facts.

Because the accident report is the product of the investigation, it should be prepared carefully and adequately justify the conclusions reached. It must be issued soon after the accident. When a report is delayed too long, employees may feel themselves in limbo. If a final report must be postponed pending detailed technical analysis or evaluation, then an interim report should be issued.

Summaries of vital information on major injury, damage, and loss incidents should be distributed to department heads. Such summaries should include information on causes and recommended action for preventing similar incidents. Incident and statistical report files should be maintained for two years or as dictated by company policy.

Supervisors should keep employees informed of significant accidents and preventive measures proposed or executed. Posting accident reports is one way to make information available.

The preceding section on inspection emphasized that hazard control benefits accrue only after the inspection report is written and disseminated. Until corrective action is initiated, recommendations—no matter how earnest, thorough, and relevant—remain "paper promises."

The same truth applies to accident investigation when used as a monitoring technique. Viewed from the perspective of hazard control, accident investigation serves a monitoring function only when it provides the impetus for corrective action.

When management and safety professionals review monthly accident reports, they exercise an essential auditing function. Management (including the CEO) can demonstrate interest in safety by requiring prompt reporting of all serious or potentially serious incidents. They use accident reports to make decisions to prevent similar accidents from occurring, and they look for answers to certain key questions. Are all significant accidents being reported? Are all parts of the organization equally committed to the hazard control effort? Are there trends or patterns in accidents or injuries? What system breakdowns predominate? What supervisors require additional training? Are employees advised of the results of accident investigation and of preventive measures being instituted? What management deficiencies are indicated?

Accident investigation is a monitoring function that occurs *after* the fact. The hazard control system already has broken down. No amount of investigation can reverse the accident. Nevertheless, accident investigation serves an important monitoring function. Past mistakes are being used to improve future operations. As George Santayana has written, "Those who cannot remember the past are condemned to repeat it."

## REFERENCES

Boggs, Richard F. "Environmental Monitoring and Control Requirements," *National Safety News,* Vol. 118, No. 2 (August 1978).

*Facility Inspection.* Philadelphia, Pa., Insurance Company of North America, 1973.

Firenze, Robert J. *The Process of Hazard Control,* Dubuque, Iowa, Kendall/Hunt Publishing Co., 1978.

"Industrial Hygiene Instrumentation," *National Safety News,* Vol. 117, No. 3 (March 1978).

Johnson, William G. *MORT Safety Assurance Systems.* New York, N.Y., Marcel Dekker, Inc., 1980. (Also available through National Safety Council.)

National Safety Council, 444 N. Michigan Ave., Chicago, Ill. 60611. *You Are the Safety and Health Committee.*

Nertney, Robert J. "Practical Applications of System Safety Concepts," *Professional Safety,* Vol. 22, No. 2 (February 1977).

Olshifski, Julian. "Air Sampling Instrumentation," *National Safety News,* Vol. 120, No. 2 (August 1979).

_____. "Selecting and Using Industrial Hygiene Consultants," *National Safety News,* Vol. 118, No. 3 (September 1978).

Plog, Barbara, ed. *Fundamentals of Industrial Hygiene,* 3rd ed. Chicago, Ill., National Safety Council, 1988.

Recht, Jack L. "Systems Safety Analysis: A Modern Approach to Safety Problems." Chicago, Ill., National Safety Council, 1966.

Scerbo, Ferdinand A., and Pritchard, James J. "Fault Tree Analysis: A Technique for Product Safety Evaluation," *Professional Safety,* Vol. 22, No. 5 (May 1977).

# Computers and Information Management Systems



## PART I
## Information Systems

THE COMPUTER HAS COME OF AGE in safety and health departments. Once the exclusive domain of personnel trained in data processing, computers have been "demystified," especially since the widespread use of personal computers (see Table 5-A). Not too many years ago, safety and health professionals had to expend considerable effort to justify the need for computer systems to support their organizations. Today, computer systems are an integral part of the business environment, and safety and health professionals are using them to help prevent work-related deaths, injuries, and illnesses; manage risks; and control losses (see Figure 5-1).

Computers are used by safety and health professionals to:

- Compile, store, analyze, and produce corporate summary reports of occupational injuries and illnesses—reports that would otherwise be time-consuming and expensive to manually prepare each month, quarter, or year.
- Produce follow-up reports pointing to hazards that were identified during inspections, but are not yet abated.
- Store and correlate employee health records with data on workplace exposures to chemicals.
- Maintain records needed to document regulatory compliance, such as training required by the OSHA hazard communications standard.

Computers are tools to help you obtain the information necessary to make decisions; computers make your job easier, your efforts more productive, and help protect your employees and organization.

The importance of information management to occupational safety and health programs did not become critical until the 1970s. The passage of the Occupational Safety and Health Act in 1970, followed by passage of environmental laws such as the Toxic Substances Control Act and the Resource Conservation and Recovery Act, contributed to a regulatory environment that quickly overwhelmed the resources of existing occupational safety and health programs. In the late 1970s and early 1980s, right-to-know laws and regulations added to the problems. All of these new regulations imposed complex procedural, reporting, and recordkeeping requirements on industry and, by executive order, on government agencies. As safety, industrial hygiene, and medical surveillance requirements increased, so did the requirements for documentation, reporting, and analysis.

As public awareness of workplace hazards grew, litigation involving occupational illnesses and deaths also began to have an impact. Employers were asked to prove in court that employees' diseases were not caused or aggravated by exposures in the workplace. Unfortunately, many employers found their records to be incomplete or unavailable. Similar situations arose with workers' compensation and disability cases. Organizations found their records were inadequate to respond to the increasing number of questionable injury claims.

These factors, combined with the need for better internal records management, led many organizations to look for ways to improve their recordkeeping systems. Most of them chose a common solution—computerized occupational safety and health information systems. Today, it is known computers can solve many of the information management problems faced by safety and health professionals.

**COMPUTERESE**

The definitions used here are meant to define some of the more basic computer terms as they are commonly used.

**Bit** A unit of measuring/SIZE, a bit is the smallest piece of information that the computer can handle. Technically, a bit represents whether an electronic impulse is turned on or off.

**Bomb** A software program "bombs" if the program does not perform its required task.

**Bug** An error or irritation in a software program—usually not so severe that it causes a program to bomb.

**Byte** This unit of measuring/SIZE is equivalent to eight bits, about the amount of information needed to store a single letter, number, or other character.

**Central Processing Unit** Part of the insides of the computer that controls all of the work the computer does.

**Data** Single pieces of information stored in a file. For example, in a file of accident records, the date of the accident would be one piece of data.

**DOS** Disk Operating System is the software that acts as the intermediary between the computer hardware and other software.

**File** Collection of information stored on a disk. For example: a letter of correspondence might be saved as a single file.

**Floppy Disk or Diskette** A flexible, round electromagnetic device that stores computer programs and data. When used in a floppy drive (the part of the computer where the diskette is inserted) the information is read into the computer's memory for processing. Often the storage capacity of the disk exceeds the memory of the computer. The information on the diskette is permanent until erased or replaced.

**Graphics** A display of information in picture form. For example: a pie chart or a bar graph on-screen are graphic capabilities of a computer.

**Hardware** All of the physical components of the computer system. This includes the monitor, printer, keyboard, central processing unit, and everything else that you can touch; not floppy disks, however.

**Hard Disk** Resembling stacked floppy diskettes, this component is actual hardware that sits inside the computer. Hard disks, then, can hold much more information than a floppy. Information on the hard disk is read into the computer's memory for processing, and it is permanent until erased or replaced.

**Hard Copy** The paper copy of information, as printed out by the computer's printer.

**K** This unit of measuring/SIZE is equal to 1,024 bytes. The term is usually used to express the memory capacity of the microcomputer. For example: 64K, 128K, 256K, 512K. If a software program requires 256K to run, and your computer has only 128K, then your computer cannot run the software program unless you purchase additional memory-expansion hardware.

**meg** This unit of measuring/SIZE equals 1,048,576 bytes or 1,024 K. It is often used to express the capacity of a hard disk. For example: 10, 20, or 30 meg.

**Memory** The part of the computer that stores programs and data currently being processed.

**Modem** The MOdulator-DEModulator or the hardware that allows computer-generated signals to be transmitted and received on telephone lines.

**Monitor** The hardware part that includes the screen.

**Printer** The hardware that produces hard copies.

**Program** A set of computer instructions on a disk.

**Software** A series of programs that, together, allow the computer to function as a word processor, data base manager, spreadsheet, or whatever.

**Table 5-A.**

## ADVANTAGES OF AUTOMATED INFORMATION SYSTEMS

### Improved availability of data

A well-designed computer system can store a huge amount of data and retrieve it easily and quickly for reporting and analysis. For example, an employer might want to know if any employees exposed to benzene have had abnormal hematology results. With a manual system, a time-consuming search through industrial hygiene and medical files for this information would be necessary. With an interactive computer system, this search should be completed within minutes. In addition, interactive systems can be used to provide emergency information, such as first aid and clean-up procedures for chemical spills.

### Elimination of duplication

The time spent maintaining duplicate records within an organization can be reduced with an automated system. For example, with a manual system, it could be necessary to keep records of occupational injuries in four offices: safety, medical, human resources, and area supervision. With a shared information system, one comprehensive file on each injury can be kept in the computer.

### Improved communications

With manual systems, important data can be lost in unfinished piles of paperwork. Often, statistics generated by one of the company's offices do not match those generated by another. Again take occupational injuries as an example; injury reports from one facility may not reach other facilities, decreasing the chance that all groups within the organization can benefit from the "lesson learned." A good data base management system makes data available to all authorized users. In addition, electronic mail systems can facilitate communication within and between facilities.

### Data standardization and accuracy

Standardizing the data input and the way in which it is organized is particularly important in organizations with many system users. When the same data is being input in the same way by

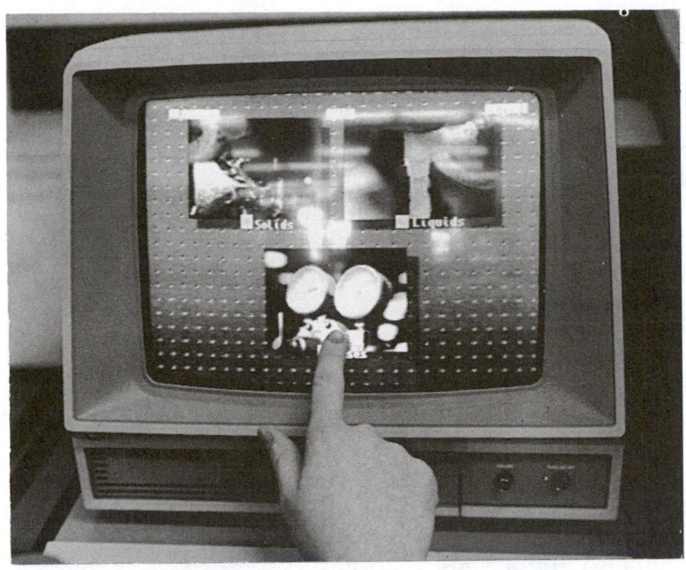

**Figure 5-1.** The microcomputer already is in use by many safety and health professionals for word processing and data base management.

each user, the analysis and decision-making capabilities of the data base are enlarged. For instance, accident and incident rates can be analyzed for all locations at the same time, rather than analyzing each location separately and then standardizing and entering the collective data for final analysis.

Accuracy of data can be vastly improved by customized entry screens, editing routines, and checks for the completeness of records. For instance, a computer system can alert the medical department that part of an OSHA-required physical examination was not conducted.

### Improved analytical capabilities

With most manual systems, compilation and analysis of data from various records consumes professional time that could be better spent on program management activities. A good automated system performs analytical tasks quickly and formats the results for reports to management. For example, the ability to correlate workplace exposure and medical records leads to improved medical surveillance. Trends in employee health can become apparent in time to act to prevent injuries and illnesses. Analyses not practical with manual systems routinely can be done using an automated system.

### Cost savings

This advantage is probably the hardest to document, but automated systems can save organizations money above and beyond the costs of system development and operation. Increased employee productivity, decreased incident rates with attendant savings, and more effective program management can lead to cost savings.

## SAFETY AND HEALTH FUNCTIONS

Today's occupational safety and health systems support single functions, such as management of training records, or a comprehensive set of functions. This section is an overview of the system functions and tools in use by safety and health professionals today.

### Incident management

Many organizations are using automated systems to store, analyze, and report on incident data. Data for all types of incidents is entered, such as property damage, occupational injury, near miss, and transportation. This data usually parallels the organization's first report and incident investigation forms. The system can produce: (1) individual incident reports; (2) summary reports that categorize incidents by location, type, rate, severity, loss, and other factors; (3) OSHA 200 logs; and (4) analytical reports that pinpoint major causes and types of accidents within organizational subsets. For example, an organization might determine through use of its automated system that a disproportionate number of accidents is occurring in a particular operation, and then arrange for the employees in that operation to have additional training.

### Workplace conditions

Many systems manage safety, industrial hygiene, health physics, environmental, and other sampling, audit, and inspection data capable of describing and quantifying workplace conditions. Industrial hygiene sampling results can be entered and used to produce exposure profiles for each workplace. Also, if safety inspection findings, recommendations, target dates for completion, and completion dates are automated, a system can produce follow-up reports highlighting identified deficiencies that have not been corrected.

In comprehensive systems, workplace data can be correlated with employee health data to support health surveillance programs.

### Environmental agents

Automated systems greatly facilitate environmental agent monitoring, including the preparation and maintenance of facility and corporate inventories of environmental agents, toxicology data, and material safety information on toxic agents within the workplace. These agents include chemical, biological, radiological, and physical hazards. Inventories of agents can be maintained by department, area, process, job, or any combination of these.

Right-to-know and hazard communications requirements make it mandatory that organizations identify agents in each workplace, train employees in their safe use, maintain material safety data sheets, and promptly respond to requests for information. Automated systems can facilitate compliance by maintaining agent inventories, training records, material safety data sheets, and information on right-to-know requests.

### Protective measures

Computer systems can maintain information on measures taken to protect employee safety and health through: (1) engineering controls such as hoods for ventilation; (2) administrative controls for health hazards; (3) personal protective equipment assignment and fit testing; and (4) individual employee training and group training programs. These functions allow an organization to maintain the information needed to facilitate regulatory compliance and document the measures taken to protect employees.

### Employee health

Monitoring employee health is a key function of comprehensive occupational safety and health systems. The protection of employees is, after all, a major goal of safety and health programs. This function primarily is used by occupational medicine specialists to manage and store records related to employee health surveillance. These records, shown in Table 5-B, result from both scheduled and unscheduled events. Scheduled health events include all physical examinations (preplacement, periodic, certification, and termination examinations and biological monitoring). Unscheduled health events are those that cannot be anticipated. It is necessary to maintain a record of both scheduled and unscheduled health events to compile a historical profile of an employee's health.

**Table 5-B.** Occupational Medicine Records

| |
|---|
| Physical Examination Records |
| Personal and Family Medical History |
| Occupational History |
| Physical Measurements (such as vision tests and blood pressure levels) |
| Blood and Urine Laboratory Tests |
| Clinical Evaluations and Diagnoses |
| Audiometry Results |
| Spirometry Results |
| Electrocardiogram Results |
| X-Ray Results |
| Other or Special Test Results |
| Immunization Records |
| Clinic Visit Records |
| Occupational Injury and Illness Records |
| Disability Data |
| Workers' Compensation Records |
| Biological Monitoring Results |
| Sickness and Absence Records |
| Mortality Records |

As with other system functions, individual components of employee health programs can be automated. Hearing conservation program records are a frequent candidate. Noise monitoring, audiogram, and employee history data can be entered and stored. These data can be used to produce monitoring schedules, threshold shift evaluations, notices to employees, and summary reports for program management.

### Employee demographics and job histories

In addition to maintaining health records, this function (or an independent but related function) must also maintain data that identifies and describes individual employees and maintain a history of job assignments and locations. The identification of employees is usually accomplished through use of one or more of the following—name, social security number, employee number, and badge number. Descriptive data includes such demographic data as sex, date of birth, race, address, identification of nearest relative, and so forth. These normally are found in personnel information systems and can be downloaded to a safety and health system.

Job assignments and location tell where an employee works and what he or she does. This data is a necessary bridge between the employee health records and the workplace and environmental records. They allow health professionals to correlate medical records with workplace records and are an integral part of comprehensive occupational safety and health surveillance systems.

## REGULATORY, ADMINISTRATIVE, AND ACTION ITEMS

Another function can provide information on regulatory requirements, including OSHA permissible exposure limits or action levels, or internally adopted standards, such as the American Conference of Governmental Industrial Hygienists (ACGIH) Threshold Limit Values (TLVs). This function also can be used to track ongoing events or problems requiring follow-up, for example, hazard abatement plans and schedules.

### Scheduling

The computerized system can be used to schedule physical examinations, workplace monitoring or inspections, fit testing of personal protective equipment, training, or any other event documented in an occupational safety and health system. The schedules can be generated by system logic, or they can be generated by users and maintained by the system.

### Standard and ad hoc reporting

Reporting is the most important function of an automated system. Standard reports are predefined and routinely used, and they are usually preformatted. They can be prepared with one or few instructions (often a choice of a menu of reports). Examples of typical standard reports are audit reports and lists of occupational injuries by workplace. All systems should have a standard reporting function.

Ad hoc reports cannot be predefined or preformatted, but are designed by users for special analyses and reporting. There are several methodologies for producing ad hoc reports. Two of them, the question-and-answer and fill-in-the-blanks approaches, prompt the system user to name the variables for study, indicate how they are to be manipulated, and specify the format for the report. Another method requires the use of a simple computer language for queries. The method chosen should be easy to learn and simple to use. Adequate query capability is one of the most important features of an occupational safety and health system, since nothing is more frustrating than to find out after years of data entry that some data cannot be retrieved in the manner required.

### Statistical analysis

Statistical analyses are necessary for many safety and health studies. Commercial statistical software packages are available for this function, but they often are separate from the occupational safety and health software. These are available on personal computers and mini and mainframe computers. It is usually an easy task to pass data from newer data base management systems to these packages for analysis. In addition, many packages are available for preparation of graphics from safety and health data systems.

## DEVELOPMENT AND IMPLEMENTATION OF A COMPUTER SYSTEM

There are five major steps to the successful development and implementation of an automated system:

1. Understand your operations and needs
2. Identify and evaluate software and hardware meeting those needs
3. Purchase and customize a system or develop your own system

4. Implement the system
5. Evaluate the system.

### Understanding needs

When you first consider buying a computer system, you should conduct a requirements study, which is a careful assessment of the needs that a system will fulfill. The requirements study must say *why* a system is needed, based on current and foreseen conditions. It must say *what* system features will satisfy the needs. It must also say *how* the system is to be constructed.

To do this, the requirements analysis must address three subjects:

1. *Needs analysis.* The reasons *why* the system should be created. How the system will solve a problem. Why certain technical, operational, and economic feasibilities are the criteria for the system.
2. *Functional specification.* A description of *what* the system will be in terms of the functions it must accomplish. How the system will be in terms of the functions it must accomplish. How the system will fulfill its role. Why certain design components should be considered.

To determine the functional specifications you need to look at your operations, recordkeeping and reporting requirements, and information management problems. Then estimate the time you spend on each activity, quantify your activities, make a "wish list," and answer questions such as:

- How much of what types of work do you do?
- How many employees are covered by your safety and health office?
- How many accident cases or employee-health related problems do you have per year?
- How many on-site chemicals do you track for employee exposure?
- How many Material Safety Data Sheets (MSDSs) are you responsible for maintaining?
- How do you keep track of employee safety training?
- How do you measure and track permissible exposure limits?
- How do you track safety equipment distribution and maintenance?
- How do you track employee medical records, such as audiograms?

3. *Design constraints.* A summary of conditions specifying *how* the system is to be constructed and implemented. What will compose the system. Why particular designs are feasible. Design constraints could include:

- Computer hardware—If your department already has IBM PCs your system must operate on them.
- Computer software standards—Does your company's data processing group support only certain data base management systems?
- Personnel—Must your system not require additional staff for data entry or operation?

Each of these subjects should be fully documented during the requirements analysis. Collectively, these analysis components should provide all of the information you need to design or evaluate a system.

### Identify and evaluate systems

After the requirements and specifications for a system have been defined, you should have enough information to identify the computer systems capable of meeting your needs. You will have two alternatives: (1) purchase commercially available software and customize it, if necessary, or (2) develop your own system.

Since 1977, occupational safety and health software has been commercially available. Today, many good packages are available, and any group planning to implement a system should find out if any of these packages meet their requirements. The advantage of purchasing software is the time and expense that can be saved. The disadvantages are that modifications and work practice adaptations are sometimes necessary when generalized packages are used. So carefully study the software features before you purchase.

The *Accident/Illness Analysis System* software package, published by the National Safety Council, is in wide use and at this writing has had few customization requests.

The major advantage of developing customized software is the exact fit between the requirements and the system. Little, if any, accommodation should be required by users. Today, new hardware and software technologies make custom software development a viable alternative.

When comparing the available alternatives, you should remember two cardinal rules. First, the alternative should be judged on the basis of criteria established by the requirements analysis. A commercially available system cannot be evaluated for use in an organization until the requirements for the system have thoroughly been defined. Second, life cycle costs should be used to compare the costs of the considered alternatives. Life cycle costs are all costs associated with system design, development, implementation, and operation over the life of the system. A system's life is the amount of time the software-hardware configuration can be expected to perform without major modification.

### Purchase and customization

If you purchase a system, you may have to modify it before it will meet all of your requirements. Naturally, when you evaluate systems, you should look at the ease with which they can be changed. Once you make your purchase, you should treat the package as a prototype. Let your system users work with it and define its deficiencies. Then make any required changes. You may have to go through several rounds of modifications before you are completely happy with the system.

### System development

If you decide to develop your own system, you must participate in the design and development process. It helps at this stage to get some assistance from a systems analyst. The intricacies of system development will depend on the size and complexity of your system and the hardware and software environment in which it will be built. Naturally, a small PC-based system for management of one or more unrelated functions will require less effort and systems expertise than a comprehensive safety and health system.

The first step is to establish the system structure. This activity, often called general design, consists of defining the subparts of the system and the interfaces between them. In the detailed design step that follows, the precise algorithms or system processes and data structures are defined. Detailed design may involve several iterations. Then, system production begins; this involves programming or use of system tools for defining file structures and report formats, testing of individual pieces, integrating various pieces, system testing, documentation, and, finally, the initial performance evaluation.

Many organizations build and implement large systems in phases. This can work well *if* an overall plan for total system design is prepared and followed during the phased development. Too often, groups have built system modules without considering total system requirements and necessary connections between modules, thus leading to project failure.

### System implementation

Implementating the new information system should be a joint effort between system developers and users. While there are many activities associated with implementation, the most important one is user training. This training should establish realistic expectations among users about the system's capabilities and requirements. Any changes required in recordkeeping procedures and potential transitional difficulties should be thoroughly explained.

User training should not be left to programmers or systems analysts. Ideally, it should be carried out by user representatives. If a system has been purchased, some training will, by necessity, have to be conducted by the supplier. However, a core of system users should be trained by the supplier, and then, if possible, these users should train others.

Training should be conducted as part of the everyday work routine. Naturally, some general training sessions must be held, but the most effective type of training is on-the-job training, with safety and health professionals learning to use terminals for data input and output in the context of their actual work.

Training should be timed so there is no gap between training and actual use of the system. When something entirely new is learned, most people need to practice it to remember it. The longer the gap between training and system use, the less learning retention there will be.

Follow-up to training also is important. Periodically, the system manager should return to the users to determine if their expectations for the system are realistic and if they are using the system correctly. Error rates should be studied. If they are too high, the reasons should be identified and the users retrained. System success depends just as much on follow-up training as it does on initial training.

An effective user's manual is important to user training. During the development process for any information system, system documentation is written. Much of this documentation is for the use of systems analysts and programmers. However, a user's manual is also prepared, and great care must be taken in its preparation. Both the system developers and user representatives must participate in writing the user's manual. It must be simple to understand and direct in its instructions. If it is properly written, it will be a permanent, valued reference for all users. If it is not well prepared, it either will not be used or will promote errors.

Every system should have a manager or data base administrator. This person has ultimate responsibility for the day-to-day operation of the system, including: (1) management of system security; (2) supervision of the data base content; (3) problem solving; (4) coordination of changes to the system; (5) archiving of data; (6) data quality; and (7) planning for future needs and applications. The system manager should know the application well and work closely with users to make sure the system meets user expectations.

### Evaluating the system

After an information system has been implemented, it should be evaluated annually. Estimating an information system's value to an occupational safety and health program can be difficult for the same reasons that conducting a cost-benefit analysis is difficult. However, evaluation is an important follow-up, and the following factors should be considered when evaluating your system:

- Completeness
- Reliability
- User acceptance
- Costs
- Improved availability of information
- New capabilities.

If problems are identified in any but the last area, corrective steps should be taken. Problems can result from human, hardware, or software elements of the system.

## SYSTEM SOFTWARE

Advancements in microcomputer technologies have made personal computers a part of our personal and professional lives. Personal computers are ideal tools for information management because they are easy to use and relatively inexpensive. Many companies are making personal computers available to managers for information management, word processing, spreadsheet manipulation, and electronic mail. They are part of an overall strategy for office automation.

**Word processing,** one of the most favored software packages because of its ease-of-use and range of office applications, is the ability to create and change text on-screen before printing it out. The text is saved, and then retrieved at a later time for possible editing before reprinting. The computer does indeed become a typewriter, but with additional features that make word processors save about 50 percent of the time it would take a typist to accomplish the same amount of work.

*Word wrap* is a word processing feature that allows you to type forever without pressing return at the end of a line: When you approach the pre-set margin, you are automatically brought to the next line. Most word processors also allow you to insert or delete characters, words, lines, or paragraphs, and to center, boldface or underline with minor keystrokes. The "new wave" of word processors includes a dictionary to check spelling and edit text and a thesaurus to help you find the right word to express yourself.

Typical word processing safety applications include quarterly reports, audit reporting, mail merges for multiple mailings, and custom notices and posters for safety commitee activities.

**Data base management systems** are an integral component of most contemporary occupational safety and health systems. It is important to understand the distinction between applications and data base management software. Today, most applications, including safety and health systems, are written with data base management systems. All data base management systems allow the storage, sorting, and retrieval of information in useful ways. They have tools for building files and reports, and they offer programming languages or commands for building complex programs. A data base management system looks like a generic computer facility until you use its tools to build your specific application.

*Applications* are programs or sets of programs that provide tailored menus, data entry screens, and reports to support specific functions—like occupational safety recordkeeping and analysis. The commercially available packages for safety and health are applications software, and most of them were built with data base management systems.

Many different data base management systems are available. Some safety and health professionals are using them to develop automated functions on personal computers. There are data base management systems available for all types of computers—some are easier to use than others. If you decide to develop your own applications software, you should look to your organization's information center or data processing department to find out what's available and if you can get development support.

## SYSTEM HARDWARE

The types and placement of computers to be used for an occupational health and safety information system must be determined during the requirements study. Today, there are many alternatives for computers, ranging from large mainframe computers to personal computers, and information processing can be accomplished on one or many of these.

Many safety and health departments are using personal computers for information management. Personal computer technologies are evolving at a rapid pace. Every year, vendors offer greater speed, more memory, and better connectivity. New operating systems permit users to run multiple tasks simultaneously, and new networking capabilities let several PCs share disks, programs, and printers.

Because rapid advances in personal computer technologies have occurred and prices have decreased, many companies have developed occupational safety and health systems relying solely on personal computers, and some use local area networks to link the PCs.

Personal computers also can be used in system configurations that dedicate the PCs to local information management and tie them into central mainframe or minicomputers for long-term storage, cross-functional analyses, and corporate data base management. This type of system configuration distributes system tools and mimics the decentralized organizational structure of many corporations. It is ideally suited to the seemingly opposing needs for local records management and control and corporate-wide data base management.

Other configurations also are being used in the development of safety and health systems. They include: (1) use of a mainframe or minicomputer for all system functions with terminals distributed to users and communication over telephone lines and (2) use of mini or maxiframe computers for regional data base management with terminals or PCs used for data input, and offering communication among the larger machines to allow corporate-wide analyses.

There are many factors that will influence an organization's choice and location of computer hardware for its information system, including currently available hardware, corporate standards for computer hardware, telecommunications, and software, and the information requirements of system users. All system options and constraints should be studied during the requirements study so an economical, efficient, and integrated system can be developed.

## SYSTEM COMMUNICATIONS

After deciding the roles of the mainframe, minicomputer, and microcomputers in your organization, as well as what the information sharing ideal would be, you should consider the communication tools available including: modems, mainframe-mini-micro links, and networking.

When considering communication tools, modems usually are mentioned first. The modem is an excellent communication link, especially if departments are located at considerable distances from each other, perhaps in another city. With a modem, two computers are always as close as the phone lines.

Most communications software packages not only provide a way to transmit information, but make possible transfers between different brands of incompatible computers. Advanced terminal emulation features of some communications software enable communications between mainframes and minicomputers, and other microcomputers.

The hardware and software to make mainframes and micros talk to each other is available, but is dependent on the hardware and software you use.

Networking is a communication form gaining rapid popularity. Local area networking, or LAN, is a team of personal computers tied together, so they can talk to each other and easily share information. The network is local in the sense that the computers are generally in the same building or within 1,000 feet of each other.

There are several species of LANs made by a number of different manufacturers. These varieties have names like *star, token ring,* and *bus,* which describe how the computers are connected.

The LAN can be connected to a minicomputer or even a mainframe so information can be transferred back and forth, thus enabling the company or department to communicate with the main company system. This way department can talk to department, and departments can talk to the main company system as well.

Because a LAN can connect two or more microcomputers, each of these computers has the power to work on its own but in addition has the ability to use information from the other computers connected to the network. Each department, then, can operate its microcomputers independently, but still remain in communication with the other computers on the network.

In addition to the benefits of information sharing, networking microcomputers permit the sharing of devices. For example: Several different stations can be connected to a printer. A laser printer could serve several different locations; the same hard disk could accommodate the storage needs. This multiple use of equipment represents some savings, especially with expensive printers.

Similarly, a network enables a number of users to access the same software. This is useful where network versions of popular programs are used by several different work stations.

This capability is even more valuable where there is customized in-house software in use by a number of different people in separate locations.

From your point of view, the LAN permits the establishment of an information network, integrating departments in useful ways, while effortlessly increasing communication.

Networking does have some additional costs associated with it. Yet, because of the associated benefits, more companies are turning away from pure mainframe systems, and are incorporating microcomputers in a networking environment. With an increased emphasis on cost control, the LAN offers a way to get operating information quickly down to the plant level to facilitate effective management and produce competitive results.

Networks have unique properties that make costs justification easier:

- *First,* you can start small and expand. A simple four-unit network can be expanded to a complex 20 terminal LAN. Later, if needed, the LAN can be connected to another LAN. This means you can seek authorization for a modest investment, and after you have demonstrated the value, then propose expansion of the system and the functions performed by the system, thereby meeting all your system communications needs.
- *Second,* the initial costs of a LAN can be substantially less than a comparable minicomputer. This is especially important when trying to justify a capital expenditure without the customary return on investment calculations.

Having decided to pursue the installation of a LAN for a defined purpose, the question arises: "What kind?"

There are several choices, and the choice depends upon the performance required and your budget. In any case, the answer requires some skilled professional assistance from your management information department, information center, or outside consultant. It is very important that your resource have actual operating experience with networks.

Initially, you will need a good deal of assistance because networks are complicated to set up and do require some training in their use. Once in operation, however, they are easy to use and have abundant advantages.

It is a good idea to assign someone the job of network administrator. This person will assign passwords, access rights, write the log-in scripts, train people to use the network, and make sure the network is up and ready to serve the users.

# PART II
# Loss Control in the Computer Room

SPURRED BY TODAY'S COMPETITIVE BUSINESS CLIMATE, the use of electronic data processing (EDP) equipment is continuing to gain popularity. Microprocessors are becoming less expensive and more versatile, allowing companies of all sizes to realize the advantages of high speed computing equipment.

The advantages of electronic data processing equipment in a business operation are many, such as high speed, more efficient information processing and reduced labor costs.

As the EDP system is expanded and tailored to meet the firm's specific needs, it becomes increasingly critical to protect it. Many firms would be in serious trouble if their EDP systems went down for any length of time. Even a company with a relatively small system could suffer devastating losses if its system became non-functional. Lost data, business interruption, failure to meet customer needs, subsequent loss of customers and replacement costs of EDP equipment are only a few examples of losses that could occur.

Just as the intelligent businessperson recognizes that EDP equipment must be used to stay competitive in today's marketplace, the potential for costly loss of that equipment must also be recognized. Positive actions must be taken to reduce this loss potential. The safety professional is the appropriate person to point out to management not only the loss potential but also safeguards that can substantially reduce the exposure.

The National Fire Protection Association (NFPA) outlines recommended safe procedures to be used when installing electronic data processing equipment. This chapter contains a summary of the more important requirements outlined by NFPA. More specific information can be obtained by reviewing the NFPA standard, *Electronic/Data Processing Equipment,* 75.

## THE ROOM

**Physical location and construction** of a computer room should be considered from both a fire and security standpoint. The interior finish, furnishings, and layout of the room should be chosen prior to equipment installation, because any necessary changes will be less costly to accomplish at this time. Naturally, the location of the room is decided during the design stage and after thoroughly reviewing the NFPA 75 standard.

To minimize exposure to fire, water, corrosive fumes, heat, and smoke from adjoining areas and activities, the computer room should not have other processes above, below, or adjacent to it. It should be located away from street-side windows and exterior walls, and above ground level for security reasons and to minimize dampness.

**Entries** to the computer room should be lockable from the outside and equipped with panic bars so operators can quickly exit from the area in an emergency. The operators will need to use keys to enter and exit the room at other times. The emergency exits should have local alarms.

**Utility lines,** including electric, gas, and phone lines and air conditioning cables, should be located so they cannot be tampered with. The floors should be raised and contain easily accessible cable raceways that are designed to have drainage adequate to handle potential domestic water, sprinkler, or coolant leakage or water and chemical runoff from firefighting operations. Cable openings should be covered or protected so debris and other combustible materials cannot fall into the openings.

**Contents.** Only the EDP equipment, approved furnishings, and auxiliary electronic equipment should be kept in the computer room. To minimize the chance of a fire starting, furniture should be of metal construction and the amount of paperwork and paper storage allowed in the room should be limited. The interior room finish should be of non-combustible materials containing no exposed cellular plastics.

## THE EQUIPMENT

After the location of the computer room has been decided, equipment requirements are considered. The arrangement and wiring of equipment, such as electric shutoff devices, air handling equipment, and smoke and heat detectors, play a substantial role in providing adequate fire protection.

The equipment should be arranged to allow for quick and easy emergency egress from the room. Provision should be made to allow operators to quickly de-energize the equipment as they exit. This is especially important when the room is equipped with an automatic sprinkler system.

The electrical system must be adequate for the expected power load, and should be provided with devices to protect against power surges, brownouts, and power failures.

Specific equipment requirements include: (1) emergency electric cutoffs both at the operators' stations and at the exitways (the electric cutoffs should be so labeled); (2) electric service rated at 125 percent of the total amperage load for the equipment; and (3) covered electric junction boxes at a 15-ft. maximum distance apart on flexible electric cords.

All air handling equipment should be connected to a separate electrical system. Dampers should be provided on all ducts; the dampers should be wired into smoke and heat detectors so they will automatically close when the presence of fire is detected.

The entire equipment complex should have an auxiliary source of electricity in case of a power failure. All electric equipment should be properly grounded and provided with power surge protection and a manual override system to protect against fluctuations in electric power. All automatic systems should have manual back-up.

It is to the company's advantage to arrange for emergency use of other facilities so data can be run even if there is an equipment failure. Before finalizing emergency equipment use arrangements, the rental expense should be carefully evaluated, and any agreements reached with other equipment owners should be obtained in writing. The best time to make emergency use arrangements is before a loss occurs.

Waste containers in the computer room should be metal with self-closing lids. If any sound-deadening materials are used, they should be non-combustible. In addition, flammable liquids should not be kept in the room, and liquids with a flash point of less than 300 F (150 C) should not be used for lubrication.

## FIRE PROTECTION

Protecting the room and its contents against fire is the next step. To make the protection system as fail-safe as possible, an automatic fire suppression system should be considered.

### Fire suppression systems

An effective fire suppression system is a Halon total flood system. Halon gas extinguishes a fire by disrupting the chemical chain reaction feeding the fire. With a Halon system, there are several advantages: no residue to clean up after agent discharge; reduced chance of equipment damage resulting from extinguishment materials being applied to sensitive printed circuits; and the non-toxicity of Halon when it is used in approved concentrations. All Halon automatic fire suppression systems should meet the requirements of the NFPA standard, *Halogenated Fire Extinguishing Agent Systems—Halon 1301, Fire Extinguishing Systems,* 12A.

The issue of using sprinklers on EDP equipment is a controversial one. According to NFPA, *Fire Protection Handbook,* 15th edition, ". . . automatic sprinkler protection and water spray fixed systems are valuable as means of reducing fire damage, even where such electrical or electronic equipment may be exposed. There should be little concern relative to the shock hazard, or of the water causing excessive damage to the equipment. Experience has proven that if a fire activates sprinklers, the sprinklers, if properly installed and maintained, provide for effective fire protection with virtually no hazard to personnel and with no measurable increase in damage to the equipment (as compared with the damage done by heat, flame, smoke, and manual hose streams)."

The philosophy that should be followed is to use a total flood system (Halon or otherwise) as primary protection, then install automatic sprinklers as secondary protection.

If the EDP equipment is de-energized before it gets wet, and if it is quickly dried out afterwards, very little damage should result. A hand-held hair dryer is a quick, convenient way to dry out printed circuit boards. Equipment should be thoroughly dried before being re-energized.

### Fire protection systems

The following items also should be considered when designing a fire protection system:

1. The fire protection of the room and the equipment can take many forms. One method is to use a Halon 1301 total flooding system in accordance with the NFPA 12A standard. The system should be interlocked into all air conditioning and air handling equipment in order to maintain the gas concentrations necessary to extinguish the fire.
2. Other forms of equipment protection are hand-held fire extinguishers of the carbon dioxide or Halon 1211 types. A Class A fire extinguisher also should be available for use on any paper or other ordinary combustible-type fire. Each extinguisher should be clearly labelled with its function and class.
3. An automatic fire detection system with UL-listed smoke and heat detectors should be provided. It should be wired to a central station supervisory service to assure quick fire department response in an emergency.
4. A dual system (mentioned previously) using both sprinklers and a Halon total flood system is a very effective method of automatic fire suppression. Using this method, primary protection is furnished by the Halon system; if the Halon system fails, the sprinkler system acts as secondary protection. It is important that the electric equipment is de-

energized before waterflow or Halon "dump." This can be done manually by operators as they exit, or automatically by means of a signal triggered by activation of the Halon system. Automatic shutdown should be provided for equipment that normally runs unattended for long periods of time. Automatic "power down" programs can be added to existing software to avoid "head crashes."

5. The sprinkler system protecting the EDP room should be valved separately from other sprinkler systems in the building. It also is necessary to provide a means for drying out the equipment to reduce further damage.
6. Waterproof covers of non-combustible materials will minimize water damage to the EDP equipment.

## RECORDS PROTECTION

The loss of vital (irreplaceable) and important (time-consuming and expensive to replace) records has caused some businesses to experience hardships ranging from lowered credit ratings to near-bankruptcy. Today these losses are unnecessary; equipment is available that can adequately protect records from extreme temperatures, impact, and other fire conditions.

A description of the available equipment and other control measures can be found in the NFPA standard, *Protection of Vital Records,* 232. *Only those records that need to be frequently used or referenced should be kept in the EDP room.* All other records should be located elsewhere, and protected according to their value. Vital and important records should not be stored in the EDP room.

When appropriate, record duplication is the best control measure. The original document is stored off the premises, and the duplicate is used and stored at the main operating location. To prevent records from being stolen or damaged, it is wise to investigate the security practices of off-premises storage areas. Both off-premises and in-house record storage areas should have automatic sprinkler systems for optimum protection against fire loss.

Periodic in-house records security checks should be conducted, and should include assessment of records in storage and in use, the EDP equipment, and records of computer time use. In addition, all personnel using the equipment should undergo security screening consisting of reference checks, previous employer checks, salary versus standard of living studies, and other relevant verifications.

The EDP equipment can be additionally protected through internal system safeguards such as error checking circuitry, redundancy checks, limiting transactions above a stated amount, bound registers, and data encrypting. These security features should periodically be checked to ensure they are in proper working condition.

Smoking in the computer room should be strictly prohibited.

### Safe Storage Practices

It is common practice to store extra computer paper in the EDP room. This can be a poor practice if the volume of unused and scrap paper is allowed to build up, since the fire risk is accordingly increased. A minimum supply of paper can be kept in the EDP room, and all scrap containers should be frequently emptied.

Another common practice is to locate fire-resistant file cabinets or media safes in the EDP room. Again, only those records needed for daily operation should be kept in the EDP room. The EDP room should not be used as a storage room but should be considered a vital area deserving high quality fire and security protection. If fire-resistant cabinets or media safes must be located in the EDP room, they should be kept closed, except when actually retrieving or replacing material.

All cabinets or safes used for records storage should be UL listed. Such equipment is classified by an interior temperature limit and a time limit (in hours). Two standard temperature and humidity limits are used:

1. 150 F (65.5 C) with 85 percent relative humidity (for photographic, magnetic, or similar non-paper records).
2. 350 F (196 C) with 100 percent relative humidity (for paper records).

Time limits are 1, 2, 3, and 4 hours. Ratings are listed by class and time, for example, *Class 150—1 hour* means the internal temperatures will be held to 150 F for 1 hour in normal fire conditions. On UL-listed equipment, the rating is shown on a metal label affixed to the cabinet or safe door.

## UTILITIES PROTECTION

Utility protection also needs to be considered, the prime concern being protection against sabotage. Electrical lines or phone lines that are exposed and accessible present a good target for anyone bent on causing serious monetary damage to a firm through business interruption.

Air conditioning intake ducts should be inspected and maintained to prevent damaging gases, vapors, fumes, or mists being drawn into the EDP room, and possibly causing equipment corrosion or damage. Particulate matter should be removed from the air by means of UL-listed filters, or by electrostatic precipitators.

Other utility protection measures for EDP rooms are:

- Make security checks of all utilities leading into the room.
- Prohibit electric transformer use in the computer room.
- Make provisions for protection against lightning surges in accordance with NFPA70—*National Electrical Code.*
- Provide an emergency lighting system to allow for safe employee egress from the area in an emergency.

## WORKERS' COMPENSATION

The NFPA encourages the following safety requirements where Halon 1301 systems are used (NFPA 12A, Section A-1-6.1.2):

a. Provision of adequate aisle-ways and routes of exits and keeping them clear at all times. . . .
b. Provision of emergency lighting and directional signs to ensure quick, safe evacuation. . . .
c. Provision of alarms within such areas that will operate immediately upon detection of the fire. . . .
d. Provision of only outward swinging self-closing doors at exits from hazardous areas and, where such doors are latched, provision of panic hardware. . . .
e. Provision of continuous alarms at entrances to such areas until the atmosphere has been restored to normal. . . .
f. Provision of warning instruction signs at entrances to and inside such areas. . . .
g. Provision for prompt discovery and rescue of persons rendered unconscious in such areas. . . .
h. Provision of instruction and drills of all personnel within or in the vicinity of such areas. . . .

i. Provision of means for prompt ventilation of such areas. . . .
j. Prohibition against smoking by persons until the atmosphere has been purged of Halon 1301. . . .
k. Provision of such other steps and safeguards that a careful study of each particular situation indicates is necessary to prevent injury or death.

## COMPETITIVE EDGE

As mentioned earlier, the EDP equipment has become a vital and commonplace tool for business, industry, government and research groups. Following the procedural steps (outlined earlier) for installation and protection of EDP equipment is necessary for any business seeking to remain viable and competitive.

Potential problems can be eliminated at the design stage when management wisely plans ahead. The computer room location, the equipment, fixed and portable fire protection systems, records and utilities protection, and employee safety must all be considered, and proper safeguards installed.

By taking positive steps to minimize exposures to loss, the intelligent businessperson will be in a better position to take full advantage of the competitive edge provided by the use of EDP equipment.

With permission of *Professional Safety Magazine,* the preceeding portion of this chapter expanded and up-dated an article appearing in that publication. (Klonicke, 1983.)

## REFERENCES

*Best's Loss Control Engineering Manual.* Oldwick, N.J.: A. M. Best Co., published annually.

Helander, M. G. (ed.) *Handbook of Human/Computer Interaction.* Amsterdam and New York: Elsevier, 1987.

Klonicke, D. W. Loss Control in the Computer Room, *Professional Safety,* (April 1983) 17-20.

Miller, Earl, and O'Hern, Carol. Microcomputers Can Make It Work for You, *Safety & Health,* 35 (Jan. 1987) 28-32.

National Fire Protection Association, Batterymarch Park, Quincy, Mass. 02269.
*Fire Protection Handbook,* 15th ed.
Standards:
*Electronic/Data Processing Equipment,* NFPA 75.
*Halogenated Fire Extinguishing Agent Systems—Halon 1301, Fire Extinguishing Systems,* NFPA 12A.
*Protection of Records,* NFPA 232.

Ross, D. T., and Schoman, K. E. "Structural Analysis for Requirements Definition," in *Software Design Techniques,* 3rd ed. P. Freeman and A. I. Wasserman, eds. Long Beach, Calif.: IEEE Computer Society, 1980.

Whyte, Adrienne. "Occupational Health and Safety Information Management Systems," in *Introduction to Occupational Health and Safety.* Joseph La Dou, ed. Chicago: National Safety Council, 1986.

# Accident Records and Incidence Rates



In this chapter, the terms *accident, incident,* and *injury* are restricted to occupational injuries and illnesses. In other chapters, accident and incident are used in their broad meanings—"unplanned events that interrupt the completion of an activity, and that may (or may not) include property damage or injury."

The Williams-Steiger Occupational Safety and Health Act of 1970 requires the majority of employers to maintain specific records of work-related employee injuries and illnesses. In addition to these records, many employers also are required to make reports to state compensation authorities. Insuring agencies also may require reports. For contest and award programs, reports based on OSHA recordkeeping requirements may be filed. Occupational injury and illness reports and records are now required of nearly every establishment by management or government.

Safety personnel are faced with two tasks—maintaining those records required by law and by their management, and maintaining records that are useful to an effective safety program. Unfortunately, the two are not always synonymous. A good recordkeeping system necessitates more data than that contained in most required forms.

This chapter deals with both aspects of recordkeeping, except specific legal requirements which are beyond the scope of this volume because they differ from industry to industry and state to state and are subject to change. The appropriate federal and state authorities need to be contacted to obtain the most current requirements. An outline of the general recordkeeping requirements under the OSHAct, as it is at this time, is presented in this chapter. The basic definitions and the method of keeping records under the American National Standards Institute Z16.1 standard are also presented, in the last section of this chapter, for employers not using the OSHA recordkeeping system.

Although this chapter covers injuries and illnesses occurring to employees while on the job, a standard for off-the-job injuries to employees (ANSI Z16.3) is briefly summarized in the off-the-job section. The standard itself should be consulted by the safety personnel concerned with that aspect of the overall safety program.

The first section explains general recordkeeping systems and contains sample forms and recommendations for establishing a good system.

## ACCIDENT RECORDS

Records of accidents and injuries are essential to efficient and successful safety programs, just as records of production, costs, sales, and profits and losses are essential to efficient and successful business operations. Records supply the information necessary to transform haphazard, costly, ineffective safety work into a planned safety program that controls both the conditions and the acts that contribute to accidents. Good recordkeeping is the foundation of a scientific approach to occupational safety.

### Uses of records

A good recordkeeping system can help the safety professional in the following ways:

1. Provide safety personnel with the means for an objective evaluation of the magnitude of their accident problems and with a measurement of the overall progress and effectiveness of their safety program.
2. Identify high-rate units, plants, or departments and problem areas so extra effort can be made in those areas.

Case No. 164 Date 2-12-

**First Aid Report**

**Name** S. D. Smith **Department** Shipping

**Male** [X] **Female** [ ] **Occupation** Packer **Foreman** Miller

**Date of Occurrence** 2-12 **Time** 10 **a.m. p.m.** **Date of First Treatment** 2-12 **Time** 10 **a.m. p.m.**

**Nature of Occurrence** Splinter in index finger of left hand

**Sent: Back to Work** [X] **Doctor** [ ] **Home** [ ] **Hospital** [ ]

**Estimated Disability** 0 **days**

**Employee's Description of Occurrence** Handling wooden crates without gloves, ran splinter into finger

**Signed** Mr. Miller
**First Aid**

**Issued by National Safety Council, Inc.**

Form IS-6 Printed in U.S.A. STOCK No. 129.26

**Figure 6-1.** This First Aid Report (4 × 6 in. or 10 × 16 cm) is prepared by the first aid attendant at the time an injured or ill person comes for treatment. A report serves as a record and permits quick tabulation of such data as department, occupation, and the key facts of the occurrence.

3. Provide data for an analysis of accidents and illnesses that can point to specific circumstances of occurrence which can then be attacked by specific countermeasures.
4. Create interest in safety among supervisors by furnishing them with information about their departments' accident experience.
5. Provide supervisors and safety committees with hard facts about their safety problems so their efforts can be concentrated.
6. Measure the effectiveness of individual countermeasures and determine if specific programs are doing the job they were designed to do.

## Recordkeeping systems

The system presented in this section is a model that can be used to provide the basic items necessary for good recordkeeping. It is designed to dovetail with the present recordkeeping requirements of the OSHAct and attempts to avoid a duplication of effort on the part of the personnel responsible for keeping records and filing reports. Provision is also made for easily entering the data necessary for those organizations using the Z16.1 standard and computing those frequency and severity rates without the necessity of a separate set of records. Some of the forms presented in this section are also constructed with modern data processing methods in mind. In general, a self-coding, check-off form can save time for both the person who fills out the report and the person who is responsible for tabulating and processing the data.

A well-designed form takes into account the person who will fill it out and the way in which the forms will be processed. It is more likely to be filled out accurately and will present fewer problems for those who process and analyze the data. Care in the choice and design of forms will pay dividends in better, more reliable data.

The recordkeeping system in this section is not the only way to keep records, but rather it is an example. The accident problems of individual establishments are unique and no one form or set of forms can provide every establishment with all the data for solving all of its individual problems. A system that does a good job of collecting the basic facts, however, makes it easier to zero-in later on data relating to a specific problem.

The following sections deal with occupational injuries and illnesses. Property-damage accidents are covered in Chapter 7. Nonemployee accidents are covered in Chapter 21. Specific recordkeeping requirements were explained in Chapter 2, Governmental Regulation and Compliance.

## Accident reports and injury records

To be effective, preventive measures must be based on complete and unbiased knowledge of the causes of accidents. The primary purpose of an accident report is to obtain such information but not to fix blame. Since the completeness and accuracy of the entire accident record system depend upon information in the individual accident reports, be sure the forms and their purpose are understood by those who must fill them out. Necessary training or instruction should be given to these personnel.

**The first aid report.** The collection of injury data generally begins in the first aid department. The first aid attendant or nurse fills out a first aid report for each new case. Copies are sent to the safety department or safety committee, the worker's supervisor, and other departments as management may wish. See Figure 6-1.

# ACCIDENT INVESTIGATION REPORT

CASE NUMBER

COMPANY ______________________ ADDRESS ______________________

DEPARTMENT ______________________ LOCATION (if different from mailing address) ______________________

1. NAME of INJURED

2. SOCIAL SECURITY NUMBER

3. SEX ☐ M ☐ F

4. AGE

5. DATE of ACCIDENT

6. HOME ADDRESS

7. EMPLOYEE'S USUAL OCCUPATION

8. OCCUPATION at TIME of ACCIDENT

9. LENGTH of EMPLOYMENT
☐ Less than 1 mo. ☐ 6 mos. to 5 yrs.
☐ 1-5 mos. ☐ More than 5 yrs.

10. TIME in OCCUP. at TIME of ACCIDENT
☐ Less than 1 mo. ☐ 6 mos. to 5 yrs.
☐ 1-5 mos. ☐ More than 5 yrs.

11. EMPLOYMENT CATEGORY
☐ Regular, full-time ☐ Temporary ☐ Nonemployee
☐ Regular, part-time ☐ Seasonal

12. CASE NUMBERS and NAMES of OTHERS INJURED in SAME ACCIDENT

13. NATURE of INJURY and PART of BODY

14. NAME and ADDRESS of PHYSICIAN

15. NAME and ADDRESS of HOSPITAL

16. TIME of INJURY
A. ______ A.M. P.M.
B. Time within shift
C. Type of shift

17. SEVERITY of INJURY
☐ Fatality
☐ Lost workdays—days away from work
☐ Lost workdays—days of restricted activity
☐ Medical treatment
☐ First aid
☐ Other, specify ______

18. SPECIFIC LOCATION of ACCIDENT

ON EMPLOYER'S PREMISES? ☐ Yes ☐ No

19. PHASE OF EMPLOYEE's WORKDAY at TIME of INJURY
☐ During rest period ☐ Entering or leaving plant
☐ During meal period ☐ Performing work duties
☐ Working overtime. ☐ Other ______

20. DESCRIBE HOW the ACCIDENT OCCURRED

21. ACCIDENT SEQUENCE. Describe in reverse order of occurrence events preceding the injury and accident. Starting with the injury and moving backward in time, reconstruct the sequence of events that led to the injury.

A. Injury Event ______

B. Accident Event ______

C. Preceding Event #1 ______

D. Preceding Event #2, #3, etc. ______

**Figure 6-2.** Accident Investigation Report (8½ × 11 in. or 22 × 28 cm) provides a record of contributing circumstances to provide a basis for specific remedial action. Users should be trained to properly fill it out.

22. TASK and ACTIVITY at TIME of ACCIDENT

A. General type of task ______

B. Specific activity ______

C. Employee was working:

☐ Alone ☐ With crew or fellow worker ☐ Other, specify ______

23. POSTURE of EMPLOYEE

______

24. SUPERVISION at TIME of ACCIDENT

☐ Directly supervised ☐ Not supervised

☐ Indirectly supervised ☐ Supervision not feasible

25. CAUSAL FACTORS. Events and conditions that contributed to the accident. Include those identified by use of the Guide for Identifying Causal Factors and Corrective Actions.

26. CORRECTIVE ACTIONS. Those that have been, or will be, taken to prevent recurrence. Include those indentified by use of the Guide for Identifying Causal Factors and Corrective Actions.

PREPARED BY ______

TITLE ______

DEPARTMENT ______ DATE ______

**Developed by the National Safety Council**

APPROVED ______

TITLE ______ DATE ______

APPROVED ______

TITLE ______ DATE ______

**Figure 6-2.** (Concluded.)

SERVICE NO. (NSC)
▶ 1-9 ______
CASE OR FILE NO.
▶ 10-15 ______

**SUPPLEMENTARY RECORD OF OCCUPATIONAL INJURIES AND ILLNESSES**

OSHA No. 101 NSC revision (Meets OSHA requirements when Instruction 1. has been followed.)

THIS REPORT IS
▶ 16, 1 ☐ First report 2 ☐ Revised report

**EMPLOYER**

1. NAME ______
2. MAIL ADDRESS ______
3. LOCATION, if different from mail address ______

**INJURED OR ILL EMPLOYEE**

4. NAME ______
SOCIAL SECURITY NO. ______
▶ EMPLOYEE NO. 17-26 ______
5. HOME ADDRESS ______
▶ 6. AGE 27-28 ______
▶ 7. SEX 29, 1 ☐ Male 2 ☐ Female
▶ 8. OCCUPATION (specify) ______
30-31, 01 ☐ Manager, official, proprietor
02 ☐ Professional, technical
03 ☐ Foreman, supervisor
04 ☐ Sales worker
05 ☐ Clerical worker
06 ☐ Craftsman—construction
07 ☐ Craftsman—other
08 ☐ Machinist
09 ☐ Mechanic
10 ☐ Operative (production worker)
11 ☐ Motor vehicle driver
12 ☐ Laborer
13 ☐ Service worker
14 ☐ Agricultural worker
15 ☐ Other
16 ☐ Unknown
9. DEPARTMENT ______
(Enter the name of department or division in which the injured person is regularly employed.)

**CLASSIFICATION OF CASE**

A. INJURY OR ILLNESS (see code on Log, OSHA No. 100)
▶ 32, 1 ☐ Injury (10)
2 ☐ Occupational skin disease or disorder (21)
3 ☐ Dust disease of the lungs (pneumoconioses) (22)
4 ☐ Respiratory conditions due to toxic agents (23)
5 ☐ Poisoning (systemic effects of toxic materials) (24)
6 ☐ Disorder due to physical agents (other than toxic materials) (25)
7 ☐ Disorder due to repeated trauma (26)
8 ☐ All other occupational illnesses (29)

B. EXTENT OF INJURY OR ILLNESS
▶ 33, 1 ☐ Fatality
2 ☐ Lost workday case
3 ☐ Nonfatal case without lost workdays

▶ C. Number of workdays lost 34-36 ______

D. Permanently transferred or terminated
▶ 37, 1 ☐ Yes 2 ☐ No

**INSTRUCTIONS**

1. *Type or print the narrative where requested.*
2. *Check the one box which most clearly describes each narrative statement.*
3. *See also original OSHA No. 101 for more details.*
4. *Complete form in duplicate. Retain original. Mail duplicate to: National Safety Council, 425 N. Michigan Ave., Chicago IL 60611.*

**THE ACCIDENT OR EXPOSURE TO OCCUPATIONAL ILLNESS**

10. PLACE OF ACCIDENT OR EXPOSURE (mail address) ______

11. WHERE DID ACCIDENT OR EXPOSURE OCCUR?
a. On employer premises
▶ 38, 1 ☐ Yes 2 ☐ No 3 ☐ Unknown
b. Place (specify) ______
▶ 39-40, 01 ☐ Office
02 ☐ Plant, mill
03 ☐ Shipping, receiving, warehouse
04 ☐ Maintenance shop
05 ☐ General or public area of employer premises (corridor, washroom, lunchroom, parking lot, etc.)
06 ☐ Retail establishment (store, restaurant, gasoline station, etc.)
07 ☐ Farm
08 ☐ Motor vehicle accident
09 ☐ Other
10 ☐ Unknown

12. WHAT WAS THE EMPLOYEE DOING WHEN INJURED? (Be specific)
______

a. Task performed at time of accident
▶ 41-42, 01 ☐ Operating machine
02 ☐ Operating hand tool (power or nonpower)
03 ☐ Materials handling
04 ☐ Maintenance & repair—machinery
05 ☐ Maintenance & repair—building & equipment
06 ☐ Motor vehicle driver, operator or passenger
07 ☐ Office and sales tasks, except above
08 ☐ Service tasks, except above
09 ☐ Other
10 ☐ Not performing task
11 ☐ Unknown

b. Activity at time of accident
▶ 43-44, 01 ☐ Climbing
02 ☐ Driving
03 ☐ Jumping
04 ☐ Kneeling
05 ☐ Lying down
06 ☐ Lifting
07 ☐ Reaching, stretching
08 ☐ Riding
09 ☐ Running
10 ☐ Sitting
11 ☐ Standing
12 ☐ Walking
13 ☐ Other
14 ☐ Unknown

**Figure 6-3.** A self-coding supplementary record of occupational injuries and illnesses.

13. **HOW DID THE ACCIDENT OCCUR?** (Describe fully the events)

a. **AGENCY.** (Object or substance involved)
**ACCIDENT AGENCY (1st column). The first object or substance involved in accident sequence.**
**INJURY AGENCY** (2nd column). The agency inflicting the injury. See also section 15.
(Example: Worker fell from ladder and struck head on machine. Check "Ladder" under accident and check "Machine" under injury.)

**ACCIDENT** **INJURY** (Check one box in each column)

| ACCIDENT | INJURY | |
|---|---|---|
| ►► 45-46, 01 ☐ | 47-48, 01 ☐ | Machine |
| 02 ☐ | 02 ☐ | Conveyor, elevator, hoist |
| 03 ☐ | 03 ☐ | Vehicle |
| 04 ☐ | 04 ☐ | Electrical apparatus |
| 05 ☐ | 05 ☐ | Hand tool |
| 06 ☐ | 06 ☐ | Chemical |
| 07 ☐ | 07 ☐ | Working surface, bench, table, etc. |
| 08 ☐ | 08 ☐ | Floor, walking surface |
| 09 ☐ | 09 ☐ | Bricks, rocks, stones |
| 10 ☐ | 10 ☐ | Box, barrel, container (empty or full) |
| 11 ☐ | 11 ☐ | Door, window, etc. |
| 12 ☐ | 12 ☐ | Ladder |
| 13 ☐ | 13 ☐ | Lumber, woodworking materials |
| 14 ☐ | 14 ☐ | Metal |
| 15 ☐ | 15 ☐ | Stairway, steps |
| 16 ☐ | 16 ☐ | Other |
| 17 ☐ | 17 ☐ | Unknown |
| 18 ☐ | 18 ☐ | None |

b. **ACCIDENT TYPE.** (First event in the accident sequence)

► 49-50, 01 ☐ Fall from elevation
02 ☐ Fall on same level
03 ☐ Struck against
04 ☐ Struck by
05 ☐ Caught in, under or between
06 ☐ Rubbed or abraded
07 ☐ Bodily reaction
08 ☐ Overexertion
09 ☐ Contact with electrical current
10 ☐ Contact with temperature extremes
11 ☐ Contact with radiations, caustics, toxic and noxious substances
12 ☐ Public transportation accident
13 ☐ Motor vehicle accident
14 ☐ Other
15 ☐ Unknown

This space may be used for additional information.

**OCCUPATIONAL INJURY OR ILLNESS**

14. **DESCRIBE THE INJURY OR ILLNESS** in detail and indicate the part of the body affected.

a. **NATURE OF INJURY OR ILLNESS.** (Check most serious one)

► 51-52, 01 ☐ Amputation
02 ☐ Burn and scald (heat)
03 ☐ Burn (chemical)
04 ☐ Concussion
05 ☐ Crushing injury
06 ☐ Cut, laceration, puncture, abrasion
07 ☐ Fracture
08 ☐ Hernia
09 ☐ Bruise, contusion
10 ☐ Occupational illness
11 ☐ Sprain, strain
12 ☐ Other

b. **PART OF BODY.** (Check most serious one)

► 53-54, 01 ☐ Eyes
02 ☐ Head, face, neck
03 ☐ Back
04 ☐ Trunk (except back, internal)
05 ☐ Arm
06 ☐ Hand and wrist
07 ☐ Fingers
08 ☐ Leg
09 ☐ Feet and ankles
10 ☐ Toes
11 ☐ Internal and other

15. **NAME THE OBJECT OR SUBSTANCE WHICH DIRECTLY INJURED THE EMPLOYEE.** Also check one box in injury column under **13a.**

16. **DATE OF INJURY OR INITIAL DIAGNOSIS OF OCCUPATIONAL ILLNESS.**

a. **MONTH**

| | | | |
|---|---|---|---|
| ► 55-56, 01 ☐ | Jan. | 07 ☐ | July |
| 02 ☐ | Feb. | 08 ☐ | Aug. |
| 03 ☐ | March | 09 ☐ | Sept. |
| 04 ☐ | April | 10 ☐ | Oct. |
| 05 ☐ | May | 11 ☐ | Nov. |
| 06 ☐ | June | 12 ☐ | Dec. |

► b. **DATE OF MONTH** 57-58 ________

17. **DID EMPLOYEE DIE?**

► 59, 1 ☐ Yes Date of Death ________
2 ☐ No

**OTHER**

18. **NAME AND ADDRESS OF PHYSICIAN** ________

19. **IF HOSPITALIZED, NAME AND ADDRESS OF HOSPITAL** ________

**DATE OF REPORT** ________

**PREPARED BY** ________

**OFFICIAL POSITION** ________

National Safety Council
Printed in U.S.A.

**Figure 6-3.** (Concluded.)

PAGE 1

**FORM 2020**
**SUPERVISOR'S INJURY / ILLNESS**
**INVESTIGATION REPORT**

## GENERAL INSTRUCTIONS

### WHEN MUST FORM 2020 BE USED?

All accidents and near-miss accidents should be investigated. Form 2020 must be completed and forwarded to Corporate Safety in Trexlertown within **48 hours** in the following circumstances:

1. When near-miss accidents, with the potential for serious injury occur.
2. All accidents involving personnel injury or illness serious enough to require a State Workmen's Compensation Insurance Report (an Employer's First Report of Injury). This includes injuries sustained in automobile accidents while the employee is on company business.
3. All suspected OSHA recordable injuries / illnesses.
4. All known injuries or illnesses to non-employees that involve our products, facilities, or employees.

ONE FORM MUST BE COMPLETED FOR EACH PERSON INJURED / ILL AS DESCRIBED IN 2, 3 & 4 ABOVE.

### WHO MUST FILL OUT FORM 2020?

Section I on Form 2020 is all coded indexing information for the computer system. This section will be completed by Corporate Safety.

Sections II through V will be filled out by the injured / ill person's supervisor.

Sections VI will be filled out by the Supervisor's Manager, after Sections II through V have been completed.
When the injured / ill person is not an employee, Sections II through V will be filled out by the person investigating the accident and Section VI by the Supervisor's Manager.
Each line on the form is keyed to the appropriate instruction by the circled numbers e.g.(5.)

PLEASE NOTE: The format of some of the entries, such as the date, etc. is very critical. PLEASE FOLLOW THE DETAILED INSTRUCTIONS IN THIS PAMPHLET CAREFULLY WRITING ONLY IN THE UNSHADED AREAS.
FILL IN ALL APPLICABLE BOXES — ONE ENTRY — ONE BOX.

### WHAT TO DO WHEN THE FORM IS COMPLETED?

When Sections II through V are completed, the form should be sent to the Supervisor's Manager who will complete Section VI.
When Section VI is completed, the form should be detached from the instructions, copied and the following distribution made:

1. The original 2020 and a copy of the State Workmen's Compensation Insurance Report (Employer's First Report of Injury) should be sent to Corporate Safety, Trexlertown. (A copy of the State Workmen's Compensation Insurance Report should also be sent to Corporate Risk Management, Trexlertown).
2. One copy should be retained in the facility's file.
3. One copy should be sent to Group / Division Safety.
4. Other copies as required by Group / Division or Local requirements.

### WHAT IS FORM 2020 USED FOR?

Form 2020 will be used as the data entry form for the computerized Safety Information System (SINS). SINS will compile accident information and allow us to determine, among other things, what kind of accidents and injuries are occurring, their causes, and how much they cost. With this information, which is available to all supervisors and managers in the company, we can develop specific programs to help you solve your particular safety problems.

### NOTICE!

**THIS FORM MUST BE SUBMITTED IN**
**ORDER FOR INSURANCE CLAIMS TO BE PAID.**

FORM 2020 (REV 1 / 81)

**Figure 6-4.** This 6-page Form 2020, Supervisor's Injury/Illness Investigation Report, is well designed to solicit additional information for causal and cost analyses. The pages are folded to open as illustrated here, making instructions easy to follow. (Reprinted with permission from Air Products and Chemicals, Inc.)

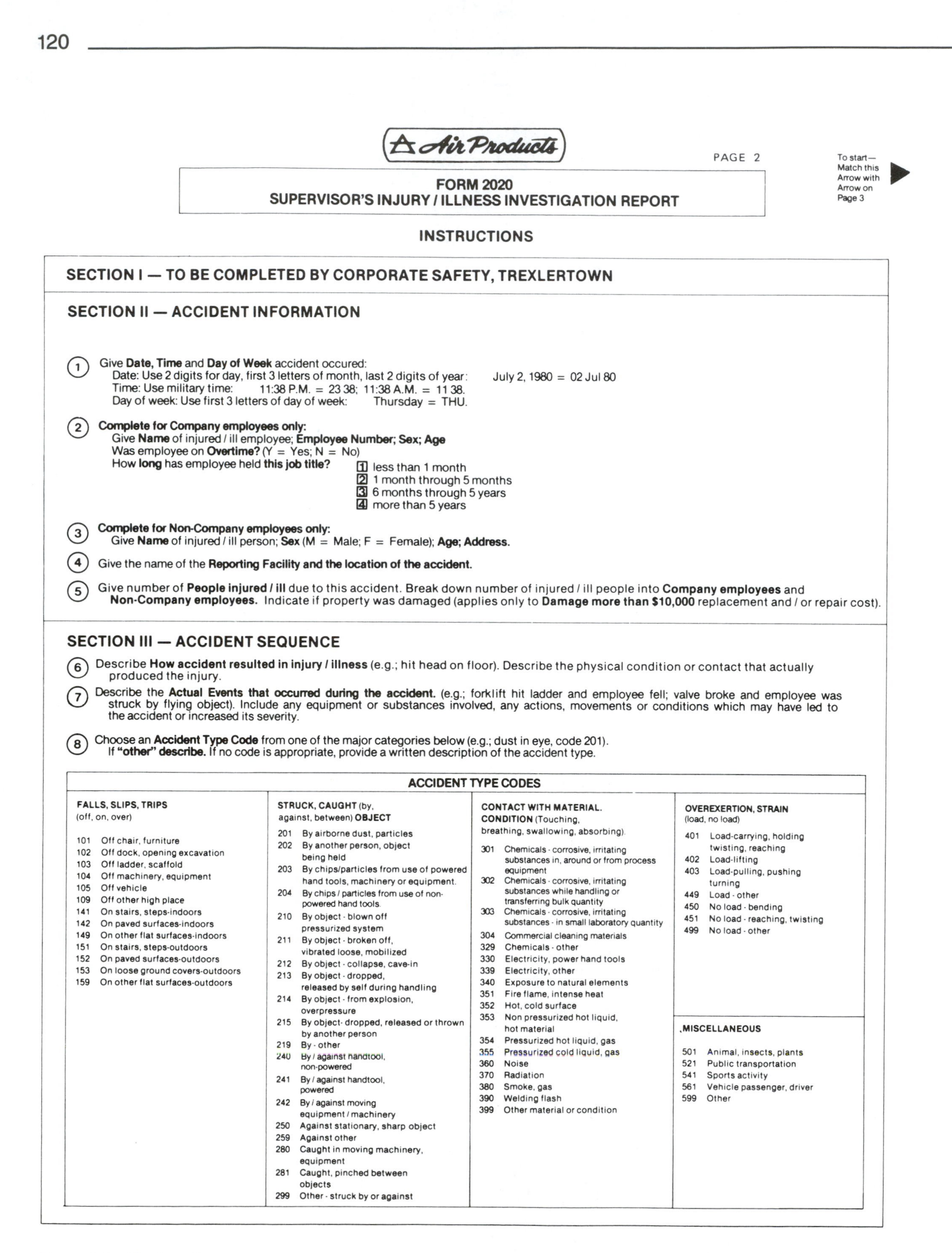

Air Products

PAGE 2

To start— Match this Arrow with Arrow on Page 3

**FORM 2020**
**SUPERVISOR'S INJURY / ILLNESS INVESTIGATION REPORT**

**INSTRUCTIONS**

**SECTION I — TO BE COMPLETED BY CORPORATE SAFETY, TREXLERTOWN**

**SECTION II — ACCIDENT INFORMATION**

(1) Give **Date, Time** and **Day of Week** accident occured:
Date: Use 2 digits for day, first 3 letters of month, last 2 digits of year: July 2, 1980 = 02 Jul 80
Time: Use military time: 11:38 P.M. = 23 38; 11:38 A.M. = 11 38.
Day of week: Use first 3 letters of day of week: Thursday = THU.

(2) **Complete for Company employees only:**
Give **Name** of injured / ill employee; **Employee Number; Sex; Age**
Was employee on **Overtime?** (Y = Yes; N = No)
How **long** has employee held **this job title?**
[1] less than 1 month
[2] 1 month through 5 months
[3] 6 months through 5 years
[4] more than 5 years

(3) **Complete for Non-Company employees only:**
Give **Name** of injured / ill person; **Sex** (M = Male; F = Female); **Age; Address.**

(4) Give the name of the **Reporting Facility and the location of the accident.**

(5) Give number of **People injured / ill** due to this accident. Break down number of injured / ill people into **Company employees** and **Non-Company employees.** Indicate if property was damaged (applies only to **Damage more than $10,000** replacement and / or repair cost).

**SECTION III — ACCIDENT SEQUENCE**

(6) Describe **How accident resulted in injury / illness** (e.g.; hit head on floor). Describe the physical condition or contact that actually produced the injury.

(7) Describe the **Actual Events that occurred during the accident.** (e.g.; forklift hit ladder and employee fell; valve broke and employee was struck by flying object). Include any equipment or substances involved, any actions, movements or conditions which may have led to the accident or increased its severity.

(8) Choose an **Accident Type Code** from one of the major categories below (e.g.; dust in eye, code 201).
If **"other" describe.** If no code is appropriate, provide a written description of the accident type.

**ACCIDENT TYPE CODES**

**FALLS, SLIPS, TRIPS** (off, on, over)

101 Off chair, furniture
102 Off dock, opening excavation
103 Off ladder, scaffold
104 Off machinery, equipment
105 Off vehicle
109 Off other high place
141 On stairs, steps-indoors
142 On paved surfaces-indoors
149 On other flat surfaces-indoors
151 On stairs, steps-outdoors
152 On paved surfaces-outdoors
153 On loose ground covers-outdoors
159 On other flat surfaces-outdoors

**STRUCK, CAUGHT** (by, against, between) **OBJECT**

201 By airborne dust, particles
202 By another person, object being held
203 By chips/particles from use of powered hand tools, machinery or equipment.
204 By chips / particles from use of non-powered hand tools.
210 By object - blown off pressurized system
211 By object - broken off, vibrated loose, mobilized
212 By object - collapse, cave-in
213 By object - dropped, released by self during handling
214 By object - from explosion, overpressure
215 By object- dropped, released or thrown by another person
219 By - other
240 By / against handtool, non-powered
241 By / against handtool, powered
242 By / against moving equipment / machinery
250 Against stationary, sharp object
259 Against other
280 Caught in moving machinery, equipment
281 Caught, pinched between objects
299 Other - struck by or against

**CONTACT WITH MATERIAL, CONDITION** (Touching, breathing, swallowing, absorbing).

301 Chemicals - corrosive, irritating substances in, around or from process equipment
302 Chemicals - corrosive, irritating substances while handling or transferring bulk quantity
303 Chemicals - corrosive, irritating substances - in small laboratory quantity
304 Commercial cleaning materials
329 Chemicals - other
330 Electricity, power hand tools
339 Electricity, other
340 Exposure to natural elements
351 Fire flame, intense heat
352 Hot, cold surface
353 Non pressurized hot liquid, hot material
354 Pressurized hot liquid, gas
355 Pressurized cold liquid, gas
360 Noise
370 Radiation
380 Smoke, gas
390 Welding flash
399 Other material or condition

**OVEREXERTION, STRAIN** (load, no load)

401 Load-carrying, holding twisting, reaching
402 Load-lifting
403 Load-pulling, pushing turning
449 Load - other
450 No load - bending
451 No load - reaching, twisting
499 No load - other

**MISCELLANEOUS**

501 Animal, insects, plants
521 Public transportation
541 Sports activity
561 Vehicle passenger, driver
599 Other

**Figure 6-4.** (Continued.)

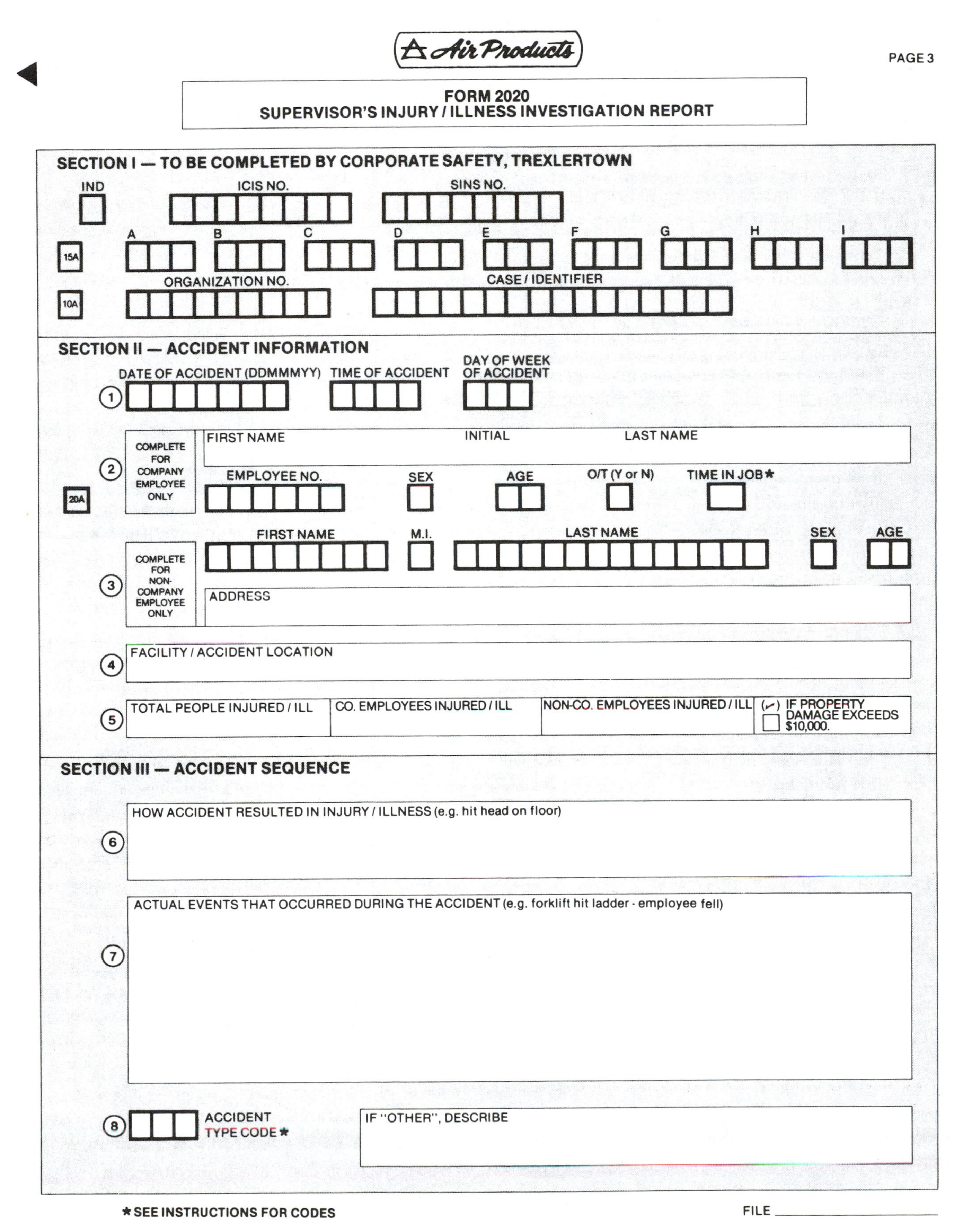

Air Products

PAGE 3

FORM 2020
SUPERVISOR'S INJURY / ILLNESS INVESTIGATION REPORT

SECTION I — TO BE COMPLETED BY CORPORATE SAFETY, TREXLERTOWN

IND | ICIS NO. | SINS NO.

15A | A | B | C | D | E | F | G | H | I

10A | ORGANIZATION NO. | CASE / IDENTIFIER

SECTION II — ACCIDENT INFORMATION

1 | DATE OF ACCIDENT (DDMMMYY) | TIME OF ACCIDENT | DAY OF WEEK OF ACCIDENT

20A

2 | COMPLETE FOR COMPANY EMPLOYEE ONLY | FIRST NAME | INITIAL | LAST NAME
EMPLOYEE NO. | SEX | AGE | O/T (Y or N) | TIME IN JOB*

3 | COMPLETE FOR NON-COMPANY EMPLOYEE ONLY | FIRST NAME | M.I. | LAST NAME | SEX | AGE
ADDRESS

4 | FACILITY / ACCIDENT LOCATION

5 | TOTAL PEOPLE INJURED / ILL | CO. EMPLOYEES INJURED / ILL | NON-CO. EMPLOYEES INJURED / ILL | (✓) IF PROPERTY DAMAGE EXCEEDS $10,000.

SECTION III — ACCIDENT SEQUENCE

6 | HOW ACCIDENT RESULTED IN INJURY / ILLNESS (e.g. hit head on floor)

7 | ACTUAL EVENTS THAT OCCURRED DURING THE ACCIDENT (e.g. forklift hit ladder - employee fell)

8 | ACCIDENT TYPE CODE* | IF "OTHER", DESCRIBE

* SEE INSTRUCTIONS FOR CODES

FILE ____________

**Figure 6-4.** (Continued.)

INSTRUCTIONS CONT'D

PAGE 4

SECTION IV — ACTIVITY INFORMATION

(9) **What was the injured / ill person doing?** (e.g. bolting a flange).

(10) **How often has the person done this activity?** (1 DAILY 2 WEEKLY 3 MONTHLY 4 LESS THAN ONCE PER MONTH 5 NEVER BEFORE)

**Was this activity a normal part of the job?** (Y = Yes; N = No).
**Was person adequately trained** for this activity in your opinion? (Y = Yes; N = No)

(11) **Do standard methods or procedures exist** for the task the person was doing? (Y = Yes; N = No) **If Yes, Were they followed?** (Y = Yes; N = No) **If procedures were not followed,**

**Did something discourage following procedures?** (Y or N)

(12) **If procedures were not followed, what was done differently** than called for by the procedure?

## SECTION V — INJURY/ILLNESS INFORMATION

(13) **Was person using equipment to protect against this injury / illness?** (Y = Yes; N = No)
**Was person supposed to be using protective equipment?** (Y = Yes; N = No)
Describe the **Type of protective equipment required and actually used.**

(14) Select the most serious and second most serious **Nature of Injury / Illness code from below.**

| NATURE OF INJURY/ ILLNESS CODES | | |
|---|---|---|
| INJURY | | OCCUPATIONAL ILLNESS |
| 101 Amputation | 110 Foreign Body, Sliver, Chip, Dust | 201 Skin Disease, Disorder |
| 102 Bite, Sting | 111 Fracture, Crush, Dislocate | 202 Lung Problem, Dust-Related |
| 103 Bruise, Contusion | 112 Internal Injury, Hernia, Heart | 203 Lung Problem, Toxic-Agent Related |
| 104 Burn, Hot, Cold, Chemical, Scald | 113 Loss of Senses, Faculties | 204 Poisoning |
| 105 Concussion, Unconscious | 114 Scrape, Scratch, Abrasion | 205 Disorders Due To Physical Agents (Other Than Toxic Agents) |
| 106 Cut, Laceration, Puncture | 115 Sprain, Strain, Torn | 206 Disorders Associated With Repeated Trauma |
| 107 Exhaustion, Heat Stroke | 116 Suffocation, Drowning | 299 All Other |
| 108 Electric Shock | 199 All Other | |

(15) Select the **Part of Body Code** from below corresponding to the most serious and second most serious injury / illness.
For 14 and 15, **Describe** each injury / illness and how each part of the body was affected.

| PART OF BODY CODES | | | | |
|---|---|---|---|---|
| HEAD / NECK | ARM / SHOULDER | TORSO | LEG | FACULTY / SYSTEM |
| 301 Scalp | 401 Shoulder | 501 Chest/Ribs | 601 Thigh | 701 Hearing |
| 302 Skull | 402 Upper Arm | 502 Back-Muscles | 602 Knee | 702 Vision |
| 303 Ears | 403 Elbow | 508 Back-Skeletal / Nervous | 603 Shin, Calf | 703 Smell |
| 304 Eyes | 404 Forearm | 503 Heart | 604 Ankle | 704 Taste |
| 305 Face | 405 Wrist | 504 Abdomen | 605 Foot | 705 Touch |
| 306 Nose | 406 Hand | 505 Groin | 606 Toe | 706 Respiratory |
| 307 Mouth / Teeth | 407 Finger | 506 Hip | 610 Whole Leg | 707 Circulatory |
| 308 Neck | 410 Whole Arm | 507 Buttocks | | 708 Digestive |
| 310 Whole Head | | 510 Whole Torso | | 710 Nervous |

(16) Describe **Treatment provided** (e.g. X-rayed and released; 6 stitches; hospitalized).

(17) Print and sign name of **Supervisor** preparing this report; give **Telephone Number at work (include area code); Date Report was prepared.**

**STOP — FORWARD TO YOUR MANAGER FOR COMPLETION**

## SECTION VI — CLASSIFICATION, PREVENTION RECOMMENDATIONS — To be completed by supervisors' manager

(18) Select the appropriate **Injury / Illness Classification Code**
- 0 First aid case, treated at plant—Form 2020 not required by Corporate Safety.
- 1 Near miss-no injury but serious potential for fatality or disability.
- 2 Injury requiring first aid, with outside professional attention - Workmen's Compensation Claim.
- 3 Medical treatment without restricted or lost workdays but OSHA recordable
- 4 Restricted and / or lost work day case.
- 5 Partial or total permanent disability
- 6 Fatality

Estimate number of **Days Away From Work;** number of **Days of Restricted Duty;** number of **Hospital Days.** If "none", leave blank.

(19) Indicate if permanent non-disciplinary T **Transfer** or D **Dismissal** (Termination) is required.
If neither, leave blank, **Give Group, Division, Department of injured / ill person.**

(20) **Was accident, injury, illness preventable?** Explain.

(21) Circle all realistic areas for corrective **Action** (By **Management**) to prevent recurrence. Circle all realistic areas for corrective **Action** (By **Worker**) to prevent recurrence.

(22) Give **Recommendations** to prevent similar problems in the future.

(23) Assign specific **Tasks, Responsibilities,** and **Completion Dates** for implementation of recommendations.

(24) Obtain **Management Approvals** in accordance with local requirements.

**Figure 6-4.** (Continued.)

## SECTION IV — ACTIVITY INFORMATION

PAGE 5

(9) WHAT WAS PERSON DOING? (e.g. bolting a flange)

30A

(10) A [ ] How often has person done this activity?* — B [ ] Was activity a normal part of job? (Y or N) — C [ ] Was person adequately trained? (Y or N)

(11) D [ ] Do standard methods / procedures exist? (Y or N) If Yes, → E [ ] Were they followed? (Y or N) If No, → F [ ] Did something discourage following procedures. (Y or N)

(12) IF PROCEDURES WERE NOT FOLLOWED, WHAT WAS DONE DIFFERENTLY?

## SECTION V — INJURY / ILLNESS INFORMATION

(13) G [ ] Was person using it? (Y or N) — EQUIPMENT TO PROTECT AGAINST THIS INJURY — Was Person supposed to be using it? (Y or N) H [ ] — TYPE REQUIRED — TYPE USED

(14) I [ ][ ][ ] NATURE OF INJURY * ILLNESS CODE — J [ ][ ][ ] — DESCRIPTION

(15) K [ ][ ][ ] PART OF BODY CODE * — [ ][ ][ ]

MOST SERIOUS — SECOND MOST SERIOUS

(16) TREATMENT PROVIDED

(17) SUPERVISOR: PRINT NAME AND SIGN — WORK TELEPHONE NO. ( ) — — DATE OF REPORT

**STOP — FORWARD TO YOUR MANAGER FOR COMPLETION**

## SECTION VI — CLASSIFICATION, PREVENTION, RECOMMENDATIONS — To be completed by Supervisor's Manager

(18) M [ ] INJURY / ILLNESS CLASSIFICATION CODE * — N [ ] DAYS AWAY FROM WORK — O [ ] DAYS OF RESTRICTED DUTY — P [ ] HOSPITAL DAYS

(19) Q [ ] (T) TRANSFER OR (D) DISMISSAL — GROUP, DIVISION, DEPARTMENT OF INJURED / ILL

(20) WAS ACCIDENT, INJURY, ILLNESS PREVENTABLE? EXPLAIN

(21) CIRCLE ALL REALISTIC ACTIONS MANAGEMENT AND WORKERS CAN TAKE TO PREVENT RECURRENCE

| MANAGEMENT (CIRCLE) | WORKER ACTIONS (CIRCLE) |
|---|---|
| A. Emergency Procedures Training Equipment | A. Follow Instructions, Work Permits |
| B. Facilities, Lighting, Ventilation | B. Follow Safe Work Practices |
| C. Guarding, Safety Devices | C. Follow Training Program Directives |
| D. Housekeeping | D. Operate Tools / Equipment Properly, Safely |
| E. Maintenance | E. Secure, Shut-off, Disconnect Systems |
| F. Methods Safety Work Practice | F. Stop Horseplay With Others |
| G. Personal Protective Equipment | G. Stop Recklessness, Inattentiveness |
| H. Process Engineering, Hazard Analysis | H. Stop Unauthorized Work |
| I. Staffing-Quality, Quantity | I. Use Common Sense, Good Judgement |
| J. Supervision | J. Use Guards, Safety Equipment Properly |
| K. Training | K. Wear Personal Protective Equipment |
| L. Working Conditions - Hours, Etc. | L. Wear Proper Clothing |

(22) DETAILED RECOMMENDATIONS TO PREVENT RECURRENCE

(23) TASK — PERSON REPONSIBLE — COMPLETION DATE

(24) LOCAL APPROVALS — FACILITY MANAGER

* SEE INSTRUCTIONS FOR CODES

**Figure 6-4.** (Continued.)

# THE BIG PICTURE
## Supervisor's Accident Investigation Checklist

PAGE 6

A. CONTROL THE ACCIDENT SITUATION - PEOPLE ARE THE FIRST PRIORITY

- ☐ Send For Help - Notify Management.
- ☐ "Safe" The Area and Administer First-aid.

**To Stop Ongoing Hazards To Rescue Personnel you may have to...**

- ☐ Shut off electrical power
- ☐ Bleed or isolate pressurized systems
- ☐ Block mechanical equipment - prevent movement
- ☐ Check air quality
- ☐ Issue personal protective equipment
- ☐ Provide emergency lighting, power, air, etc.

**Secure the Scene and Protect Evidence**

- ☐ Rope off area or Station a guard
- ☐ Issue tagouts, lockouts, permits

B. COLLECT EVIDENCE

**Identify Transient Evidence - Make notes, take pictures or provide sketches of the following...**

- ☐ Positions of tools, equipment, layout, etc.
- ☐ Note air quality, things that evaporate or melt, etc.
- ☐ Tire tracks, foot prints, loose material on floor, etc.
- ☐ Collect operating logs, charts, records
- ☐ Identification numbers of the equipment and maintenance records

Note: Put dimensions on all sketches, sign and date all photos

**Note General Conditions - Yes or No (Y or N) - Did the following factors contribute to the accident?**

- ☐ Housekeeping
- ☐ Work Environment or Layout
- ☐ Floor or Surface Condition
- ☐ Lighting or Visibility
- ☐ Noise or Distractions
- ☐ Air Quality, Temperature or Weather
- ☐ Equipment Condition or Malfunction History
- ☐ Training, Experience or Supervision
- ☐ Periodic Rule or Procedure Violations
- ☐ Employee Morale or Attitude
- ☐ Health or Safety Record
- ☐ Alcohol or Drug Abuse

C. GET THINGS BACK TO NORMAL

D. INTERVIEW WITNESSES

DO...

- Interview as soon as possible
- Interview at the accident scene
- Take notes or use a tape recorder
- Put the witness at ease
- Ask open ended questions
- Repeat the story back to the witness
- End the interview on a positive note

DON'T...

- Pressure the witness
- Blame the witness for the accident
- Interrupt an answer
- Ask questions that can be answered "yes or no"
- Ask "why" questions and "opinion" questions first

ALWAYS...

- Stress that you only want the facts
- Stress that you want to prevent the next accident
- Take the extra time to get understanding

E. ANALYSIS

- Write down the accident story
- List the facts (parts of the story) which are in dispute
- Compare the facts and dispute with the physical evidence to establish the best answer
- Finalize the story and identify accident causes with your Manager

F. REPORT

- Form 2020 for each person injured
- Form 2021 for accidental property loss in excess of $10,000 - 1 per occurrence
- Form 3175 for vehicle accidents - 1 per occurrence

**Figure 6-4.** (Concluded.)

INJURY AND ILLNESS RECORD OF EMPLOYEE

S. S. Jones — 845
(Name) (Employee Number)

Occupation Packer Department Shipping Date Employed 1-23-69

| Case Number | Injury or Illness | Date of Occurrence | Z16 Type (Fatal, Permanent, Temporary, Non-disabling) | Z16 Days Charged | Comp. and Other Costs | OSHA Type (Fatal, Lost Workday, Non-Lost Workday) | OSHA Lost Workdays |
|---|---|---|---|---|---|---|---|
| 164 | Inj. | 2-12-70 | Non-disab. | 0 | 0 | | |
| 349 | Inj. | 3-16-73 | Temporary | 3 | 0 | Lost workday | 1 |
| 766 | Ill. | 8- 3-79 | Temporary | 8 | 0 | Lost workday | 6 |
| | | | | | | | |
| | | | | | | | |
| | | | | | | | |
| | | | | | | | |
| | | | | | | | |

(Reverse side may be used for remarks)

Issued by NATIONAL SAFETY COUNCIL, 444 N. Michigan Ave., Chicago, Ill. 60611

IS3 Rev. 10M17499' Printed in U.S.A. Stock No. 129.23

**Figure 6-5.** Injury and Illness Record of Employee is a 4 × 6-in. (10 × 15-cm) card for recording injuries.

The first aid attendant or the nurse should know enough about accident analysis and investigation to be able to record the principal facts about each case. The company's occupational physician also should be informed of the basic rules for classifying cases since, at times, his opinion of the seriousness of an injury may be necessary to record the case accurately.

**Accident investigation report.** It is recommended that the supervisor make a detailed report about each accident, even when only a minor injury or no injury is the result. For purposes of OSHA or ANSI Z16.1 summaries, only those reports that meet the minimum severity level can be separated and tallied. Minor injuries occur in greater numbers than serious injuries and records of these injuries can be helpful in pinpointing problem areas. By working to alleviate these problems, serious injuries can sometimes be prevented. Furthermore, complications may arise out of the less serious injuries and their end result may be quite serious.

The supervisor's accident investigation report form should be completed as soon as possible after an accident occurs. Copies of these reports should be sent to the safety department and to other designated persons. Information concerning activities and conditions that preceded an occurrence is important in the prevention of future accidents. This information is particularly difficult to get unless it is obtained promptly after the accident occurs.

Generally, analyses of accidents are made only periodically, and often long after the accidents have occurred. Because it is often impossible to accurately recall the details of an accident, if details are not recorded accurately and completely at once, they may be lost forever.

Because all information may not be available at the time that the accident report is being filled out, items such as total time lost and dollar amount can be added later. This should not, however, prevent the other items from being answered as soon as possible after the accident occurs.

Three different supervisor's report forms are presented. They fulfill all of the information requirements of the present OSHA 101 form. The forms also include questions in addition to those contained in the present OSHA 101 form (see Figure 6-8). These questions ask for additional basic data that should be known about each accident.

The first supervisor's report form (Figure 6-2) is an open end, mostly narrative type of form. The next forms (Figures 6-3 and -4) are self-coding to allow key punching of data items directly from the form without the extra step of recoding this information for data processing equipment. By using self-coding forms, data processing equipment can easily be used to process the information and produce a variety of summary reports (such as summaries by department and by type of accident). Detailed cross tabulations thus can be produced with little effort.

For definitions of the terms regarding severity of injury, what cases are recordable, and the like, please consult the OSHA and Z16.1 sections of this chapter. For further clarification of OSHA definitions, see the guidelines section of this chapter. You also may want to consult federal or state authorities (if your state has an approved plan in effect).

**Injury and illness record of employee.** After cases are closed, the first aid report and the supervisor's report are filed by agency of injury (type of machine, tool, material, etc.), type of accident, or other factor that will facilitate use of the reports for accident prevention. Another form, therefore, must be used to record the injury experience of individual employees. (See Figure 6-5.)

This form helps supervisors remember the experience of individual employees. Particularly in large establishments or plants

**MONTHLY SUMMARY OF INJURIES AND ILLNESSES, 19___**

Company *ABC Mfg. Co.* Plant *Dayton, Ohio* Department *All*

| Period | Average Number of Employees | Number of Man-Hours Worked | Z16.1 CASES: Disabling Injuries and Illnesses: Fatal, Permanent Total | Permanent Partial | Temporary Total | Total Z16 Cases | Frequency Rate* | Time Charges: Fatal, Permanent Total | Permanent Partial | Temporary Total | Total Z16 Charges | Severity Rate* | COSTS (Compensation, Other) | OSHA CASES: Fatals | Lost Workday Cases | Non Lost Workday Cases | Total OSHA Cases | Incidence Rate† | Total Lost Workdays | FIRST AID CASES ONLY |
|---|---|---|---|---|---|---|---|---|---|---|---|---|---|---|---|---|---|---|---|---|
| Jan. | 2,060 | 345,000 | 0 | 1 | 1 | 2 | 5.80 | 0 | 150 | 18 | 168 | 487 | $ 284.50 | 0 | 3 | 18 | 21 | 12.2 | 42 | 20 |
| Feb. | 2,010 | 298,000 | 0 | 0 | 3 | 3 | 10.07 | 0 | 0 | 22 | 22 | 74 | 42.65 | 0 | 4 | 27 | 31 | 20.8 | 34 | 36 |
| Cum. | | 643,000 | 0 | 1 | 4 | 5 | 7.78 | 0 | 150 | 40 | 190 | 295 | 327.15 | 0 | 7 | 45 | 52 | 16.2 | 76 | 56 |
| Mar. | 2,080 | 353,000 | 0 | 0 | 1 | 1 | 2.83 | 0 | 0 | 42 | 42 | 119 | 77.82 | 0 | 1 | 9 | 10 | 5.7 | 12 | 10 |
| Cum. | | 996,000 | 0 | 1 | 5 | 6 | 6.02 | 0 | 150 | 78 ~~82~~ | 228 ~~232~~ | 229 | 404.97 | 0 | 8 | 54 | 62 | 12.4 | 88 | 66 |
| Apr. | 2,000 | 332,000 | 0 | 0 | 4 | 4 | 12.05 | 0 | 0 | 47 | 47 | 142 | 92.64 | 0 | 5 | 35 | 40 | 24.1 | 64 | 42 |
| Cum. | | 1,328,000 | 0 | 1 | 9 | 10 | 7.53 | 0 | 150 | 125 | 275 | 207 | 497.61 | 0 | 13 | 89 | 102 | 15.4 | 152 | 108 |
| May | 2,150 | 375,000 | 0 | 0 | 5 | 5 | 13.33 | 0 | 0 | 63 | 63 | 168 | 123.24 | 0 | 7 | 45 | 52 | 27.7 | 88 | 55 |
| Cum. | | 1,703,000 | 0 | 1 | 14 | 15 | 8.81 | 0 | 150 | 203 ~~188~~ | 353 ~~338~~ | 207 | 620.85 | 0 | 20 | 134 | 154 | 18.1 | 240 | 163 |
| June | 1,900 | 303,000 | 0 | 1 | 0 | 1 | 3.30 | 0 | 1,000 | 0 | 1,000 | 3300 | 985.56 | 0 | 1 | 13 | 14 | 9.2 | 25 | 15 |
| Cum. | | 2,006,000 | 0 | 2 | 14 | 16 | 7.98 | 0 | 1,150 | 203 | 1,353 | 674 | 1,606.41 | 0 | 21 | 147 | 168 | 16.7 | 265 | 178 |
| July | 1,825 | 295,000 | 0 | 1 | 3 | 4 | 13.56 | 0 | 250 | 23 | 273 | 925 | 368.18 | 0 | 5 | 40 | 45 | 30.5 | 91 | 51 |
| Cum. | | 2,301,000 | 0 | 3 | 17 | 20 | 8.69 | 0 | 1,400 | 226 | 1,626 | 707 | 1,974.59 | 0 | 26 | 187 | 213 | 18.5 | 356 | 229 |
| Aug. | 1,800 | 285,000 | 0 | 0 | 4 | 4 | 14.04 | 0 | 0 | 31 | 31 | 109 | 63.24 | 0 | 5 | 43 | 48 | 33.7 | 123 | 53 |
| Cum. | | 2,586,000 | 0 | 3 | 21 | 24 | 9.28 | 0 | 1,400 | 257 | 1,657 | 641 | 2,037.83 | 0 | 31 | 230 | 261 | 20.2 | 479 | 282 |
| Sept. | 1,875 | 301,000 | 0 | 0 | 0 | 0 | 0.00 | 0 | 0 | 0 | 0 | 0 | 843.66 | 0 | 0 | 9 | 9 | 6.0 | 0 | 12 |
| Cum. | | 2,887,000 | 0 | 4 ~~3~~ | 20 ~~21~~ | 24 | 8.31 | 0 | 2,300 ~~1,400~~ | 215 ~~257~~ | 2,515 ~~1,657~~ | 871 | 2,881.49 | 0 | 31 | 239 | 270 | 18.7 | 479 | 294 |
| Oct. | 1,795 | 302,000 | 0 | 0 | 1 | 1 | 3.31 | 0 | 0 | 14 | 14 | 46 | 45.60 | 0 | 1 | 5 | 6 | 4.0 | 10 | 6 |
| Cum. | | 3,189,000 | 0 | 4 | 21 | 25 | 7.84 | 0 | 2,300 | 229 | 2,529 | 793 | 2,927.09 | 0 | 32 | 244 | 276 | 17.3 | 489 | 300 |
| Nov. | 1,665 | 280,000 | 0 | 0 | 2 | 2 | 7.14 | 0 | 0 | 17 | 17 | 61 | 36.76 | 0 | 2 | 23 | 25 | 17.9 | 13 | 26 |
| Cum. | | 3,469,000 | 0 | 4 | 23 | 27 | 7.78 | 0 | 2,300 | 246 | 2,546 | 734 | 2,963.85 | 0 | 34 | 267 | 301 | 17.4 | 502 | 326 |
| Dec. | 1,620 | 275,000 | 1 | 0 | 0 | 1 | 3.64 | 6,000 | 0 | 0 | 6,000 | 21,818 | 6,785.25 | 1 | 1 | 7 | 9 | 6.5 | 2 | 8 |
| YEAR | | 3,744,000 | 1 | 4 | 23 | 28 | 7.48 | 6,000 | 2,300 | 246 | 8,546 | 2,283 | 9,749.10 | 1 | 35 | 274 | 310 | 16.6 | 504 | 334 |

*Z16 Rates: Frequency rate is the total number of Z16 cases per 1,000,000 man-hours worked. Severity rate is the total Z16 charges per 1,000,000 man-hours worked.
†OSHA Rate: Incidence rate is the total number of OSHA cases per 200,000 man-hours worked.

Issued by NATIONAL SAFETY COUNCIL, 444 N. Michigan Ave., Chicago, Illinois 60611

**Figure 6-6.** Results of the safety program can be gauged from data on this Monthly Summary of Injuries and Illnesses form (8½ × 14 in.). Rates computed for the month, year to date (cumulative), and year permit comparisons between time periods, and departments, plants, and companies during the same period. Changes in the classification of injuries and other adjustments can be easily made.

where supervisors have many people working for them, they probably cannot remember the total number of injuries—especially if the injuries are minor—that are suffered by individual employees.

The employee injury card, therefore, fills a real need. It has space for recording such factors about the injury as date, classification, days charged, costs, and OSHA lost workdays.

Because of the importance of the personal factor in accidents, much can be learned about accident causes from studying employee injury records. If certain employees or job classifications have frequent injuries, a study of employee working habits, physical and mental abilities, training, job assignments, working environment, and the instructions and supervision given them may reveal as much as a study of accident locations, agencies, or other factors.

**Filing reports.** After injury reports have been used to compile monthly summaries, the incomplete reports can be kept in a temporary file for convenient reference as later information about the injuries becomes available.

After the injury reports are complete, they should be filed in a way that will permit ready use for making special studies of accident conditions. To facilitate this work, the reports can be filed by agency of the injury, by occupation of the injured person, by department, or by some similar item.

The employee injury card should be cross referenced to the file location of the detailed accident report.

## Periodic reports

The forms discussed in the preceding paragraphs are prepared when the accidents occur; they are used to record the accidents and preserve information about contributing circumstances. Periodically, this information should be summarized and related to department or plant exposure so the safety work can be evaluated and the principal accident causes brought into proper focus.

**Monthly summary of injuries and illnesses.** A summary of injuries and illnesses should be prepared monthly to reveal the current status of accident experience. This monthly summary of injury and illness cases (Figure 6-6) allows for tabulating monthly and cumulative totals and the computation of OSHA incidence rates as well as ANSI Z16.1 frequency and severity rates (for those organizations still using that standard). Space is also provided for yearly totals and rates. This form would be filled out on the basis of the individual report forms that were processed during the month or from form OSHA No. 200, Log and Summary of Occupational Injuries and Illnesses. (See Figure 6-7.)

The monthly summary should be prepared as soon after the end of each month as the information becomes available, but not later than the 20th of the following month. Because this report is primarily prepared to reveal the current status of accident experience, it is essential that this information be determined as soon as possible.

If an accident report is still incomplete on the 20th of the following month because the employee has not returned to work,

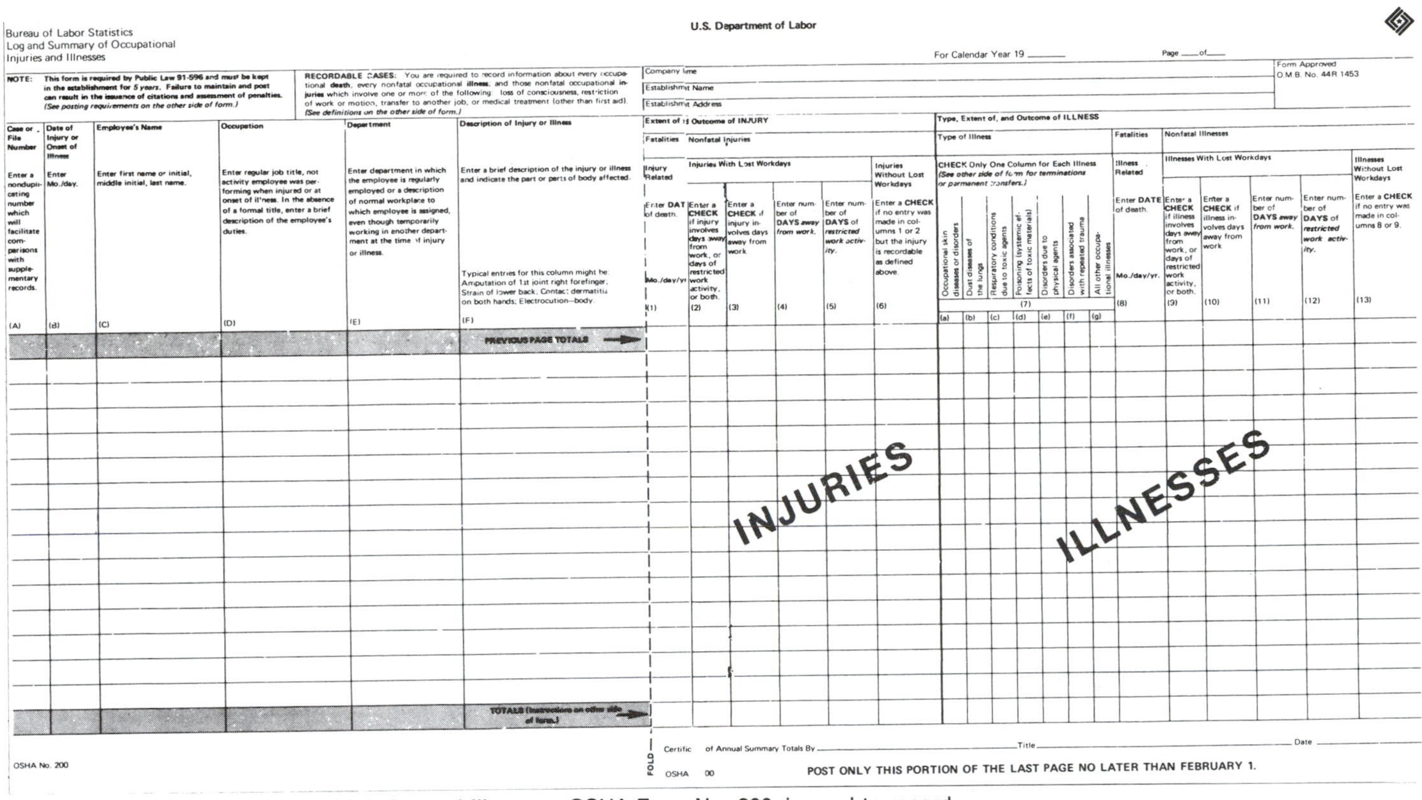

Bureau of Labor Statistics
Log and Summary of Occupational
Injuries and Illnesses

U.S. Department of Labor

For Calendar Year 19 ______ Page ___ of ___

Form Approved
O.M.B. No. 44R 1453

NOTE: This form is required by Public Law 91-596 and must be kept in the establishment for *5 years*. Failure to maintain and post can result in the issuance of citations and assessment of penalties. *(See posting requirements on the other side of form.)*

RECORDABLE CASES: You are required to record information about every occupational **death**; every nonfatal occupational **illness**; and those nonfatal occupational **injuries** which involve one or more of the following: loss of consciousness, restriction of work or motion, transfer to another job, or medical treatment (other than first aid). *(See definitions on the other side of form.)*

Company Name
Establishment Name
Establishment Address

Case or File Number — Enter a nonduplicating number which will facilitate comparisons with supplementary records. (A)

Date of Injury or Onset of Illness — Enter Mo./day. (B)

Employee's Name — Enter first name or initial, middle initial, last name. (C)

Occupation — Enter regular job title, not activity employee was performing when injured or at onset of illness. In the absence of a formal title, enter a brief description of the employee's duties. (D)

Department — Enter department in which the employee is regularly employed or a description of normal workplace to which employee is assigned, even though temporarily working in another department at the time of injury or illness. (E)

Description of Injury or Illness — Enter a brief description of the injury or illness and indicate the part or parts of body affected. Typical entries for this column might be: Amputation of 1st joint right forefinger; Strain of lower back; Contact dermatitis on both hands; Electrocution—body. (F)

Extent of and Outcome of INJURY

Fatalities — Injury Related — Enter DATE of death. Mo./day/yr. (1)

Nonfatal Injuries — Injuries With Lost Workdays — Enter a CHECK if injury involves days away from work, or days of restricted work activity, or both. (2) — Enter a CHECK if injury involves days away from work (3) — Enter number of DAYS away from work. (4) — Enter number of DAYS of restricted work activity. (5)

Injuries Without Lost Workdays — Enter a CHECK if no entry was made in columns 1 or 2 but the injury is recordable as defined above. (6)

Type, Extent of, and Outcome of ILLNESS

Type of Illness — CHECK Only One Column for Each Illness *(See other side of form for terminations or permanent transfers.)* (7)
(a) Occupational skin diseases or disorders
(b) Dust diseases of the lungs
(c) Respiratory conditions due to toxic agents
(d) Poisoning (systemic effects of toxic materials)
(e) Disorders due to physical agents
(f) Disorders associated with repeated trauma
(g) All other occupational illnesses

Fatalities — Illness Related — Enter DATE of death. Mo./day/yr. (8)

Nonfatal Illnesses — Illnesses With Lost Workdays — Enter a CHECK if illness involves days away from work, or days of restricted work activity, or both. (9) — Enter a CHECK if illness involves days away from work (10) — Enter number of DAYS away from work. (11) — Enter number of DAYS of restricted work activity. (12)

Illnesses Without Lost Workdays — Enter a CHECK if no entry was made in columns 8 or 9. (13)

PREVIOUS PAGE TOTALS →

INJURIES

ILLNESSES

TOTALS (Instructions on other side of form.) →

Certific... of Annual Summary Totals By ______ Title ______ Date ______

OSHA No. 200

POST ONLY THIS PORTION OF THE LAST PAGE NO LATER THAN FEBRUARY 1.

**Figure 6-7.** Log of Occupational Injuries and Illnesses, OSHA Form No. 200, is used to record injuries or illnesses that result in fatalities, lost workdays, require medical treatment, involve loss of consciousness, or restrict work or motion. Complete instructions for using the Log are on its reverse side (not shown here). The form measures 10½ × 20 in. (28 × 49 cm). Because forms are subject to change, be sure to ask your OSHA Area Office for the latest forms.

or if the classification of an injury is still in doubt at that time, an estimate of the outcome should be made by the company physician and the report included with the completed cases in the monthly summary.

When definite information becomes available for estimated cases, any change in classification or OSHA lost workdays should be entered in the appropriate columns in the month of closing of the case, and the adjustment included in the cumulative figures for the year through that month. This procedure provides reliable monthly data and an easy method of adjusting cumulative data.

**Annual report.** Every establishment subject to the OSHAct is obliged to post its annual report on February 1st. The cumulative totals on form OSHA No. 200, Log and Summary of Occupational Injuries and Illnesses, serve as the annual report. Those establishments designated as part of the annual Bureau of Labor Statistics sample also must send a copy of this report to Washington.

For management purposes, however, the annual report fulfills a more direct function. Whereas monthly summaries of injuries and illnesses are primarily prepared to show the trend of safety performance during the year, annual reports are prepared so comparisons for the longer period can be made with the experience of previous years, and with the experience of similar organizations and of the industry as a whole.

Especially in smaller companies, monthly injury rates often show wide variations that make it difficult to correctly evaluate safety performance. If a small company has only two or three injuries in a year, in the months in which these injuries occur, the rates will jump to extreme highs; but in the other months, the rates will be zero. These variations will be smoothed out in annual totals, however, and the rate for the longer period will have increased significance.

Annual reports should be prepared as soon after the close of the year as information becomes available, but again not later than the twentieth of the month following. If any injury report is still incomplete twenty days after the close of the year because the employee has not returned to work, or if the classification of any injury is in doubt at that time, an estimate of the outcome should be made by the company physician and the report included with the completed cases.

When annual records are closed, the estimated charges or OSHA lost workdays on cases still pending become final for that year. Any injuries or time charges reported later are considered only in rates for two or three years of which that year is a part. They need not be added to the record of that year, and they should *not* be added to the record of the succeeding year.

If time charges or OSHA lost workdays that occurred in one year were included in the record for the next year, the latter year's experience would be biased and the record would fail to give a true picture of the severity of injuries or OSHA lost workdays.

On the other hand, absolute accuracy in rates does not justify delaying the annual report more than twenty days after the end of the year in order to get exact time charges or OSHA lost workdays, or to make sure that there are no delayed cases. It

OSHA No. 101
Case or File No. 1002

Form approved
OMB No. 44R 1453

**Supplementary Record of Occupational Injuries and Illnesses**

EMPLOYER

1. Name ABC Manufacturing Co., Inc.
2. Mail address P.O. Box 123, Chicago, Illinois 60007
(No. and street) (City or town) (State)
3. Location, if different from mail address Aspinwall, Iowa 51432

INJURED OR ILL EMPLOYEE

4. Name Joseph E. Malcolm Social Security No. 807-42-0123
(First name) (Middle name) (Last name)
5. Home address 609 Eastern Blvd., Aspinwall, Iowa.
(No. and street) (City or town) (State)
6. Age 27 7. Sex: Male ✓ Female ______ (Check one)
8. Occupation Laboratory Technician
(Enter regular job title, *not* the specific activity he was performing at time of injury.)
9. Department Research
(Enter name of department or division in which the injured person is regularly employed, even though he may have been temporarily working in another department at the time of injury.)

THE ACCIDENT OR EXPOSURE TO OCCUPATIONAL ILLNESS

10. Place of accident or exposure RRI, Aspinwall, Iowa
(No. and street) (City or town) (State)
If accident or exposure occurred on employer's premises, give address of plant or establishment in which it occurred. Do not indicate department or division within the plant or establishment. If accident occurred outside employer's premises at an identifiable address, give that address. If it occurred on a public highway or at any other place which cannot be identified by number and street, please provide place references locating the place of injury as accurately as possible.
11. Was place of accident or exposure on employer's premises? Yes (Yes or No)
12. What was the employee doing when injured? Running Laboratory Tests
(Be specific. If he was using tools or equipment or handling material, name them and tell what he was doing with them.)
13. How did the accident occur? While using glacial acetic acid in laboratory test, employee dropped container which broke and acid splashed all over his hands.
(Describe fully the events which resulted in the injury or occupational illness. Tell what happened and how it happened. Name any objects or substances involved and tell how they were involved. Give full details on all factors which led or contributed to the accident. Use separate sheet for additional space.)

OCCUPATIONAL INJURY OR OCCUPATIONAL ILLNESS

14. Describe the injury or illness in detail and indicate the part of body affected. Acid burns, both hands.
(e.g.: amputation of right index finger at second joint; fracture of ribs; lead poisoning; dermatitis of left hand, etc.)
15. Name the object or substance which directly injured the employee. (For example, the machine or thing he struck against or which struck him; the vapor or poison he inhaled or swallowed; the chemical or radiation which irritated his skin; or in cases of strains, hernias, etc., the thing he was lifting, pulling, etc.)
Glacial acetic acid
16. Date of injury or initial diagnosis of occupational illness 7-10-19-
(Date)
17. Did employee die? No (Yes or No)

OTHER

18. Name and address of physician Dr. Robert Smith - Aspinwall, Iowa
19. If hospitalized, name and address of hospital Good Samaritan Hospital Aspinwall, Iowa

Date of report 7-12-19- Prepared by Morgan Jones
Official position Bookkeeper

**Figure 6-8.** Supplementary Record of Occupational Injuries and Illnesses, OSHA Form No. 101, gives details of each recordable occupational injury or illness. Records must be available for examination by representatives of the U.S. Department of Labor and the Department of Health and Human Services and states accorded jurisdiction under the OSHAct. Records must be kept at least five years following the calendar year to which they relate.

is better to omit such delayed cases from the rates entirely, unless a two- or three-year rate is calculated, or unless the company wishes to make revisions for its own information.

## Use of reports

**Reports to management.** Management is increasingly interested in the accident experience of its companies. Therefore, monthly and other periodic summary reports which show the results of the safety program should be furnished to the responsible executive. Such reports do not need to contain details or technical language and can be supplemented by simple charts or graphs to show the recent accident experience in relation to that of the preceding period and that of other companies in the industry.

In a large company, departmental data help the executive visualize accident experience in various plant operations and provide a yardstick for better evaluation of progress made in the elimination of accidents. If cost figures are obtained, comparisons of such figures for different periods are of particular interest.

**Bulletins to supervisors.** A supervisor is primarily interested in his own department and workers. One of the most effective ways to create and maintain the interest of supervisors in accident prevention is to keep them informed about the accident records of their departments. Department injury rates based on sufficient amounts of exposure reflect the effectiveness of the supervisors' safety activities.

Because interest increases with knowledge, bulletins containing analyses of the principal causes of accidents in each department not only will maintain the supervisors' interest at a high level, but will provide the supervisors with the type of information that will help them to effect further reductions in injuries.

The agenda for employee safety meetings should particularly include information about the outstanding injury and illness problems, frequent unsafe practices, hazardous types of equipment, and similar data disclosed by analysis of the accidents that have occurred in the department and plant.

**Bulletin board publicity.** Posting a variety of materials on bulletin boards is one of the best ways to maintain the interest of employees in safety. Accident records furnish many items, such as the following:

No-injury records
Unusual accidents
Frequent causes of accidents
Charts showing reductions in accidents
Simple tables comparing departmental records
Standings in contests

**Reports to National Safety Council.** The Council sponsors two recognition programs for its members requiring periodic reports of their occupational injury and illness experience.

The *Award Plan for Recognizing Good Occupational Safety Records,* open to all employer members, is a noncompetitive evaluation of a company's annual experience. Awards are given if the records meet predetermined criteria.

The annual reports are tabulated to determine injury and illness incidence rates by industry. An annual "Work Injury and Illness Rates" pamphlet containing these incidence rates by industry is published and distributed to members, so each company can compare its experience with the average experience of other companies in the same industry. These rates (along with data from the Bureau of Labor Statistics, state and national vital statistics authorities, state compensation authorities, and other federal, state, and private agencies) are shown in *Accident Facts,* the Council's annual statistical publication.

National Safety Council industrial employer members may also compete in more than 20 employee safety contests administered by the Council. Monthly bulletins are sent to contestants so that they can compare their experience with other competitors. Awards are made to the leaders of the individual contests at the end of each contest year.

### The concept of bilevel reporting

As mentioned earlier in this chapter, each company has accident problems that are different from those of establishments in other industries and, in many cases, different from those of other establishments in the same industry. No individual form or set of forms can possibly include all of the information necessary to fully investigate the causes of all accidents. With this in mind and because very long forms are rarely completed accurately and are, very often, received with much resistance by those persons who must fill them out, the concept of bilevel reporting has arisen.

The basic idea of bilevel reporting is that in addition to the general information contained in the standard report form (such as the Supervisor's Accident Report Form), further facts are necessary concerning specific types of accidents. To obtain this additional data, a supplementary form, containing a few specific questions about the accident type under investigation, is prepared and made available. This supplementary form is then filled out and attached to the regular report—only for those accidents about which the investigator is trying to obtain in-depth data.

When a sufficient number of the bilevel forms have been collected, the supplementary form is discontinued and the results can be analyzed. In this way, a minimum of time needs to be spent by the persons who fill out the forms in order to obtain useful information. Several bilevel forms, each a different color for easy handling, can be used at any one time. They can be discontinued when their job is done and replaced by other supplementary forms, while the basic report form remains unchanged and in use.

## OSHA RECORDKEEPING AND ANSI Z16

The OSHA recordkeeping system is mandatory for all establishments subject to the OSHAct. However, not all workplaces come under this Act and there are some establishments still using the ANSI Z16.1 standard for their management reports. Some do so to maintain continuity with past records of accident experience and others find parts of the standard valuable since it provides information not specified in the OSHA system. Many organizations want a separate record of permanent impairments, for example. In addition, many safety people feel that the Z16.1 severity rate provides a better measurement of severity than the lost workday incidence rate.

The National Safety Council changed from the ANSI Z16.1 standard to the OSHA Recordkeeping system, beginning January 1, 1977, for all its contests and awards. However, for the benefit of those organizations still using all or part of the Z16.1, the basic elements of the standard have been included in this chapter.

It should also be noted that in July 1977, a new standard Z16.4 was approved by the American National Standards Institute. This standard conforms to the OSHA Recordkeeping system but it provides an alternate means to compute the Z16.1 severity rate.

## A GUIDE TO OSHA RECORDKEEPING REQUIREMENTS

The information in this section explains the occupational injury and illness recording and reporting requirements of the Occupational Safety and Health Act of 1970 and Title 29 of the *Code of Federal Regulations,* Part 1904. It is directly taken from a June 1986 publication of the U.S. Department of Labor, Bureau of Labor Statistics (BLS), titled "A Brief Guide to Recordkeeping Requirements for Occupational Injuries and Illnesses."

The requirements and definitions are subject to change, so safety personnel should contact their regional OSHA or Bureau of Labor Statistics office for the latest information.

In addition to the requirements of 29 CFR 1904, many specific OSHA standards and regulations require maintenance and retention of records of medical surveillance, exposure monitoring, inspections, accidents, and other activities and incidents, and for the reporting of certain information to employees and to OSHA. These additional requirements are not covered in this section.

The remainder of this section is a direct quotation from the BLS guide.

### Employers subject to the recordkeeping requirements of the Occupational Safety and Health Act of 1970

The recordkeeping requirements of the Occupational Safety and Health Act of 1970 apply to private sector employers in all States, the District of Columbia, Puerto Rico, the Virgin Islands, American Samoa, Guam, and the Trust Territories of the Pacific Islands.

**A. Employers who must keep OSHA records.** Employers with 11 or more employees (at any one time in the previous calendar year) in the following industries must keep OSHA records. The industries are identified by name and by the appropriate Standard Industrial Classification (SIC) code:

- Agriculture, forestry, and fishing (SIC's 01-02 and 07-09)
- Oil and gas extraction (SIC's 13 and 1477)
- Construction (SIC's 15-17)
- Manufacturing (SIC's 20-39)
- Transportation and public utilities (SIC's 41-42 and 44-49)
- Wholesale trade (SIC's 50-51)
- Building materials and garden supplies (SIC 52)
- General merchandise and food stores (SIC's 53 and 54)
- Hotels and other lodging places (SIC 70)
- Repair services (SIC's 75 and 76)
- Amusement and recreation services (SIC 79), and
- Health services (SIC 80).

If employers in any of the industries listed above have more than one establishment with combined employment of 11 or more employees, records must be kept for *each* individual establishment.

**B. Employers who infrequently must keep OSHA records.** Employers in the industries listed below are normally exempt from OSHA recordkeeping. However, each year a small rotating sample of these employers is required to keep records and participate in a mandatory statistical survey of occupational injuries and illnesses. Their participation is necessary to produce national estimates of occupational injuries and illnesses for *all* employers (both exempt and nonexempt) in the private sector. If an employer who is regularly exempt is selected to maintain records and participate in the Annual Survey of Occupational Injuries and Illnesses, he or she will be notified in advance and supplied with the necessary forms and instructions. Employers who normally do not have to keep OSHA records include:

1. All employers with no more than 10 full- or part-time employees *at any one time* in the previous calendar year.
2. Employers in the following retail trade, finance, insurance and real estate, and services industries (identified by SIC codes):

- Automotive dealers and gasoline service stations (SIC 55)
- Apparel and accessory stores (SIC 56)
- Furniture, home furnishings, and equipment stores (SIC 57)
- Eating and drinking places (SIC 58)
- Miscellaneous retail (SIC 59)
- Banking (SIC 60)
- Credit agencies other than banks (SIC 61)
- Security, commodity brokers, and services (SIC 62)
- Insurance (SIC 63)
- Insurance agents, brokers, and services (SIC 64)
- Real estate (SIC 65)
- Combined real estate, insurance, etc. (SIC 66)
- Holding and other investment offices (SIC 67)
- Personal services (SIC 72)
- Business services (SIC 73)
- Motion pictures (SIC 78)
- Legal services (SIC 81)
- Educational services (SIC 82)
- Social services (SIC 83)
- Museums, botanical, zoological gardens (SIC 84)
- Membership organizations (SIC 86)
- Private households (SIC 88), and
- Miscellaneous services (SIC 89)

Even though recordkeeping requirements are reduced for employers in these industries, they, like nonexempt employers, must comply with OSHA standards, display the OSHA poster, and report to OSHA within 48 hours any accident which results in one or more fatalities or the hospitalization of five or more employees. Also, some State safety and health laws may require regularly exempt employers to keep injury and illness records, and some States have more stringent catastrophic reporting requirements.

**C. Employers and individuals who never keep OSHA records.** The following employers and individuals do not have to keep OSHA injury and illness records:

- *Self-employed individuals;*
- *Partners with no employees;*
- *Employers of domestics* in the employers' private residence for the purpose of housekeeping or child care, or both; and
- *Employers engaged in religious activities* concerning the conduct of religious services or rites. Employees engaged in such activities include clergy, choir members, organists and other musicians, ushers, and the like. However, records of injuries and illnesses occurring to employees while performing secular activities must be kept. Recordkeeping is also required for employees of private hospitals and certain commercial establishments owned or operated by religious organizations.

State and local government agencies are usually exempt from OSHA recordkeeping. However, in certain States, agencies of State and local governments are required to keep injury and illness records in accordance with State regulations.

**D. Employers subject to other Federal safety and health regulations.** Employers subject to injury and illness recordkeeping requirements of other Federal safety and health regulations are not exempt from OSHA recordkeeping. However, records used to comply with other Federal recordkeeping obligations may also be used to satisfy the OSHA recordkeeping requirements. The forms used must be equivalent to the log and summary (OSHA No. 200) and the supplementary record (OSHA No. 101).

## OSHA recordkeeping forms

Only two forms are used for OSHA recordkeeping. One form, the OSHA No. 200 [shown in Figure 6-7], serves as both the Log of Occupational Injuries and Illnesses, on which the occurrence and extent of cases are recorded during the year; and as the Summary of Occupational Injuries and Illnesses, which is used to summarize the log at the end of the year to satisfy employer posting obligations. The other form, the Supplementary Record of Occupational Injuries and Illnesses, OSHA No. 101 [shown in Figure 6-8], provides additional information on each of the cases that have been recorded on the log.

**A. The Log and Summary of Occupational Injuries and Illnesses, OSHA No. 200.**
The log is used for recording and classifying occupational injuries and illnesses, and for noting the extent of each case. The log shows when the occupational injury or illness occurred, to whom, the regular job of the injured or ill person at the time of the injury or illness exposure, the department in which the person was employed, the kind of injury or illness, how much time was lost, whether the case resulted in a fatality, etc. The log consists of three parts: A descriptive section which identifies the employee and briefly describes the injury or illness; a section covering the extent of the injuries recorded; and a section on the type and extent of illnesses.

Usually, the OSHA No. 200 form is used by employers as their record of occupational injuries and illnesses. However, a private form equivalent to the log, such as a computer printout, may be used if it contains the same detail as the OSHA No. 200 and is as readable and comprehensible as the OSHA No. 200 to a person not familiar with the equivalent form. It is important that the columns of the equivalent form have the same identifying number as the corresponding columns of the OSHA No. 200 because the instructions for completing the survey of occupational injuries and illnesses refer to log columns by number. It is advisable that employers have private equivalents of the log form reviewed by BLS to insure compliance with the regulations.

The portion of the OSHA No. 200 to the right of the dotted vertical line is used to summarize injuries and illnesses in an establishment for the calendar year. Every nonexempt employer who is required to keep OSHA records must prepare an annual summary for each establishment based on the information contained in the log for each establishment. The summary is prepared by totaling the column entries on the log (or its equivalent) and signing and dating the certification portion of the form at the bottom of the page.

**B. The Supplementary Record of Occupational Injuries and Illnesses, OSHA No. 101.**
For every injury or illness entered on the log, it is necessary to record additional information on the supplementary record, OSHA No. 101. The supplementary record describes how the accident or illness exposure occurred, lists the objects or substances involved, and indicates the nature of the injury or illness and the part(s) of the body affected.

The OSHA No. 101 is not the only form that can be used to satisfy this requirement. To eliminate duplicate recording, workers' compensation, insurance, or other reports may be used as supplementary records if they contain all of the items on the OSHA No. 101. If they do not, the missing items must be added to the substitute or included on a separate attachment.

Completed supplementary records must be present in the establishment within 6 workdays after the employer has received information that an injury or illness has occurred.

## Location, retention, and maintenance of records

Ordinarily, injury and illness records must be kept by employers for *each* of their establishments. This [section] describes what is considered to be an establishment for recordkeeping purposes, where the records must be located, how long they must be kept, and how they should be updated.

**A. Establishments.**
If an employer has more than one establishment, a separate set of records must be maintained for *each* one. The recordkeeping regulations define an establishment as "a single physical location where business is conducted or where services or industrial operations are performed." Examples include a factory, mill, store, hotel, restaurant, movie theater, farm, ranch, sales office, warehouse, or central administrative office.

The regulations specify that distinctly separate activities performed at the same physical location (for example, contract construction activities operated from the same physical location as a lumber yard) shall each be treated as a separate establishment for recordkeeping purposes. Production of dissimilar products; different kinds of operational procedures; different facilities; and separate management, personnel, payroll, or support staff are all indicative of separate activities and separate establishments.

**B. Location of records.**
Injury and illness records (the log, OSHA No. 200, and the supplementary record, OSHA No. 101) must be kept for every physical location where operations are performed. Under the regulations, the location of these records depends upon whether or not the employees are associated with a fixed establishment. The distinction between fixed and nonfixed establishments generally rests on the nature and duration of the operation and not on the type of structure in which the business is located. A nonfixed establishment usually operates at a single location for a relatively short period of time. A fixed establishment remains at a given location on a long-term or permanent basis. Generally, any operation at a given site for more than 1 year is considered a fixed establishment. Also, fixed establishments are generally places where clerical, administrative, or other business records are kept.

1. *Employees associated with fixed establishments.* Records for these employees should be located as follows:
   a. Records for employees working at fixed locations, such as factories, stores, restaurants, warehouses, etc., should be kept at the work location.
   b. Records for employees who report to a fixed location but work elsewhere should be kept at the place where the employees report each day. These employees are generally engaged in activities such as agriculture, construction, transportation, etc.

c. Records for employees whose payroll or personnel records are maintained at a fixed location, but who do not report or work at a single establishment, should be maintained at the base from which they are paid or the base of their firm's personnel operations. This category includes generally unsupervised employees such as traveling salespeople, technicians, or engineers.

2. *Employees not associated with fixed establishments.* Some employees are subject to common supervision, but do not report or work at a fixed establishment on a regular basis. These employees are engaged in physically dispersed activities that occur in construction, installation, repair, or service operations. Records for these employees should be located as follows:
   a. Records may be kept at the field office or mobile base of operations.
   b. Records may also be kept at an established central location. If the records are maintained centrally: (1) The address and telephone number of the place where records are kept must be available at the worksite; and (2) there must be someone available at the central location during normal business hours to provide information from the records.

**C. Location exception for the log (OSHA No. 200).**
Although the supplementary record and the annual summary must be located as outlined in the previous section, it is possible to prepare and *maintain the log* at an alternate location or by means of data processing equipment, or both. Two requirements must be met: (1) Sufficient information must be available at the alternate location to complete the log within 6 workdays after receipt of information that a recordable case has occurred; and (2) a copy of the log updated to within 45 calendar days must be present at all times in the establishment. This location exception applies only to the log, and not to the other OSHA records. Also, it does not affect the employer's posting obligations.

**D. Retention of OSHA records.**
The log and summary, OSHA No. 200, and the supplementary record, OSHA No. 101, must be retained in each establishment for 5 calendar years following the end of the year to which they relate. If an establishment changes ownership, the new employer must preserve the records for the remainder of the 5-year period. However, the new employer is not responsible for updating the records of the former owner.

**E. Maintenance of the log (OSHA No. 200).**
In addition to keeping the log on a calendar year basis, employers are required to update this form to include newly discovered cases and to reflect changes which occur in recorded cases after the end of the calendar year. Maintenance or updating of the log is different from the retention of records discussed in the previous section. Although all OSHA injury and illness records must be retained, only the log must be updated by the employer. If, during the 5-year retention period, there is a change in the extent or outcome of an injury or illness which affects an entry on a previous year's log, then the first entry should be lined out and a corrected entry made on that log. Also, new entries should be made for previously unrecorded cases that are discovered or for cases that initially weren't recorded but were found to be recordable after the end of the year in which the case occurred. The entire entry should be lined out for recorded cases that are later found nonrecordable. Log totals should also be modified to reflect these changes.

### Deciding whether a case should be recorded and how to classify it

This [section] presents guidelines for determining whether a case must be recorded under the OSHA recordkeeping requirements. These requirements should not be confused with recordkeeping requirements of various workers' compensation systems, internal industrial safety and health monitoring systems, the ANSI Z.16 standards for recording and measuring work injury and illness experience, and private insurance company rating systems. Reporting a case on the OSHA records should not affect recordkeeping determinations under these or other systems. Also—

*Recording an injury or illness under the OSHA system does not necessarily imply that management was at fault, that the worker was at fault, that a violation of an OSHA standard has occurred, or that the injury or illness is compensable under workers' compensation or other systems.*

**A. Employees vs. other workers on site.**
Employers must maintain injury and illness records for their own employees at each of their establishments, but they are *not* responsible for maintaining records for employees of other firms or for independent contractors, even though these individuals may be working temporarily in their establishment or on one of their jobsites at the time an injury or illness exposure occurs. Therefore, before deciding whether a case is recordable an employment relationship needs to be determined.

Employee status generally exists for recordkeeping purposes when the employer supervises not only the output, product, or result to be accomplished by the person's work, but also the details, means, methods, and processes by which the work is accomplished. This means the employer who supervises the worker's day-to-day activities is responsible for recording his injuries and illnesses. Independent contractors are not considered employees; they are primarily subject to supervision by the using firm only in regard to the result to be accomplished or end product to be delivered. Independent contractors keep their own injury and illness records.

Other factors which may be considered in determining employee status are: (1) Whom the worker considers to be his or her employer; (2) who pays the worker's wages; (3) who withholds the worker's Social Security taxes; (4) who hired the worker; and (5) who has the authority to terminate the worker's employment.

**B. Method used for case analysis.**
The decisionmaking process consists of five steps:

1. Determine whether a case occurred; that is, whether there was a death, illness, or injury;
2. Establish that the case was work related; that it resulted from an event or exposure in the work environment;
3. Decide whether the case is an injury or an illness; and
4. If the case is an illness, record it and check the appropriate illness category on the log; or
5. If the case is an injury, decide if it is recordable based on a finding of medical treatment, loss of consciousness, restriction of work or motion, or transfer to another job.

Chart 1 [Figure 6-9] presents this methodology in graphic form.

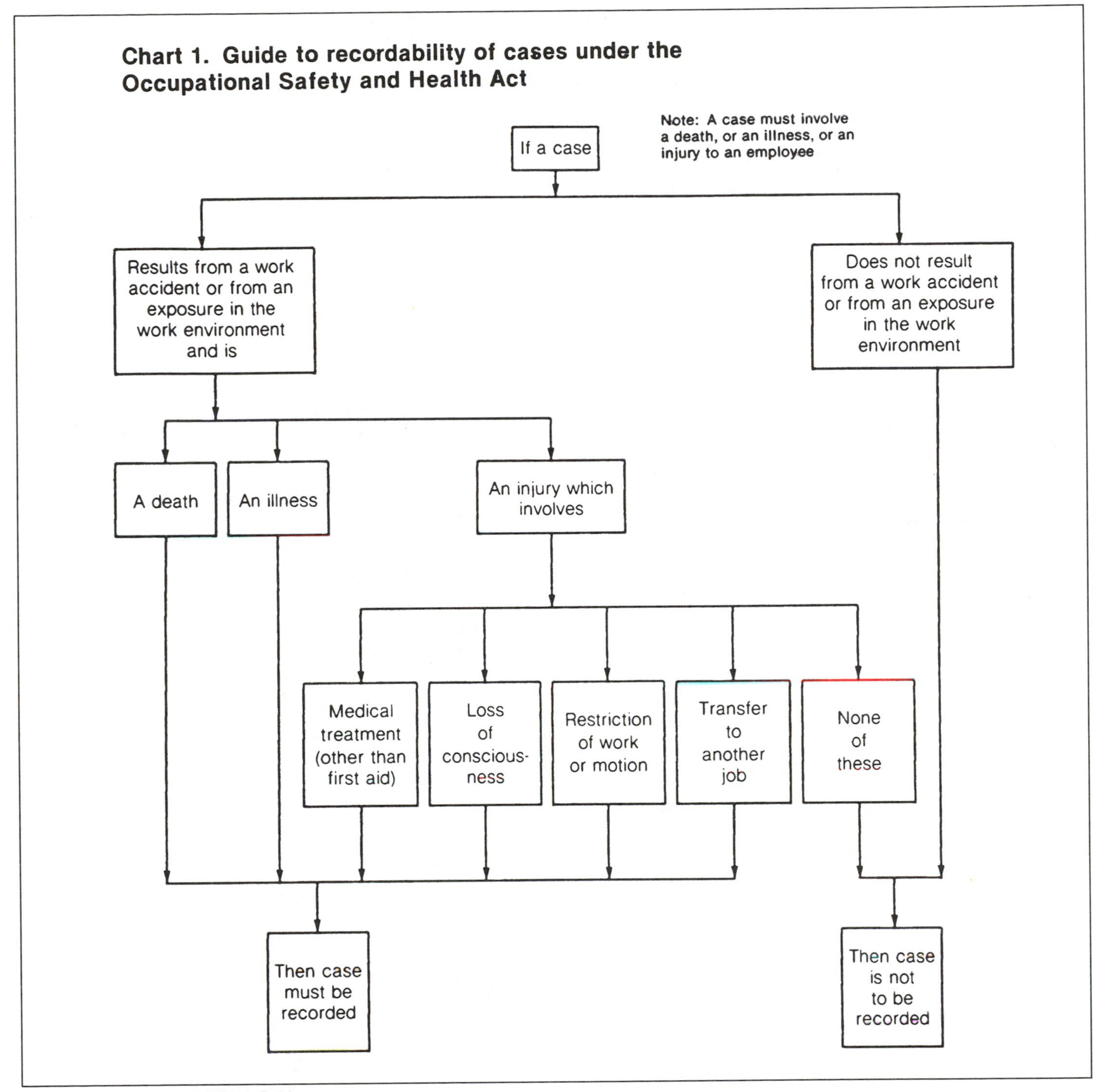


**Figure 6-9.** Chart 1 of OSHA Guide.

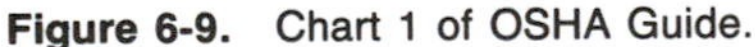

**C. Determining whether a case occurred.**
The first step in the decisionmaking process is the determination of whether or not an injury or illness has occurred. Employers have nothing to record unless an employee has experienced a work-related injury or illness. In most instances, recognition of these injuries and illnesses is a fairly simple matter. However, some situations have troubled employers over the years. Two of these are:

1. *Hospitalization for observation.* If an employee goes to or is sent to a hospital for a brief period of time for observation, it is not recordable, provided no medical treatment was given, or no illness was recognized. The determining factor is not that the employee went to the hospital, but whether the incident is recordable as a work-related illness or as an injury requiring medical treatment or involving loss of consciousness, restriction of work or motion, or transfer to another job.
2. *Differentiating a new case from the recurrence of a previous injury or illness.* Employers are required to make new entries on their OSHA forms for each new recordable injury

or illness. However, new entries should not be made for the recurrence of symptoms from previous cases, and it is sometimes difficult to decide whether or not a situation is a new case or a recurrence. The following guidelines address this problem:

a. *Injuries.* The aggravation of a previous injury almost always results from some new incident involving the employee (such as a slip, trip, fall, sharp twist, etc.). Consequently, when work related, these new incidents should be recorded as new cases.

b. *Illnesses.* Generally, each occupational illness should be recorded with a separate entry on the OSHA No. 200. However, certain illnesses, such as silicosis, may have prolonged effects which recur over time. The recurrence of these symptoms should not be recorded as new cases on the OSHA forms. The recurrence of symptoms of previous illnesses may require adjustment of entries on the log for previously recorded illnesses to reflect possible changes in the extent or outcome of the particular case.

Some occupational illnesses, such as certain dermatitis or respiratory conditions, may recur as the result of new exposures to sensitizing agents, and should be recorded as new cases.

**D. Establishing work relationship.**

The Occupational Safety and Health Act of 1970 requires employers to record only those injuries and illnesses that are work related. *Work relationship is established under the OSHA recordkeeping system when the injury or illness results from an event or exposure in the work environment. The work environment is primarily composed of: (1) The employer's premises, and (2) other locations where employees are engaged in work-related activities or are present as a condition of their employment.* When an employee is off the employer's premises, work relationship must be established; when on the premises, this relationship is presumed. The employer's premises encompass the total establishment, including not only the primary work facility, but also such areas as company storage facilities. In addition to physical locations, equipment or materials used in the course of an employee's work are also considered part of the employee's work environment.

1. *Injuries and illnesses resulting from events or exposures on the employer's premises.* Injuries and illnesses that result from an event or exposure on the employer's premises are generally considered work related. The employer's premises consist of the total establishment. They include the primary work facilities and other areas which are considered part of the employer's general work area.

The presumption of work relationship for activities on the employer's premises is rebuttable. Situations where the presumption would not apply include: (1) When a worker is on the employer's premises as a member of the general public and not as an employee, and (2) when employees have symptoms that merely surface on the employer's premises, but are the result of a nonwork-related event or exposure off the premises.

The following subjects warrant special mention:

a. Company restrooms, hallways, and cafeterias are all considered to be *part* of the employer's premises and constitute part of the work environment. Therefore, injuries occurring in these places are generally considered work related.

b. For OSHA recordkeeping purposes, the definition of work premises *excludes* all employer controlled ball fields, tennis courts, golf courses, parks, swimming pools, gyms, and other similar recreational facilities which are often apart from the workplace and used by employees on a voluntary basis for their own benefit, primarily during off-work hours. Therefore, injuries to employees in these recreational facilities are not recordable unless the employee was engaged in some work-related activity, or was required by the employer to participate.

c. Company parking facilities are generally *not* considered part of the employer's premises for OSHA recordkeeping purposes. Therefore, injuries to employees on these parking lots are not presumed to be work related, and are not recordable unless the employee was engaged in some work-related activity.

2. *Injuries and illnesses resulting from events or exposures off the employer's premises.* When an employee is off the employer's premises and suffers an injury or an illness exposure, work relationship must be established; it is not presumed. Injuries and illness exposures off premises are considered work related if the employee is engaged in a work activity or if they occur in the work environment. The work environment in these instances includes locations where employees are engaged in job tasks or work-related activities, or places where employees are present due to the nature of their job or as a condition of their employment.

Employees who travel on company business shall be considered to be engaged in work-related activities all the time they spend in the interest of the company, including, but not limited to, travel to and from customer contacts, and entertaining or being entertained for the purpose of transacting, discussing, or promoting business, etc. However, an injury/illness would not be recordable if it occurred during normal living activities (eating, sleeping, recreation); or if the employee deviates from a reasonably direct route of travel (side trip for vacation or other personal reasons). He would again be in the course of employment when he returned to the normal route of travel.

When a traveling employee checks into a hotel or motel, he establishes a "home away from home." Thereafter, his activities are evaluated in the same manner as for nontraveling employees. For example, if an employee on travel status is to report each day to a fixed worksite, then injuries sustained when traveling to this worksite would be considered off the job. The rationale is that an employee's normal commute from home to office would not be considered work related. However, there are situations where employees in travel status report to, or rotate among several different worksites after they establish their "home away from home" (such as a salesperson traveling to and from different customer contacts). In these situations, the injuries sustained when traveling to and from the sales locations would be considered job related.

Traveling sales personnel may establish only one base of operations (home or company office). A sales person with his home as an office is considered at work when he is in that office and when he leaves his premises in the interest of the company.

Chart 2 [Figure 6-10] provides a guide for establishing the work relationship of cases.

**E. Distinguishing between injuries and illnesses.**

Under the OSHAct, all work-related illnesses must be recorded, while injuries are recordable *only* when they require medical treatment (other than first aid), or involve loss of consciousness, restriction of work or motion, or transfer to another job. The distinction between injuries and illnesses, there-

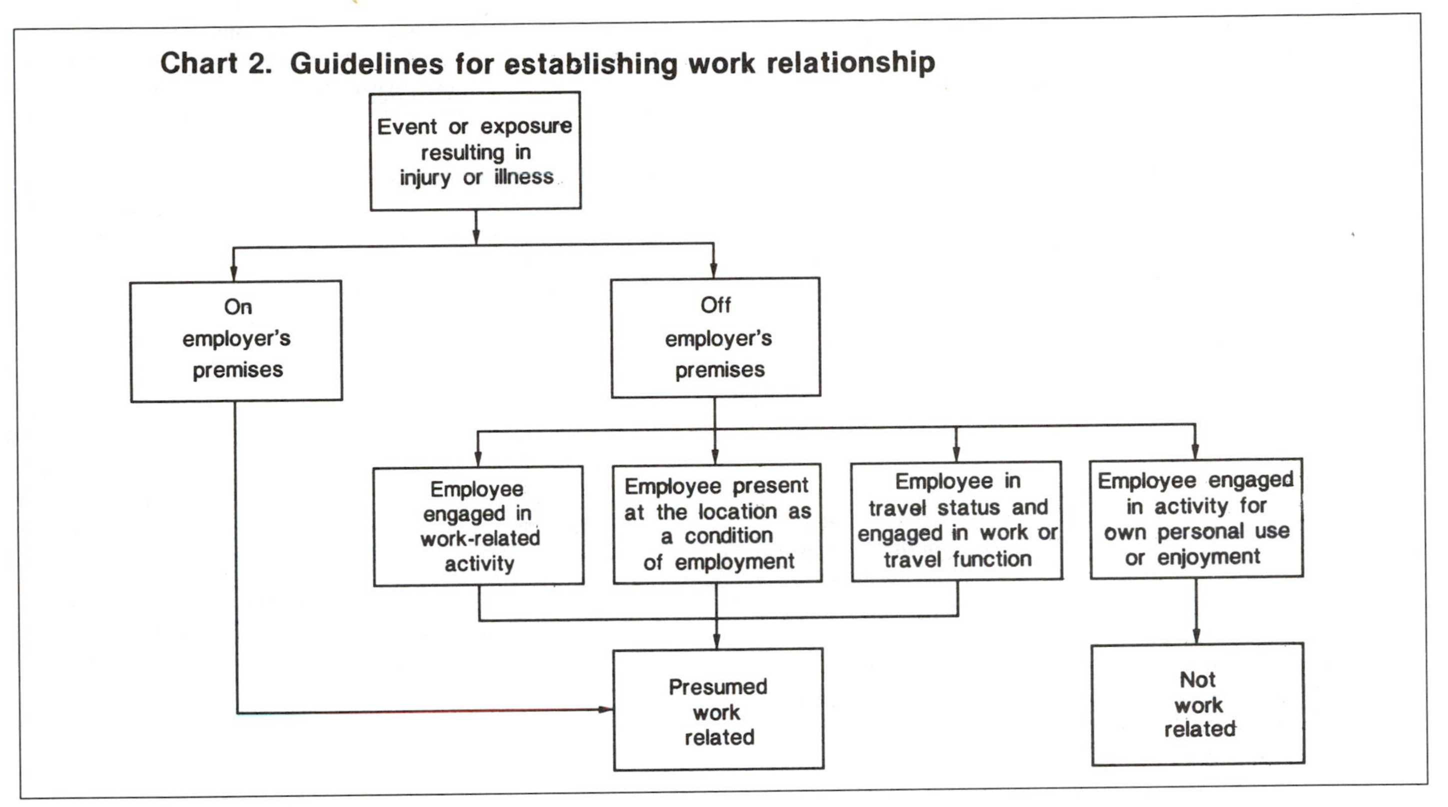


**Figure 6-10.** Chart 2 of OSHA Guide.

fore, has significant recordkeeping implications.

Whether a case involves an injury or illness is determined by the nature of the original event or exposure which caused the case, not by the resulting condition of the affected employee. Injuries are caused by instantaneous events in the work environment. Cases resulting from anything other than instantaneous events are considered illnesses. This concept of illnesses includes acute illnesses which result from exposures of relatively short duration.

Some conditions may be classified as either an injury or an illness (but not both), depending upon the nature of the event that produced the condition. For example, a loss of hearing resulting from an explosion (an instantaneous event) is classified as an injury; the same condition arising from exposure to industrial noise over a period of time would be classified as an occupational illness.

**F. Recording occupational illnesses.**
Employers are required to record the occurrence of all occupational illnesses, which are defined in the instructions of the log and summary as:

> Any abnormal condition or disorder, other than one resulting from an occupational injury, caused by exposure to environmental factors associated with employment. It includes acute and chronic illnesses or diseases which may be caused by inhalation, absorption, ingestion, or direct contact.

The instructions also refer to recording illnesses which were "diagnosed or recognized." Illness exposures ultimately result in conditions of a chemical, physical, biological, or psychological nature.

Occupational illnesses must be diagnosed to be recordable. However, they do not necessarily have to be diagnosed by a physician or other medical personnel. Diagnosis may be by a physician, registered nurse, or a person who by training or experience is capable to make such a determination. Employers, employees, and others may be able to detect some illnesses such as skin diseases or disorders without the benefit of specialized medical training. However, a case more difficult to diagnose, such as silicosis, would require evaluation by properly trained medical personnel.

In addition to recording the occurrence of occupational illnesses, employers are required to record each illness case in 1 of the 7 categories on the front of the log. The back of the log form contains a listing of types of illnesses or disorders and gives examples for each illness category. These are only examples, however, and should not be considered as a complete list of types of illnesses under each category.

Recording and classifying occupational illnesses may be difficult for employers, especially the chronic and long term latent illnesses. Many illnesses are not easily detected; and once detected, it is often difficult to determine whether an illness is work related. Also, employees may not report illnesses because the symptoms may not be readily apparent, or because they do not think their illness is serious or work related.

The following material is provided to assist in detecting occupational illnesses and in establishing their work relationship:

1. *Detection and diagnosis of occupational illnesses.* An occupational illness is defined in the instructions on the log as any work-related abnormal condition or disorder (other than an occupational injury). Detection of these abnormal conditions or disorders, the first step in recording illnesses, is often diffi-

cult. When an occupational illness is suspected, employers may want to consider the following:

a. A medical examination of the employee's physiological systems. For example:
   - Head and neck
   - Eyes, ears, nose, and throat
   - Endocrine
   - Genitourinary
   - Musculoskeletal
   - Neurological
   - Respiratory
   - Cardiovascular, and
   - Gastrointestinal;
b. Observation and evaluation of behavior related to emotional status, such as deterioration in job performance which cannot be explained;
c. Specific examination for health effects of suspected or possible disease agents by competent medical personnel;
d. Comparison of date of onset of symptoms with occupational history;
e. Evaluation of results of any past biological or medical monitoring (blood, urine, other sample analysis) and previous physical examinations;
f. Evaluation of laboratory tests: Routine (complete blood count, blood chemistry profile, urinalysis) and specific tests for suspected disease agents (e.g., blood and urine tests for specific agents, chest or other X-rays, liver function tests, pulmonary function tests); and
g. Reviewing the literature, such as Material Safety Data Sheets and other reference documents, to ascertain whether the levels to which the workers were exposed could have produced the ill effects.

2. *Determining whether the illness is occupationally related.* The instructions on the back of the log define occupational illnesses as those "caused by environmental factors associated with employment." In some cases, such as contact dermatitis, the relationship between an illness and work-related exposure is easy to recognize. In other cases, where the occupational cause is not direct and apparent, it may be difficult to determine accurately whether an employee's illness is occupational in nature. In these situations, it may help employers to ask the following questions:

a. Has an illness condition clearly been established?
b. Does it appear that the illness resulted from, or was aggravated by, suspected agents or other conditions in the work environment?
c. Are these suspected agents present (or have they been present) in the work environment?
d. Was the ill employee exposed to these agents in the work environment?
e. Was the exposure to a sufficient degree and/or duration to result in the illness condition?
f. Was the illness attributable solely to a nonoccupational exposure?

**G. Deciding if work-related injuries are recordable.** Although the OSHAct requires that all work-related deaths and illnesses be recorded, the recording of nonfatal injuries is limited to certain specific types of cases: Those which require medical treatment or involve loss of consciousness; restriction of work or motion; or transfer to another job. Minor injuries requiring only first aid treatment are *not* recordable.

1. *Medical treatment.* It is important to understand the distinction between medical treatment and first aid treatment since many work-related injuries are recordable only because medical treatment was given.

The regulations and the instructions on the back of the log and summary, OSHA No. 200, define medical treatment as any treatment, other than first aid treatment, administered to injured employees. Essentially, medical treatment involves the provision of medical or surgical care for injuries that are not minor through the application of procedures or systematic therapeutic measures.

The act also specifically states that work-related injuries which involve only first aid treatment should not be recorded. First aid is commonly thought to mean emergency treatment of injuries before regular medical care is available. However, first aid treatment has a different meaning for OSHA recordkeeping purposes. The regulations define first aid treatment as:

> . . . any one-time treatment, and any follow-up visit for the purpose of observation, of minor scratches, cuts, burns, splinters, and so forth, which do not ordinarily require medical care. Such one-time treatment, and follow-up visit for the purpose of observation, is considered first aid even though provided by a physician or registered professional personnel.

The distinction between medical treatment and first aid depends not only on the treatment provided, but also on the severity of the injury being treated. First aid is: (1) Limited to one-time treatment and subsequent observation; *and* (2) involves treatment of only minor injuries, *not* emergency treatment of serious injuries. Injuries are *not* minor if:

a. They must be treated only by a physician or licensed medical personnel;
b. They impair bodily function (i.e., normal use of senses, limbs, etc.);
c. They result in damage to the physical structure of a nonsuperficial nature (e.g., fractures); or
d. They involve complications requiring follow-up medical treatment.

Physicians or registered medical professionals, working under the standing orders of a physician, routinely treat minor injuries. Such treatment may constitute first aid. Also, some visits to a doctor do not involve treatment at all. For example, a visit to a doctor for an examination or other diagnostic procedure to determine whether the employee has an injury does not constitute medical treatment. Conversely, medical treatment can be provided to employees by lay persons; i.e., someone other than a physician or registered medical personnel.

The following classifications list certain procedures as either medical treatment or first aid treatment.

**Medical treatment:** The following are generally considered medical treatment. Work-related injuries for which this type of treatment was provided or should have been provided are almost always recordable:

- Treatment of **INFECTION**
- Application of **ANTISEPTICS during second or subsequent visit** to medical personnel
- Treatment of **SECOND OR THIRD DEGREE BURN(S)**
- Application of **SUTURES (stitches)**

- Application of **BUTTERFLY ADHESIVE DRESSING(S) or STERI STRIP(S)** in lieu of sutures
- Removal of **FOREIGN BODIES EMBEDDED IN EYE**
- Removal of **FOREIGN BODIES FROM WOUND;** if procedure is **COMPLICATED** because of depth of embedment, size, or location
- Use of **PRESCRIPTION MEDICATIONS** (except a single dose administered on first visit for minor injury or discomfort)
- Use of hot or cold **SOAKING THERAPY during second or subsequent visit** to medical personnel
- Application of hot or cold **COMPRESS(ES) during second or subsequent visit** to medical personnel
- **CUTTING AWAY DEAD SKIN** (surgical debridement)
- Application of **HEAT THERAPY during second or subsequent visit** to medical personnel
- Use of **WHIRLPOOL BATH THERAPY during second or subsequent visit** to medical personnel
- **POSITIVE X-RAY DIAGNOSIS** (fractures, broken bones, etc.)
- **ADMISSION TO A HOSPITAL** or equivalent medical facility **FOR TREATMENT.**

**First Aid Treatment:** The following are generally considered first aid treatment (e.g., one-time treatment and subsequent observation of minor injuries) and should not be recorded if the work-related injury does not involve loss of consciousness, restriction of work or motion, or transfer to another job:

- Application of **ANTISEPTICS during first visit** to medical personnel
- Treatment of **FIRST DEGREE BURN(S)**
- Application of **BANDAGE(S)** during any visit to medical personnel
- Use of **ELASTIC BANDAGE(S) during first visit** to medical personnel
- Removal of **FOREIGN BODIES NOT EMBEDDED IN EYE** if only irrigation is required
- Removal of **FOREIGN BODIES FROM WOUND;** if procedure is **UNCOMPLICATED,** and is, for example, by tweezers or other simple technique
- Use of **NONPRESCRIPTION MEDICATIONS AND** administration of **single dose of PRESCRIPTION MEDICATION** on first visit for minor injury or discomfort
- **SOAKING THERAPY on initial visit** to medical personnel or removal of bandages by **SOAKING**
- Application of hot or cold **COMPRESS(ES) during first visit** to medical personnel
- Application of **OINTMENTS** to abrasions to prevent drying or cracking
- Application of **HEAT THERAPY during first visit** to medical personnel
- Use of **WHIRLPOOL BATH THERAPY during first visit** to medical personnel
- **NEGATIVE X-RAY DIAGNOSIS**
- **OBSERVATION** of injury during visit to medical personnel.

The following procedure, by itself, is not considered medical treatment:

- Administration of **TETANUS SHOT(S) or BOOSTER(S).** However, these shots are often given in conjunction with more serious injuries; consequently, injuries requiring these shots may be recordable for other reasons.

2. *Loss of consciousness.* If an employee loses consciousness as the result of a work-related injury, the case must be recorded no matter what type of treatment was provided. The rationale behind this recording requirement is that loss of consciousness is generally associated with the more serious injuries.

3. *Restriction of work or motion.* Restricted work activity occurs when the employee, because of the impact of a job-related injury, is physically or mentally unable to perform all or any part of his or her normal assignment during all or any part of the workday or shift. The emphasis is on the employee's ability to perform normal job duties. Restriction of work or motion may result in either a lost-worktime injury or a nonlost-worktime injury, depending upon whether the restriction extended beyond the day of injury.

4. *Transfer to another job.* Injuries requiring transfer of the employee to another job are also considered serious enough to be recordable regardless of the type of treatment provided. Transfers are seldom the sole criterion for recordability because injury cases are almost always recordable on other grounds, primarily medical treatment or restriction of work or motion.

### Categories for evaluating the extent of recordable cases

Once the employer decides that a recordable injury or illness has occurred, the case must be evaluated to determine its extent or outcome. There are three categories of recordable cases: Fatalities, lost workday cases, and cases without lost workdays. Every recordable case must be placed in only one of these categories.

**A. Fatalities.**

All work-related fatalities must be recorded, regardless of the time between the injury and the death, or the length of the illness.

**B. Lost workday cases.**

Lost workday cases occur when the injured or ill employee experiences either days away from work, days of restricted work activity, or both. In these situations, the injured or ill employee is affected to such an extent that: (1) Days must be taken off from the job for medical treatment or recuperation; or (2) the employee is unable to perform his or her normal job duties over a normal work shift, even though the employee may be able to continue working.

1. Lost workday cases involving days away from work are cases resulting in days the employee would have worked but could not because of the job-related injury or illness. The focus of these cases is on the employee's inability, because of injury or illness, to be present in the work environment during his or her normal work shift.
2. Lost workday cases involving days of restricted work activity are those cases where, because of injury or illness, (1) the employee was assigned to another job on a temporary basis, or (2) the employee worked at a permanent job less than full time, or (3) the employee worked at his or her permanently assigned job but could not perform all the duties normally connected with it. Restricted work activity occurs when the employee, because of the job-related injury or illness, is physically or mentally unable to perform all or any part of his or her normal job duties over all or any part of his or her normal workday or shift. The emphasis is on the employee's inability to perform normal job duties over a normal work shift.

Injuries and illnesses are not considered lost workday cases unless they affect the employee beyond the day of injury or onset of illness. When counting the number of days away from work or days of restricted work activity, do not include the initial day of injury or onset of illness, or any days on which the employee would not have worked even though able to work (holidays, vacations, etc.).

**C. Cases not resulting in death or lost workdays.** These cases consist of the relatively less serious injuries and illnesses which satisfy the criteria for recordability but which do not result in death or require the affected employee to have days away from work or days of restricted work activities beyond the date of injury or onset of illness.

### Employer obligations for reporting occupational injuries and illnesses

This [section] focuses on the requirements of Section 8(c)(2) of the Occupational Safety and Health Act of 1970 and Title 29, Part 1904, of the *Code of Federal Regulations* for employers to make reports of occupational injuries and illnesses. It does not include the reporting requirements of other standards or regulations of the Occupational Safety and Health Administration (OSHA) or of any other State or Federal agency.

**A. The Annual Survey of Occupational Injuries and Illnesses.** The survey is conducted on a sample basis, and firms required to submit reports of their injury and illness experience are contacted by BLS or a participating State agency. A firm not contacted by its State agency or BLS need not file a report of its injury and illness experience. Employers should note, however, that even if they are not selected to participate in the annual survey for a given year, they must still comply with the recordkeeping requirements listed in the preceding [sections] as well as with the requirements for reporting fatalities and multiple hospitalization cases provided in the next section.

Participants in the annual survey consist of two categories of employers: (1) Employers who maintain OSHA records on a regular basis; and (2) a small, rotating sample of employers who are regularly exempt from OSHA recordkeeping. The survey procedure is different for these two groups of employers.

1. *Participation of firms regularly maintaining OSHA records.* When employers regularly maintaining OSHA records are selected to participate in the Annual Survey of Occupational Injuries and Illnesses, they are mailed the survey questionnaire in February of the year following the reference calendar year of the survey. (A firm selected to participate in the 1985 survey would have been contacted in February of 1986.) The survey form, the Occupational Injuries and Illnesses Survey Questionnaire, OSHA No. 200-S, requests information about the establishment(s) included in the report and the injuries and illnesses experienced during the previous year. Information for the injury and illness portion of the report form usually can be copied directly from the totals on the log and summary, OSHA No. 200, which the employer should have completed and posted in the establishment by the time the questionnaire arrives. The survey form also requests summary information about the type of business activity and number of employees and hours worked at the reporting unit during the reference year.

2. *Participation of normally exempt small employers and employers in low-hazard industries.* A few regularly exempt employers (those with fewer than 11 employees in the previous calendar year and those in designated low-hazard industries) are also required to participate in the annual survey. Their participation is necessary for the production of injury and illness statistics that are comparable in coverage to the statistics published in years prior to the exemptions. These employers are notified prior to the reference calendar year of the survey that they must maintain injury and illness records for the coming year. (A firm selected to participate in the 1985 survey would have been contacted in December 1984.) At the time of notification, they are supplied with the necessary forms and instructions. During the reference calendar year, prenotified employers make entries on the log, OSHA No. 200, but are not required to complete a Supplementary Record of Occupational Injuries and Illnesses, OSHA No. 101, or post the summary of the OSHA No. 200 the following February (regularly participating employers do both).

**B. Reporting fatalities and multiple hospitalizations.** All employers are required to report accidents resulting in one or more fatalities or the hospitalization of five or more employees. (Some States have more stringent catastrophic reporting requirements.)

The report is made to the nearest office of the Area Director of the Occupational Safety and Health Administration, U.S. Department of Labor, unless the State in which the accident occurred is administering an approved State plan under Section 18(b) of the OSHAct. Those 18(b) States designate a State agency to which the report must be made.

The report must contain three pieces of information: (1) Circumstances surrounding the accident(s), (2) the number of fatalities, and (3) the extent of any injuries. If necessary, the OSHA Area Director may require additional information on the accident.

The report should be made within 48 hours after the occurrence of the accident or within 48 hours after the occurrence of the fatality, regardless of the time lapse between the occurrence of the accident and the death of the employee.

### Access to OSHA records and penalties for failure to comply with recordkeeping obligations

The preceding [sections] describe recordkeeping and reporting requirements. This [section] covers subjects related to insuring the integrity of the OSH[A] recordkeeping process—access to OSHA records and penalties for recordkeeping violations.

**A. Access to OSHA records.** All OSHA records, which are being kept by employers for the 5-year retention period, should be available for inspection and copying by authorized Federal and State government officials. Employees, former employees, and their representatives are provided access to only the log, OSHA No. 200.

Government officials with access to the OSHA records include: Representatives of the Department of Labor, including OSHA safety and health compliance officers and BLS representatives; representatives of the Department of Health and Human Services while carrying out that department's research responsibilities; and representatives of States accorded jurisdiction for inspections or statistical compilations. "Representatives" may include Department of Labor officials inspecting a workplace or gathering information, officials of the Department of Health and Human Services, or contractors working for the agencies mentioned above, depending on the provisions of the contract under which they work.

Employee access to the log is limited to the records of the establishment in which the employee currently works or formerly worked. All current logs and those being maintained for the 5-year retention period must be made available for inspection and copying by employees, former employees, and their representatives. An employee representative can be a member of a union representing the employee, or any person designated by the employee or former employee. Access to the log is to be provided in a reasonable manner and at a reasonable time. Redress for failure to comply with the access provisions of the regulations can be obtained through a complaint to OSHA.

**B. Penalties for failure to comply with recordkeeping obligations.**
Employers committing recordkeeping and/or reporting violations are subject to the same sanctions as employers violating other OSHA requirements such as safety and health standards and regulations.

## Glossary of terms

*Annual summary.* Consists of a copy of the occupational injury and illness totals for the year from the OSHA No. 200, and the following information: The calendar year covered; company name; establishment address; certification signature, title, and date.

*Annual survey.* Each year, BLS conducts an annual survey of occupational injuries and illnesses to produce national statistics. The OSHA injury and illness records maintained by employers in their establishments serve as the basis for this survey.

*Bureau of Labor Statistics (BLS).* The Bureau of Labor Statistics is the agency responsible for administering and maintaining the OSHA recordkeeping system, and for collecting, compiling, and analyzing work injury and illness statistics.

*Certification.* The person who supervises the preparation of the Log and Summary of Occupational Injuries and Illnesses, OSHA No. 200, certifies that it is true and complete by signing the last page of, or by appending a statement to that effect to, the annual summary.

*Cooperative program.* A program jointly conducted by the States and the Federal Government to collect occupational injury and illness statistics.

*Employee.* One who is employed in the business of his or her employer affecting commerce.

*Employee representative.* Anyone designated by the employee for the purpose of gaining access to the employer's log of occupational injuries and illnesses.

*Employer.* Any person engaged in a business affecting commerce who has employees.

*Establishment.* A single physical location where business is conducted or where services or industrial operations are performed; the place where the employees report for work, operate from, or from which they are paid.

*Exposure.* The reasonable likelihood that a worker is or was subject to some effect, influence, or safety hazard; or in contact with a hazardous chemical or physical agent at a sufficient concentration and duration to produce an illness.

*Federal Register.* The official source of information and notification on OSHA's proposed rulemaking, standards, regulations, and other official matters, including amendments, corrections, insertions, or deletions.

*First aid.* Any one-time treatment and subsequent observation of minor scratches, cuts, burns, splinters, and so forth, which do not ordinarily require medical care. Such treatment and observation are considered first aid even though provided by a physician or registered professional personnel.

*First report of injury.* A workers' compensation form which may qualify as a substitute for the supplementary record, OSHA No. 101.

*Incidence rate.* The number of injuries, illnesses, or lost workdays related to a common exposure base of 100 full-time workers. The common exposure base enables one to make accurate interindustry comparisons, trend analysis over time, or comparisons among firms regardless of size. This rate is calculated as:

$$N/EH \times 200{,}000$$

where:

N = number of injuries and/or illnesses or lost workdays
EH = total hours worked by all employees during calendar year
200,000 = base for 100 full-time equivalent workers (working 40 hours per week, 50 weeks per year).

*Log and Summary (OSHA No. 200).* The OSHA recordkeeping form used to list injuries and illnesses and to note the extent of each case.

*Lost workday cases.* Cases which involve days away from work or days of restricted work activity, or both.

*Lost workdays.* The number of workdays (consecutive or not), beyond the day of injury or onset of illness, the employee was away from work or limited to restricted work activity because of an occupational injury or illness.

*(1) Lost workdays—away from work.* The number of workdays (consecutive or not) on which the employee would have worked but could not because of occupational injury or illness.

*(2) Lost workdays—restricted work activity.* The number of workdays (consecutive or not) on which, because of injury or illness: (1) The employee was assigned to another job on a temporary basis; or (2) the employee worked at a permanent job less than full time; or (3) the employee worked at a permanently assigned job but could not perform all duties normally connected with it.

The number of days away from work or days of restricted work activity does not include the day of injury or onset of illness or any days on which the employee would not have worked even though able to work.

*Low-hazard industries.* Selected industries in retail trade; finance, insurance, and real estate; and services which are regularly exempt from OSHA recordkeeping. To be included in this exemption, an industry must fall within an SIC not targeted for general schedule inspections and must have an average lost workday case injury rate for a designated 3-year measurement period at or below 75 percent of the U.S. private sector average rate.

*Medical treatment.* Includes treatment of injuries administered by physicians, registered professional personnel, or lay persons (i.e., nonmedical personnel). Medical treatment does not include first aid treatment (one-time treatment and subsequent observation of minor scratches, cuts, burns, splinters, and so forth, which do not ordinarily require medical care) even though provided by a physician or registered professional personnel.

*Occupational illness.* Any abnormal condition or disorder, other than one resulting from an occupational injury, caused by exposure to environmental factors associated with employment. It includes acute and chronic illnesses or diseases which may be caused by inhalation, absorption, ingestion, or direct contact. The following categories should be used by employers to classify recordable occupational illnesses on the log in the columns indicated:

Column 7a. *Occupational skin diseases or disorders.*
Examples: Contact dermatitis, eczema, or rash caused by primary irritants and sensitizers or poisonous plants; oil acne; chrome ulcers; chemical burns or inflammations; etc.

Column 7b. *Dust diseases of the lungs (pneumoconioses).*
Examples: Silicosis, asbestosis, and other asbestos-related diseases, coal worker's pneumoconiosis, byssinosis, siderosis, and other pneumoconioses.

Column 7c. *Respiratory conditions due to toxic agents.*
Examples: Pneumonitis, pharyngitis, rhinitis or acute congestion due to chemicals, dusts, gases, or fumes; farmer's lung, etc.

Column 7d. *Poisoning (systemic effects of toxic materials).*
Examples: Poisoning by lead, mercury, cadmium, arsenic, or other metals; poisoning by carbon monoxide, hydrogen sulfide, or other gases; poisoning by benzol, carbon tetrachloride, or other organic solvents; poisoning by insecticide sprays such as parathion, lead arsenate; poisoning by other chemicals such as formaldehyde, plastics, and resins; etc.

Column 7e. *Disorders due to physical agents (other than toxic materials).*
Examples: Heatstroke, sunstroke, heat exhaustion, and other effects of environmental heat; freezing, frostbite, and effects of exposure to low temperatures; caisson disease; effects of ionizing radiation (isotopes, X-rays, radium); effects of nonionizing radiation (welding flash, ultra-violet rays, microwaves, sunburn); etc.

Column 7f. *Disorders associated with repeated trauma.*
Examples: Noise-induced hearing loss; synovitis, tenosynovitis, and bursitis; Raynaud's phenomena; and other conditions due to repeated motion, vibration, or pressure.

Column 7g. *All other occupational illnesses.*
Examples: Anthrax, brucellosis, infectious hepatitis, malignant and benign tumors, food poisoning, histoplasmosis, coccidioidomycosis, etc.

*Occupational injury.* Any injury such as a cut, fracture, sprain, amputation, etc., which results from a work accident or from a single instantaneous exposure in the work environment.

Note: Conditions resulting from bites, such as insect or snake bites, and from one-time exposure to chemicals are considered to be injuries.

*Occupational injuries and illnesses, extent and outcome.* All recordable occupational injuries or illnesses result in either:

1. Fatalities, regardless of the time between the injury, or the length of illness, and death;
2. Lost workday cases, other than fatalities, that result in lost workdays; or
3. Nonfatal cases without lost workdays.

*Occupational Safety and Health Administration (OSHA).* OSHA is responsible for developing, implementing, and enforcing safety and health standards and regulations. OSHA works with employers and employees to foster effective safety and health programs which reduce workplace hazards.

*Posting.* The annual summary of occupational injuries and illnesses must be posted at each establishment by February 1 and remain in place until March 1 to provide employees with the record of their establishment's injury and illness experience for the previous calendar year.

*Premises, employer's.* Consist of the employer's total establishment; they include the primary work facility and other areas in the employer's domain such as company storage facilities, cafeterias, and restrooms.

*Recordable cases.* All work-related deaths and illnesses, and those work-related injuries which result in: Loss of consciousness, restriction of work or motion, transfer to another job, or require medical treatment beyond first aid.

*Recordkeeping system.* Refers to the nationwide system for recording and reporting occupational injuries and illnesses mandated by the Occupational Safety and Health Act of 1970 and implemented by Title 29, *Code of Federal Regulations,* Part 1904. This system is the only source of national statistics on job-related injuries and illnesses for the private sector.

*Regularly exempt employers.* Employers regularly exempt from OSHA recordkeeping include: (A) All employers with no more than 10 full- or part-time employees at any one time in the previous calendar year; and (B) all employers in retail trade; finance, insurance, and real estate; and service industries; i.e., SIC's 52-89 (except building materials and garden supplies, SIC 52; general merchandise and food stores, SIC's 53 and 54; hotels and other lodging places, SIC 70; repair services, SIC's 75 and 76; amusement and recreation services, SIC 79; and health services, SIC 80). (Note: Some State safety and health laws may require these employers to keep OSHA records.)

*Report form.* Refers to survey form OSHA No. 200-S which is completed and returned by the surveyed reporting unit.

*Restriction of work or motion.* Occurs when the employee, because of the result of a job-related injury or illness, is physically or mentally unable to perform all or any part of his or her normal assignment during all or any part of the workday or shift.

*Single dose (prescription medication).* The measured quantity of a therapeutic agent to be taken at one time.

*Small employers.* Employers with no more than 10 full- and/or part-time employees among all the establishments of their firm at any one time during the previous calendar year.

*Standard Industrial Classification (SIC).* A classification system developed by the Office of Management and Budget, Executive Office of the President, for use in the classification of establishments by type of activity in which engaged. Each establishment is assigned an industry code for its major activity, which is determined by the product manufactured or service rendered. Establishments may be classified in 2-, 3-, or 4-digit industries according to the degree of information available.

*State (when mentioned alone).* Refers to a State of the United States, the District of Columbia, and U.S. territories and jurisdictions.

*State agency.* State agency administering the OSHA recordkeeping and reporting system. Many States cooperate directly with BLS in administering the OSHA recordkeeping and reporting programs. Some States have their own safety and health laws which may impose additional obligations.

*Supplementary Record (OSHA No. 101).* The form (or equivalent) on which additional information is recorded for each injury and illness entered on the log.

*Title 29 of the Code of Federal Regulations, Parts 1900-1999.* The parts of the *Code of Federal Regulations* which contain OSHA regulations.

*Volunteers.* Workers who are not considered to be employees under the Act when they serve of their own free will without compensation.

*Work environment.* Consists of the employer's premises and other locations where employees are engaged in work-related activities or are present as a condition of their employment. The work environment includes not only physical locations, but also the equipment or materials used by the employee during the course of his or her work.

*Workers' compensation systems.* State systems that provide medical benefits and/or indemnity compensation to victims of work-related injuries and illnesses.

This is the end of the quotation.—Ed.

## OSHA INCIDENCE RATES

Safety performance is relative. Only when a company compares its injury experience with that of its entire industry, or with its own previous experience, can it obtain a meaningful evaluation of its safety accomplishments.

To make such comparisons, a method of measurement is needed that will adjust for the effects of certain variables contributing to differences in injury experience. For two reasons, injury totals alone cannot be used.

First, a company with many employees may be expected to have more injuries than a company with few employees. Second, if the records of one company include all the injuries treated in the first aid room, whereas the records of a similar company include only injuries serious enough to cause lost time, obviously the first total will be larger than the second.

A standard procedure for keeping records, which provides for these variables, is included in the OSHA recordkeeping requirements. First, this procedure uses incidence rates which relate injury and illness cases, and days lost as a result of them, to the number of employee-hours worked; thus these rates automatically adjust for differences in the hours of exposure to injury. Second, this procedure specifies the kinds of injuries and illnesses that should be included in the rates.

These standardized rates, which are easy to compute and to understand, have been generally accepted as procedure in industry, thus permitting the necessary and desired comparisons.

A chronological arrangement of these rates for a company will show whether its level of safety performance is improving or getting worse. Within a company, the same sort of arrangement by departments will not only show the trend of safety performance for each department, but can reveal to management other information which will make safety work more efficient.

If it is found, for example, that the trend of incidence rates in a company is up, a review of the rate trends by department may reveal that this adverse change is accounted for by the rates of just a few departments. With the source of the highest company rates isolated, safety efforts can be concentrated at the points of worst experience.

A comparison of current incidence rates with those of similar companies and with those of the industry as a whole will provide the safety professional with a more reliable evaluation of the safety performance of the company than could be obtained merely by reviewing numbers of cases.

### Formulas for rates

Incidence rates are based on the exposure of 100 full-time workers using 200,000 employee-hours as the equivalent (100 employees working 40 hours per week for 50 weeks per year). An incidence rate can be computed for each category of cases or days lost depending on what number is put in the numerator of the formula. The denominator of the formula should be the total number of hours worked by all employees during the same time period as that covered by the number of cases in the numerator. (Note that this is the same formula that was given in the previous section, which quoted from the government publication.)

$$\text{Incidence Rate} = \frac{\text{No. of injuries \& illnesses} \times 200{,}000}{\text{Total hours worked by all employees during period covered}}$$

or

$$\frac{\text{No. of lost workdays} \times 200{,}000}{\text{Total hours worked by all employees during period covered}}$$

There are two other formulas that can be used to measure the average severity of the recorded cases:

$$\text{Average lost workdays per total lost workday cases} = \frac{\text{Total lost workdays}}{\text{Total lost workday cases}}$$

$$\text{Average days away from work} = \frac{\text{Total days away from work}}{\text{Total cases involving days away from work}}$$

If these numbers are small, then it is known that the cases are relatively minor. If, however, they are large, then the cases are of greater average severity and should receive serious attention. For example, to calculate the incidence rate for total recordable cases at the end of the year, one would simply multiply the number of recordable cases by 200,000 and divide that by the number of hours worked by all employees for the whole year.

The incidence rates may also be interpreted as the percentage of employees that will suffer the degree of injury for which the rate was calculated. That is, if the incidence rate of lost workday cases is 5.1 per 100 full-time workers, then about 5 percent of the establishment's employees incurred a lost workday injury.

## SIGNIFICANCE OF CHANGES IN INJURY AND ILLNESS EXPERIENCE

The incidence rate for lost workday cases is the best measure for comparing the occupational injury and illness experience between companies of various sizes, although it is often not sufficient for determining the significance of month-to-month changes in the actual number of cases within a single company. For comparing month-to-month changes within a company, it is usually necessary to use all cases, not only the lost workday cases. This provides a more objective measure for determining the significance of month-to-month fluctuations in the number of cases, especially when there is an apparently large increase or decrease from the monthly average.

Since the average number of cases per month is calculated from numbers of cases that are larger and smaller than the average itself, variation from the average is to be expected. The variation can be either random or caused; caused variation is significant and random variation is nonsignificant. Therefore, the significance of variations can easily be determined by distinguishing between those which are random and those which are caused.

Manufacturing organizations are already faced with the task of determining the significance of variations in such things as dimensions, weight, or performance of their products. To ease this task, they frequently employ a tool known as a quality control chart. This chart identifies and distinguishes between a random variation, which is said to be "in control," and caused variation, which is said to be "out of control." Being able to distinguish between the two types of variation permits management to concentrate its efforts on those variations that are out of control.

A similar control chart can be developed to evaluate the significance of changes in injury and illness experience. The first step in developing the chart is to calculate the average number of cases per month. Because this monthly average will fluctuate from year to year, several years' experience should be used to develop a stable average. Preferably, the average should be calculated using 60 months' experience. After the average number of cases per month has been determined, the upper and lower control limits (UCL and LCL, respectively) are calculated using the following equation:

$$\text{UCL and LCL} = n \pm 2\sqrt{n}$$

where $n$ is the average number of cases per month.

Figure 6-11 shows a control chart developed for one company. Using 60 months' experience, it was determined that the company had an average of 25 cases per month. Substitution of 25 for $n$ in the equation yielded the upper and lower control limits shown below:

$$25 \pm 2\sqrt{25} = 25 \pm 10 = \begin{cases} 25 + 10 = 35 \text{ for the upper limit} \\ 25 - 10 = 15 \text{ for the lower limit} \end{cases}$$

The control chart was then constructed and the company recorded the actual number of cases each month. Note that the actual monthly number of cases, with the exception of February, fall within the upper and lower control limits. These variations are random and do not represent significant changes from the monthly average of 25 cases. The variation for February, probably, is a caused variation indicating that the experience for February is out of control and that corrective measures should be taken. Suppose, for example, an investigation revealed that an unguarded machine was the cause of the increase in cases for the month of February. It can be assumed that the installation of a mechanical guard corrected this cause and brought the injury and illness experience for March back in control.

The control chart can be a useful tool when it is used properly. As with any tool, it will depend on the skill of the user to yield good results. Follow these rules when constructing a control chart:

1. Always use several years' experience when calculating the average number of cases per month to assure that a stable average will be developed. Sixty months' experience is preferable, except as noted in (3), (4), and (5) following.
2. Count *all* cases when constructing a control chart. The use of lost workday cases alone may not provide enough data. A control chart cannot be constructed if the average number of cases per month is less than four because the lower limit will be negative. Since a plant cannot experience a negative number of cases per month, a chart based on an average of less than four would not yield meaningful results.
3. Calculate a new average each year—adding the latest year's cases and subtracting the oldest year's cases—to reflect any change in the monthly average number of cases.
4. Construct more than one chart if there are seasonal changes in employment. For example, if an operation employs one hundred people for the first 6 months of each year and four hundred people for the last 6 months, a separate chart must be constructed for each 6-month period.
5. Construct a new control chart if external factors, such as a change in the level of employment, a change in the work environment, or a change in work hazards, bring about a permanent change in injury and illness experience. When calculating a new average in such a situation, *do not* use experience prior to the change.

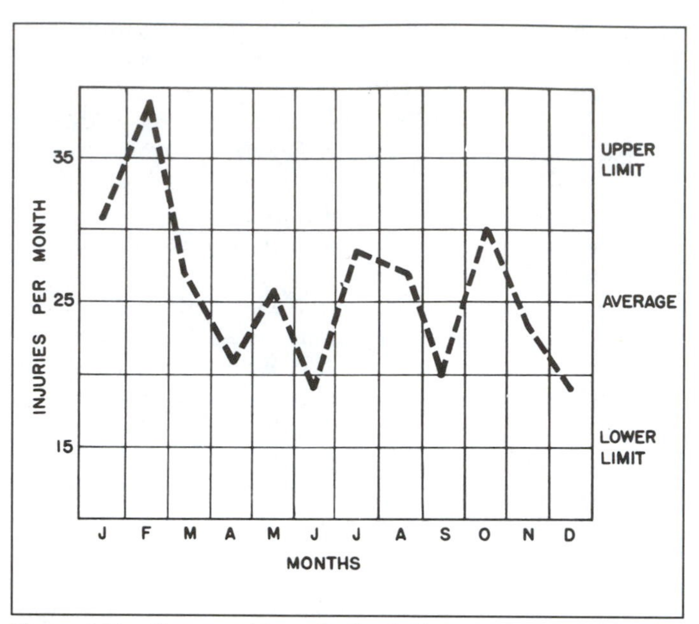


**Figure 6-11.** Upper and lower limits of this control chart were based on 60 months' experience, and show points at which the number of injuries can statistically be shown to have a contributing cause rather than be by chance. Thus, action can be taken where it will be most effective.

## OFF-THE-JOB INJURIES

In recent years, off-the-job disabling injuries of employees have exceeded on-the-job disabling injuries by 30 to 40 percent. Since the unscheduled absence of employees for any reason can cause production slowdowns and delays, costly retraining and replacement, or costly overtime by remaining employees, many safety specialists are concerned with the off-the-job injuries that occur to their employees. Moreover, activity to reduce off-the-job accidents should help promote interest in on-the-job safety.

The ANSI Z16.3 standard provides a means for recording and measuring these off-the-job injuries—those injuries suffered by an employee that do *not* arise out of and in the course of employment. Definitions and rates used under the ANSI Z16.3 standard are very similar to those used under ANSI Z16.1. Because the data on off-the-job injuries is not as easy to obtain, however, certain simplifications are introduced in ANSI Z16.3. Exposure (for use in rates per million employee-hours) is standardized at 312 employee-hours per employee per month (equal to four and one-third weeks less forty hours per week at work and 56 hours per week for sleeping). For the calculation of severity rates, each permanent partial disability is recorded at 390 days of disability (based upon Bureau of Labor Statistics averages for all permanent partial disabilities). Provision is also made in ANSI Z16.3 for recording home, public, and transportation injuries separately to allow for concentrated effort in problem areas.

## ANSI Z16.1 RECORDKEEPING AND RATES

Work on the ANSI Z16.1 standard goes back to Bulletin No. 276, "Standardization of Industrial Accident Statistics," published by the U.S. Bureau of Labor Statistics in 1920. Although developed by a body representing governmental statistical agencies, the rate provisions of this Bulletin were widely followed in whole or in part by private agencies. At a national conference on Industrial Accident Prevention, called by the U.S. Secretary of Labor in Washington in 1926, a resolution was adopted in favor of a revision of Bulletin 276 by a committee set up under the procedures of the American Standards Association (presently the American National Standards Institute). As a result of the work of this committee, the first edition of this standard was completed and approved in 1937 as Z16.1-1937. Since then this standard has been reviewed and revised.

### Dates for compiling rates

Injury rates should be determined as soon after each period (month or year, for example) as the information becomes available. A reasonable time may be allowed for completion of reports. However, absolute accuracy in rates does not justify long delays.

The ANSI Z16.1 standard suggests the following schedule for compiling injury rates:

1. Annual frequency rates should be based on all disabling injuries occurring within the year and reported within twenty days after the close of the year. Monthly frequency rates should be based on all disabling injuries occurring within the month and reported within twenty days after the close of the month.
2. Days charged for reported cases in which disability continues beyond the closing dates in (1) should be estimated on the basis of medical opinion of probable ultimate disability.
3. Cases first reported after closing dates stated in (1) need not be included in the rates for that period, or for any similar subsequent period. However, they should be included, and should replace estimates, in rates for longer periods of which that period is a part.

A disabling injury, and all days lost or charged because of it, should be charged to the date on which the injury occurred, except that for injuries, such as bursitis, tenosynovitis, pneumonconiosis (black lung), or silicosis, which do not arise out of specific accidents, the date of the injury should be the date when the injury is first reported. (Remember, OSHA calls these occupational illnesses.)

### Standard formulas for rates

The injury frequency rate and the injury severity rate are based on standard formulas as set forth in ANSI Z16.1.

**Frequency rate.** The disabling injury frequency rate relates the injuries to the hours worked during the period and expresses them in terms of a million-hour unit by use of the following formula:

$$\frac{\text{Number of disabling injuries} \times 1{,}000{,}000}{\text{Employee-hours of exposure}}$$

**Severity rate.** The disabling injury severity rate relates the days charged to the hours worked during the period and expresses them in terms of a million-hour unit by use of the following formula:

$$\frac{\text{Total days charged} \times 1{,}000{,}000}{\text{Employee-hours of exposure}}$$

**Average days charged.** The frequency and severity rates show, respectively, the rate at which disabling injuries occur and the rate at which time is charged. A third measure included in the standard procedure shows the average severity of the disabling injuries. It is called the average days charged per disabling injury and may be calculated by either of the following formulas:

I. $$\frac{\text{Total days charged}}{\text{Total disabling injuries}}$$

or

II. $$\frac{\text{Severity rate}}{\text{Frequency rate}}$$

### Calculation of employee-hours

Employee-hours used in calculating injury rates is the total number of hours worked by all employees, including those of operating, production, maintenance, transportation, clerical, administrative, sales, and other departments.

Employee-hours should be calculated from the payroll or time clock records. If this method cannot be used, they may be estimated by multiplying the total employee-days worked for the period covered by the number of hours worked each day.

The experience of a central administrative office or central sales office of a multi-location company should not be included in the experience of any one location, nor should it be prorated among the locations, but it should be included in the overall experience of the company.

In calculating hours for employees who live on company property, only those hours are counted during which the employee is on duty.

For traveling personnel, such as salesmen, executives, and others whose working hours are not defined, an average of 8 hours per day should be used in computing the hours worked.

For standby employees who are restricted to the confines of the employer's premises, including seamen aboard vessels, all standby hours should be counted, as well as all work injuries occurring during such hours.

### Disabling injuries

The standard specifies that a work injury is any injury, including occupational disease and other work-connected disability, which arises out of and in the course of employment. The following descriptions paraphrase the standard. For full details, refer to the standard.

**Occupational disease** is a disease caused by exposure to environmental factors associated with employment. Work-connected disability includes such ailments as silicosis, pneumoconiosis, tenosynovitis, bursitis, and loss of hearing. Even though there is no traumatic injury in such disabilities, if they are work connected, they are considered work injuries.

**Definitions.** To ensure uniformity in the computation of injury rates and thereby provide for comparison among rates, the standard specifies that only *disabling injuries* shall be counted in the computation of standard injury rates. In general terms, a disabling injury is one which results in death or permanent impairment or which renders the injured person unable to work for a full day on any day after the day of injury. Disabling injuries are of four classes, as follows:

1. *Death* is any fatality resulting from a work injury, regardless of the time intervening between injury and death.
2. *Permanent total disability* is an injury other than death which permanently and totally incapacitates an employee from following any gainful occupation, or which results in the loss (or the complete loss of use) of any of the following in one accident: (1) both eyes; (2) one eye and one hand, or arm, or foot, or leg; (3) any two of the following not on the same limb, hand, arm, foot, leg.
3. *Permanent partial disability* is any injury other than death or permanent total disability which results in the complete loss or loss of use of any member or part of a member of the body, or any permanent impairment of functions of the body or part thereof, regardless of any preexisting disability of the injured member or impaired body function.
4. *Temporary total disability* is any injury which does not result in death or permanent impairment, but which results in one or more days of disability.

**Temporary total disability—key points.** Of the four classes of disabling injuries used in calculating standard injury rates, temporary total disability is the most difficult to interpret uniformly. Key points in the definition are:

- *Day of disability.* A day of disability is any day on which an employee is unable, because of injury, to perform effectively throughout a full shift the essential functions of a regularly established job that is open and available. Days include Sundays, days off, plant shutdowns, and other nonwork days subsequent to the day of injury. (Note: Disability days for an injury are not counted when scheduled charge(s) apply.)

Here is how the standard describes it.

**1.5.1.1.** The day of injury and the day on which the employee was able to return to full-time employment shall not be counted as days of disability; but all intervening calendar days, or calendar days subsequent to the day of injury (including weekends, holidays, other days off, and other days on which the plant may be shut down), shall be counted as days of disability provided they meet the criteria of the preceding paragraph.

**1.5.1.2.** Time lost on a workday, or on a nonworkday, subsequent to the day of injury, ascribed solely to the unavailability of medical attention or of necessary diagnostic aids, shall be considered disability time, unless in the opinion of the physician authorized by the employer to treat the case the person was able to work on all days subsequent to the day of injury.

**1.5.1.3.** If the physician authorized by the employer to treat the case is of the opinion that the injured employee is actually capable of working a full normal shift at a regularly established job, but has prescribed certain therapeutic treatments, the employee may be excused from work for those treatments without counting the excused time as disability time, provided: (*a*) the time required to obtain the treatments does not, on any workday, prevent the employee from performing effectively the essential functions of the job assignment on that day, and (*b*) the treatments are professionally administered and constitute more than simple rest.

**1.5.1.4.** If the physician authorized by the employer to treat the case is of the opinion that the injured employee was actually capable of working a full normal shift at a regularly established job, but because of transportation problems, associated with the injury, the employee is forced to arrive at the place of work late, or to leave the workplace before the established quitting time, such lost time may be excused and not counted as disability time, provided: (*a*) that the excused time does not materially reduce working time, and (*b*) that it is clearly evident that this failure to work the full shift hours is the result of a bona fide transportation problem and not a deviation from the regularly established job.

**1.5.1.5.** If the injured employee receives medical treatment for the injury, the determination of ability to work shall rest with the physician authorized by the employer to treat the case. If the employee rejects medical attention offered by the employer, the determination may be made by the employer based upon the best information available. If the employer fails to provide medical attention, the employee's decision shall be controlling.

- *Specific disabilities.* Following are examples of specific disabilities and the standard procedures for handling them. In all these cases, the regular shift is assumed to be 8:00 A.M. to 5:00 P.M.

1. Injury occurred at 4:30 P.M. on Monday. The employee worked on Tuesday and Wednesday, but was prevented from working Thursday through the following Monday because of delay infection. The employee returned to work on Tuesday at 11:00 A.M. after seeing the doctor. This is a disabling injury with five chargeable days of disability (Thursday through Monday).
2. Injury occurred at 1:00 P.M. on Friday. The doctor certified that the employee could not have worked on Saturday if it had been a workday but could have worked on Sunday and thereafter. The employee returned to work on Monday. This is a disabling injury with one day of disability (Saturday), even though the employee lost no actual work time except Friday afternoon.
3. Injury occurred at 4:00 P.M. on Tuesday. The employee was hospitalized for treatment of the injury until noon on Wednesday, then was released by the doctor with permission to resume work on Thursday morning. This is a disabling injury with one day of disability.
4. Injury occurred at 11:00 A.M. on Tuesday. The employee was treated by the doctor and sent home with instructions not to work Wednesday, but to report to the doctor's office Wednesday afternoon. At 5:30 P.M. Wednesday the employee was seen by the doctor and discharged as fit to work immediately. This is a disabling injury with one day of disability.

5. The doctor determined that an injury would not allow an employee to return to the job for at least a week. On checking with the personnel department, the doctor learned that another job was open that the employee could handle, and the employee was cleared for this work. This is not a disabling injury because the employee could perform another regularly established job which was open and suitable.

- *Additional key points* with regard to all the classifications of disabling injuries are:

1. An injury need not be traumatic—there does not have to be a visible wound. More precisely, the standard covers work-connected *disabilities,* which may include such conditions as bursitis, tenosynovitis, loss of hearing, and similar ailments.
2. An injury need not arise out of an accident, although it must be work connected. Examples of such injuries would be dermatitis, silicosis, and bursitis. A strain resulting from overlifting would be a work injury even though there was no accident.
3. The classification of an injury is entirely independent of workers' compensation laws, and rulings of workers' compensation agencies. This provision is necessary to promote uniformity in injury classification. Otherwise, similar injuries might be classified differently in different states.

As may readily be seen, minor injuries are excluded in the calculation of the standard injury rates. The reason is that it is impossible to obtain standardization of minor cases. The standard does provide classification for minor injuries, though, as *medical treatment injuries,* and identifies them as injuries that do not result in death, permanent impairment, or temporary total disability, but which require medical treatment, including first aid.

Although these nondisabling injuries are not included in the standard injury rates, they should be given consideration and attention. The health and safety professional may well keep totals of them, and note their frequency, the departments in which they occur, the types of injuries, and the relationship between them and disabling injuries.

### Classification of special cases

**Work injuries.** In addition to the usual injuries and occupational diseases, the following are specifically identified as work injuries for statistical purposes:

1. Inguinal hernia, if it is precipitated by an impact, sudden effort, or severe strain and meets *all* of the following conditions:
   *a.* There is a clear record of an accident or an incident, such as a slip, trip, fall, sudden effort, or overexertion;
   *b.* There is actual pain in the hernia region at the time of the accident or incident;
   *c.* The immediate pain was so severe that the injured employee was forced to stop work long enough to draw the attention of the supervisor or fellow employee to the condition, or the attention of a physician was secured within twelve hours.
2. Back injury, if:
   *a.* There is a clear record of an accident or an incident, such as a slip, trip, or fall, sudden effort or blow on the back; or
   *b.* The employee was engaged in a work activity which, in the opinion of the physician authorized by the employer to treat the case, produced a physical condition resulting from overexertion.

A back condition which is revealed in the course of employment, but which does not satisfy *either* of these conditions, should not be considered a work injury.

3. Aggravation of a pre-existing physical deficiency, if the aggravation arises out of and in the course of employment.
4. Aggravation of a minor work injury, whether due to improper diagnosis or treatment or to infection either on the job or off the job.
5. Animal and insect bites incurred in the performance of duties of employment.
6. Skin irritations and infections, such as poison ivy or dermatitis, when the employee is exposed to irritants in connection with work.
7. Exposure to temperature extremes in the course of employment.
8. Muscular disability, such as bursitis or tenosynovitis, if it arises out of duties of employment.

**Nondisabling injuries.** The following cases resulting in lost time are not classified as disabling, provided that in every such case, the physician determines that the injury was in reality slight and that the injured person could have returned to work without permanent impairment or temporary total disability. In cases of exposure to ionizing radiation, the period of observation may be extended to ten days. NOTE: if *any* treatment or medication is given for a confining injury (or a suspected confining injury) after the first 24 hours of observation, the injury shall be classified as a work injury.

1. Hospitalization for observation, for a period not to exceed 48 hours from the time of injury, or of suspected injury known to have a delayed effect, from such accidents as:
   *a.* A blow on the head
   *b.* A blow to the abdomen
   *c.* The inhalation of harmful gases
2. Illness from antitoxin, provided that the illness results *solely* from the antitoxin, vaccines, or drugs used in the treatment of a nondisabling injury.
3. Disability arising solely out of a physical deficiency, provided that a worker without such physical deficiency would not have experienced the incident which resulted in the injury. (An injury which results from the work activity or the environment of the employment shall be considered a work injury, even though the employee had an existing physical deficiency.)
4. Injury from an external event of such proportions and character as to be beyond the control of the employer, such as a tornado, hurricane, earthquake, flood, conflagration, or explosion originating outside of employment, or from an immediate secondary event, such as a fire, boiler explosion, or falling electrical wire. (An injury would be reportable, though, if the victim were a policeman, firefighter, member of a disaster or emergency squad, utility line worker, or other employee who is assigned duties in connection with such events. An injury also would be reportable if it arose out of activities necessitated by an external event, such as fighting a fire, cleaning up debris, or repairing equipment.)

**TABLE 6-A.*** Scheduled Charges
(See Figure 6-12 for bone location)

A. FOR LOSS OF MEMBER—TRAUMATIC OR SURGICAL

(For loss of use of member, see footnote**)

*Fingers, Thumb, and Hand*

Amputation Involving All or Part of Bone**

| | Thumb | Fingers Index | Middle | Ring | Little |
|---|---|---|---|---|---|
| Distal phalange | 300 | 100 | 75 | 60 | 50 |
| Middle phalange | | 200 | 150 | 120 | 100 |
| Proximal phalange | 600 | 400 | 300 | 240 | 200 |
| Metacarpal | 900 | 600 | 500 | 450 | 400 |
| Hand at wrist | | | | | 3,000 |

*Toe, Foot, and Ankle*

| Amputation Involving All or Part of Bone** | Great Toe | Each of Other Toes |
|---|---|---|
| Distal phalange | 150 | 35 |
| Middle phalange | | 75 |
| Proximal phalange | 300 | 150 |
| Metatarsal | 600 | 350 |
| Foot at ankle | | 2,400 |

*Arm*

| | |
|---|---|
| Any point above*** elbow, including shoulder joint | 4,500 |
| Any point above wrist at or below elbow | 3,600 |

*Leg*

| | |
|---|---|
| Any point above*** knee | 4,500 |
| Any point above ankle and at or below knee | 3,000 |

B. IMPAIRMENT OF FUNCTION

| | |
|---|---|
| One eye (loss of sight), whether or not there is sight in the other eye | 1,800 |
| Both eyes (loss of sight), in one accident | 6,000 |
| One ear (complete industrial loss of hearing), whether or not there is hearing in the other ear | 600 |
| Both ears (complete industrial loss of hearing), in one accident | 3,000 |
| Unrepaired hernia (for repaired hernia, use actual days lost | 50 |

C. FATAL OR PERMANENT TOTAL DISABILITY ....... 6,000

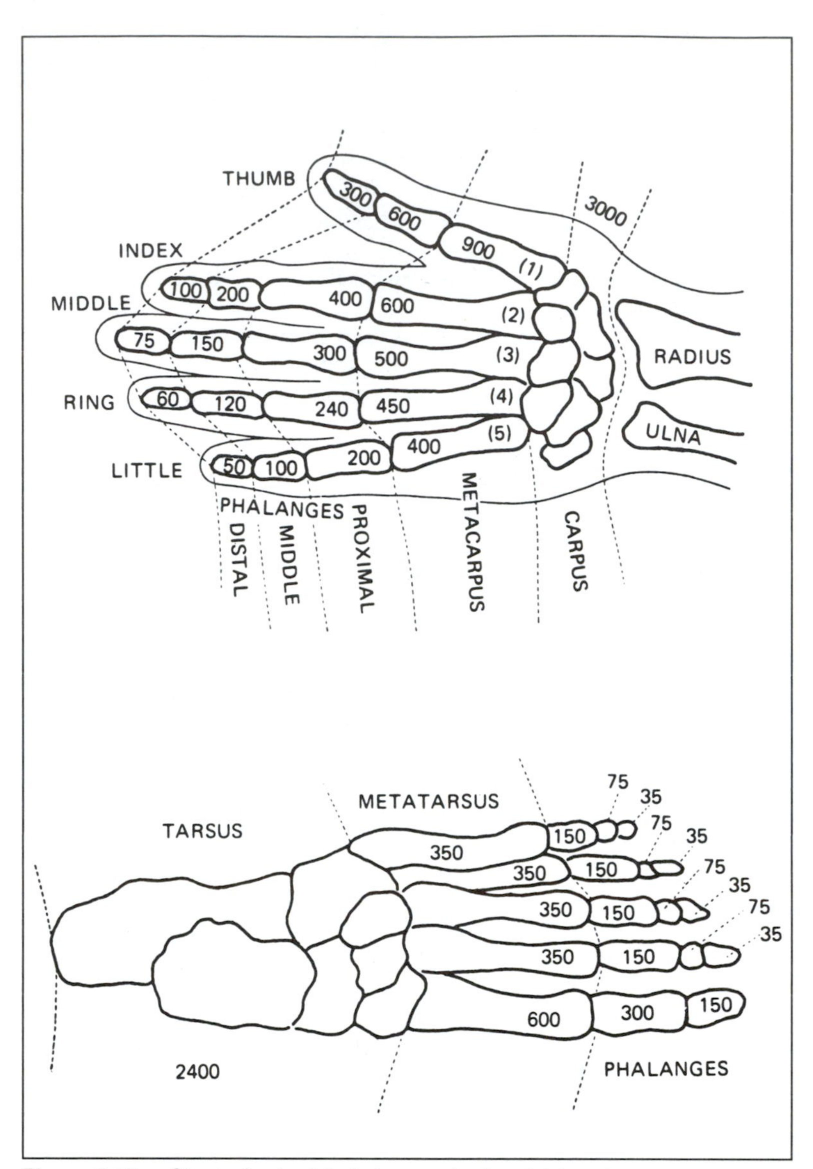


**Figure 6-12.** Chart of scheduled charges for hand *(above)* and foot *(below)*. Numbers on the bones are the charges (in days) for loss involving part or all of the bone. (Also see Table 6-A.) (Reprinted with permission from American National Standards Institute.)

**Footnotes to Table 6-A**

*Source: ANSI Z16.1, *Method of Recording and Measuring Work Injury Experience.*

**For loss of use, without amputation, use a percentage of the scheduled charge corresponding to the loss of use as determined by the physician authorized to treat the case. If the bone is not involved, use actual days lost and classify as temporary total disability.

***The term "above" when applied to the arm means toward the shoulder, and when applied to the leg means toward the hip.

### Days charged

Losses from work injuries are evaluated in terms of days of disability, or inability to produce, either actual or potential. These losses are referred to simply as "days charged." For the first three classes of injuries—death, permanent total disability, and permanent partial disability—the number of days charged is a predetermined total. For permanent partial disability, the predetermined total usually exceeds the actual time lost to reflect potential future losses of productive capacity. The predetermined totals are referred to as "scheduled charges."

This procedure is based on the philosophy of economic loss which reasons, for example, that a worker who has a hand amputated will produce less during the remaining working years than a worker who completely recovers from a hand injury, even though both injuries resulted in the same number of actual days lost at the time of the injury. If both injuries resulted in, say, 60 days lost at the time of the injury, the injury from which the victim completely recovers would be charged only the 60 days, whereas the amputation would be charged 3,000 days, the scheduled charge for this kind of injury.

**Death and permanent total disabilities.** For death and permanent total disabilities, a scheduled charge of 6,000 days is made in each case. There are no variations in this amount. If the injury is fatal, or if it results in any of the losses specified as constituting permanent total disability, the charge is the same—6,000 days.

The original basis for measuring the permanent disability of an injured worker was that, on the average, death or permanent total disability of a worker resulted in losing 20 years of productive labor at 300 days per year, or 6,000 days. Today, the death or permanent total disability of a worker would, on the average, result in losing about 24 years of productive labor at 250 days per year, giving approximately the same total loss. Time charges for permanent partial disabilities are partly related to this possible total.

**Permanent partial disabilities.** For permanent partial disabilities, the scheduled charges vary depending on the specific loss. For example, amputation of the index finger at the first joint has a scheduled charge of 100 days; at the second joint, 200 days; and at the third joint, 400 days.

Scheduled charges for permanent partial injuries are shown in Table 6-A. Figure 6-12 illustrates the charges for various hand and foot losses.

For permanent partial injuries that result in loss of use of an injured member, a percentage of the scheduled charge is used which corresponds to the percentage loss of use, as determined by the physician authorized to treat the injury.

For fatal, permanent total, and permanent partial disabilities, only the scheduled charges are used—the actual days of disability are disregarded. In some permanent partial injuries, there may be no losses at all, or there may be losses which exceed the scheduled charge. In either of these cases, though, disregard the number of days lost and *use only the scheduled charge.*

**Temporary total disabilities.** For temporary total disabilities, the fourth class of injuries, the number of days charged is the total number of full calendar days on which the injured person was unable to work as a result of the injury. The total does not include the day the injury occurred nor the day the injured person returned to work, but it does include all intervening calendar days (including Sundays, days off, or plant shutdowns). It also includes any other full days of inability to work because of the specific injury, subsequent to the injured person's return to work.

**Days charged in special cases** are indicated in the following paragraphs.

If a hernia is unrepaired (whether or not it can be repaired), it is classified as a permanent partial disability and carries a scheduled charge of 50 days. If a hernia is repaired, it is classified as a temporary total disability and the charge is only the actual number of calendar days lost.

For permanent impairments affecting more than one part of the body, the total charge is the sum of the scheduled charges for the individual body parts impaired. The total charge, however, shall not exceed 6,000 days.

If an employee suffers a permanent partial injury to one part of the body and a temporary total injury to another part in one accident, whichever charge is greater is used and determines the injury classification.

The charge for a permanent injury not identified in the schedule of charges (such as damage to internal organs, lungs, or back, or loss of speech) is a percentage of 6,000 days corresponding to the percentage of permanent total disability which results from the injury. This percentage is determined by the physician who treats the case.

## REFERENCES


American National Standards Institute, 1430 Broadway, New York, N.Y. 10018.

*Method of Recording and Measuring Work Injury Experience,* Z16.1-1967 (R1973).

*Method of Recording Basic Facts Relating to the Nature and Occurrence of Work Injuries,* Z16.2-1962 (R1969).

*Method of Recording and Measuring the Off-the-Job Disabling Accidental Injury Experience of Employees,* Z16.3-1973.

*Uniform Recordkeeping for Occupational Injuries and Illnesses,* Z16.4-1977.

National Safety Council, 444 N. Michigan Ave., Chicago, Ill. 60611. *Accident Facts* (published annually).

Recht, J. L. "Bilevel Reporting," *Journal of Safety Research,* Vol. 2, No. 2 (June 1970), pp. 51-54.

Tufte, Edward R. *The Visual Display of Quantitative Information.* Cheshire, Conn.: Graphics Press, 1983.

U.S. Dept. of Labor, Bureau of Labor Statistics, Washington, D.C. 20212. "A Brief Guide to Recordkeeping Requirements for Occupational Injuries and Illnesses." April 1986.

# 7

# Accident Investigation, Analysis and Costs



THIS CHAPTER COVERS THE INVESTIGATION of noninjury accidents as well as injury accidents. Therefore the term accident is used in its broadest sense to include occurrences that may lead to property damage or work injuries, or both.

Successful accident prevention requires a minimum of four fundamental activities:

1. A study of all working areas to detect and eliminate or control the physical or environmental hazards that contribute to accidents.
2. A study of all operating methods and practices.
3. Education, instruction, training, and discipline to minimize the human factors that contribute to accidents.
4. A thorough investigation and causal analysis of at least every accident resulting in OSHA lost workdays or a disabling injury (under ANSI Z16.1) to determine contributing circumstances. Accidents that do not result in personal injury (so-called near-accidents or near-misses) are warnings. They should not be ignored.

This fourth activity, accident investigation and analysis, is a defense against any hazards overlooked in the first three activities, those that are not obvious, or hazards that are the result of combinations of circumstances difficult to foresee.

## ACCIDENT INVESTIGATION AND ANALYSIS

Accident investigation and analysis is a means used to prevent accidents. As such, the investigation or analysis must produce information leading to corrective actions that prevent or reduce the number of accidents. The more complete the information, the easier it will be for the safety professional to take effective corrective actions. For example, knowing that 40 percent of a plant's accidents involve ladders is not as useful as knowing that 80 percent of the plant's ladder accidents involve broken rungs.

A good recordkeeping system, as discussed in Chapter 6, is essential to accident investigation because it allows the basic facts about an accident to be recorded quickly, efficiently, and uniformly.

All accidents should be investigated, regardless of severity of injury or amount of property damage. The extent of the investigation depends on the outcome or *potential* outcome of the accident. An accident involving only first aid or minor property damage is not investigated to the same degree as an accident resulting in death or extensive property damage.

For purposes of accident prevention, investigations must be for fact-finding, not fault-finding; otherwise, they can do more harm than good. This is not to say responsibility should not be fixed where personal failure has caused injury, or that such persons should be excused from the consequences of their actions. It does mean the investigation itself should be concerned only with facts. The investigating individual, board, or committee is best kept free from involvement with any punitive actions resulting from the investigation.

### Types of investigation and analysis

There are a variety of accident investigation and analysis techniques available to the investigator. Some of these techniques are more complicated than others. The choice of a particular method will depend upon the purpose and orientation of the investigation. The Failure Mode and Effect approach discussed in Chapter 4, Acquiring Hazard Information, could be very useful for investigating situations where large, complex, and interrelated machinery and procedures are involved, but may be of limited value in the investigation of accidents involving hand tools. If management procedures and communications and their relationship to accidents are of great interest, the Management Oversight and Risk Tree analysis (MORT, see references at the end of Chapter 4) could prove to be very helpful.

The accident investigation and analysis procedure outlined in the ANSI Z16.2 Standard primarily focuses on unsafe acts and unsafe conditions. Other similar techniques involve investigation within the framework of defects in man, machine, media, and management (the 4M's), or education, enforcement, and engineering (the 3 E's). For analysis purposes, these techniques involve classifying the data about a group of accidents into various categories. This has been referred to as the statistical method of analysis. Corrective actions are designed on the basis of most frequent patterns of occurrence.

Other techniques discussed in Chapter 4 come under the systems approach to safety. Systems safety stresses an enlarged viewpoint that takes into account the interrelationships between the various events that could lead to an accident. As accidents will rarely have one cause, the systems approach to safety can point to more than one place in a system where effective corrective actions can be introduced. This allows the safety professional to choose the corrective actions best meeting the criteria for effectiveness, speed of installment, cost/benefit, and the like. There are additional advantages to using systems safety techniques; they can be implemented before accidents have occurred and can be applied to new procedures and operations.

### Persons making the investigation

Depending on the nature of the accident and other conditions, the investigation should be made by the supervisor, a fellow worker familiar with the process involved, the safety and health professional or inspector, the safety and health committee, the general safety committee, or an engineer from the insurance company. If the accident involves unusual or special features, consultation with an engineer from the state labor department or U.S. Bureau of Mines, or with a union representative, may be warranted.

**The supervisor or foreman** should make an immediate report of every injury requiring medical treatment and other accidents he or she may be directed to investigate. The supervisor is on the scene, and he probably knows more about the accident than anyone else, and it is up to him or her, in most cases, to put into effect whatever measures can be adopted to prevent similar accidents. The Accident Investigation Report in the previous chapter illustrates one form that can be used to record the findings of an accident investigation.

**The safety professional.** A representative of the safety department should verify the findings of the supervisor and make an investigation of every important accident for personal information, and in most cases should make a written report to the proper official or to the general safety committee.

Nowhere are the safety professional's value and ability better shown than in the investigation of an accident. Specialized training and analytical experience enable the safety professional to search for all the facts, apparent and hidden, and to submit a report free from bias or prejudice. A professional has no interest in the investigation other than to get information that can be used to prevent a similar accident.

**Special investigative committee.** In some companies, a special committee is set up to investigate and report on all serious accidents. To be acceptable to all involved, this committee should have representatives of both management and workers. If the report of this committee is published, it would be more readily accepted by not only the workers but also management, than a report made solely by a safety and health professional.

See the discussion of safety committees in Chapter 3, Hazard Control Program Organization.

**The safety and health committee.** In many companies, especially those of small or moderate size, a number of safety activities are handled by a safety and health committee, one of which is accident investigation. Ordinarily, such investigation would be handled in a routine manner, but in important cases the chairman might call an extra meeting of the committee to conduct a special investigation.

### Cases to be investigated

An accident causing death or serious injury obviously should be thoroughly investigated. The near-accident that *might* have caused death or serious injury is equally important from the safety standpoint and should be investigated; for example, a broken crane hook or scaffold rope, or a pressure vessel explosion.

Each investigation should be made as soon after the accident as possible. A delay of only a few hours may permit important evidence to be destroyed or removed, intentionally or unintentionally. Also, the results of the inquiry should quickly be made known, inasmuch as their publicity value in the safety education of employees and supervisors is greatly increased by promptness.

Any epidemic of minor injuries demands study. A particle of emery in the eye or a scratch from handling sheet metal may seem to be a very simple case; the immediate cause is obvious, and the loss of time is small. However, if cases of this or any other type occur frequently in the plant, or in any one department, an investigation should be made to determine the underlying causes.

The chief value of such an investigation lies in uncovering contributing causes. The energetic safety professional or manager is constantly aware of the advantage of this type of accident investigation because it can prove more valuable, though less spectacular, than an inquest following a fatal injury.

Fairness and impartiality are abosolutely essential. The value of the investigation can be destroyed if there is any suspicion that its purpose is to place the blame or pass the buck. No one should be assigned to investigation work unless he or she has earned a reputation for fairness and is trained and experienced in gathering evidence. It should be made clear that accident investigations are conducted solely for the purpose of obtaining information to help prevent recurrence of accidents.

In the early years of the safety movement, accident prevention usually was a hit-or-miss activity. This approach has been replaced by more scientific techniques—see Chapter 4, Acquiring Hazard Information.

In the earlier years, a reduction in accident rates was prompted primarily by humanitarian appeal to management and workers. Although this appeal is still important, methods today are aimed at isolating and identifying accident causes in order to permit direct, positive and corrective action to prevent their recurrence.

Like other phases of modern business management, accident prevention must be based on facts clearly identifying the problem. An approach to the accident prevention problem on this basis will not only result in more effective control over accidents, but also will permit this objective to be accomplished with savings in time, effort, and money.

Accident analysis of individual cases will identify the plants, locations, or departments in which injuries most frequently occur, and will suggest necessary corrective actions to reduce accidents.

Sometimes an overall high rate is not identified with one or a few departments, but instead represents a high frequency of accidents throughout the company. Under such circumstances, it is even more important that an analysis of the accidents be made.

Similar accidents frequently may occur but at widely separated locations, so their high incidence is not apparent. Accidents may be more numerous in some machine operations than in others, or in certain procedures. Some unsafe practices that cause accidents may be committed repeatedly but at different times and in different places, so their importance as accident causes is not immediately recognized.

Analysis of the circumstances of accidents can produce these results:

1. Identify and locate the principal sources of accidents by determining, from actual experience, the materials, machines, and tools most frequently involved in accidents, and the jobs most likely to produce injuries.
2. Disclose the nature and size of the accident problems in departments and among occupations.
3. Indicate the need for engineering revision by identifying the principal hazards associated with various types of equipment and materials.
4. Disclose inefficiencies in operating processes and procedures where, for example, poor layout contributes to accidents, or where outdated physically overtaxing methods or procedures can be avoided, for instance, by using mechanical handling methods.
5. Disclose the unsafe practices which necessitate training of employees or changing work methods.
6. Enable supervisors to use the time available for safety work to the greatest advantage by providing them with information about the principal hazards and unsafe practices in their departments.
7. Permit an objective evaluation of the progress of a safety program by noting in continuing analyses the effect of corrective actions, educational techniques, and other methods adopted to prevent injuries.

### The minimum data

The purpose of an accident investigation is to identify facts about each injury and the accident that produced it and to record those facts. These records, individually and collectively, serve as guides to the areas, conditions, and circumstances to which accident prevention efforts can most profitably be directed.

The following paragraphs describe the data elements comprising the minimum amount of information that should be collected about each accident. The Accident Investigation Report Form, Figure 6-2 in the previous chapter, shows a minimum data set that was developed to improve the quality of accident investigation and analysis. This minimum data set identifies the why of some of the accident characteristics as well as the who, what, when, and where. It acknowledges the existence of multiple causes of accidents by not restricting the investigator or analyst to selecting a single causal act or condition.

The investigation also is expanded from focusing solely on the injury and accident type to include the entire sequence of events leading to the injury, as far back in time as the investigator feels is relevant. This expanded view of the accident sequence allows an employer to identify and implement a wider variety of corrective actions.

- The first of the eight groups of data elements is *employer characteristics*. This includes the type of industry and the size of the company (number of full-time equivalent employees). It is needed when data from one company is compared with that of another.
- Second is *employee characeristics*. The victim's age and sex, the department and occupation in which he or she worked, and whether a full-time, part-time, or seasonal employee. Questions about the victim's experience are also asked. How long has the victim been with the company? How long in current occupation? How often the employee repeats the activity engaged in when the accident occurred?
- The third group is about the accident itself. A *narrative description* should be prepared; it should include what the person was doing, what objects or substances were involved, and actions or movements which led to the injury. This is elaborated into a detailed accident sequence beginning with the injuring event and works backward in time through all of the preceding events that directly contributed to the accident. The data also includes a description of any product or equipment directly involved with the accident sequence and the task being performed. Any other conditions, such as temperature, light, noise, and weather, pertaining to the accident also should be noted here.
- The fourth group deals with the characteristics of *the equipment* associated with the accident. The description should include the type, brand, size, and any distinguishing features of the equipment, its condition, and the specific part involved.
- Fifth is the characteristics of *the task* being performed when the accident happened: The general task (such as repairing a conveyor) and the specific activity (such as using a wrench). The description should include the posture and location of the employee (for example, squatting under the conveyor) and whether he was working alone or with others.
- *Time factors* are the sixth group. The investigation should record the time of day and how it related to the shift the victim was working, whether first hour of the shift, second hour, or later. What type of shift—day, swing, straight, rotating, or other. And the phase of the employee's workday; performing work, rest period, meal period, overtime, entering or leaving plant.
- The seventh group, use and nature of *preventive measures*, includes the following questions: What personal protective equipment was being worn and did the employee's apparel affect the accident sequence? What kind of training did the employee have for the task he was performing? Did standards or procedures exist for the task? Were they written? Were they followed? If not followed, how did what happened differ from what should have happened? Were all guards in place and in use? What was the nature of supervision at the time of the accident? What immediate remedial actions were taken to prevent recurrence?

- The last group of questions concerns the *severity of the injury.* The nature of the injury or injuries and the parts of the body affected must be recorded as well as the OSHA severity class. If the accident resulted in some permanent impairment this should be noted.

The answers to the questions in these eight groups constitute the *minimum* information needed to proceed with an analysis. The nature of a company's operations or the interests of the analyst may suggest other questions to be answered in the investigation.

There are two types of analysis that can be done. First, the individual accident can be examined to determine the corrective action or actions to prevent future occurrences of this specific sequence of events. The other kind of analysis, the statistical analysis, examines a group of similar occurrences for patterns lending themselves to corrective actions. Over time this statistical analysis can show which corrective actions have been more effective than others.

### Identifying causal factors and selecting corrective actions

In any accident, there are many factors at work that permit the occurrence of the sequence of events leading to the injury. The idea behind the corrective action selection procedure is to identify all the factors for which a corrective action is possible and then to select the ones likely to be most effective, most cost/beneficial, most acceptable, and so on, and implement those.

Figure 7-1 shows the Guide for Identifying Causal Factors and Corrective Action. This guide is used with the Accident Investigation Report in Figure 6-2. There are four parts to the Guide: Equipment, Environment, People, and Management. Usually these elements combine to produce products and profits, but sometimes they conjoin to produce accidents. The questions in the four sections assist the investigator to systematically consider the contribution each factor made to the accident.

The structure of the Guide makes it easy to identify the causal factors. Questions are answered by placing an X in a circle or a box. An X in a circle means the item is a causal factor. When a box is marked with an X, the item was not a causal factor.

The Comment column provides space to record the specific information about the accident being investigated. The Recommended Corrective Actions column has room to enter specific corrective actions for each causal factor.

After listing all of the possible corrective actions identified on the Guide each action must be evaluated for effectiveness, cost, feasibility, reliability, acceptance, effect on productivity, time required to implement, and any other factor deemed important, before deciding which ones to implement.

This systematic approach to selecting corrective actions ensures that all major types are considered, that the analyst does not stop with familiar and favorite corrective actions, and that each corrective action chosen for implementation is well thought out.

### Classifying accident data

Even in large company operations where hundreds of accidents can occur annually, only rarely do two accidents occur in exactly the same way. Accidents do follow general patterns, however, and grouping them according to pattern is necessary for purposes of analysis.

Finding the patterns and common features of groups of cases is the statistical approach to accident analysis.

**Setting up classifications.** Before the actual analysis work is begun, classifications must be set up for grouping the various data. For each basic fact, general classifications should be established in which similar data can be grouped. Then, more specific classifications should be set up within each general classification to preserve as many of the details as possible.

For example, in ANSI Z16.2, among the general classifications recommended for the key fact Hazardous Condition are the following:

1. Defects of agencies
2. Dress or apparel hazards
3. Environmental hazards
4. Placement hazards
5. Inadequate safeguarding
6. Public hazards.

Within each one of these general classifications, more specific classifications are set up. Under Defects of agencies, for example, are listed:

1. Composed of unsuitable materials
2. Dull
3. Improperly constructed, assembled, etc.
4. Improperly designed
5. Rough
6. Sharp
7. Slippery
8. Worn, cracked, broken, etc.
9. Other.

It is not always possible to set up classifications before the analysis is begun. In this case, classifications can be developed as reports are reviewed and situations are revealed.

For example, if an analysis is being made of ladder accidents and it is found in a number of cases broken rungs caused the accidents, a specific group for Broken rungs should be set up under the classification Defects of agencies.

ANSI Z16.2 recommends general and specific classifications for most key facts. These classifications are presented principally as *suggestions* to guide the analyst in setting up classifications to fit individual problems. The point must be emphasized that for an analysis to be of maximum usefulness, classifications must be set up to encompass the situations pertinent to the particular company.

**Use of a numerical code.** Regardless of the method eventually used to sort and tabulate the various key facts, the work will be facilitated if code numbers are assigned to the different classifications. With this method, each case is read only once, at which time code numbers are assigned to the different facts, and subsequent sorting of the various facts can be quickly completed by an easy reference to the code numbers.

A numerical code is simply the assigning of numbers in sequence to a list of similar facts. For each basic fact (agency of accident, accident type, hazardous condition, etc.), there

**Figure 7-1.** The Guide for Identifying Causal Factors and Corrective Actions assists the accident investigator to systematically consider four contributing accident factors: Equipment, Environment, People, and Management.

Answer questions by placing an *X* in the "Y" circle or box for YES or in the "N" circle or box for NO.

# GUIDE for IDENTIFYING CAUSAL FACTORS and CORRECTIVE ACTIONS

CASE NUMBER

| PART 1—EQUIPMENT | | | | |
|---|---|---|---|---|
| ○ Y □ N | **1.0 WAS A HAZARDOUS CONDITION(S) A CONTRIBUTING FACTOR?** If yes, answer the following. If no, proceed to Part 2. | | | |
| | **CAUSAL FACTORS** | **COMMENT** | **POSSIBLE CORRECTIVE ACTIONS** | **RECOMMENDED CORRECTIVE ACTIONS** |
| ○ Y □ N | 1.1 Did any defect(s) in equipment/tool(s)/material contribute to hazardous condition(s)? | | Review procedure for inspecting, reporting, maintaining, repairing, replacing, or recalling defective equipment/tool(s)/material used. | |
| □ Y ○ N | 1.2 Was the hazardous condition(s) recognized? If yes, answer A and B. If no, proceed to 1.3. | | Perform job safety analysis. Improve employee ability to recognize existing or potential hazardous conditions. Provide test equipment, as required, to detect hazard. Review any change or modification of equipment/tool(s)/material. | |
| □ Y ○ N | A. Was the hazardous conditions(s) reported? | | Train employees in reporting procedures. Stress individual acceptance of responsibility. | |
| □ Y ○ N | B. Was employee(s) informed of the hazardous condition(s) and the job procedures for dealing with it as an interim measure? | | Review job procedures for hazard avoidance. Review supervisory responsibility. Improve supervisor-employee communications. Take action to remove or minimize hazard. | |
| □ Y ○ N | 1.3 Was there an equipment inspection procedure(s) to detect the hazardous condition(s)? | | Develop and adopt procedures (for example, an inspection system) to detect hazardous conditions. Conduct test. | |
| □ Y ○ N | 1.4 Did the existing equipment inspection procedure(s) detect the hazardous condition(s)? | | Review procedures. Change frequency or comprehensiveness. Provide test equipment as required. Improve employee ability to detect defects and hazardous conditions. Change job procedures as required. | |
| □ Y ○ N | 1.5 Was the correct equipment/tool(s)/material used? | | Specify correct equipment/tool(s)/material in job procedures. | |
| □ Y ○ N | 1.6 Was the correct equipment/tool(s)/material readily available? | | Provide correct equipment/tool(s)/material. Review purchasing specifications and procedures. Anticipate future requirements. | |
| □ Y ○ N | 1.7 Did employee(s) know where to obtain equipment/tool(s)/material required for the job? | | Review procedures for storage, access, delivery, or distribution. Review job procedures for obtaining equipment/tool(s)/material. | |
| ○ Y □ N | 1.8 Was substitute equipment/tool(s)/material used in place of correct one? | | Provide correct equipment/tool(s)/material. Warn against use of substitutes in job procedures and in job instruction. | |
| ○ Y □ N | 1.9 Did the design of the equipment/tool(s) create operator stress or encourage operator error? | | Review human factors engineering principles. Alter equipment/tool(s) to make it more compatible with human capability and limitations. Review purchasing procedures and specifications. Check out new equipment and job procedures involving new equipment before putting into service. Encourage employees to report potential hazardous conditions created by equipment design. | |
| ○ Y □ N | 1.10 Did the general design or quality of the equipment/tool(s) contribute to a hazardous condition? | | Review criteria in codes, standards, specifications, and regulations. Establish new criteria as required. | |
| ○ | 1.11 List other causal factors in "Comment" column. | | | |

**Figure 7-1.** (Continued.)

| | | | | |
|---|---|---|---|---|
| **PART 2—ENVIRONMENT** | | | | |
| Y N | **2.0 WAS THE LOCATION/POSITION OF EQUIPMENT/MATERIALS/EMPLOYEE(S) A CONTRIBUTING FACTOR?** If yes, answer the following. If no, proceed to Part 3. | | | |
| | **CAUSAL FACTORS** | **COMMENT** | **POSSIBLE CORRECTIVE ACTIONS** | **RECOMMENDED CORRECTIVE ACTIONS** |
| Y N | 2.1 Did the location/position of equipment/material/employee(s) contribute to a hazardous condition? | | Perform job safety analysis. Review job procedures. Change the location, position, or layout of the equipment. Change position of employee(s). Provide guardrails, barricades, barriers, warning lights, signs, or signals. | |
| Y N | 2.2 Was the hazardous condition recognized? If yes, answer A and B. If no, proceed to 2.3. | | Perform job safety analysis. Improve employee ability to recognize existing or potential hazardous conditions. Provide test equipment, as required, to detect hazard. Review any change or modification of equipment/tools/materials. | |
| Y N | A. Was the hazardous condition reported? | | Train employees in reporting procedures. Stress individual acceptance of responsibility. | |
| Y N | B. Was employee(s) informed of the job procedure for dealing with the hazardous condition as an interim action? | | Review job procedures for hazard avoidance. Review supervisory responsibility. Improve employee-supervisor communications. Take action to remove or minimize hazard. | |
| Y N | 2.3 Was employee(s) supposed to be in the vicinity of the equipment/material? | | Review job procedures and instruction. Provide guardrails, barricades, barriers, warning lights, signs, or signals. | |
| Y N | 2.4 Was the hazardous condition created by the location/position of equipment/material visible to employee(s)? | | Change lighting or layout to increase visibility of equipment. Provide guardrails, barricades, barriers, warning lights, signs or signals, floor stripes, etc. | |
| Y N | 2.5 Was there sufficient workspace? | | Review workspace requirements and modify as required. | |
| Y N | 2.6 Were environmental conditions a contributing factor (for example, illumination, noise levels, air contaminant, temperature extremes, ventilation, vibration, radiation)? | | Monitor, or periodically check, environmental conditions as required. Check results against acceptable levels. Initiate action for those found unacceptable. | |
| | 2.7 List other causal factors in "Comment" column. | | | |

| | | | | |
|---|---|---|---|---|
| **PART 3—PEOPLE** | | | | |
| Y N | **3.0 WAS THE JOB PROCEDURE(S) USED A CONTRIBUTING FACTOR?** If yes, answer the following. If no, proceed to Part 3.6. | | | |
| | **CAUSAL FACTORS** | **COMMENT** | **POSSIBLE CORRECTIVE ACTIONS** | **RECOMMENDED CORRECTIVE ACTIONS** |
| Y N | 3.1 Was there a written or known procedure (rules) for this job? If yes, answer A, B, and C. If no, proceed to 3.2. | | Perform job safety analysis and develop safe job procedures. | |
| Y N | A. Did job procedures anticipate the factors that contributed to the accident? | | Perform job safety analysis and change job procedures. | |
| Y N | B. Did employee(s) know the job procedure? | | Improve job instruction. Train employees in correct job procedures | |
| Y N | C. Did employee(s) deviate from the known job procedure? | | Determine why. Encourage all employees to report problems with an established procedure to supervision. Review job procedure and modify if necessary. Counsel or discipline employee. Provide closer supervision. | |
| Y N | 3.2 Was employee(s) mentally and physically capable of performing the job? | | Review employee requirements for the job. Improve employee selection. Remove or transfer employees who are temporarily, either mentally or physically, incapable of performing the job. | |

**Figure 7-1.** (Continued.)

| | CAUSAL FACTORS | COMMENT | POSSIBLE CORRECTIVE ACTIONS | RECOMMENDED CORRECTIVE ACTIONS |
|---|---|---|---|---|
| ○ Y □ N | 3.3 Were any tasks in the job procedure too difficult to perform (for example, excessive concentration or physical demands)? | | Change job design and procedures. | |
| ○ Y □ N | 3.4 Is the job structured to encourage or require deviation from job procedures (for example, incentive, piecework, work pace)? | | Change job design and procedures. | |
| ○ | 3.5 List other causal factors in "Comment" column. | | | |
| ○ Y □ N | **3.6 WAS LACK OF PERSONAL PROTECTIVE EQUIPMENT OR EMERGENCY EQUIPMENT A CONTRIBUTING FACTOR IN THE INJURY?** If yes, answer the following. If no, proceed to Part 4. NOTE: The following causal factors relate to the *injury*. | | | |
| | **CAUSAL FACTORS** | **COMMENT** | **POSSIBLE CORRECTIVE ACTIONS** | **RECOMMENDED CORRECTIVE ACTIONS** |
| □ Y ○ N | 3.7 Was appropriate personal protective equipment (PPE) specified for the task or job? If yes, answer A, B, and C. If no, proceed to 3.8. | | Review methods to specify PPE requirements. | |
| □ Y ○ N | A. Was appropriate PPE available? | | Provide appropriate PPE. Review purchasing and distribution procedures. | |
| □ Y ○ N | B. Did employee(s) know that wearing specified PPE was required? | | Review job procedures. Improve job instruction. | |
| □ Y ○ N | C. Did employee(s) know how to use and maintain the PPE? | | Improve job instruction. | |
| □ Y ○ N | 3.8 Was the PPE used properly when the injury occurred? | | Determine why and take appropriate action. Implement procedures to monitor and enforce use of PPE. | |
| □ Y ○ N | 3.9 Was the PPE adequate? | | Review PPE requirements. Check standards, specifications, and certification of the PPE. | |
| □ Y ○ N | 3.10 Was emergency equipment specified for this job (for example, emergency showers, eyewash fountains)? If yes, answer the following. If no, proceed to Part 4. | | Provide emergency equipment as required. | |
| □ Y ○ N | A. Was emergency equipment readily available? | | Install emergency equipment at appropriate locations. | |
| □ Y ○ N | B. Was emergency equipment properly used? | | Incorporate use of emergency equipment in job procedures. | |
| □ Y ○ N | C. Did emergency equipment function properly? | | Establish inspection/monitoring system for emergency equipment. Provide for immediate repair of defects. | |
| ○ | 3.11 List other causal factors in "Comment" column. | | | |
| | **PART 4—MANAGEMENT** | | | |
| ○ Y □ N | **4.0 WAS A MANAGEMENT SYSTEM DEFECT A CONTRIBUTING FACTOR?** If yes, answer the following. If no, STOP. Your causal factor indentification exercise is complete. | | | |
| | **CAUSAL FACTORS** | **COMMENT** | **POSSIBLE CORRECTIVE ACTIONS** | **RECOMMENDED CORRECTIVE ACTIONS** |
| ○ Y □ N | 4.1 Was there a failure by supervision to detect, anticipate, or report a hazardous condition? | | Improve supervisor capability in hazard recognition and reporting procedures. | |
| ○ Y □ N | 4.2 Was there a failure by supervision to detect or correct deviations from job procedure? | | Review job safety analysis and job procedures. Increase supervisor monitoring. Correct deviations. | |

**Figure 7-1.** (Continued.)

| | CAUSAL FACTORS | COMMENT | POSSIBLE CORRECTIVE ACTIONS | RECOMMENDED CORRECTIVE ACTIONS |
|---|---|---|---|---|
| Y N | 4.3 Was there a supervisor/employee review of hazards and job procedures for tasks performed infrequently? (Not applicable to all accidents.) | | Establish a procedure that requires a review of hazards and job procedures (preventive actions) for tasks performed infrequently. | |
| Y N | 4.4 Was supervisor responsibility and accountability adequately defined and understood? | | Define and communicate supervisor responsibility and accountability. Test for understandability and acceptance. | |
| Y N | 4.5 Was supervisor adequately trained to fulfill assigned responsibility in accident prevention? | | Train supervisors in accident prevention fundamentals. | |
| Y N | 4.6 Was there a failure to initiate corrective action for a known hazardous condition that contributed to this accident? | | Review management safety policy and level of risk acceptance. Establish priorities based on potential severity and probability of recurrence. Review procedure and responsibility to initiate and carry out corrective actions. Monitor progress. | |
| | 4.7 List other causal factors in "Comment" column. | | | |

**Developed by the National Safety Council**

**Figure 7-1.** (Concluded.)

should be no duplication of numbers; but for the different facts, the numbering series may be repeated.

After the cases have been reviewed and code numbers have been assigned to the different key facts, the reports can be easily and quickly sorted or arranged by any of the facts to reveal the principal data concerning the accidents.

Numerical codes already are assigned to all the classifications included in ANSI Z16.2, and if an analyst uses this standard, he can use the code number too. If the analyst uses the standard as a starting point and adds other classifications to cover the specific accident experience of his own company, he can code the additional items to fit into the code of the standard. Using the ANSI Z16.2 standard has the additional advantage of compatibility with data and analyses published by the Bureau of Labor Statistics, state workers' compensation authorities, and the National Safety Council.

## Making the analysis

Experience has proved that the most effective way to reduce accidents is to concentrate on one phase of the accident problem at a time rather than attempting to stop all accidents at once. There are different ways in which the problem can be approached on this basis, any one of which should prove effective.

The reports may be grouped by occupation of the injured person. Each group of reports, then, may be reviewed to determine what accident types, sources of injury, and agencies of accident are most prevalent among different occupations. Such information is particularly helpful in planning employee training and in developing educational materials and programs.

Injury incidence rates computed by departments may reveal that injuries occur at sharply higher rates in some departments than in others. If this is the case, an analysis should be made of the accidents in the high-rate departments to find the sources

**Table 7-A.** Nature of Injury versus Part of Body

| | *Part of Body* | | | | | | | | | | | |
|---|---|---|---|---|---|---|---|---|---|---|---|---|
| *Nature of Injury* | *Eyes* | *Head, Face, Neck* | *Back* | *Trunk* | *Arm* | *Hand, Wrist* | *Finger* | *Leg* | *Foot, Ankle* | *Toe* | *Internal, Other* | *Total* |
| Amputation | 0 | 0 | 0 | 0 | 0 | 0 | 1 | 0 | 0 | 0 | 0 | **1** |
| Burn & scald (heat) | 0 | 0 | 0 | 0 | 0 | 0 | 0 | 0 | 0 | 0 | 0 | **0** |
| Burn (chemical) | 2 | 0 | 0 | 0 | 0 | 0 | 0 | 0 | 0 | 0 | 0 | **2** |
| Concussion | 0 | 2 | 0 | 0 | 0 | 0 | 0 | 0 | 0 | 0 | 0 | **2** |
| Crushing | 0 | 0 | 0 | 0 | 0 | 1 | 0 | 0 | 0 | 2 | 0 | **3** |
| Cut, laceration, puncture, abrasion | 0 | 1 | 0 | 0 | 1 | 3 | 18 | 0 | 0 | 0 | 0 | **23** |
| Fracture | 0 | 0 | 0 | 0 | 0 | 2 | 5 | 0 | 1 | 0 | 0 | **8** |
| Hernia | 0 | 0 | 0 | 1 | 0 | 0 | 0 | 0 | 0 | 0 | 0 | **1** |
| Bruise, contusion | 0 | 2 | 0 | 1 | 0 | 3 | 2 | 3 | 0 | 1 | 0 | **12** |
| Occupational illness | 0 | 1 | 0 | 0 | 1 | 2 | 0 | 0 | 0 | 0 | 8 | **12** |
| Sprain, strain | 0 | 0 | 24 | 0 | 0 | 2 | 2 | 3 | 4 | 0 | 0 | **35** |
| Other | 1 | 0 | 0 | 0 | 0 | 0 | 0 | 0 | 0 | 0 | 5 | **6** |
| Total | **3** | **6** | **24** | **2** | **2** | **13** | **28** | **6** | **5** | **3** | **13** | **105** |

This crosstabulation shows how the nature of injury and the part of body interact. In this example, cuts most often affect the fingers and sprains and strains usually involve the back. Note that bruises and contusions affect several body parts.

of the accidents and their causes. This method will permit concentration of effort in the locations in which accidents occur most frequently.

If injury incidence rates reveal a high rate of occurrence in general throughout the plant, the analysis procedure usually starts with information about the injury, goes on to identify the injury-producing event, and then looks at the circumstances and causal factors. The same procedure can be followed to examine injuries occurring within a high-rate occupation or department.

A crosstabulation of the injury data can be done to show the relationship or interaction between the two categories. Table 7-A illustrates an analysis of the Nature of Injury versus the Part of Body. This crosstabulation, in addition to showing what types of personal protective equipment might be useful, also points out common injury patterns needing further investigation.

A crosstabulation can extend in several directions and cause the need for further, separate crosstabulations. For example, the categories shown in Table 7-A having the highest frequency of injuries (Cut, laceration, puncture, or abrasion to fingers and Sprain or strain to the back) would be used to construct a second (see Table 7-B) and a third crosstabulation.

As shown in Table 7-B, by the second tabulation the number of cases in a category, such as the nine finger cuts resulting from being struck by metal, usually is small enough to be practical to read the individual accident reports and find common causal factors and determine corrective actions. If the number of cases in a category is too large, then additional crosstabulations, such as Location versus activity or Activity versus occupation, can sufficiently reduce the number of cases and allow study of individual accident reports.

**Methods of tabulating.** For analyzing a small number of reports (up to about 100), hand sorting and tallying is efficient and satisfactory. The principal advantage is that the original records are being used, and all the information is available should reference to it become necessary.

A personal computer or other data processing equipment is best for large collections of cases, and is useful for smaller data sets as well because it can sort and display cases very quickly, allowing the investigator to concentrate on the accidents and test alternative hypotheses easily.

### Using the analysis

Of course, merely obtaining the information will not prevent recurrence of accidents; the contributing conditions must be corrected. Thorough analysis of groups of accident investigation reports can point to corrective actions that might not be evident when studying an individual case. In particular, inadequate policies, procedures, or management systems often are apparent after looking at the "forest" rather than the "trees."

The statistical evidence revealed in an analysis can provide the guidance to direct safety efforts along the most productive and efficient path. The analysis will provide objective support and justification for budget requests, training programs, or other safety activity.

## ESTIMATING ACCIDENT COSTS

This discussion concerns the elements of cost most likely to result from a work accident and presents a method whereby an organization can obtain an accurate estimate of the total costs of its work accidents. (This procedure for estimating costs was developed by Rollin H. Simonds, Ph.D., Professor, Michigan State College, under the direction of the Statistics Division, National Safety Council.)

Reliable cost information is one basis for decisions upon which efficiency and profit depend. Even in so obviously desirable an activity as accident prevention, some proposed measures or alternatives must be accepted or rejected on the basis of their probable effect on profits.

Although most executives want to make their company a safe place to work, they also have a responsibility to run their business profitably. Consequently, they may be reluctant to spend money for accident prevention unless they can see a prospect for saving at least as much as they spend. *Without information on the cost of accidents, it is practically impossible to estimate the savings brought about by expenditures for accident prevention.*

Annual reports stressing dollar savings are more meaningful

Table 7-B. Source of Injury versus Type of Accident

| Source of Injury | Fall from elevation | Fall on same level | Struck against | Struck by | Caught in, under, or between | Rubbed or abraded | Bodily reaction | Overexertion |
|---|---|---|---|---|---|---|---|---|
| | *Type of Accident* | | | | | | | |
| Machine | 0 | 0 | 0 | 0 | 3 | 0 | 0 | 0 |
| Conveyor, elev. hoist | 0 | 0 | 0 | 0 | 0 | 0 | 0 | 0 |
| Vehicle | 0 | 0 | 0 | 0 | 0 | 0 | 0 | 0 |
| Electrical apparatus | 0 | 0 | 0 | 0 | 0 | 0 | 0 | 0 |
| Hand tool | 0 | 0 | 0 | 4 | 0 | 0 | 0 | 0 |
| Chemical | 0 | 0 | 0 | 0 | 0 | 0 | 0 | 0 |
| Working surface, bench, etc. | 0 | 0 | 0 | 0 | 0 | 0 | 0 | 0 |
| Floor, walking surface | 0 | 0 | 0 | 0 | 0 | 0 | 0 | 0 |
| Bricks, rocks, stones | 0 | 0 | 0 | 0 | 0 | 0 | 0 | 0 |
| Box, barrel, container | 0 | 0 | 0 | 0 | 0 | 0 | 0 | 0 |
| Door, window, etc. | 0 | 0 | 0 | 0 | 0 | 0 | 0 | 0 |
| Ladder | 0 | 0 | 0 | 0 | 0 | 0 | 0 | 0 |
| Lumber, woodworking metals | 0 | 0 | 0 | 0 | 0 | 0 | 0 | 0 |
| Metal | 0 | 0 | 0 | 9 | 0 | 0 | 0 | 0 |
| Stairway, steps | 0 | 0 | 0 | 0 | 0 | 0 | 0 | 0 |
| Other | 0 | 0 | 0 | 0 | 0 | 0 | 0 | 0 |
| Unknown | 0 | 0 | 0 | 0 | 0 | 0 | 0 | 0 |
| None | 0 | 0 | 0 | 0 | 0 | 0 | 0 | 0 |
| Total | 0 | 0 | 0 | 13 | 3 | 0 | 0 | 0 |

This crosstabluation shows Source of Injury versus Type of Accident for one of the most frequent injuries identified in Table 7-A. In this example, being struck by a metal object was the source of most finger cuts. Other causes were being struck by hand tools or becoming caught in machinery.

to management than those using incidence rates. Facts about the costs of accidents also may be effectively used in securing the active cooperation of supervisors. Supervisors usually are cost conscious because they are expected to run their departments profitably. Monthly reports showing the cost of accidents or the savings resulting from good accident records are an important motivation to achieve safe operating procedures.

## Definition of work accidents for cost analysis

Work accidents, for the purpose of cost analysis, are unintended occurrences arising in the work environment. These accidents fall into two general categories: (1) incidents resulting in work injuries or illnesses and (2) accidents that cause property damage or interfere with production in such a manner that personal injury might result.

The inclusion of the no-injury accidents makes "work accidents" roughly synonymous with the type of occurrences a safety department strives to prevent.

## Method for estimating

To be of maximum usefulness, cost figures should represent as accurately as possible the specific experience of the company. A fixed ratio of indirect to direct costs developed from experience representing many different companies in many different industries does *not* serve such a purpose. Estimated costs of accidents in general do not take into account differences in hazards from one industry to another or the more important differences in safety performances from one company to another.

Since the distinctions between direct and indirect costs are difficult to maintain, they have been abandoned in favor of the more precise terms "insured" and "uninsured" costs. Using these data, a company can estimate its accident cost with reasonable accuracy.

**Insured costs.** Every organization paying compensation insurance premiums recognizes such expense as part of the cost of accidents. In some cases, medical expenses, too, may be covered by insurance. These costs are definite, and they are known. They comprise the insured element of the total accident cost.

In addition to these costs, many other costs arise in connection with accidents. The cost of damaged equipment is easily identified. Others, such as wages paid to the injured employee for hours during which he is not producing, are hidden. These items comprise the uninsured element of the total accident cost.

**Uninsured costs.** Insured costs can be determined easily from accounting records. The difficult part is determining applicable uninsured (frequently called "indirect") costs, and the method described here will serve that purpose.

The first step is to make a pilot study to ascertain approximate averages of uninsured costs for each of the following four classes of accidents:

CLASS 1 —Cases involving lost workdays, if records are kept under OSHAct, or permanent partial disabilities and temporary total disabilities, if records are based on ANSI Z16.1.

CLASS 2 —Medical treatment cases requiring the attention of a physician outside the plant.

CLASS 3 —Medical treatment cases requiring only first aid or local dispensary treatment and resulting in property damage of less than $100 or loss of less than 8 hours in work time.

CLASS 4 —Accidents that either cause no injury or cause minor injury not requiring the attention of a physician, and result in property damage of $100 or more, or loss of 8 or more employee-hours.

Once average costs have been established for each accident class, they can be used as multipliers to obtain total uninsured costs in subsequent periods. These costs then may be added to known insurance premium costs to determine the total cost of accidents.

**Table 7-B.** (Continued.)

| | Type of Accident | | | | | | | |
|---|---|---|---|---|---|---|---|---|
| | Contact with electrical current | Contact with temperature extremes | Radiations, caustics, toxic and noxious substances | Public transportation accident | Motor vehicle accident | Other | Unknown | Total |
| Machine | 0 | 0 | 0 | 0 | 0 | 0 | 0 | **3** |
| Conveyor, elev. hoist | 0 | 0 | 0 | 0 | 0 | 0 | 0 | 0 |
| Vehicle | 0 | 0 | 0 | 0 | 0 | 0 | 0 | 0 |
| Electrical apparatus | 0 | 0 | 0 | 0 | 0 | 0 | 0 | 0 |
| Hand tool | 0 | 0 | 0 | 0 | 0 | 0 | 0 | **4** |
| Chemical | 0 | 0 | 0 | 0 | 0 | 0 | 0 | 0 |
| Working surface, bench, etc. | 0 | 0 | 0 | 0 | 0 | 0 | 0 | 0 |
| Floor, walking surface | 0 | 0 | 0 | 0 | 0 | 0 | 0 | 0 |
| Bricks, rocks, stones | 0 | 0 | 0 | 0 | 0 | 0 | 0 | 0 |
| Box, barrel, container | 0 | 0 | 0 | 0 | 0 | 0 | 0 | 0 |
| Door, window, etc. | 0 | 0 | 0 | 0 | 0 | 0 | 0 | 0 |
| Ladder | 0 | 0 | 0 | 0 | 0 | 0 | 0 | 0 |
| Lumber, woodworking metals | 0 | 0 | 0 | 0 | 0 | 0 | 0 | 0 |
| Metal | 0 | 0 | 0 | 0 | 0 | 0 | 0 | **9** |
| Stairway, steps | 0 | 0 | 0 | 0 | 0 | 0 | 0 | 0 |
| Other | 0 | 0 | 0 | 0 | 0 | 0 | **2** | **2** |
| Unknown | 0 | 0 | 0 | 0 | 0 | 0 | 0 | 0 |
| None | 0 | 0 | 0 | 0 | 0 | 0 | 0 | 0 |
| Total | 0 | 0 | 0 | 0 | 0 | 0 | **2** | **18** |

## Example of a cost estimate

An estimate of costs made by one company is given in the following example. First, a pilot study was made to get the average cost of each class of accident. Included in the study were 20 Class 1 accidents, 30 Class 2 accidents, 50 Class 3 accidents, and 20 Class 4 accidents. Costs were determined and averages developed as in Table 7-C.

**Table 7-C.** Average Costs Determined by Pilot Study

| Class of Accident | Number of Accidents Reported | Average Uninsured Cost |
|---|---|---|
| Class 1 | 20 | $217.90 |
| Class 2 | 30 | 70.10 |
| Class 3 | 50 | 13.60 |
| Class 4 | 20 | 440.00 |

During the entire year, the company had 34 Class 1 accidents, 148 Class 2 accidents, and 4,000 Class 3 accidents. No record was kept of the Class 4 accidents after the pilot study was completed. Instead, the ratio of the number of Class 4 to Class 1 accidents found in the pilot study was used. This ratio was shown to be about 1 to 1, and since there were 34 Class 1 accidents during the year, it was assumed there were about 34 Class 4 accidents. (A separate record could be kept of the number of Class 4 accidents.)

The average cost for each accident class was applied to these totals to secure the results shown in Table 7-D.

Since the final total is the sum of many estimates, it should not be implied that the total figure suggests absolute accuracy. Whether $134,000 or $135,000 is chosen as the final figure depends largely on the analyst's judgment of whether the various elements may have been overestimated or underestimated. In this case, the analyst judged that the pilot study represented conservative estimates of the average costs. So he reported to the plant manager, "During the past year, accidents cost this company about $135 thousand in compensation, medical expense, lost time, and property damage."

*The average costs determined in this pilot study represent the actual experience of this particular company. Until important changes take place in this company's safety program, in the kind of machinery used or persons employed, or in other aspects affecting costs, the same average costs can be used.*

### Adjusting for inflation

The effects of inflation quickly can make obsolete the cost figures found in a pilot study. To account for this effect, the cost factors should be adjusted each year. The wage-related cost elements can be multiplied by the change in the general level of wages in the company. Other cost elements can be brought up to date by multiplying them by the general inflation rate as measured by the change in the Consumer Price Index. Because these adjustments are only approximate, the pilot study should be repeated at least every five years to establish new benchmarks.

### Items of uninsured cost

Important to a pilot study is a careful investigation of each accident to determine all the costs arising out of it. The following items of uninsured or indirect cost are clearly the result of

**Table 7-D.** Estimate of Yearly Accident Costs

| Class of Accident | Number of Accidents | Average Cost per Accident (from pilot study) | Total Uninsured Cost |
|---|---|---|---|
| Class 1 | 34 | $217.90 | $ 7,408.60 |
| Class 2 | 148 | 70.10 | 10,374.80 |
| Class 3 | 4,000 | 13.60 | 54,400.00 |
| Class 4 | 34 | 440.00 | 14,960.00 |
| Total Uninsured Cost | | | $ 87,143.40 |
| Insurance Premiums | | | 47,200.00 |
| Total Accident Cost for the Period | | | $134,343.40 |

DEPARTMENT SUPERVISOR'S ACCIDENT COST INVESTIGATION REPORT

Injury/Accident________________

Date______________ Name of Injured____________________________ Dept.________________

**TIME LOST**

1. How much time did other employees lose by talking, watching, or helping at accident? Number of employees ____________ x hours = [ ]
2. How much productive time was lost because of damaged equipment or loss of reduced output by injured worker?
   Estimate Hours = [ ]
3. How much time did injured employee lose for which he was paid on the day of the injury?
   Estimate Hours = [ ]
4. Will overtime be necessary? Estimate Hours = [ ]
5. How much of the supervisors or other managements' time was lost as a result of this accident?
   Estimate Hours = [ ]
6. Were additional costs incurred due to hiring and training or replacement?
   Training Time Estimate Hours = [ ]
7. Describe the damage to material or equipment. ________________________________

   ______________________________________________________________________________
8. If machine and/or operations were idle, can loss of production be made up?
   Yes__________ No__________
9. Will overtime be necessary? Yes__________ No__________
10. Any demurrage or other cost involved? Yes__________ No__________

ADDITIONAL ACCIDENT COSTS

To compute the total costs of the accident, it is necessary to complete the following costs. Should the supervisor have access to this information it is advised he complete as much as possible. Safety Department will develop those costs not known by supervisor.

**Figure 7-2.** This cost form (8½ × 11 in.) should be prepared by the department supervisor as soon after the accident as information becomes available on the amount of time lost by all persons and the extent of damage to product and equipment. It is sent to the safety department not later than the day after the accident.

```
11.  Estimate of demurrage or other costs.                      $

12.  Costs associated with giving medical attention, first-aid,
     ambulance costs, etc.                                      $

13.  Workers Compensation costs.                                $

14.  Hospital medical costs.                                    $

15.  Costs associated with placing injured on other work when
     unable to perform regular work.                            $

16.  Costs associated with questions 1 through 6.

                      16-1                                      $
                      16-2                                      $
                      16-3                                      $
                      16-4                                      $
                      16-5                                      $
                      16-6                                      $__________

17.  Company dollars lost on accident:                    TOTAL $

                                                     Stock No. 129.27
```

**Figure 7-2.** (Continued.)

work accidents and are subject to reasonably reliable measurement. Less tangible losses, such as the effect of accidents on public relations, employee morale, or on wage rates necessary to secure and retain employees, are not included in this method of estimating costs but can be important factors in some cases.

Information on some of the items is derived from the Department Supervisor's Accident Cost Report, National Safety Council's Form IS-7 (Figure 7-2).

The items are discussed in the order in which they appear on the Investigator's Cost Data Sheet, Form IS-8 (Figure 7-3, shown on the next page).

**1. Cost of wages paid for time lost by workers who were not injured.** These are employees who stopped work to watch or assist after the accident or to talk about it, or who lost time because they needed the equipment damaged in the accident or because they needed the output or the aid of the injured worker.

**2. Cost of damage to material or equipment.** The validity of property damage as a cost can scarcely be questioned. Occasionally, there is no property damage, but a substantial cost is incurred in reorganizing material or equipment. The charge should, however, be confined to the net cost of repairing or putting in order material or equipment that has been damaged or displaced, or to the current worth of the equipment less salvage value if it is damaged beyond repair.

An estimate of property damage should have the approval of the cost accountant, particularly if the current worth of the damaged property differs from the depreciated value established by the accounting department.

**3. Cost of wages paid for time lost by the injured worker,** other than workers' compensation payments. Payments made under workers' compensation laws for time lost after the waiting period are not included in this element of cost.

**4. Extra cost of overtime work necessitated by the accident.** The charge against an accident for overtime work is the difference between normal wages and overtime wages for the time needed to make up lost production, and the cost of extra supervision, heat, light, cleaning, and other extra services.

**5. Cost of wages paid supervisors for time spent on activities concerning the accident.** The most satisfactory way of estimating this cost is to charge the wages paid to the foreman for the time spent away from normal activities as a result of the accident.

**6. Wage cost caused by decreased output of injured worker after return to work.** If the injured worker's previous wage payments are continued despite a 40 percent reduction in his output, the accident should be charged with 40 percent of his wages during the period of low output.

**7. Cost of learning period of new worker.** If a replacement worker produces only half as much in his first two weeks as the injured worker would have produced for the same pay, then half of the new worker's wages for the two weeks' period should be considered part of the cost of the accident that made it necessary to hire him.

A wage cost for time spent by supervisors or others in training the new worker also should be attributed to the accident.

INVESTIGATOR'S COST DATA SHEET

Class 1 ______________
(Permanent partial or temporary total disability)

Class 2 ______________
(Temporary partial disability or medical treatment case requiring outside physician's care)

Class 3 ______________
(Medical treatment case requiring local dispensary care)

Class 4 ______________
(No injury)

Name ______________

Date of injury ______________ Its nature ______________

Department ______________ Operation ______________ Hourly wage ______

Hourly wage of supervisor $______________

Average hourly wage of workers in department where injury occurred $______________

1. Wage cost of time lost by workers who were not injured, if paid by employer $______
   a. Number of workers who lost time because they were talking, watching, helping ____
      Average amount of time lost per worker ______ hours ______ minutes
   b. Number of workers who lost time because they lacked equipment damaged in accident or because they needed output or aid of injured worker ________.
      Average amount of time lost per worker ________ hours ________minutes.
2. Nature of damage to material or equipment ______________

   ______________

   Net cost to repair, replace, or put in order the above material or equipment $______
3. Wage cost of time lost by injured worker while being paid by employer (other than workers compensation payments) $______
   a. Time lost on day of injury for which worker was paid _____ hours _____ min.
   b. Number of subsequent days' absence for which worker was paid ______ days.
      (Other than workers' compensation payments) _____ hours per day.
   c. Number of additional trips for medical attention on employer's time on succeeding days after worker's return to work __________
      Average time per trip ____ hours. ____ min. Total trip time ____ hrs. ____ min.
   d. Additional lost time by employee, for which he was paid by company ____ hrs. ____ min.

(over)

**Figure 7-3.** This 8½ × 11 in. form can be used to convert time losses into money losses. Initial time losses are obtained from the Department Supervisor's Accident Cost Report (Figure 7-2), and subsequent time losses are obtained from first aid and other departments as necessary. Wage rate information is obtained from the accounting department. The reverse side of the form (shown at the right) contains space for additional costs pertinent to the accident under study. See discussion under the heading, Making a pilot study.

4. If lost production was made up by overtime work, how much more did the work cost than if it had been done in regular hours? (cost items: wage rate difference, extra supervision, light, heat, cleaning for overtime.) $________

5. Cost of supervisor's time required in connection with the accident $________
   a. Supervisor's time shown on Dept. Supervisor's Report _____ hrs. _____ min.
   b. Additional supervisor's time required later _____ hrs. _____ min.

6. Wage cost due to decreased output of worker after injury if paid old rate $________
   a. Total time on light work or at reduced output ____ days ____ hours per day
   b. Worker's average percentage of normal output during this period _____________

7. If injured worker was replaced by new worker, wage cost of learning period $________
   a. Time new worker's output was below normal for his own wage ____ days _____ hours per day. His average percentage of normal output during time _________ His hourly wage $__________.
   b. Time of supervisor or others for training _______ hrs. Cost per hour $________

8. Medical cost to company (not covered by workers' compensation insurance) $________

9. Cost of time spent by higher supervision on investigation, including local processing of worker's compensation application form. (No safety or prevention activities should be included.) $________

10. Other costs are not covered above (e.g., public liability claims; cost of renting replacement equipment; loss of profit on contracts cancelled or orders lost if accident causes net reduction in total sales; loss of bonuses by company; cost of hiring new employee if the additional hiring expense is significant; cost of excessive spoilage by new employee; demurrage). $________

Explain fully:

Total uninsured cost ..........................................................$________

Name of Company ________________________________________________

Published by National Safety Council

444 North Michigan Avenue
Chicago, Illinois 60611

Stock No. 129.28

Figure 7-3. (Continued.)

**8. Uninsured medical cost borne by the company.** This cost is usually that of medical services provided at the plant dispensary. There is no great difficulty in estimating an average cost per visit for this medical attention.

The question may be raised, however, whether this expense can be considered a variable cost. That is, would a reduction in accidents result in lower expenses for operating the dispensary?

**9. Cost of time spent by management and clerical workers** on investigations or in the processing of compensation application forms. Time spent by management or supervision (other than the foreman or supervisor covered in Item 5) and by clerical employees in investigating an accident, or settling claims arising from it, is chargeable to the accident.

**10. Miscellaneous usual costs.** This category includes less typical costs, the validity of which must be clearly shown by the investigator on individual accident reports. Among such possible costs are public liability claims, cost of renting equipment, loss of profit on contracts canceled or orders lost if the accident causes a net long-run reduction in total sales, loss of bonuses by the company, cost of hiring new employees if the additional hiring expense is significant, cost of above normal spoilage by new employees, and demurrage. These cost factors and any others not suggested above would need to be well substantiated.

Miscellaneous costs were found in less than 2 percent of the cases in a group of several hundred reviewed in connection with this study.

### Making a pilot study

The purpose of the pilot study is to develop average uninsured costs, for different classes of accidents, that can be applied to future accident totals. Therefore, it is desirable not to include the costs of deaths and permanent total disabilities. Such accidents occur so seldom that the costs should be calculated individually and not estimated on the basis of averages.

Some flexibility in the grouping of classes of cases is desirable. If no distinction in made in the records between medical treatment cases requiring a physician's attention and those not requiring a physician's attention, the pilot study may combine Classes 2 and 3. (See Method for estimating earlier in this chapter.)

The following discussion assumes the study of costs will be made with the injuries grouped in the recommended classes. The discussion covers Classes 1, 2, and 4. A different method must be applied to Class 3 injuries, and it will be discussed later.

**Classes 1, 2, and 4.** To analyze uninsured costs for accidents in Classes 1, 2, and 4, the supervisor in charge of the department where an accident occurs should secure for each accident the information indicated on the Department Supervisor's Accident Cost Report form (Figure 7-2). These data can be obtained during the supervisor's regular investigation of the accident. As soon as each report form is completed, it should be sent to the safety department.

In the safety department, the information from the department supervisor's report will be transferred to the Investigator's Cost Data Sheet (Figure 7-3). The safety department then assumes responsibility for securing the supplementary information from the accounting department, industrial relations department, and other departments where records on lost time and other necessary information are kept.

As an alternative, a member of the safety department could secure all information needed on the data sheet. In this case, the supervisor's report form is not used, and the supervisor is required only to report each accident in Classes 1, 2, or 4 to the safety department as soon as it occurs.

Before he computes averages, the investigator should be certain that the pilot study has covered a sufficient number of cases of Classes 1, 2, and 4 to be representative. This number will rarely be less than 20 cases. However, more cases should be studied if the costs of the cases in a particular class vary widely. Information should be secured on enough cases of each class so the average cost per case in each class is fully representative of past experience and will, by inference, be applicable to future experience.

Once a sufficient number of cases has been accumulated, the investigation of individual cases can be discontinued. For the data thus collected, separate averages should be calculated for the cases of each class. It is recognized that these costs are averages of the uninsured costs only.

**Class 3 injuries.** These injuries are the common first cases in which no significant property damage results from the accident. They are the most difficult to analyze from the standpoint of cost, because such loss of time is likely to occur repeatedly and only for short periods, and the injuries can occur so frequently as to place an undue burden on the supervisor and safety director if a complete report form and data sheet are filled out for each case.

The points of essential information needed are the average amount of working time lost per trip to the dispensary, the average dispensary cost per treatment, the average number of visits to the dispensary per case, and the average amount of supervisor's time required per case.

The following method of developing averages for each of these items is recommended:

1. *Secure an estimate of average working time lost per trip to the dispensary* for first aid. Departmental time records should be consulted as they may show the amount of time each worker is absent from his job while receiving first aid. If so, a random sample of 50 to 100 records of persons known to have received first aid should be selected from different departments. The average time lost per dispensary visit is calculated by adding the absence time for all visits in the sample and dividing by the total number of visits.

   If departmental records do not contain this information, it will be necessary to assign an investigator to observe a random sample of 50 or more persons visiting the dispensary.

   As before, to secure the average time, all the estimated time intervals of absence are added and then divided by the total number of persons observed.
2. *Make an estimate of the average cost of providing medical attention for each visit* by dividing the total cost of operating the dispensary for a year by the total number of treatments given during the year.
3. *Calculate the average number of visits to the dispensary per case* by dividing the number of treatments of Class 3 injuries in a representative period, perhaps a month or six weeks, by the number of Class 3 injuries reported during the same period of time.

4. *Calculate the average amount of supervisor's time required per case,* where possible, by observing the activities of representative supervisors in connection with first aid cases.

   When a sufficient number of cases has been studied to be representative both of the activities of supervisors in different departments and of different types of first aid cases, the average time spent by a supervisor is computed by adding all the time intervals recorded and dividing by the number of cases.

   If it is impossible to make a time study of the supervisor's activities in connection with first aid cases, the only alternative is to secure from each supervisor an estimate of the time he spends on the usual first aid case, and to average these estimates by adding them and dividing by the number of supervisors.

   Determine the average value of this time by multiplying it by the average hourly wage of a supervisor.

The average total uninsured cost of a case in Class 3 is estimated from the data accumulated above as follows: the average amount of time lost for a trip to the dispensary (1, above) is multiplied by the plant's average wage rate, secured from the payroll department, to get the average cost per trip for the worker's time lost. To this figure is added the estimated cost of providing medical attention for a single visit (2). This figure is then multiplied by the average number of dispensary visits per medical treatment case (3), and to this result is added the average value of supervisor's time required (4).

This method of recording costs is designed to provide estimates of the average uninsured cost per case for accidents causing localized property damage or, at most, a few injuries.

The method of cost investigation for accidents resulting in deaths, permanent total disabilities, or extraordinarily extensive property damage is the same as for others, but the difference is that every one is separately investigated and should be included in the final cost estimate as a separate item. In estimating the cost of a fire, the investigator should bear in mind the company's fire insurance may cover property damage that would appear as an uninsured cost in other work accidents.

### Development of final cost estimate

Once the average for each class of case has been established, costs for any period in which a sufficiently large number of accidents has occurred to be representative can be estimated with considerable accuracy by multiplying the average uninsured cost per case for each of the four classes by the number of cases occurring in that class during the period.

If any deaths, permanent total disabilities, or extraordinarily extensive property damage accidents have occurred, the specific uninsured costs of these should be added to the estimated costs of the four classes of accidents.

To these uninsured cost totals should be added the cost of workers' compensation and insured medical expense. For companies that are self-insured, this will be the total amount paid out in settlement of claims plus all expenses of administering the insurance. For companies not carrying their own insurance, it will be the amount of their insurance premiums.

The method will have to be modified in accordance with the recordkeeping systems of different companies. For example, most self-insurers will find it impossible to separate compensated medical expense from dispensary care. In that case, these items should be combined into one, and the dispensary cost omitted from the analysis of noncompensated costs on the data sheets.

For an illustration of the development of a final cost estimate, see the example in Tables 7-C and -D.

## OFF-THE-JOB DISABLING INJURY COST

The employer loses the same services whether the employee is injured off the job or on the job, and incurs just about the same direct and indirect costs. Nevertheless the costs of off-the-job (OTJ) accidents and illnesses are, at least in part, handled differently. This section describes why the difference and takes you through some calculations.

When accidents take place off the job, a major portion of the costs are borne by employers. Some of the costs are evident, such as insurance benefits and wages paid to absent employees. Some of the costs are hidden, such as training for new or transferred workers and medical staff time demanded for workers returning to work after an accident. For example, a new worker does not produce at the same level as an experienced worker; thus the decreased productivity of the new worker indirectly increases the manufacturing overhead.

Some of the cost is more difficult to assess, although it still is very real. As accident rates in the community rise, so do insurance rates, taxes, and welfare contributions. Not all organizations are aware of all the costs that can result from off-the-job accidents and the impact that they have on operations and profits. Enough experience has been accumulated, however, to develop a simplified plan for estimating such cost.

### Categorizing OTJ disabling injury cost

The cost of off-the-job disabling injuries (OTJ DI) to an organization falls into the following two categories: insured and uninsured, just like on-job accidents that result in disabling injuries, described in the previous section.

Most uninsured costs are hidden. Aside from wage costs, most organizations do not keep records of uninsured costs. However, uninsured costs are associated with all OTJ DI accidents and, therefore, affect profit margins.

- Insured-worker productivity, cost, product loss and equipment damage.– Costs directly associated with the employee who sustained the OTJ DI injury are included in this expense subcategory.
- Noninjured-worker productivity cost, product loss, equipment damage, and administrative cost.– Costs incurred by personnel other than the employee who sustained the OTJ DI injury are included in this subcategory.
- Miscellaneous costs.–This subcategory includes loss of profit for cancelled contracts or orders and the costs of demurrage, telephone calls, transportation, or other miscellaneous expenses.

### Estimating your company's OTJ DI costs

Some experts say the ratio of insured cost to uninsured cost is 3:2, respectively. In order to estimate your company's losses from employee OTJ DIs, first determine the insured cost. Next, using the 3:2 factor, escalate the insured cost to determine the total (insured and uninsured) employee costs. The insured cost for injuries to dependents of employees is then added to the total employee cost to ascertain total losses. The following examples illustrate calculation procedures.

**Example 1. Company A** is insured by an outside carrier. Twenty percent or $225,000 of its annual premium charge was required to pay for its previous calendar year OTJ DI accident experience. Of that total, $75,000 was required for 11 employee injuries and the remaining $150,000 was required for 22 employee-dependent injuries. These figures include the administrative fee paid by the company to the carrier. The total cost for employee-dependent injuries is a conservative figure, since it does not include the administrative cost incurred by the company's insurance office staff to process claims; if this cost is known, it should be added to the employee-dependent injury expense category. When the $75,000 insured cost category is escalated to include the uninsured cost category, the total expense for employee injuries becomes $125,000.

*Company A Estimated OTJ DI Costs*

| | |
|---|---|
| Insured cost for employee injuries | $ 75,000 |
| Uninsured cost for employee injuries | 50,000 |
| Insured cost for employee-dependent injuries | 150,000 |
| Total annual estimated OTJ DI cost | $275,000 |

**Example 2. Company B** insured by an outside carrier. Its carrier stated that $850,000 was paid for 138 employee injuries and $1,800,000 was paid for 279 employee-dependent injuries for its previous calendar year OTJ DI accident experience. Their administrative fee to the carrier was 6% of the total cost, resulting in a cost of approximately $900,000 for employee injuries and $1,900,000 for employee-dependent injuries. If the cost for insurance-office staff claim processing is known, it should be added to the total employee-dependent cost. Escalation of the insured cost category for employee injuries indicated a total of $1,500,000.

*Company B Estimated OTJ DI Costs*

| | |
|---|---|
| Insured cost for employee injuries | $ 900,000 |
| Uninsured cost for employee injuries | 600,000 |
| Insured cost for employee-dependent injuries | 1,900,000 |
| Total annual estimated OTJ DI cost | $3,400,000 |

**Example 3. Company C** is self-insured. Insurance office records indicated that for the previous calendar year, OTJ DI accident experience for the amount of medical and health claims paid for 350 employee injuries was $2,280,000; $4,700,000 was paid for 690 employee-dependent injuries. Insurance office staff administrative costs for claim-processing should be added to the employee-dependent cost category if that cost is known; $2,280,000 escalated to include the uninsured cost category for employee injuries resulted in a total cost of approximately $3,800,000.

*Company C Estimated OTJ DI Costs*

| | |
|---|---|
| Insured cost for employee injuries | $2,280,000 |
| Uninsured cost for employee injuries | 1,520,000 |
| Insured cost for employee-dependent injuries | 4,700,000 |
| Total annual estimated OTJ DI cost | $8,500,000 |

### Measuring effect of OTJ programs

Calculations of average costs are useful tools when justifying initiating or accelerating off-the-job safety awareness programs. The calculations also can be used to measure the effects of safety programs. For example, 350 employees of Company C experienced OTJ injuries during the previous year for a total of approximately $3,800,000 and an average cost per accident of approximately $10,860. Based upon the OTJ DI loss experience, top management allocated $50,000 from the budget to initiate a safety awareness program. At the end of the year, the $10,860 average cost figure will be escalated to adjust for inflation and, in turn, the new figure will be used to calculate losses. For illustration purposes, we will assume that the new average cost per accident figure is $11,500 and that employee injuries were reduced from 350 to 300. Calculations (300 × $11,500) indicate losses of $3,450,000, a savings of approximately $350,000 from last year's total. The estimated net return is $300,000, which is a 600 percent return on investment. To further support justification for operating funds, savings realized from reduced employee-dependent injuries can be added to the employee savings total. This type of analysis provides a management tool to evaluate the impact of off-the-job disabling injuries to profit margins; thus, it can be used to gain management commitment in support of operating budgets for safety awareness programs.

## REFERENCES


American National Standards Institute, 1430 Broadway, New York, N.Y. 10018. *Standard Method of Recording Basic Facts Relating to the Nature and Occurrence of Work Injuries,* Z16.2-1969.

Blankenship, L.M. Nonoccupational Disabling Injury Cost Study, K/DSA-457. Oak Ridge, Tenn.: Martin Marietta Energy Systems, Inc., October 1981.

DeReamer, Russell. *Modern Safety and Health Technology.* New York, N.Y., John Wiley and Sons, Inc., 1981.

Grimaldi, John V., and Simonds, Rollin H. *Safety Management—Accident Cost and Control,* 4th ed. Homewood, Ill., Richard D. Irwin, Inc., 1984.

Kuhlman, Raymond. *Professional Accident Investigation.* Loganville, Ga., Institute Press, Div. of International Loss Control Institute, 1977.

National Safety Council, 444 N. Michigan Ave., Chicago, Ill. 60611. *Accident Investigation. . .A New Approach,* 1983. *Off-the-Job Safety,* Data Sheet 601.

# 8

# Workers' Compensation Insurance



In the mid-1980s, about 11,600 workers died each year from work-connected injuries or diseases, and more than 2 million were temporarily disabled—and these deaths and disabilities were only one-fifth of the total accidental injuries from all causes. Injured workers and their families suffered substantial economic losses as well as bodily injuries. Their employers and society also suffered sizable economic losses.

When a worker dies, is disabled, or merely requires medical attention because of work-connected injury or disease, the economic consequences affect the worker, his family, his employer, and society.

## ECONOMIC LOSSES

The worker and his family may suffer two types of economic losses (1) a loss of earnings and (2) extra expenses.

If a worker dies because of work-related injury or sickness, his survivors lose the income he would otherwise have earned, less the amount that he would have spent to maintain himself during the remainder of his working career and his retirement years. This loss can be substantial.

Total and permanent disability cause an even greater earnings loss than death because the worker must be maintained.

Permanent partial disability causes some fraction of the permanent total disability loss, depending upon the proportion of the annual earnings lost. A worker who is totally disabled for a temporary period loses his income for a specific number of weeks or months. Loss of even a month's earnings is a serious loss for the typical worker. In addition to these earnings losses, the deceased or disabled worker may no longer provide valuable household services that must now be forgone or replaced at additional cost.

Not all injured workers are disabled but almost all require some form of medical attention. For all injuries combined, medical expenses are less than the total earnings loss; but for many workers, their medical expenses exceed their earnings loss.

In addition to these losses, society loses the taxes that would have been paid by the injured employees and the products or services they would have produced. Some injured employees and families become public assistance beneficiaries and must be supported by other members of society.

### Cost details

It would be inappropriate to overlook the effect of increasing medical and hospital costs on costs of work injuries and illnesses. (More details are given later in this chapter under Cost Levels and Allocation.) For several years, these costs have increased at a rate somewhat in excess of inflation rates, and there are many reasons to believe that they will continue to do so. National Safety Council estimates, in its 1986 edition of *Accident Facts,* that the total cost of work accidents in 1985 was $37.3 billion. Cost categories that make up this loss are Visible Costs, Other Costs, and Fire Losses. Total cost per worker is estimated at $350.

Visible costs are estimated at $17.5 billion and include wage losses, insurance administrative costs, and medical costs. Other costs are also estimated at $17.5 billion and include the money value of time lost by workers other than those with disabling injuries who are directly or indirectly involved in accidents. Also included would be the time required to investigate accidents, complete accident reports, etc.

Fire losses are estimated at $2.3 billion.

It is of interest that visible costs have a one-to-one ratio to other costs. Other costs are comparable to what have been called indirect, or uninsured, costs related to the direct, or insured, costs of work injuries and illnesses.

There is a much greater interest in the total costs related to incidents that may or may not result in work injuries or illnesses. Safety practitioners have a greater opportunity than has been the case in the past to influence managements toward the adoption of more effective safety measures—that interest being related to the more important place rapidly rising costs must be given in executive decision making.

## WORKERS' COMPENSATION IN THE UNITED STATES

This discussion elaborates on the History of the U.S. Safety and Health Movement given on page 4 in Chapter 1.

In the United States, efforts to implement a system of compensation for industrial injuries lagged far behind the countries of Europe. As work-related injuries and diseases and their consequences grew less and less tolerable towards the end of the 19th century, the situation became ripe for a radical change. The first evidence of interest in workers' compensation was seen in 1893 when legislators seized upon John Graham Brooks' account of the German system as a clue to the direction of efforts at reform. This interest was further stimulated by the passage of the British Compensation Act of 1897.

### Early laws

In 1902 Maryland passed an act providing for a cooperative accident insurance fund; this represented the first legislation embodying to any degree the compensation principle. The scope of the act was restricted. Benefits, which were quite meager, were provided only for fatal accidents. Within three years, the courts declared the act unconstitutional. In 1908, a Massachusetts act authorized establishment of private plans of compensation upon approval of the state board of conciliation and arbitration. This law had no practical significance; it was a dead letter from the start.

By 1908, there was still no workers' compensation act in the United States. President Theodore Roosevelt, realizing the injustice, urged the passage of an act for federal employees in a message to Congress in January. He pointed out that the burden of an accident fell upon the helpless man, his wife, and children. The President declared that this was "an outrage." Later in 1908, Congress passed a compensation act covering certain federal employees. Though utterly inadequate, it was the first real compensation act passed in the United States.

During the next few years, agitation continued for state laws. A law passed in Montana in 1909, applying to miners and laborers in coal mines, was declared unconstitutional. Nevertheless, many states appointed commissions to investigate the feasibility of compensation acts and to propose specific legislation. The greater number of compensation acts were the result of these commissions' reports, all of which favored some form of compensation legislation, combined with recommendations from various private organizations. Widespread agreement on the need for compensation legislation unfortunately did not end all conflict over reform. Special interest groups clashed over specific bills and over questions of coverage, waiting periods, and state versus commercial insurance.

In 1910, New York adopted a workers' compensation act of general application which was compulsory for certain especially hazardous jobs, and optional for others. None of the early state compensation acts expressly covered occupational diseases. Statutes which provided compensation for "injury" were frequently interpreted to include disability from disease, but those acts which limited compensability to "injury by accident" excluded occupational disease. All except Oregon's act required uncompensated waiting periods of one to two weeks; several states provided retroactive payments after a prescribed period.

The 1911 Wisconsin workers' compensation act was the first law to become and remain effective. The laws of four other states (Nevada, New Jersey, California, and Washington) also became effective that year. In 1916, the United States Supreme Court declared workers' compensation laws to be constitutional; see page 4 for details. Although 24 jurisdictions had enacted such legislation by 1925, workers' compensation was not provided in every state until Mississippi enacted its law in 1948.

### Current acts

Today there are compensation acts in the 50 states, the District of Columbia, Guam, and Puerto Rico. In addition, the Federal Employees' Compensation Act covers the employees of the U.S. Government, and the Longshoremen's and Harbor Workers' Compensation Act covers maritime workers, other than seamen, and workers in certain other groups. The latter act provided compensation for workers in the "twilight zone" between ship and shore, since the U.S. Supreme Court had ruled they could not be covered under state compensation laws.

While economic changes and public policy have prompted increases in benefits and scope of the laws, the basic concepts have not undergone any radical changes. Employees and labor are both dissatisfied with certain aspects of workers' compensation. Labor attacks the system for inadequate benefits, coverage limitations, and exclusion of many injuries, illnesses, and disabilities that they consider job-related. Employers are critical because the system covers some injuries and diseases they do not consider job-related and is costly relative to its apparent benefits. Thus, while the early advocates of workers' compensation conceived it as a simple, speedy, efficient, equitable remedy that would reduce litigation over industrial injuries, some people have expressed doubt that their hopes have been realized.

## OBJECTIVES OF WORKERS' COMPENSATION

Workers' compensation programs can be evaluated by the extent to which they satisfy the following commonly accepted objectives:

1. Income replacement
2. Restoration of earning capacity and return to productive employment
3. Industrial accident prevention and reduction
4. Proper allocation of costs
5. Achievement of the other four objectives in the most efficient manner possible

Not all of these objectives are equally important or accepted. The first two generally are considered most important. These objectives sometimes conflict with one another, but in most ways they are linked by the design of the program.

### Income replacement

The first objective listed for workers' compensation is to replace the wages lost by workers disabled by a job-related injury or illness. According to this objective, the replacement should be adequate, equitable, prompt, and certain.

To be adequate, the program should replace lost earnings (present and projected, including fringe benefits), less those expenses such as taxes and job-related transportation costs that do not continue. The worker, however, should share a proportion of the loss in order to provide incentives for rehabilitation and accident prevention. The two-thirds replacement ratio that is found in most state statutes is generally considered acceptable, although the alternative of replacing 80 percent of "spendable earnings" has received favorable attention.

To be equitable, the program must treat all workers fairly. According to one concept of fairness, most workers should have the same proportion of their wages replaced. However, a worker with a low wage may need a high proportion of his lost wage in order to sustain himself and his family. If a guaranteed minimum income plan existed, there would be less need to favor low-income workers. A high-income worker who can afford to purchase private individual protection may have his weekly benefit limited to some reasonable maximum. If workers' compensation insurance is regarded primarily as a wage replacement program, however, relatively few persons should be affected by this maximum. An alternative philosophy would argue in favor of a more substantial welfare component with a higher minimum benefit, low maximum benefit, and extra benefits when there are dependents.

Ideally, workers would be treated the same regardless of the jurisdiction in which they are injured. This criterion, therefore, implies a minimization of interstate differences in statutory provisions and their administration.

The program should pay all disabled persons an income starting as soon after their disability commences as possible. Finally, workers should know in advance what benefits they will receive if they are injured on the job and that these benefits will be paid regardless of the continued solvency of the employers.

Under the whole-man theory, the system would be required to indemnify the worker or his family for the effect on all his personal activities, not his earning capacity alone.

### Restoration of disabled workers

The second listed objective is medical and vocational rehabilitation and return to productive employment. To achieve this objective, the worker should receive quality medical care at no cost to himself, care which will restore him as well as possible to his former physical condition. If complete restoration is impossible, he should receive vocational rehabilitation that will enable him to maximize his earning capacity. Finally the system should include incentives to disabled workers and prospective employers so the workers will return to productive employment as quickly as possible.

### Accident prevention and reduction

Occupational accident prevention and reduction is a third commonly accepted objective of workers' compensation. Those who consider this objective to be important believe that the system should and can provide significant financial and other incentives for employers to introduce measures that will decrease the frequency and severity of accidents. More specifically, the pricing of workers' compensation should reward good safety practices and penalize dangerous operations. Employees should also have some incentive to follow safe work practices by sharing some of the losses. Injured workers should have the opportunity and be encouraged to return to work as soon as they are physically able.

### Proper cost allocation

The fourth objective of workers' compensation, which has a narrower support than the first three, is to allocate the costs of the program among employers and industries according to the extent to which they are responsible for the losses to employees and other expenses. Such an allocation is considered equitable by supporters of this objective because each employer and industry pays its fair share of the cost. The economic effects are considered desirable because this allocation tends in the long run in a competitive economy to shift resources from hazardous industries to safe industries and from unsafe employers within an industry to safe employers. Higher workers' compensation costs will force employers with hazardous operations to consider raising their prices. To the extent that consumers will not accept the price increase, employer profits and their willingness to commit resources to this use will decline.

Critics of this objective argue that workers' compensation costs are such a small part of the cost of production that they have little, if any, effect on resource allocations. Consequently, they would avoid the complicated pricing practices necessary to achieve this objective.

## MAJOR CHARACTERISTICS

### Covered employment

While most of the state workers' compensation laws apply to both private and public employment, none of the laws covers all forms of employment. For various historical, political, economic, or administrative reasons, each of the laws has certain gaps. Laws that are elective rather than compulsory permit the employer to reject coverage; but in the event he does, he loses the customary common law defenses: assumed risk of the employment, negligence of a fellow servant, and contributory negligence.

A few states still restrict compulsory coverage to so-called hazardous occupations. Many laws exempt employers having fewer than a specified number of employees. The most common exception is for employers having fewer than three employees or less in eight states to fewer than four in three states. Most of the laws exclude farmwork, domestic service, and casual employment. Many laws also contain other exemptions, such as employment in charitable or religious institutions.

Federal workers are covered by the Federal Employers' Compensation Act (FECA). Employees of the District of Columbia are covered by the District of Columbia Workers' Compensation Act. It went into effect in 1982. Its provisions follow closely those of FECA.

Two other major groups outside the coverage of the compensation laws are interstate railroad workers and maritime employees. Railroad workers, any part of whose duties involve the furtherance of interstate commerce, are covered by the Federal Employers' Liability Act (FELA). Maritime workers are subject to the Jones Act, which applies provisions of the FELA to seamen.

The Federal Employers' Liability Act is not a workers' compensation law. It gives an employee an action in negligence against his employer and provides that the employer may not plead the common law defenses of fellow servant or assumption of risk; moreover, the principle of comparative negligence is substituted for the common law concept of contributory negligence.

As to the state and local employees, the actual number of these employees subject to workers' compensation or provided with such protection voluntarily is not available. All states (as well as Puerto Rico, Guam, and the District of Columbia) have some coverage of public employees but with marked variations. Some laws specify no exclusions or exclude only such groups as elected or appointed officials. Others limit coverage to employees of specified political subdivisions or to employees engaged in hazardous occupations. In still others, coverage is entirely optional with the state or with the city or political subdivision.

Certain other groups, such as the self-employed, unpaid family members, volunteers, and trainees, generally are not protected by workers' compensation.

Gradual extension of coverage over the years has been achieved by piecemeal actions: replacement of elective laws by compulsory provisions, elimination or reduction of numerical exemptions, and adoption of amendments granting protection to farm workers and other previously excluded groups. States still must strive for complete coverage.

### Covered injuries and diseases

Workers' compensation is presently intended to provide coverage only for certain work-related conditions, not all of the worker's health problems. Statutory definitions and tests have been adopted to provide the line of demarcation between those conditions which are compensable and those which are not. Because, in drafting workers' compensation laws, all jurisdictions relied to some extent on the English system (or other statutes that in turn relied upon the English model), their statutory language is remarkably similar. Nevertheless, as there are variations in language as well as differences in interpretation, a condition considered compensable in one state may be held non-compensable in others.

The statutes usually limit compensation benefits to personal injury caused by accident arising out of and in the course of the employment. Although this presents four distinct tests which must be met, in practice they are often considered in pairs: The "personal injury" and "by accident" requirements in one set, and the "arising out of" and "in the course of" requirements in the other.

**Personal injury and "by accident."** If interpreted narrowly, personal injury would deal solely with bodily harm, such as a broken leg or a cut, while the "by accident" test would refer to the cause, such as a blow to the body or an episode of excessive or improper lifting. In practice, however, the distinctions are blurred.

The "by accident" concept is a carry over from the English law. Early judicial interpretations of the English law made it quite clear that for their purposes the "by accident" requirement was intended to do little more than deny compensation to those who injured themselves intentionally. A number of U.S. jurisdictions, however, have applied the test so as to narrow the range of unintentional injuries which can be compensated.

One of the early victims of the "by accident" requirement was occupational disease coverage. As the typical judicial holding was that occupational disease and accidental injury were mutually exclusive, special legislation was required in order to provide disease coverage.

At present, occupational diseases are covered under separate legislation in six states.

**Occupational diseases.** The rest of the states incorporate occupational disease into existing workers' compensation statutes. However, controversy surrounds the specific diseases covered, the criteria for establishing that the diseases are work-related, and the time allowed for filing claims for diseases with latency periods, such as asbestos-related conditions.

Several states have provided special coverage for police and firefighters who have heart and respiratory diseases.

The difficulty of distinguishing between occupational and non-occupational hearing loss has led to enactment of special coverage provisions in many state statutes. They attempt to isolate the occupational component in the hearing loss and compensate accordingly.

One category of occupational ailment, black lung disease, is covered by a federal benefits program under the Federal Black Lung Act, part of the Coal Mine Health and Safety Act of 1969, as amended. Since enactment, $17.1 billion in benefits have been paid to victims of this respiratory disease, attributed to exposure to coal dust, and to their survivors. In 1984 alone, $1.6 billion was paid out to some 500,000 claimants.

The money to pay for these benefits comes in part from a special tax on coal production. Because the resulting fund shows a deficit—$2.8 billion in 1986—the bulk of the benefits are paid from the federal government's general revenues.

The tremendous cost of compensation for disability arising from occupational disease has led to pressure to fund such coverage in whole, or in part, through federal programs. During the first half of the 1980s, a series of measures were introduced in the U.S. Congress. They range from bills to federalize workers' compensation, i.e., remove the programs from state control and funding, to setting up special funds, along the lines of that for black lung disease, for other occupational ailments, such as asbestos-related diseases and cotton-dust disease. Because of the prospective administrative costs and open-ended benefits involved, none of these proposals has gotten beyond the committee hearing stage.

**"By accident" concept used to deny compensation.** The injury "by accident" concept was also used in many jurisdictions to deny compensation unless the injury was caused by some sort of unusual, traumatic occurrence, generally requiring the application of outside agency. Obviously this would and did drastically limit the kinds of cases which would be compensated. At present, this use of the by accident test is limited to a few narrow areas.

**Psychological difficulties.** Impairment involving psychological difficulties has been the source of much controversy based on application of the personal injury requirement. In some cases, a mental stimulus such as fear can produce a physical lesion, such as a cerebral hemorrhage. In the event of a physical lesion, the courts have not encountered much difficulty in conceding personal injuries. Compensation is usually approved also if, as a result of a clear physical injury, the patient suffers psychological disorder.

As might be expected, disagreement is most likely when it is alleged that mental stimulation has resulted in a mental illness without obvious physical change. Although many jurisdictions award compensation in such cases, others are still reluctant to compensate work-related psychological disorders.

**Work-related impairment.** The term "arising out of and in the course of employment," applied by almost every jurisdiction, is meant to define a certain level of relationship between the employment and injury of disease as a condition of eligibility for workers' compensation. The phrase obviously lacks certainty. Often it is quite difficult to determine whether a given set of facts will support an award of compensation.

The "course of the employment" aspect of this test refers primarily to the time frame of the injury. Virtually every jurisdiction holds that an employee is within the course of his employment, barring certain types of unusual circumstances or unreasonable conduct, from the moment he steps onto the employer's premises at the beginning of the work day to the moment he leaves the premises at the end of the day.

Although this test appears to be relatively simple to apply, it has not been so. One uncertain issue is, what are the premises? Injuries which clearly occur off premises but appear to deserve compensation lead to a search for exceptions and encourage courts to modify the basic rules. Many workers are not attached to particular premises. Even though an injury occurs off premises, as in travel to and from work, the employee may be compensated if a sufficient employment relationship can be found, such as payment for time or expense of travel or the provision of a company vehicle for transportation. In these circumstances, the period of travel time to and from home may be incidental in the course of employment.

The "arising out of" segment of the test is intended to provide a causal relationship between the employment and the injury. For example, it is not enough that an employee suffer a heart attack while at work. He must show that the heart attack arose out of the employment or, in other words, that it was causally related to the employment.

This means that at a minimum (some states have more stringent rules) it must be shown that it was the stress and strain or exertion of the employment that caused the heart attack, not merely a spontaneous breakdown of the cardiovascular system.

The degree of employment relationship necessary varies from state to state and has been modified as workers' compensation law has evolved. In earlier years, it was generally felt that the hazard-causing injury must be peculiar to the particular employment or be increased by the employment before the injury could be said to "arise out of the employment." This rather narrow view of compensability has been modified and to some extent abandoned in recent years.

Although it is difficult to place each jurisdiction in a particular category as to what it will hold sufficient to meet the "arising out of" test, two additional theories have been developed and followed. The first and more widespread is the "actual risk doctrine," which requires that the hazard resulting in injury be a risk of the particular employment, without regard to whether it was also a risk to which the general public is exposed. The second or "positional risk doctrine" could also be called the "but for" test. Here, if the employment places the worker in a position where he is injured ("but for" the employment the injury would not have occurred), the injury "arises out of the employment."

### Benefits

More than $17.6 billion in cash and medical benefits were received by workers in 1983—the latest year for which statistics are available—through the workers' compensation system. Benefits include medical service, cash benefit payments to the worker while totally disabled, payments for residual partial disability, burial allowances for work-related deaths and benefits to the worker's dependent survivors. Three states paid more than a billion dollars each in 1983: California, $2.24; Ohio and Texas, $1.08 billion each.

Some states provide special benefits also to cover attendants or prostheses; about three-fourths of the states provide maintenance and other services for rehabilitation. The largest proportion of benefits are in cash, either as periodic payments or as lump sums in settlement of claims. More than $12.2 billion, almost two-thirds of the $17.6 billion 1983 benefit total, were paid to workers or their survivors in cash.

Benefits are paid through three channels: commercial insurance policies; publicly operated state insurance funds; and self-insured employers. In 1983, more than $9.3 billion in workers' compensation was paid by private insurers, $5.0 billion by state funds, and $3.3 billion by self-insurers.

### Income replacement

Of the $12.2 billion benefits paid in 1983 as cash income, almost 85 percent went to disabled workers and the other 15 percent to survivors of workers killed on the job. Although 70 percent or more of recent workers' compensation cases are for temporary total disablement, such cases have accounted roughly for only one-fourth of cash benefits. At the same time, income benefits in the last few years to workers for permanent partial disabilities accounted for almost two-thirds of the total dollar amount.

**Basic features.** In general, the cash benefits provided for temporary total disability, permanent total disability, permanent partial disability, and death are payable as a wage-related benefit—the weekly amount is computed as a percentage of the worker's wage. The benefit varies by state and by type of disability but most commonly is set at two-thirds of wages, although some states provide 100 percent coverage. In some states, the statutory percentage varies with the worker's marital status and the number of dependent children, especially for survivors' benefits, which in a majority of states pay 67 percent or less of the deceased worker's wage to surviving widows without dependent children.

The benefit rate is limited to less than two-thirds of wages for many beneficiaries by another statutory provision—the maximum ceiling on the weekly benefit payable. Disabled workers whose wages are at or above the statewide average receive benefits below the statutory benefit rate in almost all states because of this ceiling, although for such individuals benefits may exceed preinjury take-home pay because benefits are tax-free.

As inflation boosts wage levels, some states have attempted to prevent the deterioration in effective benefit-wage rates by providing for future increases in the maximum without need for further legislation, for example, by automatically adjusting the maximum (for new beneficiaries only) in relation to changes in the state's average weekly wage. Much less common but gaining in acceptance are provisions that, as wage levels of workers rise, benefits rise for beneficiaries already on the rolls.

Another type of limitation on benefits sets maximum time periods or aggregate dollar amounts. Such limitations in permanent total disability and death cases may cut off benefits even though the income need continues. Only fourteen states limit the duration of total dollar benefits to widows and orphans.

In order to reduce administrative costs and to discourage malingering, benefits in all states are payable only after a waiting period following the report of disability. This delay in payment applies to the cash indemnity payments, not to medical and hospital care. The waiting period ranges from three days to seven. In all states, workers who are disabled beyond a specified minimum period of time receive payment retroactively for the waiting period. In more than three-fourths of workers' compensation laws, the minimum period before retroactive payment of disability benefits begins is two weeks.

**Benefits by type of disability.** Most compensation cases concern workers who incur temporary disability but recover completely. The maximum weekly benefit for temporary total disability is at least $98, and the median is $197. In a majority of the states, the percentage of the state's average weekly wage for temporary total disability is $100 or more.

For workers with dependents, about one-third of the states augment the weekly benefit for temporary disability, usually by some dollar amount for each dependent up to a specified total.

Benefits for permanent total disability are for disabilities that preclude any work or regular work in any well-known branch of the labor market and that can be of indefinite duration. These are similar to benefits for temporary total disability benefits. In a few states, the weekly payment for permanent disability benefits is less than for temporary.

A small number of states restrict the duration of benefits for permanent disabilities, typically to from 6 to 10 years.

Residual limitations on earning capacity after recovery (that is after a permanent partial disability) are awarded on a relatively complex basis. Partial disabilities are divided into two categories— "schedule" injuries, those listed in the law such as loss of specific bodily members; and "non-schedule" injuries, those which are of a more general nature, such as back and head injuries.

Weekly benefits for schedule injuries are a percentage of average weekly wages, usually the same as the benefit rate for permanent total disability. The maximum weekly benefit is for the most part the same as or lower than that for total disability.

Nonschedule injuries are paid at the same or similar rate but as a percentage of wage loss, the difference between wages before injury and the wages the worker is able to earn after injury.

The schedule benefits are paid for fixed periods varying according to the type and severity of the injury. For example, most state laws call for payments ranging from 200 to 300 weeks for loss of an arm and 20 to 40 weeks for loss of a great toe.

The maximum period for non-schedule injuries for each state is the same as or, more generally, less than the duration limits established for permanent total disability.

In the majority of states, compensation payable for permanent partial disability is in addition to that payable during the healing period or while the worker is temporarily and totally disabled.

Death benefits are intended to furnish income replacement for families dependent upon the earnings of an employee whose death is work-related. As is true for the other types of benefits, the amount of survivor benefits and the length of time they are paid vary considerably from state to state. Benefits computed as a percentage of the deceased worker's wage often are less than that for permanent total disability benefits if the survivor is a widow without dependents. If there are dependent children, the benefit in many states will be augmented. In most states, the duration of benefits is unlimited, although nine states retain duration limits in a range from 7 to 20 years. In a number of states, payments to widows continue usually as long as they do not remarry and to children until they are no longer dependent, usually to age 18. In many states, benefits to children continue to age 23 or 25 if in school. Benefits may be terminated earlier in the four states which also limit total dollar benefits.

In addition to benefits for widows, widowers, and children, some states pay survivor benefits to dependent invalid widowers, parents, or siblings of the dead worker. Burial expenses are payable in all states.

### Medical benefits

For many years, disbursements for medical services provided under workers' compensation have comprised about one-third of total outlay for benefits. Care includes first aid treatment, services of a physician, surgical and hospital services, nursing and drugs, supplies, and prosthetic devices. Some large employers, in addition to first aid facilities, employ staff physicians for workers. Most employers insure their medical care responsibility as they do the income benefits under workmen's compensation.

Every state law requires the employer to provide for medical care to the injured worker. In most jurisdictions, such treatment is provided without limit either through explicit statutory language or administrative interpretation. In the few states that limit the total medical care by specified maximum dollar amounts or maximum periods, the initial ceiling may be exceeded by administrative decision. Also, if specified types of injuries or disease are denied cash benefits, medical care also is denied.

A major issue concerning medical benefits for workers' compensation is the procedure for choosing the physician who is to furnish the care. Almost half of the states give the employer the right to designate the physician. In practice, the insurance company of the employer will ordinarily select the physician since it is the insurer that handles the claim for benefits. Where the doctor is chosen in this way, the medical care furnished may be more highly skilled and effective because of the selected physician's specialized experience. On the other hand, workers feel that a more important consideration is the emphasis that their personal family physician is likely to put on their health and well-being. They feel other considerations may influence a physician they do not select.

**Two other sources of benefits** for the disabled worker are Social Security and private disability programs provided as part of fringe benefit packages by larger employers, in particular.

Any disabled worker whose disability can be established to last at least 12 months or result in death may receive disability benefits if eligible under Social Security rules. The combination of Social Security benefits and workers' compensation benefits cannot exceed 80 percent of the disabled worker's earnings prior to disability.

These benefits are financed out of the Social Security tax paid by all employees.

Many applicants for Social Security benefits in the first half of the 1980s have been unsuccessful in obtaining coverage, however. Even though qualifying for workers' compensation it is not uncommon for a disabled worker not to be able to meet Social Security's controversial requirements for proving inability to perform any kind of gainful work.

Private disability insurance programs usually are coordinated with worker compensation. Frequently, a private disability program requires filing workers' compensation and Social Security disability claims before qualifying for its benefits.

### Rehabilitation

Along with industrial safety, medical care, and cash compensation, rehabilitation of workers is recognized at least theoretically as one of the primary goals of the workers' compensation system. At present the most widespread benefits offered through workers' compensation laws to restore a worker to his fullest economic capacity are the special maintenance benefits authorized in more than half of the states. These benefits usually are paid (sometimes in addition to the regular disability compensation) for various training, education, testing, and other services designed to aid the injured person to return to work. In addition, some state programs provide for travel expenses and for books and equipment needed for the training.

Ohio, Oregon, Rhode Island, Washington, and Puerto Rico directly operate rehabilitation facilities under the workers' compensation program. Some insurance companies also havc in-house facilities for rehabilitating workers.

Probably the main source of retraining and rehabilitation is the federal-state vocational program. The federal government, through the Federal Vocational Rehabilitation Act, is a major source of funds for such programs. The facilities operated by this program accept individuals with work-related disabilities as well as others. In all states, these institutions are directed by state vocational rehabilitation agencies. They provide medical care, counseling, training, and job placement. Unfortunately, not all workers' compensation cases referred for vocational rehabilitation can be accepted promptly. Many others are never brought to their attention.

One notable drawback preventing full utilization of available rehabilitation facilities is the often protracted, adversary proceedings for determining a worker's right to benefits. Because the determination of whether there should be an award for permanent partial or total disability (and how large an award should be for partial disability) is conditioned on the worker's lack of ability to work, the claim may be a strong disincentive for rehabilitation. Further, in the many compromise settlements, the employer's (or insurer's) motivation is to pay an agreed amount of money and foreclose future responsibility for medical, vocational, or other needs arising from the injury. Such settlements also work against a full-fledged effort to restore the worker to full health and productivity.

## ADMINISTRATION

The goal of workers' compensation is to provide for quick, simple, and inexpensive determination of all claims for benefits and to provide such medical care and rehabilitation services as are necessary to restore the injured worker to employment. Nearly all of the states have agencies to carry out these administrative responsibilities. The agency is either in the labor department; a separate compensation board or commission, or other department; in four, administration is left to the courts. Several states have separate, independent appeals boards to review claims when agency decisions are appealed.

### Objectives

An agency's many correlated responsibilities include close supervision over the processing of cases. The primary objective is to assure compliance with the law and to guarantee an injured worker's rights under the statute. Administration by a division within the labor department or by a board or commission has been found to be more effective in achieving the full purpose of the law than administration by the courts. The courts are not organized and equipped to render the services needed.

One criticism of state agencies concerns the delays in the first payment of compensation to the disabled worker. Although in most states insurers mail the checks, the state administrative agency has the responsibility to see that payments commence promptly. Full and prompt payment is essential because few workers can afford to wait long for benefits due. In one state, perhaps the best, the claims are paid within 15 days; in most states, however, it appears that the first payment usually arrives as much as 30 days late.

Another responsibility of the agency is to see that the injured worker gets the full benefit due. To do this, it is important to follow an injury from the first report to the final closing of the case. Some states not only check the accuracy of total payments, but also require signed receipts for every compensation payment. Some require the filing of a final receipt which itemizes the purpose of each element in the total outlay, to permit a complete audit of individual payments.

Frequently, however, the legislation itself requires a workers' compensation agency to operate on the assumption that each injured worker is responsible for securing his rights and that its primary function is to adjudicate contested claims. Even where the law does not favor this policy, lack of staff may force the agency to this restricted role.

Although it is known that many workers are not familiar with the provisions of their state's workers' compensation act, in only a few states does the administrator (as soon as possible after the injury is reported) advise the worker of his rights to benefits, medical and rehabilitation services, and assistance available at the commission's office. Too many states fail to insist on prompt reporting of accidents by employers, on prompt payment of benefits, or on final reports which spell out the amounts paid and how these amounts were computed. Although prompt reporting is usually required, sometimes no penalty is imposed for violation.

### Handling cases

Workers' compensation claims may be either uncontested or contested.

**In uncontested cases**, the two main methods followed are the direct payment system and the agreement system.

Under the direct payment system, the employer or his insurer takes the initiative and begins the payment of compensation to the worker or his dependents. The injured worker does not need to enter into an agreement and is not required to sign any papers before compensation starts. The laws prescribe the amount of benefit. If the worker fails to receive this, the administrative agency can investigate and correct any error. Jurisdictions whose laws provide the direct payment system include Arkansas, Michigan, Mississippi, New Hampshire, Wisconsin and the District of Columbia; it is also provided for in the Longshoremen's and Harbor Workers Compensation Act.

Under the agreement system, in effect in a majority of the states, the parties (that is, the employer or his insurer, and the worker) agree upon a settlement before payment is made. In some cases, the agreement must be approved by the administrative agency before payments start.

**In contested cases,** most workers' compensation laws provide for a hearing by a referee or hearing officer, with provisions for an appeal from the decision of the referees or hearing officer to the commission or appeals board and from there to the courts. As the administrative agency usually has exclusive jurisdiction over the determination of facts, appeals to the courts usually are limited to questions of law. In some states, however, the court is permitted to consider issues both of fact and law anew.

## SECURITY REQUIREMENTS

All states except Louisiana require their employers in private industry to demonstrate that they are able to pay the benefits required under the workers' compensation law. About two-thirds of the states have a similar provision for public employers.

These security provisions, in effect, require that active steps be taken by employers to guarantee that workers, when they are disabled, will receive the benefits called for by the law.

### Types of insurers

Most laws allow employers to satisfy the security requirement by insuring with private companies or to self-insure. As of January 1, 1986, only six states (Nevada, North Dakota, Ohio, Washington, West Virginia, Wyoming), Puerto Rico, and the Virgin Islands required employers to purchase protection from exclusive state operated funds. Two of these states and Puerto Rico, Guam, and the Virgin Islands also prohibit self-insurance. Besides the jurisdictions with exclusive state funds, 12 others have publicly operated programs in competition with private insurers. Regardless of the method of protection purchased, the same statutory benefits must be provided to the injured worker in that state.

### Compliance checks

In order to make sure that workers will receive benefits as intended, states need a method of checking that employers do in fact meet the security requirements. The workers' compensation agency ordinarily requires not only notice of insurance secured by employers but of cancellations of such insurance. In more than a fifth of the states, however, no formal procedures are in effect to make sure that all employers have given proper notice.

Generally, it is believed that most employers comply with security requirements. In part, a high degree of compliance may be expected because of sanctions available to the state for noncompliance. In almost four-fifths of the states, noncomplying employers become liable to worker suits with the employer's traditional common law defenses abrogated; in some states, the business may be stopped from operating. In addition some statutes call for fines against the employer or imprisonment or both.

### Regulation of insurers

Where employers are allowed to self-insure, they must generally demonstrate sound financial condition. An employer may have to make a deposit of a specified amount with the workers' compensation agency or post a surety bond. In at least a third of the states, all applicants for self-insurance must meet this requirement; in a similar proportion of states, at the discretion of the agency, this deposit may not be required. Other types of requirements imposed on self-insurers in various states are minimum payroll size, minimum number of employees, type of business, safety record, and proof of proper facilities for administering claims.

Besides restrictions imposed upon employers directly, activities of workers' compensation insurers also are regulated. Such regulation serves in part to assure that workers receive benefits when disabled. In order to write workmen's compensation insurance, insurers must conform to rules and regulation of both the state insurance department and the state agency administering the workers' compensation act (usually the industrial commission).

The insurance department primarily regulates the conditions for establishment of insurance companies in the state, their continuing solvency, and their business practices. Like self-insurers, insurance companies (in many states) must post bond or make a deposit with the state insurance department.

Generally the role of the industrial commissions in regulating insurers is limited. Few have either the authority concerning companies' rights to underwrite workers' compensation in their state or the information about financial status and operations of insurers. Further, although industrial commissions would seem to have a direct interest and concern in the claims-handling performance of insurers, few state agencies collect data on promptness of payment, amount of benefits paid, number of beneficiaries currently receiving benefits, and similar aspects of benefit operations. Generally, industrial commissions (to the extent they supervise claims operations) do so through review of individual cases, often only in the event of a dispute.

## FINANCING

The total cost of workers' compensation to employers has increased over the years and now is about two percent of covered payroll. Since insurance is the main vehicle for meeting the statutory requirements of the workers' compensation acts, the programs are financed mostly through insurance premiums.

### Financing insured benefits

For both public funds and commercial insurance, class premiums are established by an elaborate system of rates that take into account the general occupational classifications or industrial activities of the insured. About 15 percent of the employers, paying about 85 percent of the premiums, are experience-rated. That is, their premiums are modified to reflect their loss experience in the past relative to others in the same class. Also, the statistical reliability of that experience is taken into account; the larger the business, the more credible its experience. Since employers with a small number of workers are likely to experience volatile changes in injury rates from year to year, only employers of large numbers are experience rated.

Another factor in the premium-setting procedure is that discounts are given according to the size of the risk; this is an advantage to large companies. Their rates thus reflect the economies of scale which result from spreading certain fixed costs over a larger amount of premium. Finally, large companies by retrospective rating may have their premiums adjusted at the end of a policy year to match their actual experience.

Most insurers use rates developed by a rating bureau. In some states, the rates developed by the bureau are mandatory; in others, advisory only. Almost half the premiums are written on a participating basis. Participating policyholders receive periodic dividends that reflect insurer experience and sometimes their own.

### Financing self-insured benefits

Firms that cover workers' compensation risks through private insurance companies or state funds pay a premium in advance. In contrast, self-insurers have several options for financing. They may simply pay for liabilities as they are experienced, directly from operating funds, or they may provide some advance funding in one or more ways. In those states requiring deposits of funds by self-insurers, part or all of the funding for outstanding liabilities is provided for in advance mandatorily. Even if not required, a self-insurer may set aside reserves, or even formally insure its risk through a wholly owned subsidiary insurance company created for this purpose.

Such advance funding prevents severe disruptions in cash flow from unforeseen loss experience or accumulated liabilities.

### Insurer administrative costs

One of the recurring issues in evaluating workers' compensation is the financial efficiency of the insurance mechanism for providing benefits. A major part of the issue is the comparison between private and state fund insurance.

The premiums collected by private insurers are used not only to pay benefits but also for expenses associated with claims such as investigation and legal fees; for sales, supervision, and collection; for administration; for safety programs; and for taxes, licenses, and other mandatory fees as well as for earnings.

State funds have much the same costs as private insurers with these exceptions: lower (or no) taxes and fees to the state government, no margin for private profit, and lower selling costs. Consequently, although the variation among individual funds is great, expenses have averaged less than 10 percent of total premiums paid, well below the ratio for private insurers. Some state funds incur smaller expenses for administrative and legal services, which may be financed from other government funds. On the other hand, state funds in some instances may insure greater proportions of high-risk companies than private carriers and incur proportionately heavier charges for benefits.

### Other administrative costs

Another aspect of financing workers' compensation relates to the cost of supporting the public agency that administers the program. The cost of operating the industrial commission (or other administering agency) is borne either by assessments upon insurers and self-insurers or through appropriations from public funds. In the former event, the cost of administering the program is simply one more expense item in the premium charge to the employer. Where funds for the industrial commission are obtained from legislative appropriations, this part of the program is paid out of general taxes. More than one-third of state agency administrative costs nationally are funded by legislative appropriations.

In addition to that part of administrative costs financed by state general revenues, other elements in the workers' compensation system may not be financed through insurance premiums paid for by employers. For example, many employers provide medical services at their establishment or by direct payments to medical facilities. Second- or subsequent-injury funds, which bear part of the cost of injuries to handicapped workers, are financed sometimes through assessments on insurers, reflected in premiums; in some states, as direct charges upon employers; as appropriations from state funds; in a few states, as joint employee and employer contributions to the fund; or by other means. Other special funds, paid for by general revenues, have been established for such purposes as supplementing benefits depreciated by inflation or paying benefits for specified occupational diseases.

## OCCUPATIONAL SAFETY

From the beginning, the workers' compensation movement in the United States has been associated with the movement to prevent occupational injury or disease. Although some interest in this work was manifested by various employers before the enactment of workers' compensation laws, the organized safety movement, as we know it, began shortly after the first compensation laws went into effect. This movement was due in large part to an assumption on the part of industrial leaders that one of the best ways to reduce compensation costs would be to reduce the number of accidents.

The first move toward an organized effort came at the convention of the Association of Iron and Steel Electrical Engineers at Milwaukee in 1912, as discussed in Chapter 1. A session devoted to safety set up a committee on organization, which called a meeting of all interested groups and individuals in New York City the following year. This meeting resulted in the formation of the National Council for Industrial Safety, which since 1915 has been called the National Safety Council.

### Safety activities of insurers

Although the insurance business had been extended into the industrial accident field before the first workers' compensation act was passed in this country, industrial accident insurance received great impetus from this legislation. As specific schedules of payments for all work-connected injuries made the risk more definitely calculable, the business became more attractive to the underwriter.

From the first, insurance companies writing workers' compensation policies have had a large part in the movement to prevent accidents. They have developed or aided in the development of safety standards and safe practices and have contributed to the development of methods and techniques of accident prevention. Much of the basic data of safety engineering has been supplied by insurance engineers. An important motive for their accomplishment is, of course, the fact that their business thrives on a declining injury rate. Unduly large losses jeopardize the financial solvency of an insurance company. Progressively lower losses make it easier for the stock company to show a profit to its stockholders and for the mutual to pay dividends to its policyholders.

The effective work of insurers, however, has been confined mainly to large establishments. The cost of providing technical assistance in accident prevention makes it difficult for an insurer to provide service adequately to plants whose premiums are small. Owners of small plants, moreover, cannot expect to receive much reduction in premium rates either through dividends or experience rating, no matter how effective their safety program.

### Insurance price incentives

In addition to offering technical assistance, insurers have also tried, by various means, to make safety pay for the policyholder in the form of immediately reduced premium rates. This monetary benefit for accident prevention is to some extent inherent in mutual insurance, since the surplus remaining after losses and expenses is distributed among the policyholders. To offer a similar inducement to policyholders in stock companies and to give further incentive to mutual policyholders, a system of merit rating was adopted to obtain reductions in premium rates for policyholders.

The first type of merit rating used was schedule rating, under which the reduction in premium rate was computed on the basis of the policyholder's performance in providing physical safeguards. This tactic proved unsatisfactory because safeguards, while vital, are only one part of prevention. The system now in general use is experience rating, described earlier in this chapter.

Historically, it has been assumed by many authorities that merit rating provided a powerful stimulus to the safety movement. However, safety professionals do not rely solely on the merit rating system to stimulate accident prevention efforts by industry. In fact, some studies have questioned the value of merit rating as a strong impetus to safety.

## COVERED EMPLOYMENT

Although many controversies have divided students of workers' compensation, one point on which there is broad agreement is that coverage under the acts should be virtually if not completely universal. With few exceptions, employers and workers alike agree on the desirability of this basic protection. For the employer, it represents a relatively inexpensive way to protect himself against the possibility of lawsuits for injuries to his employees. For the worker it represents an important segment of his protection against income loss.

The principle of virtually universal coverage of all gainfully employed workers is basic. Yet, even today, none of the state laws meets this goal, though a few come close. Although it is believed generally that coverage has progressed and although the public has been educated on the justification for including all employment in the workers' compensation system, for the past 25 years the proportion of covered civilian wage and salary workers included has hovered around four-fifths of the potential. Recently, it has edged up to about 88 percent, largely as a result of a shift of workers to covered employment. According to the Social Security Administration, average weekly coverage under state and federal workers' compensation for 1983—latest year available—totaled 80.9 million, out of the 93.0 million civilian wage and salaried workers in the country.

### Limitations on coverage

Most of the arguments originally brought against extension of coverage have lost their force. Nevertheless, in view of the persistence of some of the exemptions or exclusions in many state laws, a review of some of the reasons behind the original limitations may be helpful. Nearly all state acts were prepared and enacted in the face of constitutional challenges and outright opposition of certain interests. Thus, each act was the result of compromises rather than the outcome of a consistent, ideal program, even if, in some instances, much weight was given to a carefully studied plan.

### Other criteria

Workers' compensation was hailed as an innovation that would introduce a great deal of certainty in the calculation and payment of benefits, in contrast to the common law system. A worker could sue under common law. If he won, he might be assured of an adequate payment; those who lost would be left with nothing but debts. To reduce uncertainty, the workers' compensation law specified the benefits that would be paid to all regardless of fault. Although the outcome of workers' compensation cases is far more certain than the ordinary suit where negligence must be shown, the law is not "automatically" applied.

In part, this uncertainty stems from the variety of the permanent partial disability cases which the schedules do not cover satisfactorily. Two factors give rise to compensation litigation. One is the uncertainty as to whether an accident did or did not arise out of and in the course of employment; the other is the extent of disability. As workers' compensation comes to encompass more and more of the ailments to which the general population might be susceptible, it becomes difficult to separate impairments that are work-connected from those that are not. In addition, it requires an exercise of legal skills and medical judgment to assess the extent of disability in occupational diseases, injuries to the soft tissue of the back, heart conditions, or cases where the only evidence before the commission may be a subjective complaint.

### Conclusion

Workers' compensation permanent partial disability benefits are the least duplicated benefit of any paid to injured workers. There is a variety of benefit levels for temporary disability, permanent total disability, and death. Nowhere is the variety more apparent than in permanent partial disability. States differ in benefit levels, in relationships among benefits paid to the various types of disabilities, in minimums, in maximums, in weeks scheduled for particular losses, and above all in benefit payment philosophies. Jurisdictions differ in the time allotted for the same partial disability as well as in assessing the degree of disability. A particular residual impairment may rate 50 percent total disability in one state and go uncompensated in another because the employee has lost no wages. In a third jurisdiction, the award may be at a different percentage because of an administrative judgment about the estimated loss in wage-earning capacity. Such variety is in addition to the differences in statutory replacement ratios and minimums and maximums on a weekly or aggregate basis.

Because of the difficulty of predicting an award for a standard impairment and because of a lack of knowledge about the relationship between a given impairment and wage loss, we have no consistent and reliable estimates of the adequacy of permanent partial awards. Some estimates were presented on the basis of a 50 percent disability, but these estimates are possible only in states where aggregate limits or rating philosophies permit one to specify the number of weeks which will be paid.

Death benefits are payable to widows and other eligible survivors at a rate that usually is the same as that paid in the permanent disability cases. Although most jurisdictions provide for payments during widowhood and until children reach a specified age, some limit the duration of payments.

Adequacy comparisons were based on a hypothetical worker making the average wage in his jurisdiction. The percentage of wage loss replaced in each jurisdiction was shown in the analysis. Certainty of payment, a prime objective of early compensation laws, has not been attained because of the litigation which persists over issues of liability and extent of disability. Promptness of payment, also an early goal, has been attained in some jurisdictions but, for most states, data on this aspect of administration are not available.

In the compromise and release cases, certainty is attained but often at the expense of closing out all possibility of future recovery should a worker's condition worsen. As most states do not maintain records on postsettlement developments, it is difficult to know how much compromise and release settlements interfere with the basic objectives of the law. In the usual case, the workers' compensation benefit is paid in periodic installments as wages are paid so that, if the employee's condition changes, the case can be reopened within the period designated in the statute of limitations. The compromise and release settlements, of course, obviate this possibility.

## REHABILITATION

Most employees injured in work accidents return to their jobs after minor medical attention with little if any worktime lost. As the effects of the injury are transient, the incident usually fades from memory. Even those who suffer days or weeks of disability and possibly endure substantial medical treatment may find the injury is not permanent. Although the loss of income and the medical expenses are distressing, eventually, when workers resume their jobs, they recover economically, too.

A minority, unfortunately as much as 10 percent of the total injured, according to one California report, suffer injuries that disrupt their lives. Even when these workers receive effective medical care so that eventually they return to productive jobs, their lives are physically and emotionally scarred. Injuries for some are so severe that prolonged medical treatment and convalescence fail to restore them completely. Residual handicaps prevent their acceptable performances in their former jobs. Only retraining and education combined with special treatment offer a prospect for future employment.

Some never return to work. If they do not die from their injuries, they live with such severe disabilities that they barely can manage for themselves. Often, the most that health services can do is to lighten the burden on those who take care of these persons.

The treatment for workers whose livelihood is threatened by work-related impairments consists of medical rehabilitation and vocational rehabilitation.

### Medical rehabilitation

It is easier to discuss individual programs than to review medical rehabilitation in the United States as a whole. Each program, whether set up by an insurance company or workers' compensation agency, contains its own requirements for treatment, qualifications for eligibility and definitions of service.

**The delivery system.** The worker who requires medical rehabilitation often receives it much as he receives other medical care. Workers' conpensation laws obligate employers and insurers to pay costs of medical care for injured workers. The worker receives whatever medical care is needed to treat the impairment and restore lost function. He may report first to the plant nurse or physician for immediate attention. If the injury is serious, he may go to a hospital. Costs are covered by having health service workers on salary, by contractual arrangement with health personnel, or by payment of hospital and doctor bills. The insurer may or may not have much influence in the selection or course of treatment.

For injuries associated with chronic disabilities, the insurer usually attempts to control the selection of the rehabilitation services, frequently by directing the worker to a particular specialist or facility with a particular expertise. Often the insurer pays for transportation to the specialist or facility as well as for rooms during treatment.

Some insurance companies operate rehabilitation facilities, under individual or joint ownership, with medical personnel on salary, at least parttime.

When insurers contract to share rehabilitation programs or facilities, they may pay expenses case by case or through a rental agreement.

Some workers' compensation agencies may be isolated from and unaware of rehabilitation procedures. Others keep relatively close tabs on the services rendered.

When informed of the potential need for rehabilitation, some agencies do little more than notify the worker and insurer that medical rehabilitation is worth considering. Other agencies conduct formal evaluations of the need for further medical care and recommend action. They seek to convince disabled workers of the wisdom of rehabilitation. When the workers agree, the insurers can be required to finance the care.

**Rehabilitation in insurance.** One study of rehabilitation programs in workers' compensation insurance coverage offered by 22 insurance companies reveals a variety of policies and practices in rehabilitation. Generalizations are thus difficult. The following comments on various rehabilitation programs, therefore, are not to be regarded as typical of the entire industry.

In the insurance industry, the concern some carriers have for both medical care and medical rehabilitation is termed "medical management." It is the attempt to minimize the total costs of compensation through emphasis on well-timed, high quality medical treatment. The concept tends to focus attention on the physical condition of the worker rather than on the monetary compensation due. The ultimate goal is to reduce the degree of disability. The insurer would like to see the worker's earning abilities fully restored rather than pay compensation indefinitely.

### Vocational rehabilitation

Vocational rehabilitation prepares the injured worker for a new occupation or for ways of continuing in his old one. Usually, vocational rehabilitation is assigned when medical treatment fails to restore the worker to the job he held when injured. The worker's injury may be so severe or his work requirements such that residual impairment prohibits effective performance. Workers with such impairment must be trained to surmount or bypass the residual limitations. Many will enter new occupations. In practice, the more effective the medical rehabilitation, the less the need for vocational rehabilitation.

This definition of vocational rehabilitation distinguishes it from medical rehabilitation more than it should. While the difference in kind of treatment seems clear enough—retraining as opposed to medical care—the categories overlap. In the public vocational rehabilitation programs in each state, services include medical diagnosis and evaluation, surgery, psychological support, the fitting of prostheses, and other health services along with education, vocational training, on-the-job training, and job placement.

The two programs blend also on the record. Recordkeeping by workers' compensation insurers does not separate claimants who receive medical rehabilitation from those who receive vocational rehabilitation, although some distinguish between medical rehabilitation and acute medical care. In contrast, records kept by workers' compensation agencies usually separate vocational rehabilitation from other benefits.

**The delivery system.** The relatively few injured workers who need vocational rehabilitation are served by several means. An employer or insurer may channel the worker to whatever sources he thinks will provide satisfactory service. Some workers are referred to the public vocational rehabilitation program where services may be financed by taxes, although insurers may reimburse the public agency. Other insurers direct workers into private facilities where vocational training is conducted by technical schools or on the job. For such services, insurers always pay the costs.

As with medical rehabilitation, some workers' compensation agencies support vocational rehabilitation so that, if the insurer does not direct the worker into a program, the agency often will. Several jurisdictions select candidates either in conjunction with screening for medical rehabilitation or separately. Workers with serious injuries, permanent disabilities, or those who receive extended compensation payments are reviewed by the agency for referral to the state's public vocational rehabilitation agency or to the insurer.

Some workers obtain vocational rehabilitation through their own efforts. If no one refers them, they may go directly to the public vocational rehabilitation office. Since 1920, the federal government and the states have cooperated financially in supporting a vocational rehabilitation program, 80 percent federal and 20 percent state, which can be utilized by anyone with a vocational handicap. Rehabilitation counselors, who usually determine a referral's acceptability, simply look for a vocational handicap without regard to the source and consider the possibilities of overcoming the handicap. If the candidate shows relatively good prospects, a plan is designed for his restoration. For those who cannot return to a paying job, the objective of vocational restoration may be to enable clients to care for themselves and to free other members of the family to earn wages.

The worker may be referred also by his physician, a friend, or a member of his family.

Once a worker is established in a vocational rehabilitation program, he is aided by whatever sources the counselors think best fit his needs. Generally, the sources are not owned and operated by the vocational rehabilitation agency but are private vendors or other public agencies. A worker may be sent to a private rehabilitation center or school or a sheltered workshop such as those run by Goodwill Industries of America, or he may be enrolled in a public institution.

## DEGREE OF DISABILITY

Determination of the extent of disability is perhaps responsible for more litigation than any other single issue in workers' compensation. It requires not only correct application of legal principles but also evaluation of facts, subjective complaints and opinions, and attempts to predict the future.

As a general proposition (some jurisdictions use different terminology and slightly different classifications), disability can be categorized in one of four classes: temporary total disability, temporary partial disability, permanent partial disability, and permanent total disability.

### Temporary disability

Temporary total disability occurs when an injured worker, incapable of gainful employment, has a possibility or probability of improvement to the degree he will be able to return to work with either no disability or merely a partial disability. Temporary partial disability is similar to temporary total in that it assumes a physical condition which has not stabilized and is expected to improve. The difference lies in the worker's current abilities. When temporarily partially disabled, the worker is capable of some employment, such as light duties or part-time work, but is expected to improve to the degree that he will attain much of his former capability.

Permanent partial disability is reached when the injured worker has attained maximum improvement without full recovery. That is, the worker has benefited from medical and rehabilitative services as much as possible and still suffers a partial disability.

Permanent total disability represents the same physical situation except that the disability is total.

The determination of temporary disability, either total or partial, is least difficult as it requires merely a determination of the employee's present physical condition in comparison with the work opportunities available. In practice, evaluation of temporary disability is concerned only with the ability of the employee to return to work for his last employer. Assuming that an employee will be able at some point to return to work for his employer, and given the difficulties involved in obtaining employment for workers still under medical care, compounded by the probability that the new employment will be temporary, most adjudicators have either expressly or in practice adopted the proposition that unless the worker can return to his last job, or can be supplied with temporary light or part-time duties with his last employer, he remains temporarily totally disabled, even though he might be able to perform another job involving duties within his temporary physical limitations.

### Permanent partial disability

The determination of the extent of permanent partial disability depends on what the jurisdiction chooses to label "permanent partial disability." Three basic theories have been discussed in conjunction with the payment of workers' compensation benefits for such disability. Their underlying philosophies differ somewhat, as do the factors to be considered in applying each to a specific situation.

**The "whole-person" theory** is concerned solely with functional limitations. Here, the only considerations are whether the worker has in fact sustained a permanent physical impairment and, if he has, to what extent does it interfere with his usual functions and abilities. Age, occupation, educational background, and other factors are not considered.

**In the "wage loss" theory,** the aim is to determine what wages the worker would have been able to earn had he not suffered a permanent impairment. When, owing to impairment, his earnings dip below the estimated wage figure, he is paid compensation equal to some percentage of the difference between the wages that he should have earned and those he is actually earning. Here the actual degree of physical impairment is of little or no importance. The only concern is the actual wage loss which has been sustained and whether it is due to the impairment.

**The "loss of wage-earning capacity" theory** requires a peek into the future. After the worker has reached his maximum physical improvement, many factors, such as impairment, occupational history, age, sex, educational background, and other elements which affect one's ability to obtain and retain employment are all considered in an effort to estimate, as a percentage, how much of the worker's eventual capacity to earn has been destroyed by his work-related impairment. The worker is awarded benefits on the basis of this computation. Benefits may be paid at the maximum weekly rate for a limited number of weeks or they may be based upon a percentage of the difference between wage-earning capacity before and after disability, to be paid until a dollar or time maximum has been reached.

**Combinations.** These three basic theories are capable of being used also in combination. For example, some states expressly or in practice provide for the use of either the "wage loss" theory or the "loss of wage-earning capacity" theory but also provide a benefit floor determined by the actual medical impairment. Thus, a worker who sustains impairment but no loss of wages or of wage-earning capacity would still receive some permanent disability benefits.

The tedious or controversial aspects of rating for a significant proportion of permanent partial disability cases have been relieved by the use of schedules. The typical schedule covers injuries to the eyes, ears, hands, arms, feet, and legs. It states that for 100 percent loss (or loss of use) of that body part, compensation at the claimant's weekly rate will be paid for a specified number of weeks. If loss or loss of use is less than total, the maximum number of weeks is reduced in proportion to the percentage of loss or loss of use. Only physical impairment is considered and the effect of the injury on wages or wage-earning capacity is ignored. If an injury is confined to a scheduled body part, the benefits provided by the schedule are exclusive, even though disability rating on a wage loss or loss of wage-earning capacity basis might result in greater benefits. Although this statement is true generally, some states provide additional benefits if use of one of the other theories does result in higher benefits being paid, if diminished wage-earning capacity continues after the scheduled amount is paid, if the scheduled injury results in permanent total disability, or if several scheduled injuries are sustained in the same accident.

Use of the schedule may also be avoided by showing that the effect of the scheduled injury, such as radiating pain, extends into other parts of the body. A few jurisdictions limit the use of schedules to amputation or 100 percent loss of use of a body part, as opposed to partial loss of use. Another group not only makes the schedule exclusive for permanent disability awards but requires that the weeks for which benefits are paid during the healing period be deducted from the number of weeks authorized by the schedule before an award is made for permanent partial disability.

Despite commentary and statutory language to the contrary, the workers' compensation systems of the United States, with a few exceptions, operate primarily on the "loss of earning capacity" theory. Even where statutory language seems to indicate clearly that only functional impairment is to be considered, the courts have managed to hold that loss of earning capacity is the real consideration. Even the use of schedules has been justified on an earning capacity basis as merely a legislative determination of presumed wage loss resulting from the impairment listed in the schedule.

It is questionable that any legislative history would back up this rationale. A consideration of its practical day-to-day application shows that, in individual cases, the presumption is without basis in fact. When one considers that a concert pianist and a laborer who have both lost two fingers would receive exactly the same compensation under a schedule award, the justification for the use of schedules, that of administrative efficiency, is questionable by all who hold equity as a basic aim of workers' compensation.

### Permanent total disability

Permanent total disability evaluation is, in most respects, merely an extension of the determination of permanent partial disability. In fact, it is a part of the same process, since the fact-finder's only additional task is to determine whether the worker's wage-earning capacity is so destroyed that he is unable to compete in the job market.

Two aspects of the permanent total disability question warrant special attention.

First, most states employ presumptions that make the fact-finder's job much easier. For example, it may be presumed that the loss of sight of both eyes or the loss of any two limbs will constitute permanent total disability and thereby relieve the fact-finder of the difficult task of evaluating all the factors previously mentioned. These presumptions in some cases may be rebutted by evidence of an established wage-earning capacity or may only apply for a limited period of time.

Second is the concept of what permanent total disability actually means. The injured employee need not be completely helpless nor unable to earn a single dollar at a job. His limitations need only prevent him from competing in a practical way in the open job market and are such that no stable job market exists for him.

### The 'exclusive remedy' doctrine

Before workers' compensation laws were enacted in the states, an employee, in order to recover damages for a work-connected injury, always was required to show some degree of fault on the part of his employer. Under what is now known as the "quid pro quo of workers' compensation law," employers accepted, or were required to accept, responsibility for injuries arising out of and in the course of employment without regard to fault. In exchange, employees gave up the right to sue employers for unlimited damages. These agreements are usually referred to in the state acts as "exclusive remedy" provisions, a term that is quite misleading.

In no state are workers' compensation benefits necessarily the only remedy available to an injured worker. Depending upon the working of the applicable statute, the worker may bring a negligence action against a fellow worker, another contractor on the same job, or some other entity or individual who caused the compensable injury. For example, the worker may sue the manufacturer of a piece of equipment which caused the injury. From the employer's viewpoint, the doctrine should be the "exclusive liability rule." As the employee sees the rule, it remains an "exclusive remedy" for obtaining compensation from the employer. Neither liability nor remedy is perfectly exclusive.

Two growing inroads into the exclusive remedy provision are (1) the expansion of the dual capacity doctrine and (2) the intentional tort exception.

An injured employee can bring a lawsuit against an employer for an injury, even if it arose out of and in the course of employment, *if the injury resulted from the employer's product or a service available to the public.* Example: A driver of a tire company delivery truck is injured when a defective tire, made by the employer, causes the truck to have an accident. Or a hospital employee is injured as the result of negligent treatment, following an on-the-job accident, by a member of the hospital medical staff. In both cases the injury did not occur as a result of the employer-employee relationship but rather through a relationship more akin to that of a supplier or service provider and the public.

*If an employer commits an intentional tort,* i.e., either deliberately causes harm to an employee or is grossly negligent or engages in reckless behavior that results in an injury, the employee has the right to sue the employer for damages. The rationale is that the exclusive remedy provision should not protect an employer against being sued for an injury resulting from a deliberate harmful action or from gross negligence and recklessness.

### Protection against termination

Court decisions and statutes have begun in the first half of the 1980s to give workers filing worker compensation claims protection against termination for making the claims. The courts and state legislatures, in precedent-setting decisions and in legislation, have clearly established that it is against public policy, and thus against the law, to fire an employee merely for exercising his or her right to apply for benefits if injured. Employees winning retaliatory discharge lawsuits have gotten reinstatement to jobs, back pay, attorney's fees, and punitive damages.

## PRIVATE INSURER PROGRAMS

Private insurers dominate the coverage. The ten leading groups wrote almost half the business; the top twenty about 68 percent, although the share of the top ten has changed little. The ten leaders wrote as much as 88 percent of the coverage in Hawaii to as little as 51 percent in Kansas and Nebraska. In four states, one insurer wrote one-fourth of the business.

Workers' compensation is the second largest property-liability insurance line; it is topped only by automobile insurance. Workers' compensation premiums in 1983—the latest year for which figures are available—were 13 percent of the total premium income.

### Classifications of insurers

Private insurers can be classified according to their legal form of organization, their marketing methods, and their pricing policies.

Legally, insurers may be classified as proprietary or cooperative insurers. Proprietary insurers have owners who bear the risk of the insurer and whose representatives manage the operations. The leading example by far is the stock insurer owned by stockholders who elect the board of directors.

Cooperative insurers have no owners other than their policyholders. The leading example is the advance premium mutual whose board of directors is elected by those policyholders who exercise their right to vote.

Unlike stock insurers, these insurers have no capital stock. Instead, retained earnings serve as a cushion against adverse experience.

Almost all of the workers' compensation insurance premiums not written by stock or mutual insurers were written by reciprocal exchanges which, in their modern form, closely resemble advance premium mutuals.

The relative importance of stock, mutuals, and reciprocal exchanges varies among states. In several states, stock insurers write over 75 percent of the business. In a few states, mutual insurers dominate the private insurance field.

### Self-insurers

In all jurisdictions (except Nevada, North Dakota, Guam, Puerto Rico, Texas, the Virgin Islands, and Wyoming), employers are permitted under certain conditions to self-insure workers' compensation. Although most jurisdictions reported 200 or fewer self-insurers, self-insured employers in 1983 paid 18.4 percent of the workers' compensation benefits.

Self-insurance is becoming popular among those who qualify. In 1960 self-insurers paid only 12.4 percent of the benefits; in 1977, 14.1 percent. Most states report an increase also in the number of self-insurers. Possible explanations are increasing cost consciousness, sales activities of agencies who seek to manage self-insurance programs, and the business merger movement which increases the size of firms and their ability to self-insure.

Self-insurance is attractive primarily because it may be less costly than insurance. An insurance premium is designed to pay the losses and expenses of the insurer and provide a margin of profit for contingencies. The self-insurer hopes to save money on the loss or the expense and profit components of the premium.

The loss component equals the average loss the insurer expects the employer and others like him to experience. If the employer is so much better than the average employer in his class that his expected loss may be less, he would save money by self-insuring. In the short run, however, the loss experience may differ substantially from the expected loss. Indeed the loss in a single year might be catastrophic. The larger the number of employees, the less the risk of fluctuation in the annual losses. For most employers, the risk is such that self-insurance is out of the question. For others, the comparison between actual and expected loss is an important consideration.

By self-insuring, the employer can save that part of the premium charged to cover selling expenses, some general administration expenses, and profits. There may also be savings on loss prevention and loss adjustment services even though these services must still be performed.

Other considerations include the relative quality of the safety and claims services provided by insurers, by management service organizations, and by employers themselves; by tax factors; and by the opportunity cost of paying an insurer a premium instead of paying losses and expenses as they occur.

## COST LEVELS AND ALLOCATION

Workers' compensation costs and other costs of industrial accidents impose a burden on industry, workers, and society generally. The magnitude of these costs and their distribution have important economic implications and pose several critical issues of policy.

### Cost levels

Workers' compensation in 1970 cost employers almost $5 billion, or more than $1.13 per $100 of payroll. The cost per $100 of payroll increased to $1.67 in 1983. This works out to close to $23 billion spent by industry in 1983 to provide workers' compensation coverage, according to the Social Security Administration.

These costs include the premiums paid to private insurers on state funds and the benefits and administrative costs paid by self-insurers. Other employer costs of industrial accidents or the losses to employees not covered by workers' compensation are not available.

**Variation among industries.** In its 1980 sample survey of employee benefits, the Chamber of Commerce of the United States found that workers' compensation costs were 1.5 percent of gross payroll. In manufacturing industries, the costs were 1.9 percent; in nonmanufacturing industries 0.7 percent. These rates changed from 0.2 percent for banks and financial institutions, to 3.0 percent for the manufacture of glass, stone, and quarry products.

### Incentives for safety

The asserted safety incentive of workers' compensation is based on the merit-rated pricing policy. "Merit rating" includes both the experience rating and retrospective rating systems. All state funds use merit rating of some sort. In most states, although private insurers are not required to rate employers on merit, they are permitted to. Under this procedure, the firm is charged a premium that is related to the dollar amount of claims for which it is liable. Consequently a merit-rated firm has an incentive to reduce the amount of its claims through loss preventive measures.

The strength of this incentive has been challenged. Only about one-fourth of insured firms, usually large ones, are eligible for merit rating.

The yearly accident record of firms with only a few employees is not a sufficiently reliable indication of their characteristic experience to be considered in establishing premium rates. On the other hand, merit-rated firms account for 85 percent of the dollar volume of premiums paid. In addition, self-insured firms which pay approximately 14 percent of all benefits are implicitly merit rated. Nevertheless, if incentive effects are inherent in experience rating, they are not available to a large number of small firms and their employees.

Firms not eligible for merit rating are class rated. Under this procedure, all employers engaged in similar business operations within a state pay the same rate per $100 of payroll. These employers have strong incentives to reduce the rates paid by their industry and may therefore exert efforts to reduce accidents within the industry. The only accident prevention incentive generated for individual employers within an industry is that, as poor risks, they land in an assigned risk pool.

Other problems, even under merit rating, moderate the incentive for safety. As benefit levels do not reflect the full costs of accidents, the premiums paid are less than adequate; consequently, any savings from safety programs are proportionally minimized. A saving in premium costs could be a significant reward for success in accident prevention if benefits were higher because premiums would then more truly measure the cost of accidents at work. The higher the benefit levels, the larger the premium costs to be avoided and the larger the incentive for prevention.

Even for the merit-rated firm, the functional relationship between injury rates and premium levels is not as direct as might be desirable. The sensitivity of premium to accident experience is dependent on the firm's payroll. As firms increase in size, the premium rate more nearly reflects the individual firm's experience. It has been suggested that more credibility be assigned to the experience of smaller merit-rated companies to increase their safety incentives.

The premium rate for firms of all sizes is more dependent on the frequency rate than the severity rate on the assumption that loss frequency is more within the control of the employer. Thus, firms are encouraged to be more concerned with the number of accidents than with the consequences of accidents. Some critics, however, believe too much emphasis has been placed on loss frequency.

Merit rating suffers the further criticism that, since premiums are related to the level of claims paid, some firms may try to reduce costs by fighting claims rather than by preventing accidents.

Finally, workers' compensation safety incentives have been questioned because the costs usually amount to little more than 1 or 2 percent of payroll and many feel that a firm is insensitive to any cost that small. Other costs of worker accidents that are not insurable provide even stronger incentives. In evaluating this criticism, it should be remembered that the costs of workers' compensation for employers in hazardous industries or with unsafe operations are much greater than 1 or 2 percent of payroll.

Unfortunately, there has been little research on this question, but there is little evidence to indicate that any substantial connection exists between merit rating and accident prevention in most states. Conceptually, however, workers' compensation should impel firms to operate at an optimal level of accident prevention activities in a least-cost fashion.

An improvement in functioning is what is needed.

## REFERENCES

Bureau of the Census, U.S. Department of Commerce, *Statistical Abstracts of the United States,* Superintendent of Documents, Washington, D.C. 20402, 1986.

Chamber of Commerce of the United States, Washington, D.C. *Analysis of Workers' Compensation Laws,* 1986.

Clifford, Joseph A. *Workmen's Compensation—New York.* Binghamton, N.Y., Gould Publications, 1979.

Gaunt, Larry, D., and McDonald, Maurice E. *Examining Employers' Financial Capacity to Self-Insure Under Workmen's Compensation.* Atlanta, Ga., Georgia State University, College of Business Administration, 1977.

La Dou, Joseph, ed. *Introduction to Occupational Health and Safety,* Ch. 17, "Workers' Compensation," National Safety Council, 444 N. Michigan Ave., Chicago Ill. 60611, 1986.

Martin, Roland A. *Occupational Disability—Causes, Prediction, Prevention.* Springfield, Ill., Charles C. Thomas, 1975.

National Council on Compensation Insurance, 200 E. 42nd St., New York, N.Y. 10017. Rate and rating plan manuals.

National Safety Council, 444 N. Michigan Ave., Chicago, Ill. 60611. *Accident Facts* (annually).

# 9

# Safety Training



An effective accident prevention and occupational health hazard control program is based on competent job performance. When people are trained to properly do their jobs, they will do them safely. This, in turn, means that supervisors must (1) know how to train an employee in the safe, efficient way of doing a job, (2) understand company policies and procedures, (3) know how to detect and control hazards, investigate accidents, and handle emergencies, and (4) know how to supervise. It means the safety professional also should be familiar with sound training techniques. He may, or may not, become directly involved in the training effort, but he should be able to recognize the elements of a sound training program.

This chapter is devoted to safety training. Although training and education cannot completely be separated, safety education is broader in scope and usually covers a number of subjects not normally included in a training program. The human relations value of education will be covered in the section on learning in Chapter 11, Human Behavior and Safety.

Training is only one way to influence human behavior. The good example of an employer who spares no effort to create safe working conditions will encourage employees to perform safely. Safe performance also is promoted by developing safe work procedures, by teaching the procedures effectively, by insisting that they be followed, and by teaching people the facts about accident causes and preventive measures.

A well-planned training program not only will train employees, but also can help alter other influences to complement the effect of the training. Firms that conduct intensive safety activities, including training, for all employees make the greatest progress in safety. Supervisors and managers, as well as unskilled and skilled workers, benefit from safety training.

## DEVELOPING THE TRAINING PROGRAM

When developing a training program, consider the program needs and objectives, course outlines and materials, and training methods. Safety training must be integrated with job training.

### Training needs

A training program is needed (1) for new and reassigned employees, (2) when new equipment or processes are introduced, (3) when procedures have been revised or updated, (4) when new information must be made available, and (5) when employee performance needs to be improved. Unless a training program is needed however, there is little justification for spending time and money just to have one.

Here are some indications of a need for a good training program:

1. Proportionately more accidents and injuries, or insurance rates higher than other companies in the same type of work, or a rising insurance rate.
2. High labor turnover.
3. Excessive waste and scrap.
4. Company expansion of plant and equipment.

### Program objectives

Training programs should be based on clearly defined objectives that determine the scope of the training and guide the selection and preparation of the training materials. Objectives should be planned carefully and written down. They should state what the trainee is expected to know or do by the end of the training period.

**Figure 9-1.** Typical training session for supervisors. Prepared course materials are being used.

To make sure the objectives cover the training needs, the duties and responsibilities of the trainees should be determined. Job descriptions and job analyses, including job hazards, should be reviewed. (See the discussion later in this chapter.) These, along with personal observations and performance tests will reveal where training is needed. For example, if a job description (or list of duties) for supervisors includes training workers, then supervisors should be trained to train. This now becomes one of the objectives of a supervisory training program.

A total safety training program is complicated; it has many different kinds of activity. Thus, a statement of safety training objectives is equally complex, and it will cover activity that may not be obvious from a quick reading of the statement. While any program must have management approval, of course, and full acceptance by the supervisors, in many companies the key level that must be reached is the group of people one level above the supervisors. All these groups must give both approval and acceptance.

There is no set of objectives that will serve all companies and all situations. Each must be developed to meet its own circumstances and its own situation. All objectives must be adapted to the needs of the company, and above all, it must be so phrased that management will fully and completely accept it. Training objectives are covered in detail later in this chapter.

All recommendations must be included in a statement of corporate policy, such as those discussed in Chapter 3, Hazard Control Program Organization. Funds for implementation should be included in the corporate budget, and placed under the control of the safety director.

### Course outlines and materials

After defining the training program objectives, the next step is developing the outline of what is to be covered. Often, an outline can meet the program objectives by using existing texts and course materials, or by combining parts of several texts or courses already in use. (A later section in this chapter, a basic course for supervisors, describes the National Safety Council's 14-session safety course outline for supervisors.)

Sometimes, a completely new program must be designed. Either way, the outline must be developed in detail. Major topics and subtopics should be logically arranged and the time allotted to each should be in proportion to the importance of each. So arranged, the course outline is the framework that determines the method of training (see Figure 9-1).

### Training methods

At this stage in planning a training program, a decision must be made as to the training method that will best reach the stated objective. For job or skill training, the best method is the four-step method called job instruction training, or JIT.

Full details of this system—ideal for training new workers or upgrading more experienced ones—are given under the section On-the-job training later in this chapter.

### The lesson plan

The safety professional and others teaching safety subjects should be familiar with lesson plans. They are blueprints for presenting material contained in course outlines. In addition to standardizing training, lesson plans help the instructor:

1. Present material in proper order.
2. Emphasize material in relation to its importance.
3. Avoid omission of essential material.
4. Run the classes on schedule.
5. Provide for trainee participation.
6. Increase the trainee's confidence, especially if he or she is new.

Names for the parts of a lesson plan and their order can vary. The following is a good example of a lesson plan arrangement:

1. **Title:** Must indicate clearly and concisely subject matter to be taught.
2. **Objectives:**
   a) Should state what the trainee should know or be able to do at the end of the training period.
   b) Should limit the subject matter.
   c) Should be specific.
   d) Should stimulate thinking about the subject.
3. **Training aids:** Should include such items as actual equipment or tools to be used, and charts, slides, films, and other aids.
4. **Introduction:**
   a) Should give the scope of the subject.
   b) Should tell the value of the subject.
   c) Should stimulate thinking about the subject.
5. **Presentation:**
   a) Should give the plan of action.
   b) Should indicate the method of teaching to be used, such as lecture, demonstration, class discussion, or a combination of these.
   c) Should contain directions for instructor activity, for example, show chart or write key words on chalkboard.
6. **Application:** Should indicate, by example, how trainees will apply this material immediately: problems may be worked; a job may be performed; or trainees may be questioned on their understanding of procedures.
7. **Summary:**
   a) Should restate main points.
   b) Should tie up loose ends.
   c) Should strengthen weak spots in instruction.
8. **Test:** Tests help determine if objectives have been reached. They should be announced to the class at the beginning of the session.
9. **Assignment:** Should give references to be checked or indicate materials to be prepared for future lessons.

The self-checklist shown in Figure 9-2 will help the instructor improve his teaching techniques.

## SAFETY TRAINING FOR SUPERVISORS

The immediate job of preventing accidents and controlling work health hazards falls upon the supervisor, not because it has been arbitrarily assigned there, but because safety and production are part of the same supervisory function.

### Responsibilities of the supervisor

Whether or not a company has a safety program, the supervisor has these principal responsibilities:

1. Establish work methods.
2. Give job instruction.
3. Assign people to jobs.
4. Supervise people at work.
5. Maintain equipment and the workplace.

**SELF-CHECK TEST FOR INSTRUCTORS**

A good instructor should be able to answer "Yes" to at least 18 of these questions. Under 15 would be below average.

| | | Yes | No |
|---|---|---|---|
| 1. | Do you check your classroom for ventilation, lighting, seating arrangement? | ☐ | ☐ |
| 2. | Do you prepare a lesson plan? | ☐ | ☐ |
| 3. | Do you use training aids when possible? | ☐ | ☐ |
| 4. | Do you preview all films before showing them? | ☐ | ☐ |
| 5. | Do you use test results to find weak points in your teaching? | ☐ | ☐ |
| 6. | Do you vary your type of presentation according to the material? | ☐ | ☐ |
| 7. | Do you stay on the subject? | ☐ | ☐ |
| 8. | Do you cover the material in each lesson? | ☐ | ☐ |
| 9. | Do you make any attempt to know your students and to learn their names? | ☐ | ☐ |
| 10. | Do you introduce yourself to each class? | ☐ | ☐ |
| 11. | Do you make clear the objectives of the lesson? | ☐ | ☐ |
| 12. | Do you summarize each lesson? | ☐ | ☐ |
| 13. | Do you introduce each new subject and explain its importance? | ☐ | ☐ |
| 14. | Do you refrain from using sarcasm in your class? | ☐ | ☐ |
| 15. | Do you start and dismiss your class on time? | ☐ | ☐ |
| 16. | Do you have all your equipment ready and tested before class begins? | ☐ | ☐ |
| 17. | Do you talk directly to the class and avoid such practices as staring at the floor, pacing the floor, juggling a piece of chalk? | ☐ | ☐ |
| 18. | Do you keep your meeting place orderly? | ☐ | ☐ |
| 19. | Do you use words that are easily understood by the class? | ☐ | ☐ |
| 20. | Do you make assignments, and are they clear? | ☐ | ☐ |

**Figure 9-2.** A self-check test can help a supervisor rate teaching methods.

These principal responsibilities of the supervisor are the very activities through which the work of preventing accidents is carried out. A brief examination of these jobs and their relation to safety will make this fact apparent.

**Establishing work methods** that are well understood and consistently followed is essential to orderly and safe operation. Many injuries and health hazards have been reported to result from unsafe methods of procedure. Often later investigation disclosed that no standard method or procedure had been set up for those jobs. The method was declared hazardous only after it resulted in an accident. Making sure safe procedures are established is a supervisory responsibility.

**Giving job instruction,** with necessary emphasis on safety aspects of the job, will help eliminate one of the most frequent causes of accidents—lack of knowledge or skill. If employees are expected to do their work safely, supervisors must demonstrate exactly how the tasks are done and make sure the employees have the knowledge and skill to duplicate the safe work procedures (see Figure 9-3).

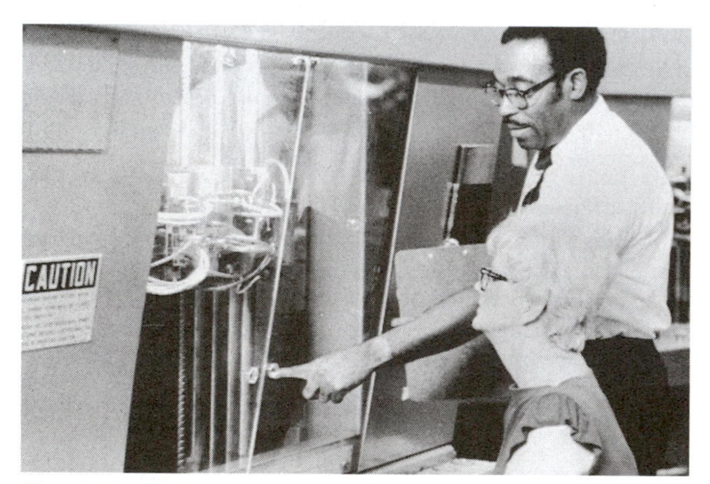


**Figure 9-3.** This supervisor is explaining an enclosed operation to one of his workers. He is instructing her in the correct and safe operation of the machine and will make frequent followups on her progress until certain of her proficiency.

**Assigning people to jobs** is closely related to job instruction. Whenever making a work assignment, the supervisor should make sure the worker is qualified to do the job and thoroughly understands the work method to insure safety and good job performance. Even an experienced worker needs some direction.

**Supervising people at work** is necessary even after a safe work method has been established and job instruction training has been given. When people deviate from established safe practices, injuries can result. Usually, it is found that the injured employees have been neglecting safe practices. In order to prevent injuries from this cause, supervisors must watch for unsafe work methods and correct them as soon as they are observed.

**Maintaining equipment and workplace** in safe condition is no different from maintaining them in efficient condition. Accidents result from tools and equipment in poor condition, from a disorderly workplace, or from makeshift tools used because the right tools are not available. The supervisor who keeps his department and equipment in top condition helps to prevent accidents while improving efficiency.

### Supervisors must accept their duties

Not only are these five functions a normal part of the supervisor's job but, unless lines of organization and authority are seriously disregarded, nobody but the supervisor can perform them. Sometimes this fact is overlooked. Since all these functions are closely allied with safety, it is only natural that safety should be integrated into the duties of the supervisor. If, however, management wants to be sure that supervisors accept this role, it should be included in a policy statement and lines of authority established.

A clear and positive executive order defining the safety duties of all persons and departments should be issued. A careful program of education should be instituted to help supervisors understand and accept their role and to give them specific help in their work of preventing accidents.

Many supervisors have acquired their present positions in organizations while some sort of safety program was in existence, and their understanding of the program as it has existed is firmly established. However, a safety professional undertaking the safety training of supervisors will almost invariably find that the first and major job is to get supervisors at all levels to understand and accept their role in accident prevention. This job cannot be done in a single meeting or by a single communication.

Simply getting supervisors to agree in theory that safety is one of their duties is not enough. They must come to understand the many ways in which they can prevent accidents, and they must become interested in improving their safety and health performance.

### Objectives of supervisor safety training

Before the safety training of supervisors is undertaken, the objectives of that training must be understood and stated specifically. Determining objectives for supervisory training is so important and involves so many kinds of activity that it should be made the subject of careful study and consultation. Members of the safety department should not attempt to set up objectives alone, but should confer with representative supervisors and members of higher management.

The following statement of objectives should be used only as a guide to be studied and modified to fit specific situations and should be accepted only to the extent that the management persons concerned agree to it. The general objectives of safety training for supervisors can be any or all of the following:

1. To involve supervisors in the company's accident prevention program
2. To establish the supervisor as the key person in preventing accidents
3. To get supervisors to understand their safety responsibilities
4. To provide supervisors with information on causes of accidents and occupational health hazards and methods of prevention
5. To give supervisors an opportunity to consider current problems of accident prevention and to develop solutions based on their own and others' experience
6. To help supervisors gain skill in accident prevention activities
7. To help supervisors keep their own departments safe.

## A BASIC COURSE FOR SUPERVISORS

The knowledge and philosophy of accident prevention is not just common sense, as some proclaim. It is a specialized body of information accumulated over a period of many years. It is the job of the safety professional to help supervisors gain whatever information is available that will make their safety efforts more productive.

The most direct way to develop the desired attitudes and to impart the necessary information about safety to supervisors is to provide a course of instructions. Courses in safety and other management subjects are conducted for supervisors by many companies. They follow a fairly well-established pattern.

### Subject outline of course

The following outline, based on the National Safety Council's 14-hour "Supervisors Development Program," shows the subjects that should usually be included in a safety course for supervisors. Visual aids are available on all the subjects and should be used at every meeting. Titles of films and other aids are not included here because new ones are being produced constantly. The person conducting a course should secure and use those best suited to his needs. (See the latest *General Materials Catalog* published by the National Safety Council.)

*Session 1*
LOSS CONTROL FOR SUPERVISORS
Accidents and incidents, areas of responsibility, the cost of accidents, and a better approach to occupational safety and health.

*Session 2*
COMMUNICATIONS
Elements of communication, methods of communication, and effective listening.

*Session 3*
HUMAN RELATIONS
Human relations concepts, leadership, workers with special problems, and the drug and alcohol problem.

*Session 4*
EMPLOYEE INVOLVEMENT IN SAFETY
Promoting safe-worker attitudes, employee recognition, safety meetings, and off-the-job accident problems.

*Session 5*
SAFETY TRAINING
New employee indoctrination, job safety analysis (JSA), job instruction training (JIT), and other methods of instruction.

*Session 6*
INDUSTRIAL HYGIENE AND NOISE CONTROL
General concepts, chemical agents, physical agents, temperature extremes, atmospheric pressures, ergonomics, biological stresses, Threshold Limit Values (TLVs), and controls.

*Session 7*
ACCIDENT INVESTIGATION
Finding causes, emergency procedures, effective use of witnesses, and reports.

*Session 8*
SAFETY INSPECTIONS
Formal inspections, inspection planning and checklists, inspecting work practices, frequency of inspections, recording hazards, and follow-up actions.

*Session 9*
PERSONAL PROTECTIVE EQUIPMENT
Controlling hazards; overcoming objections; protecting the head, eyes, and ears; respiratory protective equipment; safety belts and harnesses; protecting against radiation; safe work clothing; and protecting the hands, arms, feet, and legs.

*Session 10*
MATERIALS HANDLING AND STORAGE
Materials handling problems; materials handling equipment; ropes, chains, and slings; and material storage.

*Session 11*
MACHINE SAFEGUARDING
Principles of guarding, safeguard design, safeguarding mechanisms, and safeguard types and maintenance.

*Session 12*
HAND TOOLS AND PORTABLE POWER TOOLS
Safe working practices, use of hand tools, use of portable power tools, and maintenance and repair of tools.

*Session 13*
ELECTRICAL SAFETY
Electrical fundamentals review, branch circuits and grounding concepts, plug- and cord-connected equipment, branch circuit and equipment testing methods, ground fault circuit interrupters, hazardous locations, common electrical deficiencies, safeguards for home appliances, and safety program policy and procedures.

*Session 14*
FIRE SAFETY
Basic principles, causes of fire; fire-safe housekeeping; alarms, equipment, and evacuation; and reviewing the supervisor's fire job.

### Tips on running a course

All supervisors should be given a basic course of the type just outlined. It should be repeated from time to time for new supervisors and prospective supervisors.

The format should be formal; attendance records should be kept and certificates or diplomas issued upon completion. Some companies give a diploma that can be framed and displayed; others favor a pocket-size card, and still others present both to their graduate supervisors.

It adds to the dignity of the training program if, at the beginning of a course, a company executive meets with the group and explains the importance of the supervisor's role in the safety program and the value of the training. A formal graduation ceremony is often held upon completion of the course.

Some large companies with widespread operations personalize a course and give it throughout the organization. The preparation of such a course can be a major and costly project. It is not recommended except for very large organizations with proportionately large safety and training staffs.

Prepared courses are available, such as the National Safety Council's "Supervisor's Development" course which has a text book, the *Supervisors Safety Manual, Instructors' Guide,* student kit, certificate of completion, and visual aids. The Council's courses can be adapted to individual needs. They have been thoroughly tested by use in hundreds of companies and many kinds of operations. Use of a prepared course saves a great deal of preparation time and effort. Let the instructor give it the company slant.

### Providing instructors

One problem in giving a basic in-house safety course to supervisors is providing qualified instructors. A considerable burden is placed on the safety personnel if they must conduct every class. Many companies have found that the best instructors are general supervisors or division managers. In addition to usefully serving as instructors, these people also learn a great deal about safety during the process of teaching.

It is advisable that potential instructors take a short course on how to instruct or, at least, get together several times to plan how best to give the instruction. Each instructor should take charge of one class by preparing a lesson plan following the format outlined in this chapter and by using company situations and problems as examples.

Safety professionals may feel there is some risk of not getting good instruction by this method, but those who have tried it say that the supervisors' experience and understanding more than offset a lack of formal teaching experience.

In addition, acceptance by the group is improved because the members know that the instructor is a person who has done the things he is talking about. In the case of a general supervisor who is teaching a group of his own supervisors, the meetings can become planning sessions in which agreement is reached as to how the group is to handle situations in the future. Training linked with actual operation can be most effective.

Courses conducted locally should be planned and instructed by safety professionals experienced in training and in the subject to be covered. Criteria for such instructors might include:

- Professional member or member of the American Society of Safety Engineers
- Certified Safety Professional (CSP)
- Holder of the National Safety Council's "Advanced Safety Certificate"
- Two years' recent experience in safety training, or equivalent experience.

Regardless of experience, instructors should make an evaluation of teaching methods. A self-check test for instructors, such as the one shown in Figure 9-2, is an ideal evaluation.

### Conducting the in-house course

Training sessions should not be longer than one hour. Longer sessions are likely to become tiresome, and supervisors will become apprehensive about their teaching abilities if the trainees show signs of boredom.

It is desirable to hold sessions during working hours, if possible. Regular attendance and freedom from interruptions are of great importance and should be required by management.

If courses are held after working hours or between shifts, sessions are usually two hours with dinner or refreshments usually provided.

A supervisor's safety course produces many benefits—an increased understanding of safety, acceptance of responsibility for the prevention of accidents, and greater interest in all supervisory duties.

## SPECIAL TRAINING FOR SUPERVISORS

Success in supervision requires skill as much as it requires knowledge and understanding. Basic safety training is aimed toward opening awareness of the need for safety programs, increasing knowledge about safety, and creating a proper attitude toward safe behavior.

Many accidents, however, result from unskillful handling of some of the important tasks of supervision—giving orders, assigning jobs, correcting workers for poor performance, and the like.

To be able to do a good safety job, supervisors also need training in supervisory skills.

Among the most important of these skills are giving job instruction, supervising workers at work, determining accident causes, and building safety attitudes among workers. Special courses may be given in these subjects and others, such as accident investigation procedures.

## ON-THE-JOB TRAINING

On-the-job training (OJT) is widely used because the trainee can be producing while he is being trained. Whether the supervisor does the instructing himself or has someone else do it, the training should be carefully planned and organized. Again the safety professional should be familiar with his organization's training program so that appropriate and correct safety training is integrated into it.

In too many cases, on-the-job training is a hit-or-miss procedure where the trainee is told to follow another worker around and learn his job. In situations of this type, the worker may be too busy to do any training, or may not know good training procedures. He may even be reluctant to train another to do his work.

Remember, it is the supervisor's responsibility to make sure the person doing the training can, in fact, train. On-the-job training includes many techniques and approaches and there is no one method that will fit all situations. Job safety analysis (discussed in the next section) is one method, job instruction training is another, and over-the-shoulder coaching is still another widely used method. These methods may be used separately or in combination depending upon the complexity of the job and the time element.

### Over-the-shoulder coaching

Over-the-shoulder coaching is perhaps the most flexible and direct of the three training methods. In coaching, the trainee is expected to develop and apply his skills in typical work situations under the guidance of a qualified person. The person to whom the trainee has been assigned should be one who knows the job thoroughly, is a safe operator, and has the patience, time, and desire to help others (see Figure 9-4).

The advantages of training of this type are:

1. The worker is more likely to be highly motivated because the guidance is personal.
2. The instructor can identify specific performance deficiencies and take immediate and proper corrective action.
3. Results of the training are readily apparent since real equipment is being used and finished work can be judged by existing standards.
4. The training is practical and realistic and can be applied at the proper time.

Timing is important; not only does the trainee like to get help when needed, but also the instructor can judge the trainee's progress continually and present the next unit or phase of instruction when the trainee is ready.

To help keep track of the progress of each individual, a training chart is valuable. On the top of a typical chart are listed

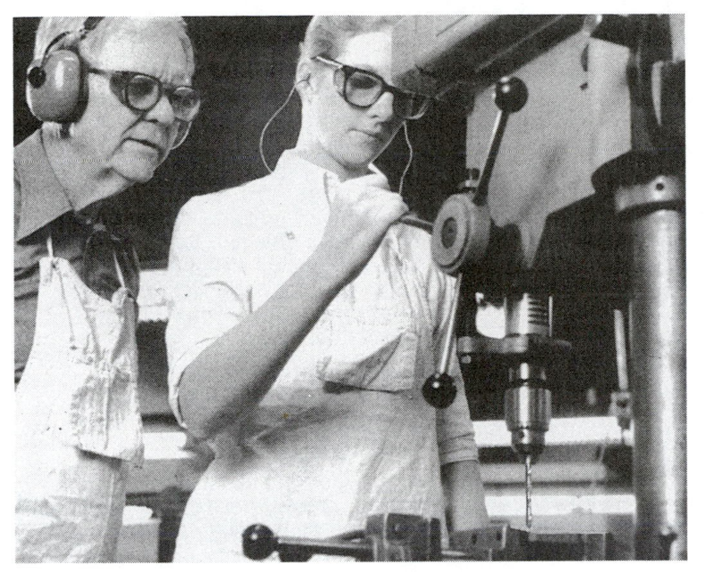

**Figure 9-4.** Over-the-shoulder coaching. This trainee is developing and applying skills in the work situation under the guidance of a qualified person.

the various tasks required of a person in a particular job classification. By observation and direct discussion, the instructor can determine whether the employee is qualified to perform the tasks necessary to fill the job. The instructor determines the trainee's degree of skill and knowledge and estimates the training needs. Notes and comments can be made on the chart. Such a chart prevents missing important training; it also prevents unnecessary training (see Figure 9-5).

## JOB SAFETY ANALYSIS

Job safety analysis (JSA) is a procedure used to review job methods and uncover hazards (1) that may have been overlooked in the layout of the plant or building and in the design of the machinery, equipment, tools, work stations, and processes, or (2) that may have developed after production started, or (3) that resulted from changes in work procedures or personnel. It is one of the first steps in hazard and accident analysis and in safety training.

Once the hazards are known, the proper solutions can be developed. Some solutions may be physical changes that eliminate or control the hazard, such as placing a safeguard over exposed moving machine parts. Others may be job procedures that eliminate or minimize the hazard, for example, safe piling of materials. These will require training and supervision

A job safety analysis can be written up in the manner shown in Figure 9-6. In the left-hand column, the basic steps of the

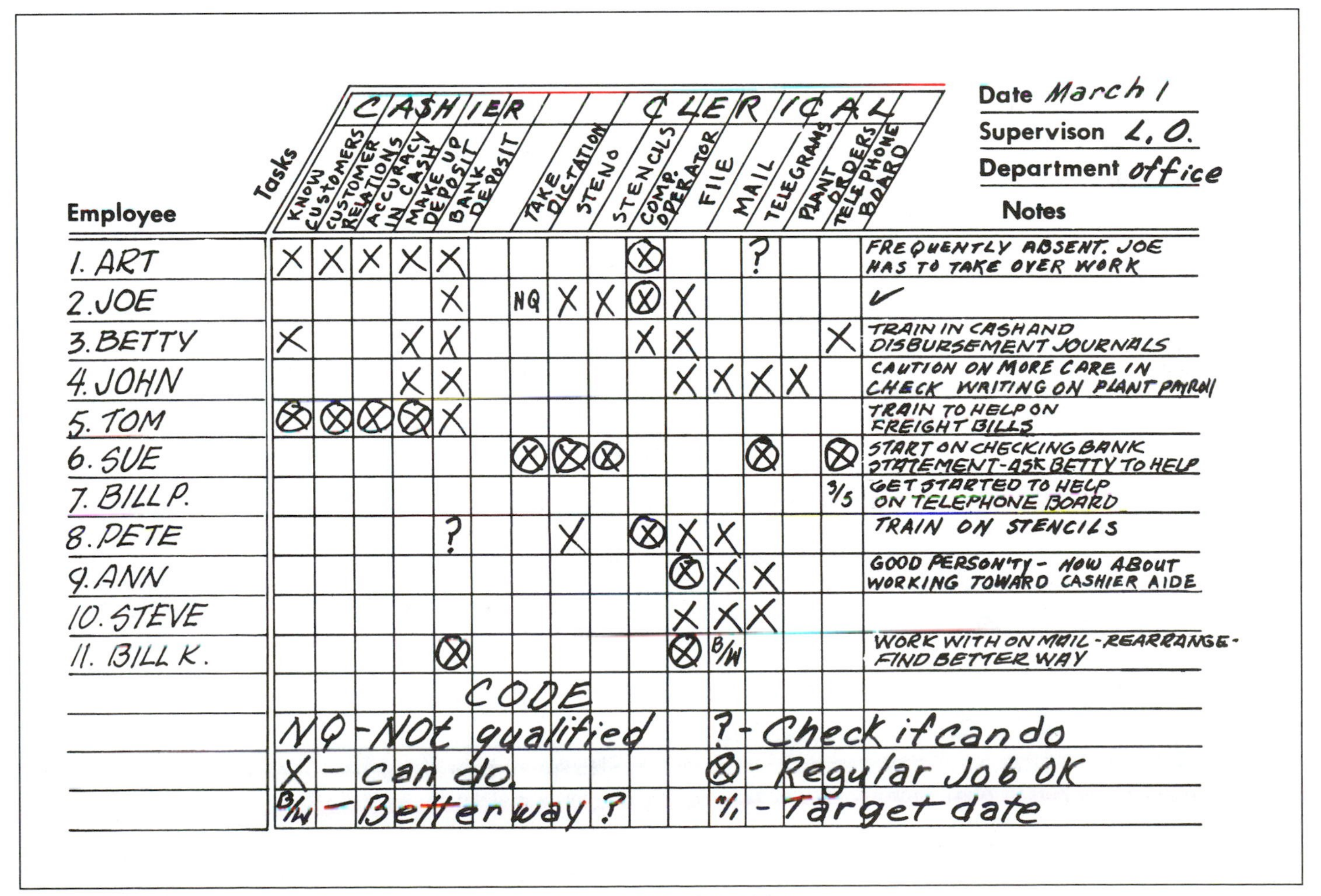

Date March 1
Supervison L. O.
Department office

| Employee | CASHIER: Know customers | Customer relations | Accuracy in cash | Make up deposit | Bank deposit | | CLERICAL: Take dictation | Steno | Stencils | Comp. operator | File | Mail | Telegrams | Plant orders | Telephone board | Notes |
|---|---|---|---|---|---|---|---|---|---|---|---|---|---|---|---|---|
| 1. ART | X | X | X | X | X | | | | | ⊗ | | | ? | | | FREQUENTLY ABSENT. JOE HAS TO TAKE OVER WORK |
| 2. JOE | | | | | X | | NQ | X | X | ⊗ | X | | | | | ✓ |
| 3. BETTY | X | | | X | X | | | | | X | X | | | | X | TRAIN IN CASH AND DISBURSEMENT JOURNALS |
| 4. JOHN | | | | X | X | | | | | | X | X | X | X | | CAUTION ON MORE CARE IN CHECK WRITING ON PLANT PAYROLL |
| 5. TOM | ⊗ | ⊗ | ⊗ | ⊗ | X | | | | | | | | | | | TRAIN TO HELP ON FREIGHT BILLS |
| 6. SUE | | | | | | | ⊗ | ⊗ | ⊗ | | | | ⊗ | | ⊗ | START ON CHECKING BANK STATEMENT-ASK BETTY TO HELP |
| 7. BILL P. | | | | | | | | | | | | | | | 3/5 | GET STARTED TO HELP ON TELEPHONE BOARD |
| 8. PETE | | | | | ? | | | X | | ⊗ | X | X | | | | TRAIN ON STENCILS |
| 9. ANN | | | | | | | | | | | ⊗ | X | X | | | GOOD PERSON'TY - HOW ABOUT WORKING TOWARD CASHIER AIDE |
| 10. STEVE | | | | | | | | | | | X | X | X | | | |
| 11. BILL K. | | | | | ⊗ | | | | | | ⊗ | B/W | | | | WORK WITH ON MAIL - REARRANGE - FIND BETTER WAY |

CODE
NQ - NOt qualified    ? - Check if can do
X - can do    ⊗ - Regular Job OK
B/W - Better way?    %, - Target date

**Figure 9-5.** Training chart helps keep track of who needs, and who has, what training

**JOB SAFETY ANALYSIS WORK SHEET**
**JOB: Using a Pressurized Water Fire Extinguisher**

| *WHAT TO DO (Steps in sequence)* | *HOW TO DO IT (Instructions) (Reverse hands for left-handed operator.)* | *KEY POINTS (Items to be emphasized. Safety is always a key point.)* |
|---|---|---|
| 1. Remove extinguisher from wall bracket. | 1. Left hand on bottom lip, fingers curled around lip, palm up. Right hand on carrying handle palm down, fingers around carrying handle only. | 1. Check air pressure to make certain extinguisher is charged. Stand close to extinguisher, pull straight out. *Have firm grip, to prevent dropping on feet.* Lower, and as you do remove left hand from lip. |
| 2. Carry to fire. | 2. Carry in right hand, upright position. | 2. Extinguisher should hang down alongside leg. (This makes it easy to carry and reduces possibility of strain.) |
| 3. Remove pin. | 3. Set extinguisher down in upright position. Place left hand on top of extinguisher, pull out pin with right hand. | 3. Hold extinguisher steady with left hand. Do not exert pressure on discharge lever as you remove pin. |
| 4. Squeeze discharge lever. | 4. Place right hand over carrying handle with fingers curled around operating lever handle while grasping discharge hose near nozzle with left hand. | 4. Have firm grip on handle to steady extinguisher. |
| 5. Apply water stream to fire. | 5. Direct water stream at base of fire. | 5. Work from side to side or around fire. After extinguishing flames, play water on smouldering or glowing surfaces. |
| 6. Return Extinguisher. Report Use. | | |

**Figure 9-6.** Here is the first step in preparing a job safety analysis. The work sheet then is broken down as shown in Figures 9-7 and 9-8.

job are listed in the order in which they occur. The middle column describes all hazards, both those produced by the environment and those connected to the job procedure. The right-hand column gives the safe procedures that should be followed to guard against the hazards and to prevent potential accidents.

For convenience, both the job safety analysis procedure and the written description are commonly referred to as JSA. Health hazards are also considered when making a JSA.

The four basic steps in making a job safety analysis are:

1. Select the job to be analyzed.
2. Break the job down into successive steps or activities and observe how these actions are performed.
3. Identify the hazards and potential accidents. (This is the critical step because only an identified problem can be eliminated.)
4. Develop safe job procedures to eliminate the hazards and prevent the potential accidents.

### Benefits of JSA

The principal benefits that arise from job safety analysis are these phases of a supervisor's work:

- Giving individual training in safe, efficient procedures
- Making employee safety contacts
- Instructing the new person on the job
- Preparing for planned safety observations
- Giving pre-job instructions on irregular jobs
- Reviewing job procedures after accidents occur
- Studying jobs for possible improvement in job methods.

The various steps in making a JSA are discussed in the following four sections.

### Select the job

A job is a sequence of separate steps or activities that together accomplish a work goal. Some jobs can be broadly defined by what is accomplished. Making paper, building a plant, mining iron ore are examples. Such broadly defined jobs are not suitable for JSA. Similarly, a job can be narrowly defined in terms of a single action. Turning a switch, tightening a screw, pushing a button are examples. Such narrowly defined jobs also are unsuitable for JSA.

Jobs suitable for JSA are those job assignments that a line supervisor may make. Operating a machine, tapping a furnace, piling lumber are good subjects for job safety analyses. They are neither too broad nor too narrow.

Jobs should not be selected at random—those with the worst accident experience should be analyzed first if JSA is to yield the quickest possible results. In fact, some companies make this the focal point of their accident prevention program.

In selecting jobs to be analyzed and in establishing the order of analysis, top supervision of a department should be guided by the following factors:

1. Frequency of accidents. A job that has repeatedly produced accidents is a candidate for a JSA. The greater number of accidents associated with the job, the greater its priority claim for a JSA.
2. Production of disabling injuries. Every job that has produced disabling injuries should be given a JSA. Subsequent

injuries prove that preventive action taken prior to their occurrence was not successful.

3. SEVERITY POTENTIAL. Some jobs may not have a history of accidents but may have the potential for severe injury.
4. NEW JOBS created by changes in equipment or in processes obviously have no history of accidents, but their accident potential may not be fully appreciated. A JSA of every new job should be made as soon as the job has been created. Analysis should not be delayed until accidents or near misses occur.

## Break the job down

Before the search for hazards begins, a job should be broken down into a sequence of steps, each describing what is being done. Avoid the two common errors: (1) making the breakdown so detailed that an unnecessarily large number of steps results, or (2) making the job breakdown so general that basic steps are not recorded.

The technique of making a job safety analysis involves these steps:

1. Selecting the right person to observe
2. Briefing him on the purpose
3. Observing him perform the job, and trying to break it into basic steps
4. Recording each step in the breakdown
5. Checking the breakdown with the person observed.

Select an experienced, capable, and cooperative person who is willing to share ideas. If the employee has never helped on a job safety analysis, explain the purpose—to make a job safe by identifying and eliminating or controlling hazards—and show him a completed JSA. Reassure the employee that he was selected because of his experience and capability.

To determine the basic job steps, ask "What step starts the job?" Then, "What is the next basic step?" and so on.

In recording the job steps, each should be completely described and the employee should be asked to verify the written job description; possible deviations from the regular procedure should be recorded because it may be this irregular activity that leads to an accident.

To record the breakdown, number the job steps consecutively as illustrated in the first column of the JSA training guide, illustrated in Figure 9-7. Each step tells what is done, not how.

### INSTRUCTIONS FOR COMPLETING JOB SAFETY ANALYSIS FORM

Job Safety Analysis (JSA) is an important accident prevention tool that works by finding hazards and eliminating or minimizing them *before* the job is performed, and *before* they have a chance to become accidents. Use your JSA for job clarification and hazard awareness, as a guide in new employee training, for periodic contacts and for retraining of senior employees, as a refresher on jobs which run infrequently, as an accident investigation tool, and for informing employees of specific job hazards and protective measures.

Set priorities for doing JSA's: jobs that have a history of many accidents, jobs that have produced disabling injuries, jobs with high potential for disabling injury or death, and new jobs with no accident history.

Here's how to do each of the three parts of a Job Safety Analysis:

#### SEQUENCE OF BASIC JOB STEPS

Break the job down into steps. Each of the steps of a job should accomplish some major task. The task will consist of a *set* of movements. Look at the first *set* of movements used to perform a task, and then determine the next logical *set* of movements. For example, the job might be to move a box from a conveyor in the receiving area to a shelf in the storage area. How does that break down into job steps? Picking up the box from the conveyor and putting it on a handtruck is one logical set of movments, so it is one job step. Everything related to that one logical set of movements is part of that job step.

The next logical *set* of movements might be pushing the loaded handtruck to the storeroom. Removing the boxes from the truck and placing them on the shelf is another logical set of movements. And finally, returning the handtruck to the receiving area might be the final step in this type of job.

Be sure to list *all* the steps in a job. Some steps might not be done each time—checking the casters on a handtruck, for example. However, that task is a part of the job as a whole, and should be listed and analyzed.

#### POTENTIAL HAZARDS

Identify the hazards associated with each step. Examine each step to find and identify hazards—actions, conditions and possibilities that could lead to an accident.

It's not enough to look at the obvious hazards. It's also important to look at the entire environment and discover every conceivable hazard that might exist.

Be sure to list health hazards as well, even though the harmful effect may not be immediate. A good example is the harmful effect of inhaling a solvent or chemical dust over a long period of time.

It's important to list *all* hazards. Hazards contribute to accidents, injuries and occupational illnesses.

In order to do part three of a JSA effectively, you must identify potential and existing *hazards*. That's why it's important to distinguish between a hazard, an accident and an injury. Each of these terms has a specific meaning:

HAZARD—A potential danger. Oil on the floor is a *hazard*.

ACCIDENT—An unintended happening that may result in injury, loss or damage. Slipping on the oil is an *accident*.

INJURY—The *result* of an accident. A sprained wrist from the fall would be an injury.

Some people find it easier to identify possible accidents and illnesses and work back from them to the hazards. If you do that, you can list the accident and illness types in parentheses following the hazard. But be sure you focus on the *hazard* for developing recommended actions and safe work procedures.

#### RECOMMENDED ACTION OR PROCEDURE

Using the first two columns as a guide, decide what actions are necessary to eliminate or minimize the hazards that could lead to an accident, injury, or occupational illness.

Among the actions that can be taken are: 1) engineering the hazard out; 2) providing personal protective equipment; 3) job instruction training; 4) good housekeeping; and 5) good ergonomics (positioning the person in relation to the machine or other elements in the environment in such a way as to eliminate stresses and strains).

List recommended safe operating procedures on the form, and also list required or recommended personal protective equipment for each step of the job.

Be specific. Say *exactly* what needs to be done to correct the hazard, such as, "lift, using your leg muscles." Avoid general statements like, "be careful."

Give a recommended action or procedure for *every* hazard.

If the hazard is a serious one, it should be corrected immediately. The JSA should then be changed to reflect the new conditions.

**Figure 9-7.** Job safety analysis training guide. Use these guidelines when preparing a JSA.

The wording for each step should begin with an action word like remove, open, or weld. The action is completed by naming the item to which the action applied, for example, "remove extinguisher," "carry to fire."

In checking the breakdown with the person observed, obtain his agreement of what is done and the order of the steps. Thank the employee for his cooperation.

### Identify hazards and potential accidents

Before filling in the next two columns of the JSA—Potential Accidents or Hazards and Recommended Safe Job Procedure—begin the search for hazards. The purpose is to identify all hazards—both those produced by the environment and those connected with the job procedure. Each step, and thus the entire job, must be made safer and more efficient. To do this, ask yourself these questions about each step:

1. Is there a danger of striking against, begin struck by, or otherwise making a harmful contact with an object?
2. Can the employee be caught in, by, or between objects?
3. Is there a potential for a slip or trip? Can he fall on the same level or to another?
4. Can he strain himself by pushing, pulling, lifting, bending, or twisting?
5. Is the environment hazardous to safety or health? For example, concentrations of toxic gas, vapor, mist, fume, or dust, heat or radiation. (See discussion in the National Safety Council book *Fundamentals of Industrial Hygiene.*)

Close observation and knowledge of the particular job are required if the JSA is to be effective. The job observation should be repeated as often as necessary until all hazards and potential accidents have been identified.

Include hazards that might result. Record the type of accident and the agent involved. To note that the employee might injure a foot by dropping a fire extinguisher, for example, write down "struck by extinguisher."

Again check with the observed employee after the hazards and potential accidents have been recorded. The experienced employee will probably suggest additional ideas. You should also check with others experienced with the job. Through observation and discussion, you will develop a reliable list of hazards and potential accidents.

### Develop solutions

The final step in a JSA is to develop a recommended safe job procedure to prevent occurrence of potential accidents. The principal solutions are:

1. Find a new way to do the job
2. Change the physical conditions that create the hazards
3. To eliminate hazards still present, change the work procedure
4. Try to reduce the necessity of doing a job, or at least the frequency that it must be performed. This is particularly helpful in maintenance and material handling.

- To find an entirely new way to do a job, determine the work goal of the job, and then analyze the various ways of reaching this goal to see which way is safest. Consider work-saving tools and equipment.
- If a new way cannot be found, then ask this question about each hazard and potential accident listed: "What change in physical condition (such as change in tools, materials, equipment, layout, or location) will eliminate the hazard or prevent the accident?"

  When a change is found, study it carefully to find what other benefits (such as greater production or time saving) will accrue. These benefits should be pointed out when proposing the change to higher management. They make good selling points.
- The third solution in solving the job-hazard problem is to investigate changes in the job procedure. Ask of each hazard and potential accident listed: "What should the employee do—or not do—to eliminate this particular hazard or prevent this potential accident?" Where appropriate, ask an additional question, "How should it be done?" Because of his experience, in most cases the supervisor can answer these questions.

  Answers must be specific and concrete if new procedures are to be any good. General precautions—"be alert," "use caution," or "be careful"—are useless. Answers should precisely state what to do and how to do it. This recommendation—"Make certain the wrench does not slip or cause loss of balance"—is incomplete. It does not tell how to prevent the wrench from slipping.

  Here, in contrast, is an example of a good recommended safe procedure that tells both what and how: "Set wrench properly and securely. Test its grip by exerting a slight pressure on it. Brace yourself against something immovable, or take a solid stance with feet wide apart, before exerting full pressure. This prevents loss of balance if the wrench slips."
- Often a repair or service job has to be frequently repeated because a condition needs correction again and again. To reduce the necessity of such a repetitive job, ask "What can be done to eliminate the cause of the condition that makes excessive repairs or service necessary?" If the cause cannot be eliminated, then ask "Can anything be done to minimize the effects of the condition?"

  Machine parts, for example, may wear out quickly and require frequent replacement. Study of the problem may reveal excessive vibration is the culprit. After reducing or eliminating the vibration, the machine parts last longer and require less maintenance.

  Reducing frequency of a job contributes to safety only in that it limits the exposure. Every effort still should be made to eliminate hazards and to prevent potential accidents through changing physical conditions or revising job procedures or both.
- A job that has been redesigned may require going beyond the immediate boundaries of the specific job—affecting other jobs and even the entire work process. Therefore, the redesign should be discussed not only with the worker involved, but also co-workers, the supervisor, the plant engineer, and others who are concerned. In all cases, however, check or test the proposed changes by reobserving the job and discussing the changes with those who do the job. Their ideas about the hazards and proposed solutions can be of considerable value. They can judge the practicality of proposed changes and perhaps suggest improvements. Actually these discussions are more than just a way to check a JSA. They are safety contacts that promote awareness of job hazards and safe procedures.

  A final version of a JSA is shown in Figure 9-8.

### Use JSA effectively

The major benefits of a job safety analysis come after its com-

| JOB SAFETY ANALYSIS<br>INSTRUCTIONS ON REVERSE SIDE | JOB TITLE (and number if applicable): Banding Pallets — PAGE 1 OF 2 — JSA NO. 105 | | DATE: 00/00/00 | ☒ NEW<br>☐ REVISED |
|---|---|---|---|---|
| | TITLE OF PERSON WHO DOES JOB: Bander | SUPERVISOR: James Smith | ANALYSIS BY: James Smith | |
| COMPANY/ORGANIZATION: XYZ Company | PLANT/LOCATION: Chicago | DEPARTMENT: Packaging | REVIEWED BY: Sharon Martin | |
| REQUIRED AND/OR RECOMMENDED PERSONAL PROTECTIVE EQUIPMENT: | Gloves - Eye Protection - Long Sleeves - Safety Shoes | | APPROVED BY: Joe Bottom | |

| SEQUENCE OF BASIC JOB STEPS | POTENTIAL HAZARDS | RECOMMENDED ACTION OR PROCEDURE |
|---|---|---|
| 1. Position portable banding cart and place strapping guard on top of boxes. | 1. Cart positioned too close to pallet (strike body & legs against cart or pallet, drop strapping gun on foot.) | 1. Leave ample space between cart and pallet to feed strapping - have firm grip on strapping gun. |
| 2. Withdraw strapping and bend end back about 3". | 2. Sharp edges of strapping (cut hands, fingers & arms). Sharp corners on pallet (strike feet against corners). | 2. Wear gloves, eye protection & long sleeves - keep firm grip on strapping - hold end between thumb & forefinger - watch where stepping. |
| 3. Walk around load while holding strapping with one hand. | 3. Projecting sharp corners on pallet (strike feet on corners). | 3. Assure a clear path between pallet and cart - pull smoothly - avoid jerking strapping. |
| 4. Pull and feed strap under pallet. | 4. Splinters on pallet (punctures to hands and fingers) Sharp strap edges (cuts to hands, fingers, and arms). | 4. Wear gloves - eye protection - long sleeves. Point strap in direction of bend - pull strap smoothly to avoid jerks. |
| 5. Walk around load. Stoop down. Bend over, grab strap, pull up to machine, straighten out strap end. | 5. Protruding corners of pallet, splinters (punctures to feet and ankles). | 5. Assure a clear path - watch where walking - face direction in which walking. |
| 6. Insert, position and tighten strap in gun. | 6. Springy and sharp strapping (strike against with hands and fingers). | 6. Keep firm grasp on strap and on gun - make sure clip is positioned properly. |

**Figure 9-8.** Final form of a JSA shows how hazards are identified and safe procedures are spelled out in order to prevent occurrences of the potential hazards.

pletion. However, benefits are also to be gained from the development work.

While making job safety analyses, supervisors learn more about the jobs they supervise. When employees are encouraged to participate in job safety analyses, their safety attitudes are improved and their safety knowledge is increased. As a JSA is worked out, safer and better job procedures and safer working conditions are developed.

But these important benefits are only a portion of the total benefits to be derived from the JSA program. The principal benefits were listed at the beginning of this discussion.

When a JSA is distributed, the supervisor's first responsibility is to explain its contents to employees and, if necessary, to give them further individual training. The entire JSA must be reviewed with the employees concerned so that they will know how the job is to be done—without accidents.

The JSA can furnish material for planned safety contacts. All steps of the JSA should be used for this purpose. The steps that present major hazards should be emphasized and reviewed again and again in subsequent safety contacts.

New employees on the job must be trained in the basic job steps. They must be taught to recognize the hazards associated with each job step and must learn the necessary precautions. There is no better guide for this training than a well-prepared JSA.

Occasionally, the supervisor should observe his employees as they perform the jobs for which analyses have been developed. The purpose of these observations is to determine whether or not the employees are doing the jobs in accordance with the safe job procedures. Before making such observations, the supervisor should prepare himself by reviewing the JSA in question so he will have firmly in mind the key points to observe.

Many jobs, such as certain repair or service jobs, are done infrequently or on an irregular basis. The employees who do them will benefit from pre-job instruction that reminds them of the hazards and the necessary precautions. Using the JSA for the particular job, the supervisor should give this instruction at the time he makes the job assignment.

Whenever an accident occurs on a job covered by a job safety analysis, the JSA should be reviewed to determine whether or not it needs revision. If the JSA is revised, all employees concerned with the job should be informed of the changes and instructed in any new procedures.

When an accident results from failure to follow JSA procedures, the facts should be discussed with all those who do the job. It should be made clear that the accident would not have occurred had the JSA procedures been followed.

All supervisors are concerned with improving job methods to increase safety and health, reduce costs, and step up production. The job safety analysis is an excellent starting point for questioning the established way of doing a job. And study of the JSA may well suggest ideas for improvement of job methods.

## JOB INSTRUCTION TRAINING

The job instruction training (JIT) method described here is often called the Four-Point Method, because the instructing job is broken into four parts, each of which is shown in detail in Figure 9-9: (1) Preparation; (2) Presentation; (3) Application; (4) Testing.

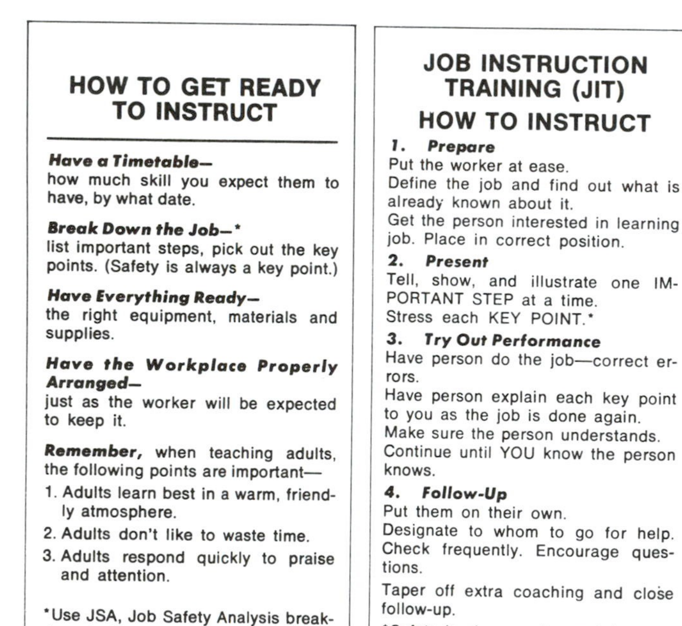

**HOW TO GET READY TO INSTRUCT**

***Have a Timetable—*** how much skill you expect them to have, by what date.

***Break Down the Job—**** list important steps, pick out the key points. (Safety is always a key point.)

***Have Everything Ready—*** the right equipment, materials and supplies.

***Have the Workplace Properly Arranged—*** just as the worker will be expected to keep it.

***Remember,*** when teaching adults, the following points are important—

1. Adults learn best in a warm, friendly atmosphere.
2. Adults don't like to waste time.
3. Adults respond quickly to praise and attention.

*Use JSA, Job Safety Analysis breakdown to locate hazards.

**JOB INSTRUCTION TRAINING (JIT)**
**HOW TO INSTRUCT**

***1. Prepare***
Put the worker at ease.
Define the job and find out what is already known about it.
Get the person interested in learning job. Place in correct position.

***2. Present***
Tell, show, and illustrate one IMPORTANT STEP at a time.
Stress each KEY POINT.*

***3. Try Out Performance***
Have person do the job—correct errors.
Have person explain each key point to you as the job is done again.
Make sure the person understands.
Continue until YOU know the person knows.

***4. Follow-Up***
Put them on their own.
Designate to whom to go for help.
Check frequently. Encourage questions.
Taper off extra coaching and close follow-up.

*Safety is always a key point.

**SAFETY TRAINING INSTITUTE**
**NATIONAL SAFETY COUNCIL**

**Figure 9-9.** Every supervisor should follow the JIT format when teaching job skills.

Selection of the JIT trainer is an important consideration. The supervisor can do the training or may choose to delegate that responsibility to a skilled person within the department. In either case, the instructor should have the following qualifications:

- Know the subject thoroughly
- Have a desire to teach
- Be friendly and cooperative
- Be a good leader
- Have a professional attitude.

The four-point method is intended to help an instructor teach a learner to do a specific job. It aims at faster learning and better learning. When combined with JSA, it becomes an excellent method for teaching safety along with job skills. (The reader will be relieved to know that it takes longer to describe some of these steps than it does to do them.)

## CONFERENCE METHOD OF TEACHING

To aid understanding and stimulate participation in instruction courses, industrial educators many years ago developed the conference method of teaching. This method is widely used in teaching management subjects to people in business and industry.

The conference leader should be skilled in this method of teaching. He should know how to draw out information and opinions from the conferees, and sum up their conclusions. As is desirable in a participation class, the number of people in a conference group should be small enough so open discussion can occur.

The leader's part is mainly that of asking questions to provoke thinking and discussion. Learning can take place in a conference if the conferees have a background of experience that enables them to discuss the subject intelligently.

### Problem-solving conferences

Since a conference is a device for getting a job done, it is particularly useful in an area like safety where many people are involved. A conference yields big returns in education for all who participate, even though its immediate purpose is to solve a current problem, not merely a hypothetical one.

Frequently, the safety professional has a problem relating to the operation of the safety program which needs to be discussed with production supervisors. The problem may have to do with the occurrence of a number of similar accidents, with some new feature of the safety program, with a new type of work to be undertaken, with a method of operation or plan to be worked out, or with any other subject in which both the supervisor and the safety department are interested. To find a solution to the problem, the safety professional may call a conference with a group of supervisors.

The conference leader should be familiar with the relationships of the members of the conference and with the general area of the subject matter.

It is of the utmost importance that the leader and members of the conference know the scope and limitations of what they are expected to do. If, for example, a conference is called to decide how to put into practice a policy or a directive, the members should understand that it is not within their authority to have the policy or directive changed. If they meet to discuss improvement of procedures for the elimination of a hazard, they should know whether or not they may consider extensive alterations of buildings and installation of new machinery. If they do not know the limitations within which they must work, then the conference becomes a source of dissatisfaction and frustration or is regarded simply as play-acting.

If the job of a conference group is to recommend action and they do recommend action, they should, of course, know what becomes of their recommendation.

Often, however, a conference group will discuss a matter that affects only the members, in which case their conclusions are drawn for their own guidance. This is probably the most usual and most satisfactory kind of conference.

### Conference leading

Every safety professional and supervisor should attempt to become a skillful conference leader. Since such skill comes mainly from practice, opportunities for holding conferences should be accepted, even if he has not had a great deal of experience. The application of good sense and an understanding of what a conference is supposed to accomplish will go far toward making the conference successful.

The sequence of conference leading is as follows:

1. State the problem
2. Break the problem into segments to keep the discussion orderly
3. Encourage free discussion
4. Make sure that members have given adequate consideration to all the significant points raised
5. Record any conclusions that are reached
6. State the final conclusion in such a way that it truly represents the findings of the group.

A person who attends many conferences has opportunities to observe examples of successful and unsuccessful conference leading. This should help him when he assumes the role of conference leader. The new leader also should seek help from training people and others who may give sound advice that can lead to productive conferences.

### Misuse of the conference method

If a conference leader, in his eagerness to get across his ideas or in order to cover ground quickly, departs from true conference procedures and steers the discussion in the way he wants it to go, then a meeting becomes a conference in name only.

Such a closely guided or controlled conference is not an effective teaching method. Worse still, because it is called a conference, it may establish in the minds of conferees and the leader a pattern of conduct that makes it difficult for them to serve usefully as members of a true conference. Conferences are almost indispensable in business, and any process that tends to make them unproductive should be avoided.

## OTHER METHODS OF SUPERVISORY TRAINING

Many methods of instruction and many teaching aids are available for training supervisors in accident prevention. Using a variety of methods in a single course or even in a single session makes for interest and understanding, so long as the methods chosen are appropriate for both the subject and the learners. Each training technique has its strengths and weaknesses, and each must be analyzed according to these criteria.

The economic factor can also be important. Taking the information that follows at its face value, one might conclude that simulation is a technique that can almost always be used. But simulators are expensive, and simulation games are time-consuming. This may not turn out to be the best choice. For example, with a small number of trainees and with many problems to be covered, the optimum selection may be a good instructor with a good understanding of the laws of learning and an ability to use good training techniques. On the other hand, if there is a large training load and relatively stable material, the best investment might be some sort of simulation.

All techniques have value as educational methods if they encourage participation.

Training techniques fall into three major categories: working with groups, working with individuals, and programmed instruction.

### Group techniques

Group training techniques encourage participants to share ideas, to evaluate information, to support each other in mutual development, and to participate in life-like situations that help in making decisions. When there are a large number of people to be trained quickly, some sort of group approach is called for.

**Brainstorming.** A form of creative thinking usually resulting from group interaction; a technique of obtaining new ideas by the free association of ideas within a group. The technique is based on four ground rules:

1. Ideas that are presented should not be criticized
2. Free wheeling, or the building of new ideas on ideas just presented, is encouraged
3. As many ideas as possible should be presented quickly
4. The combination of ideas and improvement of those already before the group are encouraged.

The function of the moderator is to cut off negative comments quickly and to encourage full participation. A recorder is responsible for jotting down the ideas as the group presents them.

Brainstorming is good for developing ideas quickly, for stirring the imagination, and for involving timid persons. The size of the group should be limited to five to ten people.

**Buzz sessions.** A group is divided into smaller groups of four to six people. A question is presented and each group discusses it for a few minutes; the group selects one of its members to chair the discussion and another to record it. At the end of the time, the recorders summarize their group's thinking for the entire audience.

This technique permits wide participation, yet screens out impractical ideas. A good briefing session should be held before the groups begin so that they can quickly start and accomplish something positive.

**Case study.** A report of a real situation or a fabricated situation—both are designed to develop basic concepts and principles. The study describes what has happened in a particular case and the events leading up to the situation, but it leaves to the group the task of deciding the nature of the problem or problems, their significance, and the probable solution.

This technique encourages participation, develops insight and ability to use problem-solving methods. Also it develops power of discrimination in drawing generalizations and conclusions. It is difficult, however, to evaluate either group or individual progress or results.

**Incident process.** A shortened case study. This method of learning presents a group with a written account of an incident. The group members ask questions concerning facts, clues, and details. The instructor supplies the answers to these questions, and the group assembles the facts, learns what happened, and arrives at a decision.

Although the technique requires less reading and preparation than the case study, the discussion can easily turn into an argument unless it is properly controlled; the discussion may be limited to a few unless all are prepared.

**Discussion.** A procedure involving an exchange of ideas and the standardizing of procedures and techniques. A discussion allows students to pool their knowledge and become active participants in a controlled program.

This procedure is very effective with small groups because the leader can fit the discussion to the background and needs of individuals. The direction of the discussion is difficult to control and can be time-consuming unless the leader is skillful.

**Role playing.** An instructional method in which incidents based on real-life situations are re-enacted by selected members of the class. The decisions made during the role playing are discussed by the entire class and the instructor to bring out and highlight behavior patterns.

This technique is excellent for developing an understanding of how people behave in specific situations; it leads to good discussions. However, it needs careful planning, it must be kept democratic with the instructor not giving the answers, and it is not effective for solving problems.

**Lecture.** A discourse or talk to encourage learning or to present new materials to a group; supplementary or background material is often passed out.

It works well with large groups with limited time. It is useful in motivating, developing attitudes, and summarizing. Students must be able to hear the lecturer.

**Panel.** A planned session consisting of two or more qualified persons, each discussing an assigned topic or subject.

Like the lecture, this technique permits a great deal of information to be quickly given; it is also one-directional in flow of information, unless a question-and-answer session follows.

**Question-and-answer sessions** can follow a lecture or a panel or be part of a discussion group.

Sessions are an excellent way to bring a group to understand new procedures or methods. It should follow an introductory session or information handout so the students are prepared to ask relevant questions.

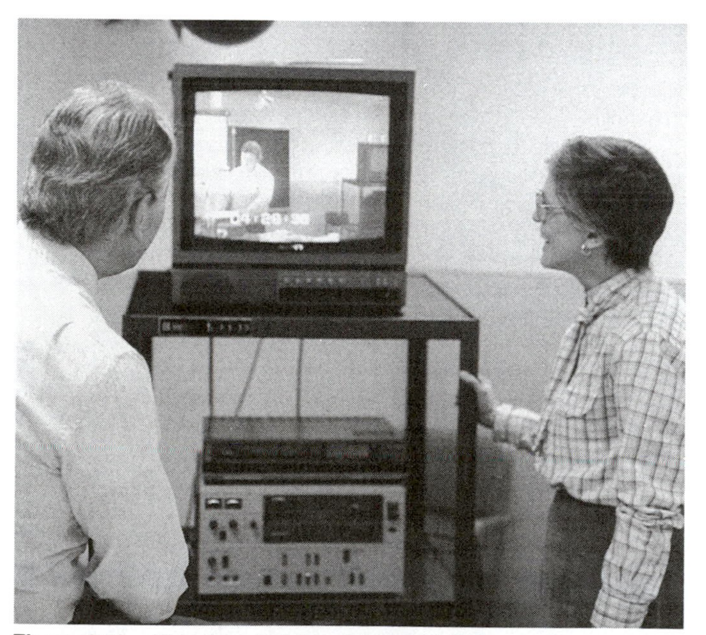

**Figure 9-10.** Videotape training uses television's instant replay features. A training session or work process or procedure can be recorded while it is happening, and then instantly replayed, and used many times over, as a training tool.

**Simulation.** Simulation often involves a simulator, such as those used for driver training and pilot training. Railroads also have developed simulators for training engineers. Simulation also is achieved by using various types of management games, such as the "in-basket technique," "war games" developed by the military, and, more recently, simulation techniques for the training of astronauts in the proper operation of space capsules. Simulation is a good method for paralleling real-life or actual conditions, thus affording opportunities for decision making without risk.

It attains intense involvement among the participants. Careful planning and attention to details are required. Participants must be carefully selected and briefed if the learning experience is to be meaningful. Initial cost is high. (It is discussed under Programmed instruction later in this section.)

### Individual techniques

**Drill.** Repetition and guided practice to develop skill. Drills are used primarily for the most important and fundamental skills of a trade, job, or task.

**Demonstration.** A widely used method to teach skills. The operation is demonstrated by the instructor and then performed by the student as in Job Instruction Training (JIT), discussed earlier in this chapter.

**Quizzes.** Usually written and often of the objective, or multiple-choice, type. This technique is good for determining achievements of objectives; it is also used as a review procedure and as a check on individual comprehension.

**Television.** Educational TV and closed circuit TV (CCTV) are widely used training media (see Figure 9-10).

Videotape training uses television's instant replay techniques. The basic technique is to record the visual procedure and the directions on a videotape and then play it back instantly by means of a monitor. Processes or manual skills, once recorded, can be replayed at will. This technique is good for training employees as well as supervisors.

Because it is also thought of as being an audiovisual, TV is discussed in Chapter 14, Audiovisual Media.

**Reading material.** Companies should provide supervisors with reading material—safety newsletters, safety magazines, magazines on supervision, booklets, and reprints. Many companies subscribe to the National Safety Council's 16-page monthly magazine *Today's Supervisor,* and some also write and distribute a monthly safety bulletin to their supervisors.

In addition to periodic material, it is desirable to have a management library for supervisory use. The library should contain book reviews and books on safety, occupational health, and related subjects.

### Programmed instruction

Programmed instruction can be used as a substitute or supplement for classroom and textbook training methods.

Programmed instruction, often called *PI,* is a self-teaching system offered in a self-contained book or through a mechanical or electronic teaching machine. In both situations, the key to the value of the system is in the program or *software.* This is the term generally adopted for the developed material, in contrast with the machinery and equipment which are termed *hardware.*

The program is presented to the learner in a series of small, sequential steps, carefully planned to progress from the simple to the complex. The student is guided through the program according to the nature of the responses. At intervals, the student must make an active response that tests comprehension of the material by writing in an answer, filling in a missing word or phrase, or choosing a correct statement or picture.

Each answer receives an immediate response. The nature of the response is determined by the answer. The student will be told to advance if the answer is correct or to reread the preceding statement if the answer is wrong or to read a related problem.

When the PI is being done through a book, the student is directed to a specific page for each answer, correct or incorrect. Or a computer can be programmed to present the desired material. Interactive video programs are an example. Another type of machine may, for example, include a filmstrip that will not advance until the correct answer has been indicated to the question just presented.

The response of a machine to those of the student can be very complex. A good example of this is the simulator, described earlier, which has been developed to train automobile and bus drivers, airline pilots, and locomotive engineers (see Figure 9-11). The trainee is placed in a realistic replica of his working position. Through his windows or on a TV screen, he sees the projection of a screen. He operates his controls in response to what he sees in the constantly changing projection. His instruments respond, and the projection changes to give him the feeling of moving through reality. In some simulators, the instructor can introduce stimuli to which the student must respond. Many of the computer-based systems have the capacity to record student responses, both in their degree of accuracy and in time taken.

PI guides the progress of the learner by giving immediate confirmation of the correctness of responses, and controls the learner's orderly process in much the same way as a tutor. It permits each student to progress at an individual rate.

PI has many of the advantages of a human tutor, although it lacks the ability of a human to adapt to new situations. It knows whether the student has selected the right response or not, but it does not know the reasons for the selection. Was it a knowledgeable answer, or simply a lucky guess? What flaw in reasoning led the student to an incorrect answer? A good instructor, on a one-to-one basis, often can determine, through experienced observation, characteristics that are difficult to build into a computer program.

Writing a programmed instructional course is a time-consuming process. The writer must make sure that each step is clear, and the reason for selecting each of the responses represents a single kind of understanding or lack of understanding. This is essential if the right kind of correctional material is to be presented. Obviously, this is an expensive operation and can be justified only if the number of trainees is large or if other training techniques are more expensive.

### School courses

**Independent study.** Courses offered through correspondence are often called home study courses or independent study courses. These have some advantages over other methods described—students can set their own pace, and can study on their own time. Home study can be an ideal training method for companies having many or scattered locations.

The National Safety Council offers three courses of this type for training supervisors—"Supervising for Safety," "Supervisors Guide to Human Relations," and "Protecting Workers' Lives."

See listings of many courses in Chapter 24, Sources of Help.

**Seminars and short courses.** A number of seminars and short courses are offered by colleges and univesities on all phases of supervisory knowhow. Insurance companies and private organizations also offer supervisory training courses. Check for what is available locally.

### Personal instruction and coaching

Personal discussion with individual supervisors often is overlooked as a method of instruction. When the health and safety professional and the supervisor meet and talk about the job, there can be a complete and free interchange of ideas that is not possible in a group. In private conversation, the supervisor often will express reservations or doubts that would not be voiced in a meeting, and the health and safety professional will have the opportunity to clear up misunderstandings.

The health and safety professional who meets with the individual supervisor not only instructs and coaches, but also learns many necessary things. Supervisors appreciate an interest in their work and are usually eager to teach about the processes of their departments. The safety professional can learn much from them about the mechanical processes and also about problems of supervision. Perhaps most importantly, the health and safety professional becomes personally acquainted with the supervisors. Every safety professional ought to allot some time to these personal contacts.

## POLICIES AND ATTITUDES IN SUPERVISORY TRAINING

The person conducting a training program for supervisors should remember a number of facts which experienced educators know.

### Teaching at the adult level

Supervisors are adults; they must be treated accordingly. The smart supervisor is aware that a portion of the job is to get the product out and do it safely. Safety is not a separate job to be done if and when other responsibilities permit time. Rather, the supervisor should be thinking of *safe production.*

In this regard, the supervisor very likely reflects the attitude of the immediate superior. If the superior lets it be known that the supervisor's value also will be measured by how safely the job is done, the proper attention will be given to that portion of the job. This fact is a strong argument for starting safety education at the upper levels of supervision.

### Integrating safety into all training

In a company that has a formal training program, safety training should fit into that program; the director of training and the safety professional should work together to plan various kinds of safety training activity.

Training supervisors is usually the first concern of the training department, because practically everything that is taught about good supervision helps to promote safety throughout the organization. Likewise, anything taught specifically for the purpose of promoting safety generally improves other areas of supervision.

**Figure 9-11.** The instructor observes while the locomotive engineer trainee runs a train in simulation. The training room is located in a specially equipped highway vehicle that travels between the railroad terminals where engineers are based. (Printed with permission from Family Lines Rail System.)

In the arrangement of formal courses, safety subjects may be integrated with other instruction on general problems of supervision, or one or two safety meetings can be included in the course.

The number of participants in a class of supervisors should be small enough to allow free discussion; fifteen is about the upper limit. The leader should carefully plan questions that encourage group discussion. There is no limit to the audiovisuals and demonstrations and other devices that are available to make the material interesting and educational, and the discussions easy to follow.

### Continuing programs get best results

Safety training for supervisors needs to be a continuing program if it is to accomplish the best results. The program should cover many subjects presented interestingly and in different ways. A limited effort, such as holding a few meetings, may cause supervisors to do a better job for a short time, but interest will lag if the initial effort is not followed up.

Industrial management must be dynamic because change is continually taking place. New methods and new ideas get attention. To hold its own in this atmosphere of change and progress, safety must not be allowed to become static.

The safety and health professional who constantly strives to help line supervisors take over direct responsibility for safety in their own operations finds the results of these efforts will be multiplied. Such health and safety professionals become true leaders who get the job done better, not by extending the scope of their own activities, but by helping supervisors assume their natural safety responsibilities.

## TRAINING NEW EMPLOYEES

Safety training begins at hiring, even before the employee starts work. An effective safety training program will include a carefully prepared and presented introduction to the company.

When a new employee comes to work, he or she immediately begins to learn. Attitudes are formed about the company, the job, the boss, and fellow employees. This happens whether or not the employer makes an effort to train the person. In order for the new employee to learn to do a safe, productive job and feel company loyalty, the employer will want to provide the right kind of start.

### Orientation

At the beginning of employment, each employee should know the company's safety policy, but the amount of information retained from the orientation procedure is limited. Natural nervousness about starting a new job, interest in many matters of seemingly more immediate concern, such as payroll procedures—all make it difficult for the employee to absorb much safety instruction. It is necessary, therefore, to consider what safety information must be given first, and then the best way to present it.

Each employee needs to learn the following things to have a good start in safety training:

1. Management is sincerely interested in preventing accidents.
2. Accidents may occur, but it is possible to prevent them.
3. Safeguarding equipment and the workplace has thoroughly been done, and management is willing to go further as needs and methods are discovered.
4. Each employee is expected to report to the supervisor any unsafe conditions encountered at work.
5. The supervisor will give job instructions. No employee is expected to undertake a job before learning how to do it and being authorized to do it by the supervisor.
6. The employee should contact the supervisor for guidance before undertaking a job that appears to be unsafe.
7. If an employee suffers an injury, even a slight one, it must be reported at once.

In addition to these points, any safety rules which are a condition of employment, such as wearing eye protection or safety hats, should be understood and enforced from the first day of employment.

### Preliminary instruction

Preliminary safety instruction is most often given to individuals or small groups by the personnel department. Sometimes the safety professional or a management executive will give the safety instruction. This method may add force and interest, but it has the practical disadvantage that safety professionals and executives are busy people and may not always be available or may be so pressed with other duties that they skimp on the job or delegate it to subordinates less able to handle it.

More important than who gives the talk is how it is given. The talk should be prepared and presented with the utmost regard for the effect it will have on the new employees. Verbal instruction should be given earnestly and with an attitude of good will and friendly cooperation.

A safety TV tape or film may do a good job of interesting and instructing new employees. It changes the pace and relieves the monotony of just listening. The presentation should be brief and limited to such essential points as those previously mentioned.

Some companies produce their own tapes or films. Although the production may seem costly, if it is shown to all new employees, the cost per employee showing may be only a few cents. Prepared tapes and films can be purchased. Other companies have had good results with sound-slide shows. (See Chapter 14, Audiovisual Media, for comparisons.)

One important advantage of an orientation film or tape is that it can present a carefully planned message in a consistent and effective manner. Management can be sure its safety message is going to be told in exactly the same way to every new employee.

Charts illustrating points presented in the orientation talk add interest and aid both understanding and conviction. Charts should be large enough and simple enough to be seen and understood easily by every member of the group (see Figure 9-12).

If a company has preplacement physical examinations, the

**Figure 9-12.** This instructor is using a chart to illustrate a safety point to his drilling rig crew trainees. This mobile classroom is a converted school bus. (Printed with permission from Parker Drilling Company.)

doctor or nurse should establish good relations with each new employee at the time of the examination. The doctor or nurse should tell about the work of the medical department as it relates to the employees and should encourage them to make use of its services. Medical or nursing personnel, as well as the person who gives general safety information, should emphasize the importance of reporting all injuries.

The final step in the personnel office safety procedure should be to emphasize the importance of the position of the supervisor in the safety program. The employee must understand that the supervisor is responsible for job training, and such training will include safe work procedures.

So there will be no gap and no contradiction between the information given in the employment office and that given later, the supervisor should know what has been covered in the orientation talks.

### Make rule books logical and enforceable

The use of rule books or manuals has long been a part of the early training given to new employees. The development, appro-

**Figure 9-13.** A job safety analysis, called a job safety write-up by some companies, can be kept in a protective plastic cover near the equipment so the operator can review it at any time. This is especially important when the job is new to the operator. Also it should be available to the supervisor when he is training a new operator for the job. (Reprinted with permission from Automotive, Tooling, Metalworking, and Associated Industries Newsletter.)

val, and distribution of printed safety rules by industrial concerns are discussed in National Safety Council Data Sheet 664, *Writing and Publishing Employee Safety Regulations.* The Council also publishes safety rule booklets and leaflets for employers to give to their employees.

Such books should be prepared so the rules are easily understood. Only logical and enforceable rules should be included. Employees cannot be expected to respect and follow rules that are illogical, unfair, or unrealistic.

Rules that supervisors have reviewed are likely to be more effective than those they have had no part in formulating.

Rule books or manuals should contain general instructions as to the employee's responsibility in safety. The rules should cover items such as first aid, personal protective equipment, work clothing, firefighting, electrical equipment, and housekeeping.

Rule books alone should not be expected to influence attitudes. Well-prepared, illustrated booklets or cards, however, can add much to the start of a good training program if the contents are briefly reviewed and discussed when the booklets are presented and if the employee is not overloaded with printed matter (see Figure 9-13).

### Departmental orientation and training

When a new employee reaches his own department, his supervisor should give him additional safety instruction. This may cover some of the points made in the employment office interview, but now specifically applied to the work the employee will be doing.

The supervisor usually repeats earlier instruction about reporting unsafe conditions, not undertaking a job without instruction and authorization, and other matters of policy. The supervisor should tell the new employee about the safety record and the department's safety and health program.

The supervisor will explain departmental general safety regulations and will see that the new employee is provided with the appropriate personal protective equipment. The supervisor also will make provisions for safe work procedure training. He will make certain that safe procedures are achieved.

Some safety and health departments have followup interviews with employees from one week to one month after employment. At this time, the safety professional reviews the points discussed at the time of employment and encourages the employee to talk about his experiences on the job.

Companies following this plan report that discussion after a few days on the job is more profitable than information given before the employee starts work. The new person has overcome his feeling of strangeness and uneasiness and can relate the discussion to what he has already experienced. If the personnel department handles the initial safety instruction, it is especially desirable for a safety department representative to talk with the employee a few days later.

Of course, any direct contact by the safety department with an employee should be with the knowledge and approval of the employee's supervisor.

This orientation procedure is the least a new employee should receive if he is to believe safety is important to his job, and if he is to understand the company's attitude toward safety.

Some companies have more elaborate orientation programs. These may include a plant tour, discussion of the company's products, viewing a film, and listening to talks by representatives of various departments. The program may take one-half to a full day.

Companies having formal orientation programs are convinced they pay off in lower labor turnover, in good employee relations, and in prevention of accidents.

In these programs, it is the job of the safety department to make certain safety is presented as interestingly and effectively as any part of the program. Safety must not be submerged in a mass of information so it will be only remembered vaguely, if at all. *The employee must carry away a deep conviction that safety is important to himself and his company.*

### Good supervision—consistent instruction and discipline

After the employee has been oriented to the new job and surroundings, it's up to the supervisor to consistently maintain safe work practices and, when necessary, discipline rule violations.

If the supervisor observes workers taking short cuts or otherwise departing from safe methods, he should correct them at once. If he does not correct them, the unsafe method soon becomes standard practice.

Much has been said about disciplining violators of safe practices. Discipline in this connection means an oral or written warning, followed by a time-off without pay penalty. Probably every employer recognizes the necessity for having penalties available to punish willful misconduct, whether the offense has to do with safety rules or other company regulations. Actually, such penalties are rarely or never needed.

Too often, violations of safety rules are overlooked until an accident occurs. If employees are corrected for every infraction of a safety rule or safe practice as soon as observed, there will be few occasions requiring discipline. In any event, if a penalty is to be assessed, it should be for the act and not for the accident.

### First aid courses

First aid courses for employees have been conducted in some industries for many years. The American National Red Cross or the Mine Safety and Health Administration have excellent courses. The need for personnel trained in first aid, if there is no infirmary, clinic, hospital, or physician in proximity or reasonably accessible, is an OSHA requirement.

The value of first aid training is greatest in companies that have night shifts or skeleton crews working when medical facilities are closed, or where working field crews are far removed from professional medical help. Public utilities, mines, oil-well drilling, and logging are a few operations where employees have become proficient in caring for seriously injured persons and in transporting them to the hospital. Providing first aid instruction is a necessity in these industries.

Even in companies with complete medical departments, trained first-aiders are a potential asset because they can stop dangerous bleeding, administer artificial respiration, and transport injured workers safely. Not only can they render these services, but their value in a disaster would be great.

In addition to possible benefit on the job, first aid training has inherent interest for employees. On the theory that first aid students learn something about the causes of accidents and acquire an accident awareness that makes them cautious, some companies believe first aid instruction makes an employee less likely to have accidents.

Companies sponsoring first aid courses usually arrange for a qualified instructor, provide a meeting place, pay for textbooks and supplies, and often reward the graduates with a dinner or other celebration.

### The so-called 'accident-prone' individual

Probably no phrase in safety causes so much disagreement than "accident proneness." Most definitions imply that a person with certain psychological characteristics or personality traits is more likely to have accidents. When a person is said to be "accident-prone," it is generally meant that he is predisposed towards having accidents.

Too often the term is applied to anyone who has more accidents than others doing the same work. A person could, however, have more than his share of accidents because he is trained improperly, needs new glasses, or simply because his working area is cramped or inefficient. Accidents also can be due to poor supervision, lax attitude of management toward accident prevention, or because the worker does not speak English and cannot understand safety instructions.

It is true that a small group of people often account for more than an expected share of accidents during a given period, but over a longer time period, the composition of the group changes—the accident repeaters of one time period do not usually show up during the next time period. Much more research has to be done in this field before any person can be positively identified as being accident prone. Above all, such a label should not be placed in the personnel file of any individual.

See the Council's book *Supervisors Guide to Human Relations* for a detailed discussion of accident- and safety-prone persons.

## OSHA AND MSHA TRAINING REQUIREMENTS

The continued importance of training is evidenced by the requirements of both the Occupational Safety and Health Administration (OSHA) and the Mine Safety and Health Administration (MSHA).

### OSHA requirements

Listed next are the major parts of the OSHA regulations (Title 29—Labor, *Code of Federal Regulations*) covering training requirements. Table 9-A gives a convenient index indicating the type of hazard and the parts of the regulations requiring training to protect against the hazard.

Part 1910, Safety and Health Training Requirements for General Industry

Part 1915-18, Safety and Health Training Requirements for Maritime Employment

Part 1926, Safety and Health Training Requirements for Construction

Part 1928, Occupational Safety and Health Requirements for Agriculture.

### MSHA regulations

The following is a summary of the training requirements under the MSHA Regulations, published in the *Federal Register,* Vol. 43, No. 199, October 13, 1978.

**Subpart B—Training and Retraining Miners Working at Surface Mines and Surface Areas of Underground Mines**

**§48.21 Scope**

Subpart B sets forth the mandatory requirements for submitting and obtaining approval of programs for training and retraining miners at surface mines and surface areas of underground mines. It also includes requirements for compensation for training and retraining.

**§48.22 Definitions**

(a) "Miner"—Any person working in a surface mine or surface area of an underground mine and who is engaged in the extraction and production process or is *regularly* exposed to mine hazards, or who is a maintenance or service worker (whether employed by operator or contractor) working at the mine for frequent or extended periods.

Short-term, specialized contract workers (drillers, blasters, etc.) who have received training under §48.26 (training of newly employed experienced miners) may be trained under §48.31 (hazard training) in lieu of other subsequent training.

Excluded from definition

(i) Construction, shaft and slope workers covered in Subpart C, Part 48.

(ii) Supervisory personnel (covered under MSHA-approved state certification requirements).

(iii) Delivery, office or scientific or short-term maintenance workers and any student engaged in *academic* projects.

(b) "Experienced miner." A person currently employed as a miner; or a person who received training *acceptable* to MSHA from an *appropriate* state agency within the preceding one month; a person with 12 months experience

**Table 9-A.** Index to OSHA Training Requirements

| Hazard | Part | Subpart | Section |
|---|---|---|---|
| Blasting or Explosives | 1910.109 | H | (d)(3)(i)(iii) |
| | 1926.901 | U | (c) |
| | 1926.902 | U | (i) |
| | 1915.10 | B | (a) thru (b) |
| | 1916.10 | B | (a) thru (b) |
| | 1917.10 | B | (a) thru (b) |
| Carcinogens | | | |
| 4-Nitrobiphenyl | 1910.1003 | Z | (e)(5)(i) thru (ii) |
| alpha-Naphthylamine | 1910.1004 | Z | (e)(5)(i) thru (ii) |
| 4, 4'-Methylene bis (2-chloroaniline) | 1910.1005 | Z | (e)(5)(i) thru (ii) |
| Methyl chloromethyl ether | 1910.1006 | Z | (e)(5)(i) thru (ii) |
| 3, 3'-Dichlorobenzidine (and its salts) | 1910.1007 | Z | (e)(5)(i) thru (ii) |
| bis-Chloromethyl ether | 1910.1008 | Z | (e)(5)(i) thru (ii) |
| beta-Naphthylamine | 1910.1009 | Z | (e)(5)(i) thru (ii) |
| Benzidine | 1910.1010 | Z | (e)(5)(i) thru (ii) |
| 4-Aminodiphenyl | 1910.1011 | Z | (e)(5)(i) thru (ii) |
| Ethyleneimine | 1910.1012 | Z | (e)(5)(i) thru (ii) |
| beta-Propiolactone | 1910.1013 | Z | (e)(5)(i) thru (ii) |
| 2-Acetylaminofluorene | 1910.1014 | Z | (e)(5)(i) thru (ii) |
| 4-Dimethylamino-azobenzene | 1910.1015 | Z | (e)(5)(i) thru (ii) |
| N-Nitrosodimethylamine | 1910.1016 | Z | (e)(5)(i) thru (ii) |
| Vinyl chloride | 1910.1017 | Z | (j)(1)(i) thru (ix) |
| Cranes and Derricks | 1910.179 | N | (m)(3)(ix) |
| | 1910.180 | N | (h)(3)(xii) |
| Decompression or Compression | 1926.803 | S | (a)(2) |
| | 1926.803 | S | (b)(10)(xii) |
| | 1926.803 | S | (e)(1) |
| Employee Responsibility | 1910.109 | H | (g)(3)(iii)(a) |
| | 1926.609 | U | (a) |
| Equipment Operations | 1910.217 | O | (f)(2) |
| | 1926.20 | C | (b)(4) |
| | 1926.53 | D | (b) |
| | 1926.54 | D | (a) |
| | 1910.252 | Q | (c)(6) |
| Fire Protection | 1916.32 | D | (e) |
| | 1917.32 | D | (b) |
| | 1926.150 | F | (a)(5) |
| | 1926.155 | F | (e) |
| | 1926.351 | J | (d)(1) thru (5) |
| | 1926.901 | U | (c) |
| Forging | 1910.218 | O | (a)(2)(i) thru (iv) |
| Gases, Fuel, Toxic Material, Explosives | 1910.109 | H | (d)(3)(i) and (iii) |
| | 1910.111 | H | (b)(13)(ii) |
| | 1910.266 | R | (c)(5)(i) thru (xi) |
| | 1910.106 | H | (b)(5)(vi)(v)(3) |
| | 1916.35 | D | (d)(1) thru (6) |
| | 1926.21 | C | (a) and (b)(2) thru (6) |
| | 1926.350 | J | (d)(1) thru (6) |
| General | 1926.21 | C | (a) |
| Hazardous Material | 1915.57 | F | (d) |
| | 1916.57 | F | (d) |
| | 1917.57 | F | (d) |
| Medical and First Aid | 1910.94 | G | (d)(9)(i) and (vi) |
| | 1910.151 | K | (a) and (b) |
| | 1915.58 | K | (a) |
| | 1917.58 | F | (a) |
| | 1926.50 | D | (c) |
| Personal Protective Equipment | 1910.94 | G | (d)(11)(v) |
| | 1910.134 | I | (a)(3) |
| | 1910.134 | I | (b)(1), (2) and (3) |
| | 1910.134 | I | (e)(2), (3) and (5) |
| | 1910.134 | I | (e)(5)(i) |
| | 1910.161 | K | (a)(2) |
| | 1915.82 | I | (a)(4) |
| | 1915.82 | I | (b)(4) |
| | 1916.57 | F | (f) |
| | 1916.58 | F | (a) |
| | 1916.82 | I | (a)(4) |
| | 1916.82 | I | (b)(4) |
| | 1917.57 | F | (f) |
| | 1918.102 | J | (a)(4) |
| | 1926.21 | C | (b)(2) thru (6) |
| | 1926.103 | E | (c)(1) |
| | 1926.800 | S | (e)(xii) |
| Pulpwood Logging | 1910.266 | R | (c)(5)(i) thru (xi) |
| | 1910.266 | R | (c)(6)(i) thru (xxi) |
| | 1910.266 | R | (c)(7) |
| | 1910.266 | R | (e)(2)(i) and (ii) |
| | 1910.266 | R | (e)(9) |
| | 1910.266 | R | (e)(1)(iii) thru (vii) |
| Powder-Actuated Tools | 1915.75 | H | (b)(1) thru (6) |
| | 1916.75 | H | (b)(1) thru (6) |
| Power Press | 1910.217 | O | (e)(3) |
| Power Trucks, Motor Vehicles, or Agricultural Tractors | 1910.109 | H | (d)(3)(iii) |
| | 1910.109 | H | (g)(3)(iii)(a) |
| | 1910.178 | N | (1) |
| | 1910.266 | R | (e)(9) |
| | 1910.266 | R | (e)(6)(viii) |
| | 1928.51 | C | (d) |
| Radioactive Material | 1916.37 | D | (b) |
| Signs—Danger, Warning, Instruction | 1910.96 | G | (f)(3)(viii) |
| | 1910.145 | J | (c)(1)(ii) |
| | 1910.145 | J | (c)(2)(ii) |
| | 1910.145 | J | (c)(3) |
| | 1910.264 | R | (d)(1)(v) |
| Tunnels and Shafts | 1926.800 | S | (e)(xiii) |
| Welding | 1910.252 | Q | (b)(1)(iii) |
| | 1910.252 | Q | (c)(1)(iii) |
| | 1915.35 | D | (d)(1) thru (6) |
| | 1915.36 | D | (d)(1) thru (4) |
| | 1916.35 | D | (d)(1) thru (6) |
| | 1917.35 | D | (d)(1) thru (6) |

working in surface operations during the preceding 3 years; a person who received new miner training (§48.25) within the past 12 months.

(c) "New miner." Not experienced.

(d) "Normal working hours." Regularly scheduled work hours.

(e) "Operator." Owner, lessee, person that controls or supervises the operation; or any contractor performing similar function.

(f) "Task." Regular work assignment which requires physical abilities and job knowledge.

(g) "Act." The Federal Mine Safety and Health Act of 1977.

**§48.23 Training Plans**

(a) Each operator shall have a MSHA-approved plan for training:

New miners

Newly employed experienced miners

Miners for new tasks
For annual refresher
For hazard.
(1) Existing mines shall submit the training plan to MSHA for approval within 150 days of the effective date (October 13, 1978). The plans must be filed by March 11, 1979.
(2) Unless extended, MSHA shall approve the operator's plan within 60 days.
(3) New mines—reopened mines. Must have an approved plan prior to (re)opening.

(b) Training plan shall be filed with the Chief of Training Center, MSHA, for the area in which the mine is located.

(c) Information to be filed:
(1) Company name
Mine name
MSHA I.D. number.
(2) Name and position of person responsible for health and safety training.
(3) List of MSHA-approved instructors along with the courses they are qualified to teach.
(4) Location of training site.
(5) Description of teaching methods and course materials which are to be used.
(6) Number of miners. Maximum number of miners to attend each session.
(7) Refresher training—a schedule of time or period of time when such training will be given. To include titles of courses, total number of instruction hours for each course, and predicted time and length of each session.
(8) New task training for miners.
(i) Submit complete list of task assignments to correspond with the definition of "task."
(ii) Titles of the instructors.
(iii) Outline training procedure for each work assignment.
(iv) The evaluation procedures used to determine the effectiveness of training.

(d) Two weeks prior to plan submission, a copy of the plan shall be given to the employees' representatives.
Should there be no employee representative, the plan shall be posted on the mine bulletin board two weeks prior to submission.
All written comments from employees must be delivered to MSHA. Miners may deliver such comments directly to MSHA.

(e) The training plan is subject to review and evaluation by MSHA. Course materials, including visual aids, handouts, etc., must be available to MSHA. At the request of MSHA, the operator must alter, change or modify the plan.
A schedule of upcoming training must be given to MSHA.

(f) A copy of the *approved* plan must be available at the mine for MSHA, miners, and miners' representatives.

(g) All courses shall be conducted by MSHA-approved instructors except as provided for in the "New Task Training of Miners" and "Hazard Training" sections.

(h) Instructors are approved in the following ways:
(1) Receive instructor training from MSHA or from a person designated qualified by MSHA.
(2) Instructors may be approved by MSHA to teach specific courses based on written evidence of qualifications and teaching experience.
(3) MSHA may approve instructors based on the performance of the instructors while teaching classes are monitored by MSHA. This program must be approved in advance by MSHA.
(4) Cooperative instructors, designated by MSHA to teach approved courses within the past 24 months, shall be considered approved.

(i) MSHA can revoke the approval of an instructor for good cause. There is a specific appeal procedure provided.

(j) MSHA shall notify the operator and miners' representative in writing the status of the MSHA approval within 60 days from the date the plan is submitted.
(1) Any revision to the plan required by MSHA in order to gain approval shall be given to the operator and employees' representative. The operator and the employees' representative have the right to discuss alternative revisions or changes with MSHA—within a specified period of time.
(2) MSHA can approve portions of a plan and withhold approval on the balance.

(k) Training shall begin within 60 days after approval of the plan.

(l) The operator shall submit any proposed changes or modifications to an approved plan to the employees' representative and MSHA prior to making such changes. MSHA must approve such changes prior to implementation.

(m) MSHA must notify, in writing, the operator and employee representative of the disapproval or recommended changes to the submitted plan. Such notification will include:
(1) State specific changes or deficiency.
(2) Action needed to bring plan into compliance.
(3) MSHA will take punitive action against the operator should remedial action to effect compliance be delayed or ignored.

(n) All MSHA-recommended changes shall be posted on the bulletin board and a copy of same delivered to the employees' representative.

**§48.24 Cooperative Training Program**

(a) Training programs may be conducted by the operator, MSHA, MSHA-approved programs conducted by state or other federal agencies, or associations of operators or miners' representatives, private associations, or educational institutions.

(b) Instructors and courses shall be approved by MSHA.

**§48.25 Training of New Miners; Minimum Courses of Instruction; Hours of Instruction**

(a) Each new miner shall receive not less than 24 hours of training. Unless otherwise stated, this training shall take place before they start work duties. At the discretion of MSHA, a new miner may receive a portion of this training after he starts his work duties. Provided that not less than 8 hours of training shall be given before the employee starts work. This first 8 hour training shall include:
(1) Introduction to work environment
(2) Hazard recognition
(3) Health and safety aspects of the tasks assigned.

The remainder of the 24 hours training or up to 16 hours will be given within 60 days. This program must be approved by MSHA. Certain conditions at a mine, such as employee turnover, mine size or safety record may cause MSHA to require the full 24 hour training prior to the start of work.

(b) New miner training program shall include:
   (1) Instruction in the statutory rights of miners and their representatives.
   Authority and responsibility of supervisors
   A review and description of the line of authority of supervisors and miners' representatives.
   Introduction to mine rules and procedures for reporting hazards.
   (2) Self-rescue and respiratory devices—instruction and demonstration in the use, care, and maintenance (where applicable).
   (3) Transportation controls and communication systems—instruction on the procedures in effect for riding on and in mine conveyances where applicable; the controls for the transportation of miners and materials; use of mine communication systems, warning signals, and directional signs.
   (4) Introduction to work environment. Includes tour of mine and a description of the entire operation.
   (5) Escape and emergency evacuation plans, fire warning and firefighting. Review mine escape system and emergency evacuation plans and instructions in fire warning signals and firefighting procedures.
   (6) Ground control; working in areas of high walls, water hazards, pits and spoil banks; illumination and night work.
   (7) Health—instruction includes the purpose of taking dust measurements and noise and other health measurements, and any health control plan at the mine shall be explained. The operator shall explain the health provisions of the act and warning labels.
   (8) Hazard recognition—course includes recognition and avoidance of hazards present in the mine.
   (9) Electrical hazards—includes recognition and avoidance of electrical hazards.
   (10) First Aid—must be a MSHA-approved course.
   (11) Explosives—includes a review and instruction on the hazards related to explosives. This course can be omitted if no explosives are used or stored at the mine.
   (12) Health and safety aspects of the tasks to which the new miner will be assigned. The course includes instruction in the health and safety aspects of the work, the safe work procedures, and the mandatory health and safety standards pertinent to the work.
   (13) Any other courses deemed necessary by MSHA based on special circumstances and conditions at the mine.

(c) The training plan shall include oral, written or practical demonstration methods to determine successful completion of the training. These methods shall be administered to the miner prior to assignment to work duties.

(d) A newly employed miner who has received the full 24 hours training within the 12 months preceding employment need not go through the operator's new miner training program. However, the miner will have to receive and complete the instruction for the "newly employed experienced miner" and "new task training of miners before commencing."

**§48.26 Training of Newly Employed Experienced Miners, Minimum Courses of Instruction**

(a) The newly employed experienced miner shall receive and complete the training listed below before being assigned to work duties.

(b) The training program includes the following:
   (1) Introduction to work environment. Includes a tour of the operation and a description of the total operation.
   (2) Mandatory health and safety standards. Includes those standards pertinent to the tasks assigned.
   (3) Authority and responsibility of supervisors and miners' representatives. Includes a review of supervisors and miners' representatives line of authority, and the responsibility of such persons. Also, an introduction to the operator's rules and procedures for reporting hazards.
   (4) Transportation controls and communication system. Includes instruction on the procedures for riding on and in mine conveyances; controls for the transportation of miners and materials; and use of the mine communication system, warning signal, and directional signs.
   (5) Escape and emergency evacuation plans; fire warning and firefighting. Includes review of the mine escape system; escape and emergency evacuation plans and instruction in the fire warning signals and firefighting procedures.
   (6) Ground controls; working in areas of high walls, water hazards, pits and spoil banks; illumination and night work. Includes introduction and instruction on the high wall and ground control plans; procedures for working near high walls, water hazards, pits and spoil banks, illuminated work areas, and procedures for working during hours of darkness.
   (7) Hazard recognition. Includes recognition and avoidance of hazards, particularly any hazards related to explosives where explosives are used or stored at the mine.
   (8) Any other courses MSHA deems necessary based on special mine circumstances and conditions.

**§48.27 Training of Miners Assigned to a Task in Which They Have Had No Previous Experience; Minimum Courses of Instruction**

(a) A miner shall be trained to safely perform any new work task prior to starting such work. The exceptions to this rule are:
   (1) If the miner received such training within the preceding 12 months and can demonstrate knowledge of the safe procedures.
   (2) If the miner performed the work within the preceding 12 months and can demonstrate knowledge of the safe procedures.
   The training program shall include the following:
      (i) Health and safety aspects and safe operating procedures for work tasks, equipment or machinery. Includes instruction in the health and safety aspects and safe operating procedures and given on-the-job.
      (ii) Supervised practice during nonproduction. Practice training in the work task will be conducted at times or places where production is not the primary objective.
      Supervised operation during production. Training will be conducted while under supervision and during production in the operation of equipment and performance of work task.

(iii) *New* or *modified* machines and equipment. Where new or different operating procedures are required as a result of new or modified equipment, the miner will be fully trained in the new procedures.

(iv) Any additional courses MSHA may deem necessary as a result of special conditions or circumstances at the mine.

(b) Miners shall not operate equipment or engage in blasting operations without direction and immediate supervision until the miner has demonstrated safe operating procedures for equipment or blasting operation.

(c) Miners assigned to a new task not covered in the paragraph shall be instructed in the safety and health aspects and safe procedures of the task prior to starting such task.

(d) All training and supervised practice shall be given by qualified trainers or experienced supervisor or other person experienced in the new task.

**§48.28 Annual Refresher; Training of Miners**

(a) Required—8 hours of annual refresher.

(b) Refresher shall include the following:

(1) Mandatory health and safety standards. The standards relating to the miner's task.

(2) Transportation controls and communication system. (Same as paragraph 48.26, b, 4.)

(3) Escape and emergency evacuation plans; fire warning and fire fighting. (Same as paragraph 48.26, b, 5.)

(4) Ground control; working in areas of high walls, water hazards, pits and spoil banks; illumination and night work. (Same as paragraph 48.26, b, 6.)

(5) First aid—method acceptable to MSHA.

(6) Electrical Hazards—recognition and avoidance of electrical hazards.

(7) Prevention of accidents—review of accidents and their causes and instruction in accident prevention in the work environment.

(8) Health—explain purpose for taking dust, noise and other health measurements, and any health control plan in effect at the mine. Further, explain warning labels and health provisions of act.

(9) Explosives—review and instruct on the hazards related to explosives: This course is not needed when explosives are not used or stored at the mine.

(10) Self-rescue and respiratory devices. Instruct and demonstrate the use, care and maintenance of self-rescue and respiratory devices.

(11) Any additional courses MSHA may deem necessary.

(c) All experienced miners will receive refresher training within 90 days after the training plan is approved by MSHA.

(d) Annual refresher training sessions shall not be less than 30 minutes of actual instruction time and miners shall be notified that the session is part of annual refresher training.

**§48.29 Records of Training**

(a) Upon completion of the MSHA-approved training, the operator shall record and certify on MSHA form 5000-23 that the miner has received the specified training. A copy of the training certificate is given to the miner. A copy of the certificate is filed at the mine site to be available to the various government agencies and the miners.

(b) False certification that training was given is punishable under Section 110 (a) and (f) of the Act.

(c) Copies of training certificates for current employees shall be retained at the mine site for two years and for 60 days after a miner terminates.

**§48.30 Compensation for Training**

(a) Training shall take place during normal working hours and the miner receives the rate of pay as though working at the work task.

(b) Should the training by given at a location other than the normal workplace, miners shall be paid for additional costs, such as mileage, meals, and lodging, they may incur in attending the training.

**§48.31 Hazard Training**

(a) All miners shall receive hazard training before starting work duties. Such training shall include the following:

(1) Hazard recognition and avoidance;

(2) Emergency and evacuation procedures;

(3) Health and safety standards, safety rules, and safe working procedures;

(4) Self-rescue and respiratory devices; and,

(5) (i) Any additional courses MSHA may deem necessary.

(ii) Miners will receive training at least once every 12 months.

(iii) The hazard training program will be submitted to MSHA along with the other training programs.

(iv) Recordkeeping and completion certification shall be maintained in the same manner as the other training plans.

**§48.32 Appeals Procedures**

The operator, miner, and miners' representative can appeal any decision of the MSHA Training Chief.

(a) Appeals to MSHA shall be in writing to:
Director of Education & Training
MSHA
4015 Wilson Blvd.
Arlington, Va. 22203
The appeal must be within 30 days after notification of a MSHA decision.

(b) The Director can request additional information from all parties.

(c) The Director shall render a decision on the appeal within 30 days after receipt of the appeal.

## REFERENCES

### Programmed instruction

"A Bibliography of Programs and Presentation Devices." Carl Hendershot, 4114 Ridgewood Dr., Bay City, Mich. 48706.
A listing of programmed instructional materials and devices with quarterly supplements.

"Library of Programmed Instruction Courses." E.I. du Pont de Nemours and Co., Inc., Education and Applied Technology Division, Wilmington, Del. 19898.
A listing of vocational training courses. Also available on request is a list of safety training courses.

National Society for Programmed Instruction, P.O. Box 137, Cardinal Station, Washington, D.C. 20017.

**Management games**

"A Catalog of Ideas for Action Oriented Training." Didactic Systems, Inc., P.O. Box 4, Cranford, N.J. 07016.
A listing of simulation games of all types, programmed instruction materials, and a listing of books on effective training.

"The In-Basket Method." Bureau of Industrial Relations, Department of Training Materials for Industry, The University of Michigan, Graduate School of Business Administration, Ann Arbor, Mich. 48104.
A series of packaged courses, each set consisting of letters, notes, memos, and reports.

"The In-Basket Kit." Allan A. Zoll, Addison-Wesley Publishing Co. Inc., Reading, Mass. 01867.
A kit of materials covering management practices that involve the learner.

"Simulation Series for Business and Industry." Science Research Associates, Inc., Department of Management Services, 155 N. Wacker Dr., Chicago, Ill, 60606.
This series includes Decision Making, Collective Bargaining, Equipment Evaluation, Supervisory Skills, Purchasing, Production, Control Inventory, and Interviewing.

**Books**

Hannaford, Earle S. *Supervisors Guide to Human Relations,* 2nd ed. Chicago, National Safety Council, 1976.

Lateiner, Alfred, and Heinrich, H.W. *Management and Controlling Employee Performance.* West New York, N.J., Lateiner Publishing, 1968.

Mager, Robert F. *Preparing Instructional Objectives,* 2nd ed. Belmont, Calif.: David S. Lake Publishers, 1984.

Mager, Robert F., and Beach, Kenneth M., Jr. *Developing Vocational Instruction.* Belmont, Calif.: David S. Lake Publishers, 1967.

Mager, Robert F., and Pipe, Peter. *Analyzing Performance Problems of 'You Really Oughta Wanna.'* 2nd. ed. Belmont, Ca.: David S. Lake Publishers, 1984.

ReVelle, Jack B. *Safety Training Methods.* New York, John Wiley and Sons, 1980.

Rose, Homer C. *The Instructor and His Job.* 2nd ed. Chicago, Ill., American Technical Society. 1966.

*Supervisors Safety Manual,* 6th ed. National Safety Council, 444 N. Michigan Ave., Chicago, Ill. 60611. 1985.

"Training Requirements of OSHA Standards," OSHA No. 2254. U.S. Department of Labor. Available from U.S. Government Printing Office, Washington, D.C. 20402, or local OSHA regional office.

# 10

# Ergonomics—Human Factors Engineering



ERGONOMICS, ALSO CALLED HUMAN FACTORS ENGINEERING, studies the interaction—both physical and behavioral—between humans and their environments. This chapter discusses the interaction between workers and their environments—the workplace, tools, and general working conditions. When workers and their environments are mismatched, injury levels rise, production is inefficient, and other incidents occur that detract from organizational efficiency and worker welfare.

## DEFINITIONS

In the U.S., the terms ergonomics, human factors, and human factors engineering have been interchangeably used. However, in other countries there has been strong distinction between the terms, and lately the U.S. seems to be following that trend. A formal definition of the term ergonomics, coined in the United Kingdom in 1950, is the study of human characteristics for the appropriate design of the living and work environment. Human factors, on the other hand, is slanted toward an understanding of the user's role in overall system performance. Human factors engineering is the application of these concepts.

For purposes of this chapter, the term ergonomics will be used. Ergonomics makes use of a number of disciplines, such as mathematics, engineering, life sciences, and behavioral sciences. Each of these sciences contributes to the identification, evaluation, and solving of man–machine problems.

## PURPOSE OF ERGONOMICS

The purpose of ergonomics is to design a system wherein the workplace layout, the work methods, the machines and equipment, and the general work environment (such as noise and illumination) are compatible with the physical and behavioral limitations of the worker(s). The better the match, the higher the level of safety and work efficiency (see Figure 10-1).

When there is a mismatch between what the worker can physically and behaviorally provide in comparison to what the job requires, then the result can be injuries, poor work efficiency, poor work quality, and other downgrading incidents.

Workers bring to their jobs certain anthropometric or physical characteristics. That is, they have specific heights, reaches, strengths, and the like. They also bring individual work methods to their jobs. That is, they may use certain lifting techniques (good or bad) and specific, sometimes unique, methods of performing a task.

On the other side is what the job brings to the worker in workplace design. For example, workstation layout is of certain dimensions and contains specific equipment. Hand tools used on the job are of definite design with fixed configurations (such as handle orientation and actuation force). There are fixed production quotas that normally must be maintained; these provide the parameter (in this case, a time element) within which the activity must be performed. If what the worker is able to physiologically or psychologically provide to the job does not meet the job requirements, a mismatch occurs. Examples of this are too high working heights (causing workers to raise their shoulders and arms for extended periods and resulting in high stress to these body parts), storage shelves or racks built too high to comfortably reach, or loads that are too heavy to lift.

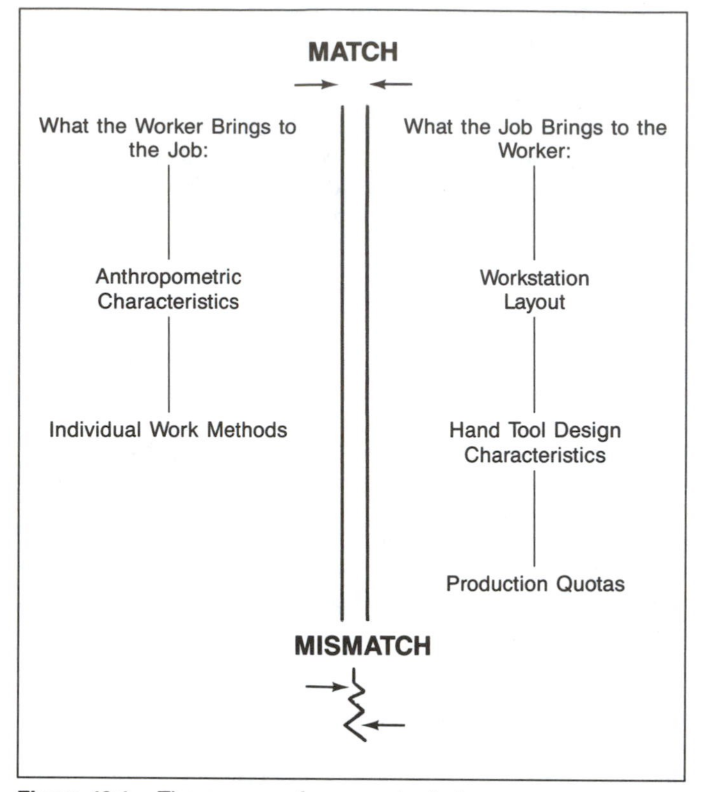


**Figure 10-1.** The purpose of ergonomics is to design a system where the general work environment is compatible with the physical and behavior limitations of the workers. The better the match, the higher the level of safety and work efficiency.

### Applications

The principles of ergonomics can be applied to safety and health problems such as overexertion injuries to the low back, carpal tunnel syndrome, and other cumulative trauma disorders. These principles can also be applied in the design of machine and equipment displays and controls to reduce response error that could have disastrous results.

Examples of disabling injuries with ergonomic implications are:

- A drop hammer operator at a forging press was using tongs to lift a heated part weighing 55 lb (25 kg) out of a forging die. Because of the size and positioning of the die, the operator was forced to lift the part straight up with his arms extended in front of him. This action resulted in a low back injury.
- A punch press operator was repeatedly lifting 4 lb (1.8 kg) metal disk blades from a pallet and flipping them prior to placement on a press stock-up table. The repetitious act resulted in wrist tendonitis.
- A worker received a low-back injury from repetitively bending over and reaching into tubs and pallet boxes to lift small light-weight parts.

In these examples, if ergonomic principles had been applied to the job design, the injuries might have been prevented.

Because low-back injuries from lifting and cumulative trauma disorders are a frequent problem and the cause of costly workers' compensation claims, this chapter will mainly discuss the ergonomic approaches to these problems. It is not possible to cover the wide area of ergonomics applications in one chapter, so readers are encouraged to consult the references listed at the end of this chapter.

## PHYSIOLOGICAL BASIS OF WORK

### Static and dynamic muscular effort

Muscles are what we use to move or exert pressure on objects. All body movements or exertions are the result of contracting (and shortening) or extending the muscles. How these muscles are used affects muscle performance. For example, dynamic muscular effort (like riding a bicycle or walking) is easier to do for long periods than static muscular effort (such as holding an object at arm's length). This is because in dynamic muscular effort, the muscles function as a pump for the blood system which, in turn, provides the muscles with a supply of energy-rich sugar and oxygen. In static muscular effort, the muscles are in a prolonged state of contraction that restricts the energy-enriched blood and allows waste products to accumulate in the muscle cells. This condition results in the muscular fatigue that we humans experience as pain. For this reason, static muscular effort cannot be maintained very long. Longer rest periods are needed to recuperate. In addition, jobs requiring long and excessive static positions (such as working while leaning forward) can lead to deterioration of the joints, ligaments, and tendons (see Figure 10-2).

**STATIC LOAD AND BODY PAINS**

| WORK POSTURE | POSSIBLE CONSEQUENCES, AFFECTING |
|---|---|
| Standing in one place | Feet and legs, possibly varicose veins |
| Trunk curved forward sitting or standing | Lumbar region, deterioration of intervertebral disks |
| Arm outstretched, sideways forward or upwards | Shoulders and upper arms; possibly periarthritis of shoulders |
| Head excessively inclined backwards or forwards | Neck, deterioration of intervertebral disks |
| Unnatural grasp of hand grip or tools | Forearm; possibly inflammation of tendons |

**Figure 10-2.** Jobs that require spending extended periods in a single posture or position can adversely affect the body. (Printed with permission from Grandjean's, *Fitting the Task to the Man; An Ergonomics Approach*, 3rd ed. London: Taylor & Francis, Ltd., 1979.)

### Energy expenditures

Strength is often thought of as the main requirement for manual jobs performance. However, if a work task is demanding (such as highly repetitive or requiring excessive static muscle holds), then the energy expended performing the task can be the limiting factor instead of available strength. "Heavy" work results in a high energy expenditure and severe stress on the heart and lungs. High energy expenditures can result in fatigue—either general or localized. As muscles fatigue, they are weakened and more likely to be injured.

Studies suggest that the human body can work for long periods of time without excessive fatigue if worked at ⅓ of its aerobic

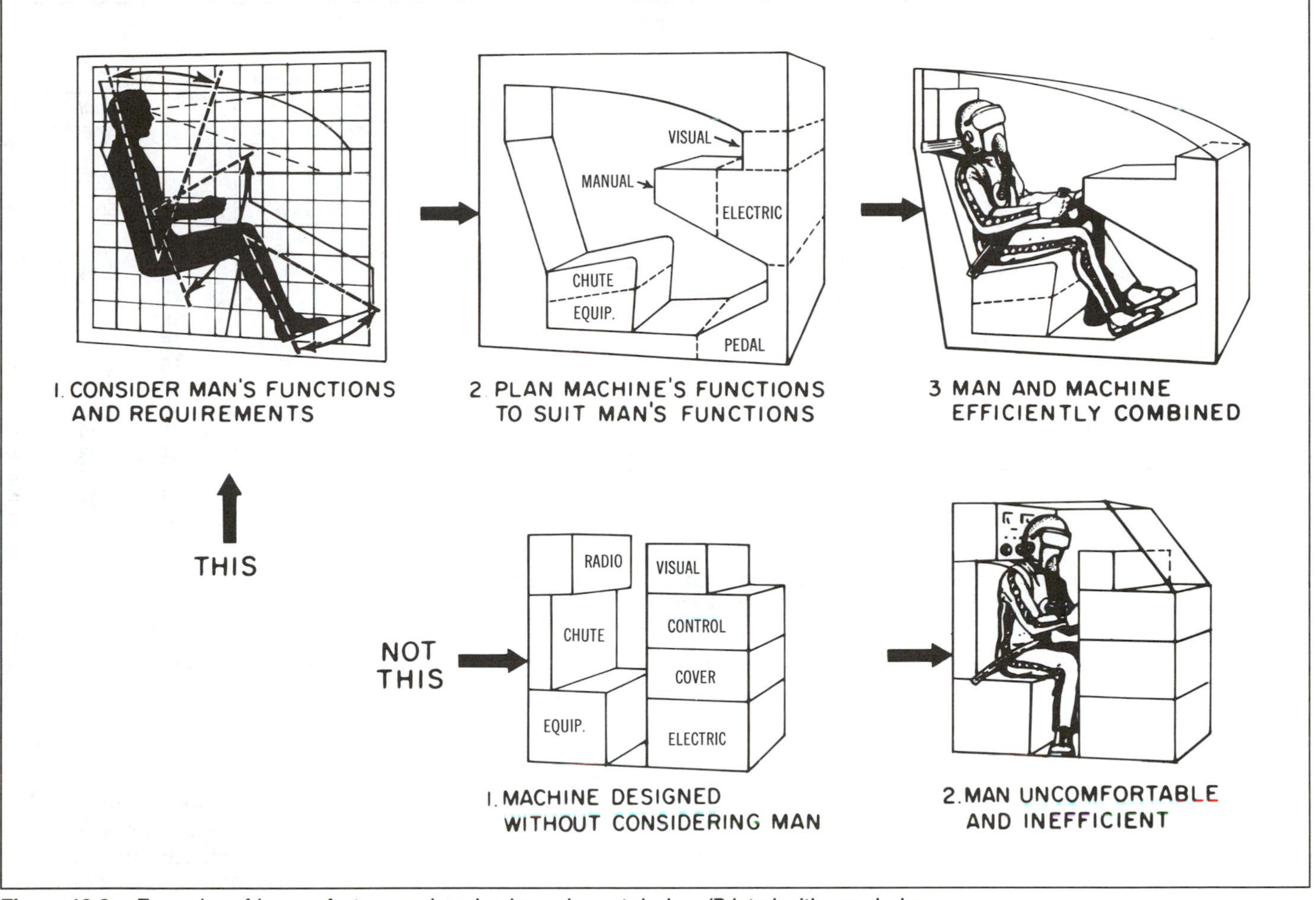


**Figure 10-3.** Examples of human factors engineering in equipment design. (Printed with permission from C. T. Morgan, et al. *Human Engineering Guide to Equipment Design.* New York: McGraw-Hill Book Company, 1963.)

capacity. This work level (4 to 5 Kcal/min) permits more work to be performed over longer periods of time. Job energy expenditures can be measured but because this requires specialized equipment, the sources listed at the end of this chapter should be consulted first to get more detailed information.

## Occupational biomechanics

Biomechanics is the study of the mechanical properties of the human musculoskeletal system during normal daily activities. It is a multidisciplinary study combining knowledge from physical and engineering sciences with knowledge from biological and behavioral sciences. Occupational biomechanics is an applied discipline concerned with the mismatching of worker physical capacities and job requirements.

Occupational biomechanical models provide a means of evaluating the physical stress of jobs to determine the degree of risk. The forces (static or dynamic) acting on any part of the body can be calculated and used as a basis for job design or redesign. In particular, biomechanical modeling of the low back and upper extremities is extremely useful in understanding the biomechanical problems of job situations that frequently create cumulative trauma disorders. (See Chaffin and Anderson, 1984, for details.)

The applications of occupational biomechanical models are: (1) to evaluate current working conditions, (2) to develop better workplaces, (3) to set guidelines for lifting activity, seated work, hand-tool design, whole-body and segment vibration, and workplace layout, and (4) to help determine worker selection and training criteria. For further information on biomechanical modeling of the specific body parts, refer to the reference list at the end of this chapter.

## Anthropometrics

Because workers come in all shapes and sizes, designing for only one segment of the population, such as "average" males, can cause problems for those not fitting this group.

The science of anthropometry takes into account human differences by measuring various characteristics, such as body dimensions, weights, and physical range of mobility. This information can be applied in workplace and equipment design to maximize worker comfort and safety and work efficiency (see Figure 10-3).

Unfortunately, an approach used by some designers is to design workstations and machinery to fit the average male. The problem with this is that half of the user population could be excluded from a workplace designed for the average. For example, at least one-fourth of the population would have to stoop to go through doorways designed for the average male

**Table 10-A.** Selected Structural Body Dimensions and Weights of Adults (Ages 18 to 79)

| Body feature (See accompanying diagrams) | *Dimensions (inches)* Male, percentile 5th | 50th | 95th | Female, percentile 5th | 50th | 95th | *Dimensions (centimeters)†* Male, percentile 5th | 50th | 95th | Female, percentile 5th | 50th | 95th |
|---|---|---|---|---|---|---|---|---|---|---|---|---|
| 1 Height | 63.6 | 68.3 | 72.8 | 59.0 | 62.9 | 67.1 | 162 | 173 | 185 | 150 | 160 | 170 |
| 2 Sitting height, erect | 33.2 | 35.7 | 38.0 | 30.9 | 33.4 | 35.7 | 84 | 91 | 97 | 79 | 85 | 91 |
| 3 Sitting height, normal | 31.6 | 34.1 | 36.6 | 29.6 | 32.3 | 34.7 | 80 | 87 | 93 | 75 | 82 | 88 |
| 4 Knee height | 19.3 | 21.4 | 23.4 | 17.9 | 19.6 | 21.5 | 49 | 54 | 59 | 46 | 50 | 55 |
| 5 Popliteal height | 15.5 | 17.3 | 19.3 | 14.0 | 15.7 | 17.5 | 39 | 44 | 49 | 36 | 40 | 45 |
| 6 Elbow-rest height | 7.4 | 9.5 | 11.6 | 7.1 | 9.2 | 11.0 | 19 | 24 | 30 | 18 | 23 | 28 |
| 7 Thigh-clearance height | 4.3 | 5.7 | 6.9 | 4.1 | 5.4 | 6.9 | 11 | 15 | 18 | 10 | 14 | 18 |
| 8 Buttock-knee length | 21.3 | 23.3 | 25.2 | 20.4 | 22.4 | 24.6 | 54 | 59 | 64 | 52 | 57 | 63 |
| 9 Buttock-popliteal length | 17.3 | 19.5 | 21.6 | 17.0 | 18.9 | 21.0 | 44 | 50 | 55 | 43 | 48 | 53 |
| 10 Elbow-to-elbow breadth | 13.7 | 16.5 | 19.9 | 12.3 | 15.1 | 19.3 | 35 | 42 | 51 | 31 | 38 | 49 |
| 10 Seat breadth | 12.2 | 14.0 | 15.9 | 12.3 | 14.3 | 17.1 | 31 | 36 | 40 | 31 | 36 | 43 |
| 12 Weight* | 120 | 166 | 217 | 104 | 137 | 199 | 58 | 75 | 98 | 47 | 62 | 90 |

*Weight given in pounds (first six columns) and kilograms (last six columns).
†Centimeter values rounded in whole numbers.
(Printed from *Weight, Height, and Selected Body Dimensions of Adults: 1960-1962.* Data from National Health Survey, USPHS Publication 1000, series 11, no. 8, June 1965.)

height of 68.3 in. (173 cm). Perhaps another one-fourth of the workers would be too short to reach a storage shelf if its height were designed to fit the average height and reach of a male worker.

Workstations and machinery should be designed to accommodate the largest variation in user population possible. Ideally, 100 percent of the user population should be accommodated but this is seldom possible. Where possible, the range between the 5th percentile and the 95th percentile should be used as this will cover about 90 percent of the user population.

The key to accommodating the various sizes and reaches of the work population is adjustability. Adjustable-height work benches, chairs, stock platforms, work fixtures, and other furnishings can be used to accommodate various size workers. Workstations and job methods should be designed where all items handled and all machine controls are positioned to eliminate excessive reaching, stooping, leaning, bending, or twisting. These postures can all produce fatigue and low back problems if they have to be frequently assumed and extensively maintained.

Table 10-A presents representative anthropometric information on structural body dimensions in fixed positions and weights for adult civilian populations. Since these dimensions and weights are for nude populations, they must be increased for clothing. For example, add 1 in. (2.5 cm) to height for shoes and 4 to 6 lb. (1.8 to 2.7 kg) for clothes. Refer to references at the end of this chapter for more specific information. Measurements, particularly weight and height, can vary with age.

### Anthropometric applications

The height at which work is being performed can be critical to job efficiency and safety. If the working height is too high, the shoulders must be frequently lifted to compensate, leading to pain in the shoulders and neck. If the working height is too low, the back must lean forward leading to backache. Work areas commonly affected are assembly and bench assembly operations.

A favorable standing work height is 2 to 4 in. (5 to 10 cm) below elbow height. For bench assembly work where there is need for more vertical space for tools and materials, the best working height is 4 to 6 in. (10-15 cm) below elbow height. Working height is the height at which the hands are working and is not necessarily the same as bench height.

Elbow height is the distance from the elbow to the standing surface when the arm is bent 90 degrees. A work height of between 37 to 41 in. (94 to 104 cm) will accommodate most males. For females, a work height between 34 to 38 in. (86 to 96 cm) will accommodate most of them. Ideally, working height should be adjustable to suit individual workers. If this cannot be done, then working height should be set to suit the tallest workers. Smaller workers can be given something to stand on, assuming this doesn't create a tripping problem.

Accommodating workers so they are comfortable on the job should not be thought of as coddling. If they are reasonably comfortable, they will be more productive and have fewer injuries. The References section at the end of this chapter lists sources of further information on how anthropometry tables can be used to increase worker comfort and productivity.

## IDENTIFYING ERGONOMIC-RELATED PROBLEMS

There are four major components in an ergonomic control-prevention program:

1. Identify existing or potential problems
2. Identify and evaluate risk factors causing the problems
3. Design and implement corrective (engineering and administrative) measures
4. Monitor and evaluate effectiveness of corrective measures introduced.

These control-prevention components apply to general safety approaches as well as to ergonomic-related problems.

In identifying existing problems, it is important to identify specific areas with a high incidence of injury or injury potential. There are two approaches to obtaining the information: (1) using passive or (2) using active data. Passive data uses existing sources

**Table 10-A (continued).** Selected Structural Body Dimensions and Weights of Adults (Ages 18 to 79)

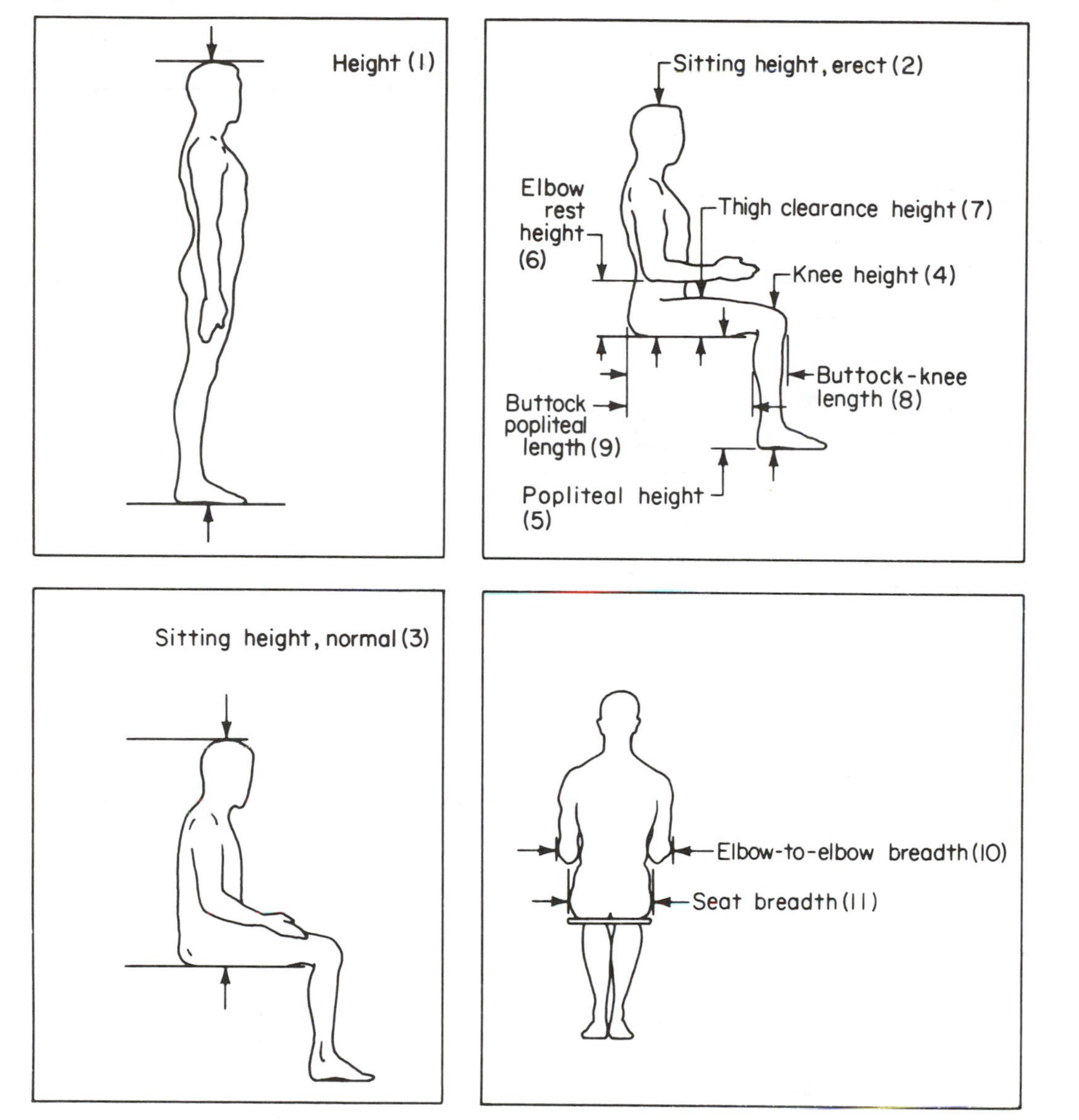


such as the Occupational Safety and Health Administration (OSHA) Form 200 Summary or internal incident investigation records to identify specific problem areas. This is the most commonly used method but is limited because all the information needed may not be available from existing data sources.

For example, accident investigation reports tend to cover sudden trauma injuries, but may not contain useable information about cumulative trauma injuries and illnesses. Some organizations overcome this limitation by either adding special sections to their accident investigation reports or developing separate reports. Figure 10-4 is an example of one company's approach to recording specific information about low back injuries caused by overexertion. This supplemental report is used in addition to their standard accident investigation report.

The second approach actively gathers on-going information about a problem. It uses a system specifically designed to collect specific information about a condition. An example is the employee who complains about wrist pain occurring on the job. In this example, efforts would be made to seek information about a potential problem even though no actual injury or illness has been recorded.

### Problem identification

Using the passive data approach, the first step in identifying the problem is to classify by job type, department, or other category, the number and types of injuries. The objective is to relate injury clusters to specific work tasks. Next, incident rates by category are developed and measured so study priorities can be decided. For this, reports of OSHA recordable injuries, first aid injuries, or other incident experience can be used.

Optionally, a form such as shown in Figure 10-5 can be used to identify the total injury and illness experience within a measured category. In this form, incident experience by job classification is being measured. Although this format is oriented toward isolating back injuries, it can be modified to include

## Supplemental Sprain/Strain Investigation Report

UNIT

INVESTIGATOR | DATE

**EMPLOYEE BACKGROUND**

1. NAME | 2. CLOCK NO. | 3. DEPT. NO. | 4. SEX | 5. AGE | 6. SHIFT | 7. OCCUPATION | 8. SEN. CLASS NO.

9. PAY STATUS ☐ Salaried ☐ Incentive ☐ Hourly | 10. BODY TYPE ☐ Rotund ☐ Muscular ☐ Thin | 11. PHYSICAL LIMITATIONS (From Medical Dept.) | 12. HEIGHT | 13. WEIGHT

14. IDENTIFY SPECIFIC AREA (Work Station, Process, Machine) WHERE INJURY WAS INCURRED | 15. DATE INJURY REPORTED | 16. DATE INJURY OCCURRED | 17. TIME OF INJURY ☐ AM ☐ PM | 18. DAY OF WEEK | 19. ONSET OF CONDITION ☐ Gradual ☐ Sudden

**INJURY DATA**

20. NATURE OF INJURY | 21. BODY PART INJURED | 22. PREVIOUS INCIDENT OF SIMILAR PAIN OR DISCOMFORT? ☐ Yes ☐ No

23. ACTIVITY PERFORMED DURING PREVIOUS INCIDENT | 24. CLASSIFICATION OF INJURY ☐ First Aid ☐ Days Lost ☐ Medical Treatment Without Lost Workdays | NO. OF DAYS LOST

**TRAINING**

25. IS THERE A JOB SAFETY ANALYSIS FOR THE JOB? ☐ Yes ☐ No | 26. IF "YES", DOES IT INCLUDE MATERIAL HANDLING TRAINING? ☐ Yes ☐ No | 27. HAS EMPLOYEE RECEIVED TRAINING IN MANUAL MATL. HANDLING PROCEDURES? ☐ Yes ☐ No | 28. APPROXIMATELY HOW LONG AGO WAS THIS TRAINING GIVEN?

**GENERAL DESCRIPTION OF INCIDENT**

29. WHAT TASK WAS EMPLOYEE PERFORMING AT TIME OF INCIDENT? (Include Part & Operation No., etc.)

30. WAS EMPLOYEE PERFORMING REGULAR JOB DUTIES AT TIME OF INCIDENT? ☐ Yes ☐ No

31. HOW DID SPRAIN/STRAIN OCCUR? (If gradual Onset, give sequence of events)

**INCIDENT DETAILS**

32. DESCRIBE OBJECT (Part, Material, Assembly, Equipment) HANDLED | 33. SIZE OF OBJECT In. | 34. WEIGHT OF OBJECT Lbs.

35. TASK DESCRIPTION: ☐ A. Lifting ☐ B. Lowering ☐ C. Pushing ☐ D. Pulling ☐ E. Carrying ☐ F. Other (Describe): ________

36. HEIGHT OF ACTIVITY: ________
☐ 1. Shoulders And Above
☐ 2. Knuckles To Shoulders
☐ 3. Floor To Knuckles

37. FREQUENCY OF HANDLING · ONCE EVERY: ________ Seconds; ________ Minutes; ________ Hours | DURATION: ☐ 8 Hours ☐ 1 Hour Or Less

38. ARM POSITION: ☐ Arms Extended ☐ Arms Next To Body

39. POSTURE: ☐ 1. Stand ☐ 2. Sit ☐ 3. Squat ☐ 4. Deep Squat ☐ 5. Stoop ☐ 6. Lean Forward (Push) ☐ 6. Lean Backward (Pull) ☐ 7. Split ☐ 8. Walk ☐ 9. Climb

40. TYPE OF CONTAINER INVOLVED IN ACCIDENT: ☐ 1. Ring Skid ☐ 2. Hairpin Tub ☐ 3. 40 x 50 Tub ☐ 4. Wood Pallet Box ☐ 5. Flat ☐ 6. Floor ☐ 7. Rack ☐ 8. Knock-down Tub ☐ 9. 40 x 50 Wire Mesh Basket ☐ 10. Other (Describe Below): ________ ☐ 11. None

41. IS BACK BEING TWISTED WHILE PERFORMING TASK? ☐ Yes ☐ No | 42. FOOTING: ☐ Wet ☐ Slippery ☐ Dry | 43. WORKPLACE SPACE: ☐ Crowded ☐ Adequate | 44. TEMPERATURE: ☐ Hot ☐ Comfortable ☐ Cold | 45. IF OBJECT CARRIED, GIVE DISTANCE CARRIED: Ft.

46. LIFT/LOWER DISTANCES: 1. Horizontal Distance: ________ In. 2. Vertical Distance (Start of Lift/Lower) ________ In. 3. Vertical Travel Distance (Distance Lifted/Lowered) ________ In. | 47. PUSH/PULL DISTANCES: 1. Distance Pushed or Pulled ________ Ft. 2. Effort Required: a. Initial Force ________ lbs. b. Sustained Force ________ lbs.

48. TOP VIEW OF WORK SITE AND POSITION(S) OF EMPLOYEE

**ANALYSIS**

1. LIFTING (2 Hand): Action Limit ________ Maximum Permissible Limit ________ Formula Factors (Decimals): H ____ V ____ D ____ F ____

2. PUSH/PULL POPULATION CAPABLE: Male | Female — 3. CARRY POPULATION CAPABLE: Male | Female — 4. LOWERING POPULATION CAPABLE: Male | Female

**CORRECTIVE ACTION** *(Use reverse side if additional space is required)*

2937-1/83

**Figure 10-4.** This Supplemental Sprain/Strain Investigation Report can be used in addition to a standard accident investigation report.

Job Classification ______________________________

| Number of Incidents | | | | | | |
|---|---|---|---|---|---|---|
| | Body Part Involved | | | | Incident Rate* | |
| *Diagnosis Classification* | *Upper Extremities* | *Shoulder and Back* | *Lower Extremities* | *Head, Neck Abdomen* | *Job Classification* | *Total Unit* |
| Non-Specific | | | | | | |
| Contact | | | | | | |
| Musculo-skeletal | | | | | | |
| Back | | | | | | |

*Per 200,000 hours worked.

**Figure 10-5.** Example of form that can be used to identify and prioritize injury and illness experience by job classification. (Adapted from NIOSH, *Preemployment Strength Testing*, DHEW Publication No. 77-163, 1977.)

other types of incidents such as carpal tunnel syndrome. Incident rates are developed for the job classification and compared to the rates for the entire plant or work site. If the job classification rates are higher than the total unit incident rates, then it is obvious a problem has been isolated. Another advantage of this type of form is that it can be used to monitor performance after countermeasures have been introduced.

### Job physical demands study

After the ergonomic-related injuries and illnesses have been identified and prioritized, the third step is to perform a job study. This means the job will be observed, manual handling tasks recorded, task variables measured, and then the job study evaluated to find the risk factors present.

A basic ergonomic job study for manual handling tasks need not be extensive nor expensive. The minimum equipment needed to perform the study includes:

- Force scale (portable spring or fish scale or equivalent)—to measure load weight and push or pull forces
- Tape measure—to measure task-variable distances
- Time piece (wrist watch or stop watch)—to measure job cycle and task times
- Pencil and paper—to record measurements and forces by task.

Performing a job study is similar to making a job safety analysis. Begin by introducing yourself to the worker, explain the purpose of the job study and solicit the worker's cooperation. Then, record the job tasks in which manual handling is performed and make the required task-variable measurements.

## OVEREXERTION LOW BACK INJURIES

### Activities leading to low back pain

Occupational low back injuries from overexertion (either sudden or cumulative trauma onset) are a major problem in industrialized societies. Traditional approaches used in this country to prevent low back injury (for example, lifting training) have not been proven to be particularly effective.

Although exercise, diet, and lifting training can contribute to safe lifting, the best approach is to apply ergonomic principles to job design. That is, design-out the physical stresses of the lift. Even when this is done, some low back injuries will occur.

Lifting is the most common task associated with low back injury. Many of these injuries do not result from a single incident but develop over a period of time. In other words, the weight lifted at the time when the low back pain first was noticed may have had little to do with the injury. The injury may have resulted from repetitive lifting, such as bending over and lifting objects out of containers, all shift long—every day.

Low back injury can come from other manual handling tasks besides lifting such as lowering, pushing, pulling, and carrying. In addition, low back injury can be the result of poor housekeeping practices: slippery floors causing tripping, slipping, and falling hazards; crowded or unorganized work conditions; or tools, materials, and debris left lying on the floor. When analyzing low back injury occurrences, incidents resulting from poor housekeeping practices should be separated from the overexertion-type of low back injuries because the countermeasures are different.

### NIOSH lifting guidelines

In 1981, the National Institute for Occupational Safety and Health (NIOSH) published its *Work Practices Guide for Manual Lifting*. Among other things, this manual contains an algebraic equation for determining the acceptability of a particular lift. This section will briefly review how to apply the equation (for further information, see References).

The NIOSH lifting guideline applies to two-hand symmetrical lifting situations in front of the body not involving trunk twisting. The lifting must be smooth (no jerking), the hand spread must be 30 in. (75 cm) or less, and the lifting posture must be unrestricted. In addition, there must be good couplings (handles, shoes, floor surfaces) and a favorable ambient environment. It is also assumed that other manual handling activities, such as holding, carrying, pushing, and pulling, are minimal. Assumptions also include a physically fit work force accustomed to physical labor. Even with all of these limitations, there still are many lifting situations where the NIOSH lifting guidelines apply.

The NIOSH lifting guidelines require that certain lifting task variables be quantified during a job study; these later are inserted in an algebraic equation and computed to determine if a particular lifting situation is acceptable.

Each manual handling task is assigned a set of variables. Each of these variables describes an aspect of the task, such as weight lifted or weight lifted × times in a minute. If any of the variables are excessive, then the result can be that (1) a high energy demand exists, which can cause weakening of the low back muscles, or (2) high compressive forces are generated between the vertebrae of the low back. Either one or both of these can result in an overexertion injury to the low back.

Following are the task variables considered in NIOSH's lifting equation:

- *Horizontal distance (H)*—the further the center of gravity of the load lifted is horizontally away from the low back, the higher the compressive force to the low back.
- *Vertical distance (V)*—the lower to the floor the load is located at the beginning of a lift, the more a worker must bend, thereby creating higher stresses to the low back.
- *Vertical travel distance (D)*—the further a load must be vertically lifted, the more energy expended.
- *Frequency of lift (F)*—the greater the lift frequency within a time period, the more energy expended.
- *Duration of lifting*—this refers to the length of the lifting cycle, such as occasional lifting for one hour, or constant lifting continuously throughout the workshift.
- *Load weight lifted (W)*—while high load weights increase the chances of low back injury, this is dependent upon the variables previously cited.

### Three regions of lifting

NIOSH's lifting guideline establishes three regions of lifting. They are: above the Maximum Permissible Limit (MPL), below the Action Limit (AL) and between the Action Limit (AL) and Maximum Permissible Limit (MPL). Figure 10-6 illustrates the three regions of lifting for infrequent lifts (less frequent than one lift per five minutes) from the floor to knuckle height and the weights that can be lifted based upon horizontal hand location. Note that the permissible weight limit becomes lower as the location of the load is moved farther from the body.

At the Maximum Permissible Limit (MPL) level, only about 25 percent of the male and less than 1 percent of the female workers would have the ability to perform lifts with reasonable safety. Lifts falling above the MPL represent hazardous lifting conditions that require engineering controls.

At the Action Limit (AL) level, more than 75 percent of the female and more than 99 percent of the male workers have the ability to perform lifts with reasonable safety. Lifts falling below the AL are thought to represent nominal lifting risk and are considered acceptable.

Lifts falling between the AL and MPL are still considered not acceptable but instead of using engineering controls, administrative controls, such as employee selection and training programs, can be used. Calculating the lifting equation for a particular lifting situation will establish the AL and MPL. Then the actual weight lifted is compared to the AL and MPL to determine where it falls.

### Lifting equation

The lifting equation established by NIOSH incorporates the major task variables that are identified and quantified during the job study. The equations for determining the Action Limit (AL) and Maximum Permissible Limit (MPL) are:

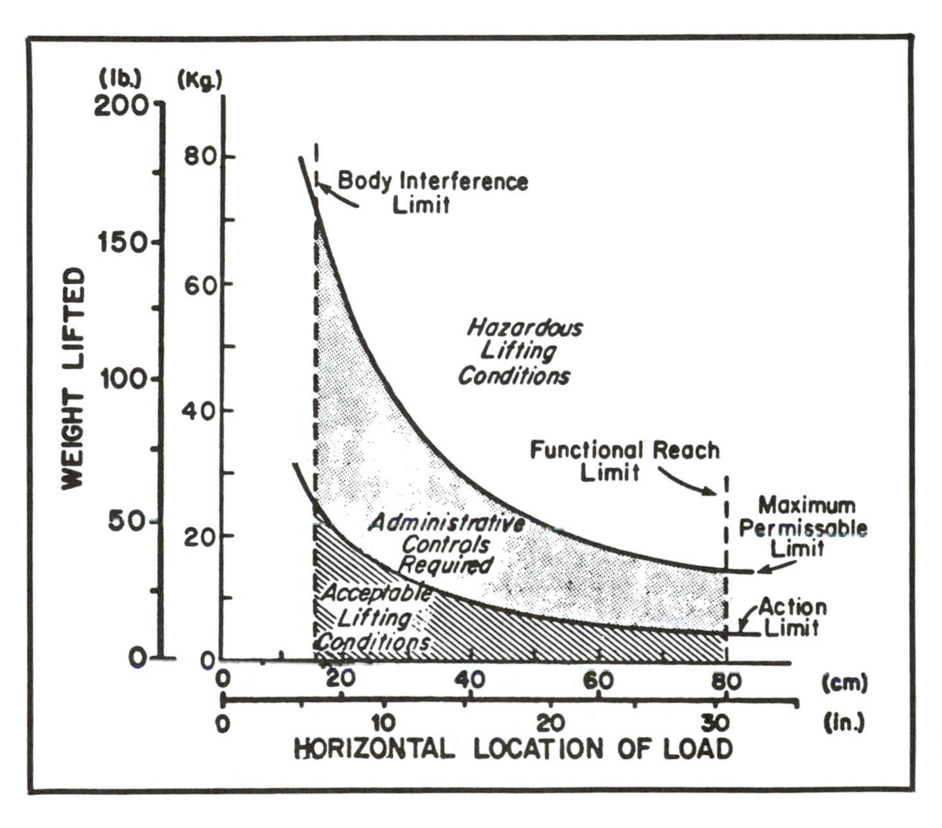


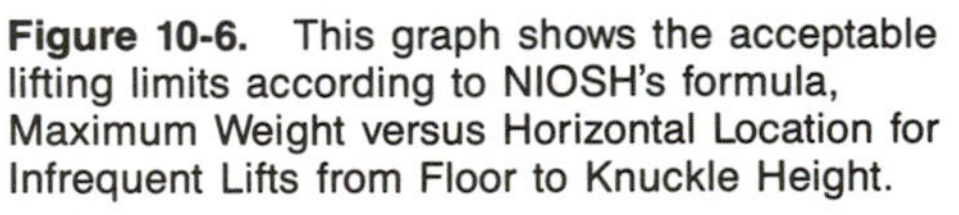
**Figure 10-6.** This graph shows the acceptable lifting limits according to NIOSH's formula, Maximum Weight versus Horizontal Location for Infrequent Lifts from Floor to Knuckle Height.

$$AL(lb)=90(6/H)(1-0.01|V-30|)(0.7+3/D)(1-F/F_{max})$$
$$MPL=3(AL)$$

Or (in metric units):

$$AL(kg)=40(15/H)(1-0.004|V-75|)(0.7+7.5/D)(1-F/F_{max})$$
$$MPL=3(AL)$$

The variables in the lifting equation are:

H=Horizontal distance measured between center of handgrasp and midpoint between ankles at the *start* of lift. It is assumed to be between 6 in. (15 cm) and 32 in. (80 cm). Generally, objects cannot be brought closer to the spine than six inches (15 cm) because of body interference nor farther than 32 in. (80 cm) because objects beyond this distance cannot be reached by most people.

V=Vertical distance measured between handgrasp and floor or standing surface at the *beginning* of the lift. This distance is assumed to be between zero and 70 in. (175 cm), which represents the range of vertical reach for most people.

D=Vertical travel distance between origin and destination of lift. This distance is assumed to be between 10 in. (25 cm) and 80 in. (200 cm). If the vertical travel distance is less than 10 in. (25 cm), set D=10 in. (25 cm) in the equation.

F =Frequency of lift is the number of lifts per minute. Frequency is assumed to be between one lift every five minutes (F=0.2) and Frequency Maximum ($F_{max}$). For lifting less frequently than once per five minutes, set F=0.

$F_{max}$ =Maximum frequency that can be sustained based on lifting duration (see Figure 10-7). To determine $F_{max}$, select column A or B depending upon the vertical distance (V) variable at the beginning of the lift. Next, select the time period (such as one hour or eight hours) that closely approximates the duration of the lifting period and then select the $F_{max}$ value located under the column previously selected. (Example: lifting a load where V=8″ and lifting duration is 1 hour. $F_{max}$=15)

A form such as Figure 10-8 can be used to record task-variable measurements during a job lifting study. This form can be modified to suit the user.

- *Object weight*—if the weight of the object lifted varies, record the average and maximum weights
- *Hand location: origin*—record the horizontal (H) and vertical (V) distance measurements at the beginning of the lift.
- *Hand location: destination*—record the horizontal (H) and vertical (V) distance measurements at the end of the lift.
- *Task frequency*—record the average number of lifts per minute. Task duration such as one hour or eight hours also should be noted.

If the values vary significantly from one lift to another, such as unloading a stack of cartons, then each lift should be analyzed separately.

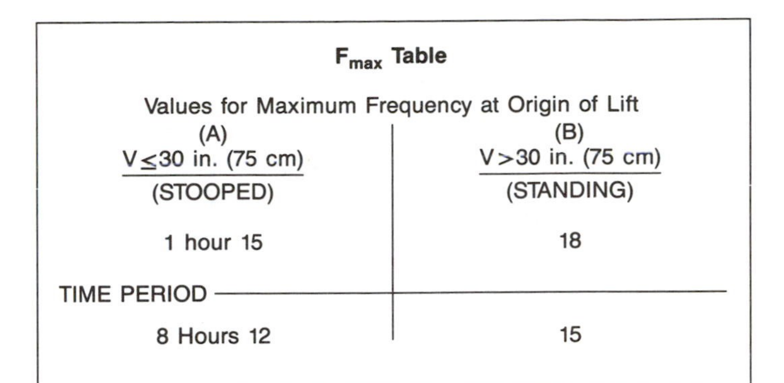

**$F_{max}$ Table**

Values for Maximum Frequency at Origin of Lift

| TIME PERIOD | (A) V≤30 in. (75 cm) (STOOPED) | (B) V>30 in. (75 cm) (STANDING) |
|---|---|---|
| 1 hour | 15 | 18 |
| 8 Hours | 12 | 15 |

**Figure 10-7.** Maximum frequency table ($F_{max}$) for selecting appropriate value when computing the NIOSH lifting equation.

### Lifting equation example

A worker is lifting 60-lb (27-kg) parts out of the bottom of a deep container prior to setting them on a machine table. The worker continuously performs this work for eight hours at a lift frequency of one per minute. Is this an acceptable lift acording to the NIOSH lifting guide? Figure 10-9 illustrates the lift and provides the task-variable measurements. After inserting the task-variable measurements into the lifting equation and solving, the results indicate that the 60-lb (27-kg) load lifted exceeds the MPL of 48 lb (21 kg). This is an unacceptable lift requiring engineering controls.

One of the main features of this lifting equation is the ability to evaluate the decimal equivalents to determine which task variable penalizes the lift. In this example, the horizontal factor (0.29) is the lowest and penalizes the lift the most. If the 21 in. (53 cm) horizontal distance at the beginning of the lift could be reduced, then this would improve the lift. For example, setting the container on a raised platform would permit the worker to lift the part without bending over thereby decreasing the horizontal distance as well as the vertical distance. This improvement might be enough to bring the lift into an accept-

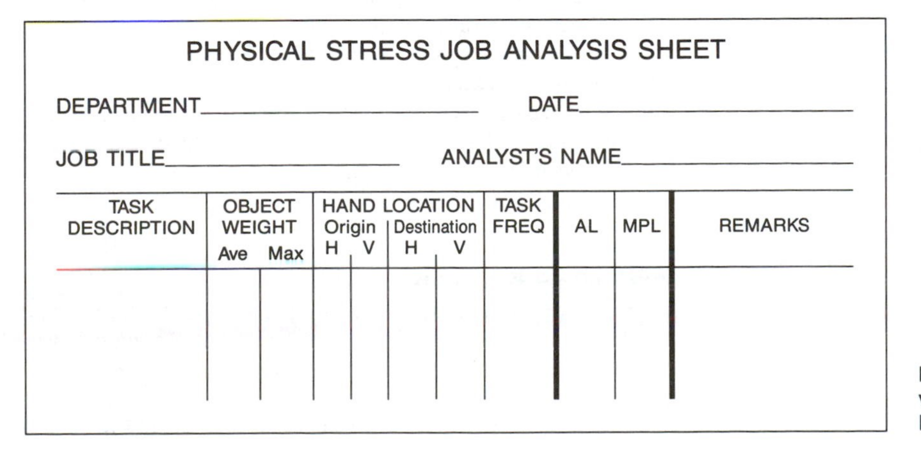

PHYSICAL STRESS JOB ANALYSIS SHEET

DEPARTMENT__________ DATE__________

JOB TITLE__________ ANALYST'S NAME__________

| TASK DESCRIPTION | OBJECT WEIGHT Ave | OBJECT WEIGHT Max | HAND LOCATION Origin H | HAND LOCATION Origin V | HAND LOCATION Destination H | HAND LOCATION Destination V | TASK FREQ | AL | MPL | REMARKS |
|---|---|---|---|---|---|---|---|---|---|---|
| | | | | | | | | | | |

**Figure 10-8.** Example of form used to record task-variable measurements when computing the NIOSH lifting equation.

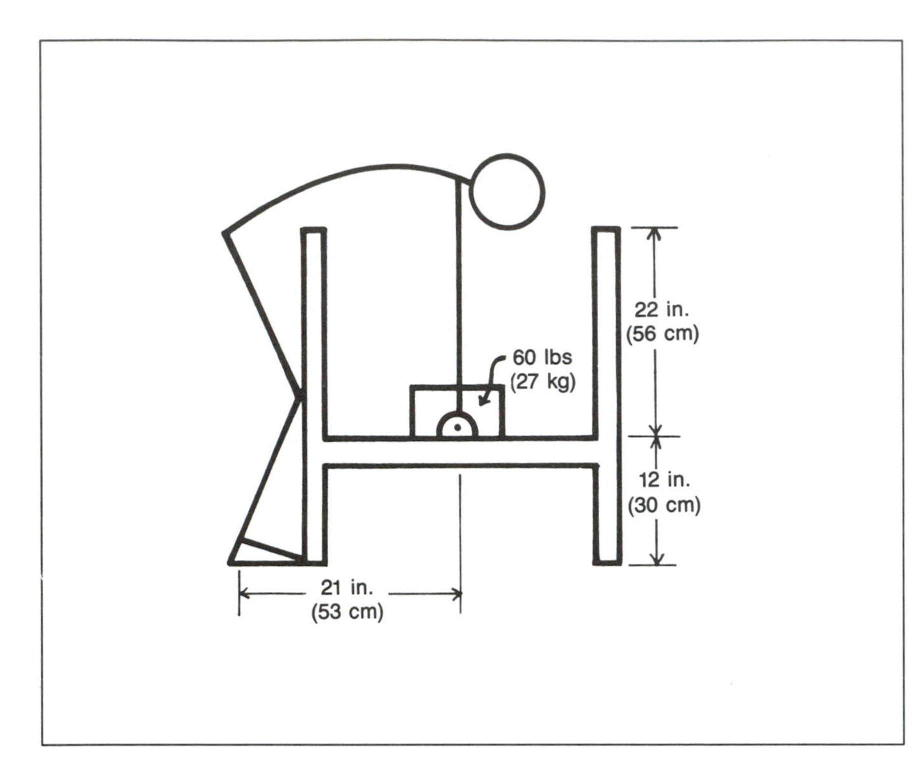


**Figure 10-9.** This is an example of how the NIOSH lifting equation is used to determine if a lifting problem exists. The 60-lb (27-kg) lift exceeds NIOSH's maximum permissible limit (MPL) and would require engineering controls.

able range below the AL. (The AL and MPL values should be used as relative values for comparison purposes—not as absolute values.)

Other engineering controls might include: raising the containers or parts off the floor; setting containers on tilt dollies; using containers with removable or drop fronts; using scissor or platform lifts or hoists; or eliminating lift by other means. There are many possible approaches to use when improving lifting tasks. The best approach is to design-out the lifting stresses.

While not all back injuries can be eliminated by using NIOSH's lifting guidelines, preventing high lifting stresses certainly will reduce the number of injuries.

## UPPER EXTREMITY CUMULATIVE TRAUMA DISORDERS

### Risk factors

Upper extremity cumulative trauma disorders are injuries caused by repeated trauma to the arms and hands—from shoulder to fingers. These disorders involve the soft tissues (tendons and muscles) or the nervous system.

Cumulative trauma disorders (CTD) can be caused by:

- Repetitive motions
- Forceful exertions
- Awkward postures
- Mechanical stress (pressure of body part on sharp edges of tools or equipment)
- Vibration level of handtool
- Exposure to cold
- Gloves.

CTD often involves multiple risk factors such as performing a repetitive task and forceful exertions while assuming an awkward posture.

### Carpal tunnel syndrome

Carpal tunnel syndrome is an impairment of the median nerve inside the wrist, and is usually caused by repeated trauma. This nerve impairment results in sensations of numbness, tingling, pain and clumsiness. Many times, the symptoms are severe enough to awaken the sufferers during bedrest.

The carpal tunnel is a structure inside the wrist formed on three sides by the wrist bones and on the fourth side by a tough ligament. The median nerve and several finger flexor tendons pass through this pathway. The finger flexor tendons move back and forth inside the carpal tunnel during use of the fingers and wrist. High rates of movement and exertions can result in injury. When the wrist is bent, these tendons rub on adjacent surfaces of the carpal tunnel. This rubbing can result in irritation and thickening of the synovial membranes surrounding and lubricating the finger flexor tendons. This thickening compresses the median nerve, which in turn results in tingling and numbness of the fingers.

While most carpal tunnel syndrome can be traced to repetitive or cumulative trauma, carpal tunnel symptoms also can occur because of a blow to the wrist, a laceration, or a burn. Women appear to be more susceptible to carpal tunnel syndrome than men, for reasons not fully understood.

The presence of carpal tunnel syndrome in a worker does not necessarily mean the injury is work-related. Outside activities, such as playing a musical instrument and gardening, can cause or contribute to the problem.

### Performing a job study

If upper extremity cumulative trauma disorders, such as carpal tunnel syndrome, tendinitis, trigger finger, and epicondylitis, have been determined to be work-related injuries, then a job study should be performed. The purpose of a job study is to determine the causal relationship between the CTD injury and the work task. This may be accomplished by evaluating the work methods, work station design, worker posture, and hand tool usage. While studying a job, look for risk factors occurring during the performance of the job. For example:

- Identify any stressful hand and arm postures assumed during the work cycle such as: bent wrist, finger pinch grasping, wrist twist, overhead reaches, arm torquing.
- Determine if the objects handled are difficult to grasp or use because of length or poor grasp area.
- Measure the frequency and duration of bent or torqued hand or arm movements during a job cycle.
- Identify any extended reaches (and holds at the end of the reach), body twisting, working with raised elbows (shoulder abduction), and bent upper body postures.
- Measure forces of exertion being applied by extremity.
- Measure working hand height and compare to elbow height. Generally, the optimum working hand height should be 2 in. (5 cm) to 6 in. (15 cm) below elbow height for most standing jobs unless arm rests are provided.
- Identify any handtool that requires bending the wrist to use. Note the handle configuration (pistol grips, barrel grip, or fluted grip).
- Measure the approximate forces required to actuate or operate the handtool. If the wrist is bent while using the handtool, the hand has significantly less gripping strength.
- Identify any powered handtools producing high levels of vibration.
- Identify any mechanical stress to the hands and fingers such as from sharp edges of tools or tooling.

In addition to the above, the amount of clothing and types of personal protective equipment being worn should be checked to determine if they restrict active movement. Poorly designed and bulky gloves can reduce hand strength up to 30 percent. Low operating temperatures can reduce dexterity.

After a job study has been performed, evaluate the information gathered. Unfortunately, at the present time there are no tables or formulas to use in identifying the specific problem beyond good judgment and common sense. Most commonly, high rates of manual task repetition combined with deviated postures and physical exertions are found to be the causes of the problem.

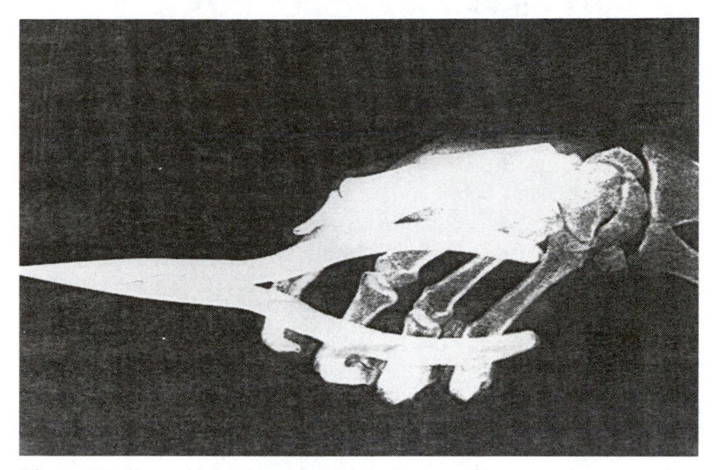

**Figure 10-10a.** Using conventional pliers requires that the wrist be bent, and contributes to worker fatigue and carpal tunnel injuries.

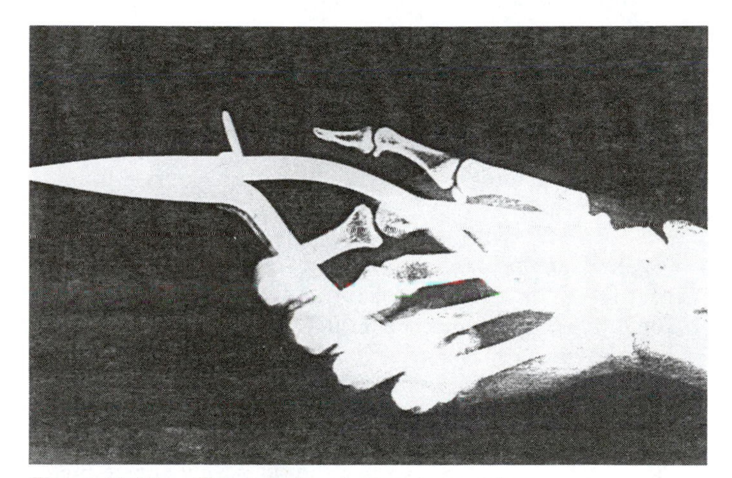

**Figure 10-10b.** The redesigned pliers has contour handles, spring, and thumb stop to reduce worker fatigue. The x-ray shows how the contour handles conform to a natural hand position. (Printed with permission from Western Electric Company.)

### Control interventions

Two types of control approaches are available for reducing or preventing problems: administrative and engineering. Of the two, engineering is the most effective.

**Administrative controls** consist of early medical intervention, worker training, matching workers to job demands, and job rotation. The disadvantages of administrative controls is that they treat the symptoms and not the cause of biomechanical stress.

**Engineering controls** are aimed at eliminating the sources of biomechanical stress through engineering and design. Such things as bending the handles on handtools to keep the wrist in a neutral position, lowering or raising work tables, reorganizing the work process, reducing hand forces, and substituting hand grips for pinch grips can successfully be used to reduce biomechanical forces of the hands and arms (see Figures 10-10a and b).

Engineering interventions specifically designed to reduce or relieve carpal tunnel syndrome include:

- Selecting hand tools with handles that spread the stress more evenly across the palm of the hand.
- Substitute a hand grip for a finger pinch, where possible. It takes 4 to 5 times as much muscle strength and tendon force to pinch an object rather than grip it.
- Avoid selecting hand tools that have fluted handle surfaces since the ridges cause stress to small areas in the palm.
- Suspend power tools on balancers to support the tool weight. Where possible, route power sources to the tool from the top to eliminate potential tripping over hoses or cables.
- Use power tools that stop when the torque setting is reached. Do not set the torque any higher than what is required to get the job done.
- Reduce power tool vibration to the hand by insulation or other means.
- Store small parts, in bench assembly operations, in bins or containers that minimize the need to bend the wrist to reach the parts. Also, orient the work so the assembly activity is within the normal range of arm movement.

## MACHINE DISPLAYS AND CONTROLS

How machines and equipment are activated, and how information is given to the operator through dials or gauges, has a definite effect on work efficiency and worker safety. For example, reaching for a machine stop button in an emergency and pushing the wrong button can result in a serious accident.

The interaction between the machine and its operator must be considered when a machine or piece of equipment is designed or purchased. This includes reaching and accessing parts of the equipment, correct working heights, properly located controls and displays, and convenient maintenance arrangements.

## DISPLAYS

There are two major types of displays: visual and auditory displays. A display can be a dial gauge, instrument, or auditory alarm that provides information to the operator about the operating status of the equipment or process. When information is displayed in a confusing format, operator error and possible disastrous consequences can be the result.

### Visual displays

Visual displays are used for one of three purposes:

- *Quantitative readings*—to determine the exact quantity involved, such as a scale.
- *Qualitative readings*—to determine the state or condition at which the machine is functioning—usually three conditions, such as above, within, or below tolerance.
- *Dichotomous (check) readings*—to check operations or to identify one or two levels, such as OFF or ON.

The purpose for which the display is to be read will dictate its best design. But, as a general principle, the simplest design is the best. The motion of the display should be compatible with (or in the same direction as) the motion of the machine and its control mechanism. A pointer that moves to the right, up, or clockwise to show an increase should have its corresponding control mechanism designed so that a rightward, upward, or clockwise movement of the control will increase the machine value and the corresponding display output value.

Multiple displays should be grouped according to their function or sequence of use. If dials are arranged in groups on a large control panel and they must all be read at the same time, the dials should be pointing in the same direction when in the desired range. This will reduce check-reading time and increase accuracy.

All displays should be labeled so the operator can immediately tell just what the display refers to, what units are being used, and what is the critical range. If the equipment is being used in a dimly lighted area, illumination must be provided.

### Auditory displays

Auditory displays should follow the principles outlined for visual displays. The question of whether an auditory or visual display should be used depends upon the situation. Figure 10-11 compares the relative advantages of auditory and visual displays. Other considerations for using auditory displays are found in the source materials at the end of this chapter.

To evaluate existing displays (visual or auditory), the following should be considered:

- Is the display intensity higher than the lower threshold level for that sense? (Because each sense has its own threshold level, energy intensities below it cannot be perceived.)
- Is the sense overloaded? What other demands are being made on this sense at the time the display in question is to be read?
- Is the display compatible with similar displays, controls, and machine movements?
- What environment factors, if any, could mask the display?

**Relative Merits of Auditory and Visual Presentations**

| *Use Auditory Presentation if:* | *Use Visual Presentation if:* |
|---|---|
| Message is simple | Message is complex |
| Message is short | Message is long |
| Message will not be referred to later | Message will be referred to later |
| Message deals with events in time | Message deals with location in space |
| Message calls for immediate action | Message does not call for immediate action |
| Receiving location is too bright | Receiving location is too noisy |
| Person's job requires him to move continually | Person's job allows him to remain in one position |
| Visual system of person is overburdened | Auditory system of person is overburdened |

**Figure 10-11.** The relative advantages of auditory and visual displays should be compared before the type of display is chosen.

## CONTROLS

A control is anything—a switch, lever, pedal, button, knob, or keyboard—used by an operator to put information into a system. Performance can be enhanced when the controls operate as the operator expects them to, if they are dimensioned to fit the human body, and their operating characteristics are within the strength and precision capabilities of the operator.

People expect things to behave in certain ways when they use operating controls. Most Americans, for example, expect a light switch to be turned on by flipping the switch up. A clockwise motion generally refers to an increase. Conversely, people expect the reverse kinds of movements to turn a system off or to decrease a function or flow. Such responses are called "population stereotypes." They are common to nearly everyone in the population. Examples of typical responses to certain situations are provided in Figure 10-12. Any control response that calls for a movement in a direction contrary to the established stereotype is likely to produce errors. The designer is asking for errors by requiring the operator to change a behavioral habit pattern.

Safety may be jeopardized if an operator misreads a poorly designed display and operates the wrong control or the right control in the wrong direction. Despite the fact that many accident reports would classify these types of accidents as unsafe acts or human error, the fact is they are design errors. Retraining the operator would probably not prevent recurrence since the operator, in an emergency, would tend to revert back to the original stereotype behavioral response. The operator should never be required to do things that are unnatural or unexpected.

### Control design principles

Through research, these principles of control design have emerged:

**Compatibility.** Just as in the design of displays, control movement should be designed to be compatible with the display and machine movement. A lift truck, for example, with lift controls that move right or left to raise or lower the lift is bound to have a number of errors associated with its operation. The correct movement would be up and down.

**Coding.** Whenever possible, all controls should be coded in some way, such as by a distinguishing shape and texture, location, or color. A good coding system can reduce many errors. The following are examples of this concept:

- *Shape and texture* can be used to code controls. This is useful where illumination is low or where a control needs to be identified and operated through touch only.
- *Location* is another way to code controls. On forklift trucks, for example, all controls pertaining to mast operation can be grouped on the right side of the control panel. Location coding can also be achieved by using a minimum distance between controls.
- *Color* also can be used as a coding technique for various controls. Color codes are useful for quick visual identification and for grouping controls for a particular operation.
- *Labels.* All controls require some type of labeling to identify their function.

### Arrangement of controls

The arrangement of controls (and displays) should support the interaction between the user and the control or display component. Consequently, the components of the system need to be arranged with these considerations in mind:

- Convenience, accuracy, speed, or other criteria determine the location of each component.
- Components having related functions should be grouped.
- Controls and displays should be grouped according to their place in a critical set of operations. The important controls should be positioned in the most convenient location for rapid and easy use.
- Controls are frequently used in a sequence or pattern relationship. In these cases, the controls should be arranged to take advantage of such patterns by locating them close to each other.
- Less frequently used controls should be placed in more distant locations than frequently used controls.

If there is conflict between any control arrangement considerations, compromises will have to be made. Generally, the frequency of use and sequence of use should be given first consideration. Avoid control or display arrangements where frequent physical transfers (of the entire body, eye, hand, or body members) from one place to another would be required.

### Control evaluations

There are other questions to consider when designing controls:

- Where are the controls placed? Can they be easily reached? Are they spaced far enough apart? Are they labeled and coded?
- What type of control is used? Is it compatible with user needs?
- Do the controls themselves present a hazard?
- What body limbs are involved? Are any of the muscles used to operate the controls being overloaded?
- Are like control operations similar in design and function? How standardized are the controls?

## ESTABLISHING AN ERGONOMICS PROGRAM

While safety professionals can individually apply ergonomic approaches to individual safety and health problems, the overall effort is not nearly as effective as when all areas within an organization include ergonomics in their efforts. When industrial and production engineers, medical, industrial relations, and other areas participate in the program, synergistic results are achieved. Instead of reacting to problems, a pro-active approach can be used to prevent problems from occurring.

The role of ergonomics within an organization can center around three areas; it can be used as:

- A means to reduce workers' compensation and associated costs, and improve worker productivity and safety.
- An enhancement to traditional approaches used by engineers, safety professionals and others within an organization.
- An opportunity to show workers that management cares and solicits their help in the program.

To establish an occupational ergonomics program, three basic approaches should be used:

- Obtain management commitment
- Establish training programs
- Encourage participation by those affected.

**Obtain management commitment.** This is mandatory in establishing and continuing a successful occupational ergonomics program. If the highest level of management is actively committed to the program, then the program becomes a priority of the organization and is reflected throughout.

**Establish training programs.** This includes creating an awareness throughout an organization of what ergonomics is and how it can be used to solve safety and health problems. Training or education programs are needed for those who will be expected to recognize ergonomic problems and for those involved in designing solutions. The program content and emphasis will depend upon the duties of those involved.

Generally, the following people will need some type of training (awareness-education):

- Management should receive awareness training covering the organization's over-all ergonomic safety and health problems to encourage its commitment to the program.
- Engineering staff, such as industrial and selected plant engineers, should receive specialized training on ergonomic problem recognition and problem solving.
- Supervisors and wage workers should receive training on problem recognition.
- Ergonomic committee members should receive training on problem recognition and problem solving.
- Other staff personnel (including those from the medical, personnel, and industrial relations departments) participating in an ergonomics program need awareness and problem recognition training.

The content and length of the training will depend upon what will be expected of them.

**Encouraging participation by those affected** by the program

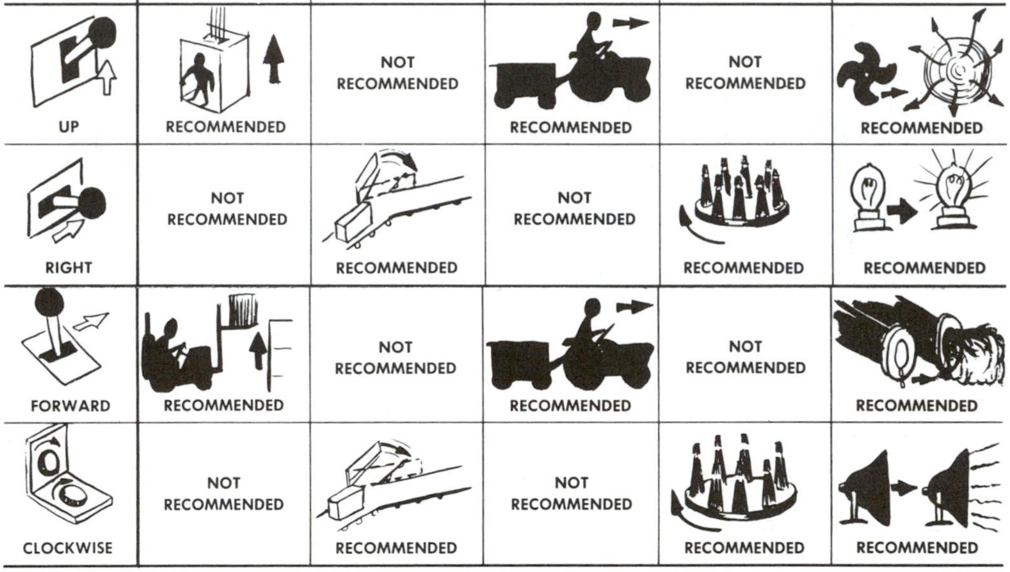

**Population Stereotypes—Behavioral Response**

**RECOMMENDED RELATIONSHIPS BETWEEN CONTROL MOVEMENT AND SYSTEM OR COMPONENT RESPONSE**

| CONTROL MOVEMENT | SYSTEM (OR EQUIPMENT COMPONENT) RESPONSE | | | | |
|---|---|---|---|---|---|
| | DIRECTIONAL | | | | NONDIRECTIONAL |
| | UP | RIGHT | FORWARD | CLOCKWISE | INCREASE* |
| UP | RECOMMENDED | NOT RECOMMENDED | RECOMMENDED | NOT RECOMMENDED | RECOMMENDED |
| RIGHT | NOT RECOMMENDED | RECOMMENDED | NOT RECOMMENDED | RECOMMENDED | RECOMMENDED |
| FORWARD | RECOMMENDED | NOT RECOMMENDED | RECOMMENDED | NOT RECOMMENDED | RECOMMENDED |
| CLOCKWISE | NOT RECOMMENDED | RECOMMENDED | NOT RECOMMENDED | RECOMMENDED | RECOMMENDED |

**Figure 10-12.** Studies show there is stereotypical expectancy for control operation among the general population. When controls operate as people expect, safety and efficiency are increased. The illustration shows the expected and recommended response from particular types of controls. (Printed with permission from C.T. Morgan, et al. *Human Engineering Guide to Equipment Design*. New York: McGraw-Hill Book Company, 1963.)

is the third approach to a successful program. People who participate in a decision tend to become committed to it.

### Organization structure and activities

There are several organizational forms which an occupational ergonomics program can take. The one discussed here is only an example. The structure actually used will depend upon management style and organizational characteristics. However, all should include the three approaches for establishing an occupational ergonomics program (covered earlier).

The first step is to appoint someone to administer and coordinate the ergonomic effort. This can be either a part- or full-time effort depending upon resources and extent of ergonomic problems. Ideally, the person appointed should have some knowledge of safety and health, industrial engineering, and medicine. Since ergonomics is a multi-disciplinary effort, it is not necessary that the appointment be out of the safety and health area. The ergonomic coordinator-administrator acts as the resident ergonomic expert and is a resource person others can consult.

Next, establish an ergonomics committee composed of representatives from management, technical staff, and wage employees. Membership might include representatives from: safety and health, industrial engineering, plant engineering, maintenance, medical, personnel, industrial relations, a wage employee representative and the ergonomic coordinator-administrator.

To make sure committee member participation is meaningful, all members should receive basic ergonomic training—not to make them experts—but so they understand the basic prin-

**Common to Nearly Every Person**

- Handles used for controlling liquids are expected to turn clockwise for off and counter-clockwise for on.
- Knobs on electrical equipment are expected to turn clockwise for on, to increase current, and counter-clockwise for off or to decrease current. (Note this is opposite to the stereotype for liquid.)
- Toggle switches are expected to turn "on" when flipped up, "off" when flipped down.
- Certain colors are associated with traffic, operation of vehicles, and safety.
- For control of vehicles in which the operator is riding, the operator expects a control motion to the right or clockwise to result in a similar motion of his vehicle, and vice versa.
- Sky-earth impressions carry over into colors and shadings. Light shades and bluish colors are related to the sky or up, whereas dark shades and greenish or brownish colors are related to the ground or down.
- Things which are further away are expected to look smaller.
- Coolness is associated with blue and blue-green colors, warmth with yellows and reds.
- Very loud sounds or sounds repeated in rapid succession, and visual displays that move rapidly or are very bright, imply urgency and excitement.
- Very large objects or dark objects imply heaviness. Small objects or light-colored objects appear light in weight. Large, heavy objects are expected to be at the bottom. Small, light objects are expected at the top.
- People expect normal speech sounds to be in front of them and approximately head height.
- Seat heights are expected to be at a certain level when a person sits down.

*Source: Woodson and Conover (1964).*

ciples. The committee should periodically meet to discuss ergonomic problems occurring in the work environment and to develop plans for eliminating or reducing them as well as to provide leadership to the effort.

Committee activities can include duties such as: developing workplace and methods design guidelines incorporating ergonomic approaches; developing ergonomic training guidelines for management, engineers, supervisors and wage employees; tracking cumulative trauma and other disorders; and developing ergonomic purchasing guidelines for machines and hand tools.

Each functional area represented on the committee contributes to the overall effort. The safety and health department usually provides injury analyses that track ergonomic problem areas. It also monitors overall performance. The medical or first aid department reviews any ergonomic-related medical problems they have treated. The ergonomics committee encourages plant workers with perceived job–related ergonomic problems to attend specific committee meetings and discuss their problems. This serves to uncover ergonomic problems that may go undetected through normal reporting procedures.

The ergonomics committee may appoint a task force committee to handle specific projects which the committee feels will require more time and specialized talent than they can offer. An example of a task force approach is to implement a program to reduce carpal tunnel syndrome problems on an assembly line. When the project is completed, the task force is disbanded.

There are two stages of growth that almost every ergonomics program goes through—the reactive stage and the pro–active stage. The reactive stage is characterized by reacting to individual problems after they occur. This approach is not necessarily bad, but it is inefficient and fragmented because time and resources are used to "put out the fire" instead of preventing it.

In the pro-active stage, planning and organization is used to prevent ergonomic–type problems from occurring. This involves developing a program approach that permanently influences the design of jobs and equipment. For example, adopting ergonomic design guidelines requiring all lifting tasks to be engineered so that 90 percent of the workforce can perform them with reasonable safety. This would permanently influence the number of overexertion injuries.

Establishing an ergonomics program is not an individual department effort—although it might start out that way. It involves the efforts of several staff departments including supervisors and wage employees. The most successful ergonomics programs utilize the efforts of many functions and encourage worker participation.

## SUMMARY

Using ergonomics–human factors engineering approaches will help solve many occupational safety and health problems. However, the lack of response to using these approaches is one of the outstanding failures in occupational and product safety efforts. This neglect is partially due to the lack of understanding of the role that ergonomics–human factors engineering can play in occupational and product safety programs.

Clearly, ergonomics–human factors engineering can be combined with traditional safety and health approaches to address these problems. Among the benefits that can be expected are:

- Fewer accidents resulting in injury or damage to property
- Minimized redesign and retrofit after the system is operational—if applied at the design phase
- Fewer performance errors
- Greater system effectiveness
- Reduced training time and cost
- More effective use of personnel with less restrictive selection requirements.

The application of ergonomics–human factors engineering principles is indispensable if optimum safety and system effectiveness are to be achieved.

## REFERENCES

Astrand, P. O., and Rodahl, K. *Textbook of Work Psychology,* 3rd ed. New York: McGraw-Hill Book Co., 1986.

Alexander, D. C., and Pulat, B. M. *Industrial Ergonomics: A Practitioner's Guide.* Norcross, Ga.: Industrial Engineering and Management Press, 1985.

Chaffin, D. B., and Anderson, G. B. J. *Occupational Biomechanics.* New York: John Wiley & Sons, 1984.

Chapanis, A. *Man-Machine Engineering.* Belmont, Calif.: Wordsworth Publishing Co., 1965.

Diffrient, N., et al. *Humanscale 1/2/3.* Cambridge, Mass.: The MIT Press, 1974.

Eastman Kodak Company: Human Factors Section. *Ergonomic Design for People at Work.* Vol. 1; Belmont, Calif.: Lifetime Learning Publ., 1983. Vol. 2; New York: Van Nostrand Reinhold Co., 1986.

Grandjean, E. *Fitting the Task to the Man; An Ergonomic Approach,* 3rd ed. London: Taylor & Francis Ltd., 1979.

Human Factors Society, P.O. Box 1369, Santa Monica, Calif. 90406. *Human Factors* (journal).

Konz, S. *Work Design: Industrial Ergonomics,* 2nd ed. Columbus, Ohio: Grid Publishing Co., 1983.

Kroemer, K. H. E. *Material Handling: Loss Control Through Ergonomics,* 2nd ed. Schaumburg, Ill.: Alliance of American Insurers, 1983.

Kroemer, K. H. E., Kroemer, H. J., and Kroemer-Elbert, K. E. *Engineering Physiology: Physiologic Bases of Human Factors/Ergonomics.* Amsterdam and New York: Elsevier Press, 1986.

McCormick, E. J. *Human Factors in Engineering and Design,* 5th ed. New York: McGraw-Hill Book Co., 1982.

Morgan, C. T., et al. *Human Engineering Guide to Equipment Design.* New York: McGraw-Hill Book Co., 1963.

National Institute for Occupational Safety and Health. *Work Practices Guide for Manual Lifting,* DHHS Publ. No. 81-122. Washington, D.C.: U.S. Government Printing Office, 1981.

Rodgers, S. H. *Working with Backache.* Fairport, N.Y.: Perinton Press, 1984.

Salvendy, G., ed. *Handbook of Human Factors.* New York: John Wiley & Sons, 1987.

Tichauer, E. R. *The Biomechanical Basis of Ergonomics.* New York: John Wiley & Sons, 1978.

Van Cott, H. P., and Kinkade, R. G. *Human Engineering Guide to Equipment Design,* revised ed. Washington, D.C.: U.S. Government Printing Office, 1972.

Woodson, W. E. *Human Factors Design Handbook.* New York: McGraw-Hill Book Co., 1981.

Woodson, W. E., and Conover, D. W. *Human Engineering Guide for Equipment Designers,* 2nd ed. Berkeley, Calif.: University of California Press, 1964.

# 11

# Human Behavior and Safety



THE INDUSTRIAL SAFETY PROFESSIONAL assists line management in achieving maximum production by preventing or mitigating work-related fatal or injury accidents. As discussed in the previous chapter, the occupational environment is composed of interacting components, such as the worker, materials, and equipment. A comprehensive safety program addresses all aspects of the work environment and recognizes that each workplace component interrelates to place safe behavior-management at the center of the program. Consequently, a major responsibility of the safety professional is directed toward changing worker behavior to facilitate safe working conditions.

This does not reduce the importance of the other facets of a sound safety program—the safety professional is also concerned with safe design and plant layouts, safety devices on machines and use of such devices by employees, the wearing of safe clothing and use of protective equipment—all of which contribute to the reduction of disabling accidents.

When, however, in spite of every precaution on the part of the manufacturer of equipment, the supplier of materials, the supervisor, and the safety professional, accidents still occur, the human element emerges as an important factor. It calls for working with supervisors and other line management, as well as individual employees, if accidents are to be reduced.

This chapter is designed to promote understanding of human behavior in the work environment. It complements the previous chapter, which emphasized designing equipment, controls, and jobs to fit the limitations of the human being. This chapter explains why some people have more limitations than others.

## PSYCHOLOGICAL FACTORS IN SAFETY

Many topics concerning the industrial psychologist are too detailed to be covered in this chapter. Some play a direct part in the success or failure of sound personnel procedures in industry, but not all are directly related to safety, nor do they fall within the assigned duties of the safety professional. They can, as part of the regular personnel procedures, indirectly contribute to safety in the shop.

Psychological factors that influence safety program success are described in the following paragraphs.

**Individual differences.** Individual difference is an ongoing industrial problem. These differences are constant and obvious; yet, there are factors common to all people, and therefore useful when dealing with work groups.

**Motivation.** Understanding what motivates people is important. To want something is motivation, but not to want something also requires motivation. To use a safety device to protect one's fingers from a saw is, perhaps, indicative of motivation for safe practices, but the desire to ignore a safety device because it might decrease production is also motivated. Conflicting motivations also should be considered in any attempt to understand human relations.

**Emotion.** Humans frequently respond to their emotions. While emotions can be constructive at times, they also can be destructive—working to the detriment of both the individual and the safety program. Emotion also can interfere with the thought processes, resulting in behavior that conflicts with a rational approach.

**Attitudes and attitude change.** Industry has recognized the effect attitudes can have on production, plant morale, turnover, absenteeism, plant safety, and the like. As a result, management has spent much time and money trying to determine workers' attitudes. Measuring, developing, and changing attitudes constitute a major problem for the personnel staff and psychologists—one of extreme importance to the safety professional.

**Learning processes.** Finally, there should be concern with learning processes. Learning starts on the first day of birth, and plays a major role in all of the topics previously mentioned. One cannot understand motivation, attitudes, emotions, or even individual differences without some consideration of the learning process involved in bringing them about.

Much of the success of a safety program depends upon its acceptance by those to whom it is directed. Program acceptance in turn is dependent upon an understanding of the psychological factors which influence program success.

The safety professional and supervisor must effectively explore and find factors common to the group that can be used to promote safe conduct on the part of all workers. The basic question is, "What factors associated with human behavior can be used to increase the effectiveness of safety programs?"

Each of the five psychological factors previously listed are important enough to discuss separately.

## INDIVIDUAL DIFFERENCES

When a chemist analyzes a chemical compound, its exact nature and composition can be accurately specified. When this compound is used, exact behavior can be expected. The action and reaction of the sample always is the same as other analyzed samples of the compound.

When studying human behavior, however, the psychologist is not dealing with the same degree of certitude as is the chemist. The psychologist does not know the composition of the agent being dealt with—in many instances, there is very little knowledge of the subject's past history.

In addition, the behavior of one person is not the same as the behavior of another person. Person A in the sample is not equivalent to Person B, in the way one cubic centimeter of distilled water equals all other cubic centimeters of distilled water.

The known fact that people differ has been referred to as the "personal equation" or, more commonly, "individual differences." The personal equation presents many problems for both the safety professional and supervisor. The case is not hopeless, for within the framework of individual differences, certain general patterns common to a group do exist.

For example, human behavior is typically motivated. Regardless of what the individual does, there is usually some purpose underlying the behavior. The purpose, for example, may be to reduce some basic tension that must be resolved. The degree and nature of this tension depends in part upon the individual's values and perceptions. More about motivation later.

Here is another example. Even though each child develops at a different rate and to different levels of efficiency, each crawls before walking and forms words before forming sentences.

The general model used to describe such modes of development is the following:

$$B = f(S,E)$$

Behavior (B) is a function (f) of the present situation (S) and all previous experiences (E).

Whether or not an employee works safely depends upon (1) the present situation—is the employee rushed? fatigued? in poor health?—and (2) past experiences—were accidents avoided in the past? What amount of training does the employee have? Under this general model, the *E* variable and the *S* variable comprise individual personality.

### The average-person fallacy

In the previous paragraphs, the point was stressed that each person differs from each other person (although there are certain common characteristics). The fact that individual differences exist should not be news. What is important is how one deals with individual differences.

Figure 11-1 presents the distribution of scores in a normal curve. Many human characteristics (for example, anthropometric characteristics or IQ) are assumed to be distributed according to the normal distribution (often referred to as the "normal curve"). Note that in such a distribution, half of the persons are below the mean (the arithmetical average) and half of the persons are above it.

Often the manager of an enterprise or director of a safety program realizes individual differences exist but that it is physically impossible to handle every person individually. The manager or director may erroneously attempt the next best thing—appeal to the "average person."

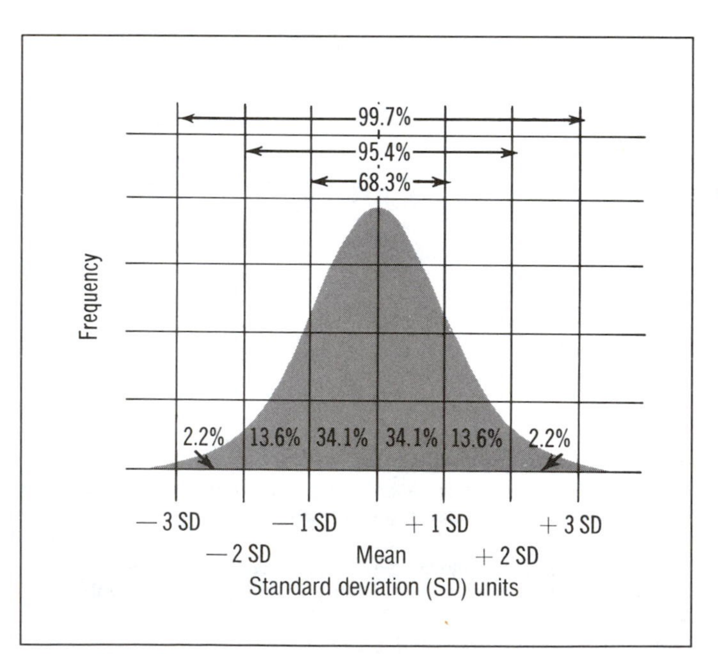


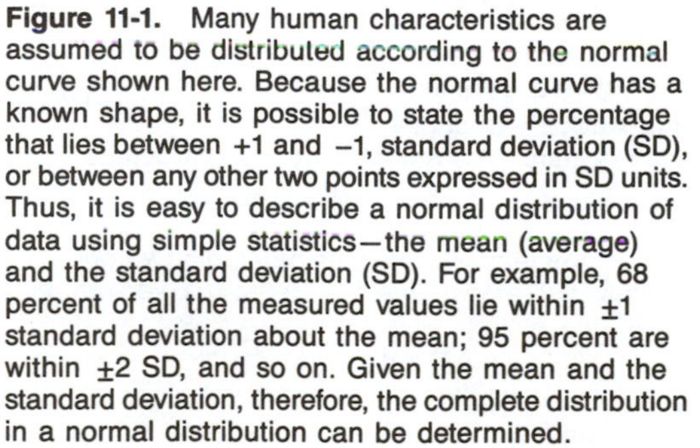

**Figure 11-1.** Many human characteristics are assumed to be distributed according to the normal curve shown here. Because the normal curve has a known shape, it is possible to state the percentage that lies between +1 and −1, standard deviation (SD), or between any other two points expressed in SD units. Thus, it is easy to describe a normal distribution of data using simple statistics—the mean (average) and the standard deviation (SD). For example, 68 percent of all the measured values lie within ±1 standard deviation about the mean; 95 percent are within ±2 SD, and so on. Given the mean and the standard deviation, therefore, the complete distribution in a normal distribution can be determined.

Unfortuntely, inept usage of the mean as the primary descriptive value of a population sample has resulted in misinterpretation and misuse of anthropometric data, culminating in a misconception called the average person. People, in reality, are highly unlikely to be average. For example, in body dimensions, which can be measured more accurately than emotional, behavioral, or intellectual characteristics (see the previous chapter), less than four percent of a test group had three common average dimensions, and less than one percent was average in five or more dimensions at the same time. Therefore, when an appeal is aimed at the average, it misses much of the population.

A better approach is to use percentiles. When designing the system, instead of setting the standards to fit the average, design it to fit all but the upper 5 percent and the lower 5 percent. The system will then fit 90 percent of the population.

The use of percentiles (for example, between the 5th and 95th percentile), rather than averages, has long been the technique for relating individual test scores to a group as a whole, and also a guiding rule in the design of man-machine relations.

### Fitting the person to the job

Very often when replacing or relocating personnel, managers strive to find the individual possessing most of the characteristics deemed necessary. Job specifications and employee requirements are given in terms of a minimum, but very seldom are psychological factors expressed within a minimum *and* maximum; see the previous chapter.

When looking for someone, often the best qualified is considered to be the one who is most intelligent, most loyal, biggest, or whatever. In other words, they look for the perfect rather than the right person. For example, a person with a below-average IQ might be more desirable for a monotonous job than a person with an average or above-average IQ. The amount of boredom and monotony may be less for the less bright, thereby eliminating carelessness, or other factors associated with poor productivity.

Another example is the person who comes to work under any condition, whether ill or not, may be as undesirable as the one who uses every excuse to take off. The worker with a high fever, bad cold, or sore back is a potential hazard, especially if under heavy medication.

**Physical characteristics of the individual.** Reaction time, psychomotor skills (for example, manual dexterity), and visual abilities seem to have at least some bearing on safe performance. While the extent to which they are directly responsible for accidents is neither clear nor constant, it appears a certain minimum degree of physical competence is required for successful, accident-free performance. Some types of jobs demand superior physical abilities while others do not. Many jobs can not be redesigned to enable a person with physical handicaps to perform efficiently and safely.

Most of the research to date has individually considered these factors. More recent studies have examined the connection between combinations of physical shortcomings and accidents. Such studies were initiated because investigators realized that physical and psychological abilities operate not as single, discrete items, but rather in interacting combination. For example, one study investigates the general perceptual skills and accidents, and not with just the separate abilities that contribute to perception.

**Individual personality.** Personnel people in industry screen candidates on the basis of relevant characteristics required by the specific job for which application is made. In many cases, both (1) physical characteristics, such as size, visual acuity, and steadiness, and (2) personality characteristics are important, so many individuals are not considered because they do not have all the necessary qualifications.

When designing equipment, human engineering experts consider physical limitations as well as other human charcteristics in an effort to make machines as nearly perfect as possible. See Chapter 10 for a discussion of human factors engineering in relation to occupational safety.

Where hazards cannot be eliminated, safeguarding provides protection against potential failure of people to correctly utilize equipment. Both safe design and safeguarding minimize the effect of individual differences on accident frequency and severity.

### Methods of measuring characteristics

Regardless of the technique used to screen, place, and motivate employees, a method of measuring program effectiveness is necessary. Techniques used to obtain feedback range from the across-company accident rates to the within-company approach of safety sampling, or critical incident technique.

Measuring techniques can be assessed by determining reliability and validity.

**Reliability** refers to consistency of measurement. The reliability of a given measurement, such as by a test or instrument, can be estimated in various ways. In general, however, the concept of reliability refers to measurement stability, whether it is (1) assessed across time or between settings, or (2) assessed using the same or a different group of individuals, or (3) assessed for internal consistency or consistency between alternate forms of the same test or instrument.

Underlying the concept of reliability is measurement error. All gages used to assess performance will have, in differing degrees, associated measurement error. Measurement error refers to the estimated fluctuations likely to occur in performance by an individual or group as a result of irrelevant or chance factors.

The degree of reliability associated with an instrument or test typically is expressed statistically by a correlation coefficient. The sign of this coefficient refers to direction of the relationship (either positive or negative); the value (for example, $r = 0.89$) is an index of the magnitude of the relationship. Therefore, the greater the magnitude of the reliability coefficient (and thus the less the measurement error), the more stable or consistent the result obtained by the instrument will tend to be (Anastasi, 1982).

**Validity** refers to the degree the test or instrument measures what it is expected to measure. The validity of a test or instrument can be assessed in various ways, such as (1) the relevance or plausibility of items (or the overall instrument) with regard to the given behavior (face validity), (2) whether all aspects within the rubrics of a given concept are adequately covered (content validity), (3) the extent to which the test or instrument can be said to measure a specific construct or trait (construct validity), and (4) a demonstrable relationship between the performance as measured by the given test or instrument and some other related behavior (such as IQ and school aptitude) (Anastasi, 1982).

Once again validity is expressed as a correlation coefficient. A negative sign, however, is as useful as a positive coefficient. An illustration might be testing for a job requiring very little mental ability—it can be negatively correlated with an IQ test so the desirable workers for the job are those with the lower test scores.

A measurement can be reliable without being valid, but a valid measurement also must be reliable. To illustrate that a reliable measure is sometimes not valid, consider a yardstick. A yardstick is very reliable—it gives a consistent measurement every time it is used. It is also valid for measuring the length of a table. But if a yardstick is used for weighing the table, it is no longer a valid measure.

In addition to reliability and validity, a measuring technique must be practical. A technique can possess high validity but be so cumbersome and intricate that it can only be used in special situations, and then only by highly skilled technicians. In spite of its statistical value, such a technique is almost worthless.

**Two sampling techniques** used for evaluating potential accident-producing behavior are (1) the critical incident technique, described in Chapter 3, and (2) behavior sampling.

- *The critical incident technique* involves the following. A random sample of employees is interviewed in order to collect accident information concerning near misses, difficulties in operations, and conditions that could have resulted in death, injury, or property loss. Those participating are asked to describe any incidents coming to their attention; see Figure 3-3 on p. 48. This technique can be useful in investigating worker-equipment relationships in past or existing systems, modifications to existing systems, or in the development of new systems.
- *The behavior sampling* or activity sampling technique involves the observation of worker behaviors at random intervals and the instantaneous classification of these behaviors according to whether they are safe or unsafe. Calculations are then made to determine either (1) the percent of time the workers are involved in unsafe acts, (2) the percent of workers involved in unsafe acts during the observation period, or (3) the percent of unsafe versus safe behavior observed. Various components of a safety program (such as safety lectures, posters, brief safety talks, safety inspections, motion picture films, supervisory training) can be applied and an immediate indication of their influence on unsafe behavior can be obtained.

Recently a series of research studies have shown that feedback can be introduced into this process with demonstrable positive effects on worker behavior and accident rates. Feedback charts are prepared that show the percent of safe behavior observed during each sample. These charts are posted in the workplace and serve as positive reinforcement for safe behavior (see Behaviors versus attitudes later in this chapter).

When screening employees or potential employees in regard to their accident potential, caution must be exercised. To date, no systematic screening procedures have been developed that meet both reliability and validity criteria and are adequate for use in all industries or even in a specific industry. While in theory such a screening procedure is possible, the state of scientific knowledge in occupational safety research is too limited to support the development of such screening measures.

## MOTIVATION

Through the interaction of hereditary and environmental factors, each worker is an individual personality. The safety professional must be continually aware of these individualities when dealing with human beings.

There are instruments by which various aspects of human behavior can be evaluated, many of which already are in use by industry. Through psychological tests, interviews, rating scales, and allied aids and techniques, personnel departments have for a long time been evaluating individual differences. Insofar as they are doing adequate jobs, personnel departments are working with safety programs to eliminate job candidates who obviously would be unsatisfactory.

Each day millions of men and women work in the manufacture and distribution of industrial products. It hardly seems logical that if all these people were individual personalities they could work together in harmony. Individual differences alone are not all there is to human behavior. There are some factors operating in all people that allow supervisors, safety professionals, and plant managers to induce work and cooperation for a common cause. The psychologists, when they try to predict and control human behavior, also are concerned with the factors all individuals have in common.

To have all personnel in a company from the president down to the lowest-paid employees productively and safely working together is one of the goals of a safety program. Such cooperative effort must be motivated as an appeal to achieve a common goal, or as a means to another goal of greater importance to the individual. In either case, the result is of value to the safety program.

It is understandable, therefore, that the question most asked by both safety professionals and supervisors is, "How do we best motivate our workers?"

### Complexity of motivation

Many theories contained in psychological literature attempt to unify the concept of human motivation. Although not all of the hypothetical concepts associated with a given theory (nor all of the theories) are equally testable on an empirical level, several general concepts associated with human motivation are suggested.

The motivational problem is perhaps the most complex one in the field of human behavior. It is not possible at the present stage of development to give clear-cut, concise answers to all the questions that might be asked about other people's motivations. Rather, attempts are made to set forth some basic factors and point out where the complexities exist. Hierarchical motivation, multiple motivation, continuing psychosocial need, and conflicting motives must be considered.

**Hierarchical motivation** simply means some needs take a higher priority—there is a hierarchy of motivational factors. (See Figure 11-2.) It has been pointed out that the psychosocial needs (for example, recognition, affection, social approval) take precedence over the biological ones (such as hunger, thirst, sex) when the latter are relatively well satisfied. This is only one aspect of the hierarchy of needs concept (Maslow, 1970).

The safety professional who carefully plans and takes each detail of the program to the boss for approval may be exhibiting an overwhelming desire for achievement and recognition, a desire much stronger than the need for affection. This is particularly true if human relations skills in dealing with workers are concomitantly ignored. The safety professional may be so concerned with personal achievement, and perhaps the accompanying recognition, plans are forced upon workers that they would not accept if given the choice.

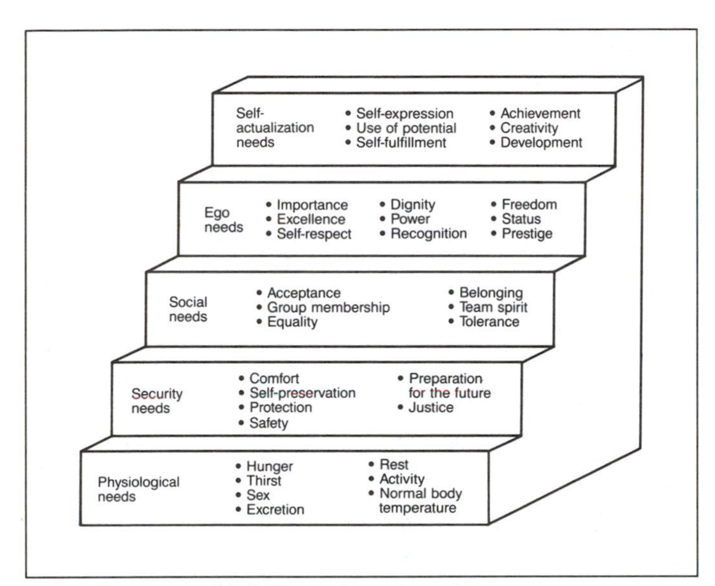


**Figure 11-2.** Abraham Maslow investigated the various needs that people have. Next he arranged them in a hierarchy; note that tissue needs are near or at the base, and that more intellectually satisfying needs are near the top.

Another example is the worker who considers recognition as the major need to be satisfied. This individual, perhaps, would go all out to become a strong militant leader of the union if he were to lose a promotion in the company. The safety director may actually be in conflict with this individual who does not care about the safety director's disapproval and instead values the affection and recognition of his fellow workers. This hierarchy can change over the years.

What is desired more early in life may assume less importance later. For example, the need for achievement and recognition perhaps is greatest in youth, thus giving a drive toward accomplishment.

Later in life the warm affection of friends or the security of belonging to groups can assume major importance if one has already had recognition for past work.

**Multiple motivation** is the second facet that complicates the behavior analysis. People are seldom motivated by one need—many forces operate at any single moment. For example at lunch time, one may eat because of the need for food, or just because it is the usual time. However, an individual might delay lunch in order to be joined by several friends. Thus the individual combined the satisfaction of a psychosocial need and a biological need.

In another case, an employee could desire recognition from fellow workers and so engage in practical jokes and harmful horseplay. At the same time, the employee might be seeking recognition from the supervisor by working hard on the safety committee. The need is the same but the means of satisfaction are in direct conflict.

This is multiple motivation. It would be to the safety professional's advantage, in the second case, to recognize what need the worker is trying to satisfy, and find ways to channel the behavior to promote recognition from both fellow employees and the supervisor.

**Continuing psychosocial need** is akin to biological need. However, people often assume that a one-time psychosocial need satisfaction should last forever and thus they need not continue to give recognition, or affection, or social approval.

For all people, these needs continue throughout life. Need satisfaction is always sought, but not always attained. It also should be apparent that the satisfactions sought by individuals will not necessarily be for the same needs. At one time, the motivation may be for social approval while at another time the need most requiring satisfaction might be affection. Thus, in dealing with people, one must recognize that what was effective yesterday may not work today, although the satisfaction-seeking behavior is to some degree similar.

Being aware that needs continually want satisfaction makes it somewhat easier for the safety professional or anyone else to deal effectively with people in the work situation. There can be daily variance and this makes it necessary for the safety professional to cultivate a personal ability, and the ability of supervisors, to provide people with the required type of satisfaction.

Since they spend so much time together and with workers, the safety professional and supervisor come to know each other and the workers very well. This should provide the key to determining what a worker's behavior at the moment means in terms of need satisfaction. This key is based on little cues in the individual's behavior that they have learned over the years to recognize, if they have taken the time and made the effort to know their people.

Careful observation is required. Safety professionals should make certain that any supervisory training program includes effective worker observation methods.

**Conflicting motives** constitute another major problem in motivation—needs can be in conflict with one another. Seeking affection could lead to behavior that might be different with fellow workers than with supervisors. It can become necessary to determine which source of affection is the most important to an employee before doing anything.

People can internalize these problems to such a degree that the resultant physical stress or even behavior is completely inconsistent with that typically expected of them. For example, an assistant safety director may strongly desire to become the safety director and yet at the same time be fearful of the duties, the responsibilities, and obligations of the job. The promotion would mean more prestige, more money—in all, a better way of life. Here then is a serious conflict. This individual's answer will depend greatly on his or her background experiences. One of the things he might do would be to quit the job and seek employment elsewhere. He might, on the other hand, seek training to better qualify for the position; there are many choices available.

The same applies to the worker on the line. He or she may strongly desire the approval of the supervisor and at the same

**Table 11-A.** Summary of Different Surveys on Job Satisfaction In Order of Importance of Different Factors

| | *Women Factory Workers* | *Union Workers* | *Nonunion Workers* | *Men* | *Women* | *Employees of Five Factories* |
|---|---|---|---|---|---|---|
| Steady work | 1 | 1 | 1 | 1 | 3 | 1 |
| Type of work | | | | 3 | 1 | 3 |
| Opportunity for advancement | 5 | 4 | 4 | 2 | 2 | 4 |
| Good working companions | 4 | | | 4 | 5 | |
| High pay | 6 | 2½ | 2 | 5½ | 8 | 2 |
| Good boss | 3 | 5½ | 5 | 5½ | 4 | 6 |
| Comfortable working conditions | 2 | 2½ | 3 | 8½ | 6 | 7 |
| Benefits | | 5½ | 6 | 8½ | 9 | 5 |
| Opportunity to learn a job | 8 | | | | | |
| Good hours | 9 | 7½ | 7 | 7 | 7 | |
| Opportunity to use one's ideas | 7 | 7½ | 8 | | | |
| Easy work | 10 | | | | | |

time desire to remain an accepted member of the work group. If the work group minimizes the importance of the safety program, this worker now is in conflict. The worker may follow the group, seek the approval of the management, or do something entirely inconsistent with either.

Some conflicts can simply be solved, for the alternatives lead to positive need satisfactions regardless of which way the individual goes. These are no problem. Others may have at one pole a positive satisfaction and at the other a negative or unwanted result. This is no problem either for the obvious choice is the one that is satisfying to the individual. Those choices that *cause* the problems are the ones that put an individual in a dilemma.

The solution depends upon how the individual has learned to work out such situations. Whether he runs away or faces up to the problem will give the safety professional or supervisor a clue as to what will occur in the future in a similar situation.

### Job satisfaction

In the interest of further understanding motivation in the work environment, many studies have been conducted to determine what constitutes job satisfaction. Generally, these studies assess the job elements workers claim contribute to their job satisfaction (or dissatisfation). The results of these investigations suggest that satisfaction of psychosocial needs rather than physiological needs may be the major motivational aspect of job satisfaction.

Table 11-A presents the results of various surveys of job satisfaction. The numbers represent rankings of the factors considered in each study. Because different language and alternatives were used in each survey, the factors have been paraphrased to represent the various elements.

The results of these surveys suggest that high pay is not in itself a primary job motivator. Although workers expect a just and equitable income, they appear to expect only what others would be paid for comparable work. While the worker might feel dissatisfied if underpaid, higher pay alone does not guarantee job satisfaction (Clark, 1958; Adams, 1965).

On the other hand, steady work or job security does appear to be a primary job motivator. Workers want the security of knowing if they perform their jobs well, they will have a job in the future. Job security, as a component of job satisfaction, can explain the willingness of a worker to maintain a low-paying, stable job instead of accepting a higher-paying, less-stable job.

Other factors appearing important as job satisfiers include—type of work, opportunities for advancement, and good working companions. Note that all of these factors seem related to the psychosocial needs of feeling important and belonging to a peer group that is acceptable to the worker. Likewise, comfortable working conditions (rated high by a number of employees) are probably associated with a desire of the worker to be humanely treated.

The results of these surveys on job satisfaction are important when considering the safety program within the context of personnel policy. Inasmuch as the safety program is designed to ensure the well-being of the employee, it helps to maintain the employee's continued ability to do the work (which, in turn, gives job security).

Likewise, the safety program represents management's interest in the working companions and working conditions of the employee. All of these aspects of safety programming should be anticipated and incorporated into the approach which is taken with both supervisory staff and the employees. Honest and sincere positioning of the safety program to enhance employee welfare makes practical sense in light of current knowledge regarding job satisfaction.

### Management theories of motivation

As previously mentioned, there are many theories in psychological literature addressing human motivation. Within this literature, specific theories have evolved with special reference to management as it exists in industrial organizations. Two such theories are presented here, although other equally cogent points of view could be discussed.

Because theories of human motivation lack sufficient data to support all their tenets, they might best be viewed as philosophies of management. They are important to the safety professional because, if accepted, they can influence the direction in which management seeks to develop and implement a safety program.

### Theory X and Theory Y

In an attempt to analyze how management personnel view human motivation, McGregor (1985) has evolved the notion that there are two basic ways in which management can view the worker. According to which view management accepts, essentially different management practices will be observed.

**Theory X,** according to McGregor, assumes the worker is essentially uninterested and unmotivated to work. In order to resolve this condition, the motivation must be instilled into the worker by adoption of a variety of external motivation agents. In effect, the worker becomes motivated to work by virtue of the external reward and punishments offered to him.

For example, in order to create motivation, a Theory X manager might use any and all of the following—introduction of rules to constrict the worker's behavior, pay incentives based on production, and threats to job security associated with performance failure.

Thus, under Theory X policy, management uses control and direction as the means of worker motivation.

**Theory Y,** according to McGregor, assumes the worker is basically interested and motivated to work. In fact, work is assumed to be as natural and desirable as other forms of human activity, such as sleep and recreation. Under such circumstances, management is confronted with the role of organizing work so the worker's job coincides with the goals and objectives of the organization. Thus, a Theory Y manager views the task as constructively using the worker's self-control and self-direction as the instruments through which work is accomplished.

By emphasizing responsibility and goal orientation, management capitalizes upon the motivation already present within the worker. If conflicts occur between the worker's goals and management's goals, they are resolved through mutual exploration and discussion. Always, under Theory Y policy, it is assumed the worker's inherent motivation is essential to completion of the organization's goals.

Both Theory X and Theory Y proponents exist, and management systems successfully operating on the basis of each of these theories can be found throughout American industry. What seems important is that whichever system is operating within an organization, it is necessary to recognize that safety programming can be initiated and implemented. While the technique of implementation may differ, Theory X and Theory Y approaches to human motivation can both be amenable to enhancing a worker's motivation to safe behavior.

### Job-enrichment theory

Another analysis of human motivation in occupational environments has been developed by Herzberg (1966). Although quite comparable to the Theory X and Theory Y distinction, Herzberg is explicit in both detail and philosophy. His concept of job enrichment, in many ways an extension of Theory Y, is a current major force in management theory.

The classic approach to motivation concerns itself with changing the environment in which a person works—the circumstances surrounding the individual while working (good or poor lighting, an agreeable or offensive supervisor), and the incentives given in exchange for work (money, a pat on the back, a writeup in the company bulletin, to name a few).

Herzberg believes the concern for environment is important—but not all-important. He says it is not sufficient in itself for effective motivation. This, he contends, requires experiences that are inherent in the work itself.

Herzberg holds there is no conflict between the classic (environmental) approach to motivation and his approach to motivation through work itself. He regards both as important. The classic approach is called hygiene whereas Herzberg's approach is called "job enrichment."

The hygiene approach may be understood by the following analogy—a person is provided with pure drinking water and waste disposal; both are necessary to keep this person healthy, but neither makes him any healthier. By extension, environmental factors always need replenishment. Good rapport may enhance an individual's job satisfaction, but job satisfaction alone will not necessarily result in safe work habits.

Figure 11-3 presents a contrast of the classic hygiene approach to motivation and the job-enrichment approach.

| Hygiene Approach (Classic) | Job-Enrichment Approach |
|---|---|
| Company policies and administration | Achievement |
| Supervision | Recognition |
| Working conditions | Work itself |
| Interpersonal relations | Responsibility |
| Money, status security | Professional growth |

**Figure 11-3.** Contrast of the hygiene approach to motivation with the job-enrichment approach.

Further, treating a person better does not enrich the job, although the individual can become unhappy if not treated well. Again, a salary increase can keep an employee from becoming dissatisfied for a time, but sooner or later another increase will be required.

Although there may be inherent hazards associated with a work environment, such as coal mining or bridge building, the worker has a right to expect controls to prevent the environment from becoming unreasonably unsafe, for example testing and removal of explosive gases, or safety nets and life lines. Such protection might not be motivating because it makes the job safer, but the worker might be very unhappy if he knew no protective effort was being made.

Herzberg's idea that work itself can be a motivator represents an important behavioral science breakthrough. Traditionally, work has been regarded as an unpleasant necessity but it has not been thought of as a potential motivator.

Although automation is helping to phase out the unstimulating aspect of many jobs, a job should provide an opportunity for personal satisfaction or growth. When it does, it becomes a powerful motivating force.

People, Herzberg further theorizes, must be given the opportunity to do work they think is meaningful. Merely telling an individual who is doing a routine job that he is happy and that he is doing something meaningful accomplishes nothing. Job rotation is not the answer, either; it does not enrich a job—it only makes it bigger.

Another point. Herzberg observes, "Resurrection is more difficult than giving birth." Obsolescence must, therefore, be eliminated by continued retraining—not just a once-in-a-while effort. Jobs should be kept up-to-date, and people doing the jobs must be kept up-to-date.

Even though a company provides the hygiene factors, it also must provide a task that has challenge, meaning, and significance. If an unchallenged individual does not quit but rather stays on—it is usually with a resigned attitude and poor morale. That, says Herzberg, is the price a company pays for not motivating people.

In summary, seven principles of job enrichment can be itemized.

- Organize the job to give each worker a complete and natural unit of work
- Provide new and more difficult tasks to each worker
- Allow the worker to perform specialized tasks in order to provide a unique contribution
- Increase the authority of the worker
- Eliminate unneeded controls on the worker while maintaining accountability
- Require increased accountability of the worker
- Provide direct feedback through periodic reports to the worker himself

## FRUSTRATION AND CONFLICT

A motivated individual is one who is attempting to reach a goal. Often, however, a barrier is placed between the goal and the one who is seeking it. When this occurs, the individual becomes frustrated. This differs from mere *lack of satisfaction* of a need, which is called deprivation. Theoretically, the *thwarting of behavior* directed toward a goal results in frustration (Miller, 1962).

- The barrier can come from within the individual. The person who sets impossible goals can become frustrated when unable to reach the goals. The way out may be an accident.
- The barrier can also arise from the environment. An example—if the person mentioned above (who sets a very high personal production goal) is unable to reach such limits because of faulty equipment.
- A third type of frustration is caused by conflict. If two motives somehow conflict, the satisfaction of one means the frustration of the other. For example, the worker who sets high personal production and safety standards, but is unable to meet both under the present system, must satisfy one and sacrifice the other.

It is to the conflict-caused frustration that we will now address ourselves. Basically, there are three types of conflicts—called "approach-approach," "avoidance-avoidance," and "approach-avoidance."

### Approach-approach

As the label implies the "approach-approach" conflict arises when an individual is faced with two goals that are equally attractive, but only one of which is obtainable at the time. An example is the college student who has been asked to go both to a dance and to a swimming party on the same evening.

The approach-approach conflict is the easiest to resolve. No matter what goal the individual selects, a need will be satisfied, without much loss to another need. Usually the individual will resolve such a conflict by satisfying one need first and then satisfying the other. If the person is both hungry and sleepy, for example, the resolution might be to eat first and then sleep.

### Avoidance-avoidance

A second type of conflict is the "avoidance-avoidance" or double-negative conflict. This conflict can arise, for example, when an employee is told to wear heavy fire-retardant outer clothing in the summer heat or else risk the possibility of clothes catching on fire. This employee is, as the saying goes, caught between the devil and the deep blue sea.

Two kinds of behavior generally result from such a conflict—vacillation and flight.

In vacillation, the individual approaches one goal; retreats, approaches the other, retreats, and so on. As the goal is approached, the unpleasant portions of the goal increase, so the individual withdraws. In the example, the worker may stall off doing the job requiring wearing the heavy clothing until the quitting bell sounds.

Another result from the avoidance-avoidance conflict is flight. The athlete may leave the field or the worker may quit his job. Quitting a job, of course, has serious consequences in lost prestige and lost income.

A more common type of fleeing, therefore, is in a figurative sense—like daydreaming and fantasy. The worker faced with an unpleasant task may repress the reality by daydreaming.

The avoidance-avoidance conflict is not as easy to resolve as the approach-approach, nor is it as difficult as this next conflict type.

### Approach-avoidance

The "approach-avoidance" conflict is the most difficult to resolve. The individual is both attracted and repelled by the same goal (Figure 11-4). The worker who is striving for top production inadvertently may take unnecessary risks to achieve that goal. One way out of such a conflict is by realizing that safety and productivity go together.

Approach-avoidance conflict refers to a single goal having both positive and negative attributes associated with it. Other things being equal, the strength of the approach or avoidance

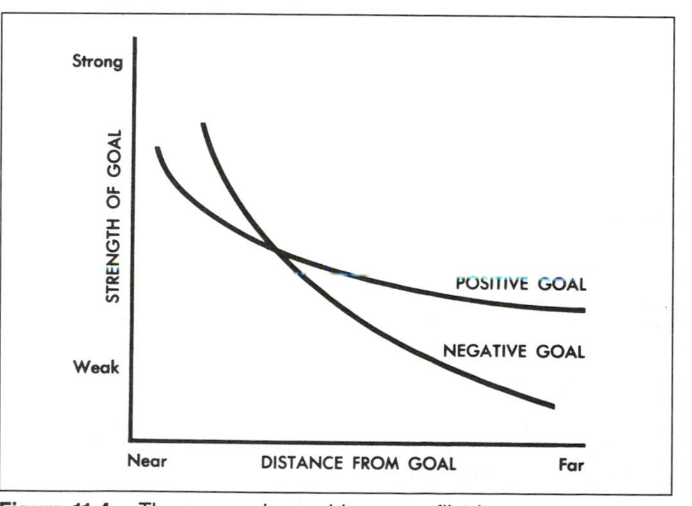


**Figure 11-4.** The approach–avoidance conflict is complicated by the different goal gradients for positive and negative goals.

tendency will depend upon the proximity of the individual to the goal. At point A, for example, the tendency to avoid (or retreat) will be greater than the tendency to approach (since, as shown, the slope of the negative goal gradient is steeper than the positive goal gradient at point A). At point B, however, the tendency to approach the goal will be greater than the tendency to avoid. According to this theory, conflict will lead to retreat, vacillation, or indecision. Theoretically, the approach-avoidance conflict, and more complex double approach-avoidance conflict situations (that is, two goals, each mutually exclusive, having positive and negative attributes associated with each), are often common situations.

### Reaction to frustration

In most cases, the conflicts described will lead to frustration. This frustration often can lead to positive, constructive resolution of the situation. For example, the individual who is facing the conflict of top production, but at increased risk, may develop a new system for processing the product at a faster *and* safer rate.

On the other hand the frustration can lead to some form of negative behavior (Miller & Dollard, 1962).

### Emotion and frustration

Negative emotions can have a disturbing influence upon a person's behavior. Anxiety is one emotional reaction that, because of its distracting influence, its stimulus to heightened reaction, and its generally upsetting effect, can make an individual more susceptible to accidents. This general upsetting feeling can spread to others working in the same situation and create an atmosphere not conducive to safe procedures.

Anxiety is, however, only one of the emotional patterns shown by individuals in the face of frustration. The anxious individual is worried, has circular thinking, and is fearful in such a way as to make behavior inadequate to reach the goal. Even when the individual realizes that this behavior is accomplishing nothing, the person is unable to find a method to solve the problem.

In frustration, some people become angry, some fearful, and some accept frustration as a challenge and attempt to solve the problem. Most people who emotionally react to frustration and threat find it extremely difficult to cope with life situations. It is not necessary to discusss the physiological pattern that develops for everyone is familiar with the feelings experienced during fear, anxiety, anger, shame, and other emotions. Rather, one needs to recognize that over the years people have developed accepted social expressions for emotions. To some degree, one individual can be aware of how another is feeling. This is not to say that one cannot be fooled or that the outward expression always truly indicates the feelings of the individual. Many people learn to mask their real feelings.

However, after day-to-day interaction over a long period of time, one can learn to tell the difference between internal feelings and those that are expressed. This is useful to the supervisor and the safety professional in dealing with individuals in the shop. As they go about the plant or establishment carrying out their normal duties, they may discover those individuals having difficulties and take some extra precautions to avoid accidents.

It seems important to make note of the problem of behavior disorders within the context of employee problems. Individuals whose normal behavior is seriously disrupted or whose ability to react to frustration is seriously impaired may require professional services beyond the scope of the normal supervisor-employee relationship. Although a supervisor should assist in directing the employee to in-plant services or outside agencies, there should not be an attempt to directly deal with psychological problems requiring professional care. While the current movement toward company-sponsored psychological services may grow, it is unlikely to be the standard in all industries in the near future. For a detailed discussion of industry programs for drug abuse and alcoholism, see Chapter 21, Occupational Health Services.

### Non-directive counseling

Although no attempt can be made here to train safety professionals or supervisors to become personnel counselors, they can do much to help reduce the immediate result of strong emotion. The supervisor is in the best position to evaluate the emotional level of an individual worker and to do something about it.

Everyone is familiar with the emotional outbursts of people who have had a trying and upsetting experience. Such expression of emotion can take many forms from an outburst of bad language, a torrent of tears, to physical assault on objects that cannot strike back. The end result is the same—relief from the intolerable tension, and emotional relaxation.

The supervisor can sometimes alleviate work-related tension by providing an employee with an opportunity to talk about whatever is bothering him or her. This may be especially helpful if the problem is work-related or concerns a minor personal problem. Even talking about nothing in particular can be helpful to the tense employee.

The stresses causing emotional upset can occur within the work environment or away from it, but each kind has its effect on the other. A simple statement, such as "How are you today?" or "What's bothering you?," may well be sufficient to set off the verbalization of pent-up emotion. After the discussion, the employee can often evaluate real or imagined problems more clearly and realistically.

To attack the problem at the roots calls for careful listening on the part of the employee's immediate supervisor. During the normal routine every supervisor has an opportunity to talk with individual employees. Through the use of open-ended questions (those that do not allow for a simple yes-or-no answer) an employee is encouraged to discuss problems. The supervisor needs to listen not only to words but the feelings they carry. The supervisor must give attention to the unspoken language of facial expressions, gestures, bodily postures, and the like. This will increase understanding of the attitudes and the strength of feeling attached to them.

Frequently it is impossible to discuss such matters at length in the work area, so the supervisor should set a time and a place away from other employees to allow the individual freedom of expression. Even though the supervisor may not have a completely private office, some place can be found for a quiet face-to-face discussion.

This relationship between the supervisor and the worker must be one of the supervisor's normal everyday functions. Diligent observation and effective and sensitive listening should be tools of every supervisor.

## ATTITUDE AND ATTITUDE CHANGE

Although many existent psychological theories address attitude and attitude change, many basic concepts are still controversial. For example, whether or not attitudes are a precursor (or

**Table 11-B.** Factors Associated with Persuasive Communication

| *Attitude Change: Situation Types* | *Communication Process* | *Attitude Change: Behavioral Step* |
|---|---|---|
| 1. Suggestive | 1. Source | 1. Attention |
| 2. Conformity | 2. Message factors | 2. Comprehension |
| 3. Group discussion | 3. Channel factors | 3. Yielding |
| 4. Persuasive messages | 4. Receiver factors | 4. Retention |
| 5. Intensive doctrination | 5. Destination factors | 5. Action |

Source: McGuire (1968).

antecedent condition) of behavior is largely an unresolved theoretical issue in social psychology. Many theories on attitude formation and change assume an active behavioral role, that is, attitudes consist of three components, knowing-feeling-acting. Thus, such a theoretical approach distinguishes, yet integrates, such concepts as knowledge, emotions, and behavior or behavioral tendencies. Other theories regard attitudes as the delineation of values toward something or someone. Most of these theories agree, however, that attitudes (at least in part) are a disposition to evaluate objects, persons, or situations favorably or unfavorably (McGuire, 1968).

What a response can be is dependent, in part, upon the previous experiences of the individual. For example, if an individual sees a person on the street who resembles a friend, the immediate response may be a smile, friendly gestures, and warm voice quality. When that person is perceived to be a stranger, there is an immediate change to another facial expression, gesture, voice quality, and so on. Anticipated reaction can be set off by a certain look from another person, a manner of speaking, a mustache or the lack of one, the color of hair, or kind of handshake. All individuals have certain feelings about these and may act in accordance to their attitudes, because they have been formed over the years.

Some of these attitudes are latent. The predisposition to respond will lie dormant within the individual until called forth. Given the appropriate stimulus, the attitude shows itself and the behavior exhibited is in accordance with the feelings of the individual. A certain word, for example, connotes different responses in different people. The words "union," "management," "labor," and even "safety" carry with them certain connotations that touch off different attitudinal reactions in individuals, depending upon the kinds of experiences they have had with the subject.

Because attitudes play such an important role in everyday relationships, consideration should be given to their development, their effects on individuals, and what can be done to change them.

### Determination of attitudes

Direct personal experience is thought to be one determinant of attitudes. In other words, experiences that individuals have had in the past, especially those involving strong associative emotions, determine attitudes. For example, the attitude of a worker toward management or the supervisor may be fearful and hostile if the worker has lost jobs for no apparent reason, as he sees it. There might have been sound reasons for the dismissals or layoffs, but to admit this would be a threat to his own pride, and so he puts the blame elsewhere. Because of the effects these dismissals have had on him and his family, such as causing financial stress, he may become very angry. Management to him now is a menacing thing, and thus the cause of his present hostile and fearful attitude.

Many people have had experiences associated with fear, sorrow, pain, or happiness. All of these will tend to make them react the same way to anything that is similar to the original emotion-provoking situation. Although there is only one positive emotion listed among those just mentioned, it is not to be concluded that most attitudes are negative, for there are many positive ones.

### Attitude change

Factors associated with communication-induced attitude change are numerous. In summarizing the large body of psychological literature on attitude formation and change, McGuire (1968) suggested three components of attitude change (see Table 11-B). These include: types of attitude-change situations (column 1), variables associated with the communication process (column 2), and behaviors associated with attitude change (column 3).

- Relevant situational factors include: suggestive situations (where the desired attitude is repeatedly presented), conformity (where social or peer pressure is used to elicit the desired response), group discussion, persuasive messages and intensive indoctrination. Each situation is associated with varying degrees of attitude change. Also, distinctive implications of each are dependent upon the type of communication variables involved.
- Source variables refer to characteristics associated with the person or represented organization presenting the message. The effect of the message on subsequent attitudes can vary as a function of the perceived credibility, attractiveness (for example, liking, familiarity), and power of the message (that is, inclusion or exclusion of basic pro and con issues), the order in which specific issues are presented, and dissimilarities between the presenter and recipient of the message. Channel factors consist of the mode of presentation, for example, direct personal experience, mass media. Receiver factors include active participation, and degree of influenceability of the audience. Destination factors refer to the degree of post-communication message decay across time, and time latency associated with delayed-action.
- According to McGuire (1968), the receiver (or audience) must proceed through the following steps in order for attitude change to occur. These are: attention, comprehension, yielding, retention, and action. According to this model, each step depends on the occurrence of the preceding one.

To add to the complexity, the communication variables, especially in daily situations, can interact with one another. And too, the sequential step process may have more intuitive than empirical support. Nevertheless, this conceptual framework should highlight the implications (such as difficulties or dangers) associated with oversimplifying or making broad unsubstantiated generalizations about the process of attitude change.

**Organization development.** Concern for the amount and rate of change within our technological society has focused on the industrial environment and its ability to withstand such change. Behavioral scientists have evolved an approach, called "organizational development," which attempts to assess a corporation's ability to adjust to such conditions as rapid and unexpected change, growth in size, increasing diversity, and management system problems. Organizational development, which generally has as its goal the implementation of an educational strategy, is designed to alter the attitude and structure of organizations so they can better adapt to a changing technology.

Generally, organizational development uses initial feedback of the employees and management to determine the climate and capacity of the organization to adapt its objectives to the technological environment. Based upon such feedback, an attempt is made to develop organic systems to replace mechanical systems within the organization. Organic systems are characterized by a preoccupation with people as they operate together whereas mechanical systems are characterized by a preoccupation with the structure that operates within a system. Figure 11-5 presents a summary of the differences between mechanical and organic systems. In the final analysis, organizational development rejects bureaucracy as an organizational model and substitutes a model based on interpersonal competence.

| Mechanical Systems | Organic Systems |
|---|---|
| Emphasis upon individual performance | Emphasis on relationships in group |
| Chain of command concepts | Confidence and trust among everyone |
| Adherence to delegated responsibility | Adherence to shared responsibility |
| Division of labor and management | Participation in multimember teams |
| Centralized management control | De-centralized sharing of control |
| Resolution of conflict through grievance procedures | Resolution of conflict through problem solving |

**Figure 11-5.** Organization development seeks to implement organic systems in place of mechanical systems.

In order to implement the ideas represented by organizational development, a number of procedures are used to effect changes in the organization. Since the changes are typically people-oriented, due consideration is given to the need to motivate acceptance of the changes within both management and work groups. Typically, education and training are used to effect reorientations within the organization's staff. These techniques have merit in implementing change only so long as top management concurs with the changes and provides incentives within the organization for their adoption.

Ultimately, organizational development seems to provide the necessary means for preparing an organization for orderly planning for the future. Included in such an effort should be the recognition of how changes in an organization will affect the current safety program. By adopting widespread involvement of all elements of the organization in safety programming, it is likely that safety programs will be able to meet the future needs of modern organizations.

### Structural change

In the previous discussion of attitude formation and attitude change, no direct relationship between attitude change and behavioral change was assumed. In practice, however, safety personnel are interested, for example, in changing an employee's attitude regarding wearing protective equipment only so long as such an attitude change ultimately results in the employee's actually wearing the equipment. While the research literature indicates that a variety of influence efforts (training and counseling for example) can change attitudes, there is much less support for a direct relationship between attitude changes and subsequent behavior changes.

In effect, it is incumbent upon the safety professional to consider techniques other than attitude change when considering changes in the employees' behavior. Evidence suggests it is possible to change behavior by the introduction of structural changes within the work environment. Structural changes are procedures designed to change the organizational constraints operating within work groups. Examples include: changing job contents, modifying the physical arrangements of work, changing worker interaction patterns, and rearranging work procedures.

In each case, other than appropriately introducing the change, it is not necessary to expend the time and effort to change attitudes prior to changing behavior directly. Rather, the introduction of the structural change can modify behavior directly and possibly attitudes will change.

While there has not been an extensive attempt to apply the structural change model in occupational safety, it warrants consideration in the future. Human behavior can be changed by eliciting the change through the very circumstances under which the individual works. Unsafe acts, for example, cannot occur where the conditions of the work and work groups preclude their occurrence.

## LEARNING

Learning underlies much of what makes for differences and similarities among people. Through learning, people have developed certain kinds of psychosocial needs, habitual patterns of behavior, ways of reacting to emotion, and the attitudes they bring with them to industry. It is important to consider learning and the laws that affect it. This is especially vital in any discussion of safety, because training is a major consideration in safety programs.

Often, to change behavior, one must substitute new learning for old habits (Hulse, Deese, and Egeth, 1980). To change behavior in need-satisfaction sequences, one must teach better ways by which the goal can be achieved. In each case, some new learning must be substituted for old.

### Motivational requirements

Repeatedly in everyday living, people recall many things they did not set out to learn. While unintentional learning does occur, it is the intentional kind of learning in which the safety professional primarily is interested.

In educational systems, great emphasis is placed upon making the individual want the knowledge that is available to him. Materials are designed to relate to practical situations. Teachers attempt to make the individual interested in the material as such. To teach a student something about angles or distances, the teacher might use a baseball diamond as an example. Teachers try to tap motivation to increase the possibility that learning will take place.

The safety professional must recognize that if workers are going to learn safe procedures, they must be motivated. To merely point out that accidents cost the company money will not motivate them. Rather, point out the risks of using unsafe work procedures—the probability of serious and painful injury, the possible loss of earning power and the effect on loved ones. The cost, not only in dollars but also psychologically to both worker and family, is of paramount, and personal, importance and will motivate safe work methods.

It is not wise to assume that, because management sees the value of safe procedures, the worker also will. Management may be motivated to start a safety program because it will reduce insurance costs, reduce the amount of waste, and increase the number of units produced. Workers cannot be expected to desire it for the same reasons. In selling a safety program to the employees, one must capitalize on the things that will motivate them. Here the safety professional can capitalize on the needs discussed previously, and probably can find many more that are consistent with the aims of the safety program. Everything that will motivate the worker to learn the right procedures should be used. (The next chapter discusses this in more detail.)

## Principles of learning

Some consideration of basic principles is valid whether the learning is to be done in a college classroom or in a work area. When training procedures utilize these principles, learning is more efficient and thorough.

**Reinforcement.** Through experimentation, psychologists have found that reinforcement can often facilitate learning. In practice, reinforcement can, for example, make work more efficient and more productive. When a worker's pay increases because of more units of production, there is a reward received for learning. This can have negative aspects as well—the worker who figures out a hazardous short cut, which produces more units, may be rewarded for an unsafe act, as was mentioned earlier in this chapter.

It is apparent that the employee must be rewarded only for safe work methods. Higher productivity because of safe work methods satisfies the need for achievement and recognition. This in itself reinforces the learning of the employee, but also spreads its effect to other employees who see this take place. A supervisor by recognizing the same needs, however, can reinforce this learning through praise of greater productivity and telling the worker how much improvement has been shown.

The publicity given to a safety record, a bonus, a promotion, or anything else that satisfies individual needs would serve as a reward to reinforce whatever learning has brought about the desired behavior.

Reinforcement tends to be more effective if it closely follows the desired behavior. Praise, for example, should be given at the time the desired behavior occurs, or if delayed, associated verbally with the desired behavior. This does not mean it must be instantaneous, but it should be within a reasonable length of time. If delay is necessary, certainly let the individual know the reason for the reward.

Reinforcement often will increase the likelihood of a reoccurrence of the desired behavior. Shortcut hazardous methods or any deliberate unsafe acts must not be rewarded. In fact, in such situations reprimands or punishment are more appropriate. In short, unsafe methods should be corrected by full and complete explanation and demonstration of safe work procedures. Such being the case, it also follows that there should be reinforcement of the new correct work pattern as soon as feasible.

This principle also applies to participation by employees in a safety program whether it be through suggestion systems, safety and health committees, discussion groups, or training sessions. In all these areas an individual gains personal worth if his or her opinions are asked for and graciously received.

A safety program that gathers the ideas of all, either individually or by representation, satisfies the need people have for being "in the know." In this way, the safety professional can create a positive atmosphere about the program and a sense of obligation and responsibility for its success. Research has demonstrated when employees feel that programs come from all, there is more chance for success.

Research on the effectiveness of punishment suggests that punishment can have diverse consequences. Punishment generally is thought to be less effective than reinforcement, perhaps because punishment provides indirect cues or information (what not to do) as opposed to positive reinforcement (reward) that provides direct information about the desired behavior (Church, 1963).

Rewarding correct procedures will lead to a more positive attitudinal response on the part of the worker than any punishment. A positive attitude toward training procedures is much to be desired. Both the positive and the negative aspects of reinforcement may generalize over the entire work situation. The supervisor's praise for a particular thing well done may spread over other aspects of the total situation, including training procedures, safety devices, and the safety program. The proper use of rewards thus can lead to efficient methods.

Much of the practice underlying programmed instruction (see Chapter 9, Safety Training) is based upon the reinforcement concept as well as the other principles of learning which follow.

**Knowledge of results.** Closely allied to reward—in fact, one aspect of reward—is knowledge of results. Everyone likes to know how the job is going. Letting employees know how well they are getting along in the training program will likely motivate them to continue training and do a better job. To train a worker and not mention improvement is defeating one's own purpose. It cannot be contended that the only one who needs to know about the effects of training is the supervisor or the safety professional. When the worker knows that this new procedure is helping increase production, reinforcement is being received for learning and effort.

One of the factors that appears in some kinds of learning is a plateau at which learning levels off for some time before again showing an increase. Often some individuals become discouraged and their learning can be retarded. If the trainer understands this phenomenon and indicates to the individual that a leveling off had been expected and that an increase will come later, the discouraging aspect of such a plateau can be avoided, with learning then proceeding more easily and efficiently. Demonstrating achievements being made through production curves, which are in effect learning curves, gives the worker knowledge of results, and this is a motivating factor for future learning.

**Practice.** The safety professional is interested in developing in an individual safe habit patterns that will become almost automatic in his work methods. Merely putting a worker through sufficient training sessions is not enough. Despite any apparent mastery, the next time through the work, one or more mistakes may be made. To make sure habit patterns are firmly entrenched, the worker must practice. The Job Instruction Training (JIT) programs include one of the important aspects of training, namely, follow-up by the supervisor. This is for no other purpose than to ensure mastery. Within a reasonable time, depending on the job complexity, this follow-up can become unnecessary.

Take, for example, a simple task such as bicycle riding. A child, in learning to ride a bicycle, needs to know how to balance, how to get on, how to get off, and how to stop, and must learn all of these things as part of the total process. Perhaps the process begins with balancing, then the start and stop, and then how to get on and off. Even after accomplishing all of these, the child should not be left alone. Several more sessions are needed to make certain the task is being correctly done. Each ride the child takes is another practice session. This is reinforcement by practice. So it is with the individual on the job.

**Whole versus part learning** has been a knotty problem for industrial trainers for many years. Whether trainers should teach the procedure as a whole, or break it down and teach it part by part, is the question. There is no best answer. Both methods have advantages and disadvantages depending upon: job complexity, the trainee, and the kind of job breakdown used. Perhaps a combination of the two is best, using the whole method, but with sufficient flexibility to emphasize meaningful parts of the task wherever necessary.

**Meaningfulness.** Studies in verbal learning have demonstrated the importance of the meaningfulness of the material. Meaningfulness is important in safety education because the worker needs to understand why a certain procedure is better than another. Adequate explanation, how a given movement or change in position can eliminate hazards with no decrease in production, gives meaningfulness to the procedure. With this understanding, workers will be motivated to learn the safe procedure. Without it, they will be inclined to use their own methods until they learn, perhaps by an accident, the inadequacy of unapproved work procedures.

The safety professional and line supervisor should not forget the advantage of workers' understanding the reason for having protective clothing, safety devices, safety meetings, and discussions, as well as the need for full and complete accident reports.

Meaningfulness to management and workers is understanding the value of a reduction of accidents, fewer disabilities, and retention of earning power.

**Selective learning.** Out of each day's many experiences, people select those they desire to retain. This probably is related to motivation more than to anything else, and for that reason motivational aspects of a training program need to be considered. Safety trainers must be sure the workers retain the most important facts. Relating subject matter to individual needs will ensure the proper selection.

**Frequency.** Everyone does best those things he or she practices most. This principle is certainly important to the safety professional for it emphasizes the necessity of frequent applications of safety rules and regulations in the training program. Frequent reference to the various kinds of problems, hazards, and procedures to eliminate accidents ensures greater effect than just a one-exposure routine. Giving the worker a copy of the safety rules and regulations in hopes these will be learned and used is not enough. Means should be adopted to bring them to his or her attention frequently and regularly.

A major railroad practiced this when its supervisors discussed with their employees a safety regulation each day before work began. Thus, each day, the staff was responsible for knowing, understanding, and applying this regulation, when appropriate or when asked by the supervisor. This company combined the principles of learning, reinforcement, and follow-up in one program activity.

A trainer must insist not only on frequent practice, but on the trainee's following the correct method. Day-after-day use of safe methods will create safe habit patterns that later will be followed almost automatically. The supervisor and safety director must make sure the work method practiced is the safe one.

**Recency.** Closely allied to the principle of frequency is that of recency. What is learned last usually can be most easily recalled. As has been indicated, handing a worker a set of safety rules and regulations does not ensure learning. Those who received printed instructions or a few training demonstrations a long time ago may not be able to recall now what the rules and regulations are. Safety professionals must devise for workers constant contacts with these regulations through continuing activities such as contests, reviews of safety regulations, and committee work.

**Primacy.** The law of primacy must be taken into consideration in two aspects of the safety program. The worker's initial contact with procedures should be one of major importance. If this initial contact is of a negative nature—such as being tossed a book of rules, accompanied by a shouted, "Make sure you learn 'em"—the worker is left with the impression that safety is unimportant. From the very beginning of employment, the worker must get the impression that not only the rules but the whole program is of great importance. This will help assure the positive response desired.

Second, in the training program, habit patterns using safe methods should assume primary importance. The supervisor must be certain the worker does not have an opportunity to work by any other than the safe method. It becomes harder to establish good patterns after having first learned the poor ones. This principle applies to the golfer who has picked up bad techniques and then has to unlearn them as well as to the worker on the job. Training must be *right* from the very beginning. Old habits are hard to break and, to the safety professional, expensive in training costs and accident costs.

**Intensity.** Those things made most vivid to the worker will be retained the longest. Safety programs already use this principle in safety publicity with eye-catching posters and the like.

It is part of the safety professional's responsibility to enhance the worker's interest in the program. In this way, it will be a long-remembered experience. Under other conditions, it might well be forgotten minutes after it is over. To some degree, this positions the safety program in the same way advertisers position their commercials to catch the public's attention.

**Transfer of training.** All new learning occurs within the context of previous learning experiences. The fact that current learning (or performance) can be influenced by previous learning is known as transfer of training. Positive transfer of training occurs when the previous learning facilitates the current learning or enhances current performance. Negative transfer of training occurs when the previous learning makes the current learning experience more difficult or in some way inhibits current performance.

Inasmuch as the safety professional desires to facilitate learning correct responses in new situations, attempt should be made to maximize the positive transfer that occurs within the industrial environment.

Generally, learning to make identical responses to new stimuli results in positive transfer. For example, learning to drive in a new (or different) automobile is facilitated by the fact that identical responses (such as accelerating or steering) are required to the new stimuli (accelerator, steering wheel). If the new stimuli are very similar to previous cars (for example, the location of the controls, their texture, and their direction of movement), positive transfer should be high. In such situations, positive transfer increases as a function of increased similarity of the stimuli in the two situations.

Learning to make new responses to identical stimuli results in negative transfer. For example, if the controls of the cars appear identical but each car requires a different response in order to be operated correctly, negative transfer can be expected. After driving a car in which the PARK position appears on the extreme right of the indicator panel, much difficulty (negative transfer) can be anticipated when driving a car in which the PARK position appears on the extreme left of the indicator panel. Or, if the OFF position of one toggle switch is the same as the ON position of another, the potential for accidents is serious.

Such negative transfer can cause errors, delayed reactions, and generally inefficient performance on the new task. In such situations the amount of negative transfer increases as a function of increases in the similarity of the stimuli in the two situations.

With an understanding of transfer phenomena, it is possible to maximize, within the industrial environment, opportunities for positive transfer and minimize those for negative transfer. This requires careful planning of machine purchases and work procedures to make certain the new tasks required of a worker make use of (and do not conflict with) previous learning experiences.

### Forgetting

This discussion of learning would not be complete without some consideration of the process of forgetting, which goes on as learning takes place. Never assume that learning and forgetting are mutually exclusive, for such is not the case. As one learns, one also forgets what was previously learned. Curves of forgetting indicate that most is lost immediately after learning has taken place. Depending on the complexity of the job, the amount of learning lost will vary after each day's training session. There should be less forgetfulness or initial mistakes in successive days of training, and more need for patient reteaching.

Consideration of the various principles of learning and the motivational aspects of the problem are essential if one is to retard this process of forgetting. The safety professional needs to consider all of them from the initial stages of employment right on through to the everyday work situations in the company.

### Behavioral management

In recent years, more and more attention and recognition have been given to the importance of behavioral management in safety. Behavioral management refers to the systematic identification, measurement, and control of safety-related behaviors. This section briefly reviews this approach.

### Behaviors versus attitudes

Attitudes refer to internal predispositions to behavior; as such, they are difficult to measure. Behavior refers to observable actions and because behaviors can be observed, they can be measured. This distinction is critically important because measurement of behaviors lays the groundwork for management.

While attitudes are difficult to directly change, changing behavior is not difficult, and this results in situational change. A good example of this is attitudes toward using seat belts. As the behavior of wearing seat belts has changed over the years, so have attitudes toward wearing them. The most efficient way to change safety-related attitudes is to change safety-related behavior.

The steps involved in changing safety-related behaviors are:

1. *Identify critical behaviors.* This means to write, in observable terms, what you want employees to do to safely perform their jobs. You can list a few critical behaviors or a complete inventory, depending on the scope and results you want.
2. *Measurement through observation.* Trained observers watch the workplace to determine if the listed behaviors are performed safely or unsafely. The total number of observed behaviors is divided into the number of safe behaviors to obtain a percentage figure for safe behaviors.
3. *Performance feedback.* The percentage figure for safe behaviors is shown on a graph displayed in the workplace. At regular intervals, behaviors are again observed and the new safe behavior figures added to the graph. Studies show this critical feedback will improve safety behaviors. Praise and recognition from supervisors is another effective way to encourage safe behaviors.

## SUMMARY

The human factor operates at all levels in industry, and is perhaps the most potent factor for success or failure of a safety program. It makes a difference whether the president of a company backs the program actively or drags his feet, whether the safety professional works hard or coasts along, whether the supervisor emphasizes safety or subordinates it to production, whether the janitor cleans well or does only the minimum. Attitudes *are* important to safety in the company, but safety-related behaviors are *critical.* Safety can be achieved only by working through all these people. The human factor must be dealt with in every area of industry.

To achieve some mark of success in dealing with people, individuals can best be considered within the framework established in this discussion. Each person is an individual, to some degree different from every other one. The differences are for the most part obvious, but there are subtle ones too that must be recognized, if only to the extent that their existence is acknowledged.

Despite great differences in people, reasons for their activities are common to all. Many needs are the same, particularly at the biological level and to a large degree at the psychosocial level. It is upon these needs that the safety professional and others in positions of leadership in industry can capitalize to most effectively promote safety.

People become frustrated when their goals cannot be achieved. There are many ways of reacting to such frustrations, but the emotional reactions are of great concern to the safety professional. These reactions, as well as the attitudes formed during the reactions, can be highly disruptive to safety precautions and procedures.

Training programs are established to teach safe work methods. These learning situations, if they are to operate efficiently, call for knowledge of the basic principles of learning that cause people to learn and act as they do. These principles should be used to make new learning more efficient.

In no way is this chapter intended to minimize or ignore the work already being carried on to reduce accidents. Safety devices, safer machines, safe work layout, and many other aids and measures all are a part of the total program. The human factor is an aspect of the whole system.

Workers, machines, and materials are still the three components of industry that can contribute to safety.

Machines and materials can be controlled, but the human factor must be guided in the interests of accident prevention.

## REFERENCES


Adams, J. S. "Inequity in Social Exchange." In L. Berkowitz (Ed.), *Advances in Experimental Social Psychology,* Vol. III. New York, N.Y., Academic Press, 1965.

Anastasi, A. *Psychological Testing,* 5th ed. Toronto, Canada, Macmillan, 1982.

Church, R. M. "The Varied Effects of Punishment on Behavior." *Psychological Review,* 70, 369-402, 1963.

Clark, J. V. *A Preliminary Investigation of Some Unconscious Assumptions Affecting Labor Efficiency in Eight Supermarkets.* Unpublished doctoral dissertation, Graduate School of Business Administration, Harvard University, 1958.

Herzberg, F. *Work and the Nature of Man.* Cleveland, Ohio, World Publishing Co., 1966.

Hulse, S. H., Deese, J., and Egeth, H. *The Psychology of Learning,* 5th rev. ed. New York, N.Y., McGraw-Hill Book Co., 1980.

Maslow, A. H. *Motivation and Personality,* 2nd ed. New York, N.Y., Harper & Row, 1970.

McGregor, D. *The Human Side of Enterprise.* New York, N.Y., McGraw-Hill Book Co., 1985.

McGuire, W. J. "The Nature of Attitudes and Attitude Change." In G. Lindzey & E. Aronson (Eds.), *The Handbook of Social Psychology,* 2nd ed., Vol. 3, Reading, Mass., Addison-Wesley, 1968.

Miller, N. E., and Dollard, J. *Social Learning and Imitation.* New Haven, Conn., Yale University Press, 1962.


### Suggested readings


Bennis, W. G. *Organization Development: Its Nature, Origins, and Prospects.* Reading, Mass., Addison-Wesley Publishing Company, 1969.

Drucker, P. F. "New Templates for Today's Organizations." *Harvard Business Review,* Jan.-Feb. 1974.

Fishbein, M. and Ajzen, I. "Attitudes and Opinions." *Annual Review of Psychology,* Vol. 23, 1972.

Hannaford, E. S. *Supervisors Guide to Human Relations.* Chicago, Ill., National Safety Council, 1976. (Course materials are available.)

Herzberg, F. "One More Time: How Do You Motivate Employees?" *Harvard Business Review,* Jan.-Feb. 1968.

Hinricks, J. R. "Psychology of Men at Work." *Annual Review of Psychology.* Vol. 21, 1970.

Shafai-Sahrai, Y. *Determinants of Occupational Injury Experience: A Study of Matched Pairs of Companies.* East Lansing, Mich., Michigan State University (Business Studies), 1973.

Zimbardo, P. G. *Psychology and Life,* 11th ed. Glenview, Ill., Scott, Foresman and Company, 1984.

# 12

# Maintaining Interest in Safety



THIS CHAPTER DEALS PRIMARILY with promoting and maintaining supervisors' and employees' interest in safety. However, management must demonstrate its interest and actively support a solid safety program. Only then can activities to promote employees' interest be undertaken.

To accomplish this, the health and safety professional must acquaint key executives and managers with the objectives of the program and how it achieves them. A well-promoted program secures the involvement of both management and employees. (See Meetings of executives, later in this chapter.)

Any lack of interest by top management should not be construed as indifference or even opposition to safety. Many times it can be traced to a lack of awareness of the basic benefits of an organized safety program.

## REASONS FOR MAINTAINING INTEREST

Why is it necessary to maintain interest in safety (1) if the work place has been designed for safety, (2) when work procedures have been made as safe as possible, and (3) after supervisors have trained their crews thoroughly and continue to enforce safe work procedures? Because even with these optimum work conditions, accident prevention basically depends upon the desire of people to work safely.

All hazardous conditions, unsafe acts, and loss-control problems cannot be anticipated. Each employee frequently must use his own imagination, common sense, and self-discipline to protect himself and to think beyond his immediate work procedures in order to act safely in questionable situations when he is "on his own."

### Indications of need for a program

Various yardsticks indicate supervisor and employee attitudes toward accident prevention. These also can point out areas with the greatest need for a program that creates and maintains interest in safety.

- An increased injury, accident, and near accident frequency may indicate that a program is needed. If such an increase cannot be explained through engineering methods, training, or supervision, the reason may be that employees are forgetting or ignoring work rules, failing to stay alert, or taking chances. A program to develop and maintain their interest in safety will help reverse this trend.
- If housekeeping is deteriorating, protective equipment is not being used, and guards are not being replaced, it is time to tighten up on supervision and to promote more supervisory interest in safety.
- Incomplete or missing accident reports also indicate a slackening of supervisory interest and, perhaps, even a failure of employees to report minor accidents and injuries. Motivation to assure better reporting is then in order.

### Program objectives and benefits

Although it cannot be expected to do everything, a well-planned program can result in workers becoming committed to safety. For example, it can:

1. Help develop safe work habits and attitudes, but it cannot compensate for unsafe conditions and unsafe procedures.
2. Focus attention on specific causes of accidents, although by itself it cannot eliminate them.

**Figure 12-1.** A sticker like this one (actual size shown here) can be placed on surfaces to remind everyone of a company's continuous efforts to work safely.

3. Supplement safety training, yet it cannot be considered a substitute for a good training program.
4. Give employees a chance to participate in accident prevention activities, such as suggesting safety improvements in job procedures.
5. Provide a channel for communication between workers and management, because accident prevention is certainly a common meeting ground.
6. Improve employee, customer, and community relations, because it is evidence of management's commitment to accident prevention. (See Figures 12-1, -2, and -3.)

**Figure 12-2.** Two youngsters participate in activities during a Safety Day/Open House at the Joliet, Illinois, Eastern Division Plant of Dow Chemical USA. (Reprinted with permission from Dow Chemical USA.)

The major objective of a program to maintain interest in safety is to involve employees in preventing accidents. Usually, though, it is as difficult to determine the degree of success such a program achieves as it is to isolate the effectiveness of an advertising campaign when separated from an entire marketing program. The reason is that, generally, companies with such programs also have sound basic safety programs: working conditions are safe, employees are well trained and safety minded, and supervision is heavily involved in the program.

However, one company that already had a good basic program attributed a reduction in its work injury rate to a stepped-up program to maintain interest. It was based on an idea submitted by an employee: Each month candy bars were distributed to injury-free employees. Wrapped with some of the candy bars were slips that could be traded for free pairs of safety shoes.

Another company gave each employee who had an injury during the month a package of gum with the slogan "Something to chew on" and a friendly safety message and wishes for an injury-free future.

A meat-packing firm was able to assess the value of its program to maintain interest. Several safety bulletin boards were installed, and posters and safety contest reports were displayed on them. These displays were credited with an impressive cut in the number of injuries and with a workers' compensation insurance refund of more than $1,200.

## SELECTION OF PROGRAM ACTIVITIES

Modern advertising and merchandising techniques have much in common with those used to "sell" safety. Just as steady and imaginative sales promotion is needed to sell most products and services, safety likewise requires constant and skillfull promotion. The basic elements of accident prevention can be made more understandable and acceptable when they are presented to the worker in this fashion.

Safety directors who undertake promotional activities with no preliminary planning or objectives in mind or use films or posters for no reason other than they happen to be available at low cost are doing their employers and coworkers a disservice. Likewise safety and health professionals who spend a disproportionate amount of time on committee work, contests, or "homemade" visual aids because of their bosses' or their own personal interests are not developing an entire safety program. Such cosmetic activities are only one piece of the entire program and are no substitute for a sound, well-planned program based on supervisory responsibility and involvement.

### Basis for the program

To be effective, a program to maintain interest in safety must be based on needs. Select activities not just because they will be popular, but because they will yield results. To develop suitable activities and promotional material, the needs of supervisors and employees must be known.

To find out what their employees really thought of their safety program and just how interested they were, one company inaugurated a safety inventory plan. Each year after the regular

**Figure 12-3.** The return on investment for public safety advertising is high, according to the Salt River Project. This billboard is one of 22 in the area that carry a new message each month and reach an estimated 76 percent of adult Valley residents. The effort costs between $50,000 and $70,000 and has been in effect 10 years. The advertising contributes to halving of the insurance premium in five years. (Reprinted with permission from Salt River Project, Arizona.)

stock inventory had been taken, safety inventory cards were distributed to all employees—salaried workers as well as hourly.

The cards were distributed by supervisors who asked each employee to take stock of his job and environment with regard to safety. The following year's safety program was planned on the basis of the questionnaire returns, which ran better than 90 percent. Many suggestions for improving the safety program were received and subsequently put into practice.

### Factors to consider

**Company policy and experience.** If a company ordinarily uses activities, such as committees, mass meetings, and contests in areas other than safety, then the safety director can consider using them in his program. It is generally unwise to spend much time on activities that are unknown to company policy and experience, unless it is believed that a new "sales pitch" is justified.

On the other hand, if supervisors and employees are overinvolved in committee work and activities such as sales promotion, quality control, and tool damage programs, similar activities for safety may be lost or burdensome. Other approaches then could prove more effective.

Once a promotional program is under way, it should not unbalance other aspects of the accident prevention program or other company activities. For example, safety meetings should not take a great deal more time than meetings for quality control, sales, or industrial relations.

**Budget and facilities.** Budget considerations always affect plans for a safety promotion program. At first the program will require extra effort, time, and money, but this expenditure can be justified because it is an investment that will produce direct as well as indirect benefits from a monetary as well as employee relations standpoint. If the program is to be successful, the budget will have to be sufficient enough to carry it out.

In selecting program activities, consider the facilities available. Sound films, for instance, require not only a sound projector and screen, but also a darkened room free of background noise.

In many companies, facilities, publications, or services may be available through the industrial or public relations departments. The latter also may be a valuable source in helping to plan promotional activities. Also the National Safety Council publishes many different types of motivational material; contact a service representative for the latest Council listings.

**Types of operations.** The nature and organization of company operations affect the choice of activities and materials for maintaining interest in safety. When operations are widely scattered and diversified, as in the construction, railroad, marine, motor transport, and air transport industries, the job of selecting and disseminating safety information becomes more complicated.

When operations are decentralized, the health and safety director must choose materials that can be easily used in the field, such as publications and films. Local supervisors must be depended upon to conduct meetings, present material, and handle posters.

The needs of employees involved in widely different kinds of

**BASIC HUMAN INTERESTS AND CORRESPONDING ACTIVITIES**

| Basic Interest Factors | Ways to Use These Factors |
|---|---|
| **Fear** of painful injury, death, loss of income, family hardship, group disapproval or ridicule, supervisory criticism. | **Visual material:** emotional or shocker posters, dramatic films, pictures and reports of serious injuries on bulletin boards, in company papers. |
| **Pride** in safe workmanship, in good records, both individual and group. | **Recognition** for individual and group achievement; trophies, personal awards, letters of appreciation. |
| **Recognition:** desire for approval of others in group and family, for praise from supervisors. | **Publicity:** photos and stories in company and community papers, on bulletin boards. |
| **Participation:** desire to be "one of the gang," "to get in the act." | **Group and individual activities:** safety committees, suggestion plans, safety stunts, campaigns. |
| **Competition:** desire to win over others, such as shown in sports. | **Contests** with attractive awards. |
| **Financial gain** through increased departmental or company profits. | **Monetary awards** through suggestion systems, profit-sharing plans, promotions, increased responsibility. |

**Figure 12-4.** Ways to put six basic human interest factors to effective use in promoting safety. Basic needs and desires that motivate people are left. The right column lists direct appeals safety promotional programs can make.

work at far-flung locations also must be considered. When a poster program is used as one means for maintaining interest in safety at each outlying location, a trustworthy employee may be designated by the local supervisor to receive posters and take care of their distribution and posting. Poster boards and display cases need to be kept clean, attractive, and free of extraneous paper.

Another method is to equip a trailer with permanent displays and transport the company safety story to far-flung locations. Thirty-five mm rear-screen projectors and video tapes (if video tape players and television monitors are available) are also well suited for use in scattered locations and in the field.

Within the same organization, the types of educational materials used for different groups of employees may vary considerably. For instance, a movie scheduled for a day shift when a large number of employees could be taken off the job would not be suitable for a small night shift when no one could leave the job site.

However, night crews and maintenance employees should not be overlooked. Their work is as vital as that of other employees to the overall accident prevention effort. Programs for them will have to be planned to tie into their specific requirements and time constraints.

**Types of employees.** The types, backgrounds, and educational levels of employees must be considered when choosing safety promotion activities. For example, migrant workers frequently do not receive sufficient job training. Material for these employees should present basic safe practices for their jobs in brief, easy-to-understand form and in languages other than English when appropriate. A similar approach should be used with temporary workers or those assigned from union halls. Employees who have difficulty understanding or reading English need visual material. Material in Spanish is available from the Inter-American Safety Council. (See Chapter 24, Sources of Help.)

**Basic human interest.** If employees seem bored or uninterested in safety activities, an extra push is needed to pep them up. One approach is to base promotional activities on employee interests, such as bowling or fishing.

The wise choice of promotional activities depends upon an understanding of basic human needs and emotions. The basic interest factors listed in Figure 12-4 are forms of motivation common to all employees. The activities suggested, therefore, should be of general appeal.

**Other considerations.** The tasteful use of employee human interest materials featuring children, animals, and cartoon figures as well as activities and contests based on a moderate amount of competition play a big part in the safety promotion programs of many companies. These elements can be as effective in safety promotion as they are in sales and advertising promotion and can be used without compromising company policy or offending anyone.

An Ohio company capitalized on the interest of many people in wagering. Employees in carpools were urged to participate in a little wager in which the rider who forgot to buckle his safety belt had to pay for lunch or dinner. The idea proved so popular the company decided to spread the word via a campaign using posters and announcements.

Ideas for maintaining interest often use humor. The "light touch" is essential, and should be good-natured. Ridicule should never be used; it only arouses resentment.

A positive, constructive approach is generally better than a negative approach. However, the latter sometimes is preferable if it is more dramatic. A picture showing the consequences of an accident, such as a fall on a slippery floor, will have a greater impact than one depicting the safe act that could have prevented the accident.

Variety is essential. Often a simple change, such as a dif-

**Figure 12-5.** The Research & Development Executive Committee of the Council's Industrial Division is addressed by Howard F. Kempsell, Sr. Staff Safety Engineer, Exxon Research and Engineering Co., during a National Safety Congress. Volunteers are the life-blood of the Council and are responsible for producing many of its publications which, in turn, help the cause of safety throughout the world.

ferent type of contest, a bulletin board redesign, or revising the format of safety meetings, can result in renewed interest. The activity itself may not be more effective, but its new form stimulates thought, discussion, and interest. Although safe work practices should become routine, their presentations should not.

Activities that require participation generate more interest than do those that involve only seeing and listening. Companies that have worked with the National Safety Council in filming movies in their plants report an upsurge in interest in safety because some employees were able to act in a film that emphasized the importance of what to them might have seemed routine.

Employees who are asked to submit suggestions for equipment guards or to help in the selection of personal protective equipment are more inclined to use the guards and the personal equipment than they would be if they had no opportunity to make their opinions known.

For the same reasons, helping to draft the safety rules encourages compliance on the part of those workers who participate in the project, and serving on a safety committee leads to increased awareness of safety responsibilities.

## STAFF FUNCTIONS

The job of creating and maintaining employee interest falls to the safety and health director, who has the basic responsibility for planning the safety program, and to the supervisors, who have the responsibility of carrying it out.

### Role of the safety and health professional

Safety and health professionals coordinate the program, supplying the ideas and inspiration, while enlisting the wholehearted support of management, supervision, and employees. Programs should be designed to involve both management and employees. Employees like to receive awards; managers like the public relations aspects of presenting them.

Safety and health directors may work with local safety councils, chapters of the American Society of Safety Engineers, and other civic or technical groups interested in accident prevention. They can gain much by attending the National Safety Congress and regional safety conferences. Here they learn what other companies are doing and how those ideas can be translated into practical activities that are useful to their own organizations. After participating in round-table discussions, listening to speakers, and meeting many people with similar interests, safety and health professionals return with renewed enthusiasm (Figure 12-5).

Because they often are called upon to address groups, safety and health professionals should be able to present their ideas clearly, effectively, and convincingly. Polishing public speaking skills will also help when dealing with people on a one-to-one basis. Safety professionals should know when and how to use visual aids.

Showmanship tactics, however, should seldom be used or used only with discretion as they could reflect unfavorably on the safety and health professional, the company, or the safety program.

Safety and health professionals should help educate supervisors so that they see that working conditions are kept as safe as possible and insist that their workers follow safe procedures consistently, simply as a part of good job performance.

A vast amount of program material of use to the safety professional is available through the National Safety Council publications. For example, many of the illustrations in this chapter were selected from Industrial Sectional *Newsletters* and the monthly feature "Ideas That Worked" in the Council's *Safety and Health,* formerly *National Safety News.*

Safety and health professionals also are invited to submit interesting data, difficult problems, or "gimmicks" of any nature to the Council for help in solving problems or in order to share ideas and problem-solutions with others. The Council has information on every phase of safety, gleaned from the experience of members in various industries. One company's solution to a problem may be of real help to many others.

### Role of the supervisor

The supervisor is the key person in any program designed to create and maintain interest in safety because he is responsible for translating management's policies into action and for promoting safety activities directly among the employees. How well this responsibility is met will determine to a large extent how favorably employees will view safety activities.

Supervisors must be made aware that under existing laws it is they who are directly accountable to both management and society for their employees' safety. It is the responsibility of management through the safety professional to see that supervisors receive adequate safety training.

The supervisor's attitude toward safety is a significant factor in the success, not only of specific promotional activities, but also of the entire safety program, because his views will be reflected by his employees.

Supervisors who are sincere and enthusiastic about accident prevention can actually do more than the safety director to maintain interest. Conversely, if supervisors only give lip service to the program or ridicule any part of it, their attitude offsets any good that might be done by the health and safety professional.

Many supervisors are reluctant to change their modes of operation or to accept new safety engineering ideas, much less to enthusiastically support contests, safety stunts, committee projects, and other activities used to promote and maintain interest in safety. It is the health and safety director's task to sell these supervisors on the benefits of accident prevention, to convince them that promotional activities are not "frills," but rather projects that can help prevent injuries, and to persuade them that their wholehearted cooperation is essential to the success of the entire program. And, having a successful program makes their jobs as supervisors easier and less time consuming.

One way supervisors can support safety is by setting a good example. By wearing safety glasses and other personal protective equipment whenever they are required supervisors can promote safety through example.

Teaching safety is an important function of supervisors. To be successful in this area, they cannot depend upon safety posters, a few warning signs, or even general rules to do their teaching for them. Supervisors themselves must first be trained to teach if they are to be competent in this area. (See Chapter 9, Safety Training, for details on training courses and techniques.)

The supervisor should not have to suddenly adopt a "get tough" approach to enforce safety rules. He should be consistently firm and fair. If workers have the impression that the supervisor either cannot recognize unsafe conditions and unsafe acts or does not care whether or not they exist, they too will become lax.

Supervisors should be encouraged to take every opportunity to exchange ideas on accident prevention with workers, to commend them for their efforts to do the job safely, and to invite them to submit safety suggestions.

They can be most effective in relaying personal reminders on safety to employees. This procedure is particularly necessary in the transportation and utility industries, where crews are on their own from terminal to terminal.

Supervisors are entitled to receive all the help the safety department can give through correspondence, and supplies of educational material for distribution, and by receiving as frequent visits as circumstances permit. They should also receive adequate recognition for independent and original activity.

## SAFETY COMMITTEES AND OBSERVERS

There are many types of safety and health committees having many different functions. (For further details, see the Council publication, "You are the Safety and Health Committee.") However, the basic function of every safety committee is to create and maintain interest in safety and health and thereby help reduce accidents.

Some organizations prefer other types of employee participation to formal safety committees because they feel safety committees require a disproportionate amount of administrative time, that they generally tend to pass the buck, that they sometimes stir up more trouble than they are worth, and that some supervisors try to unload their responsibilities onto the safety committee.

The answer to these objections is not to abolish the committees but rather to reexamine their duties, responsibilities, and methods of operation. Such analysis often can lead to constructive changes that will enable a committee to fulfill its original objective—that of stimulating and maintaining interest in safety.

Committee membership should be rotated periodically. This insures a fresh look and also compounds the number of employees who are trained to look at operations through the eyes of safety.

Involving employees in safety inspections, either alone, as observers, or as part of a formal safety and health committee has the same basic objective: to get more employees actively involved and interested in the safety and health program. Planning, publicizing, and following definite procedures will streamline the work of both committees and observers and help ensure effective results.

## QUALITY CIRCLES AND SAFETY CIRCLES

Hazard recognition and control is one area where employee involvement produces substantial results. The following discussion is taken from the Indiana Labor & Management Council's booklet "Worker Involvement in Hazard Control," see References. Used with permission.

Two reasons for the popularity of involving employees in hazard recognition and control is the desire of management to use all available resources to increase productivity and quality in the face of growing competition, and the understanding by management that employees want to accept new challenges and participate in activities that affect their work-life.

Popular forms of worker involvement programs are quality and safety circles. A quality circle is a group of employees, performing similar work or sometimes varied work, who meet weekly to learn about and apply basic techniques to identify problems within their area(s), analyze them, and recommend solutions to management. In some instances circle members discover hazards during analysis of other plant problems and make excellent recommendations for their control.

Safety circles are a type of quality circle used by some companies to reduce the number of accidents and injuries by keeping safety and all its important features foremost in the minds of the employees. This implies a change in the employee's role from passive to active, while management's role becomes less negative (fewer don'ts) and more positive.

In many situations, safety circles are established on a departmental level. Safety circle meetings are held monthly for approximately one-half to one hour in duration. Each circle meeting is usually preceded by a presentation by a department supervisor explaining the successes and problems that the safety circle team experienced during the preceding month. In addition, the circle reviews a breakdown of all injuries, including first-aid cases as a means of measuring the department's progress and as a means of pinpointing trouble areas.

Companies employing the safety circle concept indicate an improvement in their accident and injury experience.

## MEETINGS

Health and safety meetings may be conducted for supervisors, employees, or other groups, but in every case the purpose is to stimulate and maintain interest. If meetings fail to achieve this, their format or content should be changed sufficiently to make them effective or they should be stopped and a new approach taken.

### Types of meetings

The following types of health and safety meetings commonly arouse and maintain interest in accident prevention. (Also see Chapter 9.)

1. Meetings of operating executives and supervisors to formulate policies, initiate a health and safety program, or plan special activities.
2. Mass meetings of all employees, sometimes including families, or even the entire community to serve special purposes such as launching a major new program or contest and "selling" safety to everyone affected by accidents and injuries.
3. Departmental meetings to discuss special problems, plan campaigns, or analyze accidents.
4. Small group meetings to plan the day's or week's work so that it can be done safely, to discuss specific accidents, or to review safety instructions.

**Meetings of executives.** When a health and safety program is inaugurated, it is especially important that the chief executive officer of the company or top plant manager should call a meeting to announce the general accident prevention plans and policies to all foremen, supervisors, superintendents, and other operating executives.

If these persons meet at regular intervals to discuss operating problems, this announcement can be made at one of these regular meetings. Otherwise, the CEO or manager should call a special meeting.

After this first meeting, the group may hold sessions periodically to evaluate the safety program, to check on the progress being made in accident prevention, and to appraise proposed activities. It is also desirable for this group to review and/or investigate all fatalities, multiple amputations, and other serious injuries.

**Departmental meetings** serve many health and safety purposes. They may be used to discuss the company safety program so that employees will better understand what is going on, to provide information about accident causes and accident types, or they may be purely inspirational to create an awareness of hazards and a desire to prevent accidents. It is also desirable for departmental meetings to review and/or investigate all injuries involving lost work days or restricted work.

Many departmental safety meetings are held monthly and most are conducted by the supervisor. The safety department usually assists in planning and provides materials, such as visual aids.

The program for a departmental meeting may include:

1. Reports on injuries in the department since the last meeting; a safety inspection in the department; and the department's standing in a contest. (The total time spent on reports should be limited so that this part of the meeting does not become tiresome.)
2. Discussion by the supervisor of where observance of safe practices or unsafe conditions needs to be improved.
3. Talk, demonstration, or audiovisual presentation on an appropriate accident prevention subject. The speaker may be the supervisor, a department employee, the company safety director, an outside expert, or a company executive.

Departmental meetings give the supervisor an opportunity to point out the dangers of particular unsafe practices or conditions. By condemning those practices, the supervisor makes it clear that he will set a good example and expect his workers to follow. In addition, most workers welcome an opportunity to share their safety ideas.

At the conclusion of departmental meetings, the supervisor may prepare written reports for the plant (or company) safety committees and review by the managers.

**Small group meetings** with people doing similar work can be held at or near the work place (Figure 12-6). The supervisor may discuss the causes of a recent accident of which the workers have personal knowledge. Employees should be encouraged to join in the discussion, and a conclusion should be reached as to how the accident might have been prevented.

The supervisor may present a problem that has developed because of new work or new equipment. Again, all should participate and offer their views.

At times the supervisor may present a film or chart talk on a subject related to the work of the group members. Other audio-visuals such as models or exhibits may be used. Safety devices or pieces of equipment or material may be brought in and discussed.

"Production huddles" are instruction sessions about a specific job which include safety information. Such meetings are particularly useful with maintenance crews when an unusual job is about to start. The plans for doing the job safely and efficiently are discussed and a procedure is agreed upon. Public utility line crews who use this type of meeting call it a tail board conference. Before starting a job, the crew gathers around the truck and discusses the job, laying out the tools and materials they will need and agreeing upon the part each person is to do.

**Figure 12-6.** Informal safety meetings at General Electric Co., Cleveland, Ohio, help determine ways to prevent accidents.

A particular advantage of small group meetings is that they provide excellent opportunities for presenting all types of information, including safety information, directly to employees and they stimulate an exchange of ideas that can benefit the accident prevention program. To be successful, each health and safety meeting must include a tangible message, originality of presentation, opportunity for audience participation, and a conclusion that spurs action toward an attainable goal.

**Mass meetings.** Large mass meetings are held for special purposes, such as the launching of a contest, the presentation of awards, the introduction of interesting new equipment, the explanation of a change in company policy, or the celebration of an exceptionally fine safety record with an event such as "safety day" (Figure 12–2).

In companies with plants in different cities, a top executive may call a meeting of employees during a plant visit. The talk may cover safety as well as other subjects. One company president makes an annual round of plants with the safety director and speaks at a safety rally of all employees at each plant.

Under certain conditions, particularly in smaller communities, large meetings can be held in a local theater or public hall. Such circumstances necessitate making fairly elaborate arrangements and giving the meetings considerable publicity to assure good attendance.

A mass meeting in a public hall using the "family safety night" theme, makes it possible for not only employees to attend but also their wives or husbands, families, and friends. In addition to a presentation or speech targeted at the program theme, there should be some other entertainment. Often good talent can be found right in the plant or shop.

Mass meetings afford an excellent opportunity for using an outside speaker who can speak with authority and in an audience-pleasing manner on general accident prevention work. Such speakers are found in a nearby plant, an insurance company, the city administration, an automobile club, or a community safety council.

If movies, video cassettes, or slide shows relating to accident prevention are desired, a suitable selection can be made from those available through the National Safety Council, or regional film service organizations, whose locations may be obtained from the Council.

### Planning programs

Making the safety meeting interesting is of the utmost importance. Speakers should not complain or scold. Talks should be limited in time and should start and end on time. The subject matter should be studied in advance to make sure that it is pertinent and is not a repeat of other recent talks.

Large occasional meetings need even more preparation and timing than do small meetings. Speakers, including company executives, should review their remarks with the meeting planner to assure that their remarks will add to the desired purpose. Films and other visuals should be checked in advance.

Persons responsible for employee meetings should study them critically to be sure they are accomplishing their purpose. There is always a danger meetings will become a dull routine. Only continual effort and planning will prevent this.

A plan of action to develop a successful health and safety meeting includes these points:

- Prepare in advance. The preliminary arrangement determines the results. Do not conduct a meeting without preparation.
- Select a major topic. Make it timely and practical—one that the group can discuss.
- Obtain facts and figures. Be sure they are correct and complete. Prepare a visual, such as a simple chart or table, whenever possible.
- Map the presentation. Decide on the best way to present the subject of the meeting. Try to anticipate the group's reaction and questions. Outline results you hope to accomplish.
- Set a timetable. Allow adequate time, but set a reasonable limit.
- Be sincere. Your sincerity and your interest in the employees' welfare must be unmistakable.
- Introduce the topic. Tell in simple terms what the meeting is all about. Use a punch line or other short and to-the-point lead-in.
- Present facts, arouse interest. State pertinent facts in an interesting manner.
- Promote group discussion. Ask questions that cannot be answered "Yes" or "No." Prompt members of the group to think individually and collectively. Let *them* talk.
- Agree on doing something. Try for group agreement on methods of correction and improvement. Write these down.
- Summarize the meeting. Review briefly what has been discussed and decided . . . follow up in the various departments in writing, if possible.

## CONTESTS STIMULATE INTEREST

Contests (such as housekeeping contests, interdepartment contests, etc.) are not substitutes for management interest, safe procedures, and "built in" safety. While a successful accident prevention program is found where good management, good training, and efficient operation are in place, some *special* effort may be needed to maintain interest in good housekeeping, reporting hazards, and the like. Moreover, the interest value of contests has direct bonus values in good publicity and improved employee morale.

Therefore, a competition is held only after the basic steps in a safety program have been taken—a policy statement made, a record system adopted, equipment safeguarded, a first aid department installed. Such substantial demonstrations of management's interest, sincerity, and responsibility greatly help win the active participation of supervisors and workers in a contest.

### Purpose and principles

Safety contests are operated purely for their interest-creating value. A contest that creates favorable interest is valuable; one that does not create interest is worthless. In most contests, the competing groups are departments of the same plant or divisions of the same company. This type of contest is useful when investigations of injury accidents reveal that most are a result of the employee's unsafe act. Should they be the result of unsafe conditions or a combination of unsafe acts/unsafe conditions, they might well focus negative attention on one of your most conscientious safety-minded employees injured through no fault of his own.

Generally, contests are based on accident experience and are operated over a stated period, with a prize for the group attaining the best record according to the contest rules.

Contests have been important almost from the time of the first safety programs, and a fairly well-established group of operating principles has been developed:

1. A contest should be planned and conducted by a committee representative from the competing groups.
2. Competing groups should be natural units, not arbitrary divisions.
3. Methods of grading must be simple and easily understood.
4. The grading system must be fair to all groups.
5. Awards must be worth winning and of the sort that create interest.
6. Good publicity and enthusiasm are important.

Contests may run for various periods—from a few months to a year. Those who recommend longer periods believe that if workers are kept on their toes for a longer time, safe working is more likely to become a habit. Some safety professionals, however, believe that greater interest can be aroused and maintained during a short period and therefore prefer short and more frequent competitions. Contests of different duration can be tried to see which is most effective.

**A safety contest stock certificate idea** was developed by the Maxwell House Division of General Foods Corp. and ran for one year.

For each week a department worked without a disabling work injury, the department received a stock certificate worth 50 cents. Dividends were paid on this stock at the rate of: 10 cents for the first 1,000 consecutive safe hours worked; 25 cents for the first 10,000 consecutive safe hours worked; 50 cents for the first 50,000 consecutive safe hours worked; and one dollar for the first 100,000 consecutive safe hours worked. This meant that if a department worked 100,000 consecutive workhours without a disabling injury it received a total of $1.84 in dividends for each share of stock held.

There are penalties, however. If a disabling injury occurred in a department, that department was penalized one month's stock earnings. This meant that during that particular month the department could not be awarded any stock or dividends. It also meant that if an accident should occur in the latter few days of the month the department would lose all dividends and stock certificates previously issued for that month.

### Injury rate contests

In a contest based on injury rates, the measure of safety performance is the OSHA incidence rate, which was described in detail in Chapter 6, Accident Records and Incidence Rates.

Injury rate contests carry with them an inherent risk of possible abuse and the *need* to be *closely monitored.* All injuries must be reported to give notice of unsafe acts or conditions, which if not corrected, could cause a serious disabling injury. These contests have the tendency to put peer pressure on the employee not to report an injury, administrative pressure on those doing the recording to rationalize whether or not to count it, and management to focus attention on the contest rather than the safety program. Once this should happen, the entire safety program is discredited.

Contests should not be based upon severity, because severity data cannot be determined promptly and because severity frequently is a matter of luck and contributes little to knowing how to prevent the accident in the future. Using a combination of frequency and severity is not good for the same reason. Contests should not be based upon reduction of reported first aid cases because people, therefore, may fail to report such injuries.

**In-plant contests.** The most effective contests generally do not run continuously, with a new one beginning as the old one ends. They are more in the nature of special campaigns to run for a specified time and are launched with advance publicity and fanfare. Often the president or other high official makes the original announcement, presents awards, and otherwise lends prestige to the contest.

Competition, if properly organized, can do much to develop teamwork. Some workers who apparently give no thought to their own work habits can be influenced to cooperate with their fellow workers if they know that their unsafe acts and resulting accidents will discredit their department or "team."

**Council contests and awards.** National Safety Council members firmly believe in the value of contests for maintaining interest. The Council has about two dozen Industrial Section contests. All are open to Council members, and several include nonmember participants through specially arranged contests cosponsored with trade associations.

In Council contests, companies are grouped according to size and operation so that competing units will be comparable with one another. The definitions and rules are established by the Council. OSHA incidence rates are compared.

Many Council members consider the sectional contest to be one of the most important features of Council service. Awards are presented at sectional meetings, local safety council meetings, and occasionally at trade association conventions.

Most of the Council's sectional contests are integrated with the Council's Award Plan. Under this plan, four award levels are set up to provide some recognition for every good safety record of a member company or unit. In order of importance, the awards are:

AWARD OF HONOR
AWARD OF MERIT

**RULES FOR THE 'XYZ COMPANY' SAFETY CONTEST**

**Rule 1.** The contest shall begin January 1, and end December 31, 19. . . .

**Rule 2.** The contest shall consist of two divisions.
a. Fabricating units shall participate in Division I.
b. Field erection departments under the direction of a superintendent shall participate as separate units in Division II, which shall be divided on the basis of size into two groups: Group A shall consist of the five erection units working the largest number of man-hours, and Group B shall consist of all other units. Units shall be tentatively grouped by size during the first three months, and the final classification shall be made on the basis of total man-hours worked at the end of four months. No further changes in size groups will be made after April 30, 19. . . .

**Rule 3.** Recognition awards shall be:
a. Trophies to the winners in Division I and Groups A and B of Division II.
b. Engraved certificates to plants and erection units ranking second and third in Division I and in Groups A and B of Division II.

**Rule 4.** a. The winners shall be the contestants having the lowest weighted frequency rates.
b. In the event that two or more contestants in any classification have had no chargeable injuries during the contest period, the winner shall be the contestant who has worked the largest number of man-hours since the last chargeable injury.

**Rule 5.** All injuries and illnesses resulting in death or days away from work shall be counted, as defined by OSHA recordkeeping requirements.

**Rule 6.** A sum of $50.00 shall be presented to the units that have had no disabling injuries during the first six months of the contest or have reduced their average frequency rates for the first six months 50 per cent in comparison with the average rate for the preceding six months' period. The award shall be divided into prizes of $12, 10, 8, 6, 4, 2, and eight $1 prizes and raffled to employees. No employee may win more than one prize.

**Rule 7.** Standings shall be compiled monthly and published in a bulletin that will be distributed to all managers, superintendents, and foremen.

**Rule 8.** All questions pertaining to the definitions of injuries and rules shall be referred to the Contest Committee, whose decisions shall be final.

**Rule 9.** Awards shall be presented at an appropriate ceremony to be announced at the end of the contest.

**Figure 12-7.** A typical set of rules for a safety contest. See text for a point-by-point discussion.

CERTIFICATE OF COMMENDATION
PRESIDENT'S LETTER
The Award Plan recognizes perfect records (no disabling injuries) covering an entire calendar year.

**Associate contests.** A number of associations conduct their own contests. Statistics from association contests are submitted to the National Safety Council, which uses them to give the corresponding industries' injury rates in its annual *Accident Facts* booklet. In addition to stimulating the interest of the associations' members, such contests give the Council a broad and reliable accident reporting base.

### The 'XYZ Company' contest explained

The "XYZ Company" is engaged in steel fabrication and erection. Rules for one of its safety contests are given in Figure 12–7, but some explanation is needed.

RULE 1. Although contests can run for any period of time, the company chose a one year period to allow development of safe working habits—the useful objective of the contest—and, because some of the departments were relatively small, to eliminate random (chance) factors from influencing an individual department's experience record.

RULE 2a. Hazards in steel fabrication and erection vary greatly; therefore, fabricating units compete in one division and erection units in another. Operations are similar enough to have a common basis for determining standings—see Rule 4.

Other companies may find that hazards differ sharply from one plant or operation to another, and the similar units may be too few to group. Such plants or operations may compete on an equitable basis in several ways.

- The participant achieving the largest percentage reduction in its frequency rate in comparison with a base period, such as the previous year, may be the winner. Because each unit competes against its past record as par, this method provides a fair basis for comparison of rates.
- The compensation insurance rates for different types of units in the same state have been used to establish a handicap factor to compensate for differences in hazards. If the rates per $100 payroll are $3.00 for Plant A, $2.00 for Plant B, and $1.50 for Plant C, factors for the units are 3, 2, and 1.5, respectively. The frequency rate of each plant is adjusted by dividing the rate by its factor. Plants are ranked from the lowest to the highest on the basis of the adjusted rates.
- The national average rates for different types of units may be used similarly for determining standings. If Plant A

achieved a frequency rate of 10.0 and the national average for units in this industry was 12.0, Plant A's rate is 0.83 of the national average. Plant B's rate in comparison with its national average rate is 0.75. Therefore, Plant B would rank nearer the top than Plant A. Average rates for most industries are published annually in the National Safety Council's *Accident Facts.*

RULE 2b. Separation of large and small units is essential because a small unit finds it easier to go through an entire contest period without a disabling injury than does a larger unit. If the number of units is sufficient, three size classifications may be set up. Unequal size groups can be competitive if they include units that can attain no-injury records. For this reason, the erection departments of the company in this example were divided into Group A—the five largest departments—and Group B—the remaining units. Five contestants in a group are usually considered minimum.

RULE 3. Awards should be specified. "XYZ Company" follows the general practice of giving first, second, and third place awards.

RULE 4a. The frequency rate is most often used as the basis for determining standings as it can be computed promptly and easily.

RULE 4b. There should be a satisfactory method of determining the winner between two or more units having perfect records in order to give the smallest contestant a fair chance. Selecting the winner on the basis of the largest number of man-hours worked since the last chargeable injury is fair to all units, regardless of size.

RULE 5. A standard method of counting injuries is essential in order to avoid controversies and maintain confidence in standards. Use the incidence rate described in Chapter 6, Accident Records and Incidence Rates.

First aid and other minor injuries should be excluded because, if they are counted, workers may fail to obtain treatment so that cases will not be put on the record. The result could be an increase in infections.

RULE 6. Various "special rules" may be used in a contest that runs six or more months in order to stimulate and maintain employee interest.

RULE 7. Frequent contest bulletins keep everyone informed about standings. Bulletins can be posted and standings discussed at safety meetings. Contest results can be announced in plant publications and in other ways to build and maintain interest.

RULE 8. Questions about rule interpretation will arise and must be settled fairly. This is an important function of the contest committee, which, in turn, may refer decisions about disputable injuries to outside judges. A committee of the American National Standards Institute has been set up to interpret standard definitions.

RULE 9. (See the section Meaningful awards, later in this chapter, for award ideas.)

### Interdepartmental contests

Interdepartmental competitions get "close to home" and place responsibility for a good showing on supervisors. Because workers have a greater personal interest in the standing of their department than in the record of the entire plant or other unit, this type of competition has proved popular for creating interest among both supervisors and workers.

An interplant contest plan often may be adapted to an interdepartmental competition. A company operating a number of similar plants may take advantage of workers' interest in their departmental records by conducting a competition among the same departments in various plants. Public utilities may conduct a contest among districts, and other nonmanufacturing companies may follow a similar plan. Because hazards in similar operating units are about the same, the basis of standings may be the frequency rate. Departments or other units may be grouped according to size.

Most departmental contests, however, are conducted among dissimilar departments in one plant, and the departments often differ greatly in size. The difficulties due to variations in hazards and numbers of employees from one department to another may be overcome in the same manner just discussed with regard to differences between plants.

One plan that produced excellent results was aimed particularly at supervisors in charge of departments having five to sixty employees. The principal rules were:

1. This contest will cover the period from July 1 to December 31.
2. Each department will compete against its own previous six months' accident record. Only disabling injuries are counted.
3. The department's accident experience will be judged by the frequency rate developed during the contest as compared with its frequency rate of the previous six months.
4. A suitable prize will be given each supervisor who reduces his department's frequency rate in the contest months by 50 percent or more over his frequency rate in the previous six months.
5. If a supervisor had no disabling injuries in his department during the previous six months, the prize will be given if he meets his previous six months' record.

A bronze trophy was given to each supervisor and department head who met the requirements for an award, and a dinner party was held. When the rates of the contest period were compared with the rates of the previous six months, the results were reductions of 30 percent in frequency and 62 percent in severity.

It is important that competing plants or groups be kept posted on the latest standings—Figure 12–8 shows a typical contest bulletin.

### Intergroup competitions

Intergroup contests are particularly suitable for units that employ fewer than 400 people and have small departments in which hazards vary sharply. Employees are divided into teams of from 20 to 50 workers. To equalize factors of size and difference in hazards, each team has a proportionate number of employees from the most and least hazardous occupations. Each group is led by a captain, whose principal duty is to contact members of the team and create interest in winning. Team members may be members of contest committees.

Interest is promoted by naming teams after prominent baseball, football, or other outstanding sports teams, and the entire competition may be named after a league or other sports organization. Team names can be drawn from a hat and membership can be shown by colored pin-on buttons. Signs are placed in competing departments.

Identification of workers by colored buttons helps to overcome difficulties in scoring. Because the members of different teams work together, they make sure that every case involving

**STANDINGS IN THE 'XYZ COMPANY' SAFETY CONTEST**
**JANUARY-JUNE**

Oakland leads at the halfway mark!

Tulsa moved into second place.

Chicago slipped from second to third place in June.

Corbin, in last place, had no disabling injuries during June. Good work!

Tulsa still leads for the President's Award for largest improvement over the previous year's record.

| *Plant* | *Rank* | *January–June Incidence rates** | *% Increase + or decrease – over last year* |
|---|---|---|---|
| Oakland | 1 | 2.8 | –30% |
| Tulsa | 2 | 2.9 | –40% |
| Chicago | 3 | 3.5 | + 5% |
| Cincinnati | 4 | 4.4 | –20% |
| Corbin | 5 | 5.0 | +22% |

*Incidence rate is number of cases resulting in death or days away from work per 200,000 employee-hours.

**Tips on How to Win, No. 6**

**One-sixth of our disabling injuries occur in the use of cranes and hoists. Have foremen hold a safety meeting on safe practices in hitching loads and other crane operations for shop men. We'll furnish a film. Use posters on the subject. Require safe methods. See the enclosed bulletin for suggestions on how to solve this major accident problem.**

**Figure 12-8.** Simple monthly contest bulletin gives essential information about current standings and also includes a suggestion for improvement in "Tips on How to Win."

a competitor is charged against that group's record.

### Intraplant or intradepartmental contests

By setting standards of performance, an intraplant or intradepartmental safety contest puts particular emphasis on the responsibilities of supervision and employees for avoiding accidents. These standards are often no-injury records for varying periods (Figures 12–9a and b), achievement of lower injury rates in comparison with a previous period, and improvement over the average injury rates of similar units or of the industry. In this type of contest, units of an organization do not compete with one another; rather, each unit attempts to match or surpass established standards.

### Personalized contests

Some plants single out for special recognition employees who have safe records. Certificates are given to those who have worked one, two, five, and ten years without an injury. Holders of certificates have found them useful in obtaining promotion and even in seeking employment with other organizations.

Periodic raffles of merchandise or cash, or the use of a new car for three or four months, regardless of the injury record of a department or plant, also have been used successfully to encourage and acknowledge the efforts of safety-conscious workers. Employees who are involved in disabling accidents (or drivers who have had a "preventable accident") during the period become ineligible for drawings.

Various sweepstakes plans have proved popular and effective in maintaining interest in a good record from month to month. One such plan was operated successfully by a branch plant of a well-known paper company. Here's how it worked.

On the payday prior to the beginning of each month all hourly employees received cards with serial numbers at the pay office window. The workers wrote their names and departments on the cards, tore off the stubs and put them in a box, and retained their portions of the cards. Names of employees in departments having no disabling injuries during the month then were drawn for prizes ranging from $5 to $25. The company contributed a total of $75 per month.

Since eligibility for the drawing depended on a perfect departmental record, each worker had to be careful about his actions. Workers frequently corrected others for unsafe practices that might spoil chances for prizes.

If no accidents occurred during a three-month period, supervisors participated in a drawing for prizes ranging from $5 to $15. This feature proved helpful in enlisting their cooperation.

Some companies give awards like first aid kits or trading stamps to those in every department who have worked during a given period without an accident. This approach is effective only if going a month, for instance, without an accident is unusual.

### Overcoming difficulties

Several difficulties in operating contests are rather easily overcome. One is that some departments are inherently more hazardous than others.

As previously suggested to overcome this problem, handicaps can be established, based on annual rates of insurance compa-

**Figure 12-9a.** Signs such as this one at DuPont's Cape Fear site, Wilmington, North Carolina, recognize the accomplishments of Daniel International Corp. employees while encouraging them to set new goals.

**Figure 12-9b.** Employees of the joint forces of Daniel International, DuPont, and Davis Electrical Constructors celebrated 13 million safe work hours without a disabling injury. The key to this safety achievement is the attitude of the employees. "Everyone is motivated, they pay attention to what they're doing, and they use safety equipment," states Charlie Garrett, Daniel's project manager.

nies or on average accident frequency for the different kinds of work.

Another method is to base standings on improvement over past records. Thus, a department with a past average frequency of 20 and a current frequency of 15 would be rated as having made 25 percent improvement and would win over a department with a past frequency of 15 and a current frequency of 12—a 20 percent improvement.

Usually, in both methods an average of rates over three to five years is used as the base.

Another problem is that a department may have so many accidents at the beginning of the contest period that it is out of the running and loses interest. Having shorter contest periods helps to overcome this difficulty. Another remedy is to have different awards for different achievements. An award for the department having the longest run of injury-free workhours is one example.

Division of responsibility for the cause of an accident may become a problem. For example, the unsafe act of one supervisor's worker (or an unsafe condition in this supervisor's department) can result in an accident involving an employee from another department. This would affect another supervisor's record. These situations must be anticipated and the contest planned to deal with them fairly.

### Noninjury rate contests

The use of noninjury rate contests is gaining favor among safety professionals to motivate program performance. They are used to zero-in on problem areas and are not subject to the same potential misuse as are injury rate contests.

For example, an employee contest can be based on the safe worker. The contest uses either a monetary award or a status award, such as a drawing for a television set or a private parking space. All employee names are included in the competition.

During the contest period each worker detected performing an unsafe act or not wearing required protective equipment has his name removed. To maintain credibility those experiencing recordable injuries would also be eliminated.

Supervisory contests, such as a "Supervisor's Safety Club," generally are based on how well they perform program elements. Elements range from the percentage of a supervisor's employees receiving safety talks to the number of self inspections conducted with deficiencies promptly corrected. Here again to maintain credibility, the recordable injury of an employee would remove the supervisor from the contest. Awards range from steak dinners and sports events to a Supervisor's Safety Club jacket. Management of several supervisors or departments have similar competitions and similar awards.

Contests at the plant level generally cover multiplant performance in a corporation. In addition to how well corporate safety program elements are met, injury experience should be included. Such contests usually have a corporate trophy to be displayed at the winning plant rather than providing any monetary consideration.

Other noninjury rate contests are safety slogan, poster, housekeeping, and community contests. But, no matter what the contest, the objective is to get the maximum number of people talking, thinking, and participating in safety.

**Slogan, limerick, and poster contests.** Safety slogan contests vary. One can be for the best safety slogan submitted by an employee. Another may ask the employees or their spouses if they can repeat the "slogan of the week" or the message on a certain safety poster.

Company magazines may conduct contests to "finish the limerick" or "write a rhyme" or write "twenty-five words on the best way to be safe." Often these are open to both employees and family members.

The value of homemade posters is in their special application to a particular industry or company. If the posters are the result of an employee contest, their interest-creating value will be increased, possibly exceeding that of the tailor-made variety of posters. An important ingredient of such a contest is to get employees and their families participating in the planning and judging stages too.

Frequently, employee poster ideas—aside from the quality of the art work—are so good that companies even submit the winning contest entries to the National Safety Council for possible conversion into printed safety posters.

**A housekeeping contest** often is conducted among departments. This type of competition is fundamental because it is aimed at accident causes and usually tries to eliminate unsafe practices and conditions.

Housekeeping contest plans differ from one company to another. The following plan has been used successfully by a metals firm.

Once a week a committee of three management representatives inspects each department and reports unsafe conditions to the superintendent. A copy of the report is furnished to the works manager, and another is kept for the use of later inspection committees. A demerit for each unsatisfactory condition is charged to the department. If the condition is not rectified within one week, an additional demerit is added.

At the end of the month the demerits are totaled, and departments are rated on the basis of the proportion of demerits to the total number of employees in the department. If Department A employed 175 people and had 25 demerits, its rating would be 85.7. This figure is obtained by dividing 25 (number of demerits) by 175 (number of employees), multiplying by 100, and subtracting the product from 100. Standings are posted monthly on the bulletin boards in each department.

Awards are made at a mass meeting held after the lunch period. Names of employees in winning departments are placed in a box from which is drawn the name of the winner for the month. The winner's picture is posted on a special bulletin board, and a short talk on safety by a representative of management is broadcast throughout the plant. (A general rule prohibits an employee from winning more than one award during the contest.)

The name of the winning department is inscribed on a plaque each month. The head of the winning department receives the plaque from the previous month's winner at the mass meeting, and at the end of the year the department that has won the plaque the greatest number of times receives it permanently.

**Community and family contests.** Many companies have stimulated interest by sponsoring safety essay or poster contests for children of employees, local school children, or young art students. The publicity before and after such contests plus the interest generated by the posters themselves and the judging not only stimulates the interest of employees, but also promotes the company's community and public relations.

More than one company has launched a safety poster or essay contest for the children of employees with the full knowledge that the employees would give their children considerable help and that there would be much favorable discussion about the contest in locker rooms, lunchrooms, and car pools.

To promote greater safety at railroad crossings, Texas railroads sponsored a contest at the Texas Press Association Convention. The journalist who made the closest guess of the total number of grade crossings in Texas won a prize, and safety at grade crossings received much attention (Figure 12–10).

**Miscellaneous contests.** There is an endless number of different types of contest possibilities. Often they can be combined with injury reduction contests. Contests can be held for attending safety meetings, for wearing safety shoes, for reporting unsafe conditions or unsafe acts, or for off-the-job or public safety activities of individuals, departments, or branch plants.

Although contests are popular with both management and employees, the safety and health director always should attempt to determine before starting one whether or not it will take time and effort away from providing safer equipment or better training for supervisors and employees.

### Contest and other publicity

All stages of a contest should be publicized as much as possible. Placards and stories in employee newsletters should announce it. Standings should be published at frequent intervals. Special signs, banners, and posters can be used for this purpose. Bulletins can be handed out to employees urging care in keeping the record perfect. Trade journals and National Safety Council newsletters are other publicity outlets. In a smaller community, outstanding safety performance by a well-known company deserves—and usually gets—excellent publicity in the local newspaper and on the radio station.

The publicity value of a successful contest is considerable,

**Figure 12-10.** Journalist F. H. Ryan receives a contest prize from Santa Fe Assistant Regional Public Relations Manager Tom Murphy. The contest, Operation Lifesaver, sponsored by Texas railroads, focused attention on railroad crossing safety.

although difficult to estimate. It should be commensurate with the significance of the occasion.

The presentation of an award to an employee who has gone 25 years without a disabling injury has human interest value for both internal and external communications resources. Recognition of an exceptionally fine "no injury" record by a plant or company or presentation of a National Safety Council award would also call for widely disseminated publicity.

Some companies purchase radio or television time to announce the results of a contest. One company had large campaign-type buttons made, and photos of children wearing them appeared in the local press.

Publicity (including photographs) can be sent to the local newspaper. The information should be prepared as a press release, indicating the nature of the contest or award. See Chapter 13, Publicizing Safety, for details of preparing a news release.

### Meaningful awards

An award serves several purposes: an inducement, a good will builder, a continuing reminder, and a publicity tool. To serve these purposes, however, an award must be meaningful (Figure 12–11).

Employees sense when awards are given only for "sales" or publicity purposes and are based on little or no safety effort. When a multiplicity of awards are given the large number can detract from the true value of the program.

The value of awards lies in their appeal to basic human interest factors, such as pride, need for recognition, urge to compete, and desire for financial gain (Figure 12–4). Monetary awards should be presented as a bonus for making an extra safety effort. The distribution of U.S. Savings Bonds or trade stamps for safe records gets away from the direct monetary nature of a financial award.

The originality or cleverness of an award or of its presentation is an important factor. Select awards that are worthy of good publicity, photograph well, and provoke employee conversation. Refreshments for all employees in a department after the completion of a set number of injury-free hours probably will create more favorable comment than would the presentation of a fancy plaque to the department supervisor. Drawing for a small cash prize or a grab bag prize would attract more interest than a routine presentation of the same award. An award to an employee's spouse for completing a home safety checklist, for writing a safety slogan, or for contributing to the com-

**Figure 12-11.** The Campus Safety Association offers two scholarships annually to assist students in their education in environmental health and safety. Each scholarship winner receives $300 and a plaque. One of the 1985 winners, Henry R. Moore, Jr. (center) of Latham, NY, a student at Oklahoma State University, receives a plaque from Richard Giles, director of OSU's safety department. At left is Larry Borgelt, head of OSU's fire protection and safety program.

pany paper would create more interest than the same award given on the job.

- One Council member reported an award idea that received an unusual amount of publicity, both within the plant and locally. A local automobile agency loaned the company a new car that was driven for a week by an injury-free employee whose name was drawn from the hat. The employee had a special "reserved for John Doe" parking space in the company lot. The only cost to the employer was a few dollars to cover special insurance. Even a special parking space awarded on a rotating basis can be effective.
- Another way to gain interest is to let the employees participate in selecting the award, planning its presentation, and helping with publicity. Frequently, they will suggest a humorous or novel award or publicity approach that may attract more interest than one planned by management.
- Payment of bonuses as awards for good safety records evokes considerable differences of opinion. Some managements and safety people feel that this approach is unwarranted as all employees are paid to work safely. Others believe it can enhance an already successful program.
- Some companies raffle household or sports merchandise. Interest in safety among supervisors and workers often is developed by personal awards like wallets, knives, or key cases, often suitably inscribed.
- Many employees value attractive pins or engraved cards commending them for years of employment without an accident. One company places a safety record sticker on the employee's hard hat. Others provide special badges, pins or shoulder patches to recognize safety achievement or service on a safety committee.

In addition to contest awards, recognition should be given to those who have saved lives, served on safety committees, submitted valuable safety suggestions, or made other significant contributions to accident prevention.

### Award presentations

To make an award presentation something special, one company rented an auditorium and invited civic and labor leaders, company executives, other dignitaries, and employees' families to a large celebration party.

Another approach is for the president of the company, or some other high official, to present the awards at a general meeting, a picnic, or a dinner (or breakfast) that may even include entertainment. The reason for inviting VIP's is not only to add prestige to the presentation, but also to promote their interest and commitment to safety.

Such presentations require planning, and must be in keeping with the importance of the occasion. The location chosen for the event should be appropriate and comfortable, not noisy or crowded. An award to an individual might be made in an executive office; a group award might be presented in a conference room, private dining room, or in a company lounge or cafeteria (during a nonrush period).

Brief participants on the agenda. Familiarize those who make the presentation with the significance of the award, the achievement it recognizes, and the background of the individual(s) who earned it. Arrange for reporting and photographing the event. Photographs can be posed after the actual presentation to take advantage of desirable backgrounds, such as the plant or company name, some prominent trademark, or another interest-catching effect. Feature the award itself prominently in the picture. (See the next chapter and the Council Data Sheet 619, *Photography for Safety*, for other ideas.)

For a group award—such as a company, a plant, or a department's completing an injury-free year—free refreshments, such as coffee and cold drinks, can be offered for a specified time—ranging from one coffee break (per shift) to a full 24-hour period.

Another, more elaborate award was given by a company employing about 300. At the end of a year in which there had been no disabling injuries, the president took the group to a major league baseball game. The following year, after the company maintained its no-injury record, the employees, plus their families, were invited to an all-day picnic and cruise.

Such activities help build better employee relations, as they

**Figure 12-12a-c.** Posters convey safety messages forcefully and often humorously, and poster contests involve many people in thinking about safety. These winning posters promote campaigns for good housekeeping (a) and elimination of slip and fall hazards (b, c).

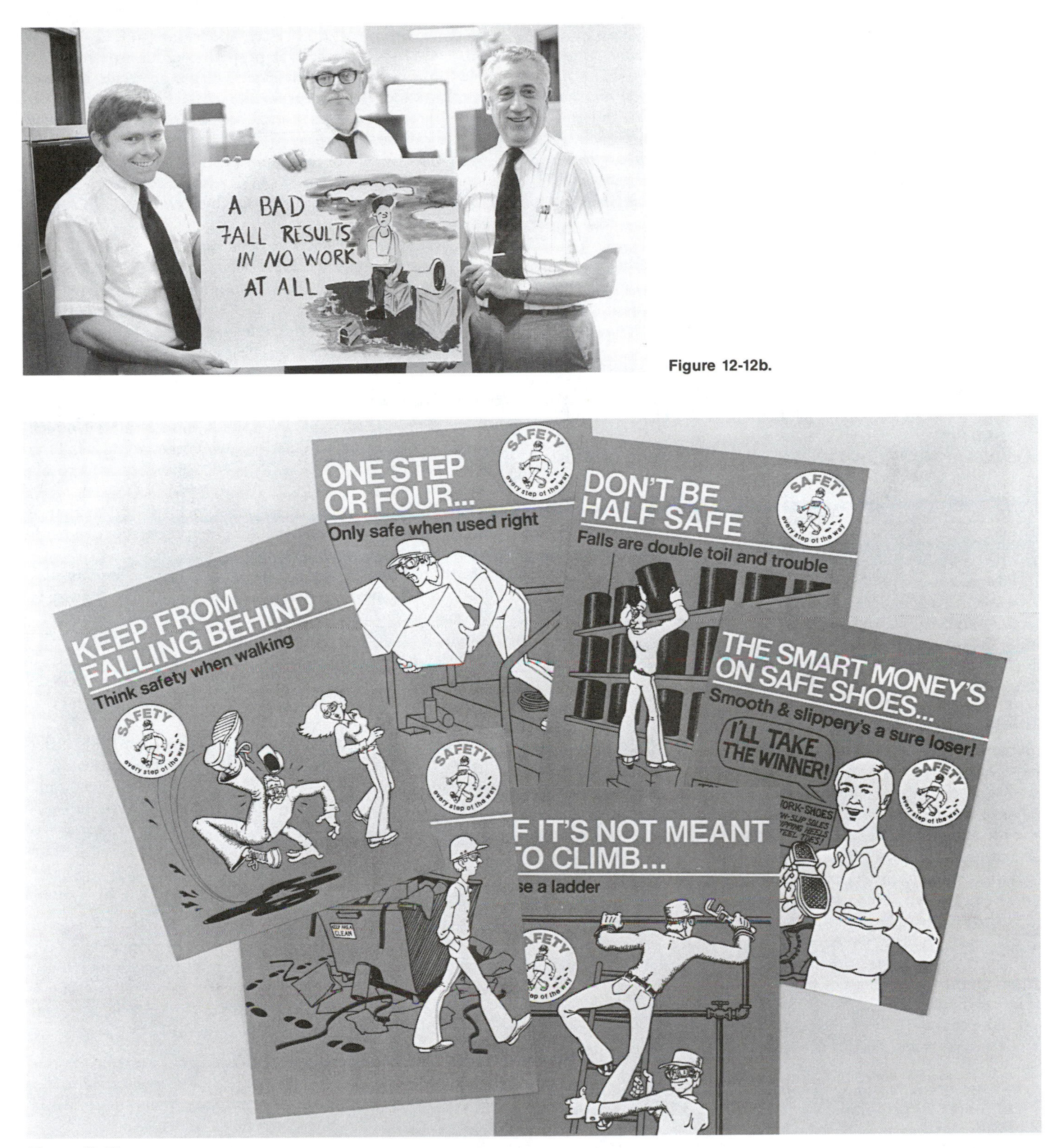


Figure 12-12b.

Figure 12-12c.

promote more interest in safety on the job and within their communities.

## POSTERS AND DISPLAYS

Posters and displays are meant to reach large numbers of people on the move with brief, simple messages, designed to accomplish one or more missions—to convey information, to change attitudes, and/or to change behavior. They are designed to communicate with people who are going about their normal activities and consequently the audience must be seized, the message conveyed, and the contact finished in a very brief span of time (Figure 12–12 a–c).

Safety posters are one of the most visible evidences of accident prevention work. Because of this, perhaps, some companies have mistakenly assumed that posters alone would do the safety job and have neglected such essentials as real management support, guarding, and job instruction. In fact, hit-and-miss use of posters in a plant where no other safety work is done is likely to have a negative influence, making employees feel that the company is not sincere. (See National Safety Council Data Sheet 616, *Posters, Bulletin Boards, and Safety Displays.*)

### Purposes of posters

Posters properly used have great value in a safety program through their influence on attitudes and behavior. One has only to see the efforts that commercial advertisers make to acquire space in business areas or near factory gates in order to appreciate the value of posters inside the workplace.

When selecting posters it is well to have their specific purposes in mind:

1. To remind employees of common human traits that cause accidents;
2. To impress people with the good sense of working safely;
3. To suggest behavior patterns that help prevent accidents;
4. To inspire a friendly interest in the company's safety efforts;
5. To foster the attitude that accidents are mistakes and safety is a mark of skill;
6. To remind employees of specific hazards.

Posters are useful also in supporting special campaigns, for instance, using guards, wearing eye protection, maintaining good housekeeping, offering safety suggestions, or driving carefully (Figures 12–12a–c). They promote traffic, home, and even pedestrian safety by encouraging safe habits both on and off the job.

### Effectiveness of posters

A number of studies have been made on the effectiveness of posters for training and motivating.

- One study, conducted by the British Iron and Steel Research Association, used three posters that reminded workers to hook cable slings. The posters were displayed in six plants over a period of six weeks. A seventh plant was used as a control. Tallies made in the six plants before and after display of the posters showed about an eight percent increase in compliance with the rule. The seventh plant, in which the posters were not used, showed a very slight decrease in compliance.

Although use of the posters merely supplemented previous training, the plants that originally had the lowest rates of compliance showed some of the best gains. Using the three test posters separately, on a biweekly basis, proved slightly more effective than simultaneous use of all three during the entire six-week test.

- In a survey conducted by a prominent casualty insurance company, over 200 employees were interviewed in depth on the effectiveness of safety posters, films, and leaflets. Results indicated that all the media were instrumental in bringing workers to a high level of safety awareness and that all were effective in sustaining that awareness. Employees were found to prefer posters to leaflets, although they acknowledged the value of leaflets for more detailed coverage of, for example, off-the-job safety.
- A survey of Council members indicated that about three-fourths of the nearly 800 respondents use a variety of poster subjects with one-third preferring cartoons and all-industry posters. Horror or shocker posters were least preferred.

Fifty-four percent of the respondents use posters to influence general attitudes; 27 percent to cover special operations; 18 percent, to meet special or seasonal problems; and 14 percent, to promote off-the-job safety.

Sixty-nine percent of the respondents in the Council's poster survey preferred posters of 8½ by 11½ in. and 17 by 23 in. sizes (21.5×28 cm and 43×58½ cm). Of these two sizes, the smaller was preferred six-to-one over the larger.

Frequently, an unusually striking photo or overly elaborate (and expensive) artwork actually detracts from the safety message. This is not to say that "eye-popping" illustrations should never be used; occasional use may serve to attract attention to the more conservative, serious messages on other posters. Be sure photos are related to the message being presented.

### Types of posters

Industrial posters available from the National Safety Council, insurance companies, associations, and other sources fall into three broad categories: general, special industry, and special hazard.

- General posters are concerned with such subjects as chance-taking, disregarding safety rules, forgetting to replace guards, and other human failures.
- Special industry posters, as the term indicates, have application only in specific industries, such as mining or logging.
- Special hazard posters, for example on lifting, ladders, the storage or handling of flammable liquids, are useful in every industry where the particular hazards are encountered. In some cases, a hazard is so serious that a special poster is developed because of the severity and not the frequency of exposure.

The National Safety Council carries more than 1,000 different safety posters in stock at any one time. These range from pocket-size pressure-sensitive stickers to billboard-size jumbo posters. About nine new posters are added each month.

Subject matter is roughly in proportion to the occurrence of certain types of accidents. For example, more posters are concerned with material handling than with chemicals and gases. (For illustrations of the many posters available from the National Safety Council, see the latest *Poster Directory.*)

**Other locations.** Posters and stickers can be mounted on delivery trucks, buses, industrial trucks, mail trucks and carts, in elevators, and even on doors. Pocket cards or plastic pocket protectors, such as those available from the National Safety Council, might be called "walking safety posters."

**Other materials.** Safety messages need not be limited to printing and artwork. They can be very effective when used in illuminated or changeable signs. One company paints a safety message on a plywood welding screen.

**Homemade posters.** A company can develop its own posters to deal with special hazards not covered by posters available from outside sources. Even the smallest company can make an occasional special poster inexpensively, using colored paper, crayons, or felt marking pens, to call attention to a special hazard, to commemorate the winning of a safety award, or to point up a problem not likely to be covered by a commercial poster. See Chapter 14, Audiovisual Media, for details. Even more information is given in the Council book, *Communications for the Safety Professional.*

Effective posters can be made using photographs of local conditions or accidents, even if the situations must be posed. A common type of homemade poster is the "testimonial" featuring a photo of an employee and a close-up of damaged safety glasses or safety shoes, with a brief statement explaining how this equipment protected the wearer.

Homemade posters on new processes, new guards, or new rules personalize the safety program and augment even the best selection of commercial posters. An excellent source of this type of poster is an employee poster contest. It is easily administered at a nominal cost.

### Changing and mounting posters

No specific rule can be given for the frequency of changing posters because of varying definitions of the term "poster."

Some types of posters may well be mounted permanently. For example, a poster on artificial respiration can be kept in the first aid room, or one on the use of a certain kind of fire extinguisher can be posted near it.

Most companies change general interest posters at definite intervals, usually weekly, perhaps rotating them from one area to another or filing them and then reusing them after a year or so.

The type of posters displayed should be varied. Consecutive posting of several infection posters, for instance, or of machinery posters is not desirable unless a special campaign is being conducted. To secure proper balance, it is better to use an eye poster, then a machinery poster, next an infection poster, and so on.

For maximum effectiveness, posters not only must be selected carefully and changed on a definite schedule, but also displayed attractively in well-lighted locations where they will be seen by the greatest number of people. They should be placed on safety bulletin boards, near time clocks, in cafeterias, and at points of special hazard, such as paint storage rooms, rubbish cans, hazardous machinery, or dangerous intersections.

### Bulletin boards

Bulletin boards should permit convenient change of posters and should be placed where employees can see them when they are momentarily at leisure, such as near drinking fountains. They should be centered at eye level, about 63 in. (1.6 m) from the floor. They should be in a well-lighted place; if more light is needed, they can be specially lighted. A good size for a bulletin board is about 22 in. wide by 30 in. long (56×76 cm).

Boards should be painted attractively and glass-covered. One board at a location in the work place is usually desirable, but in lunchrooms or locker rooms several panels may be used effectively. Flashing lights, sometimes desirable in nonproduction areas, are likely to be objectionable in workplaces.

A bulletin board should be used for only one display at a time, but need not carry safety posters exclusively. Any program of mutual interest to company and employees may legitimately use the bulletin boards. In fact, safety posters may have a stronger appeal if they appear on a board on which employees occasionally see displays on other subjects.

Bulletin boards in the same company may range from large, enclosed, illuminated boards with special sections for posters, safety bulletins, and other messages to a number of small frames or other inexpensive poster mounts installed at strategic points.

Black enameled poster frames to which clip-on literature racks can be added are available from the National Safety Council. This permits convenient distribution of leaflets and other pickup literature which support the safety message.

### Displays and exhibits

Displays and exhibits can also be used to promote off-the-job safety as well as work place safety. For example, many companies try to motivate their employees to understand that a happy vacation must also be a safe vacation.

Personal protective devices, tools, and pieces of firefighting equipment can be used to make up displays or exhibits, with or without corresponding posters. (See Figure 12–13.) Another good interest-catcher is a seasonal exhibit combining a Council poster and a safety display featuring such items as proper footwear for winter walking or skin and eye protection items in the summer to guard against sunburn and eye strain.

Signs with changeable letters, electric tape messages, or eye-catching lighting can be used for safety displays.

Many simple and attractive displays have been devised for presenting statistical data to workers. One is a safety clock, the face of which is marked off to indicate the frequency of disabling injuries. Twin clocks or dials often are used, one recording the present rate and the other the rate for the corresponding period of the past month or year.

One company used large thermometer-like boards, placed at every gate and clock-house. Arrows indicated the present and previous month's records. The comparative standings of departments were shown below.

An auto race was the theme of another display; each car represented a department. The cars moved daily to denote progress being made. Airplanes can be used similarly. Another exhibit featured race horses participating in a "Safety Derby" and named after items of personal protective equipment.

A display of highway signs with photographs of accidents below them was placed near the plant entrance of another company.

## OTHER PROMOTIONAL METHODS

Other methods that can be used effectively to arouse and maintain interest in safety are campaigns, safety stunts, courses and demonstrations, publications, public address systems, and suggestion systems.

### Campaigns

Campaigns focus attention on one specific accident problem. They are additions to, and not substitutes for, continuous accident prevention efforts the year round.

Campaigns may promote home safety, vacation safety, fire safety, or the use of safety equipment such as safety shoes. A "Clean-Up Week" may be held, or a "Stop Accidents" campaign may be run to promote the development of safe attitudes both on and off the job.

The National Safety Council's nationwide campaigns, "The Big Plus" and "Team Safety," (and others) included posters, films, booklets, and specialty items to promote and maintain employees' interest. In addition, special materials have been developed in collaboration with trade associations and to support special campaigns such as "Stop Shock," and "Fight Falls."

**Figure 12-13.** Personal protective equipment is the theme of this display at NRPC's Beech Grove, Indiana, maintenance shop. The display is lighted at night and is in Amtrak's impressive 15-foot display case.

Information on current campaign material is found in Council catalogs and other publications.

Many large corporations have conducted extensive campaigns to promote safety on and off the job with much of their safety awareness material aimed at employee families, local citizens, and even groups outside the U.S.

Suitable publicity should be planned for the campaign from kickoff to conclusion, similar to that discussed earlier in this chapter for safety contests. Signs, flags, desktop symbols, and other items can be used to dramatize the campaign. To wind it up, schedule a special event such as giving each employee an inexpensive novelty item, free coffee, or a free breakfast or dinner.

Many of the same promotion ideas that help maintain interest in contests also can be effective in special campaigns. For example, a first aid drill or a demonstration of cardiopulmonary resuscitation (CPR) may be given. Some companies use safety parades, exhibits of unsafe and safe tools and equipment, pledge cards, and other such features.

Timeliness may be an important factor in the way employees respond to a campaign. Successful safety campaigns have been linked to elections, the World Series, the football season, Thanksgiving, and other special events.

See Council Data Sheet 616, *Posters, Bulletin Boards, and Safety Displays,* for more ideas.

### Off-beat safety ideas

Off-beat safety ideas or "stunts" capitalize on all the effective aspects of showmanship and can be developed as separate devices for maintaining interest or can be used to supplement contests and campaigns. *Safety and Health* and other publications regularly give details on various stunts.

Most companies agree that constructive stunts help inspire employees to high standards while stunts that ridicule usually do more harm than good. Moreover, employees and supervisors who are the objects of ridicule may have just cause to blame management for not setting up safe procedures or providing safe facilities and equipment.

Safety stunts can involve an entire company, a department, a small group, or just the individual. A stunt may be humorous, novel, or dramatic, and occasionally even shocking.

A simple stunt is often most effective. A pivoted hammer, mounted over a pair of safety glasses in a display case, can be operated by a string to demonstrate the impact resistance of the glasses. To dramatize the importance of eye protection, the "let's pretend" test can be used. Several volunteers are blindfolded and then asked to eat, write, and move around.

Stunts developed for the company safety program often can

**Figure 12-14.** Local school children are led on a tour of the Canadian Pacific Railroad station in Winnipeg, Canada, by clerk Jim Fisher. This excellent community public relations effort includes distribution of a safety booklet and the viewing of a safety film.

be used at company open houses or safety picnics and in community safety projects as well. Such stunts, when supported by visual aids, signs, and printed material, demonstrate the company's interest in accident prevention and give the employees a chance to participate in programs that help create safer attitudes on the job, too.

A somewhat "off-beat" poster or card serves as an attention getter. One card is used to alert a co-worker that he has exposed himself to an accident.

### Courses and demonstrations

Most health and safety professionals agree that courses in first aid, lifesaving, water safety, civil defense, and disaster control have bonus values that help prevent work injuries, too.

The worker who has completed a course in first aid and has learned to do CPR will be more aware of the hazards of electric shock and more likely to help maintain electrical equipment in safe condition. Likewise, the employee who learns how to stop arterial bleeding better appreciates the consequences of using a saw or a power press or cutter without the guard.

Home study and extension courses also serve to stimulate and maintain interest by giving the employee a better understanding of the job as they dispel unsafe attitudes. Most safety training courses, in fact, are designed specifically to improve the attitudes of both supervisors and employees. Using appropriate video tapes and other visual aids will enhance the effectiveness of the courses. (See Chapter 14, Audiovisual Media.)

Public participation in courses involving employees promotes community good will. Many industrial safety people are doing an excellent job of promoting safety and fire prevention by arranging or teaching courses on these subjects for the Boy Scouts and Explorer Scouts, the Girl Scouts, Camp Fire groups, Junior Achievement companies, and other youth or school groups (Figures 12–14 and 12–15).

The National Safety Council's "Defensive Driving Course" provides an excellent means of promoting good employee and public relations. It stimulates safer attitudes both on the job as well as off.

Fire equipment demonstrations have a practical value beyond that of teaching employees how to react in an emergency. The fact that the equipment is provided for their use reminds them of management's concern about their welfare. Moreover, the demonstrations make employees more aware of the dangers of fire and point up the need for obeying fire prevention rules.

Such demonstrations are easily arranged through local fire departments or fire equipment distributors. Many companies conduct their own demonstrations, using extinguishers that require recharging, or "not-in-service" extinguishers kept specifically for this purpose.

**Figure 12-15.** Mattel Inc.'s Rainbow Brite® and Twink® characters participated in a children's program during a National Safety Congress joint session of the Educational Resources and Traffic Safety Divisions. Using nationally known characters helps gain wide local and national publicity for a safety effort.

## Publications

**Reports.** Safety program progess reports should be written in an interesting and concise manner for superiors and supervisors. Visuals can be used effectively where appropriate. (See Chapter 14, Audiovisual Media.)

The cost of accidents and, perhaps, the cost of prevention should be presented in terms that are significant to management, such as medical and compensation costs, production losses, sales losses, increased maintenance costs, and the less tangible, but perhaps more important, hidden costs involved in administrative problems and in impaired public, customer, or employee relations. Reports need not be dull. Photographs, for example, can pin-point a company's major sources of disabling work accidents.

A statement of accident losses and safety achievements may be included in the company's annual report or in a special annual or monthly safety report issued to top executives and supervisors. If departmental accident losses, like incidence rates, can be charged on an equitable basis, such as "per hundred thousand dollars of sales" or "per one thousand employee-days of production," the comparative standings of departments and improvement in departments or units are easy to evaluate. (See the discussion in Chapter 6, Accident Records and Incidence Rates.)

The fact that such information is recorded and publicized is in itself an incentive to supervisors. It reminds everyone concerned that accident costs are just as much an integral part of profit and loss as production, sales, maintenance, distribution, and advertising.

Special charts, graphs, and statistical reports also can visually show facts about accidents. One chart can track the number of disabling injuries, others the number of days lost, injury causes, accident causes, or body location of injuries. It cannot be too strongly emphasized that unless such charts are kept up to date they can do more harm than good.

**Annual reports.** In recent years, companies have worked hard to make their annual reports to stockholders interesting and clearly understandable. In many cases, annual reports are also distributed to employees so that they can become better acquainted with the company's purposes and problems. A section on aims and accomplishments in accident prevention attracts employee interest and further serves to emphasize the interest of management in the safety of its employees.

**Newsletters.** Monthly or weekly newsletters are especially important as a means of maintaining interest. They keep employees and supervisors informed, particularly in decentralized or field operations where bulletin boards are not practical. Newsletters can give detailed information on standings in a safety contest and publicize unusual accidents or serious hazards. They can help explain safety rules, remind employees of safe work practices, and support the safety program in general. If workers can serve as "reporters" or help produce such a newsletter, so much the better. (See details in the next chapter.)

A case history of a particularly unusual or spectacular accident can sometimes be featured.

**Booklets, leaflets,** and personalized messages take many forms: safety rule booklets, special "one-shot" leaflets, monthly publications such as the National Safety Council's *Today's Supervisor* magazine and its *Safe Worker* and *Safe Driver* for employees, and letters from management.

The content of an employee rule booklet, except for material involving company policy, may be developed with the help of safety committees or selected workers as a means of stimulating interest and helping ensure compliance with the rules.

Larger companies may have their own editors and artists, even their own printing facilities. Smaller companies, however, also can issue attractive booklets, leaflets, and personalized messages, and at negligible expense by using local consultants and printers.

The National Safety Council, trade associations, and professional organizations publish a wide variety of booklets and leaflets that are authoritative, attractive, and relatively inexpensive. They cover a wide range of subjects—material handling, first aid, housekeeping, fire prevention, vacation safety, safe driving, and the like. Such materials, carefully selected and regularly distributed, effectively supplement company-prepared publications.

Letters commending meritorious service, signed by the manager and addressed to individuals, make an excellent impression upon workers.

Safety calendars, published by the National Safety Council, together with Christmas letters from the manager have a direct appeal that reaches the workers' homes. Such mailings should include all employees. *Family Safety and Health,* a quarterly publication of the National Safety Council, is sent to more than two million homes by managements who are interested in the welfare of both employees and their families.

**Buttons, blotters,** book matches, pencils, and other small novelties, all conveying a safety message, also may be used. For example, booklets of silicone tissues for cleaning glasses, imprinted with brief messages or safety rules, serve to remind employees of the rules and to encourage proper use of safety equipment.

### Public address systems

Public address systems often are used to broadcast announcements and page employees. Many companies have taken advantage of these installations to broadcast safety information.

Such messages should be planned carefully. Employees might readily lose interest in long speeches or too-frequent safety reminders. When the public address system is used for broadcasting music, safety announcements can be made between numbers.

### Suggestion systems

Because accident prevention is closely associated with efficient operation, many suggestions not only help prevent accidents, but also lower production costs; not only improve manufacturing conditions and methods, but also change the outlook of workers.

If a company does not have a general suggestion system, it is probably better not to establish one for safety suggestions alone. Setting safety apart from ordinary operating procedures may de-emphasize its importance.

Getting good suggestions is important and must be encouraged by all levels of management. Posters, contests, campaigns, merchandise incentives, direct mail, printed handouts, personal appeal, supervisory training, safety clubs, and press releases are employed to motivate employees to submit safety suggestions.

Effective suggestion systems, like other things in life, cost money. Many companies pay considerable sums for employee suggestions—one firm awards more than $10 million annually, but feels the money is well spent because of the value of the suggestions.

To merit an award, a safety suggestion, like a production suggestion, should be substantial, be practical, and be a real solution. Changing a method or material, guarding a hazard, inventing a safety tool or device are examples of suggestions worthy of awards. Erecting a sign, cautioning workers, or publishing slogans are examples of ideas usually not considered eligible for an award.

Safety suggestions generally are regarded as a highly desirable way to avoid safety grievances. Improved employee interest and personal involvement are additional benefits.

**Suggestion awards.** It is easy to measure the monetary value of suggestions that result in greater efficiency, lower material cost, decreased labor cost, or reduced waste. Usually awards for suggestions in these categories are in proportion to the savings derived by the company. Although some safety suggestions also have a monetary value, they are hard to evaluate; hence, payments for suggestions that contribute to the welfare of the employees, but result in no direct savings to the company, are most often estimates or composite judgments. Some firms have developed guidelines which consider such factors as the degree of hazard, originality, extent of application, etc. One company has also developed an award guide based upon disability cost experience.

Distinguishing between "safety awards" and others is a mistake that can result only in a feeling that safety is regarded by the company as a sideline of no great importance. Payment for a safety suggestion must be on the same basis as that for other suggestions—it should be based upon its real worth if it can be determined. If a suggested safety device enables an operation to be run at a speed that would be dangerous without the device, a saving may be measured. If a number of accidents have occurred in an operation and a suggested device will eliminate them, the cost of those accidents can be projected and a saving calculated.

Awards should reflect the merit of the suggestions. Most companies award cash and/or bonds. Some award merchandise, all-expense-paid trips, company stock, certificates of merit, medals, gifts for the suggester's spouse and family, or attendance at a recognition luncheon.

Some companies exclude superintendents, supervisors, designers, methods and systems personnel, and other supervisory or technical personnel from receiving awards, so that the other workers will have someone to whom they can go for assistance. Some companies feel supervisory personnel should not be excluded and several firms have separate plans and award schedules for salaried, supervisory, technical, and management personnel.

**Suggestion committee.** If necessary, a special subcommittee can be set up to determine the monetary value of safety suggestions so that employees will be rewarded for them exactly as they would be for other money-saving suggestions. However, some firms believe such special treatment sets safety ideas apart from other ideas.

Many established suggestion systems now in operation are producing excellent results in monetary and "people" savings. No company should start a suggestion plan or decide not to start one without first studying carefully the plans now in existence. Further information can be obtained by contacting the Executive Director, National Association of Suggestion Systems. (See References.)

**Boxes and forms.** Suggestion boxes should be attractive and well placed, and stocked with special blank submission forms. Commercial suggestion forms also are available. To increase the interest of the employees and establish a spirit of cooperation and importance, it is essential that management acknowledge and resolve all suggestions promptly.

### Recognition organizations

Some organizations recognize people in the United States and Canada who have eliminated serious injuries, or who have minimized them by using certain articles of personal protective equipment. The award provides an excellent opportunity for publicity.

Three of these organizations are:

Wise Owl Club. Founded in 1947, this club honors industrial employees and students who have saved their eyesight by wearing eye protection. Address inquiries to: Coordinator, Wise Owl Club, National Society to Prevent Blindness, 79 Madison Avenue, New York, NY 10016.

The Golden Shoe Club. Awards are made to employees who have avoided serious injury because they were wearing safety shoes. Address inquiries to: Golden Shoe Club, 2001 Walton Road, P.O. Box 36, St. Louis, MO 63166.

The Turtle Club. Founded in 1946, the Turtle Club recognizes and honors workers who have escaped serious injury because they were wearing a hard hat at the time of an industrial accident. Although OSHA mandated the use of head protection in many U.S. industries in 1971, the Club's goal—to save lives and prevent injuries through promoting the use and acceptance of hard hats—and its purpose—to assist industry in creating awareness of the benefits of wearing industrial head protection—remain important today. In 1984 the Club's sponsor, E.D. Bullard, reopened membership and has admitted over 200 new members (Figure 12–16). Address inquiries to: Mr. E.D. Bullard, International Sponsor, The Turtle Club, P.O. Box 9707, San Rafael, CA 94912-9707.

**Figure 12-16.** The Turtle Club honors people who have escaped serious injury because they were wearing a hard hat. Displayed here are the items new members receive: a hard hat with the club insignia, a membership certificate, a wallet card, and a lapel pin.

## REFERENCES


Alliance of American Insurers, 1501 Woodfield Rd., Schaumburg, Ill. 60195.
*Fire Prevention and Control,* A/V.

Hannaford, Earle S. *Supervisors Guide to Human Relations,* 2nd ed. Chicago, Ill., National Safety Council, 1976.

Indiana Labor and Management Council, Inc., 2780 Waterfront Parkway, Indianapolis, Ind. 46214. *Worker Involvement in Hazard Control,* 1985.

Konikow, Robert B., and McElroy, Frank E. *Communications for the Safety Professional.* Chicago, Ill., National Safety Council, 1975.

National Association of Suggestion Systems, 230 N. Michigan Ave., Chicago, Ill. 60601.

National Safety Council, 444 N. Michigan Ave., Chicago, Ill. 60611.
*Accident Facts* (annual).
Catalog and Poster Directories.
*Family Safety and Health* Magazine.
Industrial Data Sheets
*Motion Pictures for Safety,* 556.
*Nonprojected Visual Aids,* 564.
*Photography for Safety,* 619.
*Posters, Bulletin Boards, Displays,* 616.
*Projected Still Pictures,* 574.
*Writing and Publishing Employee Safety Regulations,* 664.
*Industrial "Newsletter."*
*Today's Supervisor* Magazine.
*Safety and Health* Magazine.
*101 More Ideas that Worked.*
*Safe Driver* Magazine.
*Safe Worker* Magazine.
*You Are the Safety and Health Committee,* rev. 1986.

# 13

# Publicizing Safety



The preceding chapter covered "internal" publicity. This chapter discusses how to influence the way a company looks to people on the outside. Favorable publicity is an unmistakable bonus to a good safety program. Why it is so often left uncashed is difficult to understand.

Any company likes to have someone—especially a prospective customer—say, "I like what I hear about this company. I understand that it really takes care of its employees. So I figure it must treat its customers right; therefore, I'll be treated right."

One good way for a company to get a reputation for taking care of its employees is to be known as a really safe place to work. Yet an amazing number of companies do little or nothing to let their public—customers, stockholders, the community—know that the safety and health of their employees are important to them. That is what this chapter is all about. It is an effort to present a simple and sensible formula for letting people know that "at my company, the health and safety of the workers are important."

Most companies have a professional public relations department which handles the communication program. In smaller companies, the safety director may have to handle public communication. In both cases, the information in the following pages should prove useful. The safety professional should be aware of overall company policies and programs and know when to turn to specialists, and when to ask for creative help.

## PUBLIC RELATIONS

Public relations is the "management function which evaluates public attitudes, identifies the policies and procedures of an individual or an organization with the public interest, and plans and executes a program of action to earn public understanding and acceptance," according to the magazine *Public Relations News*.

Abraham Lincoln knew about public relations. Speaking at Ottawa, Illinois in 1858, Lincoln said: "Public sentiment is everything. With public sentiment nothing can fail; without it, nothing can succeed. He who molds public sentiment goes deeper than he who enacts statutes or pronounces decisions. He makes statutes or decisions possible or impossible to execute."

Lincoln's classic quotation on public sentiment can be traced to Jean Jacques Rousseau, the eighteenth century French philosopher who is generally credited with developing the term *public opinion* as we use it today. At its simplest, public relations is the promotion of good public opinion about a person, a company, a government or other entity. Thus, every employee, every activity, every facility of a company contributes in many ways to the overall opinion that persons outside the company have about that company. This is true public relations.

Anyone concerned with accident prevention in any way—safety and health professional, supervisor, member of the plant safety committee, or officer of the school or community safety council—should realize that any time there is communication with someone outside the committee, department, or company, public relations is involved. Even a family picnic can strengthen public relations.

Any public relations program has to be backed up by a sound organization. Public relations reflects the quality of an organization—but it cannot create that quality. Successful safety achievement merits and can result in good publicity, but canned publicity or publicity based upon inflated facts or specious statistics will be recognized for what it is—and can do more harm than good.

Public relations information, to "click", need not always be red-hot news, but it must have an element of spot news, or human interest, or self-help. Then it will have feature value.

Activity—real, honest, legitimate activity—makes news. Of course, urgent need or dramatic circumstances help make news, too. Until they turn up, however, genuine effort will go a long way toward giving a program news and publicity value.

### Basis for success

The basis of a successful public relations program is a successful management—management that makes sure that staff and employees produce good products safely and efficiently, that they cooperate with each other and with the customers, and that all give the best and friendliest service humanly possible—and give it at all times.

The plain fact is that poor public relations is costing individuals and organizations in this country millions of dollars each year. The remedy is simple: a better understanding and use of fundamental public relations on the part of everyone—and a sincere effort to put it into practical use. For lack of good public relations, many a worthy cause has failed to get the support it deserves, and many an organization has failed. A good public relations program need not cost a great deal of money. But it is worth time, effort, and a reasonable budget.

## THE VOICE OF SAFETY

In any genuine, effective public relations (PR) program, emphasizing safety can be a real help. In fact, it is hard to imagine a PR program where sincere and effective concern for the protection of employees from accidents is not a top priority.

If a company does not have a safety program, it misses vital opportunities for good public relations and dramatically increases the chance for adverse public attitudes. A safety program, properly managed and communicated to internal and external audiences, can help offset news of such incidents as employee deaths or company accidents; the safety program is long-range and sustained whereas "news" is here today and replaced by other news tomorrow. The time to implement a safety PR program is not when trouble strikes, however. The program must be in place and functioning, and the public must know about it.

The health and safety professional should not only welcome publicity for safety efforts, but should energetically seek it.

### Working within the company

First, it is essential to talk a little about the basic facts of public relations. There are two questions to be answered:

1. Does the company have a public relations department?
2. Is there an employee publication in the company?

If both answers are "yes," the safety professional should get in touch with both these units before doing anything about publicizing the safety program. This step is important. It not only assures professional skill and consistency of efforts to publicize safety activities, but it will save confusion, avoid duplication, and possibly prevent misunderstanding.

Why does such an obvious procedure have to be mentioned? The reason too often is that there is little, if any, communication between a safety and health professional and the public relations department and publications editor.

Communication between the health and safety professional and the PR staff and editor is indispensable, for these three must work together, or safety is not going to get the attention it deserves and needs.

The health and safety professional should explain to the PR staff and the editor, if this has not been done before, that accident reduction at the company is a priority and that their communications help and advice are needed to reach this goal. They need to understand the necessity of employee and public acceptance of the safety program and how much the health and safety professional is depending upon them to help earn acceptance.

There is a wealth of real news in safety, and there is a strong possibility that everyone in the safety business has been too quick in assuming that safety must by its very nature be on the dull side. In recent times, there has been more and more recognition by more and more writers and others that safety can be made interesting. It just takes the combined efforts of health and safety professionals, PR people, and editors to do the job.

### Sixteen ways to make safety news

It might be useful to list some of the things that can make safety news in an organization and that editors and PR staff ought to know.

1. No-accident records for the entire company or for any one unit—in terms of either days or worker-hours (Figure 13-1).
2. Improved safety records for the company or for any one unit, even if no prolonged no-accident period is involved.
3. An interplant safety contest, or an intercompany contest—especially if anyone has dreamed up an unusual angle or prize.
4. Any unusual safety record for safety performance by an officer or employee of the company—either in length of time or character of the job done.
5. Innovations in safety programs of the company that will prevent accidents. An invention, too, has special news value if the company has been plagued with accidents the new gadget may prevent.
6. An unusual or highly valuable safety suggestion by an employee.
7. Safety conventions or meetings, either those held by the company or those held elsewhere, to which company representatives will go. A digest of such meetings should be publicized.
8. Other special safety events besides conventions—a safety banquet, a safety training course, fire and first-aid demonstration, a special meeting, or an award ceremony.
9. Some unusual event intended to get the employees to take their safety training home to their families, or something the company is doing directly with the families of workers in an effort to promote around-the-clock safety. Open-house tours, local water safety shows or public showings of safety films are examples.
10. Some pronouncement or statement by the president or other high officer of the company on some unusual or new safety device or company safety service, such as free inspection of employees' cars.
11. A speech by the head of the company or the health and safety professional at a local, regional, state, or national safety convention or conference. The editors and public relations staff should have advance copies of it. The per-

**Figure 13-1.** A seven million worker-hour record was publicized by featuring the record itself. (Reprinted with permission from du Pont of Canada, Ltd, Shawinigan Works.)

son making the speech should be sure to say something worthy of public attention.

12. The company's annual report is the foundation for corporate communications. Stockholders *do* read these reports. A good paragraph or two on the safety record for the past year will go a long way in achieving sound publicity within the corporate family, as well as inform the analysts, who recommend stocks, what the company is doing beyond its financial performance.
13. Any act of heroism by someone in the company. This is a sure-fire story for local papers as well as company publications. Maybe this type of news is not pure safety, but news media regard it as part of safety, and it can always be tied in with an indirect safety message (Figure 13-2).
14. A survey or study of some phase of accident prevention in the company. If the investigators discover that married people who own their own homes are safer than their single counterparts, they have provided a ready-made story.
15. Election or appointment of a company officer or safety professional to an important post as a volunteer officer of the National Safety Council, American Society of Safety Engineers, Board of Certified Safety Professionals, local safety organization, or governmental agency.
16. A company or employee winning an award in a National Safety Council contest. Winners, not losers, are publicized.

A good rule of thumb is to stress the positive, rather than the negative side of an event. For example, instead of a story that says "single people are less safe," it could say that "a company study shows a need for special safety efforts by singles."

## SOME BASICS

If a company does not have a public relations department, this fact need not prevent its getting information into local papers or on the air.

It hurts, of course, because PR people are more experienced and naturally know their way around in media circles better than the health and safety professional does. However, even in the absence of a company public relations department, the health and safety professional can get publicity for safety activities by going directly to the newspapers, magazines, and radio and TV. It helps to solicit and secure PR advice from local safety organizations, or even business or trade associations with such service. The health and safety professional should always keep the company management informed of what is going on.

The health and safety professional should not pretend to be a public relations expert. Instead, he or she should be sincere in approaching editors and TV/radio program mangers with sound, worthwhile information and not worry if it is packaged perfectly for media acceptance. A good story is a good story. Most newspersons are glad to have the material and often will use it. News value and reader interest are what count the most.

Editors and program people are usually not difficult to approach—provided that the health and safety professional is

**Figure 13-2.** Rescuers carry a concrete worker whose small boat capsized, throwing him into the chilly, turbulent water of the Mississippi River. He went over two spillways, one of them 30 feet long, before he was rescued. He was wearing a U.S. Coast Guard approved Type III work vest, which also had OSHA-required safety features and hypothermia protection. (Reprinted with permission from the *Minneapolis Tribune.)*

courteous and friendly and admits to a lack of specialized knowledge of PR techniques. Naturally, no editor or anyone else likes to have someone come charging in and claim to be doing a big favor by delivering the story the world has been waiting for.

Although the common sense and salesmanship needed for success in the safety field certainly are enough to enable the safety professional to present his or her case clearly and effectively to the paper or radio or TV station, there should be willingness to accept advice from PR professionals on how to best tell the story.

### Select the audience

"Who must be reached with safety information and is it of interest to them?" These seem like fundamental questions, but many companies never try to answer them. Industrial or manufacturing companies, of course, would scarcely mind if the whole populace insisted on reading or hearing or looking at the company's public relations information and taking it deeply to heart.

Because publicity cannot reach everyone, the audience must always be chosen carefully, especially if the budget is tight or time is limited. In that case, it is logical to assume that—in addition to the in-plant (company) audience—the company would prefer to reach people who might be in a position to buy the product, or help the company in some other direct and profitable manner.

### Use humor and human interest

It is worthwhile to try to brighten safety, to make it positive, rather than ponderous and dreary. It is even possible to evoke a chuckle now and then.

Editors are familiar with the solemn pronouncement that "Safety is a serious subject, and must be taken seriously—safety is no laughing matter." No one can argue with that position.

**Figure 13-3.** Awards can be displayed against meaningful backgrounds. Clyde Nyquist, a senior warehouse specialist at Abbott Laboratories, North Chicago, Ill., poses with the trophies he won at the 10th annual International Materials Mangagement Society Fork Lift Truck Rally. The news release that Abbott sent out stressed the company's lift truck operator training programs, refresher courses, and good safety and production records. (Reprinted with permission of Abbott Laboratories.)

Of course, safety is a serious subject. Of course, an accident is no laughing matter. But does it follow, therefore, that no one can put into safety—the enemy of accidents—some of the same techniques, the same sales appeal, the same sparkle that are used so successfully to sell all the things people need to keep them shipshape? (See Figure 13-3.)

If those techniques can sell shampoo or a personal care product or an automobile, is it unreasonable to expect they can also sell safety?

Or how about a cartoon treatment? This just may brighten what might otherwise be a slightly dull and drab presentation.

The health and safety professional should not be too disturbed if someone points out that a cartoon has treated safety negatively. It may well have done just that. This is the very thing that gives a cartoon its punch. A "prat-fall" cartoon will draw attention to the slippery, icy sidewalk in a way that cannot be shown by a person walking and *not* falling. There is no need to dread being negative now and then. However, care should be taken to avoid ridiculing or negatively portraying victims of accidents.

It is even possible to get a cute child or baby into the act, or even a faithful, shaggy dog, in order to get that spark, that punch, that human touch that lifts safety activities out of an impersonal rut.

### Names are news

Remember that facts and figures about injuries and their frequency and severity are made more interesting by a good-sized injection of human interest in safety news. Human interest means people. The health and safety professional who wants to communicate must talk more about people and safety, and less about things and safety.

### Friendly rivalry

Safety awards, safety records, safety contests, safety inventions, and "gimmicks"—these are only a few of the many things that make good safety news.

If the comapany has reached a new injury-free record in its industry, it is headed for headlines. The editor and the PR department must be kept informed all along the way. They will help arouse public interest in the performance, and also stimulate greater interest and greater effort among the employees themselves.

An award is worthless if kept a secret. It is worth only what is made out of it. Photos of award presentations are sometimes used but they should be interesting and even unusual to attract special attention. Editors dislike the typical "grip and grin" award photos, so be inventive in trying to get a picture worthy of publication.

In some instances, top safety awards have been accepted by some companies as if they were a "dime a dozen." On the other hand, other companies have made similar awards the occasion for some of the biggest, bell-ringing celebrations ever seen—and "safety stock" took a big rise as a result.

A public utility company in Michigan, for example, made so much of its award-winning safety records of the various units throughout the state that an outsider might think the company had won the World Series.

At one event marking the celebration of such a victory, more than 1,000 employees, from the top brass on down, jammed the closed-off street in front of company headquarters for presentations, followed by an all-company picnic.

This celebration got coverage in the papers and on the air throughout the state of Michigan. Here was public relations that any organization would welcome. Safety had made news.

### Techniques

These pointers are offered to the safety professional who wants to make the most of public relations opportunities:

1. Be honest in what you say. Never exaggerate.
2. Deliver what you promise. If you say to the media that something is going to happen, make certain it happens—and as you said it would. This often calls for a "runthrough" in advance.
3. If for any reason there is a change in plans from what you have announced, notify the papers and radio and TV stations at once.
4. Be scrupulously accurate in your names, places, and other facts. There is no such thing as being too careful in this respect. If an editor misspells names, the only thing to do is to resubmit the names correctly spelled again and hope for the best.
5. Offer ideas but do not ask for specific space or time. Complaining about your company PR department, or complaining that the local paper or station has treated your company shabbily will not only accomplish exactly nothing, but will make for a bad relationship.
6. Do not alert your PR department or publications editor (or put out anything yourself) unless you have real news or features to offer. You must not issue material just to be issuing it. Be reasonable with the amount of material you send out. You can wear out your welcome.
7. Tip off your local or industry association, safety councils, and your PR department (or if you do not have one, the papers or radio and TV stations) to anything worthwhile you run across that might make an item or program for them, even though it has no relation to you or your company. They will appreciate it.
8. Above all, do things that make news. Almost every routine safety item can, with a little extra effort by the health and safety professional and the editor, become a more readable, more constructive piece. News will be published only if something is being done for safety that makes news. News can always be heightened by intelligent, imaginative treatment, but it must be there in the first place to be worth telling. Advertising space can even be purchased for special items.
9. Use good sense and an honest approach if the news is bad. Prove to press representatives and the public that you can roll with the punches. (Check with legal counsel and public relations officials on how far you need go, however.)

### Communication by the safety office

If a health and safety professional must handle his or her own publicity with the local media, here are some tips. These hints might seem unnecessary, but many stories have died because someone failed to observe them.

1. Editors and news directors are busy people. Unless a situation is an emergency, media people prefer to receive possible story ideas in writing. Send them a brief release or a letter outlining what you have to offer. Include facts and by all means make your material interesting. Timing is vital: Don't notify the media today of something you are planning for tomorrow. And, never suggest that a reporter or news crew be sent out—this is the editor's or news director's province, not yours.
2. Generally, the person to write to is the city editor of the paper or the news director of the radio or TV station. Of course, if an item is specifically written for a certain columnist or commentator, it is better to make direct contact. If it applies only to a specialized area (finance or sports, for example), it should be brought to the attention of that editor. Know the publication, whether it is general news or a trade journal. If you are working with TV or radio, don't waste their time on information or items they'll never use. (Example: broadcast media *seldom* mention personnel changes.)
3. Write not for your boss, but for your reader or listener. Answer objectively the questions: who, what, when, where, how and why? Do not load your releases with propaganda for the company. There is no surer way to kill your positive relations with the media.
4. Make your releases just as professional in style, appearance, and general quality as you possibly can.
5. In writing a release, be brief and to the point. Newspaper space is limited, and costly. Try to "hold down" the piece to a page. Papers receive thousands of releases each month. These are skimmed, and only the best get into print. Likewise, remember that TV and radio broadcasts usually are measured in seconds, not minutes. Keep your stuff short.
6. If you are sending a picture with the release, the caption should be typed on a piece of plain white paper and taped to the back or bottom of the photo. *Never* use paper clips, and *never* write with a pen or pencil on the back of the picture. Either of these will likely damage the photo and make it difficult to reproduce clearly. Think *visuals* when working with TV—what can they show that will help to tell or illustrate your story?
7. Leave script writing to the professionals but check facts. If a radio or TV station requests material, send them the facts, figures, and whatever narrative is necessary. The people at the station will put it into the proper form.

### Hints for TV interviews

The health and safety professional must often be the spokesperson for his or her company, not only for newspaper coverage, but also for radio and television. Although getting the facts correct may be adequate for a newspaper interview, a television or radio inteview reflects more of the company than merely what the facts show. It projects a company image through the company spokesperson. If you do not feel that you project a good image over the radio or television, pick someone who will, in your department, or in the public relations department.

Here are some tips that will help you give a better radio or TV appearance.

1. Remember that you are being interviewed for your knowledge, not for your personality, entertainment value, or good looks. Be yourself. Don't put on a special voice or worry how the lavaliere microphone looks with your clothes. Don't wave your hands or touch your face or hair. Don't jingle coins or play with jewelry. Keep your hands down.
2. If your TV appearance is preplanned, dress conservatively. Avoid loud clothing or busy pattern that can affect the camera or overpower your message.
3. Go over with the inteviewer in advance just what areas will be discussed. You can steer away from areas you cannot discuss, and you can get help on questions that you might not be able to answer.
4. Be natural and cordial. Smile, if the situation is friendly. Be sure to maintain eye contact with the interviewer or camera. Do not memorize a statement and rattle it off. You will waste everyone's time. Better, be well-prepared and let the on-air material be a question and answer or discussion between you and the interviewer. Don't rush and don't try to cram a lot of information into a TV slot. Keep it short and light. Avoid statistics.

### Handling an accident story

In any public relations program, it is just as important to know what not to do as to know what to do. In fact, it can be even more important.

The foremost warning is this: do not cover up bad news. Good media relations are of utmost importance. It is at such a time that a sound public relations program "pays off."

Every health and safety professional hopes the day will never come that an accident—a bad one—occurs and damages a company's safety record, but it has happened. In some instances, the repercussions of the way the accident was handled have been even more tragic than the accident itself—at least, to the company as a whole.

Here is an example of how *not* to handle a press representative: In a midwestern city some time ago, two workers were killed by a crane. This company enjoyed a first-rate relationship with the newpapers and radio and TV people in that city. It worked hard at safety and at public relations. It was good to its employees and had a fine reputation for playing its cards fairly and on the table.

On this particular occasion, however, someone in the company's higher echelons got "buck fever." So, when a reporter came out to the plant to get what was to her paper a routine story of the accident, she ran into censorship at the plant.

The health and safety professional shoved the reporter off to the personnel manager. This person switched her to the general manager, who gave her some "double talk" and deferred to the company doctor. The doctor said the health and safety director was the person to talk to.

By this time the reporter's righteous wrath was rising. She knew she was getting the treatment, and what had started out to be just a routine assignment now had become a challenge to dig up something that, for some reason, appeared to be covered up by the company.

The reporter could not lose in a contest like this. Since the workers had been killed, the coroner would have all the facts. If they had been badly hurt, one of the hospitals would have the information. If the workers had not been killed or hurt badly, it was no story in the first place.

So the reporter got the facts from the coroner's office, and wrote a story that was just as nasty toward the company as it could be without committing libel.

The story was edited, headed, set in type, and lay in the composing room, awaiting its turn to get into the paper.

Now at this paper, as at every other paper, there is usually more news set in type than the paper can print. Each day dozens of items get left out—the "overset," as it is called in newspaper parlance.

The story of the accident might well have ended up as "overset" and, if printed, might not have been played up. These no longer were normal circumstances, unfortunately. The coverup and runaround the reporter had received at the plant had changed all that.

This little story had been marked "must" when it was sent to the composing room. It thereupon became something very special—a story that was now given front-page prominence.

When there is bad news to report, the health and safety director will just have to swallow hard, and back up the public relations department 100 percent in giving out the news as straight and fast and completely as if the tidings were all in the company's favor. Unless directed to deal with the media, the safety professional should stay in the background and provide the proper information to management and public relations.

Along with the grief, a mention of the good things—that this is the first accident in months or years, that the company has a safety record far better than the national average for its type of operation, and that it has won a number of safety awards—will help take the curse off the story. Reporters are usually willing to include these facts, too.

It is not only fair and honorable, but downright smart to "lay it on the line" for press representatives whenever there is news, regardless of whether it is pleasant or unpleasant news. This principle is vital to a good public relations program.

It would be wise for a health and safety professional to anticipate that some day he or she may have to serve as a company spokesperson at an accident or disaster scene. It is imperative, therefore, to seek legal counsel to make certain how much can be said in a press interview.

News media can actually help during a big emergency. Families, friends, and neighbors will be clamoring for news and the media can get it to them fast. Details of any casualties must first be given to next-of-kin.

## WORKING WITH COMPANY PUBLICATIONS

If the health and safety professional thinks the company publication has neglected safety, it is time to correct the situation. The editor should be asked how more news value and human interest can be worked into safety stories. The health and safety professional should tell the editor he or she wants the program to be just as newsworthy as possible and offer to provide details and descriptions of events. The idea is to give the editor plenty of good, current information.

The editor is just as eager as anyone to publish interesting news and features, and will go more than halfway to think up ways to put news value and reader interest into safety doings.

Here is an example of how the health and safety professional and the editor can team up to make a routine safety happening more newsworthy.

Suppose one of the employees, Oscar B—, reaches his twenty-fifth anniversary of steady work without a day's lost time due to an injury. This achievement probably entitles Oscar to a button or a badge or a plaque or something.

The public relations-conscious health and safety professional asks, "Well, instead of just pinning this button on Oscar with a hearty handclasp and a few words of commendation, why not make a real thing out of it? Take the occasion to tell Oscar—and all the world—that at this plant there is nothing more important than recognizing the contribution to a safer, better way of work that Oscar has made through his personal example of safe practices over the years."

Spurred by the talk the health and safety professional and the editor have had recently about perking up safety news, the editor does not merely publish a picture of Oscar and his award along with one flabby little item. The editor finds Oscar, sits down with him over a cup of coffee, and asks him a few questions about his career, about his opinions on safety "way back yonder" and now, and about any ideas he may have for making things even safer at the plant.

Now this episode is only one little example of what can happen when the health and safety professional and the editor of the company publication get together to do a more imaginative and energetic job of publicizing the safety program.

A system can be used when one department of the company wins an interdepartmental safety contest. Instead of merely recording the results of the contest, the editor can dig into the program of the winning department, interview the people responsible for its success, and, perhaps come up with a magazine story that will give every department some hints on how to improve its safety activities.

### Producing a publication

Materials, such as safety newsletters, instruction cards, bulletins, broadsides, booklets and manuals for communicating safety rules, information, and ideas in print, require careful planning and preparation. Among steps to be taken in planning both internal (to a company audience) and external (to the general public or other out-of-company groups) publications are:

1. Clearly define the objectives of the publication. Consider the type of audience to be reached by those objectives.
2. Determine how general or how restricted the message is to be.
3. Decide what form of publication will best convey the message.
4. Estimate cost of preparing and printing the publication in whatever forms, sizes, and quantities needed. An expenditure for a new publication must, of course, be provided for in the budget, whether or not the item is produced "in house."

If the objective is to place in the hands of the worker the specific rules to be followed in doing a job safely and efficiently, an instruction card may be suitable. To stimulate general safety-consciousness, a broadside (single sheet printed on one side) may be effective. If a series of short reminders, for example, on fire prevention, is needed, posters may be the answer. To treat a topic of general interest, such as methods of materials handling, a leaflet may be used. Here, posters or leaflets from the National Safety Council, insurance company, or other organizations may be more effective, and more economical than "in house" produced material. For highly technical jobs or for more thorough coverage of a plant's safety policies and rules, manuals may be required. Even a company-wide (or plant-wide) public-address system or closed-circuit TV network would be appropriate.

When the form of the publication is being decided, it should be remembered that there is a direct relationship between the appearance of a printed piece and the degree of interest which it arouses. Most readers will react unfavorably to a bulletin, newsletter, or booklet with text in very small type, few or no illustrations, narrow margins, and long paragraphs.

Reasonably large type (10 point or 12 point), selected to fit the size of the page and, of course, to accommodate the volume of material, will help readability. For comparison, this column is set in 9-point type. Elite typewriter type is 10-point size. The *Safety Newsletters,* published by the various divisions of the National Safety Council, are set in 10-point type in 2⅛-inch wide columns. In addition, judicious use of white space and variety in size and placement of illustrations help make a publication both pleasing to the eye and easy to read. In safety, as in other fields, ideas conveyed in print are best received and best absorbed if they are well organized and attractively presented.

**Illustrations** break up the text and help to get points across to the reader. Photographs which show action described in the copy add realism in instructional materials such as manuals. Human interest photos are desirable in newsletters. Line drawings and sketches are valuable to clarify technical points on instruction cards, in manuals, and in other training materials. Awards can also be publicized.

If the printing process permits reproduction of photos and other illustrations, pictures of award winners, safety devices, and safe and unsafe practices can be used. To avoid embarassing or ridiculing employees who have been injured or caught in an unsafe act, their features can be blocked out, or pictures specially posed (and so identified) by other employees can be taken of similar situations.

Since some states have laws that forbid publication of a person's photograph without their written permission, a signed release should be obtained from every person who appears in recognizable form in any picture. Often having a new employee sign a photo release is part of the employment routine. Asking for a photo release is just good manners. Details on illustrations are given in Chapter 14, Audiovisual Media.

**Preparation of material.** Once the objectives, scope, and form of a publication are determined, the person preparing it should make an outline of the subject or subjects to be covered. For most types of material, the outline need not be elaborate, but it should be logical and complete, showing how each topic is a part of the overall plan.

Before gathering material, the writer might well spend some time studying the people for whom the message is intended so that he will know something of the knowledge and comprehension of the readers-to-be. In the interests of accuracy, completeness, and balance, material should be gathered from several sources—including articles, books, and especially supervisors, workers, and others in the company who have had experience in the matters to be treated. To ensure technical accuracy, it may be necessary to solicit help from specialists in specific areas.

No matter what the form of the publication, the writer should keep in mind certain basic rules of good writing. To get ideas across quickly and easily, short sentences, simple words, and brief paragraphs are recommended.

In a piece of some length, such as a booklet, a system of headings, kept as informal as possible, will both arouse the reader's interest and guide his thinking as he reads. In a piece designed to instruct, numbered lists of job steps, for instance, will prove helpful. In any case, the writer should follow closely the line of logical thought developed in the outline.

Copy should be written in a positive, constructive style. When the nature of the material and the form of publication permit, a friendly—but never condescending—tone can be used effectively. Personal references and names, as in a newsletter, will increase readership. Humor tied to the message and pitched to the employees' sense of what is funny can add a great deal to some types of publications. For instance, cartoon illustrations and a light touch in copy may be particularly effective in a rule booklet.

Readability of the proposed publication can be gauged by having a few of the people to whom it will be addressed test-read it for understanding.

**Production of publications.** For the technical details of printing, the advertising department or experts in the publishing field can be consulted. In layout and typography, readability should be the first consideration.

How the piece is to be used will determine its size, paper, binding, cover and similar details. For materials that are to be filed or for insertion of revised pages, loose-leaf binders may be used.

The in-company or outside editor or printer who will handle the job should be asked for technical advice.

**Getting ideas.** Everyone in the public relations business runs dry of ideas now and then. Anyone who is suffering from this affliction should not hesitate to call on others for help. Employee publications do not compete with one another, so ideas can be borrowed freely from them.

Some national agencies produce and supply safety material. See listings in Chapter 4.

The National Safety Council publishes in *Safety and Health* a "Safetyclips" page, which contains stories and illustrations for editors of employee publications. Council Sectional Newsletters and other of its publications contain a wealth of interesting and informative material. The Council publication, "How to Run a Newsletter," single copies available on request, has additional ideas. Council poster miniatures and other Council materials usually are released for general use if the customary credit is given.

Volunteering to serve as an editor of a National Safety Council Industrial Division Newsletter is good practice. At each fall Safety Congress, a training session is held for incoming *Industrial Newsletter* editors.

## REFERENCES

Culligan, Matthew J., and Dolph Crewe. *Getting Back to the Basics of Public Relations and Publicity.* Crown Publishers, 1982.

Forrestal, Dan. *Public Relations Handbook,* 2nd ed. Chicago, Ill., The Dartnell Corp., 1979.

Lesly, Philip. *Lesly's Public Relations Handbook.* Englewood Cliffs, NJ, 1978.

Moore, H. Frazier, and Frank B. Kalupa. *Public Relations Principles, Cases and Problems.* Richard D. Irwin, 1985.

National Safety Council, Chicago, Ill. *Photography for the Safety Professional,* Industrial Data Sheet 619 (1982).

Nolte, Lawrence W., and Wilcox, Dennis L., eds. *Fundamentals of Public Relations: Professional Guidelines, Concepts, and Integrations,* 2nd ed. Elmsford, N.Y., Pergamon Press, Inc., 1979.

Parkhurst, William. *How to Get Publicity.* Times Books, 1985.

# 14

# Audiovisual Media



Audiovisual (AV) is a term used to describe instructional materials and equipment designed to facilitate teaching and learning by making use of both hearing and sight.

Commonly used audiovisuals include actual equipment and models, flip charts and posters, slides and other projected transparencies, recordings, filmstrips, videotapes, and motion picture films. Teaching aids include chalkboards, flannel boards, hook and loop boards, bulletin boards, and display cases. AV equipment (hardware) includes simulators, projectors of all kinds, tape and cassette recorders for sound and video, cameras, monitors, the full range of television studio equipment and cable television hardware, and computer-interactive training equipment. Activities such as demonstrations and experiments usually are considered part of audiovisual programs.

## EFFECTIVENESS OF AUDIOVISUAL MEDIA

Audiovisuals assist the communications–learning process by presenting information concisely and with impact. People become more involved when visuals accompany the verbal teaching experience. Sometimes words, whether spoken or written, may be misconstrued by learners.

Figure 14-1 rates various communications media on a relative scale of concrete to abstract experiences—the more concrete the communication, the more effective it will be.

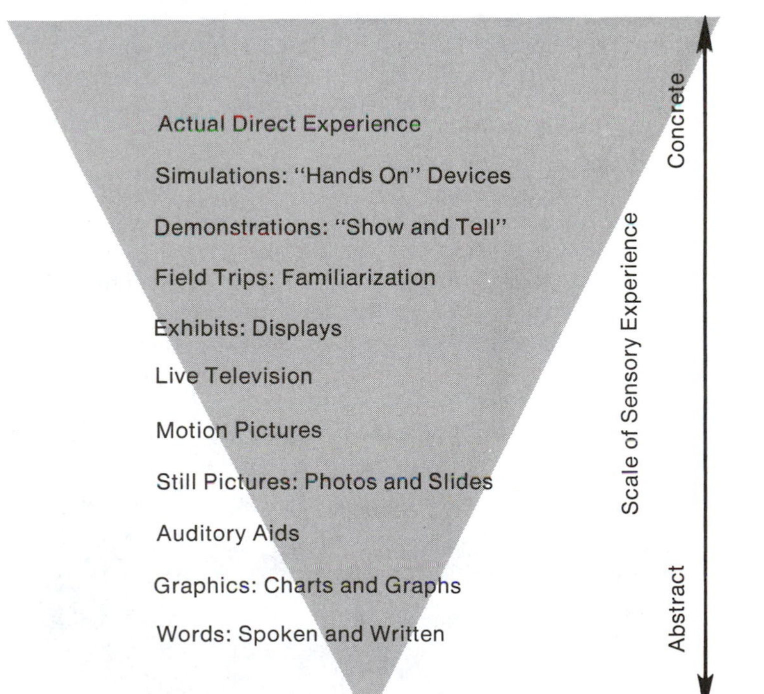


**Figure 14-1.** Experience is said to be the best teacher. In measuring the relative value of an audiovisual or other means of communication, remember that the more concrete the medium of communication, the more effective it is.

In recent years, there have been many studies made to determine the most effective ways to reach people's minds with either teaching or marketing messages. Some of these studies have been publicly funded and conducted by universities; others have been initiated and supported by businesses interested in learning the effects of various advertising media on product sales. The following figures are representative of these studies:

- How much of the total learning experience comes through each of the senses?
  - 1% through taste
  - 1.5% through touch
  - 3.5% through smell
  - 11% through hearing
  - 83% through sight.
- How much do people remember?
  - 10% of what is read
  - 20% of what is heard
  - 30% of what is seen
  - 50% of what is seen and heard
  - 70% of what they say
  - 90% of what they both say and do.

Designing a communication or training program to take advantage of concrete resources and natural learning capabilities takes more ability than merely knowing the subject matter; it demands an ability to use audiovisuals effectively, and to know when each can be most productive with the intended audience. With properly defined objectives, AV uses are endless. Only imagination—and budget—define the limits.

Safety and health professionals are using audiovisuals (AVs) as tools in presenting information for:

New employee indoctrination
Initial and remedial training
Training supervision in its role in accident prevention
Specific safety procedures, such as hot work permits, vessel entry, electrical and other power lockouts
Basic fire prevention techniques
Job safety analysis programs
Fire brigade training
Safety and health programs, such as OSHA and Right-to-Know.

Safety and health professionals should be aware of the range of audiovisual resources and their appropriateness both to transmitting the message content and to audience appeal.

Often more than one visual is necessary for the communications job. For this reason, many safety departments have numerous types of audiovisual products available.

Some of the most popular audiovisuals used in safety education include slide presentations, demonstrations, flip charts, overhead projection, chalkboards, posters, films (16 mm), videotape and closed circuit television. Units controlling more than one projector (multimedia) are in demand.

### Use in training and motivating

Audiovisual materials are important conveyors of information. Photos of hazards or of safe conditions provide visual evidence useful in reports to supervisors and to management. These, plus graphic presentation, can make routine reports and statistical analyses interesting and easy to understand. More than one safety professional has found a simple bar chart or colored graph to be more meaningful to the audience than pages of detailed statistics.

A visual can emphasize the points of information in a safety talk; it can even provide a convenient outline for the speaker. Visual materials can be used to organize group thinking and to summarize safety committee action.

As a motivational tool, visuals that appeal to the emotions can help change attitudes, encourage safe work habits and compliance with safety rules, and remind employees of special rules or hazards. Photos or displays of safety equipment that saved workers from serious injury are especially effective tools (see Figure 14-2).

Audiovisuals of all types are widely used to promote interest and obtain cooperation in special campaigns, safety contests, and similar activities.

### Selection of media

To be most effective, audiovisuals must be selected with care, after considering many factors (see Table 14-A).

- What is the purpose of the communication—motivating, training, reporting, fact finding, entertaining? What result is wanted? Which medium, or combination, will serve this purpose best, within budget limits?

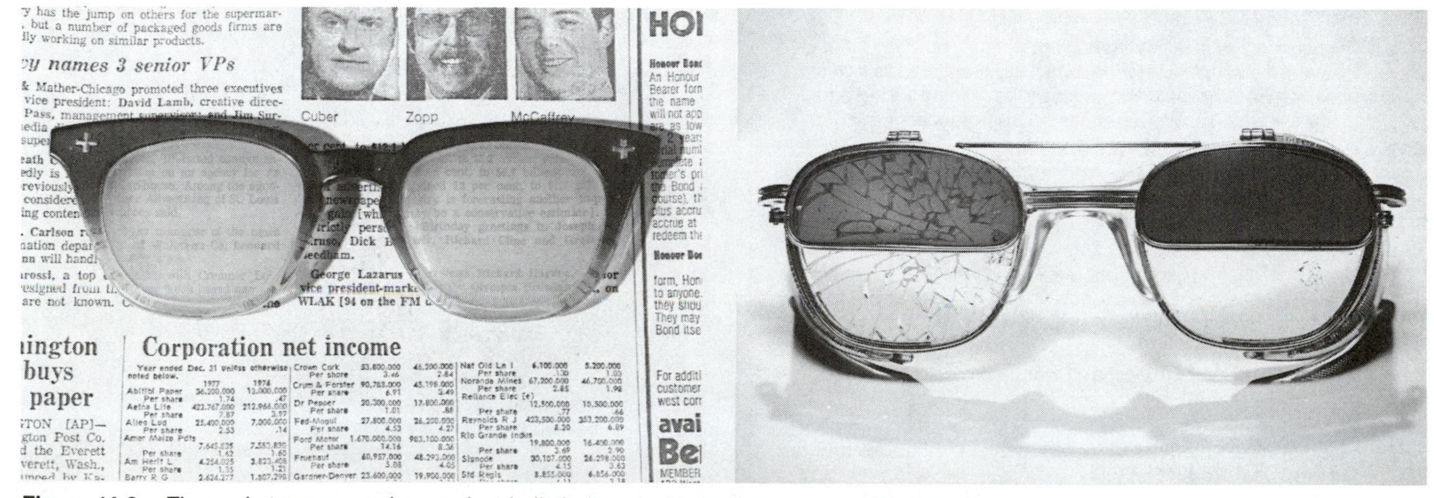


**Figure 14-2.** These photos appeared on a plant bulletin board with the following caption: "These safety glasses saved the sight of two workers. The left pair resisted heat that might have caused blinding burns to the eyes. The right pair resisted the impact of a white-hot fragment that could have penetrated the eye. The next time you feel like taking off your safety glasses—think about this picture—then make up your mind."

**Table 14-A.** Major Features and Limitations of Various Audiovisuals

| *Type and Popular Size* | *Audience Size* | *Limitations* | *Strong Points* | *Comments* |
|---|---|---|---|---|
| *MOTION PICTURES 16 mm* | M/L | Camera and projector expensive; require trained operator, except for self-threading models. Film not easily changed or updated. | Effective for training and motivating. Uniform professional message. Optical sound nonerasable. Sharper image than 8 mm for given projection size. Single-frame, stop-motion projectors are available. | Silent verson less costly, but less effective. |
| *SLIDES 2x2 in. (35 mm, 126, or 127) 2¼x2¼ in. (120 film) 3¼x4 in. (super-slides) (theater projector)* | S/L | Slides may get out of sequence, reversed, etc. Cardboard mounts not durable. | Effective for training and motivating. Less of a "canned" show since slides may be rearranged. Slides can be made and processed quickly. Color inexpensive. | Taped message or reading script easily added or changed. Remote control and multiple projection possible. |
| *FILMSTRIPS 35 mm sound* | S/L | Strips and records not easily updated. Message might not be effective or suitably paced for user's needs. | Effective for training and motivating. Message uniform. Sounds easily added to tape or disk. | Silent strips with scripts less expensive, but still effective. Seldom used any more. |
| *OVERHEAD PROJECTORS 10x10 in. 7x7 in.* | S/L | Transparencies positioned by hand. Projector close to screen; it or user may block view unless screen is raised or set at an angle. Ready-made material not widely available. | Effective for training. User can write on transparency while facing audience. No need to darken room Transparencies easily made and filed. Presentation informal and flexible. | Color transparencies or overlays easily made. |
| *OPAQUE PROJECTORS 10x10 in. max.* | S/M | Projectors require manual operation. Material in books may be difficult to store or ship. Room must be darkened. Copy may be too small. | Effective for training. No transparencies required; small objects, printed material, drawings, and photographs used "as is." | Copies or originals can be hinged or put on rolls to maintain sequence. |
| *CLOSED-CIRCUIT TELEVISION VIDEOTAPE CASSETTE* | S/M | Initial investment expensive. Requires adequate lighting. In color or black and white. Copies must be be made one at a time unless duped by lab. | Instant replay. Excellent for training situation where trainee must "see himself in action." Has relatively low operating cost. Can be shown in lighter room. | Small number of people can view screen. TV is a culturally natural transmission medium. |
| *COMPUTER-BASED INTERACTIVE TRAINING* | S | High initial expense. May be difficult to copy. Difficult to produce in-house, requires VCR or laser disk equipment to generate some images. | Low operating cost. Does not require a darkened room. High retention of material presented, user paced. Can automatically record and verify training and test scores for individual employees. Presentation and message uniform, does not require presence of instructor. | Procedure without VCR-Laser, disk effective; but less interesting to student. |
| *FLANNEL, HOOK AND LOOP, MAGNETIC 12x36 in. to 48x72 in.* | S/M | Presentation requires advance preparation. Few ready-made presentations available. Flannel board material may fall off if not applied correctly or if board too nearly vertical. | Effective for training. Message easily changed, yet can be filed and reused. Permits informal presentation with desirable audience contact. Dramatic, "slap-on" effect builds interest. | Boards suitable for heavier displays; cost slightly higher than cards or pads. |
| *FLIP CHARTS AND CARDS 38x48 in. 18x24 in.* | S/M | Limited to small groups. Limited as to amount of copy. Good lighting necessary. | Effective for training and informing. Prepared material can be arranged in sequence. Good audience contact. Material easily prepared; can be added during talk and can be saved. | Ready-made letters, color, sketches cut-outs easily added. Colored paper effective. |
| *PAPER SHEETS AND PADS 28x36 in.* | S | Speaker must print legibly. Good lighting necessary. Ink from felt markers may bleed onto adjacent sheets. | Effective for training or discussion; informal. Permits reference to other sheets both during the discussion and for later writing of minutes. Low-cost pads easily obtainable. | Used in place of chalk boards, no erasing. |
| *CHALKBOARDS (portable and wall mounted) 36x 48 in., larger for wall mounted* | S | Board must be erased before reuse and recall not possible. Good lighting necessary. Ordinary chalk marks hard to see. Dust from chalk and erasers annoying. | Effective for training or for discussing a limited number of points. Presentation informal. Portable chalkboards also useful for holding charts or displays. | Colored or fluorescent chalk adds life to talk. Magnetic boards available. |
| *POSTERS AND BANNERS 8½x11½ in., 17x23 in., and larger* | S/L | Only one or two ideas can be presented at a time; considerable time needed for changing. | Effective for motivating; support training. Specific messages can be posted at points of hazard or to meet timely situations. Ample posters available. | Homemade posters supplement general posters. |
| *WORKING MODELS, EXHIBITS, AND DEMONSTRATIONS* | S/L | May require special training to use. Live action is subject to errors. | Action can closely simulate actual conditions. Permit group participation. | |

- Which medium will best convey the content of the message? For example, detailed technical figures can be communicated by a chart that can be held up or projected in front of the audience, while it is explained. A tape recorder or a movie projector would be of little help.
- What is the size and type of audience? What is their attitude toward you and toward your subject? How knowledgeable are they? How good are their communications abilities?
- How capable are the communicators? Do they need special training in either the subject matter or in the effective use of the audiovisual? Do they need other help?
- Where is the audiovisual to be used—in a training room, at a meeting, in the office or plant, in the field, or at home? Use of audiovisuals requires scheduling, preparation or purchase, distribution, and storage. Suitable facilities must be made available.
- How flexible or how formal must the audiovisual be? In some cases, a flexible type, which each speaker can adapt to his particular use, may be desirable. In other cases, a formal aid which offers conformity of message with company policy and uniformity of presentation may be preferable.

  When a formal visual is being considered, a number of questions should be asked: Will the entire message apply to many different audiences, even though they are located in different geographical areas or are confronted with different hazards? Will the material become dated, or can it be used almost indefinitely? Is it likely, for instance, that changes will be made in machines, processes, job layouts, or even personal protective equipment that are illustrated?

  In some cases, a combination of flexible and formal aids is desirable. For example, some speakers use a carrying case containing three-dimensional exhibits, charts, flannel boards, and other nonprojected material, as well as slides and a portable projector, to give road show presentations that can be changed to meet specific needs.
- How do the costs of the various audiovisual media compare? In the selection of audiovisual equipment, this point is especially important.

  Whether or not the cost of an audiovisual is justified must be considered in the light of what it will buy. An investment of $50,000 or more in a well-planned sound movie might be justified for a long-range training program or public relations campaign. One of the advantages of such a visual is that it can be used over a relatively long period of time as a means of communicating the same message to many people. The repeated showing of a $50,000 film to large audiences over several years might well bring the cost per viewer down to a few pennies.

  In contrast, the apparently modest expenditure of $500 on a homemade videotape cassette developed without sufficient planning and applicable to only a handful of employees could be excessively high and perhaps ineffective. Moreover, the same amount of time, money, and effort devoted to a training program, individual job instruction, or perhaps production of an inexpensive safety rule booklet might get better results.

  Cost is only one of a number of factors to be weighed in the selection of a visual. An expensive visual is not necessarily the best one. For example, a simple paper pad or chalkboard may be more effective than an elaborate printed brochure for presenting a safety report to a group of executives or for training employees in safe practices.
- How is the message to be supported? Not only should there be follow-up, but other people should know what was communicated to whom, in order that they can reinforce the message, or at least not accidentally contradict it.

### Commercial versus homemade visuals

To determine whether it is more economical and practical to make audiovisual aids than it is to buy them, the same factors that affect selection of aids must be considered: the purpose for which the aid is to be used, the type and size of audience, and the degree of flexibility desired.

For an informal supervisors' meeting, for a report to a safety committee, or for a presentation to company officials, a homemade aid, such as a chalkboard or hand-lettered flip chart, would be appropriate. For more formal talks or for a number of meetings at decentralized locations, a commercially prepared aid—a videotape, a set of slides with a script, a filmstrip with a record or a tape, or a set of commercially painted charts—might be a wise investment.

Other points to be examined when a choice is being made between a commercial aid and a homemade aid are the costs involved and the availability of facilities, talent, and time.

It is a good idea to contact manufacturers of audiovisual and projection equipment; many will provide instruction booklets and other materials on how to make and use audiovisuals. See References at the end of this chapter.

Still another factor is that participation of individuals or of committees in the planning and development of a visual may be highly desirable as a means of arousing and maintaining interest in accident prevention. The net effect of employee participation, in fact, may compensate for lack of the professional touch, provided that the quality of the finished product is not seriously affected.

When a final decision is being made, the various factors must be considered in terms of one another. For example, slides may be selected as the type of visual to be produced. The speaker may have a suitable camera and lights, plus the ability and the time to do the job himself. However, the number of showings, the size of the audience, or the importance of the message may warrant the expense of professional input.

The same comparison can be made between professional and homemade charts, signs, and other visuals.

Even if a safety department alone could not justify a professional visual media staff, the needs of the department, when added to those in sales, training, and other communications areas, might justify such in-company, or in-plant, facilities and staff. The combined benefits and savings could make this worthwhile.

## PREPARATION OF AUDIOVISUAL MEDIA

Once a suitable audiovisual medium has been selected, the details of production must be worked out. The general procedure is to make an outline, develop script and picture descriptions, have pictures taken or art work made, and check the content for accuracy. Often the length of the presentation must fit an instructional time frame. Realistic deadlines must be set.

Many types of audiovisuals include written or spoken words making coordination between the illustrations and the message essential. They should complement each other for maximum effectiveness. Sometimes the art work also can be used in

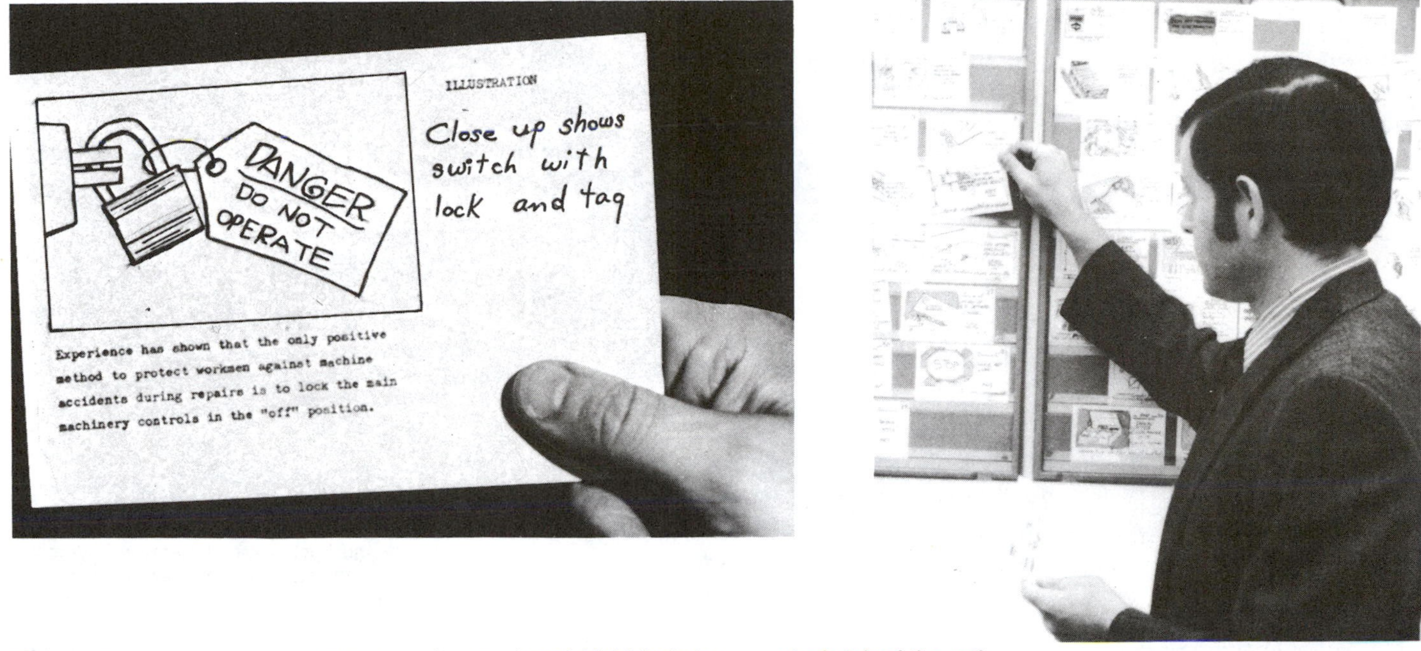


**Figure 14-3.** This typical card (*left*) from the storyboard (*right*) includes a rough sketch of the action and other information relevant to making the finished illustration. The storyboard technique permits easy rearrangement until the most effective sequence is determined.

printed brochures or magazine articles, as well as for handout instructional materials.

Preparation of AV media also can be done commercially, but if you do it yourself, here are some tips.

### Preparing a script

A video cassette, a film, a set of slides, or other type of presentation requires good organization and a script suited to the audience and the desired response. Here are some guidelines:

1. Identify the target audience. Decide what needs to be taught (task analysis), the knowledge level of the audience, the detail needed, and the level and method of presentation. Write down the objective of the visual; perhaps discuss it with colleagues or with a committee assigned to help with the project.
2. Develop a simple outline of the subjects to be covered, indicating the approach (humorous or serious, for example) to be used, and the props, illustrations, and shooting locations needed. Check to see if these are available or must be obtained or made.
3. Conduct a formal planning session. Using a script or script outline, review in detail all information to be presented; concentrate on logical sequence, technical accuracy, and possible problems. Avoid the temptation to crowd too many ideas into the outline. Concentrate on the chief objective.
4. Determine how to open and close the story. In a training script, main points should be repeated and summarized.
5. Following the outline, write a rough draft of the script and the picture descriptions, and then a final draft. Technical and management approval usually are required.
6. Develop a shooting list. Based on script or script outline, identify shots by location and approximate length of time. Identify supplementary materials needed, such as titles, cartoons, drawings, diagrams of equipment parts, and sound effects.

For a script to have maximum effectiveness, short words and simple sentences are usually recommended. The script should be kept brief and to the point. If a shot requires lengthy narration, variations or different views of the scene or subject could be developed. This makes the scene appear to be shorter than it actually is.

**Storyboard technique.** In preparing a script for a set of slides or for a motion picture or videotape, the storyboard (planning board) technique is recommended.

Usually, the copy is typed double space, down the right side of the page or illustration board, frame by frame or scene by scene, with the corresponding illustration or picture descriptions placed opposite.

Or each frame can be represented by a 4 × 6-in. (10 by 15-cm) card with the illustration on the left side and the script and production instructions on the other side. The cards can be mounted on a large piece of cardboard, laid out on a desk top, or placed in a planning board rack. This technique makes it easy to rearrange the sequence and to visualize the entire finished product (see Figure 14-3).

When a set of charts or a chalk-talk is being prepared, rough sketeches or notes can be made on a paper pad and scaled to size. Even in miniature, a rough sketch will give a good idea of the amount and size of lettering that can be used, the effect of color, and other aspects.

If the script is to be reviewed by safety committee members, company officials, or other persons, it can be typed and duplicated. Cards can be grouped on pages and duplicated. Deadlines for reviews must be set and followed. Of course, important points should be approved by key executives or other authorities.

### Lettering

The most common complaint regarding both nonprojected and projected visuals is that lettering is difficult to read—too small, too thin, too crowded—or even illegible. The best rule-of-thumb

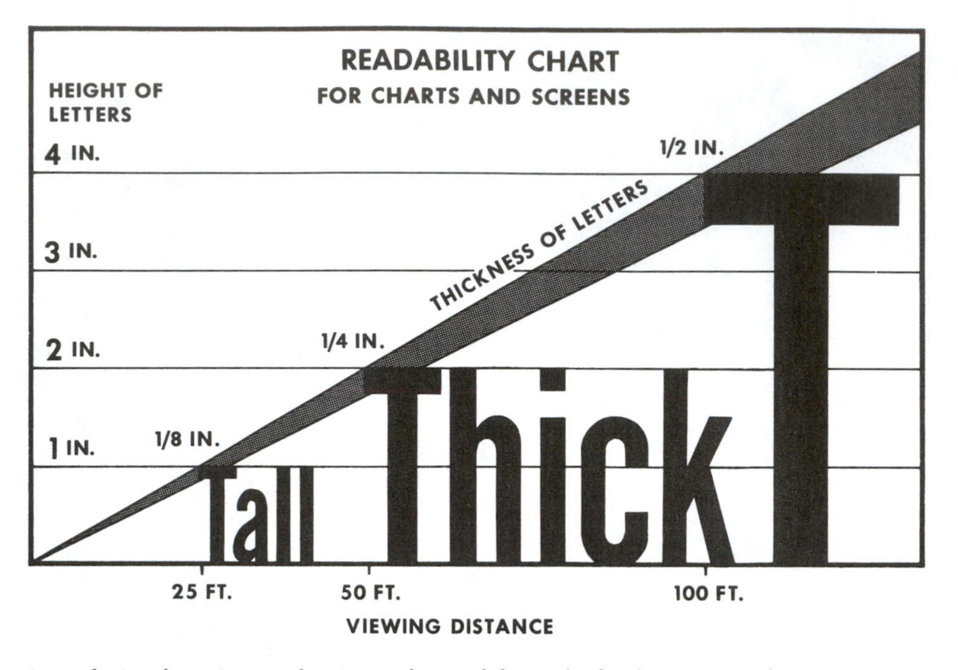


**Figure 14-4a.** This chart shows what heights and thicknesses letters and symbols should be for easy viewing of projected and nonprojected visuals.

is to design lettering so that it can be read from the back row of the audience. Simplicity is the keynote.

Block letters show up better than handwritten copy. To be easily read at a distance of 50 ft (15 m) printed or projected letters should be 2 in. (5 cm) high and ¼ in. (8 mm) thick (see Figures 14-4a and -4b).

For use with overhead or opaque projectors, material typed with characters at least ¼ in. high, with spacing of ¼ in. between lines, will give a letter height of 2 in. on a screen 6 ft (1.8 m) wide and will be clearly visible at a distance of six to eight times the screen width.

The space between rows of letters should be at least one-half the height of the letters, preferably the same as the full height. For example, there should be at least ½ in. (and preferably 1 in.) spacing between letters 1 in. (2.5 cm) high.

Printed or typed materials on 8½ by 11 in. (20 × 25 cm) sheets, such as record forms, will require large lettering or typing on a machine with oversize (¼ in.) characters. Material typed all-caps in an area 3 in. high by 4½ in. (7.5 × 11.5 cm) wide will be legible when converted to a 2-in. square slide. If possible, all illustrations and titles (art and lettering) should fit the horizontal format of the screen. That is, the width should be 1½ times the height (see Figure 14-5.)

Material on a visual should not be crowded, should be well organized, and kept simple. A simple rule for the amount of copy is: no more than six or seven lines with three to six words of typed or lettered material per line. Typewriters which permit half-spacing (technically, "one-and-a-half spacing") are of special value here; use of a gothic typeface improves legibility.

A growing variety of ready-made lettering material is available in camera and art supply stores. Examples are plastic stick-on letters, rub-on transfer letters, gummed letters, and ceramic, cork, cardboard, or other letters which give a three-dimensional effect when lighting is from one side.

Lettering guides are available for use with special lettering pens and felt-tipped marking pens. Many styles of lettering often can be made with one lettering set.

Lettering machines, often used by graphic artists, can easily and inexpensively help create professional-appearing transparencies for use in overhead projectors. More expensive tools, such as computers with CAD (computer-assisted drawing) software packages, offer infinite lettering, graph, and illustration possibilities. Having lettering done by a professional typesetter is another option.

For an informal visual to be shown to a small group, a large black or colored crayon may be used. For a visual intended for larger groups, instructor's chalk, broad-tip felt marking pens, stencils, cut-out letters, or brush-painted letters are preferable. Background colors can be varied also. Too much of any one color can be tiring. Try white on black or black on yellow.

### Drawings and graphs

The same rules that apply to lettering apply to drawings: the line work should be broad and opaque. Frequently a complex item can be constructed from cutouts of colored paper. Colored paper is also effective for bars of a bar chart or areas under a curve. Graphs may be simply constructed from colored tapes (⅛ to 1 in. wide). In presentations, graphs frequently are used to show a trend rather than specific points; hence the grid background should be omitted or should consist of only a few fine lines. Where accuracy is important, the actual numbers should be used.

Another inexpensive do-it-yourself technique is to draw a cartoon or chart. Illustrations can easily be transferred to flip charts or large sheets of paper by using an opaque projector to duplicate the image on the paper and then drawing the outline (see Figure 14-6). The illustration can be completed by using varicolored chalk on a contrasting background; this lends an interesting touch that often is lacking in conventional black-and-white material.

Try to show only one point at a time when building your overall story.

Depending upon the size of type and the amount of copy, a page of printed material may become illegible when reproduced on film. If such material must be used, it should be converted to a more suitable form. The information given on a page of statistics, for instance, might be expressed in a few simple charts or graphs. Reduce the details to those required to illustrate the point. Then do it effectively (see Figure 14-7).

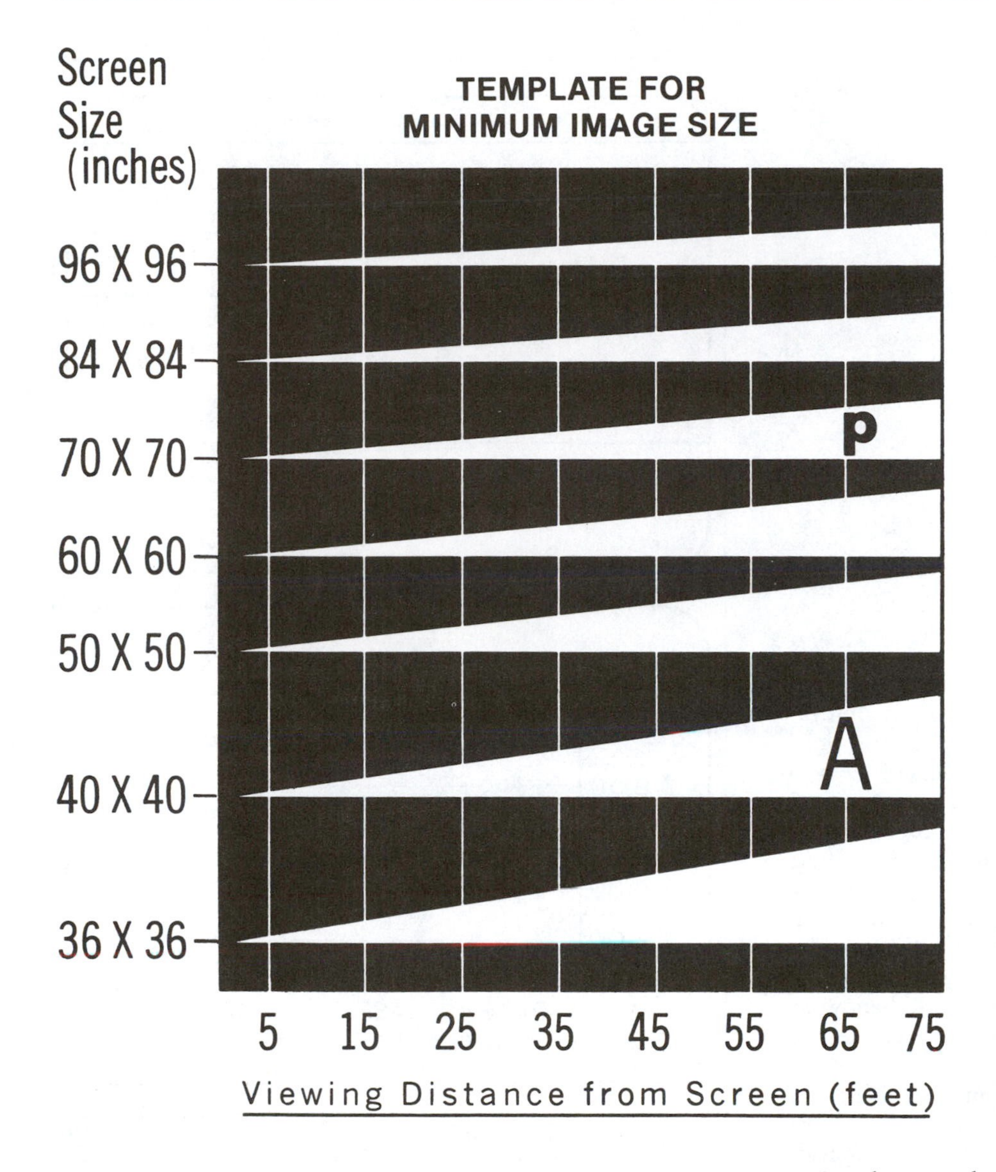


**Figure 14-4b.** The correct type size for overhead projection can be scaled by placing the transparency over this template. For example, lettering for a transparency to be projected on a 70- by 70-in. screen must be at least as large as the "p" (¼ in. high) if it is to be viewed from a distance of 65 ft. For this same viewing distance, but with a 40- by 40-in. screen, a letter of minimum size "A" (½ in. high) should be used when making the transparency.

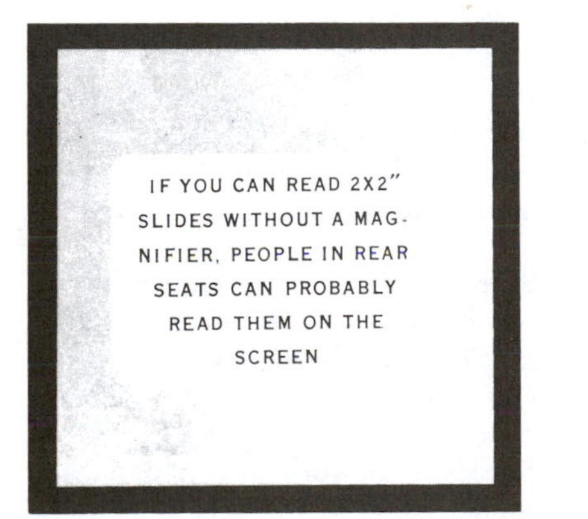


**Figure 14-5.** Rule-of-thumb for lettering to be shown on a 2- by 2-in. slide. (Printed with permission from Eastman Kodak Company.)

### Use of color

Color enhances both nonprojected and projected visuals. Contrasting colors always should be used. For example, black lettering shows up best on a white, yellow, or light-orange background, and worst on a dark blue.

With projected visuals, color adds realism, provides contrast values that can bring out important points, and gives a professional look to the completed visual. Full-color motion pictures are expensive and not always necessary. When emphasis is on the action, black and white may even be more effective.

With color film or special titling film, for instance, title frames can be made attractive and closeups can be shot against black or colored backgrounds for good contrast. Textured backgrounds and colored lighting effects lend a professional touch.

For posters, dramatic effects can be obtained with the high-visibility fluorescent paints, chalks, and papers, particularly if "black" light is used.

Color can be added with large blocks of instructor's colored chalk or with colored felt-tipped pens that make a broad, heavy line. Colored designs can be made by spraying through stencils or simple cutouts. Spray paints can be used to give the overall color, and powdered colored chalk can be daubed lightly over lettered material to give a tint.

Colored tape and ready-made arrows, circles, and other stock designs can be used to make charts and graphs, and also to mark important parts in equipment photographs.

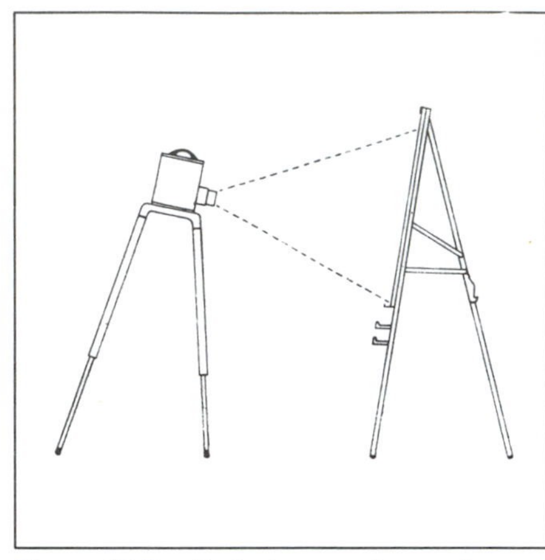

Figure 14-6. Projectors can be used to project images on large sheets of paper when making flip charts.

ANNUAL SAFETY REPORT

Accident Cost Factors by Departments

(Incurred losses per 1,000 hours)

| Group | Base Period 3 Years | Last Year | This Year |
|---|---|---|---|
| Production Division | $5.40 | $7.10 | $2.50 |
| Sales Division | 7.20 | 3.25 | 6.70 |
| Company Average | 8.50 | 5.10 | 3.60 |

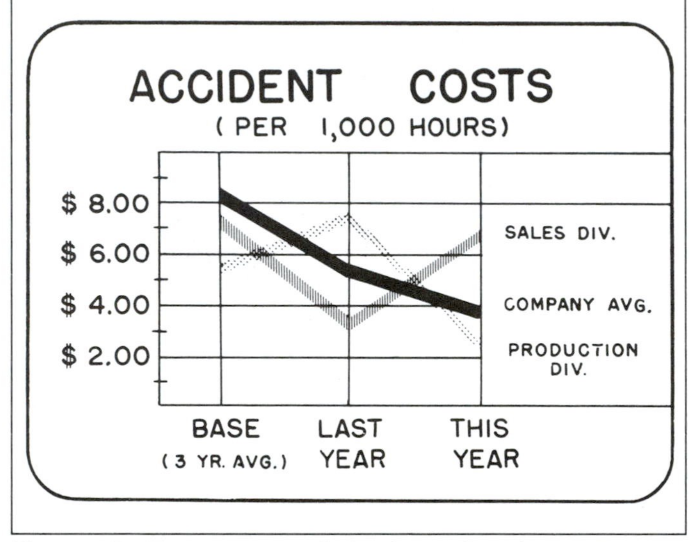


Figure 14-7. For use as an audiovisual, complicated statistics should be converted into a simple graph. Typed slide (*top*) is unclear and has poor spacing—the top part is crammed while the bottom portion is left blank. The bottom slide is better balanced and the facts more clearly presented.

## Photographic illustrations

The value of a visual depends to a considerable extent upon selecting the right illustrations and the quality of the illustrations. So far as photographs are concerned, the general principles of good photography are the same regardless of the type of camera or film used. (For a detailed discussion, see National Safety Council Industrial Data Sheet 619, *Photography for Safety.*) Here are a few suggestions for taking good pictures:

- The important part of the picture should be highlighted by means of a closeup, a supplementary sketch, a contrasting background, or by an arrow or sign placed by the item. For example, if a guard or a piece of safety equipment is being photographed, it can be painted (spray cans are handy) or shot against a colored background that provides effective contrast.
- Sometimes it may be important to include an object of known size such as a pencil, automobile, or person to given an indication of the relative size of the object being photographed. Be sure to use a late model car so the picture does not too soon become dated.
- With motion, videotape, and with still pictures, use of long shots, then medium shots, followed by close-ups help establish the scene or situation.
- If background material must be shown in detail, extra lighting should be used. A single light source will work. If background detail is not important, it can be kept out of the original picture, cropped out of the negative or the finished photo, or touched out of the print. Off-to-the-side lighting will give a pleasant three-dimensional effect and will keep light off the background, thus playing it down.
- If there is doubt about the possible result, it is a good idea to take two or three different exposures at the time of the original shooting. Probably, there then will be no need to come back later to get a better picture.
- Use of an exposure meter will help ensure good results with both natural or existing light and with floodlights or spotlights. A special exposure meter can be used for electronic flash. If a meter isn't used, exposure can be calculated by the guide numbers provided on the camera flash unit or found in the film instructions. Numbers are for average rooms. Large industrial areas require lower numbers (more light).

  Where possible, photos should be made outdoors, to take advantage of natural lighting, or in a studio under controlled light. If actual job situations are wanted, then shooting on location is called for.
- Instant-type cameras can be used to capture current activities, such as safety practices or newly discovered hazards. These instant photos can be taken and used in a visual presentation on the same day at a relatively low cost. Videotape offers this same advantage.
- By using a macro (close up) lens, artwork can be copied and made into a slide. If floodlights are used, be sure to use tungsten (Type B) color film (see Figure 14-8).

## Combining sound and images

With a set of slides, sound can be added by a tape recorder on which a prepared sound track is played. The slide or filmstrip projector operator can follow a marked reading script; a beep or other audible signal can be used to signal when to move

**Figure 14-8.** A 35mm single-lens reflex (SLR) camera is used to copy artwork in order to make a 2- by 2-in. mounted slide. When 3,200 K lights are used (as shown here), the camera must be loaded with Type B color film. Make certain the lights are equidistant from the artwork and that no glare or bright reflections off the art will spoil the slide.

the next picture into position. Using special equipment, inaudible signals can be inserted on the record or the tape to automatically advance the slides or the strip.

Videos made in-house, or professionally produced, can give added dimension to safety presentations. The sight and sound of familiar people in a familiar location can bring the safety message home through video made in the plant. Safe practices are more likely to be identified with and the danger of unsafe practices will be more memorable.

Compared to some other media, videos offer definite advantages: the cost of video equipment is lower than film equipment; video is flexible—can easily be edited and updated; it has all the advantages of film at less cost and offers more options for use; video's instant replay feature lets a trainee practice his new job then instantly see his efforts; and candid video shots can be taken of actual working conditions and practices.

Videos can also be used to advantage by having line supervisors rehearse safety sessions.

The supervisors can give the presentations in front of the video camera. The presentation is recorded and immediately played back so the supervisor can see his imperfections and take steps to correct them (see Figure 14-9).

Other advantages of training via video include:

- *Unsafe acts* can be recorded and played back instantly to someone being trained for a new job;
- *In-house programs* allow review of important job procedures and job methods that will be consistently presented;
- *Videotapes* can be the core of formal or informal discussions of safe job sequence or review of special safety precautions that ought to be considered;
- *People find it easier* to relate to pictures of equipment and people that they know;
- *Equipment,* such as the recorder-playback unit and the monitor, can be used to show commercial training tapes.

In-plant video productions can be simple or sophisticated. When a one-time use, quick review of a subject is needed, a hand-held, black and white camera can be used. This results in little need for special lighting, editing, or scripting.

When the need is to reach more people and to make more enduring points, a more sophisticated approach is called for. A script should be written and scenes should be formally blocked out. Use plant personnel for actors whenever possible, but a script that needs to make a point dramatically or entertainingly may call for professionals. The objective or point of the film must also be pin-pointed and the means of making the point decided upon.

The raw result can then be edited into a safety and health presentation that is comparable to commercial TV productions if professional scriptwriters and AV personnel are used.

Some companies make one to three tapes a year at the corporate level that pin-point special hazards, such as hazardous materials handling. Such special tapes can supplement AV presentations from outside sources.

Safe practices can also be shown in tapes on other training subjects.

In addition, videos can be used as company "newsletters" or to demonstrate the safety tip for the month (other uses are shown in Table 14-B).

### Safety considerations

Photographers, whether professional or amateur, should observe certain precautions when taking pictures on location.

**Safe background.** Nothing can ruin the effect of a safety presentation more than unsafe or extraneous material in the background. For example, if the workers demonstrating a guarded grinding wheel are not wearing safety glasses, the photograph or movie sequence is likely to do more harm than good. Or, if the background shows poor housekeeping, fire hazards, or other employees not wearing the required personal protective equipment, it detracts from the safety message in the foreground.

**Safety equipment.** All photographers and their helpers should use the safety equipment required in the area. This means they should have safety glasses, safety shoes, and safety hats where required, not only for their own protection but to demonstrate a safe attitude.

**Observe safe practices.** These include not smoking and avoiding the use of flash equipment or the wrong type of electrical equipment in hazardous areas. Clearance should be obtained for work in hazardous areas.

**Electric lighting equipment.** If photo floods are used, temporary wiring should be in good condition and of the three-wire, automatic-grounded type. Where possible, it should be strung over aisles or otherwise kept out of the way of truck and pedestrian traffic. If there could be any danger from operating machines, they should be locked out or have main switches pulled to prevent accidental starting. The services of a qualified maintenance worker or electrician to assist the photographer can provide an extra safeguard. Any electrical equip-

**Figure 14-9.** This instructor is videotaping his defensive driving training presentation. When done, he will play it back and either edit out the imperfections or re-record the spots needing correction. (Printed with permission from NL Industries.)

**Table 14-B.** Some Safety Uses for Videotape

Videotape is a versatile medium. Here are a few ways it is currently being used to put across the industrial safety message:

- **Security surveillance**—Mounted cameras observe distant gates, loading docks, and payroll departments.
- **Job review**—Used in conjunction with Job Safety Analysis, Task Safety Analysis, and Step Safety Analysis for observation and review of both good and bad procedures.
- **Management training and development procedures**—First train the trainer, then have him train employees down the line. Playbacks assure that procedures are practiced correctly by the instructor before others are taught to do as he does.
- **Incident investigation**—When brought to the accident scene, a recording is made of the physical set-up, personnel present, time, weather, lighting, and other conditions pertinent to the accident.
- **Motivation and enforcement**—Showing the employee how his performance can lead to an accident is an effective way of gaining his cooperation. OSHAct violations can be spotted and corrected. Before-and-after scenes can prove dramatic.
- **Training in sophisticated equipment, complex procedures**—Videotapes can show and repeat processes a step at a time, permitting interruptions for questions.
- **Informing distant audiences of a procedure or announcement affecting all units**—Duplicates of a tape can be made and mailed anywhere so that all personnel are informed simultaneously.

ment that the photographer uses should be UL-listed (or approved by another certified agency).

**Ladders.** Ladders or work stands should be available so camera operators and helpers do not have to use makeshifts or climb on machines, tables, or other equipment to get unusual angles. If they must work from heights, they should use safety belts and lifelines.

### Legal aspects

There are certain legal considerations in photographic production. It is customary to use model releases, discussed in the previous chapter, which permit employees and others to give written permission for the company to use their photographs. Check with a legal counsel, industrial relations department, and local photographers for practical advice on using employees or the public in a film.

Consideration should also be given to copyrighting a slide show, videotape, or film. This can be handled through a company's legal department. (For details, contact U.S. Copyright Office, Library of Congress, Washington, D.C. 20559; ask for form PA, Application for Copyright Registration for a Work of the Performing Arts.)

## PRESENTATION OF AUDIOVISUAL MEDIA

Even the best planned audiovisual will miss its mark unless the necessary facilities are at hand and unless the speaker checks them and rehearses with them.

If a company is planning on building a training room, it is best that the safety professional work with the designers and AV professionals at the earliest stages to make sure that the room incorporates all the necessary AV features.

### Room lighting

Nonprojected visuals require good general lighting. If a room only has indirect illumination, a portable floodlight or spotlight can be used on charts, exhibits, chalk boards, and other visuals. The light can be clamped to a chair or to a portable stand immediately in front of the visual, but placed so it will not interfere with the audience's view.

Colored spotlights can heighten the visual's dramatic effect. Revolving colored lights, like those used for Christmas displays, are suitable for more permanent exhibits, signs, or displays. Black or ultraviolet lighting used with fluorescent paint, paper, chalk, or ink gives a vivid effect.

If lighting must be dimmed for showing a projected visual, preparations must be made for darkening the room at the proper time, but killing house lights should not cut off power to the projector and reading lights. The location of the light switches must be noted, and someone should be asked to darken the room at a given signal. A shielded, reading light will be needed when the house lights are turned off for the presentation. A small flashlight may come in handy if the reading light is not operative.

Window blinds or drapes should permit shutting out daylight.

Well in advance of the meeting, locate the electric outlets and check electric equipment for safe working condition. Having spare bulbs and extra extension cords on hand may save embarrassment or prevent delay. Extension cords should be marked with high-visibility tape or placed so they do not create tripping hazards.

### Use of pointers

A pointer is a necessary item when a speaker is using a chalk board or charts and may be helpful with other types of aids as well. A pencil or a finger is a poor substitute for a pointer.

Use of a pointer enables the speaker to face the audience and at the same time easily relate his words to the visual material. The speaker should not play with the pointer—this distracts from the presentation. He should not touch the visual or the projection screen with the pointer—it may move the visual or screen, or even mar it.

If visibility is a problem, a pointer with a fluorescent painted tip will be helpful. In a darkened or semidarkened room, a battery-operated or 110-volt flashlight-type pointer can be used to project a spot of light or a bright arrow onto a screen from a considerable distance.

A high visibility electric pointer is useful even in a fully lighted room when the speaker must stand at some distance from his chart or screen.

A telescoping, pocket-size pointer is useful for speakers who must carry a pointer with them.

### Amplifying and recording systems

If the acoustics in the room are bad or if outside noise makes hearing difficult, an amplifying system will be needed for successful presentation. This is particularly important if there are a number of speakers and some cannot be heard in far corners of the room.

If the speaker must move around, a lavaliere (chest-type), wireless, or lapel microphone is necessary. (Be sure the lavaliere microphone does not rub against a tie clip as this makes a lot of noise.) If the speaker can remain in one place, a pedestal or lectern microphone is satisfactory.

If audience participation is desired from a large group, floor microphones placed in the aisles or roving microphones carried by assistants will enable members of the audience to be heard throughout the room. Otherwise, the speaker must repeat questions and comments from the floor, using his microphone, so the entire audience will know what has been said.

Before each use, an amplifying system should be checked for good operating condition and to be sure reception is satisfactory throughout the room. At all times while the system is being used, it should be supervised by an individual familiar with electronic equipment to assure control of volume and to eliminate annoying acoustic feedback.

Also see the discussion of public address systems at the end of this chapter.

### TelePrompters

TelePrompters or other cueing devices can be used for dramatic or formal presentations in which actors, technicians, or executives are required to follow a prepared script.

Of course, technical help is needed to set up the TelePrompter, and those using it must be familiar with the technique so their presentation will have the desired natural effect.

### Screens

There are several types of screens used for projection:

*Glass-beaded screens.* Usually portable, but often of the large pull-down variety, these screens have a high reflectance value but within a narrow angle of projection (45 degrees).

*Matte finish screens* do not give quite as bright an image as beaded screens. However, since they have a wider viewing angle (60 degrees), they are more suitable for larger audiences, and for any room where some of the audience must sit at a considerable angle to the screen.

*Lenticular-surfaced screens* have embossed surfaces that reflect a high percentage of projected light on a viewing angle wider than that of beaded screens (70 degrees). Also they reject stray incident light.

*Permanently mounted aluminum foil screens* are the brightest obtainable. They cannot be rolled up because they have a solid backing and are slightly curved. The viewing angle is only 30 degrees maximum. The screen is designed for use with room light on; it is too bright for use in a darkened room, unless projector illumination is reduced.

*Rear projection screens* can be used in partially lighted rooms and have a wide viewing angle. However, any stray light behind the rear screen must be held to an absolute minimum.

The image on the screen should be neither dazzlingly bright nor dim. Generally, a 500-watt projector bulb is satisfactory for small and medium groups in either a darkened or partially darkened room, if the film, slides, or transparencies have good color and well defined material. A 1,000-watt bulb in a 16mm movie projector is better for medium and larger groups. Special, longer-burning bulbs are available.

Where possible, screens should be set slightly above the heads of the audience for maximum visibility. With some projectors, principally the overhead type set close to the screen, a keystone effect can be created, whereby the top of the image is noticeably wider than the bottom. This distortion can be reduced by slanting the screen slightly so its plane is more nearly perpendicular to the projected light (see Figure 14-10). Some screens now are equipped with a clip or bar that permits this adjustment. Such bars can also be purchased separately and attached to portable or wall screens.

If possible, the area of the projected image should fill the screen, but not extend over on the background.

Although modern projectors equipped with powerful bulbs do not require complete room darkening, room lights should be dimmed (or some lights switched off) and windows shaded as needed. This leaves sufficient light for note-taking or script reading. There is no need to completely darken a room for showing noncontinuous-tone images (such as graphs, diagrams, or lettering). A reverse-projection unit or a shadow box permits showing filmstrips, slides, or movies to a small group without dimming lights.

Slide projectors that must be located in or behind the audience should have a remote-control device, or be operated by someone other than the speaker (if the presentation does not have a recorded narration). There should be a prearranged, subtle signal for changing slides. When the speaker must say "Next

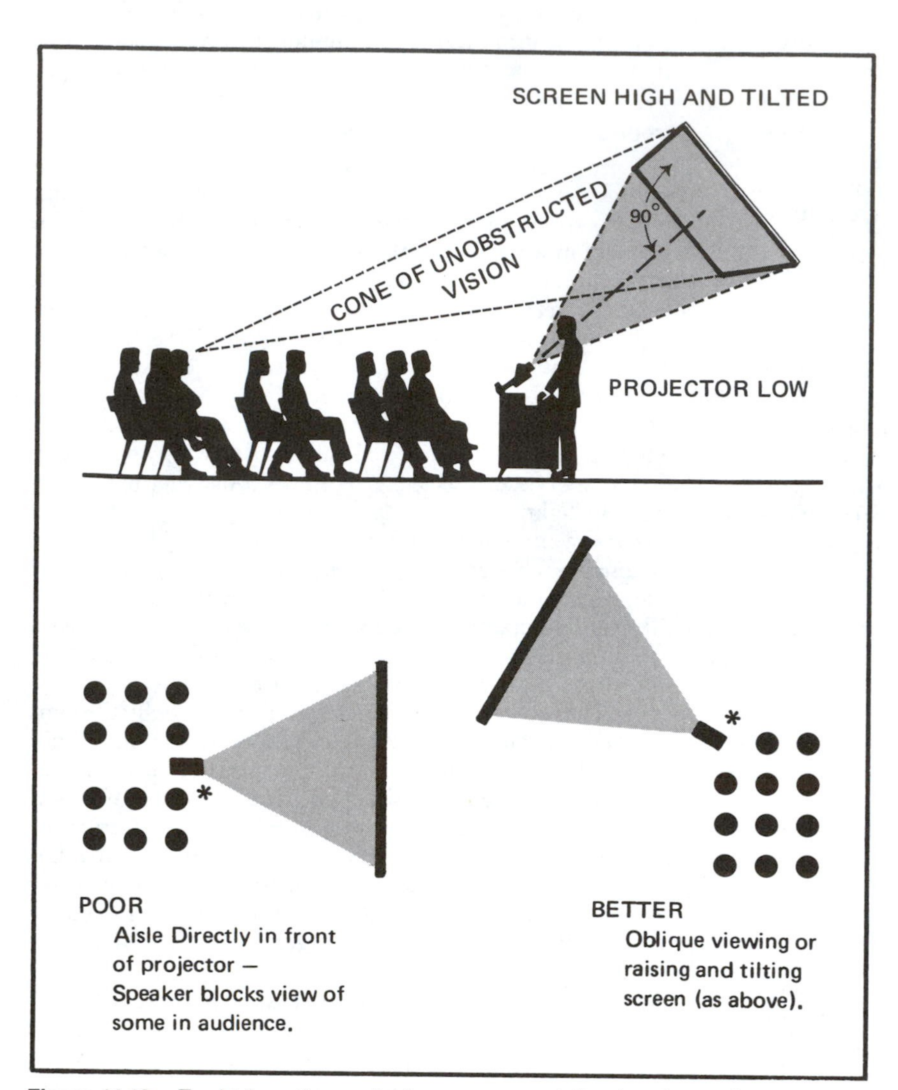


**Figure 14-10.** To obtain optimum viewing area, especially when instructing large groups, tilt the overhead screen (*top*) or place it toward the corner of the room (*lower right*). Both placements reduce the possibility of the instructor and equipment blocking the vision of some viewers, as might be the case when the projector is located in front of the group (*lower left*).

slide please," he distracts the audience. Some offhand, but clear, signal with the pointer or flashlight should be arranged.

## Seating

The presentation room should be checked for safety features and for seating arrangement. Aisle space should be adequate, exits ready for emergency use, and ash trays provided if smoking is permitted.

The seating arrangement should be planned so every member of the audience will have an unobstructed view of the visual (see Figure 14-10). For meeting room or auditorium seating, allow 21 to 24 in. (53 to 61 cm) width for each chair and 36 in. (91 cm) for each row of chairs, or about 5 to 6 sq ft (0.46 to 0.56m$^2$) per viewer. In classrooms or conference rooms, allow twice as much space.

For viewing a projected visual, the most desirable seating area is within a 30-degree angle from the projection axis with a matte screen and within a 20-degree angle from the axis with a beaded screen (see Figure 14-11).

The recommended minimum viewing distance is at least twice the screen width, and the maximum no more than eight times the screen width, although a distance no more than six times the screen width is desirable. A 6-foot-wide (1.8 m) image, therefore, could be viewed at a maximum distance of about 50 ft (15 m) by approximately 100 persons.

For extremely wide rooms, and if the seats cannot be arranged so the audience sits within the recommended viewing angle, it is possible to project images from slide projectors or over-head projectors simultaneously on two or three screens, separated by at least 25 ft (7 m).

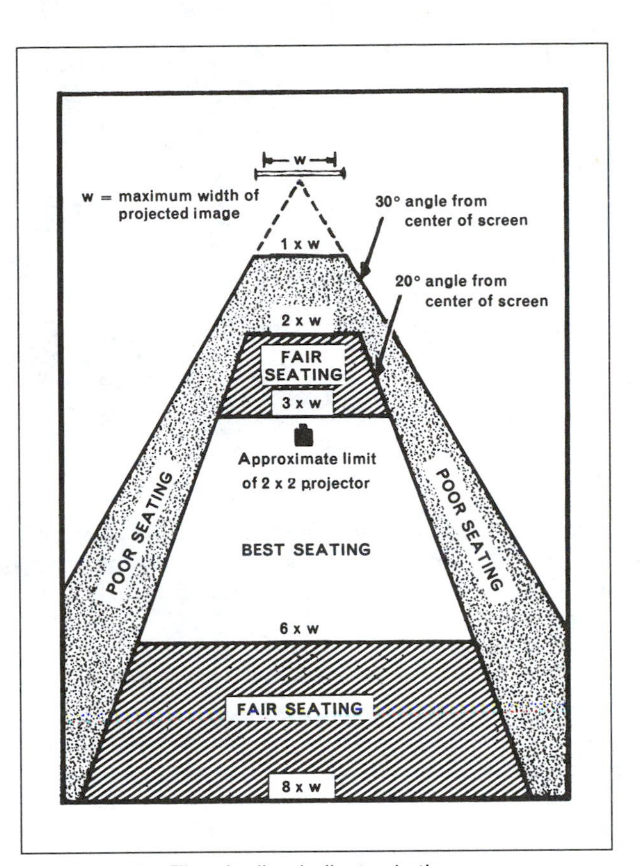


**Figure 14-11.** The shading indicates both good and poor seating areas when ordinary (beaded) projection screens are used. A lenticular screen would widen the angle, which is measured from the centerline. The best viewing area is within 20 degrees of the projection axis for a beaded screen, and within 30 degrees for a matte screen.

## Mobile presentations

Where good projection facilities are lacking, such as in the field or at branch terminals or plants, mobile presentations should be considered. Flip charts, flannel boards, and demonstrations can be used outdoors or in quarters not well suited to projected visuals.

Trucks, trailers, converted buses (Figure 14-12a), and other large vehicles can be used as mobile classrooms. Rear screen projectors, mounted in station wagons or van wagons, facilitate the use of projected visuals in the field. For such small quarters, air conditioning should be used.

Videos work well for mobile presentations where space and size of audience are limited. As a rule of thumb, a television screen can be viewed by one person per diagonal inch of screen (see Figure 14-12b).

## Rehearsal

With even the simplest visuals, practice before use is imperative. It will help prevent the speaker from running overtime and will help assure smoothness of presentation.

Unrehearsed use of a visual may reduce its effectiveness considerably. A set of charts, for instance, may be well prepared, easy to understand, and attractive, but if they are shown in random fashion or must be fumbled with by the speaker, much of their impact will be lost.

**Figure 14-12a.** The instructor aboard Trainer One, a 35-ft mobile classroom, is teaching his crew traffic control techniques. (Printed with permission from SET Trust Fund.)

**Figure 14-12b.** All the instructors and trainees on Trainer One could easily view a video training tape on a 23-in screen. (Printed with permission from SET Trust Fund.)

Moreover, if the speaker wanders from the subject, the set of charts or transparencies, instead of serving as an aid, may even prove distracting. If material other than that illustrated must be discussed, the speaker would be wise to cover the charts (or turn off the projector) until ready to return to them.

When items such as chalk, an eraser, and a pointer are needed, the speaker should make sure in advance they are at hand. A person who needs a marking pen should carry two pens in case one runs dry. Some training rooms have white chalkboards; if using a felt-tipped pen, be sure the ink can be removed by a damp cloth.

Training and rehearsal is particularly important with more complicated equipment such as a movie projector. Well in advance of the showing, a trained operator should make sure that the film is not broken, the equipment is ready to roll, and extra bulbs are on hand. The projector should be threaded or have slides inserted in correct projection position, the motion picture header should be run off, the sound adjusted, picture focused, and the screen positioned at the proper distance. When a tape or a record is used with a film or slides, the sound and picture should be synchronized. Arrangements for turning lights on and off should always be checked.

A film or a set of slides should be previewed and checked for good condition, and, in the case of slides, for proper sequence and right side up. When slides are in proper sequence, draw a diagonal line across the top of the pack—an out-of-place or mis-turned slide shows up at once.

The need for introductory remarks, discussion questions, and recall or follow-up materials should be considered.

If a script is to be read, the speaker should not indulge in lengthy ad-libs. He should stick to his plan of presentation, giving each chart or slide the time and attention it deserves, but no more. It is a good idea to preview all required visuals and have them on hand before the meeting starts.

The speaker should face the audience as much of the time as possible, particularly if he is using charts or a chalkboard. He should not talk when he is moving about or not facing his audience. Not only would the distraction be bad, but the audience would have difficulty hearing him.

If the speaker wishes to face the audience the entire time, he can arrange for an assistant to write on the chalkboard or turn the flip charts. Be sure this doesn't become more distracting than useful. If the speaker is using a projected visual, he can operate the projector with a remote control cord or have another person run it for him.

It is more effective if the speaker has the full information of the slide in front of him so that he does not have to turn continually toward the screen and away from the audience.

## NONPROJECTED VISUALS

Nonprojected visuals include graphics, three-dimensional exhibits and models, and live presentations. Among the various types of graphics are chalkboards (formerly called blackboards), paper pads, flip charts, display cards, flannel boards, magnetic boards, and hook and loop boards.

### Chalkboards and paper pads

Chalkboards are a basic visual. They come in several colors and charcoal, but light green is considered standard. A dustless chalk should be used, preferably a yellow or other bright color for maximum visibility.

A large block of instructors' chalk or the side of a stick chalk gives a heavier and wider line. Different colors of chalk, perticularly the fluorescent chalks with black light, are very effective.

Paper pads (usually 2 by 3 ft [0.5×0.76 m]) provide inexpensive visuals for small groups (see Figure 14-13). Pads, or a number of sheets of paper, clamped on light-weight wooden or aluminum folding easels (as flip charts are) are portable and always ready for use. The speaker can keep training material on the pad so he can refer to it, or throw it away, as wished. Some pads have faint lines to facilitate lettering and layout of artwork.

Pads are of real value to those who must lead discussions or run brainstorming sessions. As each chart is filled, it can be tacked to a corkboard or clipped to a wire running along one side of the conference room. This (1) provides the group with a continuous record of what has been discussed, (2) lets latecomers catch up to the discussion, without having to stop the discussion and have a review, and (3) helps in writing a good report of the meeting—the notes are right on the sheets.

Inexpensive rolls of white paper can be used for charts. Even brown wrapping paper will do if material such as lettering is added in a strongly contrasting color. Material can be written and read as the chart is unrolled like a scroll.

**Figure 14-13.** Informal visuals written on a paper pad hold the interest of this small, informal group.

### Flip charts and posters

**Flip charts.** Flip charts are a refinement of paper pads. Usually charts are on heavier paper, prepared in advance, and used in more formal meetings; frequently, blank pages are also provided for on-the-spot additions. Flip charts might combine specially made material mounted on large sheets or hinged in briefcase-size easel-binders that are used for desk-top or bench-top discussion. Portable units are easy to make and are easy to take to different locations.

To help in making the presentation, brief notes can be written lightly in blue pencil on the face of the card or flip chart. Notes are written small enough so the instructor can read them, but they are invisible to anyone more than 6 or 7 ft (1.8 or 2 m) away.

**Posters** are desirable to a successful safety program. They serve as reminders, warnings, and motivators. They deal with many accident problems, both occupational and nonoccupational, and they employ many types of art styles and formats. They are not used like a flip chart, but rather are displayed at various locations throughout an establishment or plant.

Selected carefully and used discriminately, posters can help employees avoid hazards and unsafe acts and can help the company or plant have a more effective safety program. For more details, see the discussion of Posters and displays in Chapter 12, Maintaining Interest in Safety, and in Industrial Data Sheet No. 616.

### Flannel boards

A flannel board is a plywood board, commonly 3 by 4 ft (0.9×1.2 m) covered with dark flannel. A number of frames can be fastened together or placed close together to form multiples of the original size.

Roll flannel can be tacked, tied, or otherwise stretched over any large, flat, slightly inclined surface, including a chalk board. Art work or lettering can be made on a special velour-backed paper or on light-weight cardboard or heavy-weight paper to which flocking paper with felt adhesive or strips is affixed. In an emergency, sand paper can be used instead of the special flocking paper. The flocking or sandpaper grips the long-napped flannel with sufficient strength to hold light-weight cards.

These individual cardboards, with words, designs, or messages on them, can be attached to create a dramatic effect not possible with other visuals (see Figure 14-14).

**Figure 14-14.** Dramatic effects not possible with other visuals can be created using a flannel board.

### Hook and loop boards

A hook and loop board is a heavy-duty "slap board." The material covering the board contains countless nylon loops. These loops are caught by the almost invisible nylon hooks on the small pieces of tape mounted on the back of signs or other display objects. A tiny patch of the hook material fastened to a heavy or large object will hold it securely on the board.

Hook and loop boards are available commercially in sizes ranging from 18 by 24 in. (45×60 cm) to 48 by 72 in. (1.2×1.8 m).

### Magnetic boards

A magnetic board can be made of either a spray-painted sheet metal plate or a steel-backed chalk board. Small objects or cutouts mounted on small magnets or on magnetic tape can be placed on the board and then moved at will.

This type of visual often is used for training operators of vehicles such as forklift trucks. The mobility of the objects—toy vehicles or cutouts of trucks, together with the cutouts of aisles and loads—enables the instructor to give a realistic demonstration of safe practices.

### Photographs

Photographs need not be projected to be useful. Candid shots of safe and unsafe practices or conditions, photos of award presentations, meetings, new equipment, and the like, have excellent news value on bulletin boards (see Figure 14-2), in company or plant papers, house organs, or safety magazines, and are even welcomed by national trade and professional magazines.

The instant photograph can be another excellent tool. It provides an almost immediate record of hazards and is useful in accident investigation. See Council Data Sheet No. 619.

### Exhibits and demonstrations

Exhibits and models make very effective three-dimensional displays for use in instruction. Such exhibits can be made for the purpose of demonstrating the safe working of a machine or process. Examples of first aid equipment, protective clothing, rescue equipment, respiratory protective equipment, and fire protection appliances can also be featured.

A small wooden dummy made with articulated joints and spine is often used for showing the correct and safe method of lifting and carrying heavy loads (see Figure 14-15a for lifting instructions).

Demonstrations of firefighting can sometimes be organized with the assistance of the local fire service (see Figures 14-16a and -16b).

Demonstrations of good and bad lighting can be easily arranged in a lecture room. The effect of an impact on safety hats, safety shoes, and eye protection devices can be shown. The teaching of splinting by demonstration plays an important part in first aid training.

## PROJECTED VISUALS

Projected visuals include slides, transparencies for overhead projectors, objects used in opaque projectors, motion pictures, television and video.

This chapter will not go into great detail regarding slides and movies, because excellent information is available from the manufacturers of film and projectors, camera stores, libraries, schools, and publications. The *National Safety News,* for instance, frequently carries detailed articles on producing slide shows, using slides, and related topics. The National Safety Council also publishes a series of Data Sheets on the subject (see References).

### Slides

Slide presentations have become increasingly popular since the addition of sound. Programs can be recorded and sound synchronized to slides on a cassette recorder, which can automatically advance the slides with the narration.

Learner response is encouraged when cassette recorders are equipped with a stop button on the visual synchronizing mechanisms. The slide presentation can pose questions, be stopped for class response (oral or written), and then started again for correct answers and the explanations or discussions. To add professionalism, background music can be added.

One international organization has designed basic safety presentations available in half a dozen languages. For each presentation, locals receive a package consisting of the tape cassette, slides, and a printed copy of the narration for whoever will conduct the safety training session.

- Specific advantages include:
  Slides can be easily updated
  They are inexpensive
  Almost anyone can make a slide
  They can be geared to the needs of the audience by substituting specific slides
  Audio tape for automatic advance is available
  Can be coupled with other projectors both slide and film
  Are easily stopped for discussion.
- Potential disadvantages include:
  Slides can easily get out of order
  They can stick in the holding tray
  Slides can easily be projected upside down or backwards
  Poorly made slides are distracting.

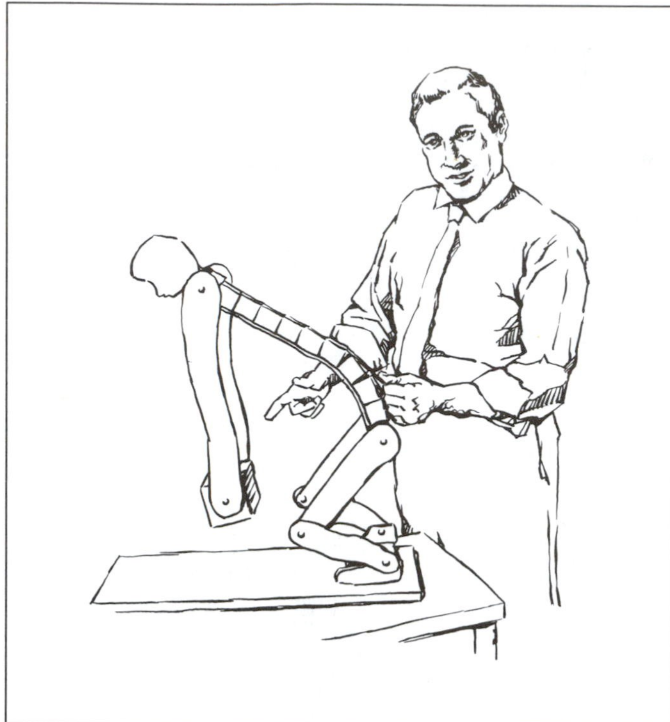

Figure 14-15a. A model is used to demonstrate correct lifting techniques.

**Here Is How to Use It**

There is a correct and an incorrect way to pick up a heavy object. The correct way is to keep the back straight, the knees bent and spread, and the load close to the body. The incorrect way is to reach way over and lift. (Twisting the back complicates the bad effect.)

To demonstrate this effectively, a special model can be used. If used with its block spine locked, the model simulates lifting with strong leg muscles; the ribbon on the spine remains limp, indicating very little tension of the back muscles. It demonstrates that the back cannot be kept straight without bending the knees.

To demonstrate improper lifting, the model is used as is shown in the sketch. The legs are bent only slightly (or held straight). One hand lifts the handle just ahead of the fulcrum at the hips in order to lift the weight in the hands. The back arches under the strain and visibly pulls each block apart.

The model can also be made to show proper foot placement (see drawing at left).

The simplicity of producing slides can give a sense of false security and result in poorly organized presentations.

Slide projectors range from inexpensive single slide viewers that can be held in the hand and looked at, to remote control or automatic-change projectors that have trays or magazines holding up to 140 slides. A table top-sized, rear projection portable machine is suitable for very small groups. With a taped or recorded message, this type is well suited to training one or two employees.

## Overhead projection

Of the many ways to get a visual message to people, overhead projection offers special advantages:

The speaker can present to any size group in a fully lighted room.

He faces his audience at all times.

The material can be revealed point by point.

The audience is not distracted by the machine.

Overhead projection equipment is easy to operate and is readily available.

Overhead transparencies can be made quickly and inexpensively.

Figure 14-16a. Railroad safety officers are receiving hands-on training in firefighting. (Printed with permission from Association of American Railroads.)

Figure 14-16b. As part of hazardous material response training, these students extinguish a liquified petroleum gas fire at the AAR Transportation Test Center. (Printed with permission from Association of American Railroads.)

**Construction Specifications**

The spine is made with 9 blocks, each 2 in. square and 1⅝ in. high. Each is drilled at the center to accommodate a standard screen door spring. The T-shaped head and shoulder piece is about 8 in. long, 2 in. thick, and supports the arms on shoulders about 5 in. apart. The hip block is 2 in. square, with sloping sides so the legs will spread open in front. Add a ⅜-in. spacer between the hip block and each leg.

Assemble the body by using wood screws to attach ends of spring to the head and hip pieces. Tack a piece of 2-in. wide belting to the *front* of the spine blocks to hold them in alignment.

The arms and legs can be shaped from ¼-in. plywood.

A block, approximately 5 in. square and 3½ in. high, represents the lifting weight. Elastic tape or multiple rubber bands are stretched from the shoulder to the hip, along the *back* of the spine blocks to represent the spine muscles.

A metal handle can be secured to the lower end of the hip block, as shown in the drawing.

The model can be mounted on a 1x12-in. board, 18 to 24 in. long.

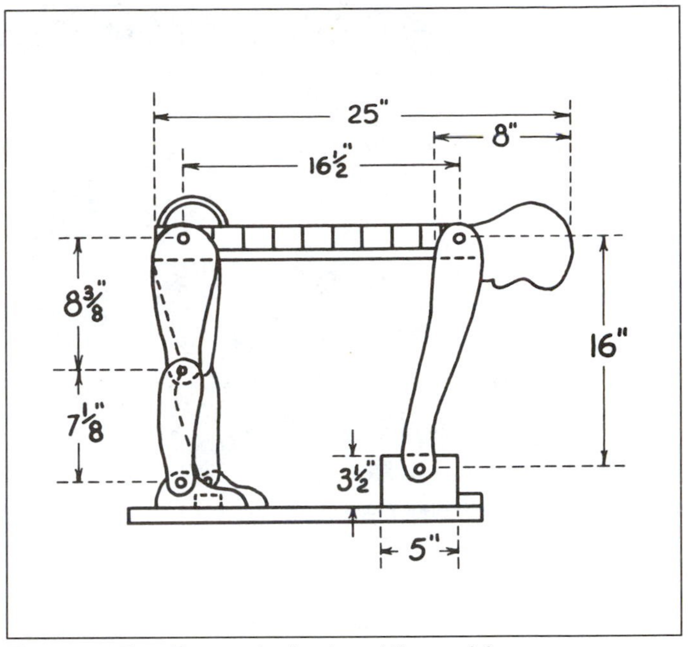


**Figure 14-15b.** Ergonomic drawing of the model.

Lettering is a key to the effectiveness of a visual for overheads. Careless lettering can detract from even the best illustration while neat, well-planned lettering can be effective by itself. Letters should be at least ¼ in. high (if the original can be read from a distance of 10 ft (3 m), the transparency should project well). The letters can be applied to transparencies by hand, stencils, tracing, transfer letters and symbols, and lettering tape which is made from an imprinting machine.

It may be helpful to use color for clarity and emphasis of certain points. Color attracts attention, is eye-pleasing and provides variety.

Where possible, limit each original to one point or comparison. Break paragraphs into sentences, and sentences into phrases and key words. Use a maximum of six lines, and six or seven words per line.

Reproduction equipment is available for making transparencies from almost any original.

To add impact to projected transparencies, overlays can be used. This method helps simplify difficult concepts and also lets the presentor build the visual's story in a meaningful way. It involves two or more imaged transparencies used in sequence over each other. Using no more than two overlays and different colors for each makes this a very effective way to present step-by-step information. Overlay visuals are hinged to the frame on one side with tape to allow the base visual to be presented first, then the overlay flapped over it to complete the message.

On overhead projection the visual can be projected unframed, but there are good reasons for using frames: The frame blocks light around the edge of the visual, adds rigidity for handling and storage; and provides a convenient border for writing notes.

If a blank transparent sheet or roll of clear plastic film is used on the overhead, the speaker can use it as a pad to gather ideas or lead a discussion. Not only are ideas written down for all to see on the screen, but the sheets can be saved and used to keep a record of the meeting.

This visual technique has a number of important advantages. The speaker needs no assistant; he can face the audience, and the room can be either partially or fully lighted. He can use any opaque material as a cover sheet, which can be withdrawn to reveal data point by point, and he can point to items on the transparency—and therefore on the screen—with a pencil or other object (see Figure 14-17).

### Opaque projectors

Opaque projectors are useful for showing printed material and even three-dimensional objects. Printed material up to 10 in. (25 cm) square or an object up to 2½ in. (6.5 cm) thick is placed in the machine, and the image is projected onto the screen by means of a powerful light and a mirror. The room must be darkened for effective viewing.

With an opaque projector, material that cannot conveniently be transferred for overhead projectors or photographed for 2-in. slide projectors can be quickly shown. An example of such material would be a safety catalog or a safety poster in full color. Of course, to be suitable for use in an opaque projector, printed material must have type large enough to be legible to the entire audience when the material is projected.

Another use of the opaque projector is to project a picture, map, or other shape on to a pad or chalkboard so it can be traced. The size of the projection can be varied to the exact size required. This results in a neat, professional-looking drawing that also can be used as a flip chart (see Figure 14-6).

### Motion pictures

Movies are rated excellent for training and motivating. Because of their higher cost compared with that of other visuals, they should be planned with special care. Usually, movies are important enough to justify commercial production or at least professional advice before and during production. Some homemade movies, however, have proved to be effective. See previous section on Making Motion Pictures.

A wide selection and variety of 16mm safety films are available

**Figure 14-17.** An overhead projector permits the instructor to face his audience because he can see the visual without turning.

from insurance companies, local safety organizations, commercial film libraries, industrial producers, and the National Safety Council.

**Polavision.** This Polaroid product consists of a camera, rear-screen playback unit, and other accessories. After the picture is made, the film cartridge is inserted into the playback unit and is fully developed in about 90 seconds. Then it is ready for projection. The cartridge runs 2½ minutes in length and is available only in 16mm color. Sound is not available. Equipment is easily operated, but the projection area is limited because the screen's diagonal measurement is less than 12 in. (30 cm).

### Multi-image

Multi-image refers to using two or more projectors simultaneously. Actually it is an expansion of the basic sound slide program format that has gained tremendous popularity. The presentation can be of the same medium (such as two or more slide projectors), or mixed media (slide/motion picture).

A programmer allows you to automatically turn slide and/or movie projectors on and off, advance slides, and change slide projector lamp currents from off to full on, at nearly any rate of change. These functions may be performed in any sequence you desire.

- Advantages include:
  - Visual information can be presented both linearly and spatially, enabling the viewer to see not only the order but also the relationship of information.
  - These programs compress the time needed to create an impression.
  - They tend to develop greater viewer involvement and enthusiasm.
- Potential disadvantages are:
  - Operator training is necessary.
  - They can require sizeable budgets.
  - They take time to develop.

In summary, it should be remembered that each audiovisual system has particular properties that represent a possible advantage or disadvantage when compared with another type of audiovisual system. The audiovisual decision-maker must weigh needs and objectives against all the other factors, including budget and audiovisual capabilities.

## VIDEOS

The word video may be used to describe any type of television equipment, such as video cameras, tape recorders, cassette players, recording tape, monitors, and a full range of television studio equipment and video television hardware.

Today a greater effort is being made to remove communication barriers and to assure that information is properly learned and used. For these reasons, video is being recognized as one of the most effective communications and training methods available, both as a production medium and as a transmission medium. It can also be used as a means of self-analysis (Figure 14-18). The following discussion is taken from "Using Videotape to Conduct Safety Training," by Gerald H. Kaiz. See References.

### Videotape advantages

The advantages of videotape lie in two areas: lower cost of production and a culturally natural transmission medium.

The lower cost of video compared to motion pictures is due to two factors. First, video has the ability to instantly play back the field-recorded sequence to assure the technical accuracy of the visual material, and to verify that the proper focus, lighting, viewing angle, and composition have been achieved. If there is any problem, the activities can be immediately repeated, taped, and reviewed until the proper content has been recorded.

If motion picture or slide film is used, the image content is not known until the film is developed and reviewed, which may be days later. Any problems requiring return to the shooting location several times can greatly increase costs. In addition, much more footage may be shot than will actually be used to provide insurance in case some shots are unusable.

The second reason for the lower cost of video is the structuring and editing of the program material. The video program is structured with all video footage taken for specific uses, and reviewed in the field prior to studio assembly. No time is spent choosing from alternate scenes later because they can be reshot right away if the first effort doesn't turn out as desired. Sometimes the audio can be rewritten to match the action, although the picture should be selected to meet script demands, especially if the script has already been approved by technical and management people. Compared to motion picture film that must be edited from the processed film, videotape is quickly edited electronically and does not require splicing. Also, video can be easily duplicated for quantity distribution, and topical subjects can help round-out a company's library of safety presentations.

Adding sound to slide programs increases the cost. The quality of a slide is not known until it is developed; selection of the best slide from the many that are taken is time-consuming because 5 to 15 slides are required per minute of program time.

Another consideration is that the television set has become the focus of household communication, bringing in the nightly news, entertainment, and cultural programming. As a result, many people have moved from a reading-oriented culture to a verbally and visually stimulated culture; people passively interact with the video screen. For the majority of people, viewing a television set to obtain information is more natural than listening to a live presentation, watching a motion picture, or viewing a slide presentation.

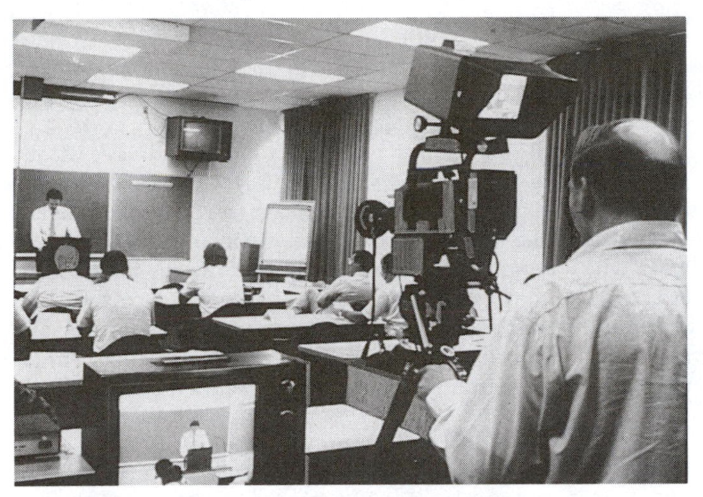

**Figure 14-18.** While this student of National Safety Council's Safety Training Institute makes a presentation, the talk is taped on closed-circuit TV. Later, the presentation will be replayed and critiqued by both the class and the student.

### Videotape players

Although professional producers of television programs prefer the reel-to-reel recorder, primarily because the tape in this format is easier to edit, there is widespread distribution of already-produced subjects for the popular cassette tape player. These tapes come in several types and sizes: ¾ in. or ½ in., VHS or Beta, or 8 mm tapes.

The two words, cassette and cartridge, are often used interchangeably, referring to any container holding tape that can be used in a player without having to be threaded. Technically, a distinction is recognized. A cassette holds both the original reel and the take-up reel in the same case; a cartridge has only one reel on which the tape is wound. Its take-up reel is incorporated into the player, with the threading being done automatically.

But in operation, they are much alike. The plastic case with the tape is put into position, a switch or two is flipped, and the player goes to work. It picks up the signal and feeds it to the television set. They are not alike, however, in physical format. Be sure to use the cartridge or the cassette that is designed for the equipment on hand. There is a limited interchangeability.

The most commonly used systems are based on magnetic tape, but several types of presentations start out with film.

### In-house and out-of-house production

In deciding whether to create such programs within company or plant facilities or to use an outside firm specializing in such work, the quality of the presentation must be considered.

Because of familiarity with the household television set, the student using television as a learning tool is conscious of the quality of production. In order to hold the student's attention, the quality must be comparable to what the student has been watching, as opposed to home movies or merely a talking face appearing on the screen. Home movies usually are amateur productions that will not convey proper images either of the company or its safety practices. The talking face may as well be a live lecture, because the real advantage of the video medium—its ability to bring the job site directly into the classroom—is not being utilized.

To achieve the quality desired, two factors must be available: people to produce the programs, and equipment to record the program materials.

As a rule of thumb, if less than one video program is prepared each month, it may be advisable to consider an out-of-house training organization to produce the videotapes, under the direction and control of an in-house coordinator.

The one program monthly is the production load that can be used to justify three full-time people: a training program developer, a video director, and an equipment operator. If these positions are filled by part-time personnel, schedules will not be met, costs will increase, and production will probably fall into the home movie classification.

The necessary video equipment for production includes a video camera, video recorder, and an editing system. If the equipment is going to be stored in a closet and used only occasionally, consider the out-of-house training or production company.

A caution is in order here. A training company is different from a video production house. The training company will perform all the steps listed next. The video production house performs only steps 4, 5, and 7, and then only under close direction.

In addition to the six steps listed early in this chapter under Preparing a script, the following steps must be followed:

1. Develop program's scope. Identify the information to be covered, its sequence, and what is to be shown to take advantage of the dynamics of the medium; identify shooting locations, equipment, and personnel requirements. Get budget approval.
2. Write narrative script and submit for review.
3. Rehearse activities and script in the field. This permits identifying video camera and sound recording equipment locations and performer knowledge, and assures proper equipment availability and operation.
4. Shoot field footage. Also, review technical accuracy and visual content of each piece of field video. If live sound is part of the presentation, it is recorded during the field shoot.
5. Studio assembly and editing. This will be performed by recording the audio track from an approved script, and integrating it with the field video, live sound recording, and the additional visual aids. This is the most important and critical aspect of videotape production.
6. Review completed program. This is done with all persons who attended the formal planning session (step three under Preparing a script, earlier in this Chapter). Identify corrections of either the visual or audio portion of the program. If all prior steps were properly performed, these should be minimal.
7. Make corrections. This will be followed by the development of written materials for students and the instructor.
8. Train the instructor to properly use the program.
9. Run the program in the field and get feedback from the class on presentation level, meeting objectives, and training effectiveness. If the target audience was correctly pinpointed and the program scope correctly defined, little modification should be needed. If modifications are needed, make them after pilot programs are complete.
10. Continue to monitor training effectiveness and document employee performance and program benefits.

## AUDIO AIDS

Purely audio aids include tape recorders, radios, commercial recordings, and public address systems.

### Tape recorders

A tape recorder can be used many ways:

- One safety professional, using a battery-operated recorder, dictates a running commentary while photographing safe or unsafe operating conditions, new processes or equipment that require subsequent study. This method proves easier than writing notes. After being edited, the tape is used with the slides to make up a training tool.
- A tape recorder can record minutes in safety meetings, valuable discussion in training sessions, and on-the-job interviews which may prove useful later for bulletins, newsletters, and future meetings.
- Tape recorders are frequently used to record speeches and conferences. Pedestal or table microphones generally pick up extraneous sounds, as well as the desired ones; also it is sometimes difficult to identify a speaker. Those planning such discussion recordings should arrange for one-at-a-time discussion or for placement of recorders at different parts of the discussion table. It is a good idea to mention names frequently as one addresses a conferee, and to rephrase discussion points that may be unclear, either to meaning or to clarity (someone else talking at the same time, for example).
- A safe, efficient job procedure can be taped. Sufficient time is allowed for performing each task. The trainee, with the tape player nearby, follows the instructions as he performs the work.

  A tape recording can be used with slides or other visual training material and used individually to train new employees at a remote location—for example, one restaurant of a chain of restaurants.

  Foreign languages can be recorded on a tape for instruction of non-English-speaking employees. If audiovisuals are to accompany presentations that are given in two or more languages, it is best that they be strictly pictorial, having no languages written or printed on them.
- A polished speaker can even use a tape recorder to have dialogue with himself. This requires close timing.

### Radio

Radio can serve as a medium to promote accident prevention through scheduled programs on various aspects of home, traffic, and community safety. Spot announcements concerned with traffic safety, for instance, commonly are broadcast at frequent intervals by local stations during long holiday periods such as the Labor Day weekend.

Radio was covered in detail in the previous chapter on publicizing safety.

### Commercial recordings

Music, special sound effects, and even dramatic episodes on commercial recordings can be added to safety talks, slide scripts, and the like, to give a professional touch.

### Public address systems

A public address system is a useful tool for making safety announcements, directing emergency evacuations, and perhaps even publicizing unusual safety achievements.

There is some question as to whether public address systems should be used to broadcast safety messages on the job. Some administrators feel that music or messages might prove dangerously distracting to workers at moving machinery or on other work which requires full attention. This technique may not be effective and may even draw complaints if used excessively.

Portable public address systems have been used by supervisors and safety professionals as an aid in on-the-job meetings, either to help overcome extraneous noise, or to create a dramatic effect.

## REFERENCES


Caiati, Carl. *Video Production the Professional Way.* Blue Ridge Summit, Pa.: Tab Books Inc., 1985.

Kaiz, Gerald H. "Using Videotape To Conduct Safety Training." *National Safety News,* January 1980, pp. 46-50.

National Safety Council, 444 N. Michigan Ave., Chicago, Ill. 60611.
  "Guide for Audiovisual Aid Development," 1981.
  Industrial Data Sheets
    *Motion Pictures for Safety,* 556.
    *Nonprojected Visual Aids,* 564.
    *Photography for Safety,* 619.
    *Posters, Bulletin Boards, and Safety Displays,* 616.
    *Projected Still Pictures,* 574.
  *Poster Directory.*

Shefter, Harry. *How to Prepare Talks and Oral Reports.* Pocket Books, Inc., 1230 Ave. of the Americas, New York, N.Y. 10020.


### Associations and government agencies


Association of Audio-Visual Technicians, P.O. Box 9716, Denver, Colo. 80209.

Association of Visual Communicators, 900 Palm Ave., Suite B, South Pasadena, Calif. 91030.

International Communications Industries Association, 3150 Spring St., Fairfax, Va. 22031. *Audio-Visual Equipment Directory.*

International Television Association, 6311 N. O'Connor Rd., Suite 110-LB 51, Irving, Texas 75039.

National Archives and Record Service, National Audiovisual Center, Washington, D.C. 20409. "List of Audiovisual Materials Produced by the United States Government for Industrial Safety." (latest edition).

National Education Association, Dept. of Audio-Visual Instruction, 1201 16th St. NW., Washington, D.C. 20036

U.S. Office of Education, Bureau of Adult and Vocational Education, 400 Maryland Ave. SW., Washington, D.C. 20202.

(Other data books and pamphlets are available from distributors and manufacturers of cameras, films, and other visual media equipment. Trade journals in the fields of photography, education, sales mangement, advertising, and training also contain excellent information.)

# 15

# Office Safety



MANY LARGE ORGANIZATIONS TODAY consist almost entirely of office workers (insurance companies and banks are good examples). Accidental injuries are just as painful, severe, and expensive to office workers as to production workers. Unless an organization has an effective office safety program, accidental injuries are far more likely to occur.

Yes, the risk of occupation-related accident or injury to the office worker is lower than the risk to those employees involved in manufacturing or transportation. Office risks often go unrecognized and unmanaged, however, and some could eventually lead to individual or plant disaster.

An injury to an office worker can break the safety record of a manufacturing plant just as effectively as can an injury to a production worker. In the aerospace industry, for example, the injury frequency rate for office workers sometimes has been greater than for production workers.

A company safety program cannot be fully effective if there is only partial participation. A safety program that is not vigorously pursued in company offices probably will not be vigorously pursued in the factory, shop, or plant. If office workers are exempt, then production workers often feel that following rules to avoid hazards is an unnecessary burden, and, perhaps, an unfair exercise of authority by management. Exempt office workers seldom understand the importance of safety and may scoff at or criticize production-oriented safety activities, especially in front of production employees.

To emphasize what was written in Chapter 3, the health and safety professional who expects to sell safety to management must get management involved in a total safety program—get them involved in preventing office accidents as well as production accidents. Any management that preaches safety must also practice it.

## SERIOUSNESS OF OFFICE INJURIES

One reason office safety programs are not more widespread is that many people believe office injuries are inconsequential.

This is an erroneous belief. One aerospace firm, for example, paid out $102,000 over eight years at one facility just for injuries incurred by people falling out of chairs. Approximately 25,000 people worked at the plant, and more than half were office workers. Of the 14 chair accidents that occurred, the two worst totaled $97,000. Not only are medical and wage replacement benefits costly for disabling accidents, but there also are hidden costs such as the loss of productivity. (See Chapter 7, Accident Investigation, Analysis, and Costs.)

Studies made by the State of California Department of Industrial Relations, and the Equitable Life Assurance Society of the U.S. (see References), show that this is not an isolated incident. (ure 15–1).

The California State Department of Industrial Relations analyzed reports filed by more than 3,000 California employers on disabling injuries to employees. These employers together employed more than one million office workers. When the results of the study were extrapolated to nationwide scale, on-the-job office accidents annually amounted to about 40,000 disabling injuries at a *direct cost* (indemnity benefits and medical expenses) of about $100 million. This figure does not include any indirect costs for employers, workers, or the nation.

Of the many accidental fatalities occurring to office workers, approximately half are work-connected automobile accidents.

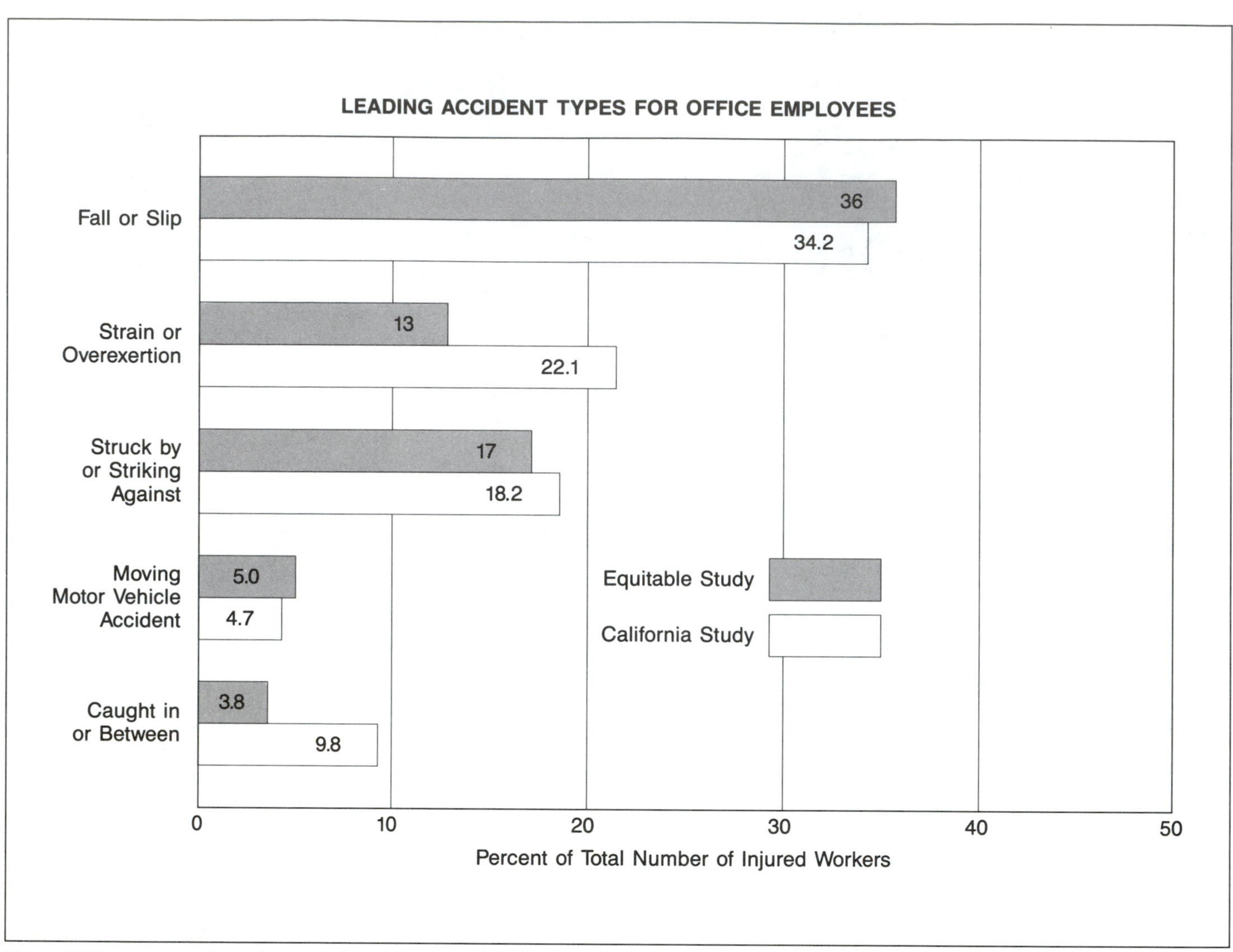


**Figure 15-1.** A comparison of two studies of accidents occurring in the office worker population.

**Table 15-A.** Disabling Accidents From Falls in Office Work

| *Equitable Life—8,000 employees (an eight-year study)* | *Disabling Accidents* | *Days Lost* |
|---|---|---|
| In hallways and work areas, caused by running, slipping, tripping over wires, desk drawers, file cabinet drawers, etc. | 53 | 553 |
| From chairs | 21 | 120 |
| Stairs | 16 | 117 |
| Escalators or elevators | 8 | 55 |
| Total | 98 | 845 |

| *California Survey—1,000,000 employees (a one-year study)* | *Disabling Accidents Totals* |
|---|---|
| Falls (All categories) | 4,360 |
| Falls or Slips on Stairs or Steps | 752 |
| Falls from Other Elevations | 370 |
| Falls on the Sale Level | 3,238 |

However, this amount can be significantly reduced by using a defensive driving program, (DDC), such as that developed by the National Safety Council.

"Office worker" is defined in the California study as "a person primarily engaged in performing clerical, administrative, or professional tasks indoors in an office at the employer's place of business." This definition does *not* cover salespersons, claims adjusters, social workers, medical and teaching personnel (other

Table 15-B. Percentage of Disabling Work Injuries* and Illnesses to Office Workers by Occupation and Length of Service

| Occupation | Total Disabling Work Injuries | Length of Service | | | | | |
|---|---|---|---|---|---|---|---|
| | | 1 mo. | 2 mo. | 4–6 mo. | 3–5 yrs. | 6–10 yrs. | 11–20 yrs. |
| Clerical & Kindred | 12,858 | 2.1 | 4.3 | 4.0 | 17.1 | 15.7 | 9.2 |
| Professional, Technical, & Kindred | 1,418 | 1.5 | 2.5 | 7.5 | 17.7 | 15.7 | 12.1 |
| Managers & Administrators | 2,000 | 1.4 | 2.4 | 3.4 | 18.1 | 16.0 | 18.9 |
| Totals | 16,276 | 2.0 | 3.9 | 8.2 | 17.2 | 15.7 | 11.0 |

*These figures are true only for the California Study.

**Figure 15-2.** Falling from a chair is extremely dangerous because a person is totally unprepared. A fall backward may cause serious injury to the head, neck, or back.

than clerical or administrative), and certain stock, order, and inventory clerks. The California study did not cover employees of the federal government, maritime workers, and railroad workers in interstate commerce. The Equitable Life Assurance Society study, on the other hand, *did* include salespeople, claims adjusters, medical personnel, and supply and warehouse personnel. The insurance company study covered approximately 8,000 employees of one company working in one building, about 5 percent of whom were maintenance personnel. During the eight year study period, the injury frequency rate was 2.3, and severity rate was 15 for the office workers, using ANSI Z16.1 method of calculation; see Chapter 6, Accident Records and Incidence Rates. The average days charged per disabling injury was 5.9. See Tables 15–A and –B for details of the studies.

### Complacency—prime cause of injury

Office injuries may seem to be inconsequential because they often lack dramatic impact. For example, the person who is injured falling backward from a chair (Figure 15–2) seems to merit little sympathy or attention. The image of an office worker slipping and crashing to the floor on his back seems amusing to some people.

Industrial accidents are commonly considered to occur more frequently and with greater severity—amputations, lost eyes, and broken bones—than office accidents. But who is likely to be absent from work the longer: the office worker who sustains a severe compression fracture of one or more lumbar vertebrae from a bad fall, or the production worker whose hand is amputated? Probably, it would be the office worker.

Complacency—the attitude that office accidents do not amount to much—is one of the prime causes of office accidents. The average office worker gives little thought to safety because office work is not perceived as being hazardous.

On the other hand, the well-instructed worker in a plant manufacturing flammable, toxic chemicals knows that it is a risky business—and understands why safe procedures must be used and safety equipment worn. The worker's safety behavior and training is the best defense. As a result, the production department may have fewer accidents than the plant office.

The office worker, therefore, has to be informed of the hazards and the safe work procedures that must be used. The employee must also be shown that management provides a safe environment and safe equipment to encourage proper safety behavior. The worker must be willing to adopt safe procedures, and be encouraged to do so. Even more important, office supervisors must understand the nature of office hazards and unsafe practices and take necessary measures to reduce and manage these hazards.

### Who gets injured?

The California and the Equitable Life studies pointed to whom most injuries occurred, and how they occurred.

**New surroundings.** These studies showed the importance of teaching office workers to look for new hazards and to correct them. It was also found that there was a substantial increase in the number of injuries in the first year a company moved into a new office building. The change upset established routines and presented unknown hazards. Even going to and from work became more hazardous as employees had to explore new driving and transportation routes.

**New and young employees.** The California study found that the new and younger employee does not have a higher accident rate than the longer-employed and older worker—at least this was true for those employed in California (see Table 15–B). In California, office employees who had been on the job for one and two months had an injury percentage rate of 2.0 and 3.9, respectively. At the same time, those that had been employed for three to five years had the highest accident percentage rate (17.2 percent). As for age, the study showed that only three percent of the injuries occurred in the 18 to 19 years bracket.

**Sex of employee.** In the California study, 70 percent of the disabling injuries occurred to women, who comprised 68 percent of the office labor force.

The study also shows that for office occupations, the estimated disabling work injury rate for women was very close to the injury rate for men. The rate was approximately 7.8 disabling injuries per 1,000 women employed in office work compared with 7.1 for the same number of men.

The injury statistics compiled from the California study were similar to the Equitable study. The rate of injury accidents per thousand male employees was about the same as it was for female employees. However, the rate of total days lost from disabling injuries was two and one-half to three times higher for the men than for the women.

### Types of disabling accidents

**Falls** are the most common office accident, and account for the most disabling injuries, according to both surveys, Figure 15–1. They cause from two to two and one-half times as high a disabling injury rate among office as among nonoffice employees. Falls were the most severe office accident and were responsible for 55 percent of the total days lost because of injuries.

The most common accidents resulting in injury occurring to women office workers were slips and falls. The California survey showed that 39 percent of all disabling injuries to women were caused by falls and slips; men sustained only 22 percent of their injuries from this source.

- Most chair falls came when a person was sitting down, rising, or moving about on a chair. A few were caused by people leaning back and tilting their chairs in the office or cafeteria, or putting their feet up on the desk. Although stairs seem more hazardous than chairs, people seem to recognize the hazard and are more cautious. Furthermore, people are not as often exposed to the stair hazard as they are to a chair.
- Another major accident category is falls on the same level. This includes slips on wet or slippery floor and tripping over equipment, cords, or litter left on the floor. Good housekeeping procedures can reduce accidents in this category.
- A final category was falls from elevations, caused by standing on chairs or other office furniture and by falls from ladders, loading docks, or other miscellaneous elevations. These falls (not including stairs) accounted for approximately two percent of the disabling injuries in the California study, whereas falls on stairs accounted for almost five percent.

**Over-exertion.** Almost three-fourths of the strain or exertion mishaps occurred while employees were trying to move objects—carrying or otherwise moving office machines, supplies, file drawers and trays, office furniture, heavy books, or other loads (Figure 15–3). Often, the employees were moving office equipment or furniture without authorization from their supervisors.

A significant number resulted when the employee made a sudden or awkward movement, and did not involve any outside agency. Reaching, stretching, twisting, bending down, straightening-up, and cumulative trauma were often associated with these injuries.

**Objects striking** workers accounted for about 11 percent of injuries to office workers in both studies. Most of these injuries were sustained when the employee was struck by a falling object—file cabinets that became overbalanced when two or more drawers were open at the same time, file drawers that fell out when pulled too far, (Figure 15–4), office machines and other objects that employees dropped on their feet when attempting a move, or typewriters that fell from a folding pedestal or rolling stand.

In addition, a number of employees were struck by doors being opened from the other side. Office supplies or other material and equipment sliding from shelves or cabinet tops caused a few injuries in this classification.

**Striking against** objects caused approximately seven percent

**Figure 15-3.** Moving office equipment or materials should be left to the designated personnel. Improperly lifting and carrying heavy objects may cause muscular and/or back injuries. Even if light, a large object can distract attention and result in collisions.

of the office injuries in both studies discussed here. Two out of three of these injuries were the result of bumping into doors, desks, file cabinets, open drawers, and even other people (Figure 15–5) while walking. Hitting open desk drawers or the desk itself, while seated at a desk, or striking open file drawers while bending down or straightening-up caused most of the rest of these injuries.

Other incidents of striking against objects included bumping against sharp objects such as office machines, spindle files, staples, and pins. Also included are infected cuts incurred while

**Figure 15-4.** Pulling a file drawer out can be hazardous if lower drawers are empty and the cabinet becomes top heavy. (Reprinted with permission from "Play It Safe in the Office," *Lewis News,* Lewis Research Center, NASA, Cleveland, Ohio.)

handling paper, file drawers, and supplies.

**Caught in or between.** The final major classification was accidents where the worker was caught in or between machinery or equipment. Mostly, this was getting caught in a drawer, door, or window. However, a number of employees got caught in duplicating machines, copying machines, addressing machines, and fans. Several got their fingers under the knife edge of a cutter.

**Miscellaneous office accidents** included foreign substances in the eye, spilled hot coffee or other hot liquid, burns from fire, insect bites, and electric shocks.

Eye strain or muscle aches from extended CRT use and cumulative trauma injuries also are to be considered hazards to office workers.

## CONTROLLING OFFICE HAZARDS

Office accidents can be controlled by eliminating hazards or reducing exposure to them. Hazards are most easily eliminated or reduced when the office is in the planning stage, when equipment is purchased, or when new office procedures are set up.

### Layout and ambience

Offices should be laid out for efficiency, convenience, and safety. The principles of work flow apply to offices as well as to factories. Location of the office should take into account the inherent hazards of adjoining operations.

**Stairways and exits** (including access and discharge) should comply with NFPA 101, *Life Safety Code;* floor and wall openings should comply with ANSI *Safety Requirements for Floor*

**Figure 15-5.** Convex mirrors help prevent collisions at blind intersections of aisles.

*and Wall Openings, Railings, and Toe Boards,* A12.1. Handrails, not less than 30 in. or more than 34 in. (.8 and .9 m) above the upper surface of the tread, are specified for one side of stairs up to 44 in. (1.1 m) wide, and both sides for stairs wider than 44 in. For stairs wider than 88 in. (2.2 m) add an intermediate (center) rail.

Exits should be frequently checked to be sure that stairways are unobstructed and well illuminated. Exit hallways or paths should have emergency lighting. Exit doors, if locked, shall not require the use of a key for operation from inside the building.

**Doors** are another frequent source of accidents in offices. Glass doors should have some conspicuous design, either painted or decal, about 4½ ft above the floor and centered on the door so that people will not walk into it. (See Chapter 21 for details.)

Safety glass complying with ANSI standard Z97.1 should be installed, particularly in doors, rather than plate glass. Sometimes local codes specify the type that must be installed.

Frosted safety glass windows in doors provide vision to prevent accidents while preserving privacy. Solid doors present a hazard because they can be approached from both sides at the same time, and one person can be struck when the door opens. Employees should be warned of this hazard and instructed (1) to approach a solid door in the proper manner, that is, away from the path of the opening door; (2) to reach for the door knob so that, if the door is suddenly opened from the other side, the hand receives the force of the impact rather than the face; and (3) to open the door slowly, especially if the door opens outward.

Another hazard is the door that opens directly onto a passageway. If the door opens directly into the path of on-coming traffic, somebody might bump into the edge of the door. If doors that open onto hallways cannot be recessed, they should be protected with short-angled, deflector rails or U-shaped guardrails that protrude about 18 in. (46 cm) into the passageway, or the area they swing over can be marked as suggested in the following paragraph. Another procedure is to place storage lockers or benches along the wall near the door, which provides the safety of a recessed door.

Some offices having tile floors will paint white or yellow stripes or apply tape to mark traffic flows or to guide people away from a rapidly opening door. The floor in front of a swinging door also can be marked or painted as a warning. A warning sign also could be posted. As a final precaution, it is good practice to have the door hinges on the upstream side of the traffic; that is, on the right hand side as one faces the door from the hallway.

Doors are covered in the *Life Safety Code,* NFPA 101.

**Lighting.** Adequate light, ventilation, and other employee services have an important influence on employee morale. Business growth often requires installing more desks and other equipment than originally planned. Overcrowding is bad from the standpoints of the appearance and the physiological effect on employees, especially if it overtaxes ventilation facilities. Smaller offices can be made to appear larger and less crowded, if walls, woodwork, and furniture placed against the walls are the same color.

Illumination levels recommended by the Illuminating Engineering Society for an office are listed in Table 15–C. Also see especially Chapter 1, Industrial Buildings and Plant Layout, of the *Engineering and Technology* volume of this Manual.

Some accidents can be attributed to poor lighting. However, many less-tangible factors associated with poor illumination are contributing causes of office accidents. Some of these are direct glare, reflected glare from the work, and harsh shadows, all of which hamper sight.

Excessive visual fatigue can be an accident-causing element. Accidents also can be prompted by delayed eye adaptation when moving from bright surroundings into dark ones and vice versa. Some accidents attributed to an individual's carelessness can actually be traced to difficulty in seeing.

Office design can facilitate good lighting. If offices depend primarily on daylight, for example, employees engaged in visual tasks should be located near windows. North light is preferred by draftsmen and artists. However, employees generally should not face windows, unshielded lamps, or other sources of glare. Indirect, shielded fluorescent lamps can produce high levels of illumination without glare. Furthermore, walls and other surfaces should conserve light while avoiding annoying reflections. Ceiling, walls, and floor act as secondary large area light sources and, if finished with the recommended reflectances, will increase light and reduce shadows. Finally, modern office lighting must be designed to counteract potential stress and eye fatigue in CRT users.

As a uniform means of evaluation, a standard procedure, entitled "How to Make a Lighting Survey" has been developed in cooperation with the U.S. Public Health Service. (This publication is not part of the standard, *Practice for Office Lighting,* ANSI/IES, RP1–1982, but is presented as background material

**Table 15-C.** Levels of Illumination for Offices

| | *Recommended Illumination* (Footcandles)* |
|---|---|
| Cartography, designing, detailed drafting | 200 |
| Accounting, auditing, tabulating, bookkeeping, business machine operation, reading poor reproductions, rough layout drafting | 150 |
| Regular office work, reading good reproductions, reading or transcribing handwriting in hard pencil or on poor paper, active filing, index references, mail sorting | 100 |
| Reading or transcribing handwriting in ink or medium pencil on good quality paper, intermittent filing | 70 |
| Reading high-contrast or well-printed material, tasks and areas not involving critical or prolonged seeing such as conferring, interviewing, inactive files, and washrooms | 30 |
| Corridors, elevators, escalators, stairways | 20 (or not less than 1/5 level in adjacent areas) |

*Minimum on task at any time.

for the use of the standard.) Table 15–C shows the illumination levels recommended by the Illuminating Engineering Society and approved by ANSI.

**Ventilation.** Where there are large interior spaces, forced ventilation is needed if the space is to be used for office purposes. All mechanical ventilation and comfort conditioning systems require careful planning and installation by qualified specialists. Private offices installed around the outer walls of a large office space should not cut off light and ventilation from the other employees. If fans are used in an office, they should be guarded, secured, and placed where they cannot fall.

The designs of many office buildings seal in office air—odors, smoke, office solvents and chemicals, molds, fungus, and other contaminants. In these buildings, it is particularly important to design and install adequate ventilation systems. The systems should be regularly inspected and updated if problems appear.

**Electrical.** Workers should be protected against electrical equipment hazards in an office. In some cases, the hazard can be avoided completely, such as not using electric-key switches. In other cases, hazards can be reduced by using UL-listed equipment, installing sufficient well-located outlets (receptacles), and arranging cords and outlets to avoid tripping hazards.

Employee use of poorly maintained, unsafe, or poor quality non-UL-listed electrical equipment, such as coffee makers, radios, and lamps, must be prohibited. Such appliances can create fire and shock hazards.

A sufficient number of outlets should be installed to eliminate the need for extension cords. Cords that are necessary should be clipped to backs of desks or taped down. If cords cannot be dropped from overhead, and must cross the floor, cover them with rubber channels designed for this purpose. Cords should not rest on steam pipes or other hot or sharp metallic surfaces.

Outlets should accommodate 3-wire grounding plugs to help prevent electric shock to operators. Floor outlets should be located, if possible, so they are not tripping hazards, and should be placed where they will not be accidentally kicked or used as a foot rest. A floor outlet protruding above floor level is frequently shielded by a desk or some other piece of furniture. However, when the desk is moved, the outlet becomes an immediate tripping hazard unless it is appropriately covered. Such floor design is not common any longer; underfloor or cellular floor raceways are usually used in new construction.

The National Electrical Code requires ground fault circuit interrupters in restroom areas. In all areas, wall receptacles should be so designed and installed that no current-carrying parts will be exposed, and outlet plates should be kept tight to eliminate possibility of shock or collision injury.

Cords for electrically operated office machines, fans, lamps, and other equipment should be properly installed and frequently inspected to see if there are any defects that can cause shocks or burns. Switches should be provided, either in the equipment or in the cords, so it is not necessary to pull the plugs to shut off the power. The fact that office equipment is operated on 110-volt circuits is no assurance that serious injury will not occur. Fatalities can be caused by current as low as 100 milliamperes if it passes through the vital organs. (See Chapter 15, Electrical Hazards, in the *Engineering and Technology* volume.)

Office electrical service should be designed to accommodate change. Electrical, coaxial, computer, and telephone cables are easily serviced and movable when run through modular channels dropped from the ceiling.

Installation or repair of any electrical equipment should be done by qualified workers using only approved materials. Because defective wiring may constitute both shock and fire hazards, all recommendations of the *National Electrical Code,* NFPA 70, should be observed.

**Equipment.** An office machine should not be placed on the edge of a table or desk. Machines that tend to creep during operation should be secured either directly to the desk or table or placed on a nonslip pad. And, particularly, typewriters on folding pedestals should be fastened to the pedestal.

Heavy equipment and files should be placed against walls or columns; files also can be placed against railings. File cabinets should be bolted together or fastened to the floor or wall so they cannot be tipped over (see Figure 15–4).

**Floor surfaces** are one of the major causes of office accidents. They should be as durable and maintenance free as possible.

Floor finishes should be selected for slip-resistant qualities. Well-maintained carpet provides good protection against slips and falls. Defective tiles or boards or carpet should be immediately repaired. Worn or warped mats under office chairs and rubber or plastic floor mats with curled edges or tears should be replaced or repaired as these conditions can create tripping hazards.

Slipperiness is characteristic of highly polished and extremely hard but unwaxed surfaces such as marble, terrazzo, and steel plates. Slip-resistant floor wax can give these materials a higher coefficient of friction than they already have and can reduce their slipping hazard. However, wax must not be applied so thickly that a smeary coating results. Also, an oil mop should not be used on a wax floor because a soft and smeary coating will result (see Chapter 21, Nonemployee Accident Prevention).

Special slip-resistant protection should be used on stairways and at lobby or elevator entrances, and these specially hazardous areas should always be maintained in the best possible condition. Floor mats and runners often provide a better, more slip-resistant walking surface. Their use is discussed in the National Safety Council's Industrial Data Sheet 595, *Floor Mats and Runners*. A good, routine maintenance program is needed to keep entrance, cafeteria, and vending area floors dry. Refer to Chapter 21, Nonemployee Accident Prevention.

**Parking lots and sidewalks** are major problem areas. Many slipping and falling accidents occur in the company parking lot. To reduce the hazards, the lot must be maintained, debris removed, potholes filled, and uneven surfaces corrected. In colder climates, effective snow and ice removal controls should be used during the winter months. Again, Chapter 21 covers the subject in greater detail.

**Aisles and stairs.** A suggested minimum width for aisles is four feet. Passages through the work area should be unobstructed. Waste baskets should be kept where people do not trip over them. Telephone cables and electrical outlets should not be placed so that when in use, the wire creates a tripping hazard in passages that people use. These and other obstructions, such as low tables and office equipment should be placed against walls or partitions, under desks or in corners. Stepoffs from one level to another in an office should be avoided. If one exists it should be well marked and guarded with a railing.

File drawers should not open into aisles, particularly narrow ones, unless extra space is provided. Pencil sharpeners and typewriter carriages must not jut out into aisles.

**Storage.** Materials stored in offices sometimes cause problems. In general, materials should be stored in areas specifically set aside for the purpose (Figure 15–6). The storage area should be located so general traffic patterns do not have to be crossed to reach the stored items. Also, nothing should be stored or left on the floor in a passageway where it could become a tripping hazard.

Materials should be neatly stacked or piled in stable piles that will not fall over. The heaviest and largest pieces should be on the bottom of the pile. Where materials are stored on shelves, the heavy objects should be on the lower shelves. Objects should not be stacked on window sills if there is a danger of breaking or falling through the window.

Storage should be planned to make items accessible. Appropriate step ladders should be provided where necessary. Office falls can occur when workers stand on chairs, counters, or shelves to reach inconveniently stored items. Rolling ladders are discussed later in the Safe office equipment section.

**Figure 15-6.** Store boxes, records, equipment, and other items in designated storage areas. Do not store anything on top of lockers or in aisles or walkways. (Reprinted with permission from "Play It Safe in the Office," *Lewis News*, Lewis Research Center, NASA, Cleveland, Ohio.)

Smoking should be forbidden in mailing, shipping, print shops, or receiving rooms, and in other areas where there may be large quantities of loose paper and other combustible material, and in areas where flammable fluids are used, such as duplicating rooms or artists' supply areas.

Flammable and combustible fluids and similar materials should be stored in safety cans, preferably in locked and identified cabinets. Only minor quantities should be left in the office and bulk storage should be in properly constructed fireproof vaults. (See Chapter 16, Flammable and Combustible Liquids, in the *Engineering and Technology* volume of this Manual.)

### Safe office equipment

A good quality of office furniture not only contributes to the safety of the office but also to its appearance. This, in turn, improves the attitudes of both employees and visitors.

**Chairs,** especially, should be comfortable and sturdily built with a wide enough base to prevent easy tipping. Five-legged chairs are more stable and discourage employees from tilting back on their chairs. The casters on swivel chairs should be on at least a 20-in. (0.5 m) diameter base, but a 22-in. base is preferred. The casters should be securely fixed to the base of the chair and well constructed because loose or broken casters are a frequent cause of chair falls (Figure 15–2). About 20 percent of the chair falls in the California study were due to chair defects.

Chairs should be purchased with easy-to-adjust seat heights and back supports. Employees should be shown how to properly adjust their chairs. The correct fit will make the employee more comfortable and reduce acute and chronic strain potential—making the office environment safer and more productive.

**Desks and files.** Spring-loaded typing desks should be carefully selected. If some models are opened without due care, the typewriter table will snap out and cause a bruise or a cut.

Also, even if good quality desk and file cabinets are purchased, it is still possible that occasionally one will have a sharp burr or corner on it. Office furniture should be inspected when received, and such burrs or corners should be removed immediately.

Drawers on desks and file cabinets should have safety stops.

Purchase office machines, such as rotary files, copying machines, paper cutters, and paper shredders, with well-designed guards.

Glass tops on desks and tables can crack and cause safety hazards. Durable synthetic surfaces are safer.

A sufficient number of noncombustible waste baskets should be furnished. Also enough safety-type ash trays should be available, if smoking is permitted in the area, and they should be large enough and stable enough to safely contain smoking materials.

**Office fans** should have substantial bases and convenient attachments for moving and carrying. If located less than 7 ft (2.1 m) from the floor, they must be well-guarded, front and back, with mesh to prevent the fingers from getting inside the guard.

Many cut fingers result when people try to move fans by grasping the guard, or try to catch falling fans. Fans should not be handled until the power is turned off and the blades stop turning.

**Computers.** If a computer is to be installed in a building that has overhead sprinklers, keep sprinkler protection in service throughout the area, but get advice on necessary protection against both fire and water damage. Actually, water damage is not to be feared as much as previously thought. For one thing, most new computers are less susceptible because of their solid state circuitry. In addition, tests have shown that water does not harm magnetic tape. Most of the damage suffered by computers in a fire results from the heat. One of the best ways to prevent a damaging fire is to keep combustible materials such as paper, tapes, and cards at an absolute minimum in the room with the computer. When safeguarding such an investment, call in a fire protection adviser, as well as a computer installation expert (see Chapter 5, Part II).

**Rolling ladders** and stands used for reaching high storage should have brakes that operate automatically when weight is applied to them. All step ladders should have nonskid feet.

**Chemical products.** The number and hazards of office chemicals are often underestimated. Assess all chemical products used in copying and duplicating machines, print shops, and assess all adhesives and cleaning materials. Workers must be informed of any dangers and instructed in the safe use of hazardous chemical products.

If possible, substitute nontoxic and nonflammable solvents for those used in printing and duplicating or other operations. (Details are given in *Fundamentals of Industrial Hygiene,* part of this Occupational Health Series.) If chlorinated bleaches are purchased for cleaning purposes, make sure that they will not be mixed with strongly acidic or easily oxidized materials. Purchase a good grade of slip-resistant floor wax.

**Purchasing equipment.** As discussed in Chapter 3 and Chapter 5, the company health and safety professional should work with the purchasing agent in buying office furniture and equipment. Both should be aware that although advertisers sometimes stress the safety features of office equipment, the machines may be delivered without these important safeguards. Mechanical hazards of heavy office equipment can be ascertained by careful, expert inspection before purchase. These hazards can almost always be eliminated or minimized, although sometimes at substantial expense.

The purchasing department also must be informed of precautions to be taken in connection with chemicals, dyes, inks, and other supply items. Particular attention should be paid to toxic, irritant or flammable properties. Where hazards are unavoidable, labels and specific instructions for careful handling should be supplied or issued when the material is received.

The purchasing department should gather all pertinent information from the manufacturer on equipment design and electrical and space requirements, and should try to determine the composition of proprietary compounds. This they can forward to the safety professional (or safety department) for an opinion concerning inherent safety hazards before purchasing new equipment or supplies.

**Office machines.** All machines having external moving parts that could be hazardous should have enforced safety procedures; constant training and retraining of operators is necessary.

Employees should be told that if any office machine gives a shock, appears defective, or if it sparks or smokes—turn it off, pull the plug, and tell the supervisor.

Some office machines are noisy, especially the telex, computer printers, and printing equipment. This noise is usually more of an annoyance than a health hazard, but may need to be evaluated.

## Printing services

**Larger offset presses.** Check the operation of offset presses. Is the operator putting his fingers on the blanket while the press is in motion? One offset press department had seven finger injury accidents in the first two weeks of operation, all caused by press operators who put their fingers in the running press to remove dirt or other particles from the plate.

Presses should meet all guarding regulations imposed by local, state or provincial, and federal agencies.

The area around the presses should be free from clutter and well lighted, the flooring should be resilient or rubber mats should be provided to minimize operator fatigue and to prevent slipping.

Only qualified operators should operate presses. Loose clothing and long hair are hazardous around these machines. Low flashpoint flammable liquids or toxic solvents should not be used to clean the presses; office supervisors and press operators should understand the fire and health hazards involved and follow all instructions for safe use, storage, and disposal of flammable and toxic substances.

**Gathering and stitching machines.** Guards should be installed on open sprockets and collector chain drives of gathering and wire stitching (or stapling) machines to protect employees from hand and body injury. The operating arm on the end of the gathering machine should be guarded.

Hinged drop-guards should be installed to cover any exposed operating mechanism that creates nipping hazards under the machine and along the working area where operators fill the pockets. The floors and work platform at this area should be covered with nonskid material.

Operators should be trained to open signatures in the middle and place them on the saddle or rod between the hooks on the moving chain. If the hook is not put on the rod or chain correctly, *no attempt should be made to straighten it out until the machine has been shut off.* The machine also should be shut down when threading stitcher heads, making any adjustments or removing jams.

**Folding machines.** Here are points to be stressed for safe operation of folding machines:

1. Before jammed paper is pulled from the machine, shut the motor off to avoid getting hands in the feed rollers.
2. Finger clearance at the folding knife should be checked before pulling out paper, putting tape on rollers, or adjusting plates and roller pressure.
3. Workers should walk down the steps of folder feeder platforms facing forward, never backward.
4. On large-size folders, all steps and platforms should be protected by railings.

Defective staples protruding from reports or booklets should be removed to avoid cuts from them while books are being jogged, trimmed, or wrapped. Workers should be trained to cup their hands over the work when removing defective wire staples.

Employees engaged in this operation should wear eye or face protection, and passers-by should be protected against flying staples by screens or by isolation of this work.

### Enforced safety procedures

Because the major category of office accidents is slips and falls, running in offices, for whatever reason, must be prohibited.

A number of office accidents can be prevented if everybody walking in passageways would keep to the right. Convex mirrors should be placed—and used—at corners and other blind intersections. As discussed under Doors earlier in this chapter, collisions at doors also can be prevented if people do not stand directly in front of the door, but away from the path of its swing when they go to open it.

People carrying material must be sure they can see over and around it when walking through the office. They should not carry stacks of materials on stairs; they should use the elevator, if available if not, make two trips. People should not have both arms loaded when using stairs; one hand should be free to use the handrail.

When using stairs, outside at night or in a dimly lit area, workers should be instructed to go single file, to keep to the right, and to always hold the handrail. People should not crowd or push on stairways; they should pay attention to where they are going. Commonly, falls on stairs occur when the person is talking, laughing, and turning to friends while going downstairs.

Other safety rules for stairs include: do not congregate on stairs or landings, and do not stand near doors at the head or foot of stairways.

Good housekeeping is essential to prevent falls. Littering should not be allowed. Spilled liquids must be wiped up immediately, and pieces of paper, paper clips, rubber bands, pencils, and other loose objects must be picked up as soon as they are spotted.

Broken glass should immediately be swept up. It should not be placed loose in a waste container but it should be wrapped in heavy paper and marked BROKEN GLASS. Glass that shatters into fine pieces can be picked up with damp paper towels.

Tripping hazards, such as defective floors, rugs, floor mats, should be reported to the maintenance department and immediately repaired.

Many falls could be prevented by choosing supportive footwear with non-slip soles; shoes with high heels are undesirable.

**Chair falls.** Habits that lead to chair falls (Figure 15–2) must be discouraged. Scooting across the floor while sitting on a chair should be forbidden. Leaning sideways from the chair to pick up objects on the floor is dangerous and should be discouraged. Leaning back in the chair and placing the feet on the desk should be discouraged.

People should properly seat themselves in their chairs. They should form the habit of placing a hand behind them to make sure the chair is in place. Sitting down on the edge of the seat rather than in the center, or backing too far without looking, or kicking the chair out from under can result in a sudden fall to the floor. Standing on a chair to reach an overhead object is particularly dangerous and must be forbidden.

**Filing cabinets,** as discussed earlier in this chapter, are a major cause of injuries including bumped heads from getting up too quickly under open drawers, mashed fingers from improperly closing drawers, and hand injuries and strains from moving the cabinets.

Some precautions are necessary against these accidents.

- First, people should never close file drawers with their feet or any other part of their body. They should use the drawer handle to close the cabinet, making sure their fingers are not curled over the edge when the drawer closes. All file drawers should immediately be closed after use.
- Second, only one file drawer in the cabinet should be opened at one time in order to prevent the cabinet from toppling over. As previously indicated, where possible, file cabinets should be bolted together or otherwise secured to a stationary object to safeguard against this chance of human failure.
- Third, don't open a file drawer if someone is in immediate danger or underneath. Do not leave an open drawer unattended—not even for a minute. When one person has a file drawer open, he should warn other persons working in the area so they do not turn around or straighten up quickly and bump or trip over an open drawer.
- Fourth, climbing on open file drawers must be forbidden.
- Fifth, small stools used in filing areas are tripping hazards when left in passageways. Any person who sees one out of place should put it where it cannot cause a fall.
- Sixth, filing personnel should wear rubber finger guards to eliminate cut fingers from metal fasteners or paper edges.

Desks or files should never be moved by office personnel; they should be moved by maintenance workers preferably using special dollies or trucks made for such moving. In general, furniture should not be rearranged without authorization or checking with office management. When desks or cabinets are moved, thought should be given to floor obstructions and necessary aisle space before making the move. When a telephone terminal box on the floor or electrical outlet box is exposed after moving furniture, the box should be marked with a tripping hazard sign until it is removed. The outlet must be removed and, if it is needed, relocate it; it is far cheaper to do this than to pay for a fall.

Do not run electric cords under rugs, they sometimes come out because of traffic movement and form tripping hazards. They also are fire hazards. New outlets should be installed to eliminate the necessity for extension cords.

**Materials storage.** There are a number of precautions to be taken when storing materials. Neat storage makes it easier to find and recover materials without dropping or knocking over other materials. Supervisors must keep employees from stacking boxes, papers, and other heavy objects on file cabinets, desks, and window ledges, or from placing these materials carelessly on shelves where they could spill in an avalanche (Figure 15-6). If heavy objects spill toward a window, the glass might break and cause a serious accident.

Card index files, dictionaries, or other heavy objects should not be kept on top of file cabinets and other high furniture. Movable objects such as flower pots, vases, and bottles should not be allowed on window sills or ledges.

Razor blades, thumb tacks, and other sharp objects should not be loosely thrown into drawers. They should be carefully boxed. Blades and points should be stuck in foamed-polystyrene blocks.

**Lifting.** Occasionally it is necessary for office personnel to lift light or heavy objects, such as files, books, boxes, and computer tapes. For these times, office workes should be taught proper lifting techniques.

**Other hazards.** Some additional hazards are as follows: (1) never allow a spindle (spike) file in the office, (2) never store pencils in a glass on the desk with points outward, (3) never leave a knife or scissors on a desk with the point toward the user, and never hand sharp-pointed objects to anyone, point first, (4) paper cutters should be equipped with a guard that affords maximum protection (bar guards or single-rod barriers found in some cutters are not considered full protection), and (5) do not leave glass objects on the edge of desks or tables where they can easily be pushed off.

Office machinery should be operated only by authorized persons. This was discussed in an earlier section.

Some offices have an employee lounge or eating area with a hot plate for brewing coffee or a microwave oven for warming lunches. In these areas, spilled beverages can be a burning and a tripping hazard. Microwave ovens must be properly used. (See the discussion in Chapter 18 under Food Service.)

Where the sales department is part of the office staff, or where employees travel on company business, a safe driving program should be part of the company's safety program.

**Office supervisors** should make sure all materials are kept in their proper places. All litter must be placed in waste baskets, and all drawers not being used must be kept closed.

Supervisors are just as responsible for training their people in safe procedure as they are in training them for efficiency.

Supervisors should encourage employees to report all broken chairs, or missing casters, stuck drawers, cracked glass, and other hazards for correction. A policy of immediate correction of these defects should be followed. A formal program should be established requiring quarterly office safety inspections.

## Fire protection

**Fire hazards.** Solvent-soaked or oily rags used for cleaning duplicating equipment should be kept in a metal safety container. Smoking should be prohibited within 10 ft (3 m) of where flammable solvents are used in duplicating or any other office operation. Solvents should not be carelessly handled because there is danger of splashing in the eyes.

Never allow smoking on elevators.

Do not throw matches or cigarettes into waste baskets; the contents are usually very combustible (Figure 15-7).

Procedures shoud be established so cleaning and maintenance personnel avoid collecting possible smoldering combustible material from ash trays in other combustible containers, such as cardboard boxes or cloth bags.

Some waste containers made from plastic or other flame-resistant material may actually be combustible if subjected to fire or intense heat. If such combustion occurs, dangerous toxic gases and dense smoke may be generated that can easily endanger a whole office. In order to avert this hazard, metal or fire-safe tested materials, designed to contain the fire should be used.

**Fire extinguishers.** Portable fire extinguishers in a fully charged operable condition should be kept in their designated places at all times when they are not used (Figure 15-8). (See Chapter 17, Fire Protection, in the *Engineering and Technology* volume of this Manual, for correct type of extinguishers for specific office hazard areas.)

Employees, in general, should know what to do in case of fire.

It is important that employees be trained to operate extinguishers, and fire hoses if provided, and know how to react in case of fire or other emergency. (Panic and confusion can be as dangerous as flame and smoke.)

When a fire is discovered, an employee should do three things: (1) turn in the alarm (no matter how small the fire is), (2) alert fellow workers, and (3) if trained to do so, use the proper fire-fighting equipment, but only if the employee always has a safe path of egress while fighting the fire.

Office employees should receive annual fire and other emergency training. The training should include use of portable fire extinguishers, procedures for reporting emergencies, and location of escape routes and shelters.

**Emergency plan.** Every office should have a written emergency plan. Supervisors should be appointed in every area to safely guide people out of the building. Every department should be assigned a specific route—and also an alternate in case the exit is blocked. A surprise fire drill may save lives in an emergency situation in the future.

When the alarm sounds, supervisors should direct the show, but every employee must play his part. The group should move calmly along in an businesslike manner, without hurrying or pushing, and wait on a different floor or outside the building for the signal to return. In a real emergency, the officials in charge would authorize return to the building. (See Chapter 16, Planning for Emergencies.)

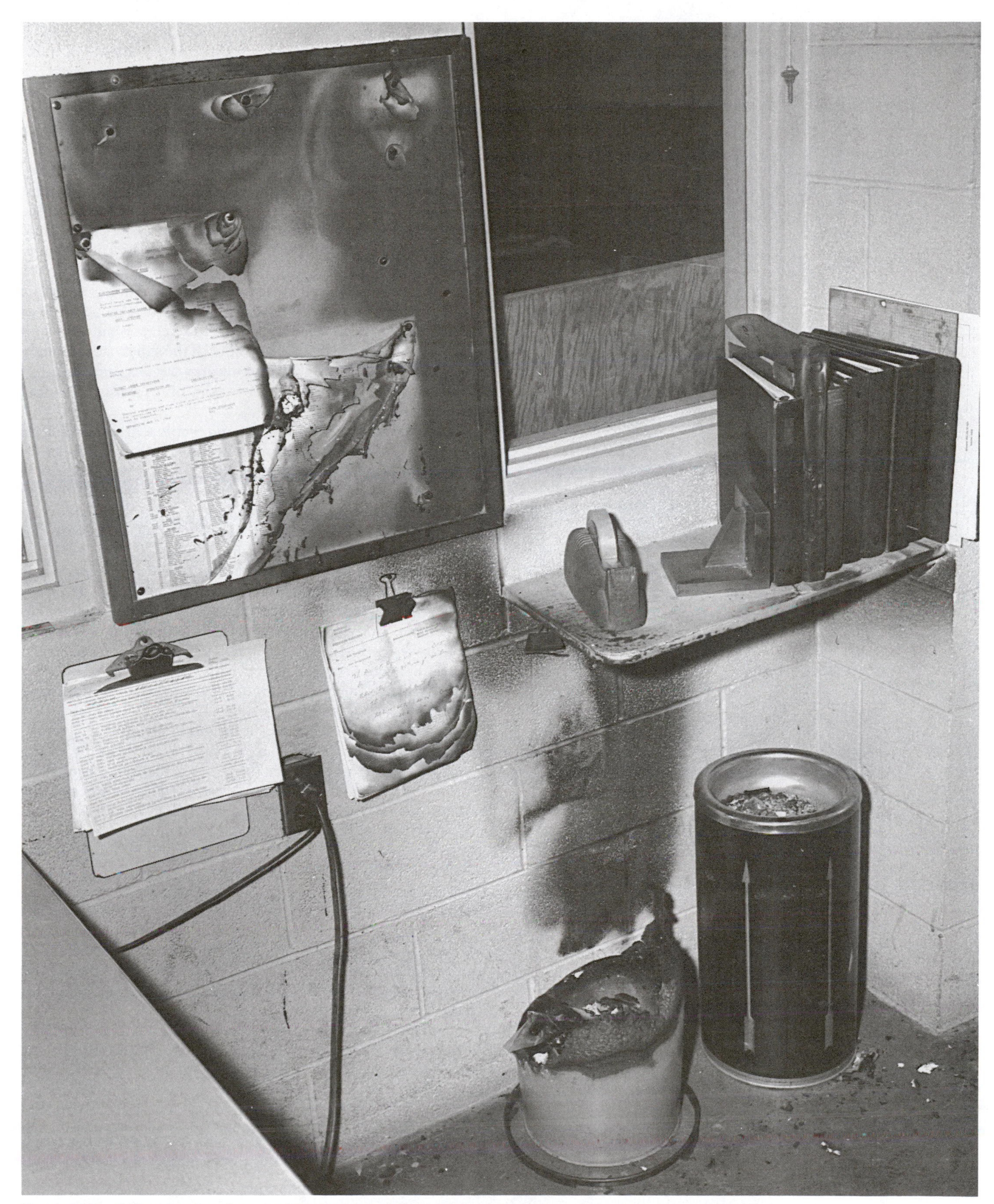

**Figure 15-7.** Still-warm ashes from smoking materials were dumped into an office wastebasket that contained paper. A fire started, causing the damage shown here. (Reprinted with permission from *Printing and Publishing Newsletter*.)

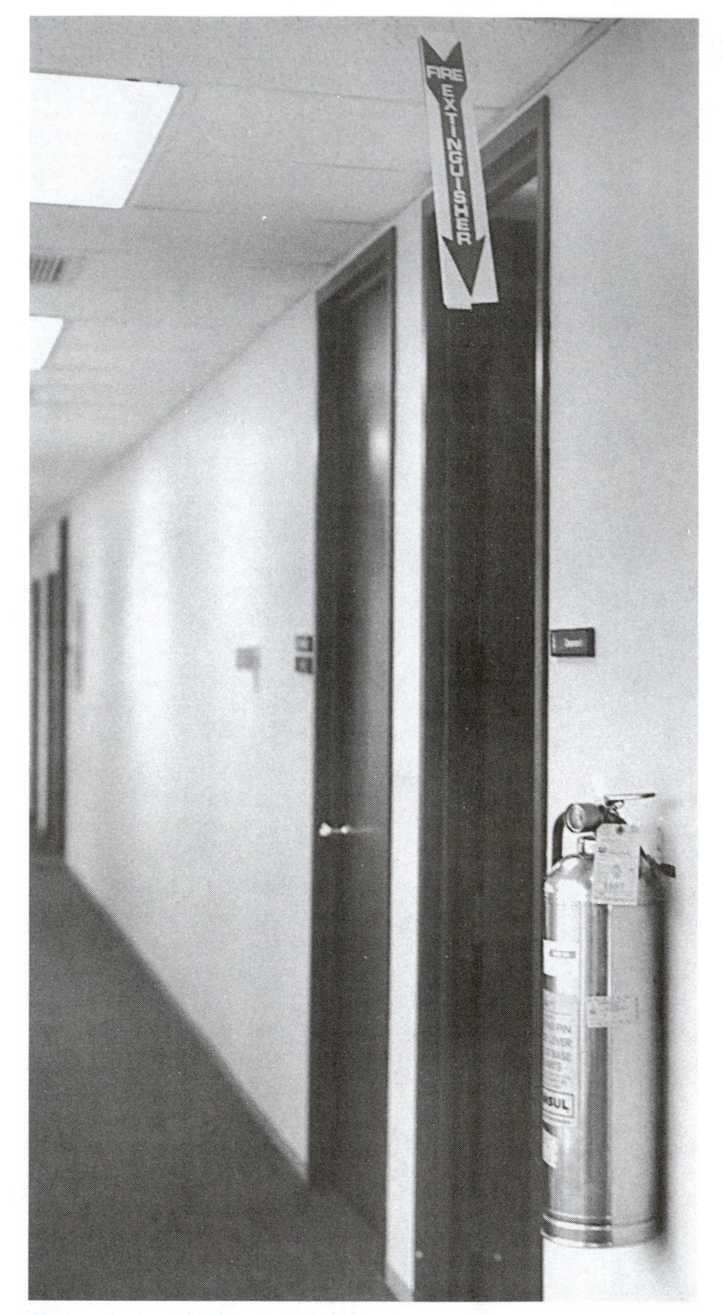


**Figure 15-8.** Fire extinguishers in office areas should be identified by a bright red triangular sign suspended from the ceiling above the extinguisher, a sign posted on the wall above it, or a bright red stripe painted above its location.

## SAFETY ORGANIZATION IN THE OFFICE

The supervisor is, of course, the key person in the office safety program. However, even the hardest-working supervisor will have difficulty maintaining full-time interest in safety all alone. The office safety committee can serve as a work horse in maintaining interest in the accident prevention program, but it cannot substitute for good management.

### Training

In order to develop proper safety behavior, safety instructions should be properly given to all new office employees. The personnel or industrial relations department can provide an accident prevention brochure or a set of printed rules. They also should arrange for all explanations of procedures as quickly as possible during the employee's early work days. Motivation and training were covered in Chapters 10 and 11 and 9, respectively.

Unfamiliar surroundings, new equipment, or altered work tasks increase the likelihood of accidents, even among veteran employees. Therefore, these people also should be instructed when beginning a new job. There should be specific instructions for each piece of equipment. No one should ever be permitted to use a machine unless fully instructed in its operation, and told the location and how to use fire equipment as well as how to summon medical aid.

Instruction for supervisors in safe office operation is necessary because they are likely to have the same lack of recognition of accident potentials as the employees. However, prevention of accidents requires the dedicated vigilance of the supervisor throughout every working day. If he fails to carry out this function, accidents due to unsafe acts by employees will continue, undiminished.

Routine training should exist, and employees should receive continuing information. Training meetings are recommended on such topics as slip and fall prevention, proper lifting, fire safety, emergency procedures, office chemical safety, and off-the-job safety. Safe attitudes and behavior are not merely put on when an employee enters the office, and taken off when he walks out the front door. (See Chapter 11 for details.)

All office employees who must enter production areas where safety hats, eye protection, and hearing protection are necessary should be provided with these items and should be required to wear them. Every employee who visits the plant should have a card of the general safety rules that apply to the plant, and should be familiar with them. The same requirement should be enforced for all visitors. Safety rules should apply to everyone if the program is to be successful.

Finally, to repeat what has been said in previous chapters, a company safety program cannot succeed unless it has the wholehearted backing of its top management. The supervisor must know that his accident prevention performance is watched and that good performance is appreciated. The Council's *Supervisors Safety Manual* has some good information that can be used.

### Safety committee

In planning an office safety program, representation on the safety committee should be on the same basis as that of any other company department or division. The office should be on the inspection itinerary of the company's safety person. The office supervisor should accompany the safety inspector on every inspection, along with an office safety committee member. (See Chapter 4 for inspection procedures.)

The committee assists all supervisors in maintaining safety and reporting directly to the safety director or whoever is in charge of the program. Along with the department head, it can make periodic inspections of the office to look for accident or fire hazards. It makes recommendations, many of them based on suggestions from supervisors and other employees. It also can help prepare and revise company safety rules.

Often the committee is in charge of office-wide communication, training, and incentive programs designed to maintain peak interest in safety, using means such as posters, bulletins, and contests.

The organization of the office safety committee can be the same as that of the company and joint safety committees discussed in Chapter 3.

### Accident records

If a safety program is to succeed accurate accident records must be kept. Not only do accident investigations and analysis of records spotlight problems that must be corrected, but the records show if progress is being made in accident prevention.

Office employees, like plant workers, should report every accident, no matter how minor the injury. The reports should be detailed, and made as quickly as possible following the accident or near accident. Unsafe conditions or procedures indicated in the reports should be corrected as quickly as possible, because near-miss accidents are warnings of worse accidents to come.

Records are the concrete foundation of the safety structure. They tell the "who, what, when, why, and how" of accidents in your office—and help you to prevent repeat performances. Accurate records also provide guidelines on which company insurance rates are based.

The average office will not have enough major injuries to warrant extensive investigation and analysis. However, it's urgent that records be kept to pinpoint problems and prevent future accidents.

If, for instance, a large number of falls are injuring workers, supervisors can double-check possible hazards and devote special attention to the problem in meetings and other communications.

Standard report forms are available. They were discussed in Chapter 6, Accident Records and Incidence Rates. Accident investigation was discussed in Chapters 4 and 7.

## REFERENCES

American National Standards Institute, 1430 Broadway, New York, N.Y. 10018.

*Safety Performance Specifications and Methods of Test for Glazing Materials Used in Buildings,* Z97.1.

*Practice for Office Lighting,* ANSI/IES, RP1-1982.

*Safety Requirements for Floor and Wall Openings, Railings, and Toe Boards,* A12.1.

Baldwin, Doris. "Caution: Office Zone." *Job Safety and Health,* Vol. 4, No. 2 (Feb. 1976).

"How Safe Is Your Office," *National Safety News,* Vol. 112, No. 4 (Oct. 1975).

Kiefer, Norvin C. "Office Safety," *Journal of Occupational Medicine,* Vol. 9, No. 11 (Nov. 1967).

"Industry's Orphan—Office Safety," *Environmental Control Management* (Dec. 1969).

National Fire Protection Assn., Batterymarch Park, Quincy, Mass. 02269.

*Life Safety Code,* NFPA 101.

*National Electrical Code* NFPA 70.

National Safety Council, 444 N. Michigan Ave., Chicago, Ill. 60611.

*Escalators* (1980), 516.

*Evacuation System for High-Rise Buildings* (1985), 656.

*Falls on Floors* (1981), 495.

*Flammable and Combustible Liquids in Small Containers* (1984), 532.

*Floor Mats and Runners* (1981), 595.

*Off-the-Job Safety* (1986), 601.

*Power Lawn Mowers* (1981), 464.

*Motor Fleet Safety Manual,* 3rd ed., 1986.

*Safety Handbook for Office Supervisors.*

*Supervisors Safety Manual,* 6th ed.

*Visit to Office Falls.*

*What You Should Know About Office Safety,* 1985.

State of California, Dept. of Industrial Relations, Div. of Labor Statistics and Research, 525 Golden Gate Ave., San Francisco, 94102.

"Disabling Work Injuries to Office Employees," 1963 and 1978 editions.

"Work Injuries and Illness in California," Annually.

*Statistical Abstracts of the United States,* U.S. Department of Commerce, 1987.

# 16

# Planning for Emergencies



No industrial, commercial, or mercantile organization is immune from disaster. Emergencies can arise at any time and from many causes, but the potential loss is the same—people and property. Advance planning for emergencies is the only way to minimize this potential loss.

Planning in advance is necessary—it is not a luxury, rather it is good insurance. Where professionally trained emergency help and assistance may not be available, the need for emergency planning is intensified. A comprehensive management plan is intended to take care of all expected emergency situations. This includes both the spectacular (such as a tornado) and the common accident situation. Quite often emergency planning is assigned to the safety and health professional. This is fine, but there is a real need for the corporate management to be fully involved in the many decisions that must be made.

The safety of employees, visitors, and customers must be the first concern in planning for an emergency. Care for the injured must be available immediately. In some disasters, evacuation may be necessary.

Next, consideration should be given to protecting the property and the operation. In a new plant, consideration should be given to arranging and locating certain facilities and operations to provide greater inherent safety to the entire operation. In general, all emergency plans will include cleanup details necessary for the situation.

Finally, planning may be concerned with restoring business to normal. In emergencies likely to damage or wipe out a unit or plant, the question of resuming operations under conditions of temporary wiring, lack of heating, or repair and construction work should be considered.

Regardless of the size or type of organization, management is responsible for developing and operating a program, which is designed to meet these eventualities. An effective plan requires the same good organization and administration as any business undertaking. There is no one emergency plan that will do all things for all organizations. Each company must therefore decide on a plan that fits its needs and can be afforded.

Emergency plans involve organizing and training of small groups of people to perform specialized services, such as fire fighting or first aid. Small, well-trained groups can serve as a nucleus to be expanded to any size needed to meet any kind of emergency. Even with outside help available, a self-help plan is the best assurance that losses will be kept to a minimum.

An organization will need to develop several plans to control different types of emergencies. Although certain basic elements would be common to all plans, the same complete plan could not, for example, be used for both a tornado and a nuclear attack.

Before an organization initiates an emergency plan, it is necessary to evaluate the potential disasters that might occur. The next section, Types of Emergencies, discusses these in detail.

The next step is to assess the potential harm to people and property. Again some adjustments, perhaps a range in the extremes of most likely and most unlikely, could be compiled. The time of day and the event itself are other factors that should be considered in assessing the potential damage. Planning should encompass all shifts and catastrophes that might occur during weekends or holidays when no one or only a skeleton staff may be on hand.

In trying to estimate potential damage to property one would follow the same general procedure. A building may be strong enough to resist a tornado, but a sudden 7-in. (18 cm) rain storm might cause dangerous flooding. On the other hand, an exploding boiler located in an adjacent building would probably not harm the main plant.

Next, probable warning time should be considered. For example, a flood may build up over a period of several days while a "bomb scare" affords only a few minutes warning from a telephone call. This warning time should permit some chance to alert personnel and mobilize the plan. It may be desirable to have two different plans, depending upon the actual time available.

The amount of change that must be made in the operations is another factor. For example, in anticipation of a heavy snow storm, it may be necessary to send employees home early. Some equipment may be left turned on or idling, instead of being shut down completely.

Finally, consider what power supplies and utilities may be needed, particularly those used for fire protection, lighting, ventilation, and communications.

A basic emergency preparedness plan will usually include a chain of command, an alarm system, medical treatment plans, a communications system, and shutdown and evacuation procedures.

This chapter points out the various elements involved in developing emergency and personnel-protective plans. Not every element discussed will apply to every organization. Also several of the functions may well be combined and handled by one person, particularly in a smaller company. Generally the text is directed to the more elaborate and expanded type of organization and planning.

## TYPES OF EMERGENCIES

Before a company begins an extensive planning, organizing, and training, it is necessary to determine just what disasters are most likely to occur. In some areas, floods are no problem; in other regions, hurricanes or earthquakes are of little concern. However, work accidents, fire or explosion, sudden shutdowns, or acts of aggression might occur anywhere.

There are many sources of information to help determine the possibility that such an event might occur in your locality—weather records, accident and fire statistics, and industry or local authorities. After listing the potential disasters, some reasonable assessment must be made of the likelihood of occurrence. Some adjustments should be made to allow for the seasonal nature of certain events.

### Fire and explosion

Except where fires result from large-scale explosions, warfare, or civil strife, the fire emergency usually allows a short time for marshaling of firefighters and organizing an evacuation if necessary. Many conflagrations originate as small fires, that is, fires that could positively be controlled by inhouse personnel. Therefore, prompt action by a small, trained group can usually handle the situation. However, plans should include the marshaling of extensive fire fighting forces upon the first indication of any fire growing beyond the "small fire" stage.

The main point is this: *small fires must be checked as soon as they start.* The first five minutes are considered the most important. Good housekeeping, prompt action by trained people, proper equipment, and common-sense precautions will prevent a small fire from becoming a disaster.

Specific information on fire extinguishing and control is in Chapter 17, Fire Protection, of the *Engineering and Technology* volume.

### Floods

When a company or plant is located in an area that can flood, it should have the protection of dikes of earth, concrete, or brick construction. The probable high-water mark can be obtained from the U.S. Weather Bureau or the U.S. Army Corps of Engineers. The latter group also provides valuable assistance in planning flood water control.

Floods—except "flash" floods caused by torrential cloudbursts, or bursting of a storage tank, dam, or water main—do not strike suddenly. Ordinarily, there is enough time to take protective measures when a flood seems imminent.

### Hurricanes and tornados

Areas most frequently exposed to winds of destructive hurricane force are the Atlantic and Gulf coasts. However, inland locations are not immune to this type of disaster.

The U.S. Weather Bureau and other agencies have developed improved methods of detecting and tracking hurricanes; thus ample warning can be given for maximum protection of property and evacuation of personnel from threatened areas.

Companies regularly exposed to this hazard have developed a system of tracking the hurricanes on a map. At predetermined locations, a specified alert condition becomes effective and each supervisor completes a checklist for that alert. As the hurricane progresses through the 100-mile (160 km) circle, 50-mile (80 km) circle, etc., the plant is shut down in an orderly manner.

Buildings constructed in areas where hurricanes occur should be built strong enough to withstand these destructive winds and tides. Basic preventive measures include equipping with storm shutters or battens which can be promptly applied, at least on the side from which the storm is expected to approach. If this is not done, shattered windows may result in the roof being lifted and the building destroyed. If the roof is lost or damaged, building contents are drenched by the heavy rain that accompanies the storm and by water from broken sprinkler pipes. To prevent this, roofs should be securely anchored and tall structures (such as chimneys, water towers, and flag poles) designed to withstand high wind velocities.

Although the central Mississippi Valley is considered the tornado area of the country, almost every state has experienced them. The damage is inflicted quickly and is usually restricted to a small area, but the destruction can be massive.

Although the U.S. Weather Bureau has effectively increased its forecasting of tornado conditions and determining possible areas of danger, it is not possible for them to give as much advance warning or to pinpoint the strike area as accurately as they can with hurricanes. Therefore, a company must be prepared to protect its personnel on short notice and to take corrective action to protect and restore undamaged equipment and materials.

U.S. National Weather Service radio bands can be monitored during likely days. In one Midwest city, several large companies have set up a cooperative warning network. A lookout is stationed atop the city's tallest building. Through a central network, not only the member companies, but also the city's radio stations and civil defense are alerted in the event of an approaching tornado.

Tornado and hurricane experience indicates that emergency plans should include:

1. Procedure for getting personnel to a safe place. If the building is not constructed to withstand the forces, emergency shelters should be located close to the work area. All personnel should be instructed in the procedure to follow, with and without advance warning.
2. Assigned trained personnel to take care of power lines—dangling wires are a serious hazard.
3. Assigned trained people to remove wreckage to prevent injury to salvage and repair workers.
4. Scheduled regular meals and rest for the repair crews.

### Earthquakes

Most seismic areas in the United States are around the Pacific Coast. Earthquakes generally occur without warning and affect the entire community or large areas, thereby making community services unavailable for assistance.

"Earthquake-resistant" construction consists of building a structure so that it "floats" above the bedrock and ballasting it as a ship is ballasted, by making lower stories heavy and upper stories light. Utility lines and water mains should be flexible and laid in trenches that are free of the building, rising in open shafts and connected to fixtures by flexible joints. Lockers, cabinets, shelves, etc. should be securely installed with seismic bracing and safety restraining strips on shelves containing bottles of chemicals.

The principal dangers from earthquakes are the collapse of buildings, fire originating from broken gas mains, and lack of water to fight any fire. Water reservoirs or emergency water sources should be provided for fighting fire if the municipal supply mains are broken or water pressure is disrupted.

### Civil strife and sabotage

Riot or civil strife is a recent addition to the list of emergencies for which a company should plan in advance.

**Civil strife.** An emergency involving civil strife raises the questions of the right to protect property and the individual's legal right to assemble. A company should obtain from an elected legal authority in the community (the district attorney, for example) a statement explaining the company's rights in protecting its property and the company's legal responsibility for the safety of employees and other people—such as customers, supplier salesmen, and visitors—who may be on the company property. A company's legal department can be helpful in determining such a position, but its opinion does not have the force of law.

Some of the problems involved include disruption of business when an office or plant area is invaded by outsiders, protection against a mob intent on destroying company property, requests from neighboring companies for assistance of your personnel during a riot, and the rights and responsibilities of armed company guards.

This type of emergency can be just as disastrous as any other type and should receive advance planning by manufacturing, mercantile, or commercial establishments.

**Sabotage.** Protection against sabotage is also an important consideration. The saboteur may be a highly trained professional or an amateur. He may be anyone—usually one of the least-suspected members of the organization. Because physical sabotage is frequently an inside job or requires the assistance—knowingly or unknowingly—of someone inside the plant, the principal measures of defense must be against entry of persons bent on sabotage. Evidence of sabotage should be reported to the FBI and if defense work is involved, to the Department of Defense.

### Work accidents and rumors

The "chain reaction" from a so-called "routine" work accident can result in an emergency situation. For example, a break in a chemical line or toxic vapors from outside the plant entering the ventilating system may create an emergency. Panic caused by a rumor or lack of knowledge can also create an emergency. Plans for such situations should include establishment of auxiliary areas in the building to be used for medical treatment, a method of notifying employees of the actual situation, a method of quickly taking a head count, and sources of oxygen supplies available on short notice.

### Shutdowns

Although a shutdown is not an emergency per se, it can result from an unscheduled action, such as a disaster or strike; hence a fast shutdown procedure should be covered under an emergency plan. This plan should be based on a priority checklist. That is, all of the tasks to be assigned and functions to be performed should be arranged in order of importance so that if time is short, at least the most vital precautions are completed. This "crash" procedure is usually an adaptation of the routine procedure used for scheduled shutdowns, such as for vacation or renovation. Naturally the amount of warning time controls the speed of shutdown. Whenever a plant or other unit or building must be shut down, safeguards against fire take on added importance. Few employees, if any, will be present to discover and deal with a fire which might start. The extent of these measures will vary with the size and purpose of the plant. It is important to organize a formal program for instructing personnel. Examples of items which need attention include removal of lint, dirt, and rubbish; draining and cleaning of dip and mixing tanks and other equipment where flammables have been used; cleaning spray booths, ducts, and flammable liquid storage; closing gas and fuel line valves; opening switches on power circuits which will be out of service; checking serviceable condition of sprinkler systems, fire extinguishers, hydrants, alarms, and other protective apparatus; and anchoring cranes.

Prior to the closing, employees are alerted by special instructions to keep their work stations clean and fire-safe.

During the shutdown, continuous inspection of any maintenance or special operations, such as remodeling, must be maintained. Gas cutting and welding should be carefully supervised. Employees who remain on duty—the plant protection force, watchmen, maintenance workers, supervisors, or executives—should be briefed in effective countermeasures in case a fire breaks out.

If there has not been sufficient notice to effect a normal shutdown, it may become necessary to allow personnel into the area to perform necessary functions.

Company management should designate someone to authorize the admittance of personnel necessary to handle emergencies arising within the area. The chief of protection or the fire chief should arrange with local police and fire department officials for assistance if an emergency gets beyond local control. It is especially important that arrangements be completed for expediting the admittance of firefighters and their equipment.

Some companies use plant protection service agencies to prevent loss from theft, fire, and accident hazards during shutdowns. Similar plans should be worked out with these people so that police and fire assistance is expedited when it is needed.

### Industrial civil defense

One of the main differences between planning for peace-time emergencies and planning for emergencies resulting from warfare conditions is that war may cripple an entire community. This difference makes it more important that emergency plans take into account self-sufficiency because the outside sources of help—fire and police departments, hospitals and doctors, regular sources of supply for material and equipment—would not be so readily available.

Industrial civil defense consists of the plans and preparations of business and industry managements to achieve a state of readiness. This would enable their plants, facilities, and employees to cope with the effects of nuclear attack.

Even if a particular area is not attacked, plants in an area that has been spared may be requested to furnish transportation to evacuate the injured from damaged areas and to house and feed the evacuees. Plant emergency squads may also be required to go to the assistance of stricken plants. In a major catastrophe, there would probably not be enough hospital space available, making it necessary to keep the injured in temporary shelter for a considerable time. In such cases, employees with the proper training might be required to administer sedatives and plasma and to treat minor and major injuries.

The Office of the President would coordinate agencies that would handle the investigation of areas dangerously contaminated with radioactivity, if this were the cause of the trouble. However, it is advisable that plant personnel with responsibility for health and decontamination have some knowledge of this work. Company personnel can be trained in this subject, as well as many others, by the Emergency Management Institute, National Emergency Training Center, Emmitsburg, MD 21727.

This chapter cannot give detailed survival plans for a nuclear attack or for protection against chemical or biological warfare. However, it is suggested that the disaster program director consult with the city or state emergency services agency or department of civil defense and emergency management, which has material and trained personnel to assist a company in formulating its own program. Top management must initiate and actively support the program. (Additional program details are given in the next section, Radioactive materials. Specific hazards of ionizing radiation are discussed in the National Safety Council's *Fundamentals of Industrial Hygiene.*)

It is incumbent upon industry to engage in preparations to protect itself and its employees in event of nuclear attack so as to ensure continued economic production or early resumption of that production.

### Hazardous materials

Because there are many chemical substances being used today, there must be concern with potential usage and handling problems. There are many rules and procedures to be observed, but again ask the question, what if a safeguard fails? What if the container cracks and substances leak out?

In addition to normal hazards, are there potential chemical reactions with other substances that cause still further dangers to people and property? See requirements of OSHA–Reg. 29CFR 1910.1200 Hazard Communication.

Chemical hazards are discussed in the Council's *Fundamentals of Industrial Hygiene.*

### Radioactive materials

Fires and other emergencies involving radioactive materials are becoming more common as the peaceful use of isotopes becomes more widespread.

**Radioactive elements and fire.** Giraud (1973, see References) makes the following observations.

Radioactivity cannot by itself cause fires, but neither can it be destroyed or modified by fire. A fire may, however, change the state of a radioactive substance and render it more dangerous by causing it to spread in the form of a gas, aerosol, smoke, or ash.

Furthermore, fires can cause structural disruptions in stocks of fissile materials and in the special equipment for their treatment or use. Such disruptions may, at worse, result in a nuclear chain reaction, and a criticality accident may then occur.

Radioactive elements are found in various forms, depending on their uses. The human eye can detect no difference between an inactive element and the same element when rendered radioactive. Both appear equally harmless. A fundamental distinction must, however, be made between so-called "sealed" and "unsealed" sources.

- In the case of sealed sources, the radioactive substance is *not* accessible. The container has sufficient mechanical strength to prevent the substance from spreading during normal conditions of use. The capsule is made of stainless steel. The sources are of small dimension—approximately one centimeter.
- In unsealed sources, however, the radioactive substance is accessible. In normal conditions of use, there is no means of preventing it from spreading. Solid substances are kept in aluminum tubes, liquids are kept in flasks, and gases in glass ampules.

The fact that a substance is radioactive does not affect its general physical properties nor its behavior when heated to an abnormally high temperature—as, for instance, during a fire. The substance will, on contact with fire, undergo the normal transformations, depending on its initial form—i.e. solid, liquid, or gas. Melting, boiling, and sublimation can be expected—with the formation of combustion products corresponding to the chemical properties of the substance, in the form of slag, ash, powder, dust, mists, aerosols, fumes, or gases.

These combustion products are generally finer and less dense than the original substance, so they disperse more easily. Although the change in the physical state of the substance will not have affected its radioactivity, the radiation hazard will be more difficult to control.

The protective containers currently in use have a widely varying resistance to fire. The protection afforded to the contents will, therefore, depend on the type of container used. In general, sealed sources have a good fire-resistance and radioactive elements thus contained are well-protected.

Unsealed sources, however, and solutions or gases in fragile containers easily fall victim to fire. The urgency of the action

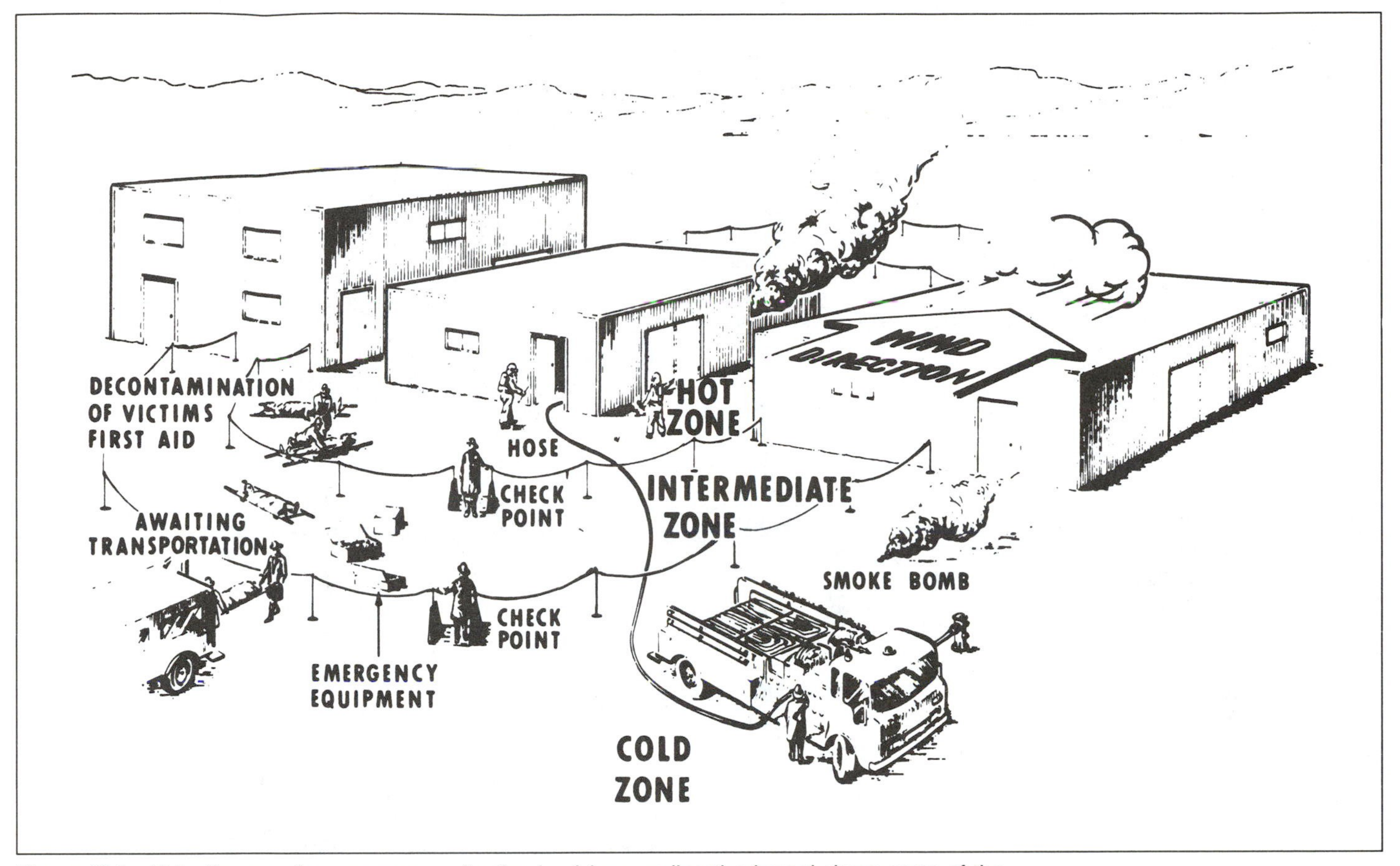


**Figure 16-1.** This diagram of an emergency situation involving a radioactive hazard shows some of the necessary precautions. Note the barricades and check points separating the "hot" and "intermediate" zones from the "cold" zone, and the air masks and extra protective clothing on the firefighters operating in the hot zone. The smoke grenade is a far more sensitive indicator of wind direction than is a windsock, and wind is extremely important. It will blow dangerous particles toward emergency personnel or away from them, depending upon how well organized is the operation. (Reprinted with permission from National Fire Protection Association. See References, Kerr, 1977.)

to be taken in the event of an accident with radioactive materials can be determined by the firefighting staff once the type of container is known, and the nature of such action will depend on the properties of the radioactive substance. See *OSHA* regulations 29*CFR* 1910.1200 (f) Hazard Communication, 'Labels and Other Forms of Warning' and 29*CFR* 1910.38 Employee Emergency Plans and Fire Preventing Plans.

When, as the direct or indirect result of a fire, the protective container has been broken, the radiation hazards for rescue workers at the point of the fire, or for personnel in the vicinity, are likely to be more serious than the danger connected with the spreading of the fire to parts of the building presenting conventional fire risks. Accordingly, the person in charge of the rescue work will sometimes be obliged to override the normal firefighting procedures to assure the protection and confinement of the radioactive elements threatened. If they are already affected by the fire, further hazards may arise.

The release of radioactive elements may result in contamination of surface areas. This may be caused by the spilling or splashing of radioactive substances or by the spreading of solid radioactive substances in paste, powder, or dust form. All possible precautions must be taken to prevent any further spread of the contamination. The means to be used, however, will differ with each case. In the first (spilling or splashing) absorbent materials should be used—such as powder, earth, sand, etc. In the case of spreading, the substances should be slightly dampened with a spray of water—unless it is otherwise specified on the container. See OSHA regulation 29CFR 1910.1200 (g), Material Safety Data Sheets.

Liquids can be prevented from spreading by the methods normally used by the firefighting brigade. The contaminated area will be clearly marked and roped off to prevent the entry of unauthorized personnel. (See Figure 16–1.)

Contamination of the atmosphere is caused by radioactive elements in the form of dust, aerosols, fumes, and gases. The spreading of such contamination is determined mainly by the prevailing weather conditions, and it is difficult to control. Such atmospheric contamination may lead to other toxic or corrosive hazards associated with the particular chemical. The most serious danger is that of inhaling the substance when it is suspended in the air. Firefighters, accordingly, should wear self-contained breathing apparatus.

The danger of internal irradiation is always present whenever there is contamination by a source of penetrating radiation. It may also occur by accidental release of an alpha or beta emitter from its protective container, or by the destruction (even partial) of the protective container.

The following material was adapted from "Preplanning for

a Nuclear Incident," by James W. Kerr, FPE, which appeared in the April 1977 issue of *Fire Command®* magazine. Copyright© by National Fire Protection Association, Boston, Mass. Reprinted with permission.

**Hazards.** How serious is the radiation problem? Although everyone is constantly exposed to radioactivity from cosmic rays and natural sources, the levels are very low. Only exposures that go much higher are of concern. The effects of radiation on living organisms—specifically people—seem to be most noteworthy when a large dose is received in a short time. If the same dose is spread over a longer time, as with workers in a factory using radioactive materials, the physical problems may be similar, but the effects show up differently.

Basically, overdoses of nuclear radiation start by causing simple symptoms that could be due to anything—even seasickness or the flu. By the time enough radiation has hit a person to cause such obvious problems as nausea and vomiting or diarrhea, some less visible changes in blood or nerve tissue also have happened. The tables in the book, *Nuclear Hazard Management for the Fire Service* (see References), make it obvious that it is vital to keep the radiation doses of those involved as low as possible.

On the fireground, do not expect immediate incapacitation of any firefighters from the radiation. The physiological effects do not appear for some time. Even lethal exposures do not cause nausea until after a few hours.

**Planning.** Regardless of the size of the nuclear event, the basic rules are always the same:

Notify the proper authorities,
Identify the hazards,
Find the limits of the area involved,
Reduce risk of exposure to people.

These steps are always required, whether for a small spill at a laboratory, a train wreck, a reactor explosion, or fallout from a nuclear testing or war. Only the degree differs, and perhaps the "proper authority" to be notified. A major help for preparing preplans relating to medical matters is NFPA 99, *Health Care Facilities* (see References).

The experts in radiation usually are found in the Department of Defense, the Department of Energy (DOE), or the Nuclear Regulatory Commission. DOE provides regional maps showing the proper telephone number for emergencies. The Chemical Emergency Center (CHEMTREC) can provide response/action information through a nationwide telephone number—800/424–9300. (See the description later in this chapter and in Chapter 24, Sources of Help.) For the wartime case, Civil Defense is the obvious contact.

Plan to keep people—including your own forces—away from the immediate area. Except for essential suppression and rescue work, 500 yards (460 m) is a good safety distance for planning purposes. Any sightseers should be kept even farther away.

Plan to hold for proper medical evaluation anyone who may have been exposed or contaminated. Reentry to the restricted area must be controlled. Be sure to keep masks on everyone involved. Firefighters must avoid smoke, dust, and vapors, insofar as possible. (See Figure 16–1.)

The first warning of a radiation problem could be the radioactive placard shown in Figure 16–2. It features a black or purple "propeller" on a yellow background. Always be alert for it. All areas in the plant or laboratories that use or store isotopes should be known to the health and safety professional and to firebrigade personnel. See OSHA regulations 29CFR 1910.38, Employee Emergency Plans and Fire Preventing Plans; 29CFR 1910.96, Ionizing Radiation; and 29CFR 1910.97, Nonionizing Radiation.

**Figure 16-2.** "Radiation Yellow—III" label, which is affixed to each package of highly radioactive material. Different labels are required for different intensities and quantities of radioactive material. For details see Title 49—Transportation, Code of Federal Regulations, Part 172, Hazardous Materials Table and Hazardous Materials Communications Regulations.

Decontamination is like any other technical operation: get expert advice. Otherwise, the radioactive material may be washed somewhere where it will do more harm than if left in place. Finally, obtain some knowledge of radiation instruments. Check with the scientific or engineering people who are using these materials. The local civil preparedness office is a prime source for instruments and for training in their use.

**Operations.** There is nothing magic about nuclear radiation, nor about the means to cope with it in an accident situation. The accompanying drawing (Figure 16–1) shows some of the necessary precautions. Invisible in the background is a pattern of fallout extending downwind; it has the same cigar shape as projections of fallout from nuclear war. It is similar to the smoke and combustion byproducts that are generated by every fire. Use common sense and experience, and keep upwind.

A few factors demand special attention. First, rescue requires extra speed. Radiation hazards, even more than the typical fire or toxic threats, require fast approach and fast exit. Victims need respirators, or at least eight thicknesses of gauze, to filter their air. Radiation monitors, using standard Civil Defense Preparedness Agency instruments or the equivalent, must check every person and item exiting from hotter to cooler zones.

Risks must be assessed mathematically. For example, criticality

## SAMPLE CALCULATION FOR A NUCLEAR INCIDENT PREPLAN

Problem: A research laboratory within your response area houses Cobalt-60. The fire inspector determines that the maximum source strength that will be on the premises is 10,000 curies, and that the cobalt is located 24 ft (7.5 m) from the entrance to its compartment.

If a fire starts in the room, how long can a firefighter stay at the doorway fighting it?

(Use a maximum allowable dosage of 25 Roentgens per man for any one incident.)

Procedure:

1. Determine the dose rate using Table 16-A. The table indicates a dose rate of 1.47 roentgens per hour per curie at a distance of 3 feet.
2. Calculate the total dose rate at 3 feet: 1.47 Roentgen/hour/curie × 10,000 curies = 14,700 roentgens/hour.
3. Calculate the dose rate at the specified distance, using the inverse square formula:

$$\frac{\text{Radiation at distance 1}}{\text{Radiation at distance 2}} = \left(\frac{\text{Distance 2}}{\text{Distance 1}}\right)^2$$

Radiation at 24 ft =

$$\left(\frac{3 \text{ ft}}{24 \text{ ft}}\right)^2 \times (\text{radiation at 3 ft})$$

Radiation at 24 ft =

$$\left(\frac{1}{8}\right)^2 (14{,}700 \text{ roentgens/hour})$$

$$= \frac{14{,}700}{64} = 230 \text{ roentgens/hour}$$

$$= \frac{230}{60} = 3.8 \text{ roentgens/minute}$$

4. Calculate the allowable exposure time for firefighters:

$$\frac{25 \text{ Roentgens/person}}{3.8 \text{ Roentgens/minute}} = 6.6 \text{ minutes/person}$$

Assume a maximum exposure time at the door of 6 minutes per person for safety. If firefighters enter the room, the allowable exposure time must be recalculated.

**Figure 16-3.** Sample calculation for a nuclear incident preplan. (Reprinted with permission from Kerr, 1977; see References.)

(the likelihood of a nuclear explosion) depends upon the amount of fissionable material on hand. But, because laws forbid having this much on a single transport, most emergency services need not be concerned about performing the calculations for this situation. Yet hazard from radiation—as distinguished from a nuclear explosion—can be calculated quite easily if the source strength is known. In a building emergency, a preplan inspection usually would have recorded the source strength. For transportation incidents, the label, the bill of lading, or both will indicate the source strength. Of course, fire or other effects of the accident could eliminate these sources of information, or render them illegible or inaccessible.

Calculating the allowable exposure time for emergency personnel is simple, provided that the source strength is known. (Table 16-A). The example shown in Figure 16–3 is conservative, in that it ignores shielding, as by walls, containers, etc. It also assumes a point source rather than a more diffuse source. Note that it follows a rule of thumb limiting each person to 25 roentgens exposure per accident. Volumes could be written on this limitation. [Check latest exposure level limit.]

**Training and prevention.** Medical aspects of the problems resulting from exposure to radiation are complex, but early symptoms are simple. The real solution is not cure, but prevention, although decontamination can help to reduce total exposures.

Two important factors are obvious:

- Preplans must identify hazardous places and define the potential problem.

**Table 16-A.** Radiation Characteristics of Common Isotopes

| *Isotope* | *Half Life* | *Dose Rate** |
|---|---|---|
| Sodium-22 | 2.6 years | 1.34 |
| Sodium-24 | 15.0 hours | 2.15 |
| Manganese-52 | 5.7 days | 2.14 |
| Manganese-54 | 300.0 days | 0.54 |
| Iron-59 | 45.1 days | 0.71 |
| Cobalt-58 | 72.0 days | 0.62 |
| Cobalt-60 | 5.3 years | 1.47 |
| Copper-64 | 12.9 hours | 0.13 |
| Zinc-65 | 245.0 days | 0.31 |
| Iodine-130 | 12.5 hours | 1.37 |
| Iodine-131 | 8.1 days | 0.25 |
| Cesium-137 | 30.0 days | 0.36 |
| Iridium-192 | 74.5 days | 0.61 |
| Gold-198 | 2.7 days | 0.27 |
| Radium-226 | 1622.0 years | 1.005 |

*Roentgens per hour per curie of source strength, measured at a distance of 3 ft (0.9 m) from a point source.

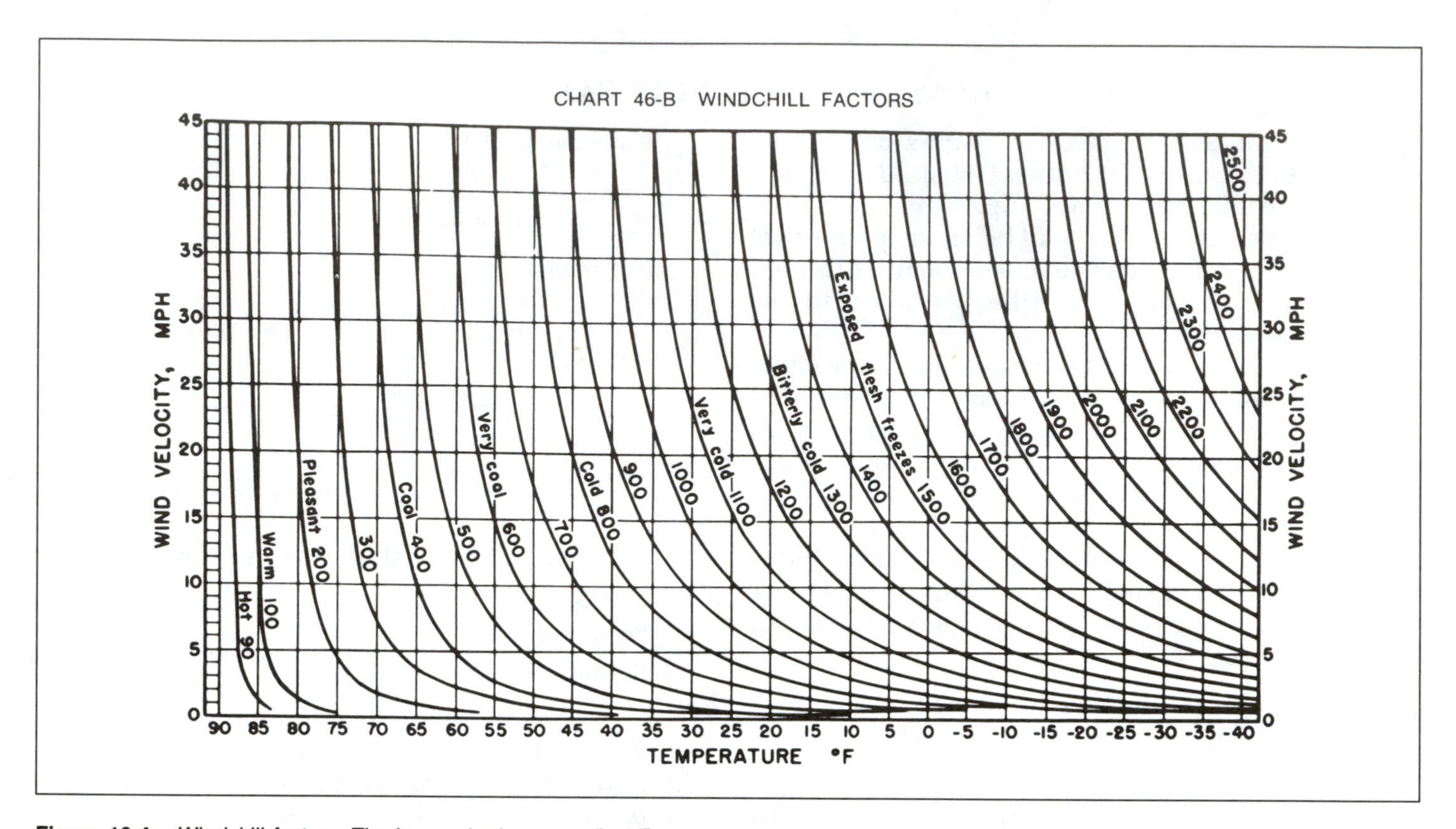


**Figure 16-4.** Windchill factors. The human body senses "cold" as a result of both temperature and wind velocity. The numerical factor that combines the effect of these two is called "the windchill factor," shown by the curves in the above nomogram of dry-shade atmosphere cooling. Take the line marked "Very cold 1100," for example. This shows that a person can feel just as cold when the temperature is 35 F and the wind velocity is 45 mph as at −35 F with a wind velocity of 1.5 mph. When the windchill factor reaches 1400, flesh that is exposed to the wind will freeze in only a few seconds; outdoor exposure is not recommended at all. Because of the extra clothing that people wear in cold weather, their physical size is greater than it is in warm weather. Be sure that equipment and controls are of adequate size and simplicity so that they can be run effectively and safely by persons wearing heavy clothing. (Reprinted from Engineer and Development Laboratories, The Engineer Center at Fort Belvoir.)

- Emergency personnel must understand the threat, and the principles of survival time and distance.

Training in the use of radiation instruments is neither complex nor expensive, and local civil preparedness offices are ready to help with it. Most other technical aspects of dealing with radiation incidents are largely covered in standard emergency procedure drills. Rules for prevention of casualties to emergency personnel after the incident also are well defined. (See References.)

Prevention of the accident depends on human behavior—adherence to rules and procedures, and possession of basic skills. It obviously helps if management enforces laws and codes.

To avoid contamination from radioactive materials, follow these recommendations.

Avoid direct beams of X ray or other machines.
Keep masked, to prevent ingestion of radioactive dust.
Avoid extreme fields of nuclear radiation.
Rotate personnel on an accident scene to prevent overdoses of a single person or team.
Know extinguishment procedures for the problem materials.
Know where to locate guidance and help.
Allow the "decay" of radiation intensity over a period of time to work for you.

### Weather extremes

Throughout a year, there may be some unusually severe and unexpected weather events that may require some changes in normal operations. For example, in North Dakota, the temperature may occasionally drop to 35 degrees below zero (−30 C), yet most activities and travel are not normally affected. But, if the wind increases in strength or the temperature drops suddenly, there may be a need to assist people in travel or other outside activity. (See Figure 16–4 for a windchill chart.) Employees could be alerted prior to leaving work. They should also know when or how they are to be notified about the company opening in the morning.

On the other hand, in the event of extremely heavy snowfall, what changes might be made in operations? What should employees be told prior to leaving for home?

Or suppose an unusually heavy rain strands hundreds of customers in a store just a few minutes before closing. Are supervisors and clerks prepared to handle the situation? May they allow telephone calls in and out? How do they control the crowd?

Hail or wind may start breaking glass windows while customers are shopping. What is the immediate action?

Suppose that adverse weather caused a power failure or someone suddenly shut off all power and lights while crowds were shopping? The emergency lighting system may operate as intended, but employees, particularly key supervisors, must understand emergency plans and be prepared to act responsibly.

## PLAN-OF-ACTION CONSIDERATIONS

Following the assessment of potential emergencies, the next step is to translate these needs into a plan of action. Management should be in charge of drafting a policy and getting the plan underway. It will usually be necessary that the union leaders (if any) be involved in the planning process. Generally, someone should be appointed emergency planning director or coordinator, perhaps with help from an advisory committee. Usually because of their experience and training, the health and safety, medical, fire, and security departments will be involved. Of course, because production and maintenance will be affected, these departments must be consulted. Also, the legal staff needs to be aware of the plan. And finally, contacts with local law-enforcement agencies, fire, and civil defense are necessary.

The cost and effort involved in giving immediate attention to emergency planning can be justified by weighing the cost of preparedness against the possibility of yearly losses from accidents, fires, floods and other catastrophes.

### Program considerations

The preliminary aspects of emergency planning have now been discussed—the need for advance planning and an evaluation of the type of emergencies and their potential harm to people and property. The next step, then, is to translate this need into a working plan within the organizational structure. In some cases, this requires working with other local agencies to most fully protect your operations.

Advance planning is the key. It is necessary to develop a written set of plans for action. The plans should be developed locally within the company (and corporate structure) and be in cooperation with other neighboring or similar organizations and with governmental agencies. It may not always be possible for them to fully cooperate or participate, but through planned action each organization should be aware of certain available assistances. The company may need to plan to be largely dependent upon its own resources to provide the internal safety.

Often an emergency manual or handbook will be developed for the plant or organization. The following outline covers many of the items that might be included, but other items may be needed as dictated by the expected emergencies and the available resources.

1. Company policy, purposes, authority, principal control measures, and emergency organization chart showing positions and functions.
2. Some description of the expected disasters with a risk statement.
3. A map of the plant, office, or store showing equipment, medical and first aid, fire control apparatus, shelters, command center, and evacuation routes.
4. A list (which may also be posted) of cooperating agencies and how to reach them.
5. A plant warning system.
6. A central communications center, including home contacts of employees.
7. A shutdown procedure, including security guard.
8. How to handle visitors and customers.
9. Locally related and necessary items.

Some of these items will be discussed in more detail on the following pages.

The plans should be rehearsed. Realistic conditions should be used to further learn the effectiveness of the plan. For example, maybe the emergency lights failed when needed, or the telephone service failed; but these are also the conditions that might occur in a real disaster. Therefore planning should include all possible, as well as probable contingencies.

### Chain of command

Once the decision has been made to establish a disaster plan, a director or coordinator should be appointed and an advisory committee, representing various departments, established.

**The director** should be a member of top management, whether it be a one-building or one plant company or a national corporation because he or she will have to be able to delegate authority and speak for the company. The head of the disaster-control organization must be a cool, quick-thinking person and should be sufficiently robust to withstand the arduous duties involved in an emergency. The emergency director's regular duties should be such that the greater part of his time will normally be spent at the unit he is responsible for. However, an alternate is always named in the plan. The alternate should be a person who has authority and qualifications similar to the director and should be trained with the director.

The director (and alternate) should be the first to be trained. Continuous liaison should be maintained with local Civil Defense authorities, if possible, to make sure that the company plans are coordinated with those of the community and to keep the company informed on new developments. See OSHA regulation 29CFR 1910.156, Fire Brigades, and 29CFR 1910.1200, Hazard Communication.

The director may be responsible for:

Communications
Firefighting
Rescue service
Guard service and warden service
First aid and medical service
Demolition and repair
Transportation
Investigation
Public relations.

All of these functions are likely to be essential although some may be combined. The person (and alternates) responsible for each function should be selected with great care and trained by the director. These chiefs should be familiar with all parts of the plan and should have experience in the fields in which they are to serve.

**Assigned personnel** must be trained to carry out their duties in accordance with the overall emergency plan. In small operations, where there may be no regular guards or firefighters, the operating personnel will be the people trained to take care of these duties. Of course, the number of members on each of the teams depends on the circumstances of each plant. Each team captain should select personnel from the available volunteers, supervise their training, and procure their equipment. Stronger people can be assigned to service in rescue squads because the

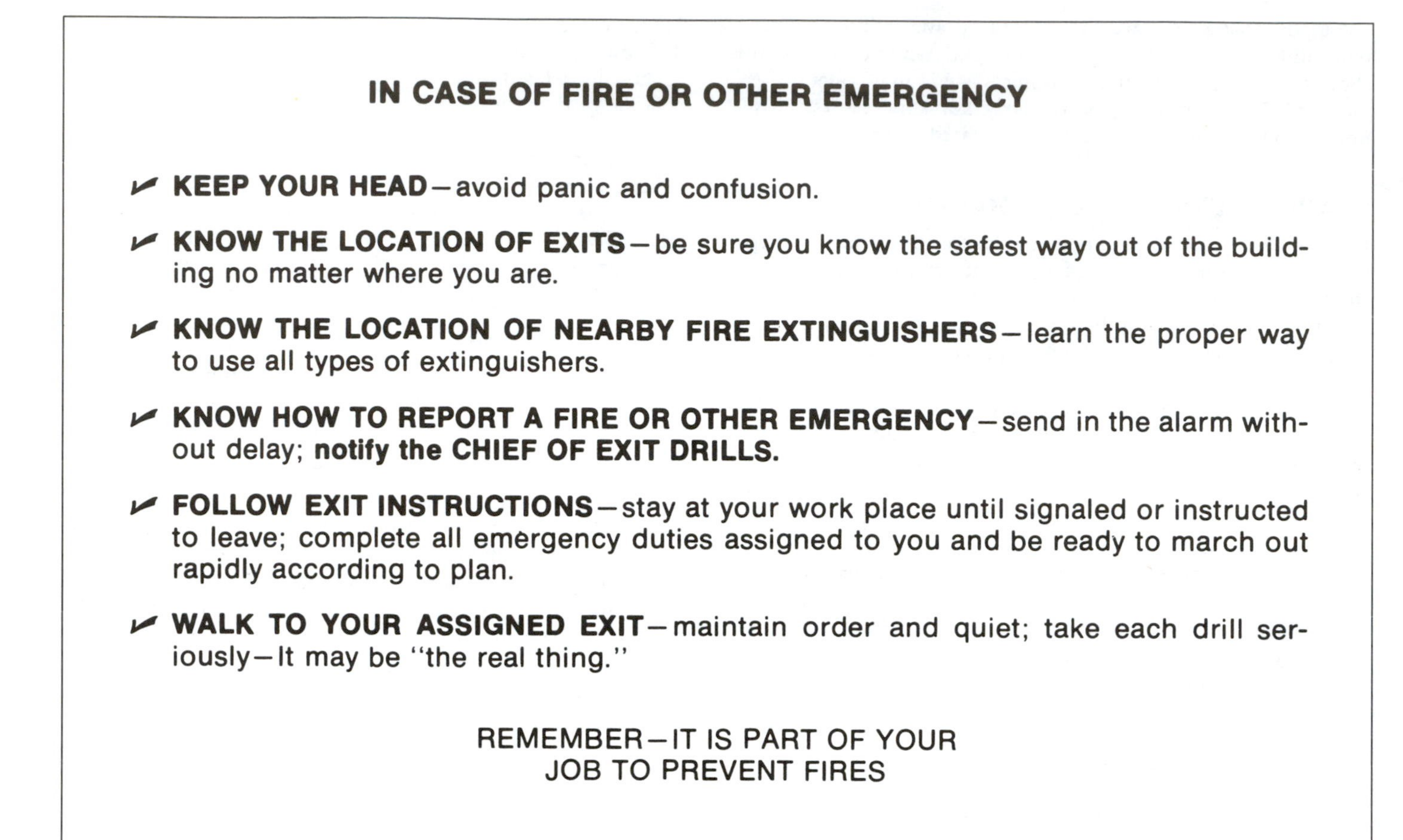


**Figure 16-5a.** Sample emergency exit notice for general posting.

work usually demands strenuous physical effort; those who are not so strong could be useful in light salvage operations.

Because wholehearted cooperation of personnel is necessary to the successful operation of an emergency plan, shop stewards or other employee representatives should take part in the planning. They should be aware that whatever measures are taken are for the protection of the lives and jobs of the workers as well as for protecting property.

Provisions should be made for emergency reporting centers so that employees will know where to report should the disaster occur while the employees are away from the plant. Reporting centers give employees a feeling of security and continuity, and aid the company in taking a "roll call." To facilitate these arrangements, each employee should carry an identification card which gives specific instructions on where to report, list of other reporting centers, basic employment record, and designation of the employee's next of kin in case they must be contacted or receive money due. The reporting center will keep a duplicate record for each employee assigned to report there. A one-plant company or a small company can consider using the home of a member of management, a supervisor, or an employee.

## Training

One of the most important functions of the director and staff, on both the corporate and plant levels, is training. Training for each type of disaster is essential in developing a disaster-control plan and keeping it functioning. Employees must be taught to realize that an emergency plan is vital and real—it cannot exist usefully if it remains a remote idea. Training and rehearsals are time consuming, but they keep the program in good working order.

Training of key people will be of little value unless it reaches all employees. The better informed and prepared the work force is, the less chance of panic and confusion during the emergency.

Practice alerts should be conducted to make sure that the employees know where to report and what their duties are. Even the most carefully prepared plans can develop flaws when put into practice, and only rehearsals can reveal them. The first one or two practices should be announced—lest there be a panic—but there should be no warning of succeeding ones. Officials should determine to their complete satisfaction that the disaster plan will work under emergency conditions. Once this has been determined, the plan should be maintained with periodic tests, staff discussions, and an occasional disaster problem. If this is not done, all the planning effort will have been wasted (see Figure 16-5a and b).

Management should assure employees that the company is doing everything possible to prevent injury to them, that every employee is an essential and necessary part of the team, and that the disaster-control organization is ready for any emergency. Such assurance will go a long way toward developing a state of mind that will not panic. Then when disaster strikes, emergency forces snap into action, workers file quietly into their shelters or other designated areas, firefighters are ready with hoses and equipment, and first-aid squads stand by ready to aid the wounded. Such planning is further evidence of management's

**EMERGENCY EXIT INSTRUCTIONS**
**MACHINE SHOP—DAY SHIFT**

**Read Carefully**

The following persons will be in command in any emergency, and their instructions must be followed:

CHIEF OF EXIT DRILL—H. C. Gordon, General Sup't.
MACHINE SHOP EXIT DRILL CAPTAIN—R. L. Jones, Foreman
MACHINE SHOP MONITORS—Dave Thomas and A. L. Smith

**In event of FIRE in machine shop**

- ✓ **NOTIFY THE GENERAL SUPERINTENDENT'S OFFICE**
- ✓ **PUT OUT THE FIRE, IF POSSIBLE**—If the fire cannot quickly be controlled, follow instructions given by Exit Drill Captain R. L. Jones or by the shop monitors. Leave by the exit door at the south end of the shop; if it is blocked by fire, use the door through the toolroom to the outside stairway.

**In event of FIRE or EMERGENCY in other sections of building**

The general alarm gong will ring for two 10-second periods as an "alert" signal. Continue work, but be on the alert for the "evacuation" signal, which will be a series of three short rings. At the evacuation signal:

- ✓ SHUT OFF ALL POWER TO MACHINES AND FANS
- ✓ TURN OFF GAS UNDER HEAT TREATING OVENS
- ✓ CLOSE WINDOWS AND CLEAR THE AISLES
- ✓ FORM A DOUBLE LINE IN THE CENTER AISLE AND FOLLOW MONITORS AND EXIT DRILL CAPTAIN TO EXIT—Walk rapidly, but do not run or crowd; do not talk, push, or cause confusion!

After leaving the building, do not interfere with the work of the plant fire brigade or the city fire department. Await instructions from the General Superintendent or your foreman.

**Returning to the building**

Return-to-work instructions will be given over the loudspeaker system or by telephone from the Superintendent's office.

**Figure 16-5b.** Sample individual instruction notice for general posting.

concern for employees.

### Command headquarters

The average command headquarters will not withstand a direct nuclear attack, but it should still be planned for any of the other emergencies which may occur.

Coordination of the disaster control organization should come from a well-equipped and well-protected control room. The headquarters should be equipped with telephones, sound-powered phones, public address system, maps of the plant, emergency lighting and electric power, sanitary facilities, a second exit, and two-way radios for communication both locally and with Civil Defense authorities.

Good communications are necessary for effective control and flexibility in a disaster situation. Communications include the telephone, radio, messengers, and the plant's alarm system (discussed separately later in this chapter). The disaster plan should provide for adequate telephones in emergency headquarters to handle both incoming and outgoing calls. Panic and disintegration of the organization will develop quickly if these calls are not handled with dispatch. An accurate log is kept of all incoming and outgoing messages.

Some means of communication independent of normal telephone service must be available during an emergency, such as provided by a battery-operated radio. The disaster plan must anticipate the possibility of losing normal telephone communications and electric power.

### Emergency equipment

An emergency checklist should include equipment and material to be ordered as well as shutdown actions to be performed.

- For example, where it is not feasible to keep on hand the necessary emergency equipment and materials, a list should be maintained of sources from which these items can be obtained on short notice. These sources must be outside of the immediate area. If the emergency were a flood alert, for example, this equipment and material would consist of sandbags, battens for windows and doorways, boats, tarpaulins, fuel-driven generating equipment (such as gasoline-powered arc welding machines or motor-generator sets), standby pumping equipment, a supply of gasoline in safety containers to fuel this equipment, lubricating oil and grease, rope, life belts, portable battery-operated radio equipment, and audio speakers.
- Some of the items on the shutdown part of the checklist would include: closing valves; protecting equipment that cannot be moved; closing and battening doors, windows, and ventilators to keep out looters as well as water; plugging vents and breather pipes. Included with the checklist should be a list of telephone numbers of supervisors and key employees to be notified.

Provision should be made for moving tank cars to higher ground and anchoring them. Portable containers should also be moved above the high-water mark, as should buoyant materials and chemicals that are soluble in water.

Storage tanks under the probable high-water mark (including underground tanks) should be specially anchored to prevent floating. Auxiliary dikes of sandbags or dirt should be built around key areas.

Other procedures must include shutting off electric and gas utility services at the *main line* before any water reaches them. Hot equipment should be cooled before water reaches it. All machine surfaces should be coated liberally with heavy grease, especially around openings to bearings. (This step applies even to machines which may not be under water, because dampness affects equipment.) Open flames should be eliminated so that any flammable liquid floating on the flood waters will not be ignited.

- If possible, a salvage crew should remain at the site to continue preventive operations after the plant or facility has been shut down to take further necessary steps if the flood shows signs of exceeding the estimated high-water level.

### Personnel shelter areas

While there could be times during natural disasters when employees would be moved to shelter areas, shelters are more often associated with a bombing or nuclear attack.

The average company would find it impractical to build a shelter to provide protection against a direct hit or "near miss," but consideration should be given to shelters for protection against fallout or other disasters. A shelter designed and equipped to provide protection against radioactive fallout will also serve as a personnel refuge for other types of emergencies. Shelters are designated by the sign shown in Figure 16-6.

Basements, tunnels (if they cannot be flooded), and inside areas of multistory buildings of concrete construction are examples of possible shelter areas in existing buildings. The denser the shield, the better it protects.

**Figure 16-6.** Civil Defense Fallout Shelter designation used throughout the United States to mark shelters in both public and private buildings. Shelters are surveyed and approved by the Federal Emergency Management Agency.

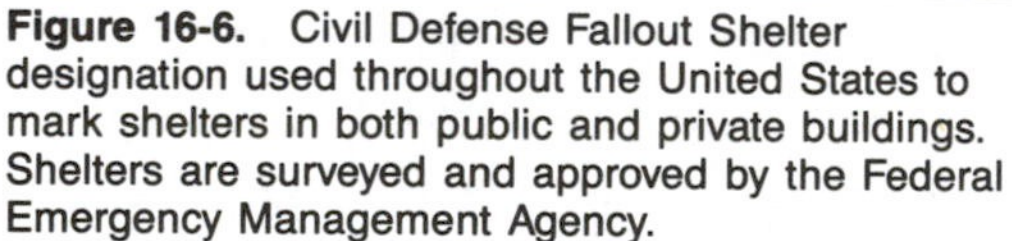

Many companies have not seriously considered constructing employee shelters because they do not think that the expense could be justified by the limited use. However, this is not necessarily the case; the shelter can have daily use as a locker room, training room, meeting room, cafeteria, or employee lounge. There might also be a psychological value in having employees use the shelter area on a daily basis; the familiarity tends to minimize any depressing effect which use of the shelter might have in an actual emergency. Although most industrial employee shelters are designed to give protection only from fallout, companies that have incorporated shelters in the construction of new buildings have found that they can also protect against blast effects.

Each employee shelter will require one person, with assistants, designated as the shelter manager. These managers must be carefully trained because they will be faced with all the psychological problems of life in close quarters, plus mass feeding, distribution of water, arranging sleeping accommodations, assigning duties, control of supplies, bolstering morale, and handling the personal problems that each individual will bring with him. The shelter manager may also be faced with emergency medical problems so training in first-aid procedures is necessary.

### Alarm systems

In most industrial operations a definite fire alarm system is set

**Figure 16-7.** Spacing of alarm systems is important for notification of personnel in remote areas of large plants.

up, using existing signaling systems such as a plant whistle; however, to avoid confusion with the regularly used signals, some plants have special codes or other signaling devices. This type of signal also may indicate the location of the fire, or separate signaling devices may be used for each building or working area within the company property.

The alarm system that activates the emergency plan may or may not go through the communications center, but should be touched off in the emergency headquarters office. Alarm systems should be provided in all buildings (Figure 16-7).

In hospitals, or other locations where both employees and nonemployees can hear an audible page system, a code name can be used to announce a fire and its location, such as "Doctor Red wanted in. . . ." Employees must be trained to be alert for this subtle signal.

Electric alarms are preferred to mechanical ones except in a shop having one large open area where there is only one alarm-summons station and one alarm-sending device, such as a manually operated gong. Manually operated alarms should supplement electric alarms. Closed circuit systems of the type specified by National Fire Protection Association (NFPA) standards are recommended; see References at the end of this chapter.

Companies in areas where municipal fire departments are available usually have a municipal alarm box close to the firm's entrance or in one of the buildings. Others may have auxiliary alarm box areas, connected to the municipal fire alarm system, at various points on the premises. Another system often used is a direct connection to the nearest fire station which may register by a water alarm in the sprinkler system or be set off manually. If possible, the fire alarm system should be connected with the local firefighting alarm, and have an independent power supply.

In large cities private central station services are available and provide excellent protection. These central stations receive signals from plant fire alarm boxes, watchmen, sprinkler head operations, and other hazard control points in the plant. Being able to give undivided attention to matters of plant security, they can relay information to fire or police departments without delay. The signal received at the fire department or assistance agency should locate specifically the site of the fire, or at least the building or area, so the fire can be found quickly (Figure 16-8).

Automatic sending stations (thermostatic detectors) may be used, but should not interfere with the sending of the manual alarm.

Regular checks should be made on the alarm system. All stations should be inspected on a monthly basis by a responsible person, and the overall system should have a daily test to ensure it is in proper operating condition. These daily tests should be conducted at a prearranged time and under a variety of wind

**Figure 16-8.** A command console for computer-based supervision and monitoring of large buildings or building complexes for loss-producing hazards including intrusion, fire, and other emergencies. Based on a central processing unit, the console includes computer display screen, high speed printer, and zoned annunciator panels.

and weather conditions to determine whether the signal can be heard in all parts of the plant at all times.

### Fire and emergency brigades

Because a fire can start from so many causes, fire prevention and fire protection must receive major attention in any emergency program. Advance planning is important. See OSHA regulation 29CFR 1910.156, Fire Brigades.

The company fire chief must be able to command people as well as have special training in fire prevention and protection. A person who has had experience in city or volunteer fire department work, or a military service veteran with experience in firefighting is a good choice. In a smaller company or plant, a master mechanic, maintenance department head, or other employee with mechanical experience can be a good part-time fire chief.

The fire chief should have one or more assistants who should have complete knowledge of the plant and equipment, command the respect and obedience of those under them, and are qualified to perform the duties of the chief officer if necessary.

The size of the plant and the fire potential presented by the occupancy determine the kind of plant firefighting brigade. The majority of plants may require only first aid firefighters under the direction of departmental foremen or managers. In larger plants, the fire brigade organization, directed by a full-time fire chief, is composed of full-time and emergency members. The full-time members maintain fire brigade equipment and are responsible for the permanent fire protection of the plant. Emergency members report for fire duty when the alarm is sounded.

The fire brigade apparatus should be selected only after a study has been made of plant conditions, to thus make sure it will be adequate for any emergency. Advice and assistance can be obtained from the local fire department, the NFPA, Office of Civil Defense, insurance companies, or perhaps a neighboring plant.

The large plant fire brigade is usually organized into squads, each with specific duties. One company's organization chart is shown in Figure 16-9.

**The evacuation squad** evacuates all employees from the emergency area as quickly and orderly as possible, without injury. They search closed areas, such as washrooms, to determine that everyone has been evacuated.

**Utility control squad** members are usually maintenance personnel, who are familiar with plant piping systems and the control of process gases, flammable liquids, and electricity.

**Sprinkler control squad** members must understand the automatic sprinkler system—the direction of rotation of the valve they

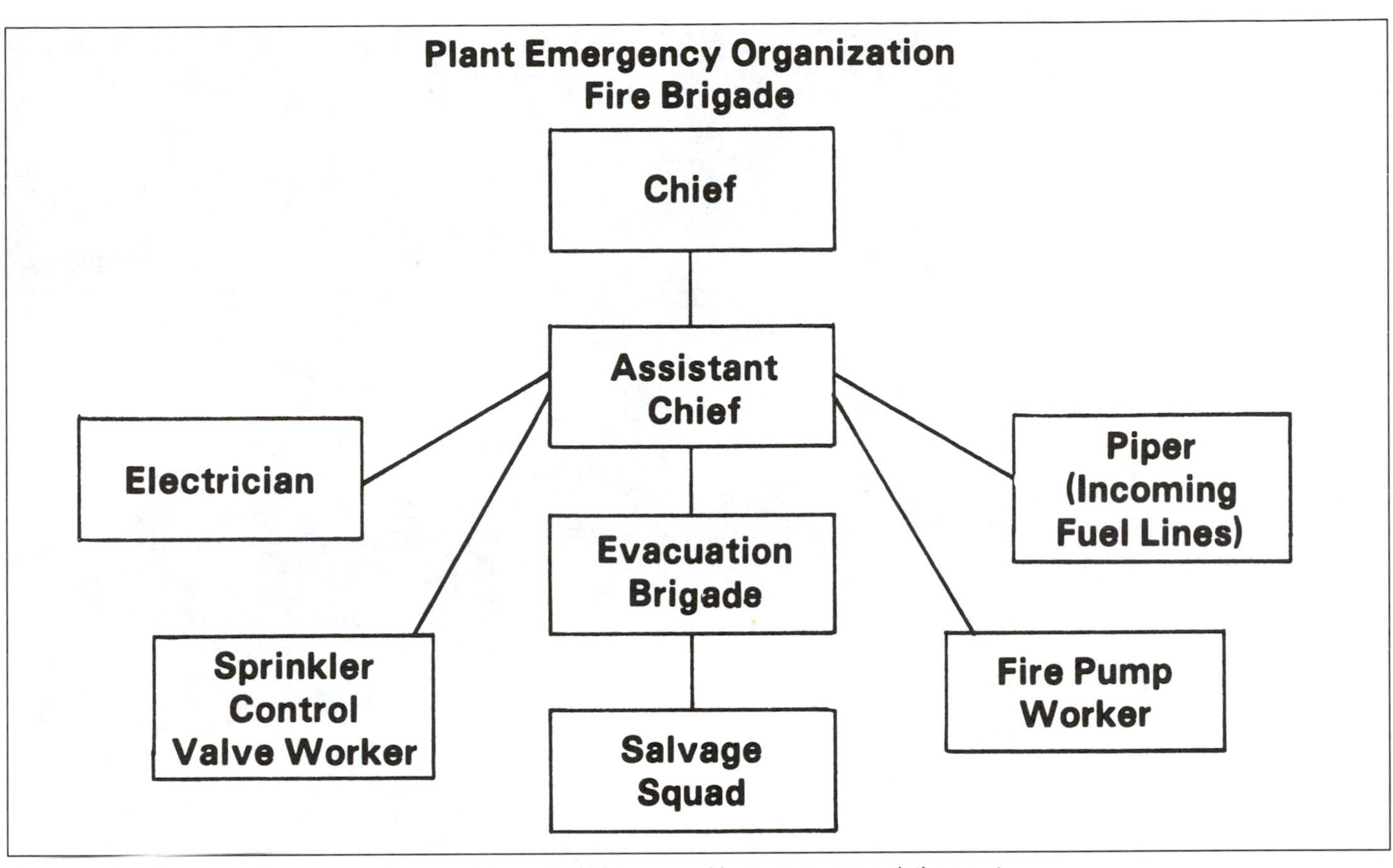


**Figure 16-9.** Because of the complexity of response needed to cope with an emergency, duties must be divided among plant emergency organization members to avoid confusion. Here is one plant's organizational chart. (Reprinted with permission from Textile Section Newsletter.)

are to operate, the use of sprinkler stops, and the replacement of sprinkler heads, if this is not a maintenance department function.

**Extinguisher squad.** Portable fire extinguishers are frequently operated by designated employees who work in the vicinity. However, as the size of the plant increases, it is advisable that special squads be selected for handling fire extinguishers. (See Figure 16-10.)

**Hose squad** members are trained to operate fire hydrants and hoses. They should drill frequently with wet hose lines so they have the feel of a charged hose. After they become proficient in handling the hose, drills once or twice a year may be ample (Figure 16-11).

**The salvage squad** is trained to protect as much stock and equipment as possible by controlling the directional flow of water and by covering stock with tarpaulins. Training should include proper methods of throwing tarpaulins and planned use of them in directing the flow of water. Members should also be familiar with the location of sawdust or other absorbent material and be aware of its value in controlling water on floors.

**The brigade-at-large or rescue team** consists of maintenance personnel or specially trained people. The main functions of this unit are to extricate casualties and eliminate hazards to other workers involved in the control of the emergency. Members respond to all alarms with a utility truck containing such rescue

**Figure 16-10.** Fire extinguisher training is very important. Underwriters Laboratories Inc. considers any fire extinguisher in the hand of a trained operator to have 2½ times more firefighting capacity than when used by a novice. Here, the fire chief instructs an office worker in the use of a pressurized water fire extinguisher. (Reprinted with permission from Paulsboro Laboratory, Mobile Research and Development Corp.)

**Figure 16-11.** Under direction of an instructor (white helmet), an emergency response drill team approaches a 300-gallon oil tank filled with burning fuel oil. Note the shower spray above the lead firefighters.

equipment as ropes, chains, block and tackle, ladders, cutting torches, saws, axes, and jacks. The amount of equipment, of course, will depend upon the size of the plant and the hazards involved, but every effort should be made to anticipate possible problems. Under the direction of the brigade officer, personnel in this unit also control utilities, ventilating fans, and blowers. They close fire doors, windows, and other openings in division walls. They open windows and doors leading to fire exits. Where escapes are the swinging section type, they should be the first to operate the escapes and to secure the steps to the ground.

Depending upon the size and inherent hazards of the plant facility, it may be necessary to train squads in handling and erecting ladders, using foam lines, recharging foam generators, and specialized rescue techniques.

**Fire pump team.** There should be at least two competent persons for pump duty in the main pump room, and a person assigned to each pump located elsewhere.

**Firefighter training.** The newly organized brigade should go through complete drills, preferably weekly. Later, less elaborate drills may be held at less frequent intervals. Drills should be held at unannounced times. They should be thorough in every respect, closely approximating fire conditions. (See Figure 16-11). See OSHA training regulations 19CFR 1910.156 (c)(1-4), Fire Brigades; 29CFR 1910.157 (g)(1-2), Employees involved with fighting incipient stage fires with portable fire extinguishers; 29CFR 1910.157 (g)(3-4), Employees designated to use firefighting equipment as part of an emergency action plan; and 29CFR 1910.155 (c)(4), Definition of Training.

No mater how thoroughly the industrial fire brigade is trained, there still must be close cooperation between it and adjoining or nearby plants and the public fire department. As stated before, it is recommended that the municipal fire department be called immediately upon discovery of every fire.

In fire emergencies, watchmen should be instructed to open yard gates and be ready to direct fire apparatus. Where plants have railroad tracks, cars should not be permitted to block crossings that may be needed in an emergency.

The brigade chief will be in full charge at a fire until the officer in charge of the public fire department takes over. The brigade chief then serves as an advisor on plant processes and special hazards.

**The fire station** itself should be centrally located, but not exposed to possible fires. If should be of fire-resistant material or located in a sprinklered part of the plant and protected with portable extinguishers. A larger plant may require mobile units, such as light hand-drawn trucks outfitted for the special hazards of the plant.

### Plant protection and security

Industrial security is management's responsibility. Government agencies can provide assistance and advice in establishing a policy. Since the basic problems of security are protection of property and control of persons, a company does not have to establish a new department to handle this function. A company's emergency security force could be built around the present security force.

Personnel need training in maintaining order, handling crowds, and coping with the threat of panic. They should also be prepared to prevent looting. They map emergency routes to shelters, both inside and outside the plant grounds.

Fires have often been caused by watchmen smoking while on duty, or overlooking fire hazards. Their failure to discover fires promptly, their shutting off sprinklers without ascertaining whether the fires have been extinguished, or their ignorance of the proper sprinkler valves to close after a fire is extinguished is often responsible for heavy loss.

As a supplement to automatic alarm and signal systems, the employment of an intelligent, well trained, and physically fit watchman can prevent or minimize fire loss. Because fires which start when the plant is idle produce more damage, the watchman becomes an important part of the fire prevention and detection organization.

The guard or watchman functions include protection against pilferage, burglary, vandalism, and espionage. The time of the inspection rounds should occur irregularly. The entire inspection should not create a detectable pattern.

The first round immediately after the plant closes is the most important. Most fires are likely to start just after employees have left, from machines or processes running unattended, or from careless smoking.

The guard or watchman should have enough time on his rounds to make a thorough inspection of the premises. His route should require no more than 40 minutes and should take him close to all hazardous occupancies. He should be provided with an approved flashlight or other illuminating device, and, where practical, plant lights along his route should be left on.

The guard can look for violations of smoking rules, improper storage of flammable material, leaking oil, gasoline, gas, or other flammable materials, and report unsatisfactory conditions to the management. The guard should be physically capable of turning in a fire alarm, dealing immediately with small fires, and with such matters as shutting off gas and closing fire doors. Failure of a guard to report on schedule at the end of his patrol should be immediately investigated.

Additional protection and security measures call for closing off certain windows and other openings in plants which are not vital to operation and limiting the number of plant entrances and exits. (All measures must be consistent with good fire prevention practices.) Installing protective wire mesh over windows along public thoroughfares is recommended. Floodlighting critical parts of plants at night is necessary for good protection.

### First aid and medical

A helpful publication in establishing a disaster medical service is the *Guide to Developing an Industrial Disaster Medical Service* compiled by the American Medical Association's Council on Occupational Health. This guide can also assist a company in evaluating its readiness and identifying weak areas. Examples of forms and casualty tags are included. Although the guide is designed to combat peacetime disasters, it provides a basis on which to expand the medical services to meet the devastation following a nuclear attack.

First-aid and medical service should be headed by the company doctor, if available, as discussed in Chapter 19, Occupational Health Services. In the organization of the medical phase of the emergency plan, those responsible must select and train personnel; decide what measures, equipment, and supplies are needed; and establish first-aid stations and a treatment center.

All employees should be encouraged to enroll in a first-aid course. People assigned to first-aid and medical units should pass standard and advanced first-aid courses. Local chapters of the American National Red Cross provide excellent training in this regard. See OSHA regulation 29CFR 1910.151, Medical Services and First Aid. This regulation says: "(a) The employer shall ensure the ready availability of medical personnel for advice and consultation on matters of plant health. (b) In the absence of an infirmary, clinic, or hospital in near proximity to the workplace which is used for the treatment of all injured employees, a person or persons shall be adequately trained to render first aid. First-aid supplies approved by the consulting physician shall be readily available. (c) Where the eyes or body of any person may be exposed to injurious corrosive materials, suitable facilities for quick drenching or flushing of the eyes and body shall be provided within the work area for immediate emergency use." (See also ANSI recommendations for eyewash and shower.)

If a major disaster occurs, there may not be enough trained doctors and nurses available. In such an eventuality, care beyond the first-aid level will have to be provided by nonmedical people who have received additional training. Such a medical team can be developed by recruiting volunteer medical aides, preferably with some experience. In addition to first-aid training, more advanced instruction by regular medical personnel or local hospitals should be provided.

A major consideration to keep in mind when establishing an emergency medical program, is that while most of the medical aid will be given in a central medical station, some of it may be given on the job site, possibly under hazardous conditions.

Plans should be made for representatives of the medical team to check all personnel at the disaster scene for trauma and to provide a written clearance for them to leave the plant when they are able to leave. In this duty, representatives of the investigation team may interview personnel before they leave to be sure that any needed eyewitness information is recorded.

Welfare and medical service includes the investigation of needs for prevention of epidemics, food inspection, and sanitation inspection. In the planning stages, company trucks, if any, should be designated as ambulances and the necessary equipment for them supplied. Two-way radio communication for such ambulance service is essential. Provision should be made for nonperishable food and water rations. There should be close coordination of plant first-aid measures with the local civil defense, health, and medical services.

The chemical service responsibility of this unit requires that gas masks be provided in case tear gas is used as part of a sabotage effort and that trained personnel, equipment, and supplies for chemical defense and decontamination be on hand. There should be a plan for priority sequence in decontamination of the plant—that is, water supply, power plant, machinery areas, warehouse areas, etc. Medical team personnel will also be responsible for any radiological monitoring thought to be necessary after a nuclear attack. The mere knowledge that such monitoring equipment is available is a morale-builder.

After a disaster, no one should be permitted to drink water until it has been examined.

### Warden service and evacuation

The warden service is responsible for maintaining employee control during emergencies, including (1) guiding employees to shelter, (2) directing employees away from hazardous areas, and (3) averting panic. In smaller companies, the wardens could also have the responsibility of taking charge of shelters. In some cases, the warden service may be responsible for seeing that shutdown of processes and equipment is carried out smoothly.

This type of service was devised primarily for areas of high population densities, such as commercial structures, factories, and residential areas. Some plants have a high concentration of employees. For these, the use of warden teams is certainly a good idea. In some plants, however, the concentration is very low. In such plants or departments with low personnel concentration, warden service is not necessary. In these cases, the operators themselves will have to be trained in shutdown details.

Management, in checking and providing for safe exits and evacuation drills, should refer to the National Fire Protection Association's standards and to local codes. Smooth, safe functioning of an evacuation plan requires a thorough knowledge of all plant operations and types of employees, number and types of exits available, width of exits, proper location of exits, possible alternate exits, and location of hazards, as well as a knowledge of warning and evacuation facilities. The subject of building exits is covered in Chapter 15, Office Safety; in Chapter 1, Industrial Buildings and Plant Layout, in the *Engineering and Technology* volume; and in the *Life Safety Code,* NFPA 101. See also OSHA regulations 29CFR 1910.157 (a)(2) and 29 CFR 1910.38.

Most plants have a rigid rule that only especially appointed people on the fire brigade shall go to the vicinity of the fire, and that everyone else shall proceed on signal to a refuge location in accordance with the organized evacuation plan.

### Transportation

Disrupted transportation facilities or restrictive traffic regulations could make it impossible for many employees to get to work. The company may need to provide transportation with company trucks and cars. Advance planning for car pools and pick up stops will greatly facilitate such a procedure.

The transportation responsibility includes arrangements for ambulance service, preparation for transportation of employees to and from work, and movement of emergency service crews as needed. Planning for adequate transportation service and traffic control requires liaison with the public police department, civil defense authorities, and possibly the military.

The transportation unit should consist of a group of regularly assigned drivers. Station wagons, from which the seats can be removed, and company trucks can be used when it becomes necessary to handle stretcher cases in evacuating any injured. The unit will be the means of getting auxiliary firefighters, first-aid teams, and salvage and rescue workers to the scene of the disaster at the earliest possible instant. The unit will also be used to deliver needed equipment and material from outside suppliers.

A source of motor fuel will have to be anticipated. One company used oversized underground gasoline storage tanks for the company service pump. Its emergency electric generating equipment also used gasoline-run engines because, in this case, gasoline is more readily available than is diesel fuel.

## SECURITY FROM PERSONAL ATTACK

Protecting the employee from muggings, rapes, and robberies is a concern of the health and safety professional when such attacks occur on company property. Even if these attacks occur elsewhere, the side effects are brought into the workplace.

### The crime problem

Health and safety personnel are not professional crime fighters. However, the on-the-job welfare of the work force is their responsibility, as is often the security of the physical property of a company, building, or plant. If a woman is attacked while waiting in her car for her husband to clock-out of the late shift, or if a man is mugged in the company parking lot on payday, or if some part of the office or plant is burglarized, the safety of an employee can be jeopardized.

When an incident occurs to a coworker, employee attitudes and emotions can range from alarm and anger to apathy. Distraction or preoccupation of any kind, if only momentary, can result in a serious accident.

How should the health and safety director cope with security and crime problems that chip away at the safe environment established within the confines of the plant or building? How does a health and safety organization incorporate this area of concern with other phases of the profession? Finally, how does the director reconcile any ensuing economic expenditures?

There are no pat answers to these questions. Each must be answered in terms of a particular industry, location, and workforce. All should be answered, however, with an attitude of commonsense and practicality that addresses itself to the welfare of the employee.

### Building and premises survey

What does the building or plant look like at night? Is it swathed in a bath of floodlights? Does an armed guard open electronic gates to the plant yard? Or, is it a small building set back from the street in an open, landscaped industrial park? Whatever the physical setup, know it, and know it well. Is there unguarded access to the parking area from a busy street or highway? Is there a viaduct or catwalk that allows easy entrance to a restricted area?

Note that not all crimes of the mugging-burglary nature take place in dark secluded areas, but a large percentage are in those locations simply because discovery is less likely. Thus a security survey of an industrial complex should probably center on secluded or remote areas. Tunnels between buildings, street underpasses, poorly lighted stairwells, dock areas, and freight elevators are all possible covert hiding places for unauthorized persons. Don't overlook washrooms and locker room facilities. Burglars have been known to hide in toilet stalls until after hours and then shop through the plant and offices for valuables.

### Implementation of the survey

Once a list of security problems has been drawn up, priorities must be established and plans for implementing them must be made.

Installing lights and a fence around the parking lot may deter a car thief from the area or a sex offender from hiding in the back seat of an unlocked car. In addition, the fence and gate may slow up employees leaving the area and make shift-change traffic more organized and safer. Adequate lighting in a stairwell would reveal a lurking figure waiting to accost an employee. Also, slipping and tripping would be less likely.

Often the plant or building survey reveals security gaps that can be closed by the maintenance department. For example, broken windows can be replaced or boarded up. Trash barrels or any container an intruder could stand on to squeeze through a window or trap door should be placed so as not to be an invitation. For small business or industrial operations that may be housed in one building, lights around the periphery of the structure are a relatively low cost method of deterrence. Decals prominently displayed in a window warning of an on-premises security alarm may be enough to dissuade the young or novice intruder. (Many establishments display such a decal whether or not an alarm system is actually installed.)

### Professional assistance

The experience of local law enforcement agencies can be a major source of help in resolving a company's security needs. In addition they can advise of any local laws or codes that require compliance.

Some police departments, particularly in suburban or rural areas, patrol industrial areas throughout the evening hours. Find out what the local force does. At the same time investigate types of crime in the area and recommend special procedures suited to a particular setup.

For instance, maybe an adjacent race car track brings large crowds of people to the track across the street from your plant, which has a large fleet. Burglars may use the diversion of the sporting event to help themselves to tools and parts in the plant's garage complex.

Even the unpredictability of weather should be considered in a security program. Do heavy rains, snow, or fog render some security procedures inoperable? If so, the procedures may have to be changed. The police (and fire department) may advise of the special operations that they are required to take in the event of unusual weather conditions that would nullify some security procedures. For example, find out if bus routes or traffic must be rerouted behind your plant or building if the viaduct floods or the prairie catches fire.

Based on the discussion with the law enforcement agencies, it may be necessary to work with additional security professionals. This can include watch service companies, insurance firms, burglar alarm and detector manufacturers, and lighting companies. Naturally, any security equipment selected will be based on needs.

Just as multi-plant industrial complexes share core medical facilities, cooperative security plans are worth investigating. Expenses for additional lighting, fences, and watch patrols may be shared by several companies. However, this should be done only with the approval of the insurance carriers.

### Personnel protection

Once the "bricks and mortar" of a plant are secure under the supervision of watchmen and guards or detector systems, much of the personal security of employees is also provided. Not all unauthorized persons questioned by a plant guard or detected on a surveillance system are there to steal. Some are on the premises (or in the area) to harm an employee, often just a person who is there by chance.

A well-organized employee identification system is basic to company and plant security. All too often it is assumed that the person sauntering through an area, be it restricted or not, has a legitimate reason for being there. By challenging the stranger's presence in an area and asking him for a company identification card or, if cards are not used, asking for verification from another employee, this problem can be stopped. At first, employees may object to "police overtones" of "identify or else," but selling security to them on the basis of, "It's for your own protection," meets with little resistance. Supporting this approach with examples of problems resulting from little or no security measures usually quells even the strongest resisters.

Wearing photo-identification cards clipped to a shirt pocket or collar can be made a condition of employment. In addition, asking the employees themselves about improving security can often bring forth ideas that fit in well with a specific plant's set up. The "we want you on the job safely" approach is a more positive way of selling security to employees than the "No-admittance-beyond-this-point" or "Don't-do-this" approach.

### Female employees

Rape is a hideous crime. It can occur at any place, at any time. It is a crime in which the victim is left with serious emotional scars and/or physical injuries.

Seen as a power crime by a male expressing his hatred for women, a rape case can trigger strong emotional reactions amongst coworkers. There has been much dialogue on the subject from all areas of society—law, sociology, psychology, security—all with a goal of improving the lot of the victim and understanding the rationale, or lack of rationale, of the accused.

Keeping in mind that the presence of other persons and the "light" of easy detection are obvious deterrents, security measures should include adequate lighting, visible presence of guards, and a guard escort service, if necessary. Women employees should know that rape victims are often the victims by chance, not necessarily because they know or recognize the assaulter.

Tell female employees about self-protection during a safety meeting set up just for the subject of security. Set the tone of the program with, again, "we want you on the job safely." Encourage comments from the women, particularly suggestions on ways they will feel more secure. Don't let them leave the meeting empty-handed. Distribute National Safety Council's *Just Another Statistic* (see References) or similar literature that may be available from your local law enforcement agency, and perhaps a whistle. A number of cities have met with success after the establishment of whistle campaigns. While stopping rape is one of the reasons for using the whistle, it can be used by men and women in other emergency situations. The whistle is to be used when a person sees something suspicious happening or is personally threatened, and when it is physically safe to sound the whistle.

Sometimes to guarantee further the safety of a workforce,

additional security measures must be taken for the in-transit employee. In-transit security is a sensible procedure for preventing muggings and robberies as well as rapes. Cabs and minibus rides from the plant gate to public transportation are standard procedures in many work situations. Guard escort from the plant gate to the employee's car is provided in some high-crime areas. For the most part, these efforts have been set up to protect female employees who work evening shifts. Similar protection is afforded the female office worker by the policy of "safety in numbers," or by requiring the presence of a supervisor or several other people during over-time hours.

### Mutual assistance

Cooperation with surrounding plants, local law enforcement authorities, and transportation systems can expedite in-transit security. The cost of operating a minibus service from several plants or establishments to the local bus or train can be shared by the participating companies. Or, a similar agreement can be made with a local taxi company. Some companies located in high-crime areas have borne the expense of taxi rides to the home of the employee in an effort to maintain a safe and qualified workforce.

## OUTSIDE HELP

A company's chance of survival and recovery is greater when knowledge, equipment, and personnel are pooled with its neighbors. Therefore, emergency plans should include a provision for exchanging aid with other plants in the industrial community.

### Mutual aid plans

A number of industrial communities are organized to assist their members in the event of an emergency or disaster. These organizations include manufacturing plants, large offices, stores, hotels, utility companies, chemical plants, law enforcement organizations, hospitals, newspapers, and radio and television stations. They operate independently of or as supplements to any civil defense groups.

It is impossible to have adequate supplies available for a really large disaster. The best defense is to have adequate supplies in other areas committed for standby use, with communication channels and a plan for rapid transportation of supplies to the stricken area. Especially important are adequate medical supplies and firefighting equipment. A plan for rapid and accurate communication is necessary, as discussed earlier in this chapter.

Mutual aid plans with neighboring companies and community agencies should include establishing an organizational structure and communication system, standardizing an identification system and procedures and equipment (such as fire hose couplings), formulating a list of available equipment, stockpiling medical supplies, sharing facilities in an emergency, and cooperating in test exercises and training.

Frequently these "cooperatives" establish a task force composed of personnel from each member company. Training is supplemented by detailed written instructions. Bulldozers, floodlights, and tools are marked by each plant for emergency use. Training on a community basis might include instruction by members of the public fire department to plant fire brigade members and also some actual training by members of a construction or wrecking company to show the salvage and rescue teams how to handle heavy weights and to work safely among debris.

### Contracting for disaster services

Some companies contract for disaster service. The service is paid for by a fixed annual retainer, plus additional pay for the actual hours worked. For example, a wrecking company can be engaged to supply the men and equipment necessary to clear debris created by a disaster. Contracting for such a service removes the burden of providing trained pesonnel and maintaining a great deal of idle emergency equipment that could easily be damaged in the very disaster for which it was designed.

### Municipal fire and police departments

Firefighters from the station most likely to respond to an alarm should be fully acquainted with all fire hazards in the plant. Cooperation may be encouraged by inviting local fire officials to inspect the company area. As a result, they can become familiar with the location, construction, and arrangement of all buildings, as well as all special hazards, such as flammable gases, liquids, and materials. Because public fire department rescue equipment can supplement plant rescue units, they can make sure that company equipment is compatible with the municipal equipment. Therefore, the local fire department can formulate an efficient plan of attack before a fire occurs. Such procedure is far better than waiting for fire to break out, and then running the risk of misunderstanding the situation and initiating improper firefighting methods. See OSHA regulations 29CFR 1910.38 and subpart L, 29CFR 1910.155-165. Also see subpart L, appendix A-E and 29CFR 1910.1200, Hazard Communication.

The public police force can aid in putting down large-scale disturbances and in assisting with evacuation from the plant premises in the event of a major disaster. Planning for this outside help should include arrangements for traffic control, particularly where a plant parking lot empties immediately onto a public highway.

### Industry and medical agencies

Details of the following services are given in Chapter 24, Sources of Help, under Emergency and Specialized Information.

- In 1970, the Chemical Manufacturers Association created the Chemical Transportation Emergency Center (CHEMTREC). Under this program a national center located at 2501 M Street, NW., Washington, DC 20037 (CMA headquarters) can relay upon request emergency information concerning specific chemicals, the hazards, and steps to take to control the emergency. A phone is available for 24 hour service—800/424-9300. It is intended primarily for use by those who transport chemicals, but others may need the information.
- The Toxicology Information On-Line Network (TOXLINE) has been designed to provide current and prompt information on the toxicity of substances. It is intended to be used by health professionals and other scientists working with pollution, safety, drug, health, and other disciplines. The service is under the auspices of the National Library of Medicine, 8600 Rockville Pike, Bethesda, Md. 20014. The service is accessible via terminals on-line through a national telephone-based network. A fee is required for use.

There are a number of other emergency and specialized information sources listed in Chapter 24, Sources of Help.

### Governmental and community agencies

During a community-wide disaster, a large number of governmental and private agencies are available to assist industries; these include the Office of Civil Defense, the U.S. Army Corps of Engineers, the Salvation Army, the American Red Cross, the U.S. Public Health Service, and the U.S. Weather Bureau. To be effective in coping with an industrial community disaster, the efforts of all of these groups must be coordinated and directed toward a common end. Therefore each plant should have an up-to-date listing of all cooperating agencies; the administrator's name, address, and telephone number; and the task assignment of the agency. If possible, these people should meet periodically to discuss mutual problems and disaster control techniques.

The company's emergency planning director should become thoroughly familiar with the authority, organization, and emergency procedures that are established by law and which will become effective upon declaration of a civil defense emergency.

In wartime, the federal, state, and local governments are responsible for relief measures. The Red Cross has offered to assist the government in providing food, clothing, and temporary shelter on a mass-care basis during the emergency period immediately following enemy attack. In many communities, local Civil Defense officials have requested Red Cross chapters to assume all or part of this responsibility, acting under Civil Defense authorities.

In natural disasters, the American Red Cross is responsible for assisting families and individuals to meet disaster-caused needs that cannot be met through their own resources. These relief operations are coordinated with the activities of the local, state, and federal governments. When a disaster occurs, the local chapter of the American Red Cross aids disaster victims. The resources of the national organization are available to supplement chapter assistance.

## REFERENCES


American Insurance Association, Engineering and Safety Service, 85 John St., New York, N.Y. 10038.

*Fire Hazards and Safeguards for Metalworking Industries,* Technical Survey No. 2.

*Fire Safeguarding Warehouses,* Technical Survey No. 1.

American Medical Association, Council on Occupational Health, 535 N. Dearborn St., Chicago, Ill. 60610. *Guide to Developing an Industrial Disaster Medical Service.*

The Conference Board, 845 Third Ave., New York, N.Y. 10022. *Studies in Business Policy,* No. 55, "Protecting Personnel in Wartime."

Factory Mutual Engineering and Research, 1151 Boston-Providence Turnpike, Norwood, Mass. 02062.

*Handbook of Industrial Loss Prevention.*

*Loss Prevention Data.*

Federal Emergency Management Agency, 500 C St., S.W. Washington, D.C. 20472. (Available through Superintendent of Documents, U.S. Government Printing Office, Washington, D.C. 20402.)

*Attack Environment Manual,* in process, June 1987.

*In Time of Emergency,* H-14, Oct. 1985.

*Mass Casualty Planning: A Model for In-Hospital Disaster Response,* Aug. 1986.

Giraud, Raymond, "Radioactive Elements," *National Safety News,* June 1973. Adapted from author's article in *Revue Technique du Feu,* Entreprise moderne d'edition, 4 rue Cambon, 75 Paris 1, France.

International Association of Fire Chiefs, 1329 18th St. NW., Washington, D.C. 20036. *Nuclear Hazard Management for the Fire Service,* 1975.

Kerr, James W. "Preplanning for a Nuclear Incident." *Fire Command!* April 1977.

National Fire Protection Association, Batterymarch Park, Quincy, Mass. 02269.

*Installation, Maintenance, and Use of Auxiliary Protective Signaling Systems,* NFPA 72B.

*Installation, Maintenance, and Use of Central Station Signaling Systems,* NFPA 71.

*Explosion Prevention Systems,* NFPA 69.

*Facilities Handling Radioactive Materials,* NFPA 801.

*Fire Protection Handbook,* latest ed.

*Guard Operations in Fire Loss Prevention,* NFPA 602.

*Guard Service in Fire Loss Prevention,* NFPA 601.

*Health Care Facilities,* NFPA 99.

*Industrial Fire Hazards Handbook,* SPP–57A.

*Life Safety Code,* NFPA 101.

*Installation, Maintenance, and Use of Local Protective Signaling Systems,* NFPA 72A.

*Installation, Maintenance, and Use of Remote Station Protective Signaling Systems,* NFPA 72C.

*Installation of Sprinkler Systems,* NFPA 13.

*Inspection, Testing and Maintenance of Sprinkler Systems,* NFPA 13A.

National Petroleum Council, 1625 K St. NW., Washington, D.C. 20006.

*Disaster Planning for the Oil and Gas Industries.*

*Security Principles for the Petroleum and Gas Industries.*

National Safety Council, 444 North Michigan Ave., Chicago, Ill. 60611.

*Fire Protection Guide.*

*Fundamentals of Industrial Hygiene,* 3rd ed.

Industrial Data Sheets

*Fire Prevention and Control at Construction Sites,* 491.

*Fire Prevention in Stores,* 549.

*Just Another Statistic.*

(See other appropriate topics treated in both volumes of this Manual, especially those pertaining to organization, training, medical and nursing services, fire extinguishment and control.)

Emergency Management Institute, National Emergency Training Center, Emmitsburg, MD 21727.

"Security from Personal Attack," *National Safety News,* July 1973.

Underwriters Laboratories Inc., 333 Pfingsten Rd., Northbrook, Ill. 60062.

*Classification of Fire-Resistance Record-Protection Equipment.*

*Gas Shutoff Valves—Earthquake.*

# 17

# Personal Protective Equipment



THE PRIMARY APPROACH in any safety effort is to maintain or change the physical environment so that accidents cannot occur. However, it is sometimes necessary for economic reasons or in temporary or changing conditions to safeguard personnel by equipping them individually with specialized personal protective equipment (PPE). Although the use of personal protective equipment is an important and necessary consideration in the development of a safety and health program, it should not be used permanently instead of engineering out or otherwise maintaining a safe and healthy work environment. In general, governmental regulations list the use of PPE as a case of last resort. Analyze accident situations to determine whether PPE can prevent a recurrence. When work conditions cannot be made safer, clearly PPE is necessary. Other chapters in this manual outline methods of creating, maintaining, and managing a safe work environment.

Clearly, the first step of management is to design a safe work environment. Commonly, engineering of the conditions is necessary. This means looking at everything in the work environment, seeing what can potentially harm the workers, and changing those items so harm cannot occur. It means considering the worst possible analysis of the conditions—not the best.

The next step is to manage conditions so that employees' exposure is controlled or reduced. Limiting the time an employee is exposed to the condition is the usual method. For those work environment conditions that can't be eliminated through engineering or by limiting work exposure, PPE becomes the basic protection device. It's important to take a very strong positive attitude toward the proper use of PPE.

## A PROGRAM TO INTRODUCE PPE

Once it is decided that personal protective equipment is going to be used, then do the following:

1. Write a policy on usage of the PPE and communicate it to employees and visitors as needed.
2. Select the proper type of equipment.
3. Implement a thorough training program.
4. Make certain the employee knows the correct use and maintenance of the equipment.
5. Enforce its use.

### Policy

The policy should simply state the need for and use of PPE. It may also contain exceptions or limitations on use of PPE. Some policies or safety rules will detail the kind, use, work condition, etc., expected. The management staff must follow the same rules.

Here is an example of one firm's policy on wearing of personal protective equipment devices:

> For safe use of any personal protective device, it is essential the user be properly instructed in its selection, use, and maintenance. Both supervisors and workers shall be so instructed by competent persons.

### Selection of proper equipment

After the need for personal protective equipment has been established, the next step is to select the proper type. The degree of protection that a particular piece of equipment affords under various conditions is the most important criterion.

Except for respiratory protective devices, few items of personal protective equipment available commercially are tested according to published and generally accepted performance specifications and approved by an impartial examiner. Although satisfactory performance specifications exist for certain types of personal protective equipment (notably protective helmets, devices to protect the eyes from impact and from harmful radiations, and rubber insulating gloves), there are no approving laboratories to test equipment regularly according to these specifications. (See the latest NIOSH Certified Equipment List, which can be obtained from the U.S. Superintendent of Documents, Washington, D.C.)

The Safety Equipment Institute (SEI) has formulated fair and objective policies for third-party certification of safety equipment. SEI voluntary certification programs involve both product testing and an on-going program of quality assurance audits. Participating manufacturers are required to submit a specific number of product models to undergo demanding performance tests in SEI authorized independent laboratories. When the lab has completed the test, SEI receives a pass or fail notification. For the quality assurance program, an audit is conducted on location at a manufacturer's production facilities. SEI wants to ensure that products coming off the assembly line are made to the same exacting specifications as the product model actually tested for certification. SEI's existing certification programs include (1) eye and face protection, such as goggles, faceshields, spectacles, and welding helmets; (2) emergency eyewash and shower equipment; (3) fire-fighter's helmets; and (4) protective headwear, such as helmets. (The latest edition of the list of *SEI Certified Products* may be obtained by writing SEI, 1901 N. Moore, Suite 501, Arlington, VA 22209.)

### Proper training

The next step is to obtain the workers' complete compliance with requirements to wear the personal protective equipment. Several factors influence compliance; among them are: (1) the extent to which the personnel who must wear the equipment understand its necessity, (2) the ease and comfort with which it can be worn with a minimum of interference with normal work procedures, and (3) the available economic, social, and disciplinary sanctions which can be used to influence the attitudes of the workers.

In organizations where workers are accustomed to wearing personal protective equipment as a condition of employment, this problem may be minor. People are simply issued equipment that meets the requirements of the job and are taught how and why it must be used. Thereafter, periodic checks are made until use of the issued equipment has become a matter of habit.

When a group of workers are issued personal protective equipment for the first time or when new devices are introduced, the problem may be more difficult. A clear and reasonable explanation as to why the equipment must be worn must be given. Traditional work procedures may have to be changed. If such changes are required, a good deal of resistance, justifiable or not, may be generated. Also, workers may be reluctant to use the equipment because of bravado or vanity.

The practice of having supervisors and foremen try out new protective equipment and devices prior to actual adoption, and getting their comments and discussing the advantages, has been successfully used in many operations.

A good deal of the resistance to change can be overcome if the persons who are going to use the PPE are allowed to choose the particular style of equipment they will wear from a group of different styles which have been preselected to meet the job requirements. In some situations, it may be advisable to have a committee from the work force help select suitable devices. Management's desire to purchase one standard style of equipment may not be realized immediately, and several styles may need to be stocked. In the latter case, the cost, though higher than the cost of stocking only one style, will be small compared to the potential cost of accidents resulting from failure to use the equipment.

For the convenience of their employees, some companies maintain equipment stores on the plant premises.

A training program outline might include:

1. Describing what hazard and/or condition is in the work environment.
2. Telling what has/can be/cannot be done about it.
3. Explaining why a certain type of PPE has been selected.
4. Discussing the capabilities and/or limitation of the PPE.
5. Demonstrating how to use, adjust, or fit the PPE.
6. Practicing using the PPE.
7. Explaining company policy and its enforcement.
8. Discussing how to deal with emergencies.
9. Discussing how PPE will be paid for, maintained, repaired, cleaned, etc.

### Use and maintenance

All equipment must be inspected periodically before and after each use. A record should be kept of all inspections by date, with the results tabulated. The recommendations of the manufacturer for inspection should be closely followed, as should the recommendations of the manufacturer for the maintenance of the device, and the repair and replacement of parts supplied by the manufacturer of the product.

### Enforcement

Employees need to know how the use of PPE will be enforced. Many companies have some kind of progressive disciplinary action, such as unpaid time off, and finally, termination. The enforcement of the use of PPE is critical to a successful program.

### Recognition clubs

Several organizations sponsor recognition awards for those who have escaped or minimized injury by wearing personal protective equipment. These organizations include the Wise Owl Club, sponsored by the National Society to Prevent Blindness, The Golden Shoe Club, and The Turtle Club. Details of these clubs and other recognition clubs are given in Chapter 12, Maintaining Interest in Safety.

### Who pays for PPE?

Policies differ regarding who pays for personal protective equipment. It is difficult to determine when the company should pay the cost of personal protective equipment and when workers should. Factors to consider are the probability and expected severity of injury, the willingness of workers to wear the equipment, its length of life, the degree to which it may be depreciated by nonoccupational use, and provisions of collective bargaining agreements.

For example, the welder's helmet is almost universally supplied by management, for the job could not be performed at all without it. Work gloves sometimes must be purchased by the user. However, welder's or other special purpose gloves are usually considered as being a necessary part of the job, like work tools, and are issued free. In other companies the cost of eye protection equipment that has prescription-ground lenses, such as safety spectacles, is often shared by worker and employers. Many industrial firms maintain shoe stores to make safety shoes convenient to purchase. In some instances, to encourage purchases, safety shoes are offered at a partial reimbursement rate. Finally, in some areas, a shoemobile service is even available. A vendor comes into the plant with a trailer completely equipped and stocked to fit and sell safety shoes.

The succeeding pages discuss seven major areas of personal protective equipment—head, eyes and face, ear, respiratory, hands, feet, and trunk. Each section will provide information on the standards available or proposed, some details about the equipment available, and suggestions for selecting equipment to meet the job hazard.

## HEAD PROTECTION

Although workers should be encouraged to use their heads to absorb knowledge, they should not use them to absorb blows. Those who are exposed to head hazards must be provided with protective headwear (Figure 17-1). Some operations in which head protection may be necessary include tree trimming, construction work, shipbuilding, logging, mining, overhead line construction or maintenance, and basic metal or chemical production.

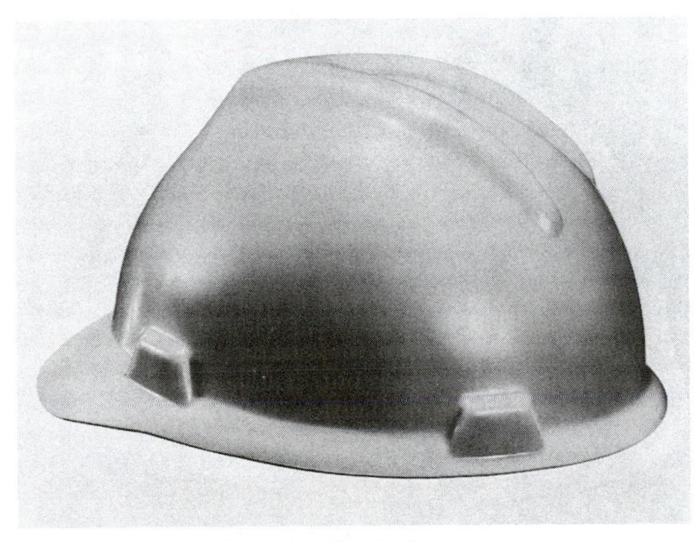

**Figure 17-1.** A standard type of protective headwear.

### Protective headwear

Protective headwear is designed to protect the wearer's head from impact and penetration of falling objects and, in some cases, from high-voltage electric shock and burns. The hazards may require protection of all three types. Protective headwear can also help shield the head and hair from entanglement in machinery or exposure to environments. Safety professionals should be aware of changes in operations that may create a need for protective headwear. For example, a firm undergoing a slack season might transfer some employees to duties requiring protective headwear. In addition, construction, maintenance, and odd jobs requiring head protection often occur in the normal operations of many companies.

The American National Standard Institute establishes specifications for helmets (ANSI Z89.1-1986). The standard is intended to be used in its entirety on a product. Identification inside the helmet shell must list the manufacturer's name, ANSI designation, and either class A, B, or C. (ANSI standards are designated by a year, and the helmet identification should be no more than 5 years old. The latest designation should be observed.)

In standard ANSI Z89.1–1986, helmet is defined as "a device that is worn to provide protection for the head, or portion thereof, against impact, flying particles, electrical shock, or any combination thereof and that includes a suitable harness" (see Figure 17-2a through e). The harness is a complete assembly

**Figure 17-2a.** Protective headwear adapted to meet the needs of particular work situations. (a) A knit helmet liner is added for cold weather protection. (b) A universal adaptor with chemical goggles is added to assure that workers will wear their goggles whenever safety conditions require protective headwear. (Courtesy Uvex Winter Optical, Inc.) (c) Protective headwear with a movable faceshield. (d) A wire mesh faceshield and muff-type ear protection are added. (Courtesy Bilsom International, Inc.) (e) Protective headwear with attached welding face mask and earmuffs. (Courtesy Mine Safety Appliances Company.)

Figure 17-2b.

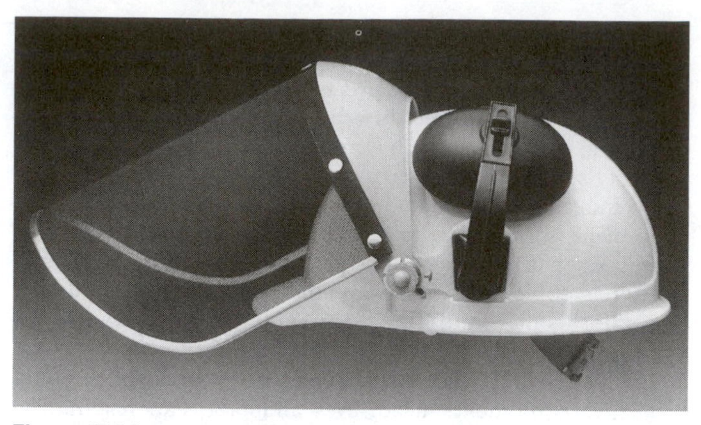

Figure 17-2d.

Figure 17-2c.

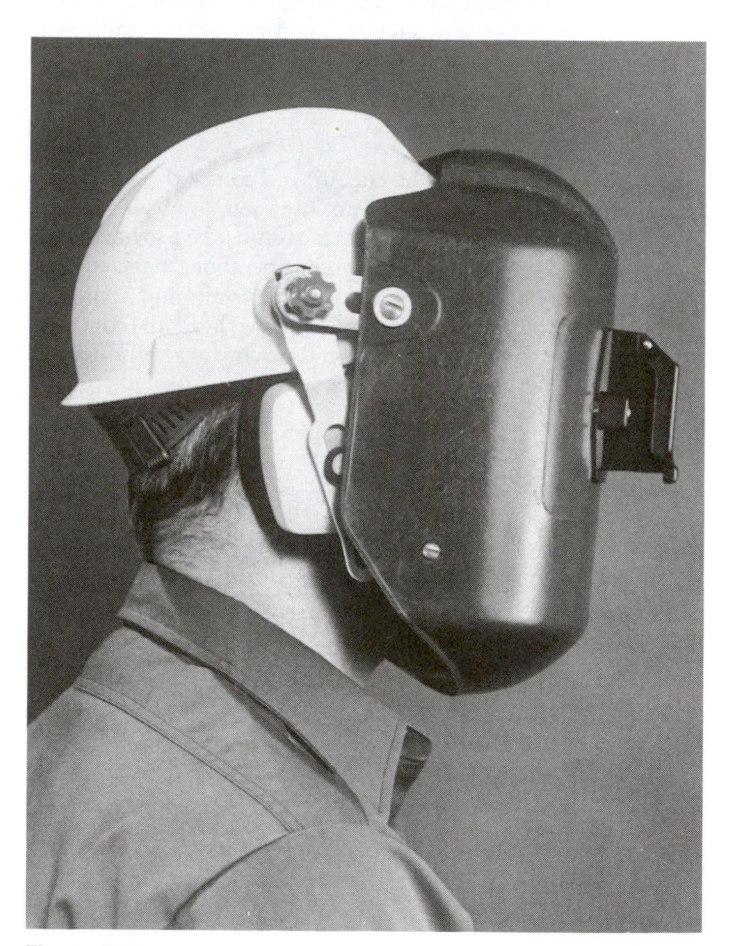

Figure 17-2e.

by means of which the helmet is monitored in position on the wearer's head. Protective helmets are commonly and incorrectly called safety helmets, safety hats, and hard hats. ANSI standard Z89.1-1986 defines three classes of helmets:

Class A—Helmets are intended to protect the head from the force of impact of falling objects and from electric shock during contact with exposed low voltage conductors.

Class B—Helmets are intended to protect the head from the force of impact of falling objects and from electric shock during contact with exposed high voltage conductors.

Class C—Helmets are intended to protect the head from the force of impact of falling objects.

### Bump caps

A bump cap is not a helmet (hard hat). There is no standard that covers bump caps, except for each manufacturer's specification. But the bump cap has its place in some work environments. When the impact hazard is represented by stationary objects, such as low slung pipes or catwalks, floor works, well-protected machinery, cleaning tight spaces, and not overhead operations, the severity of a potential injury is limited by the comparatively restricted movement of the worker's head. In these cases, the bump cap is sufficient protection (Figure 17-3).

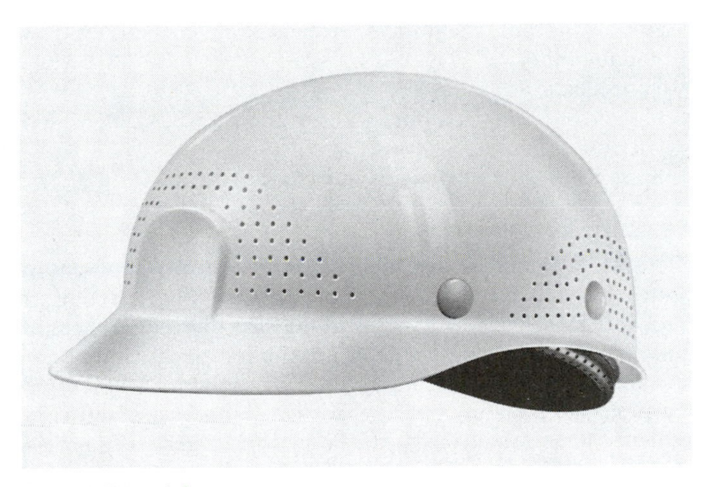

**Figure 17-3.** A bump cap.

Workers who utilize the bump cap and/or other protective headgear must be trained and supervised to ensure correct usage. Bump caps should never be used where ANSI-approved Class A, B, or C protective helmets are required.

### Hair protection

It is important that persons with long hair or beards who work around chains, belts, or other machines protect their hair from contact with moving parts. Besides the danger of direct contact with the machine, which may occur when they lean over, they are exposed to the hazard of having their hair lifted into moving belts or rolls that develop heavy charges of electricity. Since it is difficult to remove this hazard completely by mechanical means, people with long hair should be required to wear protective hair covering.

Hair nets, bandannas, and turbans are frequently unsatisfactory for hair protection because they do not cover the hair completely. Protective caps should completely cover the hair. If the wearer is exposed to sparks and hot metals, as in spot welding, the cap should be made of flame-resistant material. Disposable flame-proof caps are provided in some chemical plants.

No standards have been accepted for flexible caps, but they should be made of a durable fabric to withstand regular laundering and disinfecting.

In order to encourage its use, the cap should be as attractive as possible. Design should be simple, and the cap should be available in a variety of head sizes or should be adjustable to fit all wearers. It should be cool and lightweight. If dust protection is not required, the cap should be made of open weave material for better ventilation. Finally, it should have a visor, and it should be worn with visor in front.

After a suitable cap has been chosen, its use should be required and enforced. It is common practice among workers, for reasons of vanity, to wear the cap on the back of the head so that part of the hair over the forehead is exposed. Sometimes this practice can be discouraged by a realistic demonstration of what may happen when the hair comes in contact with a revolving spindle.

Convincing workers that caps preserve hair from the effects of dust, oils, and other shop conditions has resulted in some success in getting them to wear protective hair covering.

### Maintenance

Before each use, helmets should be inspected for cracks, even hairline cracks, signs of impact or rough treatment, and wear that might reduce the degree of safety originally provided. Prolonged exposure to ultraviolet rays (sunlight) and chemicals can shorten the life expectancy of thermoplastic helmets. Helmets that exhibit chalking, cracking, or less surface gloss should be discarded.

Protective helmets should not be stored or carried on the rear window shelf of a vehicle because sunlight and extreme heat may adversely affect the degree of protection. Another good reason not to carry hats there is that in case of an emergency stop or accident, the helmet might become a hazardous missile.

If the helmet has sustained an impact or shows any signs of damage, it should be discarded. Alterations of any sort impair the performance of the headgear.

At least every 30 days, protective helmets (in particular their sweatbands and cradles) should be washed in warm, soapy water or a suitable detergent solution recommended by the manufacturer and then rinsed thoroughly.

Before reissuing used helmets to other employees, the helmets should be scrubbed and disinfected. Solutions and powders are available which combine both cleaning and disinfecting. Helmets should be thoroughly rinsed with clean water and then dried. Keep the wash solution and rinse water temperature at approximately 140 F (60 C). Do not use steam, except on aluminum helmets (see Figure 17-4).

Removal of tar, paint, oil, and other materials may require the use of a solvent. Because some solvents can damage the shell, the helmet manufacturer should be consulted as to what solvent should be used.

Pay particular attention to the condition of the suspension because of the important part it plays in absorbing the shock of a blow. Look for loose or torn cradle straps, broken sewing lines, loose rivets, defective lugs, and other defects. Sweatbands are easily replaced. Disposable helmet liners made of plastic or paper are available for hats used by many people (such as visitors).

An adequate number of crowns, sweatbands, and cradles should be stocked as replacement parts. Some companies replace the complete suspension at least once a year.

### Color coded protective helmets

Many companies use color coded protective helmets to identify different working crews. Many colors are available; some colors are painted on and others have the color molded in. It is not recommended that paint be applied after manufacture because paint has solvents that may reduce the dielectric prop-

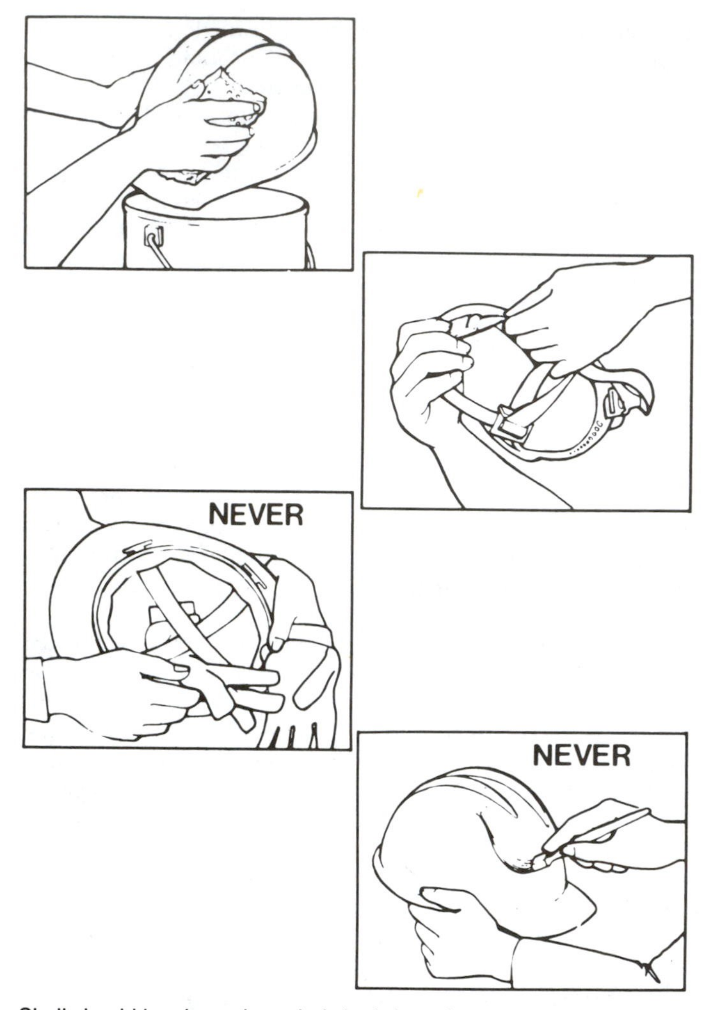


Shell should be cleaned regularly both for safety and appearance. Dirt or stains may hide hairline cracks, a reason to replace the helmet. Regular inspection of suspension system is important. Wearers should look closely for cracking, tearing or fraying of suspension materials. Never carry anything inside the helmet. A clearance must be maintained inside the helmet for the protection system to work. Never paint your helmet. Paint contains solvents which can make the shell brittle. Reflective tape is recommended for numbers or symbols.

**Figure 17-4.** Maintenance of protective headwear.

erties or affect the shell. Lighter colored hats are cooler to wear in the sun or under infrared energy sources.

## FACE AND EYE PROTECTION

Protection of the eyes and face from injury by physical and chemical agents or by radiation is vital in any occupational health and safety program. In fact, this type of protection has the widest use and the widest range of styles, models, and types.

The amount of money spent to acquire and fit eye protective devices is small when measured against the savings afforded by the protection given. For example, the purchase and fitting of a pair of impact-resistant spectacles may cost about $10; compensation costs for eye injury can exceed $5,000, according to a 1985 study in Ohio. The same study stated that over 70 percent of all eye injuries resulted from flying or falling objects. Contact with harmful substances, chemicals, etc., caused over 20 percent of injuries. Foreign bodies in the eye occurred in about 60 percent of the cases.

The eye and face protection standard, ANSI Z87.1–1979, *Practice for Occupational and Educational Eye and Face Protection,* is a fairly comprehensive document. It sets performance standards, including detailed tests, for a broad area of hazards—excluding only X rays, gamma rays, high-energy particulate radiations, lasers, and masers.

Besides general requirements, applying to "all occupations and educational processes," the standard provides requirements and limitations on the following:

- Rigid welding helmets
- Welding hand shields
- Nonrigid welding helmets
- Attachments and auxiliary equipment—lift fronts, chin rests, snoods, aprons, magnifiers, etc.
- Flammability
- Faceshields
- Goggles—eyecup (chipper's), dust and splash, welder's and cutter's
- Spectacles—metal, plastic, and combination.

### Selection of eyewear

Factors that should be considered in the selection of eyewear to protect from impact include the protection afforded, the comfort with which they can be worn, and the ease of keeping them in good repair. Styles now available are similar to regular eyewear and are cosmetically pleasing (Figure 17-5a and b). Flexible types are preferred by many because of their light weight and convenience. One drawback to the latter is they generally have a shorter wear-life than the sturdier frame and glass lens eyewear.

Proper eye protection devices should be selected, and their use should be impartially enforced so as to give maximum protection to the user for the degree of hazard involved (Figure 17-6). On certain jobs and in some locations 100 percent eye protection must be prescribed.

Faceshields are not recommended by ANSI Z87.1 as basic eye protection against impact. To get impact protection, faceshields must be used in combination with basic eye protection. Faceshields have their purpose, and are discussed later in this section.

Cup goggles should have cups large enough to protect the eye socket and to distribute the impact over a wide area of the facial bones. Cups should be flame-resistant, corrosion resistant, and nonirritating to the skin (Figure 17-7a and b).

If lenses are to be exposed to pitting from grinding wheel sparks, a transparent and durable coating may be applied to them. Welding lenses should be protected by a cover lens of glass or plastic.

Lenses must not have appreciable distortion or prism effect. ANSI Z87.1 limits the nonparallelism between the two faces to 1/16 prism diopter (4 minutes of arc) and both the refraction in any meridian and the difference in refraction between any two meridians to 1/16 diopter. Additional information on eye protection can be found in the National Safety Council book, *Fundamentals of Industrial Hygiene.*

### Contact lenses—rumor versus facts

For more than 20 years a rumor has persisted that welding or other electric flashes make contact lenses stick to the eyeball. This rumor has been proven false. Before basing an action on this

**Figure 17-5a.**

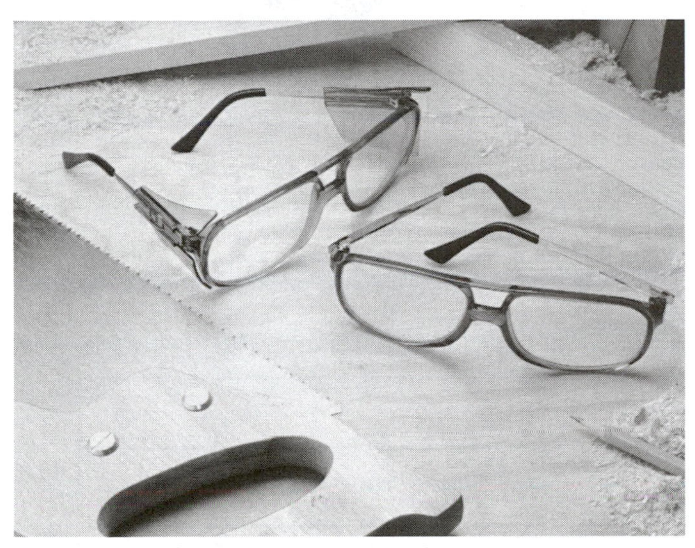

**Figure 17-5b.** Styles of safety glasses. Note side shields in Figure b. (Printed with permission from H. L. Bouton Co., Inc., Figure a, and Fendall Company, Figure b.)

rumor, it would be good to contact a reputable manufacturer, optometrist, or the National Safety Council for the current information about the subject.

Accident data and studies suggest that contact lens wearers do not appear to have problems when their eyes are properly protected in the workplace. The National Society to Prevent Blindness publishes the latest findings as a service to both business and health and safety professionals. Their purpose is to help contact lens wearers keep their eyes in good condition. The following guidelines and recommendations for contact lens use in industry are reprinted with permission from the National Society to Prevent Blindness:

> Contact lenses sometimes provide a superior means of visual rehabilitation for employees who have had a cataract removed from one or both eyes, who are highly nearsighted, or who have irregular astigmatism from corneal scars or keratoconus. Except for situations in which there exists significant risks of ocular injury, individuals may be allowed to wear contact lenses in the workplace. Generally speaking, contact lens wearers who have experienced long-term success with contacts can judge for themselves whether or not they will be able to wear contact lenses in their occupational work environment. However, contact lens wearers must conform to the prerogatives and directions of management regarding contact lens use. When the work environment entails exposures to chemicals, vapors, splashes, radiant or intense heat, molten metals, or a highly particulate atmosphere, contact lens use should be restricted accordingly. (Contact lens use considerations should be made on a case-by-case basis in conjunction with the guidelines of the Occupational Safety and Health Administration and the National Institute for Occupational Safety and Health.)

### Recommendations

The National Society to Prevent Blindness makes the following recommendations as a service to management who must direct contact lens use and the employees who must wear them:

1. Occupational safety eyewear meeting or exceeding ANSI Z87.1 standards should be worn at all times by individuals in designated areas.
2. Employees and visitors should be advised of defined areas where contacts are allowed.
3. At work stations where contacts are allowed, the type of eye protection required should be specified.
4. A specific written management policy on contact lens use should be developed with employee consultation and involvement.
5. Restrictions on contact lens wear do not apply to usual office or secretarial employees.
6. A directory should be developed which lists all employees who wear contacts. This list should be maintained in the plant medical facility for easy access to trained first-aid personnel. Foremen or supervisors should be informed of individual employees wearing contact lenses.
7. Medical and first-aid pesonnel should be trained in the proper procedures and equipment for removing both hard and soft contacts from conscious and unconscious workers.
8. Employees should be required to keep a spare pair of contacts and/or a pair of up-to-date prescription spectacles in their possession. This action will allow the employees to perform their job functions, should they damage or lose a lens while working.
9. Employees who wear contact lenses should be instructed to remove contacts immediately if redness of the eye, blurring of vision, or pain in the eye associated with contact lens use occurs.

### Comfort and fit

To be comfortable, eye-protective equipment must be properly fitted. Corrective spectacles should be fitted only by members of the ophthalmologic profession. An employee can be trained to fit, adjust, and maintain eye-protective equipment, however, and each employee can be taught the proper care of the device used.

To give the widest possible field of vision, goggles should be fitted as close to the eyes as possible, without bringing the eyelashes in contact with the lenses.

In areas where goggles or other types of eye protection are used extensively, goggle-cleaning stations can be conveniently located. Defogging materials and wiping tissues can be provided there along with a receptacle for discarding them. Various defogging materials are available. Before a selection is made, test to determine the most effective type for a specific application.

Use of sweatbands helps prevent eye irritation, aids visibility, and eliminates work interruptions for face mopping. Sweatbands

## Selection Chart for Eye and Face Protectors for Use in Industry, Schools, and Colleges

This Selection Chart offers general recommendations only. Final selection of eye and face protective devices is the responsibility of management and safety specialists. (For laser protection, refer to American National Standard for Safe Use of Lasers, ANSI Z136.1-1976.)

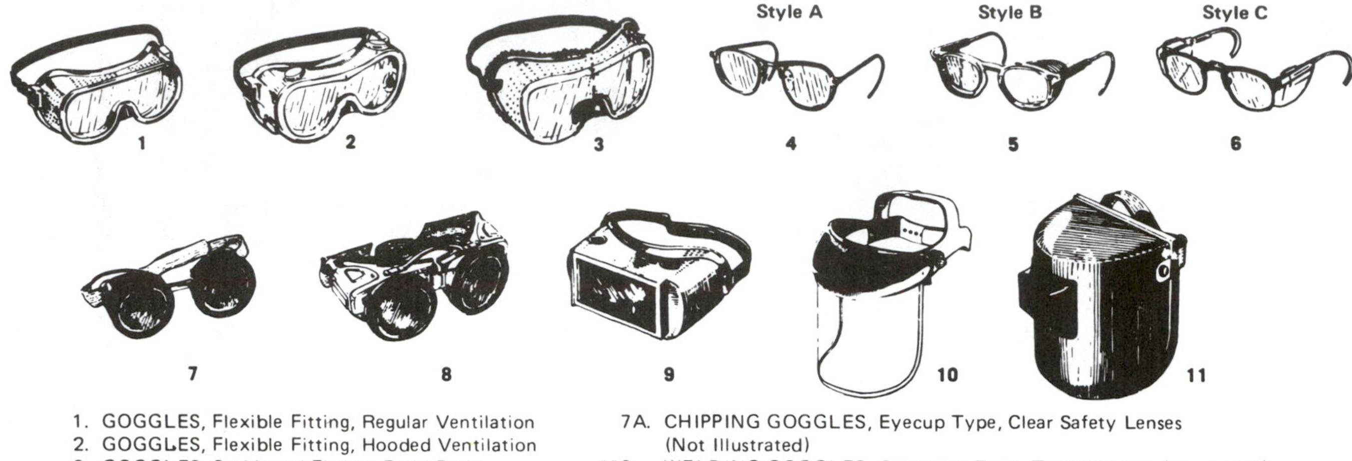


1. GOGGLES, Flexible Fitting, Regular Ventilation
2. GOGGLES, Flexible Fitting, Hooded Ventilation
3. GOGGLES, Cushioned Fitting, Rigid Body
*4. SPECTACLES, without Sideshields
5. SPECTACLES, Eyecup Type Sideshields
6. SPECTACLES, Semi-/Flat-Fold Sideshields
**7. WELDING GOGGLES, Eyecup Type, Tinted Lenses (Illustrated)
7A. CHIPPING GOGGLES, Eyecup Type, Clear Safety Lenses (Not Illustrated)
**8. WELDING GOGGLES, Coverspec Type, Tinted Lenses (Illustrated)
8A. CHIPPING GOGGLES, Coverspec Type, Clear Safety Lenses (Not Illustrated)
**9. WELDING GOGGLES, Coverspec Type, Tinted Plate Lens
10. FACE SHIELD, Plastic or Mesh Window (see caution note)
* 11. WELDING HELMET

*Non-sideshield spectacles are available for limited hazard use requiring only frontal protection.
**See Table A1, "Selection of Shade Numbers for Welding Filters," in Section A2 of the Appendix.

| APPLICATIONS | | |
|---|---|---|
| OPERATION | HAZARDS | PROTECTORS |
| ACETYLENE–BURNING<br>ACETYLENE–CUTTING<br>ACETYLENE–WELDING | SPARKS, HARMFUL RAYS, MOLTEN METAL, FLYING PARTICLES | 7, 8, 9 |
| CHEMICAL HANDLING | SPLASH, ACID BURNS, FUMES | 2 (For severe exposure add 10) |
| CHIPPING | FLYING PARTICLES | 1, 3, 4, 5, 6, 7A, 8A |
| ELECTRIC (ARC) WELDING | SPARKS, INTENSE RAYS, MOLTEN METAL | 11 (In combination with 4, 5, 6, in tinted lenses, advisable) |
| FURNACE OPERATIONS | GLARE, HEAT, MOLTEN METAL | 7, 8, 9 (For severe exposure add 10) |
| GRINDING–LIGHT | FLYING PARTICLES | 1, 3, 5, 6 (For severe exposure add 10) |
| GRINDING–HEAVY | FLYING PARTICLES | 1, 3, 7A, 8A (For severe exposure add 10) |
| LABORATORY | CHEMICAL SPLASH, GLASS BREAKAGE | 2 (10 when in combination with 5, 6) |
| MACHINING | FLYING PARTICLES | 1, 3, 5, 6 (For severe exposure add 10) |
| MOLTEN METALS | HEAT, GLARE, SPARKS, SPLASH | 7, 8 (10 in combination with 5, 6, in tinted lenses) |
| SPOT WELDING | FLYING PARTICLES, SPARKS | 1, 3, 4, 5, 6 (Tinted lenses advisable; for severe exposure add 10) |

CAUTION:
- Face shields alone do not provide adequate protection.
- Plastic lenses are advised for protection against molten metal splash.
- Contact lenses, of themselves, do not provide eye protection in the industrial sense and shall not be worn in a hazardous environment without appropriate covering safety eyewear.

**Figure 17-6.** Selection chart for eye and face protectors. See also Table 17-A. (Courtesy American National Standards Institute.)

Figure 17-7a. Two types of protective goggles. (a) Fog ban safety goggles with high impact resistant polycarbonate lenses. (Courtesy General Bandages, Inc.) (b) Dust goggles protect against fine dusts. (Courtesy INCO Safety Products.)

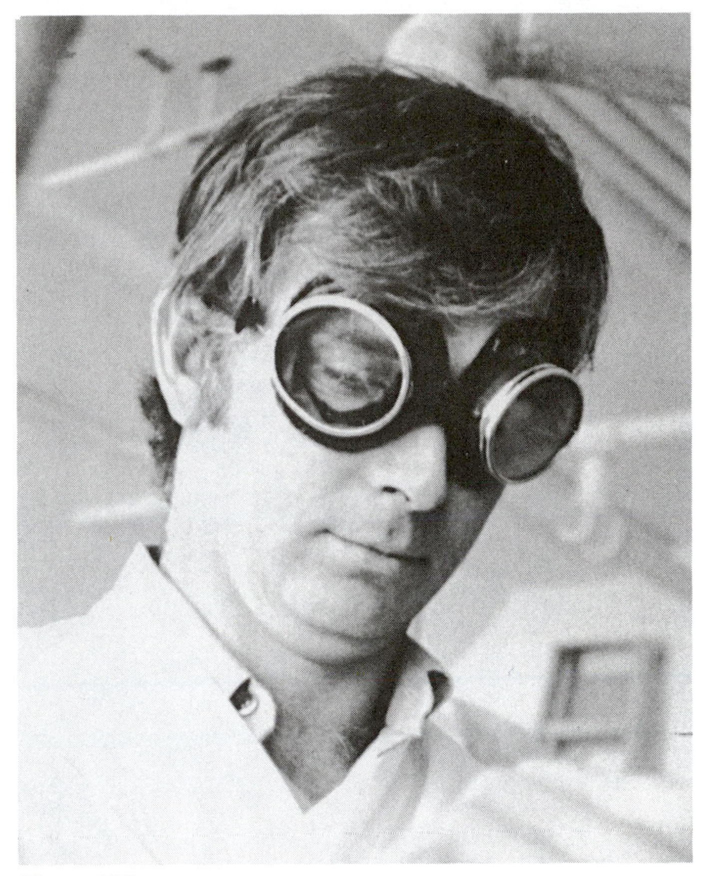

Figure 17-7b.

are usually made of a soft, light, highly absorbent cellulose sponge. An elastic band holds the sweatband in place on the wearer's forehead so that it does not interfere with glasses or goggles. Evaporation from the exposed surface produces a cooling effect which adds to comfort.

### Face protection

As a general rule, faceshields should be worn over suitable basic eye protection. Faceshields are available in a wide variety of types to protect the face and neck from flying particles, sprays of hazardous liquids, splashes of molten metal, and hot solutions (see Figures 17-8 and 17-2c, d, and e). In addition, they may be used to provide antiglare protection where required.

Three basic styles of faceshields include headgear without crown protectors, with crown protectors, and with crown and chin protectors. Each of the three is available with one of these replaceable window styles:

Clear transparent
Tinted transparent
Wire screen
Combination of plastic and screen
Fiber window with a filter plate mounting.

The materials used in faceshields should combine mechanical strength, light weight, nonirritation to skin, and the capability of withstanding frequent disinfecting operations. Metals should be noncorrosive and plastics should be of the slow-burning type. Only optical grade (clear or tinted) plastic, which is free from flaws or distortions, should be used for the windows. And plastic windows should not be used in welding operations unless they conform to the standards on transmittance of absorptive lenses, filter lenses, and plates.

On some jobs, such as the pouring of low-melting metals, protection against radiation is not necessary, but it is desirable to protect the head and face against splashes of metal.

A faceshield similar to an arc welder's faceshield, but made of wire screen (which provides much better ventilation than a solid shield), can be used. This shield is commonly used without a window because the plain wire will not fog under high temperature and high humidity. A metallized plastic shield that reflects a substantial percentage of heat has been developed for use where there is exposure to radiant heat.

### Acid hoods and chemical goggles

Head and face protection from splashes of acids, alkalis, or other hazardous liquids or chemicals may be provided in a variety of ways, depending upon the hazard. Good protection is given by a hood made of chemical-resistant material with a glass or plastic window (Figure 17-9). Some manufacturers provide a hood with replaceable inner and outer windows. In all cases, there should be a secure joint between the window and the hood materials.

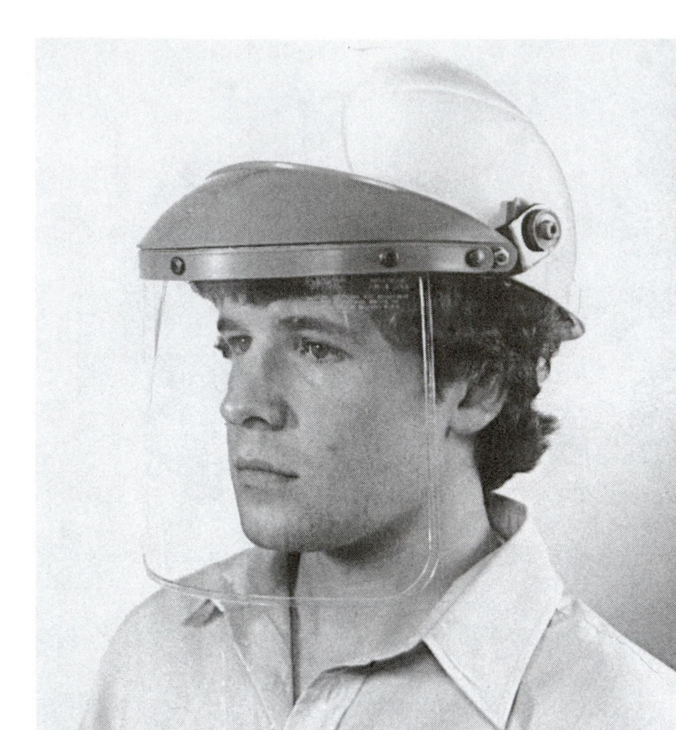

**Figure 17-8.** A faceshield protects eyes, face, and neck from flying particles and splashes.

Hoods are extremely hot to wear, but can be obtained with air lines for the wearer's comfort. When a hood is so supplied, the wearer should have a harness or belt like that on an air line respirator for support of the hose (Figure 17-17).

If protection is necessary only from limited direct splashes, the wearer can don a faceshield made of a material unaffected by the liquid or a flexible-fitting chemical goggle with baffled ventilation, if the eyes are not exposed to irritating vapor.

For severe exposures a faceshield should be worn in connection with the flexible-fitting chemical goggles.

Faceshields should be shaped to cover the whole face. They should be supported from a headband or harness, so they can be tipped back and clear the face easily. Any shield should be easily removed in case it becomes wet with corrosive liquid.

If goggles worn under the shield are nonventilated for protection against vapor as well as against splashing, they should also be nonfogging (described earlier).

### Laser beam protection

No one type of glass or plastic offers protection from all laser wavelengths. Consequently, most laser-using firms don't depend on safety glasses to protect an employee's eyes from laser burns. Some point out that laser goggles or glasses might give a false sense of security, tempting the wearer to unnecessary exposures.

Nevertheless, researchers and laser technicians do frequently need eye protection. Both spectacles and goggles are available—and glass or plastic for protection against nearly all the known lasers can be had on special order from eyewear manufacturers. Typically, the eyewear will have maximum attenuation at a specific laser wavelength—with protection falling off rather rapidly at other wavelengths.

Laser protective goggles or spectacles or an "anti-laser eyeshield" attenuate the He-Ne laser light (wavelength 6328 A) by factors of 10 (O.D. = 1), 100 (O.D. = 2), 1000 (O.D. = 3), or more. An optical density (O.D.) of three of four still renders the beam visible in bright sunlight. The goggle style of protective eyewear used in the laboratory is often unsuitable in the field because of fogging.

The American Conference of Governmental Industrial Hygienists cautions that laser safety glasses or goggles should be evaluated periodically to make sure that maintenance of adequate optical density is kept at the desired laser wavelength. There should be assurance that laser glasses or goggles designed for protection from specific laser wavelengths are not mistakenly used with different wavelengths of laser radiation. The optical density values and wavelengths should be shown on the eyewear. Eyewear storage shelves can also carry this notation.

Laser safety glasses or goggles exposed to very intense energy or power density levels may lose effectiveness and should be discarded.

Technical details, uses, hazards, and exposure criteria for lasers are given in *Fundamentals of Industrial Hygiene,* 3rd ed., National Safety Council, Chicago, 1988. Also see ANSI Z136.1-1986, *The Safe Use of Lasers.*

### Eye protection for welding

In addition to damage from physical and chemical agents, the eyes are subject to the effects of radiant energies. Ultraviolet, visible, and infrared bands of the spectrum are all able to produce harmful effects upon the eyes, and therefore require special attention.

Ultraviolet rays can produce cumulative destructive changes in the structure of the cornea and lens of the eye. Short exposures of intense ultraviolet radiation or prolonged exposures to ultraviolet radiations of low intensity will produce painful, but ordinarily self-repairing corneal damage.

Radiations in the visible light band, if too intense, can cause eyestrain and headache, and can destroy the tissue of the retina.

Infrared radiations transmit large amounts of heat energy to the eye, causing discomfort. The damage produced is superficial.

The filtering properties of filter lenses have been established by the National Bureau of Standards. The percentage transmittance of radiant energies in the three bands—ultraviolet, visible, and infrared—is established for 16 different filter lens shades (Table 17-A). Both absorptive and filter lenses are available in polycarbonate.

Welding processes (see Chapter 13, Welding and Cutting in the *Engineering and Technology* volume) emit radiations in three spectral bands. Depending upon the flux used and the size and temperature of the pool of melted metal, welding processes will emit more or less visible and infrared radiation—the proportion of the energy emitted in the visible range increases as the temperature rises. At least one manufacturer produces an aluminized cover for the usual black welding helmet. Its purpose is to reduce infrared absorption and the resulting heat stress to the wearer.

All welding presents problems, mostly in the control of infrared and visible radiations. Heavy gas welding and cutting oper-

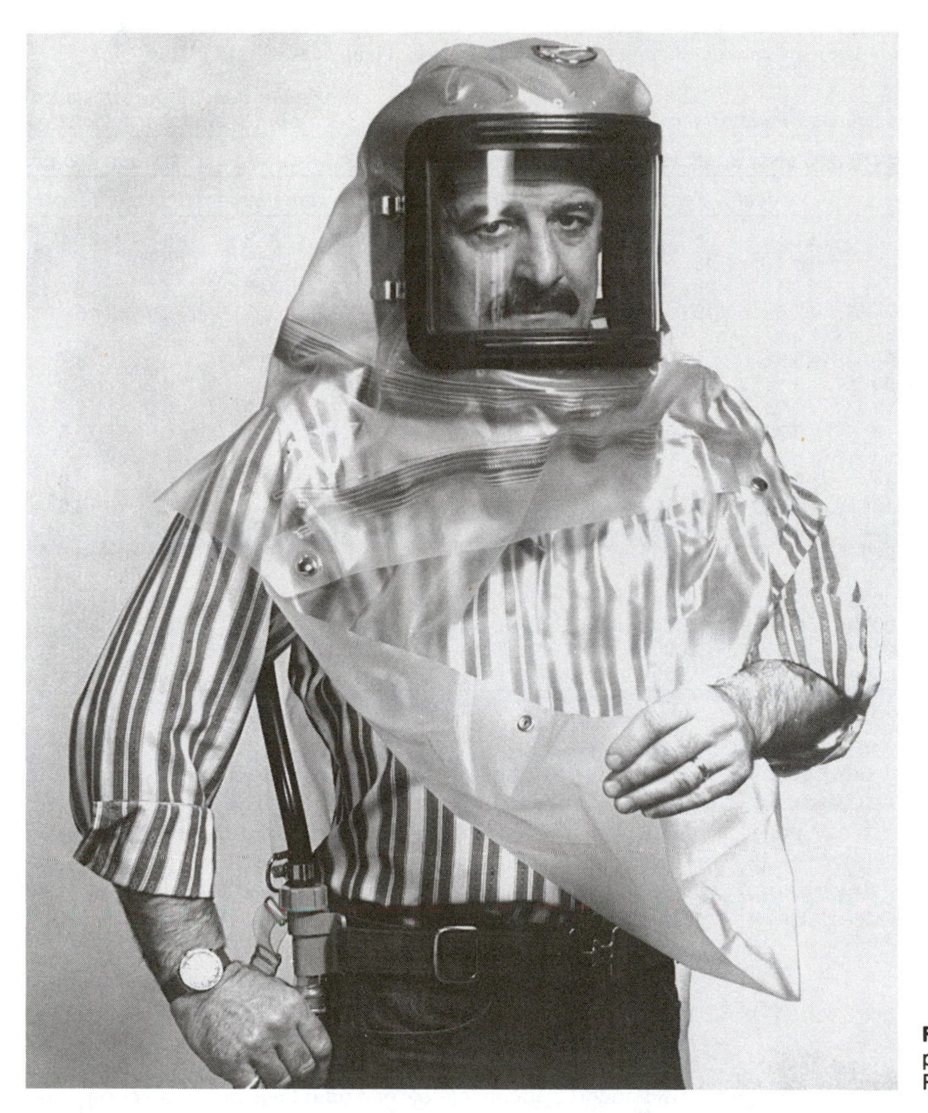

**Figure 17-9.** An air-supplied hood with a plastic window. (Courtesy AO Safety Products.)

ations, and arc cutting and welding exceeding 30 amperes, present additional problems in control of ultraviolet. Welding helmets must be used to provide head and face protection (Figures 17-2e and 17-6).

Welders may choose the shade of lenses they prefer within one or two shade numbers. Following are shades commonly used:

Shades No. 1.5 to No. 3.0 are intended for glare from snow, ice, and reflecting surfaces; and for stray flashes and reflected radiation from cutting and welding operations in the immediate vicinity (for goggles or spectacles with side shields worn under helmets in arc welding operations, particulary gas-shielded arc welding operations).

Shade No. 4, the same uses as shades 1.5 to 3.0, but for greater radiation intensity.

For welding, cutting, brazing, or soldering operations, use the guide for the selection of proper shade numbers of filter lenses or windows in Chapter 13 of the *Engineering and Technology* volume. (Recommendations are also in ANSI Z87.1, *Practice for Occupational and Educational Eye and Face Protection.*)

To protect the filter lenses against pitting, they should be worn with a replaceable plastic or glass cover plate.

Eye protection having mild filter shade lenses or polarizing lenses and opaque side shields are adequate for protection against glare only. For conditions where hot metal may spatter and where visible glare must be reduced, a plastic faceshield worn over mild filter shade spectacles with opaque side shields should be specified.

It is permissible to combine the shade of the plate in a welder's helmet with that of the shade of the goggle worn underneath to produce the desired total shade. This procedure has the added advantage of protecting the eyes from other welding operations or from an accidental arc when the helmet is raised.

To protect against ultraviolet and infrared radiation as well as against visible glare in inspection operations, protective lenses should be installed in a hand shield or welder's helmet. The shield should be made of a nonflammable material, which is opaque to dangerous radiation and a poor conductor of heat. A metal shield is not desirable, because it heats under infrared radiation.

Some tinted lenses used in special work afford no protection from infrared and ultraviolet radiations. For instance, most melters' blue glass used in open-hearth furnaces and the lenses used at Bessemer converters afford no protection against either type of harmful radiation. Probably no harm will come from

**Table 17-A.** Transmittances and Tolerances in Transmittance of Various Shades of Absorptive Lenses, Filter Lenses, and Plates

| Shade Number | Optical Density | | | Luminous Transmittance | | | Maximum Infrared Transmittance | Maximum Spectral Transmittance in the Ultraviolet and Violet | | | |
|---|---|---|---|---|---|---|---|---|---|---|---|
| | Maximum | Standard | Minimum | Maximum | Standard | Minimum | | 313 nm | 334 nm | 365 nm | 405 nm |
| | | | | Percent | Percent | Percent | Percent | Percent | Percent | Percent | Percent |
| 1.5 | 0.26 | 0.214 | 0.17 | 67 | 61.5 | 55 | 25 | 0.2 | 0.8 | 25 | 65 |
| 1.7 | 0.36 | 0.300 | 0.26 | 55 | 50.1 | 43 | 20 | 0.2 | 0.7 | 20 | 50 |
| 2.0 | 0.54 | 0.429 | 0.36 | 43 | 37.3 | 29 | 15 | 0.2 | 0.5 | 14 | 35 |
| 2.5 | 0.75 | 0.643 | 0.54 | 29 | 22.8 | 18.0 | 12 | 0.2 | 0.3 | 5 | 15 |
| 3.0 | 1.07 | 0.857 | 0.75 | 18.0 | 13.9 | 8.50 | 9.0 | 0.2 | 0.2 | 0.5 | 6 |
| 4.0 | 1.50 | 1.286 | 1.07 | 8.50 | 5.18 | 3.16 | 5.0 | 0.2 | 0.2 | 0.5 | 1.0 |
| 5.0 | 1.93 | 1.714 | 1.50 | 3.16 | 1.93 | 1.18 | 2.5 | 0.2 | 0.2 | 0.2 | 0.5 |
| 6.0 | 2.36 | 2.143 | 1.93 | 1.18 | 0.72 | 0.44 | 1.5 | 0.1 | 0.1 | 0.1 | 0.5 |
| 7.0 | 2.79 | 2.571 | 2.36 | 0.44 | 0.27 | 0.164 | 1.3 | 0.1 | 0.1 | 0.1 | 0.5 |
| 8.0 | 3.21 | 3.000 | 2.79 | 0.164 | 0.100 | 0.061 | 1.0 | 0.1 | 0.1 | 0.1 | 0.5 |
| 9.0 | 3.64 | 3.429 | 3.21 | 0.061 | 0.037 | 0.023 | 0.8 | 0.1 | 0.1 | 0.1 | 0.5 |
| 10.0 | 4.07 | 3.854 | 3.64 | 0.023 | 0.0139 | 0.0085 | 0.6 | 0.1 | 0.1 | 0.1 | 0.5 |
| 11.0 | 4.50 | 4.286 | 4.07 | 0.0085 | 0.0052 | 0.0032 | 0.5 | 0.05 | 0.05 | 0.05 | 0.1 |
| 12.0 | 4.93 | 4.714 | 4.50 | 0.0032 | 0.0019 | 0.0012 | 0.5 | 0.05 | 0.05 | 0.05 | 0.1 |
| 13.0 | 5.36 | 5.143 | 4.93 | 0.0012 | 0.00072 | 0.00044 | 0.4 | 0.05 | 0.05 | 0.05 | 0.1 |
| 14.0 | 5.79 | 5.571 | 5.36 | 0.00044 | 0.00027 | 0.00016 | 0.3 | 0.05 | 0.05 | 0.05 | 0.1 |

(Reprinted with permission from ANSI standard Z87.1-1979.)

continued use of these lenses if the exposures are of short duration. However, new personnel learning these flame reading skills should be provided with lenses that protect in these two portions of the spectrum.

The chemical composition of the lens rather than its color provides the filtering effect; this factor must be considered when selecting a filtering lens.

## HEARING PROTECTION

Medical professions have long been aware of the problem of noise-induced hearing loss in industry. It is incontrovertible that certain levels of noise can produce irreversible damage to an individual's hearing. Usually, such loss is in the high-frequency range—about 4,000–6,000 Hz. If exposure to excessive noise levels continues, the individual eventually will also experience deterioration in hearing at the lower frequency levels.

A number of factors contribute to hearing loss or hearing damage in industrial settings. Foremost, of course, is the loudness of the sound in an individual's work area. But the duration of exposure is also crucial. Working daily in steady noise of more than 85 dB for full 8-hour shifts is considered hazardous by most experts. However, individual susceptibility varies. Some people will be harmed by noise well below the 85 dB level; others can tolerate much higher levels with no damage whatsoever.

Some state regulations and portions of the OSHA legislation require audiometric testing of employees exposed to excessive noise. It is a requirement that an audiometric testing program be initiated and maintained for all U.S. employees who are exposed to noise levels in excess of 85 dBA for 8-hour time weighted average. In fact, it is a good idea to test and maintain a record on all employees. A properly carried out audiometric testing program may determine whether the hearing protective devices worn by the employees are in fact protecting their hearing from noise damage.

Before requiring any employee to wear hearing protection, the noise in the workplace needs to be surveyed and evaluated. Measuring the physical characteristics of noise in occupational environments serves to: (1) provide the physical evidence of individual exposures; (2) identify areas where controls need to be established; (3) help to prioritize noise-control and noise-reduction efforts, including administrative controls; (4) document exposures in the work environment for medical–legal purposes; (5) establish documentation for state, federal, or insurance compliance requirements; (6) provide a basis for cause–effect relationships between accuracy of auditory-risk assessments, adequacy of personal hearing protection, and changes, if any, in the status of hearing of those included in the program; and (7) provide insights for improving education and motivation indoctrinations among workers, supervisors, and managers.

When translating noise measurements into exposure estimates, remember that there is no precise safe–unsafe line of differentiation. Any unprotected encounters with steady-state or intermittent noises that exceed about 85 dBA or with impulse or impact noises that exceed about 120 dB (peak) are suspect of overtaxing the mechanisms of hearing. (See Gasaway in the References.)

When noise measurement is completed, and other possible noise control efforts are unsuccessful, then the need for hearing protection is clearly established. For explanation of noise measurement, evaluation and control, see the Council's *Fundamentals of Industrial Hygiene.*

### Selection of hearing protection

**Attenuation.** Hearing protection devices have different abilities to release (attenuate) noise levels, and also reflect differences at different frequencies of noise. Absolutely necessary for a good protection program is an accurate knowledge of the noise levels (and frequencies) that must be protected against. From the data obtained in the noise survey described above, the proper selection of hearing protection devices can be made.

An aid in the selection of hearing protection is the Environmental Protection Act requirement that calls for all protectors to have a label indicating its Noise Reduction Rating (NRR). This single number provides an estimate of effectiveness, and can be generally subtracted from a dBA value in the workplace. This value indicates the noise level being received in the workers' ear. Some caution must be observed in doing so because (1) NRR is derived in an ideal laboratory test, (2) wearing the device on the job will be less ideal, and (3) noise frequencies and level will not be equal across the spectrum and will change other factors. Thus, it is necessary that the NRR (or any other similar value) be used at something less than full value. The real world value may be as little as 50 percent of the published NRR value.

**Muff ear protectors.** When selecting a hearing-protection device, consider the work area in which the employee must use it. For example, a large-volume earmuff would not be practical for an individual who must work in confined areas with very little head clearance. In such instances, a very small or flat earcup or insert protector would be more desirable.

When using muff protectors in special-hazard areas (such as power-generating stations where there are electrical hazards), it may be desirable to use nonconductive suspension systems in connection with muff protectors. Also, if the wearing of other personal protective equipment, such as safety hats or safety spectacles, must be considered, the degree of hearing protection required must not be compromised. Furthermore, the efficiency of muff protectors is reduced when they are worn over the frames of eye-protective devices. The reduction in attenuation will depend upon the type of glasses being worn as well as the head configuration of the individual wearer. When eye protection is required, it is recommended that cable-type temples be used. Cable temples will give the smallest possible opening between the seal and the head.

Other considerations when selecting a hearing-protective device include the frequency of exposure to excess noise (once a day, once a week, or very infrequently). For such cases, possibly an insert or plug device will satisfy the requirement. If the noise exposure is relatively frequent and the employee must wear the protective device for an extended period of time, the muff protector might be preferable. If the noise exposures are intermittent, the muff protector is probably more desirable since it is somewhat more difficult to remove and reinsert earplugs.

In determining the suitability of a hearing protective device for a given application, the manufacturer's reported test data must be examined carefully. It is necessary to correlate that information to the specific noise exposure involved. The attenuation characteristics of the individual hearing protective devices are compared at different frequencies.

**Formable aural inserts** fit all ears. Many of the formable types are designed to be used once, then thrown away. Materials from which these disposable plugs are made include very fine glass fiber, wax-impregnated cotton, and expandable plastic foam.

Variations from one model to another will provide varying degrees of noise reduction. Manufacturers supply attenuation data for their products so the health and safety professional can evaluate their effectiveness for use in a given situation. The ANSI Standard S3.19–R1979, *Method for the Measurement of Real-Ear Protection of Hearing Protectors and Physical Attenuation of Earmuffs,* describes how to determine the efficiency of a specific device for a given noise exposure.

### Types

Hearing protectors in general use fall into four types: enclosure (helmets), aural (ear insert), superaural (canal caps), and circumaural (ear muffs) (see Figure 17-10).

Prior to issuing any ear insert it is important to take certain measures: (1) the ear canals should be examined. Certain diseases may not allow use of earplugs. (2) The employees must be taught proper insertation techniques. (3) They must be taught proper sanitation and checking techniques.

**Enclosure.** The enclosure hearing protector completely surrounds the head, such as the astronaut helmet. Attenuation of sound is achieved through the acoustical properties of the helmet. Additional attenuation can be achieved by wearing inserts with the enclosure helmet. Cost, and heat, as well as bulk, normally preclude general use of the enclosure hearing protector, but there may be specific needs for it.

**Aural insert.** Commonly called inserts or earplugs, the aural insert is generally inexpensive and has a limited service life. The plug or insert is generally classified in three broad categories: (1) formable, (2) custom molded, and (3) molded.

Custom-molded hearing protectors, as the name indicates, are made for a specific individual. A prepared mixture is carefully placed in the person's outer ear with a small portion of it in the ear canal; as the material sets, it takes the shape of the individual's ear and external ear canal. Only trained personnel should attempt the process of forming these hearing protectors.

Molded (or premolded) aural inserts are usually made from a soft silicone rubber or plastic. The most important aspect of this type is to get a good fit. The hearing protector must fit snugly in order to be effective. For some persons, this may cause some discomfort because of the irregular shape of the ear.

**Superaural.** The superaural or canal cap hearing protector depends on sealing the external edge of the ear canal in order to achieve sound reduction. A soft rubber-like material is used to make the cap. The caps are held in place against the edges of the ear canal by a spring band or a head suspension.

**Circumaural.** Cup (or earmuff) devices cover the external ear to provide an acoustic barrier. The attenuation provided by earmuffs varies widely due to differences in size, shape, seal material, shell mass, and type of suspension. Head size and shape also influence the attenuation characteristics of these protectors. Temple pieces of safety spectacles can cause noise leakage; to minimize this leakage employees can use the cable temple pieces. The type of cushion used between the shell and the head also has a great deal to do with attenuation efficiency. Liquid or grease-filled cushions give better noise suppression than plastic or foam rubber types, but may present leakage problems.

## FALL PROTECTION SYSTEMS

Lifelines, safety belts, and associated equipment are often used in construction and general industry when fall hazards cannot be eliminated by railings, floors, or other means. Safety belts or fall protection body support are generic in nature. The term body belt is intended for a strap worn snugly around the waist.

**Figure 17-10.** A wide variety of hearing protection devices including ear muffs, ear plugs, and Swedish wool. Properly worn and cared for, these protective devices will prevent hearing loss.

A body harness is comprised of straps worn around parts other than the soft tissue areas of the body. Users should refer to specific services and uses and to both the OSHA General Industry Regulations and appropriate ANSI Standards. Manufacturers offer specific information or applications of equipment. (Linemen's body belts, pole straps, and ladder-climbing protection belts are not covered in this discussion.)

### Personal lifelines

Personal lifeline systems are usually rope systems that provide flexibility for worker freedom of movement, yet can arrest a fall and help absorb the shock. These systems always have some type of belt or harness that is worn around the body to which a lanyard or rope-grabbing or fall-arrest device is attached.

Body belts should be used only where very short free falls of less than 2 ft (0.6 m) are anticipated. The body belt D-ring should be arranged at the back of the worker. A body harness should be used where longer free falls, up to 6 ft (1.8 m) are anticipated. A harness can spread the shock load over the shoulders, thighs, and seat area.

A combination body harness has pelvic and chest belts that work together to distribute impact force evenly and guard against slipping out of the harness during the fall. The chest belt is worn loosely to allow freedom of breathing and movement; the pelvic belt should be adjusted to fit snugly. The two belts are attached by a strong web band at back, long enough to keep them independent of each other during normal work, but short enough for the pelvic belt to catch the chest belt in case of fall.

The lifeline is defined as a vertical line suspended from a fixed anchorage or a line between two separate horizontal fixed anchorages, independent of the work surface, to which the fall-arrest device is secured. The horizontal lifeline must be capable of supporting a dead weight of 5,400 lb (2,450 kg) per person, applied at the center of the lifeline. Horizontal lifelines are constructed of wire rope at least ½ in. (12.5 mm) in diameter, and attached to at least two fixed anchorages. (See also Chapter 5, Ropes, Chains, and Slings, in the *Engineering and Technology* volume of this Manual.)

Retracting lifelines are portable self-contained devices that are attached to a fixed anchorage point above the work area. The lifeline extends from the device and is connected to the user's body support. A tension is maintained up to the maximum extension, and the lifeline freely retracts when the user moves closer to the anchorage and extends as the worker moves away from it. The result is a short free fall or a controlled descent.

The lanyard is a short piece of flexible line up to 6 ft (1.8 m) in length which can be used to secure the wearer of a body support to a lifeline, or fixed anchorage. Lanyards may be constructed of any fibrous or metallic material. A lanyard should have as little slack as possible to limit free fall distance. Shock-absorber lanyards are designed to absorb up to 80 percent of the stopping force of a regular lanyard and should be used whenever possible. Lanyards can be spliced permanently to the safety belt.

Be sure that the lanyard is attached to a fixed anchorage by means that will not reduce its required strength. A bowline knot will reduce the strength of a rope lanyard by up to 50 percent. (See Table 5–D in Chapter 5 of the *Engineering and Technology* volume of this Manual.) The free ends of lanyards must also be seared or otherwise tightened. In addition, wire rope or rope-covered wire lanyards must not be used where impact loads or electrical hazards are anticipated.

Lanyards and associated hardware also require attention as to construction and use. Lanyards must not be lengthened by connecting two snap hooks together. Snap hooks and D-rings must fit together, to minimize the chance of accidental disengagement. Locking snaps are preferred. Snap hooks must be kept in good or "as new" condition.

### Body support

Body support is used for securing, suspending, or retrieving a worker in or from a hazardous work area. This device can be secured around the waist, chest, or a large portion (torso, thighs, chest, buttocks, and shoulders) of the body to distribute the stopping force.

Strength members of belts may be made of any material, except leather, that in turn will meet the minimum performance test. Hardware, buckles, hooks, and rings are also subject to determined specifications. The buckle must be the quick release type and must withstand a tensile test of 4,000 lb (1,815 kg) without failure.

Safety belts, harnesses, and lanyards are classified according to their intended use. (1) Body belts (work belts) are used to restrain a person in a hazardous work position and to reduce the probability of falls (Figure 17-11a and b). (2) Chest harnesses are used where there are only limited fall hazards (no vertical free-fall hazard) and for retrieval purposes, such as removal of a person from a tank or bin (see Figure 17-12). (3) Body harnesses are used to arrest the most severe free falls (see Figure 17-13). (4) Suspension belts are independent work supports used to suspend or support the worker.

Belts and lanyards should always be scrutinized for weak points that may cause the piece to fail under impact. The assemblies must be inspected according to the manufacturer's recommendations not less than twice annually. The date of each inspection is then recorded on an inspection tag that should be permanently attached to the belt. Chest harness and suspension belts must not be used for stopping falls and do not have to meet impact requirements. Belts and lanyards that have been subjected to impact loading shall be removed from service and destroyed.

### Window cleaner's belts

Employers of window washing personnel should provide safety equipment and devices conforming with the requirements of ANSI A39.1 and A120.1. Specified applications should be referenced to these standards.

A window cleaner's belt is subject to a moderate static load most of the time it is in use, as the worker leans back in the belt and is held in position by it while working. It will, however, be subjected to a severe loading in case of a fall with only the terminal of the belt attached to the window anchor. This is the most common kind of fall in window cleaning and also the most hazardous one with this type of belt, because the worker will slide to the end of the safety line, transmitting impact to the single window anchor which may be broken off or pulled out, particularly in older buildings.

The amount of impact force developed in arresting a fall depends chiefly on three elements: the weight of the person, the distance of the fall, and the suddenness of stopping. Of these three elements, the suddenness of stopping is of far greater importance than the other two factors.

The maximum possible fall in a window cleaner's belt is limited to the length of the safety rope or strap, 8 ft (2.5 m). In other types of belts, the fall may be much longer than the rope length. For example, the maximum fall in a construction worker's belt may be as much as twice the length of lifeline in

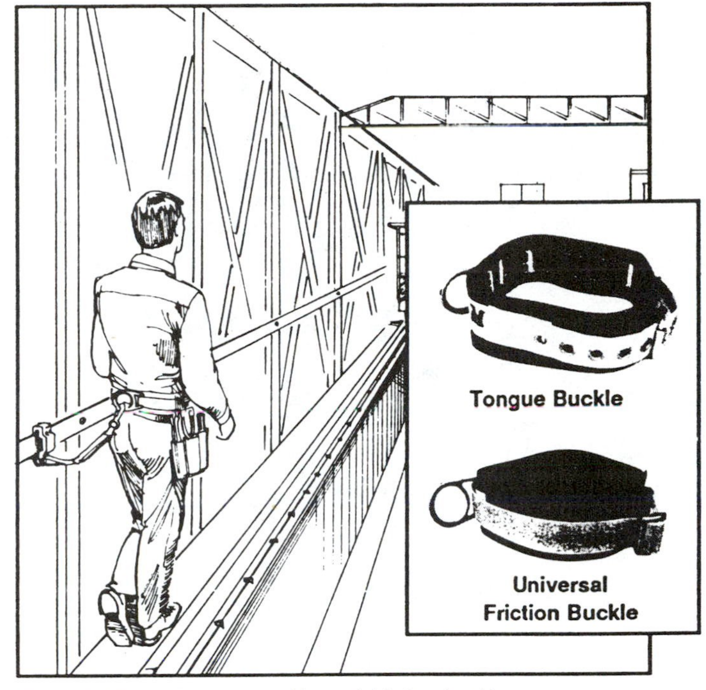


**Figure 17-11a.** Tongue buckle and friction buckle showing pad (right) and belt with lanyard (left). This belt design allows the wearer to use it as a tool belt or as a restraint in a hazardous work position.

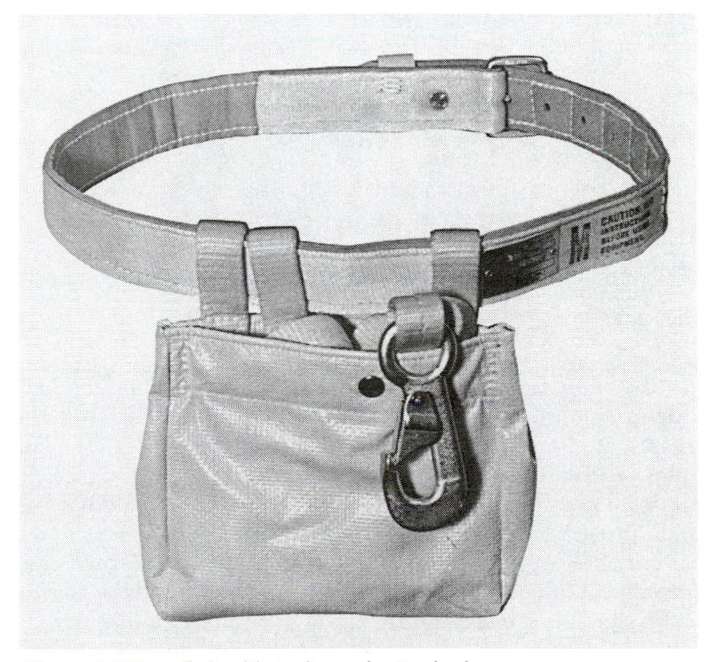

**Figure 17-11b.** Belt with tool pouch attached.

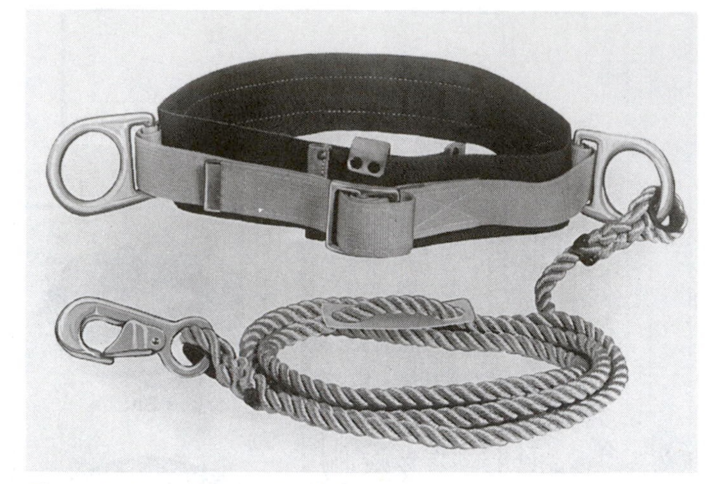

Figure 17-11c. Nylon belt with lanyard.

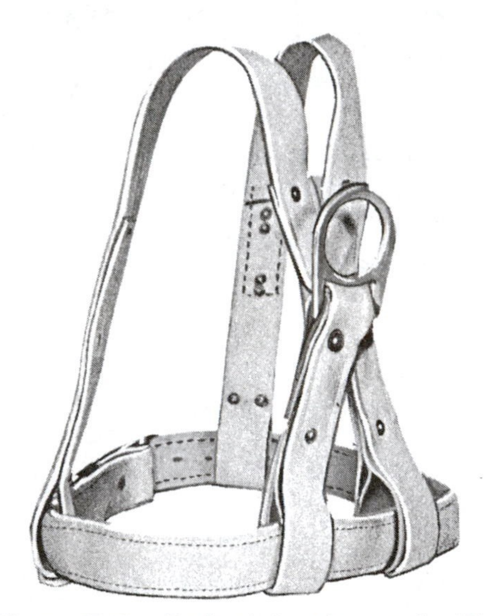

Figure 17-12. Retrieval chest harness for lifting workers from enclosed places such as quarries, tanks, and bins.

case of falls from above lifeline anchorage. In all cases where the lifeline is attached to the support on which the worker is standing, the free fall will be usually the full length of the lifeline plus the distance from the worker's feet to the D-ring of the belt. If at all possible, the wearer of the safety belt should not "tie off" (attach to an anchorage point) below waist level.

For exterior maintenance use, as in a window cleaner's rig, separate lines must be employed for the loadline and the lifeline. The loadline, which supports the worker, must be shorter than the lifeline so that the former will be taut and the lifeline hangs under its own weight. Thus, wear and tear should occur to the loadline, and the lifeline should retain its full strength for use in stopping an accidental fall if things go wrong.

The lifeline system should incorporate a shock absorber which will reduce the loading in the event of a fall, thus relieving strain on the building anchor, waist belt, and person.

Nylon rope lifelines may not be suitable because their stretch may cause collision with the ground during a fall at a lower level. Means for rescue of suspended workers should be developed.

### Construction of body support

Belts should be scrutinized for weak points which might cause the belt to fail under a heavy impact. For instance, the waist belt should always be inserted through the D-rings or other attaching devices and never riveted to them in such a way that the D-ring or lifeline could be separated from the belt through the failure of the rivets.

Buckles should hold securely without slippage or other failure, and this holding power should be achieved by only a single insertion of the strap through the buckle in the normal or natural way. If the buckle is one that requires that the free end of the webbing be turned back and inserted again through the buckle in order to achieve its full holding power, the management, through safety instruction and inspection, must make every effort to see that the workers always use this method of fastening their belts. Otherwise, the belt may slip under very minor strains. Where there is danger of falling into water or of being trapped by fire, a quick-release buckle should be used.

In general, synthetic webbing is the only material used for safety belts which may be called upon to take impact loads. Webbing has three to four times as much resistance to impact loading as leather of the same size. Leather belts are used for positioning. Webbing belts can use friction buckles which avoid loss of strength at buckle holes. It also provides a snugger fit of the belt to the body.

Cotton webbing is not seriously affected by dryness, moisture, or heat, short of actual scorching or burning. However, to avoid mildew, untreated cotton webbing should not be subjected to moisture for long periods. It should also be protected from corrosive chemicals unless the material has been prepared for such conditions.

If the only purpose of the belt is to stop a worker in case of a fall hazard, then a waist belt with a single D-ring may be satisfactory. This single D-ring may be attached for positioning at any point on the belt, for the belt can be turned to different positions after being buckled on the wearer. The preference is to attach the D-ring to the back of the belt. Where possible, a body harness is preferable for absorbing impact and keeping a person upright in case of a free fall. If the worker is suspended after a fall, harnesses are better.

If a belt is to furnish support to a person while working, it should have adequate means for such support. A belt for a person who will lean back in it while working on a sloping roof or hillside should have a wide back pad and two D-rings, one on each side of the belt, to which a suitable lanyard sliding positioning device can be attached and which, in turn, is connected to a suitable anchorage.

If the belt is used to support the person's entire weight vertically while working, as in being raised and lowered along the wall of a building, a boatswain's suspension harness may be preferred. In this type of belt, one strap is used as a seat. Sometimes a board is added to make it more comfortable. Leg straps and a strap around the worker's waist prevent falling out of the seat.

An industrial body support used to hoist a person out of a tank is usually subjected only to a static load. Such a belt should be either the shoulder harness, chest-waist, or parachute type with the D-ring so placed as to hold the person in a relatively erect position, especially for narrow openings.

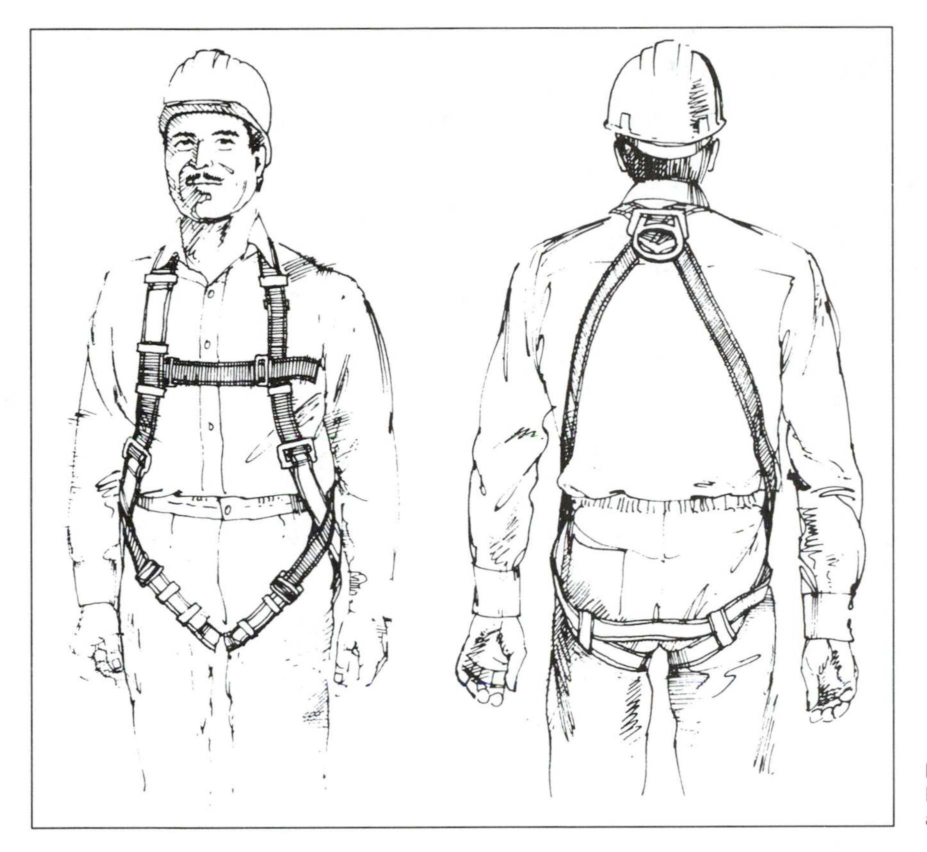

**Figure 17-13.** Full-support body harness. This body harness is ideal for workers on elevated sites. During a fall it distributes the fall impact over the body.

### Belt care

The belt or harness should be washed with warm water and domestic soap. After a rinse in clean, warm water, it should be dried at room temperature and stored away from strong sunlight or ultraviolet glare.

Leather belts should not be exposed to excessive heat, such as from a radiator, because they may be permanently damaged by a temperature as low as 150 F (66 C). Any heat painful to a person will damage leather.

Cotton or linen webbing belts should be washed in soapy water, rinsed, and dried by moderate heat. They are not damaged by temperatures up to 212 F (100 C). If a belt is to be subjected to unusual conditions of use, the manufacturer should be consulted as to its care. Synthetic fiber belts should not be exposed to excessive heat that might soften or melt the fibers, or to chemicals that might affect the composition of the fiber. Changes in appearance, color, etc. need to be carefully evaluated.

### Inspection and testing

Wearers of full protection belts should inspect them before each use. At least every one to three months they should be examined by a trained inspector, and a record of the examination should be kept.

Belt hardware should be examined and worn parts should be replaced. Each belt rivet should be examined to be certain that it is secure. No rivet should be used in a web belt which may be subjected to impact loading.

Belts in service should not be tested. Any service test to prove whether the belt could take the maximum impact loading which might be required of it would very likely so damage the belt as to render it unsafe for use. Therefore, only sample belts and worn or doubtful belts should be tested, and these should always be tested to destruction to determine their safety. They should be kept as samples and used only to help judge the safety of other belts. Belts subjected to an impact in an accidental fall should not be reused.

In judging the reliability of a system, the following vital, and to some extent conflicting, factors must be considered:

1. Sufficient strength to stop the wearer after a maximum free fall. The lanyard must be a minimum of ½-in. (12.7-mm) nylon (or equivalent), with a maximum length to provide for a free fall of no greater than 6 ft (1.8 m). The rope shall have a minimum breaking strength of 5,400 pounds (2,450 kg).
2. A shock absorber to limit the impact loading and prevent injury to the wearer or failure of the anchorage, lifeline, and belt.
3. Short enough stopping distance to prevent the wearer from striking a dangerous obstruction below, including swinging into an object.
4. Sufficient margin on all these factors, so far as possible, to cover all unknowns such as weight of the wearer, distance of fall, physical fitness, the distance to any damaging obstruction, variations in the strength or elasticity of materials, and deterioration of materials due to wear or other causes.

The primary caution regarding the use of fall protection, as with other personal protective equipment, is to see that it is worn and used correctly. A belt or harness is worthless unless it is being worn at the time that a fall is possible and attached to a lanyard or fall arrestor with an adequate overhead anchorage. It should also be securely buckled and worn snugly enough to reduce the possibility of the worker slipping out of it.

### Lifelines

Vertical lifelines should be secured above the point of operation to an anchorage or structural member capable of supporting a minimum dead weight of 5,400 lb (2,450 kg).

For most lifelines, synthetic ropes of ½ in. (12.7 mm) minimum (nylon) and ⅝ or ¾ in. (16 or 19 mm) diameter are recommended. Nylon and polyester are more resistant to wear or abrasion than manila and is also more resistant to some chemicals. Ropes made of other synthetic fibers such as polypropylene have characteristics that make them very good performers in certain applications, especially where weight is a factor. In a lifeline where strength and high energy absorption is important, nylon is still superior; however, stretch could be dangerous in some applications. In all cases, the specific rope grab must be approved for use with the chosen lifeline by the manufacturer.

The rigidity of wire ropes greatly magnifies the impact loading; therefore, they should not be used as lifelines where a free fall is possible unless a shock-absorbing device is also used. Also, wire ropes must not be used where electrical hazards are present.

If rope is spliced into snaps and D-rings instead of knotted, the splice will retain approximately 90 percent of the breaking strength of the rope. Breaking strength is always measured in a straight line pull on the rope.

Lifelines should be tied so they permit as little slack as possible and can stop a person with the minimum free fall. Serious injury or death may result if the total free fall plus the total stretch or elongation of the lifeline and shock absorber will allow the worker to strike some damaging object. Even when there is net protection, fall protection equipment may be necessary to avoid striking cross braces and beams between the worker and the net.

If there is a clear space in which a person may be stopped during an accidental fall, a low-force shock absorber (600-800 lb) may be used with far less discomfort to the worker than a rope lanyard. If space is closely limited and a shorter stop is imperative, retractable lifelines can be a great help in providing a practical answer.

Retracting lifelines are portable self-contained devices that are attached to a fixed anchorage point above the work area. The lifeline extends from the device and is connected to the user's body support. A tension is maintained up to the maximum extension, and the lifeline freely retracts when the user moves closer to the anchorage and pulls out as the worker moves away from it. In the event of a fall, retracting lifelines use a centrifugal mechanism which causes pawls to move outward, engaging a brake. The result is a short cushioned fall or controlled descent, depending on the particular device and intended application.

### Retrieval after a fall

After a fall up to six feet, or a fall in a confined space, retrieval methods are essential to avoid prolonged suspension which could aggravate an injury to a worker. Retractable lifelines can limit falls to inches so that regaining balance or recovering a worker can be relatively easy. Furthermore, winch systems designed for personnel lifting can be made part of such lifelines to assist the speedy raising or lowering process in some applications.

### Horizontal lifelines

These are essential for horizontal mobility while using lanyard systems or fall arrest devices. Methods to pass intermediate anchorage support without unhooking are available and ground level installations or bucket truck installations without fall hazard can allow reasonable safety for a variety of applications in both general industry and construction.

### Care of lifelines

Rope lines should be washed with mild soap and water and dried in circulating air. They should not be exposed to high temperatures and strong ultraviolet light.

Wire ropes should be kept clean and dry and should be frequently lubricated. Before use in acid atmospheres, they should be coated with oil. After such use, they should be thoroughly washed and again coated with oil.

## RESPIRATORY PROTECTION

The protection of employees from airborne health hazards in the workplace is the responsibility of management. At times when the hazards cannot be eliminated, the proper use of respirators becomes part of that responsibility. However, when respirators are required to provide the proper protection for employee health and safety, a complete respiratory protection program must be established that includes proper selection, use, maintenance, evaluation, training, and other factors.

Respiratory protective devices vary in design, equipment specifications, application, and protective capabilities. Proper selection depends on the toxic substance encountered, conditions of exposure, human capabilities, and equipment fit.

The National Institute for Occupational Safety and Health, under authorization of the Federal Mine Safety and Health Act of 1977 and the OSHAct of 1970, provides a testing, approval, and certification program assuring commercial availability of safe personal protective devices and reliable industrial hazard-measuring instruments. Use of the terms "approved" and "certified" reflects these applicable federal regulations. There will be an approval label on the equipment. Manufacturers may be consulted for specific applications, equipment needs, repairs, and the like which may be necessary.

### Protective factors

The term, "the protective factor," is used to describe the overall effectiveness of a respirator. The effectiveness of each kind of respiratory equipment has limitations. All air purifying equipment requires replacement of cartridges and/or filters as necessary to maintain effectiveness. Furthermore, workers must be trained in the proper use of respiratory equipment. Finally, workers should be aware that some substances are dangerous to other body systems in addition to respiratory entrance routes. In fact, some are so active that they can penetrate the skin.

### Selecting respiratory protection

Given the hundreds of toxic substances workers may encounter and the wide variety of respiratory protection equipment available, making the right choice of breathing equipment can be a difficult task.

The remainder of this section is taken from a NIOSH booklet, *Respiratory Protection—An Employer's Manual* (see References).

The proper selection of respiratory protective equipment involves three steps: (1) the identification of the hazard, (2) the evaluation of the hazard, and (3) finally, the selection of the appropriate approved respiratory equipment based on the first two considerations. (Refer to Figure 17-14 for final selections.)

### Identification of the hazard

Identification (and evaluation) of the hazard forms the basis for a decision on the need for the respirator program. If a survey of operations and work environments indicates that no employees are being exposed to contaminant concentrations exceeding established limits (in the U.S. the OSHA standards), then a respirator program is not required. However, your company may be using unregulated substance(s) for which there is no standard. An "in-house" evaluation may have indicated the need for respiratory protection equipment for this substance.

When a survey to determine the need for respirators is to be undertaken, it is important, initially, to know something about different kinds of hazardous atmospheres which may require the use of respirators.

**Gaseous contaminants.** Gaseous contaminants add another invisible material to what is already a mixture of invisible gases—the air we breathe. These contaminants are of two types.

1. Gases are the *normal* form of some substances, e.g., carbon dioxide.
2. Vapors are like gases except that they are formed by the evaporation of substances, such as acetone or trichloroethylene, which ordinarily occur as liquids.

**Particulate contaminants.** Particulate contaminants are made up of tiny particles or droplets of a substance. Many of these particles are so small that they float around in the air indefinitely and are easily inhaled. There are three types of particulates. Dusts are solid particles produced by such processes as grinding, crushing, and mixing of powder compounds. Mists are tiny liquid droplets given off whenever a liquid is sprayed, vigorously mixed, or otherwise agitated. Fumes are solid condensation particles of extremely small particle size.

**Combination contaminants.** The two basic forms—gaseous and particulate—frequently occur together. Paint spraying operations, for example, produce both paint mist (particulate) and solvent vapors (gaseous).

**Oxygen-deficient atmospheres.** In an oxygen-deficient atmosphere, the problem is not the presence of something harmful, but the absence of something essential. These atmospheres are most commonly found in confined and usually poorly ventilated spaces. Oxygen-deficient atmospheres are classified as immediately dangerous to life. Immediately dangerous to life or health is a term used to describe very hazardous atmospheres where employee exposure can: (1) cause serious injury or death within a short time (such as exposure to high concentrations of carbon monoxide or hydrogen sulfide), or (2) cause serious delayed effects (such as exposure to low concentrations of radioactive materials or cancer-causing agents).

### Evaluation of the hazard

A walk-through survey of the plant to identify employee groups or processes, or worker environments where the use of respiratory protective equipment may be required, is the next step in the respirator selection process. The survey can be facilitated by use of a hazard evaluation form (Figure 17-15). There are also instruments used to determine the concentration of airborne contaminants. However, only qualified individuals must use these instruments and interpret the results. If the facility does not have in-house qualified personnel, outside consultation will be required. Oxygen-deficient atmospheres may require special analysis. The maximum allowable concentrations of each identified hazard need to be known.

### Types of respirators

There are various types of respirators available to protect against hazards. However, before any respirator can be used safely, it is essential to know its functions and limitations thoroughly. Respirators are classified according to the respiratory hazard, such as an oxygen-deficient or air-contaminated environment (gases, particulate, etc.) or both. The two main categories of respirators are: (1) air-supplying respirators, and (2) air-purifying respirators.

### Air-supplying respirators

Air-supplying respirators are those which provide a supply of breathable air different from the workplace air. These respirators include the following:

1. Self-contained respirators provide a transportable supply of breathing air and afford complete protection against toxic gases and oxygen deficiency. An example is the Self-Contained Breathing Apparatus (SCBA), which has an air tank that can be strapped to the back (Figure 17-16).
2. Supplied-air respirators receive air through an air line or air hose (Figure 17-17a and b). The air may be supplied from a compressor or through large diameter tubing with its inlet placed in uncontaminated air. These are used where air-purifying respirators are not sufficient and where the atmosphere is not immediately dangerous to life or health (OSHA definition). There are limits prescribed for the length of the air line hose. A source of breathable air is required.

Combination self-contained and supplied-air respirators have both an air-purifying attachment and an air hose for breathing. The air hose permits long-term work use, and the short-term air purifying device provides emergency action by the worker. It should be used in situations in which potentially toxic atmosphere may occur.

**Abrasive blasting respirators.** Abrasive blasting respirators are used to protect personnel engaged in shot, sand, or other abrasive blasting operations, which involve air contaminated with high concentrations of rapidly moving abrasive particles.

The requirements for abrasive blasting respirators are the same as those for an air line respirator of the continuous flow type, with the addition that mechanical protection from the abrasive particles is needed for the head and neck (Figure 17-18). NIOSH-MSHA tests and certifies such equipment.

There are two forms of abrasive blasting respirators that cover the head and neck, and even the shoulder and chest.

These units use a rigid helmet to encase the user's head. The eyepiece is of impact-resistant safety glass or of plastic covered by a metal screen. An adjustable knitted fabric collar, covered with rubber- or plastic-coated fabric, fits over the metal helmet and down over the use's neck, shoulders, and chest in order to give additional protection. A flexible tube brings air to the helmet. Air, exhausted from the helmet, flows between the collar and the user's neck.

In both units, the eyepiece (window) and the protective screen should be easily replaceable, preferably without the use of tools.

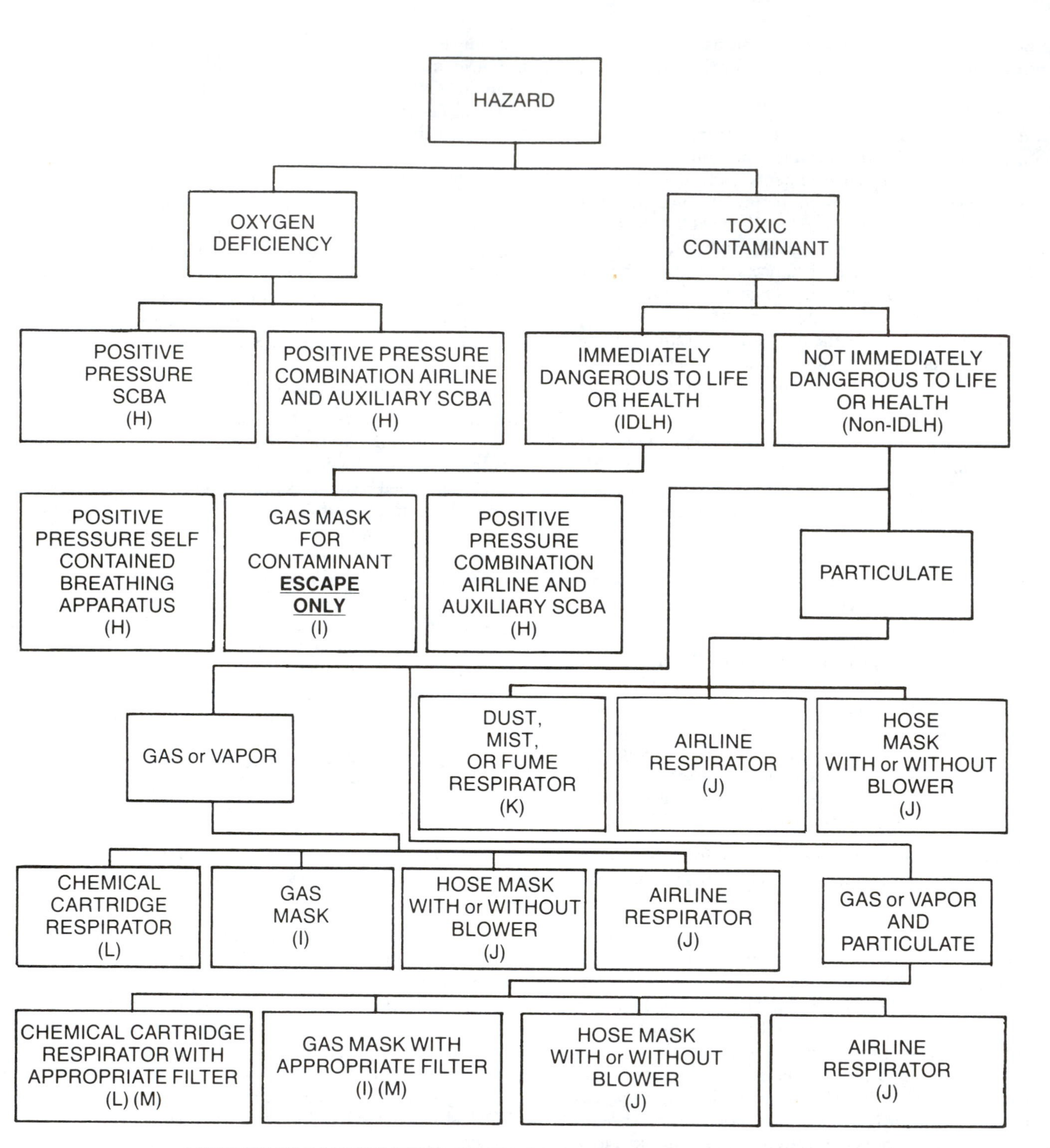


**Figure 17-14.** Suggested outline for selecting respiratory protective devices. Letters in parentheses (in shaded boxes) refer to Subparts of Title 30, CFR, Chapter I, Part 11, which discuss the items.

**Air supplied hoods.** For some long-term operations where a completely enclosed suit is not necessary, an air supplied hood may be used. These are particularly useful in hot, dusty situations (Figure 17-19). A vortex tube may also be used to reduce the ambient air temperature by up to 50 F (10 C).

Respirable air under suitable pressure should be delivered to a hood at a volume of at least 6 cfm (0.0028 $m^3/s$).

**Air supplied suits.** The most extreme condition requiring respiratory equipment is that in which rescue or emergency

Respiratory Protective Equipment Hazard Evaluation Form

Company ____________ Date ____________

Division ____________ By ____________

Department ____________ Page ____________

| Employee | Job Description | Limits | Respiratory Protection Required | | Respiratory Equipment Type-SCP | Remarks |
|---|---|---|---|---|---|---|
| | | | Yes | No | | |
| | | | | | | |

**Figure 17-15.** Respiratory protective equipment hazard evaluation form.

**Figure 17-16.** Self-Contained Breathing Apparatus (SCBA). (Courtesy Bordas.)

**Figure 17-17a.** Air line respirator. (Courtesy MSA.) Figure 17-17b shows parts and connections.

repair work must be done in atmospheres both extremely corrosive to the skin and mucous membranes and acutely poisonous and immediately hazardous to life, such as atmospheres containing ammonia, hydrofluoric, or hydrochloric acid vapors. For these conditions, a complete suit of impervious clothing, with a respirable air supply, is available (Figure 17-20 a and b).

There are no generally accepted specifications for such suits; therefore, considerable dependence must be placed upon the manufacturer. The material should have sufficient mechanical strength to resist rough handling and considerable abuse without tearing.

The hose line supplying the air should be connected to the suit itself, as well as to the helmet, since it is not only extremely fatiguing but also dangerous to wear such a suit for a long period unless it is well ventilated.

Personal air conditioning devices utilizing a vortex tube are available for air supplied suits or hoods. These cooling devices are desirable to reduce fatigue where high ambient temperatures may be encountered (as in heat-protective clothing) or where body heat may build up (as under impermeable chemical protective clothing).

The vortex device works by taking an air stream under pressure and dividing it. One portion loses heat; the other gains heat. The cold portion passes into the suit or hood; the warm portion is vented to the atmosphere, or vice versa in cold weather.

### Air-purifying respirators

Air-purifying respirators can purify the air of gases, vapor, and particulates, but do not supply clean breathing air (Figure 17-21). They must *never* be used in oxygen-deficient atmospheres. The useful life of the air purifying device is limited by (1) the concentration of the air contaminant, (2) breathing demand of the wearer, and (3) the removal capacity of the air purifying medium (cartridge). They have a facepiece and an attached cartridge which contains specific material needed for the containment. They are classified as either gas and vapor respirators or particulate respirators.

Gas and vapor respirators (also known as chemical cartridge respirators) remove gases and/or vapors by passsing the contaminated air through cartridges containing charcoal or other special material that traps these contaminants (Figure 17-22a and b). Cartridges should be matched to the contaminants (Table 17-B). These cartridges are used to protect against con-

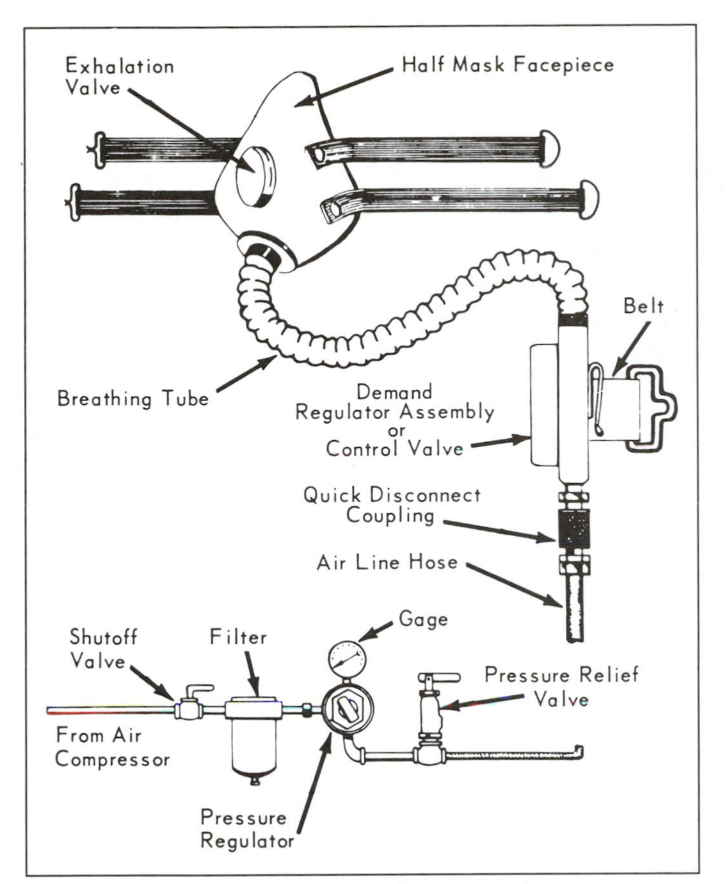


**Figure 17-17b.** Air line respirator showing parts and connections.

taminants (not exceeding certain concentrations) that have adequate warning properties of smell or irritation. This allows the wearer to judge when a cartridge is no longer usable. Some cartridges are dated as well, and should not be used after the expiration date.

Four major negative rules apply to chemical cartridge respirators:

1. Do not use chemical cartridge respirators for protection against gaseous material which is extremely toxic in very small concentrations.
2. Chemical cartridge respirators should not be used for exposure to harmful gaseous matter which cannot clearly be detected by odor.

   Example: methyl chloride and hydrogen sulfide. The former is odorless; and the latter, although foul smelling, paralyzes the olfactory nerves so quickly that detection by odor is unreliable.
3. Chemical cartridge respirators should not be used against any gaseous material in concentrations which are highly irritating to the eyes without satisfactory eye protection.
4. Chemical cartridge respirators cannot be used for protection against gaseous material which is not effectively stopped by the chemical fills utilized, regardless of concentrations.

The second type of air-purifying respirator, the particulate respirator, is also known as a mechanical filter respirator. Depending upon the design, the filters can screen out dust, fog, fume, mist, spray, or smoke by passing the contaminated air through a pad or filter. (They do not provide protection against gases, vapors, or oxygen deficiency.) They consist of a facepiece

**Table 17-B.** Color Code for Cartridges and Gas Mask Canisters

| *Atmospheric Contaminants to be Protected Against* | *Color Assigned* |
|---|---|
| Acid gases | White |
| Organic vapors | Black |
| Ammonia gas | Green |
| Carbon monoxide gas | Blue |
| Acid gases and organic vapors | Yellow |
| Acid gases, ammonia, and organic vapors | Brown |
| Acid gases, ammonia, carbon monoxide, and organic vapors | Red |
| Other vapors and gases not listed above | Olive |
| Radioactive materials (except tritium and noble gases) | Purple |
| Dusts, fumes, and mists (other than radioactive materials) | Orange |

*Notes:*
1. A purple stripe shall be used to identify radioactive materials in combination with any vapor or gas.
2. An orange stripe shall be used to identify dusts, fumes, and mists in combination with any vapor or gas.
3. Where labels only are colored to conform with this table, the canister or cartridge body shall be gray or a metal canister or cartridge body may be left in its natural metallic color.
4. The user shall refer to the wording of the label to determine the type and degree of protection the canister or cartridge will afford.

to which is attached a mechanical filter, papers, or similar filter substance. There are many types of filters capable of trapping a wide variety of classes of airborne particles. Filters should be changed at frequent intervals, when they become clogged, or when it becomes difficult to breathe through them.

Powered air-purifying respirators usc a blower to pass contaminated air through an element that removes the contaminant

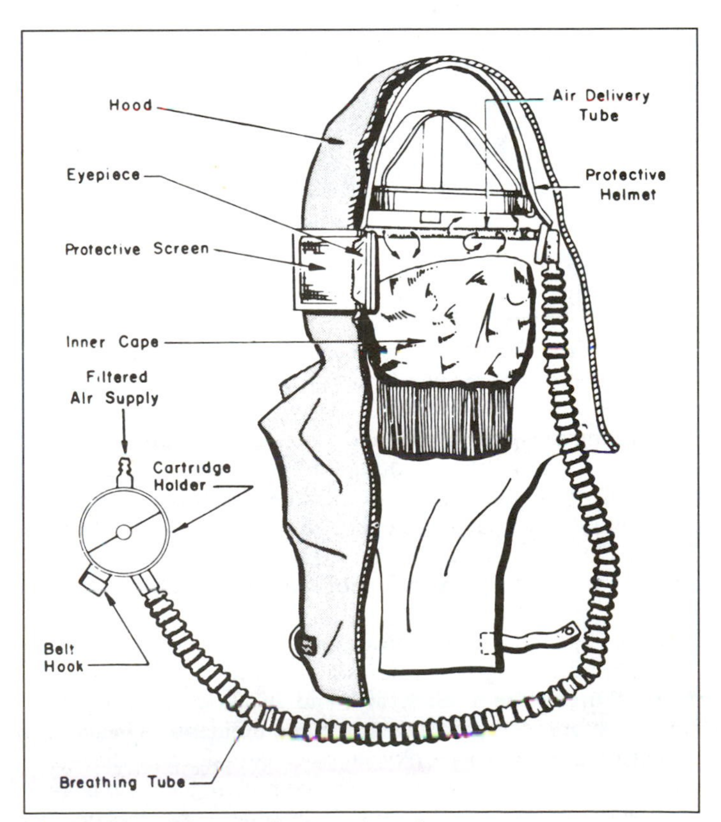


**Figure 17-18.** Diagram shows parts and connections for lightweight hood designed for use by persons doing abrasive blasting.

**Figure 17-19.** Air-supplied hood. (Courtesy Personal Environments Systems, Inc.)

and to supply purified air to a facepiece, helmet, or hood. The purifying element may be either a filter, a cartridge, or a combination of the two. A blower is used to force the contaminated air through the element and to the respirator facepiece. The unit supplies clean air to the worker at positive pressure, so that contaminated air does not leak to the worker.

Combination gas, vapor, and particulate respirators (also known as gas masks) filter out gases, vapors, and particulates by passing the contaminated air through a cartridge or canister containing both a particulate filter and a gas/vapor absorbing device. They are to be used in atmosphere containing sufficient oxygen for breathing. A variety of types is available.

**The self-generating apparatus** consists of a chemical canister, a breathing bag that acts as a reservoir, a facepiece with tube assembly, and a relief valve and check valves to regulate flow in accordance with respiratory requirements. The chemical in the canister evolves oxygen when contacted by the moisture and carbon dioxide in the exhaled breath, and also retains the carbon dioxide and moisture (moisture retention is important as it permits fog-free lenses). The apparatus is light in weight and contains no intricate parts.

**The compressed or liquid oxygen (cylinder) recirculating apparatus.** This equipment is supplied with either a full facepiece or a mouthpiece and nose clip. The seal around the facepiece must be kept absolutely tight. Such equipment consists, essentially, of a high-pressure oxygen cylinder with reducing and regulating valves, a lung-governed admission valve which supplies oxygen from the cylinder only during inhalation, a carbon dioxide scrubber and cooler, and a reservoir breathing bag connected by tubes to the mouthpiece or facepiece. Check valves direct the flow through the circuit so that the exhaled oxygen is purified of carbon dioxide, and rebreathed from the bag, with replenishment from the cylinder as required. The liquid oxygen type operates similarly.

The rebreathing principle permits the most efficient utilization of the oxygen supply. The exhaled breath contains both oxygen and carbon dioxide because the body consumes only a small part of the inhaled oxygen.

Because this equipment is quite complicated to use and maintain, personnel should be thoroughly trained.

### Storage of respirators

Respirators should be stored to protect them from dust, sun-

**Figure 17-20a.** Totally encapsulating suits with a respirable air supply provide protection from hazardous chemicals. (Courtesy MSA.)

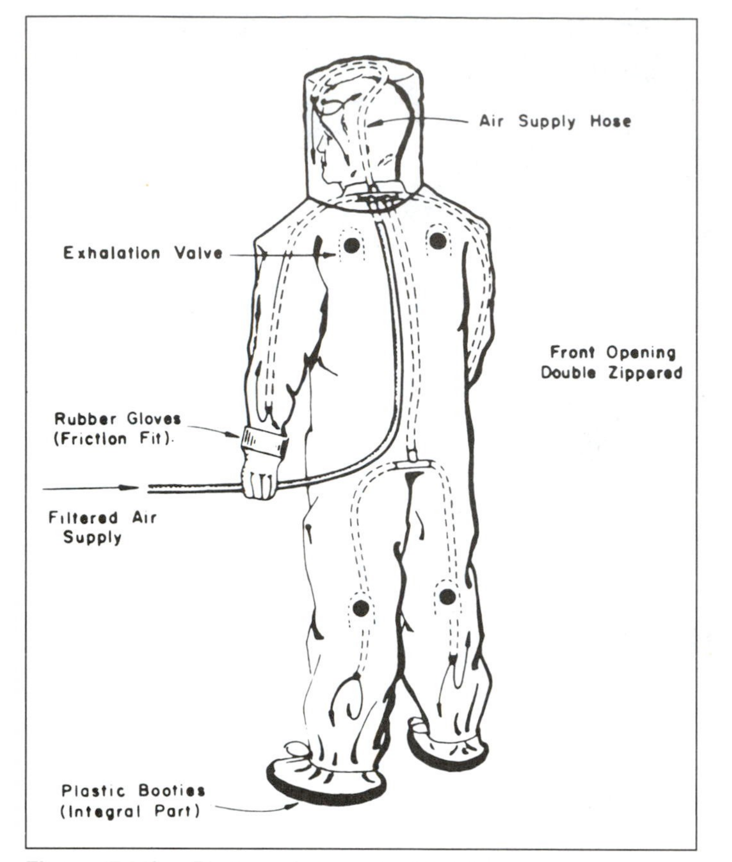


**Figure 17-20b.** Diagram of air-supplied suit for use in corrosive chemical atmospheres.

light, heat, extreme cold, excessive moisture, and damaging chemicals. Unprotected respirators can sustain damaged parts or facepiece distortion that make them ineffective.

Before being stored, a respirator should be carefully wiped with a damp cloth and dried. It should be stored without sharp folds or creases. It should never be hung by the elastic headband or put down in a position which will stretch the facepiece.

Since heat, air, light, and oil cause rubber to deteriorate, respirators should be stored in a cool, dry place and protected from light and air as much as possible. Wood, fiber, or metal cases are provided with many respirators. Respirators should be sealed in clean plastic bags.

Respirators should not be thrown into tool boxes or left on work benches where they may be exposed to dust and damage by oil or other harmful materials.

After cleaning and inspection, place respirators in individual, sealable plastic bags. Then store them in one layer with the facepiece and exhaustion valve in normal position. Respirators should not be stored in lockers unless they are protected from contamination, distortion, and damage.

Standard steel storage cabinets or steel wall-mounted cabinets with compartments are good solutions for the storage problem for air-purifying respirators. Special storage cabinets may be purchased from the manufacturer for SCBA's. The cabinets should be located in uncontaminated but readily accessible areas.

## Maintenance of respirators

The ongoing maintenance of the respirators themselves is an essential part of the respiratory protection program. If the equipment malfunctions through poor maintenance, the employee may unknowingly be exposed to a hazard which could be fatal.

The maintenance program should incorporate the manufacturer's instructions and should include provisions for disassembly, including the removal of the respirator's purifying elements, cleaning, sanitizing, inspection for defects, repair (if necessary), installation of purifying elements, reassembly, packaging, and storage. (See section above.)

The air-purifying elements (chemical cartridges or filters) should not be cleaned or exposed to excess moisture, including high humidity. Discard the elements if there is any question about them.

Supervisors should be responsible for making daily inspection, particularly of functional parts such as exhalation valves and filter elements. They should see that the edges of the valves are not curled and that valve seats are smooth and clean. Inhalation and exhalation valves should be replaced periodically.

In addition to the daily check, respirators should be inspected weekly by trained persons. During the weekly inspection, rubber parts should be stretched slightly for detection of fine cracks. The rubber should be worked occasionally to prevent setting (one of the causes of cracking), and the headband should be checked to be sure that the wearer has not stretched it in an attempt to secure a snug fit (see Figure 17-23).

Emergency respirators must be checked at least monthly, and a written record must be kept.

Sometimes, in an effort to reduce resistance to breathing, workers will punch holes in the filter, the rubber facepiece, or other parts. Underlying causes of this mistreatment should be discovered and corrected. For instance, it may be found that in the interest of economy, filters are allowed to become completely plugged before they are replaced.

### Cleaning and sanitizing

Each time an employee uses a respirator, it should be in a cleaned and sanitized condition. Respirators to be used in emergencies, or on a nonroutine basis should be cleaned and sanitized after each use.

The actual cleaning may be done in a variety of ways:

1. The respiratory protection equipment should be dismantled and washed with detergent in warm water using a brush, thoroughly rinsed in clean water, and then air dried in a clean place. Care should be taken to prevent damage from rough handling. This method is an accepted procedure for a small respirator program or where each worker cleans his or her own respirator.
2. A standard domestic clothes washer may be used if a rack is installed to hold the facepieces in a fixed position. (If the facepieces are placed loose in a washer, the agitator may damage them.) This method is especially useful in large programs where respirator usage is extensive.

If possible, detergents containing a bactericide should be used. Organic solvents should not be used, as they can deteriorate the elastomeric (rubber or silicone) facepiece. If these detergents are not available, a disinfectant may be necessary. Check with the manufacturer for disinfectants that will not damage the respirators.

Be sure that the cleaned and sanitized respirators are rinsed thoroughly in clean water no hotter than 50 C (120 F) so that all traces of detergents and cleaners are removed. Otherwise,

**Figure 17-21.** Powered air-purifying respirator designed for use by employees working in the vicinity of coke oven emissions or foundry operations. (Courtesy MSA.)

skin irritation or dermatitis may result when the employee wears the respirator. Allow the respirators to air dry by themselves on a clean surface. They may also be hung carefully on a line.

If management is unwilling or unable to run the maintenance progam, usually it is possible to contract with an outside firm that will maintain respirators.

#### Inspection of respirators

After cleaning and sanitizing, each respirator should be reassembled and inspected for proper working condition and repair or replacement of parts. Respirators should also be inspected routinely by the user immediately before each use. This is to ensure that the respirator is in proper working condition.

**Figure 17-22a.** Full-face cartridge respirator. (Courtesy Ultra-Twin.)

An inspection checklist compiled by NIOSH follows (see Figure 17-23).

**Disposable respirators** should be checked for:

- holes in the filter (obtain new disposable respirator)
- straps for elasticity and deterioration (replace straps, contact manufacturer)
- metal nose clip for deterioration, if applicable (obtain new disposable respirator).

**Air-purifying respirators** (including quarter-mask, half-mask, full facepiece, and gas mask) should be checked for:

1. The facepiece
   - excessive dirt (clean all dirt from facepiece)
   - cracks, tears, or holes (obtain new facepiece)
   - distortion (allow facepiece to sit free from any constraints and see if distortion disappears—if not, obtain new facepiece)
   - cracked, scratched, or loose-fitting lenses (contact respirator manufacturer to see if replacement is possible—otherwise, obtain new facepiece).

**Figure 17-22b.** Canister type gas mask. (Courtesy MSA.)



**Figure 17-23.** Maintenance check points for air-purifying respirators.

2. The headstraps
   - breaks or tears (replace headstraps)
   - loss of elasticity (replace headstraps)
   - broken or malfunctioning buckles or attachments (obtain new buckles)
   - excessively worn serrations on the head harness that might allow the facepiece to slip (replace headstrap).
3. The inhalation and exhalation valves
   - detergent residue, dust particles or dirt on valve or valve seat (clean residue with soap and water)
   - cracks, tears, or distortion in the valve material or valve seat (contact manufacturer for instructions)
   - missing or defective valve cover (obtain valve cover from manufacturer).
4. The filter elements
   - proper filter for the hazard
   - missing or worn gaskets (contact manufacturer for replacement)
   - worn threads—both filter threads and facepiece threads (replace filter or facepiece, whichever is applicable)
   - cracks or dents in filter housing (replace filter)
   - deterioration of gas mask canister harness (replace harness)

- service life indicator, expiration date, or end-of-service date.

5. The gas mask
   - cracks or holes (replace tube)
   - missing or loose hose clamps (obtain new connectors)
   - service-life indicator on canister (or contact manufacturer to find out what indicates the end-of-service date for the canister).

**Air-supplying respirators** should be checked for:

1. The hood, helmet, blouse, or full suit (if applicable)
   - rips and torn seams (if unable to repair the tear adequately, replace)
   - headgear suspension (adjust properly for wearer)
   - cracks or breaks in faceshield (replace faceshield)
   - protective screen to see that it is intact and fits correctly over the faceshield, abrasive blasting hoods, and blouses (obtain new screen).
2. The air supply system
   - breathing air quality
   - breaks or kinks in air supply hoses and end fitting attachments (replace hose and/or fitting)
   - tightness of connections
   - proper setting of regulators and valves (consult manufacturer's recommendations)
   - correct operation of air-purifying elements
   - proper operation of carbon monoxide alarms or high-temperature alarms.
3. The facepiece, headstraps, valves, and breathing tube inspection checks are the same as for the air-purifying respirators.

**Self-Contained Breathing Apparatus (SCBA).**

- consult manufacturer's literature.

In some plants, maintenance service for respirators, as well as for other kinds of personal protective equipment, can be effectively provided by traveling service carts (Figure 17-24).

Where a number of respirators are in regular use, a central station is often set up for their care and maintenance, as well as for the care and maintenance of other items of personal protective equipment. Each employee is then provided with two respirators and either a locker or a hook at the central station.

Some plants have found that if two respirators are assigned to each worker, equipment lasts more than twice as long. This plan is most desirable when cleaning cannot be done between shifts or before the next scheduled shift.

Under such a plan, marked respirators are turned in daily or weekly, depending upon use, for cleaning, disinfection, inspection, and repair. The worker then uses the second respirator until the first can be serviced.

Another plan is to keep quantities of disinfected respirators on hand for groups. This plan works where individual needs vary, but in such a plan the user is not so easily charged with responsibility. If the same respirator is used by several persons, it is always best to clean and disinfect it after each use.

Respirators should be marked to indicate to whom they are assigned. The method of identification should be permanent enough so that the marking cannot be changed inadvertently or without effort.

### Training

Once the respirator has been determined, the wearer must be trained in its use and care. This is important for every type of respirator. Each respirator wearer should be trained initially in the use of respirators and should be retrained periodically.

Training should include the following:

1. Reasons for respiratory protection, explanation of why other controls and methods are not being used, and explanation of efforts being made to reduce the hazards.
2. Explanation of the respirator selection procedure used by the health and safety professional, including identification and evaluation of the hazard.
3. Proper fitting, donning, wearing, and removing of the respirator.
4. Limitations, capabilities, and operation of the respirator.
5. Proper maintenance and storage procedures.
6. Wearing of the respirator in a safe atmosphere to allow the user to become familiar with its characteristics.
7. Wearing of the respirator in a test atmosphere under close supervision of the trainer—one in which the wearer can simulate work activities and detect respirator leakage or malfunction.
8. How to recognize and cope with an emergency situation.
9. Instructions for special use as needed.
10. Explanation of any regulations governing the use of respirators.

The instructor should be a qualified person, such as an industrial hygienist, safety professional, or the respirator manufacturer's representative.

## PROTECTIVE FOOTWEAR

Specifications for various kinds of footwear have been standardized by ANSI Z41-1983, *Safety-Toe Footwear.* Safety-toe footwear has been divided into three classifications—75, 50, and 30—based on its ability to meet the minimum requirements for both compression and impact shown in Table 17-C.

**Table 17-C.** Minimum Requirements of American Standard Z41-1983

| *Classification* | *Compression (pounds)* | *Impact (pounds)* | *Clearance (inches)* | |
|---|---|---|---|---|
| | | | *Men* | *Women* |
| 75 | 2500 | 75 | 16/32 | 15/32 |
| 50 | 1750 | 50 | 16/32 | 15/32 |
| 30 | 1000 | 30 | 16/32 | 15/32 |

1 lb mass = 0.45 kg
1 lb force (avoirdupois) = 4.4 N
1 in. = 2.54 cm

Steel, reinforced plastics, and hard rubber are used for safety toes with the choice depending on the protective level desired and the shoe design. The test requirements are identical for both women's and men's shoes (Figure 17-25a and b).

Toe boxes used in shoes may be conductive, nonconductive, or spark resistant. For work under wet conditions, rubber boots or rubber shoes may be obtained with a steel toe box to protect against impact. Puncture-resistant soles are another optional feature.

Regulations may require that the safety-toe shoe be used for work requiring the handling of heavy materials. Safety shoes also afford good protection against rolling objects, such as

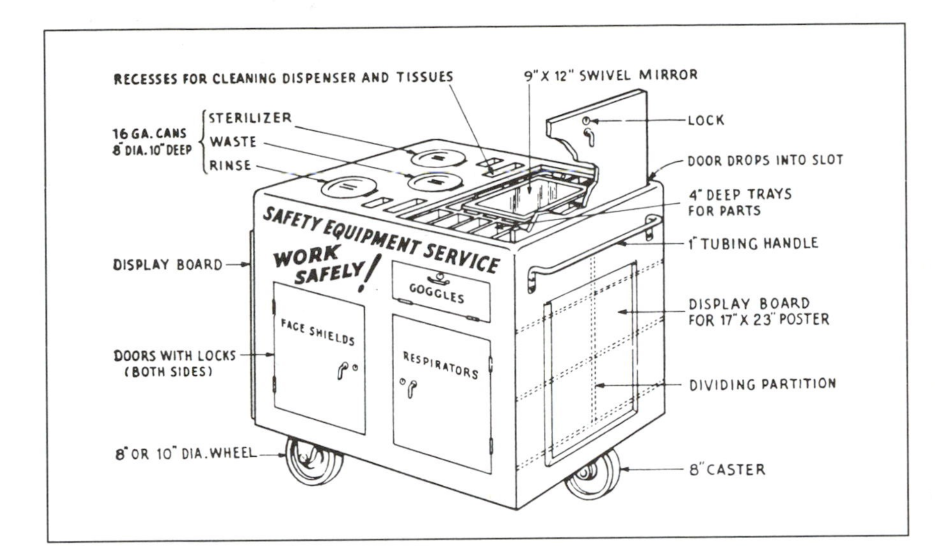


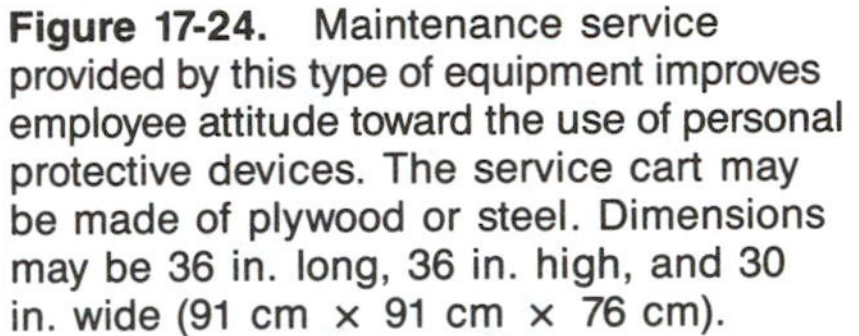
**Figure 17-24.** Maintenance service provided by this type of equipment improves employee attitude toward the use of personal protective devices. The service cart may be made of plywood or steel. Dimensions may be 36 in. long, 36 in. high, and 30 in. wide (91 cm × 91 cm × 76 cm).

**Figure 17-25a.** Protective footwear shown with the regular safety toe shoe on the left, and the metatarsal guard on the right. (Courtesy Iron-Age Shoe Company.)

barrels, heavy pipe, rolls, or truck wheels, and against the hazard of accidentally kicking sharp objects.

The toe box adds a little to the weight and cost of the shoe. A well-made and properly fitted safety shoe should be comfortable. Comfort is an important factor in any shoe, but particularly when safety footwear is required. Great care should be exercised in selecting the correct type and size of safety shoe.

Many companies set up shoe departments, with the aid of shoe manufacturers, in their plants and provide trained employees to see that individuals are properly fitted with the correct types for the hazards involved. Some retail organizations provide the same type of service. Shoemobiles are also available for in-plant customers.

To protect the upper foot area from impacts, integral metatarsal (or over-foot) guards should be worn in addition to safety-toe shoes. Heavy-gage, flanged, and corrugated sheet metal footguards help protect the feet. There are no standards developed for these items.

### Conductive shoes

Safety shoes, boots, and rubber overshoes may be obtained with a conductive construction to allow a drain off of static charges, and with a nonferrous construction to reduce the possibility of friction sparks in locations with a fire or explosion hazard. Initial and subsequent periodic tests should be made on conductive footwear in order to make sure that the maximum allowable resistance of 450,000 ohms is not exceeded. A special design is offered for use in munitions plants.

Conduction shoes are used (1) in hazardous locations where the floors are nonconductive and grounded, such as in the manufacture of certain explosive compounds, or (2) when cleaning tanks that have contained gasoline or other volatile hydrocar-

**Figure 17-25b.** Protective footwear is also available in women's styles and sizes.

bons. These shoes have conductive soles and nonferrous exposed metal parts.

### Foundry shoes

Safety shoes of the "congress" or gaiter type are used in some plants where employees are exposed to splashes of molten metal. Having no fasteners, such shoes are easily and rapidly removable in an emergency. Some National Safety Council members engaged in foundry and steel mill operations have reported that serious burns have occurred to workers who are unable to remove ordinary work shoes in an emergency. As in all such occupations, the tops of the shoes should be covered by the trouser leg, spats, or leggings, to keep out molten metal.

### Electrical hazard shoes

Electrical hazard shoes are intended to minimize hazards resulting from contacts with electric current where the path of the current would be from the point of contact to the ground. There are two general types available. If damp or badly worn, they cannot be depended on for protection.

### Special shoes

In some industries, such as construction, where there may exist an increased hazard from protruding nails and where contact with energized electric equipment is remote, shoes or boots are equipped with flexible metal-reinforced soles or inner soles.

Special shoes made with a stitched and cemented construction are available for electricians. These shoes serve as good insulators, so when they are repaired, avoid the use of nails which will destroy their insulating value.

For wet conditions such as are found in dairies and breweries, leather shoes with wood soles, or wood-soled sandals worn over shoes, are effective. Some companies offer a special sole designed to increase traction on damp, wet floors. If dirt accumulates in these shoes, a sanitation problem may develop.

Wood soles provide good foot protection on jobs which require walking upon hot surfaces which are not hot enough to char the wood. Wood soles have been so generally used by men handling hot asphalt that they are sometimes called paver's sandals or paver's shoes. They are, however, equally satisfactory for other work which requires soles that do not conduct heat.

Plastic shoe covers can be worn to protect a product from contamination.

Where shower baths are used, many organizations provide paper slippers or wooden sandals for each individual, to reduce the possibility of foot infection. Paper slippers are discarded after a single use, while the sandals are disinfected at frequent intervals, particularly before being assigned to others.

### Cleaning rubber boots

If rubber boots are reused by people on the next shift or job, great care should be exercised to disinfect boots after each shift or job. First, the boots are washed inside and outside with a hose under water pressure. Then they are dipped into a tub containing a solution of 1 part sodium hypochlorite and 19 parts water. The hose is again used for rinsing, after which the boots are ready for drying. Other disinfecting agents can be used, but this one has been satisfactory and is easily obtainable.

One company has a drying rack consisting of a tank with low-pressure steam coils having upright steel-pipe boot holders which permit circulation of hot air inside the boots. After the boots are washed thoroughly and dipped in the disinfecting solution, they are completely dried in about 12 minutes.

If much work of this type is necessary, the rack with water jets could be rearranged so that a number of boots could be cleaned and rinsed at one time.

## SPECIAL WORK CLOTHING

In our modern industrial environment, exposure to fire, extreme heat, molten metal, corrosive chemicals, cold temperature, body impact, cuts from materials which are handled, and other highly specialized hazards are often part of what is known as "job exposure." Special protective clothing is available to minimize the effects of all of these hazards. Sample swatches of materials used can usually be obtained from the manufacturers for testing.

### Protection against heat and hot metal

**Leather clothing** is one of the more common forms of body protection against heat and splashes of hot metal. It also provides protection against limited impact forces and infrared and ultraviolet radiation.

The garments should be of good quality leather, solidly constructed, and provided with fastenings to prevent gaping during body movement. Fastenings should be so designed that the wearer can rapidly and easily remove the garment. There should be no turned-up cuffs or other projections to catch and hold hot metal. The garments should have no pockets or if pockets are necessary, they should have flaps which can be fastened shut.

For ordinary protection against hot metal, radiant heat, or flame hazards of somewhat more intensity than those represented by welding operations, wool and leather clothing is used. Specially treated clothing has been developed which is impervious to metal splash up to 3,000 F (1,650 C).

**Wool garment** requirements are in general the same as those for leather except that metal fastenings should be covered with flaps to keep them from becoming dangerously hot.

**Asbestos substitutes,** including fiberglass or other special high-temperature resistant materials, are available. These materials are effective in the leggings and aprons usually worn by foundry personnel working with molten metal. Such leggings should completely encircle the leg from knee to ankle, with a flare at the bottom to cover the instep. The design of the leggings should permit rapid removal in emergencies.

The front part of the legging may be reinforced to provide impact protection when it is required. The most common material for this reinforcement is fiberboard.

**Aluminized clothing.** Where people must work in extremely high temperatures up to 2,000 F (1,090 C)—such as in furnace and oven repair, coking, slagging, firefighting, and fire rescue work—aluminized fabrics are essential. The aluminized coating reflects much of the radiant heat and the underlying material insulates against the remainder. Some of these suits consist of separate units of trousers, coats, gloves, boots, and hoods. Others are one-piece from head to foot. Some suits used in industrial operations are airfed to reduce heat and increase comfort.

Aluminized heat-resistant clothing generally falls into two classes:

- Emergency suits may be used where the temperatures may exceed 1,000 F (540 C), as in a kiln or furnace, or where workers must move through burning areas for firefighting or rescue operations. These suits are constructed of aluminized glass fiber, with layers of quilted glass fibers and a wool lining on the inside.
- Fire proximity suits are used in the proximity of high temperature, such as slagging, coking, furnace repair work with hot ingots, and firefighting where the flame area is not entered. These suits are seldom of one-piece construction. They depend primarily on the reflective ability of an aluminized coating on a base cloth of glass fiber or synthetic fiber. *Never use fire proximity clothing where fire entry is required.*

### Flame-retardant work clothes

Cotton work clothing can be protected against flame or small sparks by flameproofing. One of the commercial preparations can be applied in ordinary laundry machinery after the garment is washed. Treating the material will make it highly flame-resistant and will not add much to the weight or stiffness of the cloth.

Durable flame-retardant work clothes are readily available. Cotton treated with tetrakis (hydroxymethyl)-phosphonium chloride (THPC), developed by the U.S. Department of Agriculture Research Laboratories, gives good flame retardancy and will withstand many launderings. A high-temperature-resistant nylon fabric that chars rather than melts is available for the most severe situations. Modacrylic fabrics resembling cotton fabrics have permanent fire-retardant properties and are light in weight.

Flameproofed clothing should be marked or otherwise made distinctive to reduce the chance that untreated garments may be used by mistake.

### Cleaning work clothing

Manufacturer's recommendations should be followed in laundering and cleaning work clothes. Excessive water temperatures or use of certain washing preparations can cause deterioration of the fabric or affect its properties. Spot cleaning with organic solvents will soften or dissolve some synthetics. Chlorine bleaches will remove most flame-retardant treatment from cotton.

Compressed air used for dusting of clothing must not exceed 30 psig (200 kPa) pressure. It is recommended that a vacuum system be used; this will prevent dust from being spread into the air where it could get into someone's eyes or lungs.

Many industrial laundries and industrial clothing rental agencies can advise on cleaning and maintenance of work clothing.

### Protection against impact and cuts

It is necessary to protect the body from cuts, bruises, and abrasions on most jobs where heavy, sharp, or rough material is handled. Special protectors have been devised for almost all parts of the body and are available from suppliers of safety equipment.

**Pads** of cushioned or padded duck will protect the shoulders and back from bruises when men carry heavy loads or objects with rough edges.

**Aprons** of padded leather, fabric, plastic, hard fiber, or metal will protect the abdomen against blows. Similar devices of metal, hard fiber, or leather with metal reinforcements provide protection against sharp blows with edged tools. For jobs requiring ease of movement, aprons may be split and equipped with fasteners to draw them snugly around the legs.

**Guards** of hard fiber or metal are also widely used to protect the shins against impact.

**Knee pads** should be worn by mold loftsmen and others whose task requires continual kneeling.

No one type of personal protective equipment for the extremities is suitable for the many different work situations involved in any business or industrial operation, from the laboratory to the loading dock. Thus, proper protection for the hands, fingers, arms, and the skin must be selected on a job-rated basis. The specific type of protection and its material depends upon the type of material being handled and the work atmosphere.

**Gloves.** The material to be used for gloves depends largely upon what is being handled (Figure 17-26a and b). For most light work, a cotton or canvas glove is both satisfactory and cheap. For rough or abrasive material, leather or leather reinforced with metal stitching will be required. Leather reinforced by metal stitching or metal mesh also provides good protection from edged tools, as in butchering and similar occupations. Metal mesh or highly cut-resistant plastic gloves are also effective.

There are many plastic and plastic-coated gloves available in materials such as neoprene, latex, and nitrile. They are designed to give protection from a variety of hazards. Careful consideration must be given to actual permeation tests against hazardous chemicals. Some surpass leather in wearing ability. Others have granules or rough materials incorporated in the plastic for better gripping ability. Some are disposable.

Where the use of a complete glove is not necessary, finger stalls or cots may be used. These are available in combinations of one or more fingers. Some of the more common materials used are rubber, duck, leather, plastic, and metal mesh. The construction of the cot depends on the degree or type of hazard to be confronted.

Gloves should not be used while working on moving machinery such as drills, saws, grinders, or other rotating and moving equipment that might catch the glove and pull it and the worker's hand into hazardous areas.

In addition to gloves there are also available mittens (including one-finger and reversible types), pads, thumb guards, finger cots, wrist and forearm protectors, elbow guards, sleeves, and capes, all of varying materials and lengths.

**Hand leathers and arm protectors.** Where the problem is protection from heat or from extremely abrasive or splintery

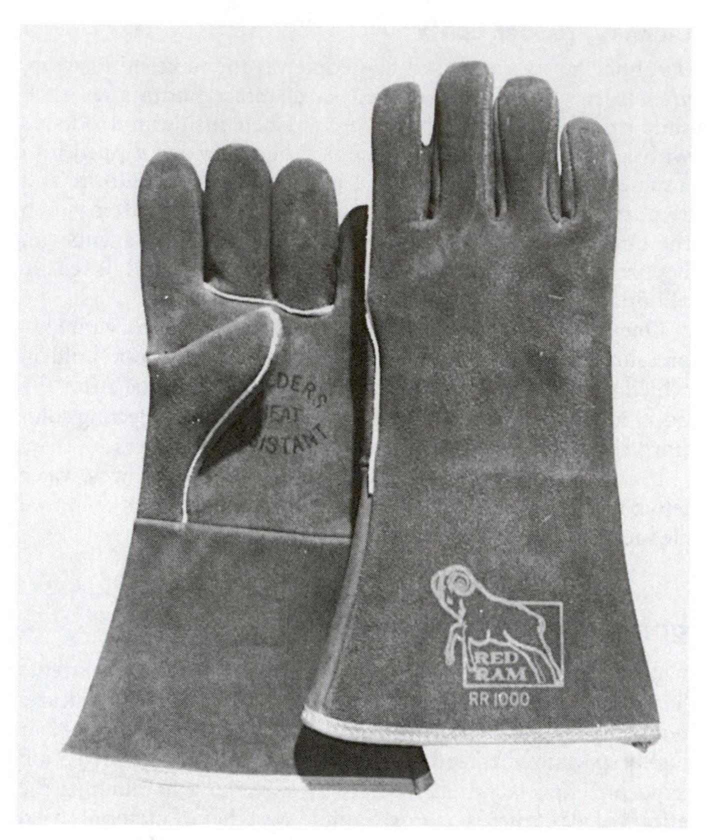


**Figure 17-26a.** Welding gloves with self-extinguishing fleece lining. (Courtesy Elliot Glove Company.)

material, such as rough lumber, hand leathers or hand pads are likely to be more satisfactory than gloves, since they can be made heavier and less flexible without discomfort.

Since hand leathers or pads are primarily for heavy materials handling, they should not be used around moving machinery. They should at all times be sufficiently loose to release the hands and fingers if caught on a rough edge or nail.

For protection against heat, hand and arm protectors should be of wool, terry, or glass fiber. Leather can be used, too, but will not stand a temperature over 150 F (65 C).

Wristlets or arm protectors may be obtained in any of the materials of which gloves are made.

### Impervious clothing

For protection against dusts, vapors, and moisture of hazardous substances and corrosive liquids, there are many types of impervious or impermeable materials available. These are fabricated into clothing of all descriptions, depending on the hazards involved. They range from aprons and bibs of sheet plastic to garments which completely enclose the body from head to foot and contain their own air supply.

Materials used include natural rubber, olefin, synthetic rubber, neoprene, vinyl, polypropylene, and polyethylene films and fabrics coated with them. Natural rubber is not suited for use with oils, greases, and many organic solvents and chemicals. Make sure that the clothing selected will protect against the hazards involved and that it has been field tested prior to actual use.

Some synthetic fabrics used for regular work clothing in chemical plants, where daily contact with acids and caustic solutions would cause rapid deterioration of regular cotton

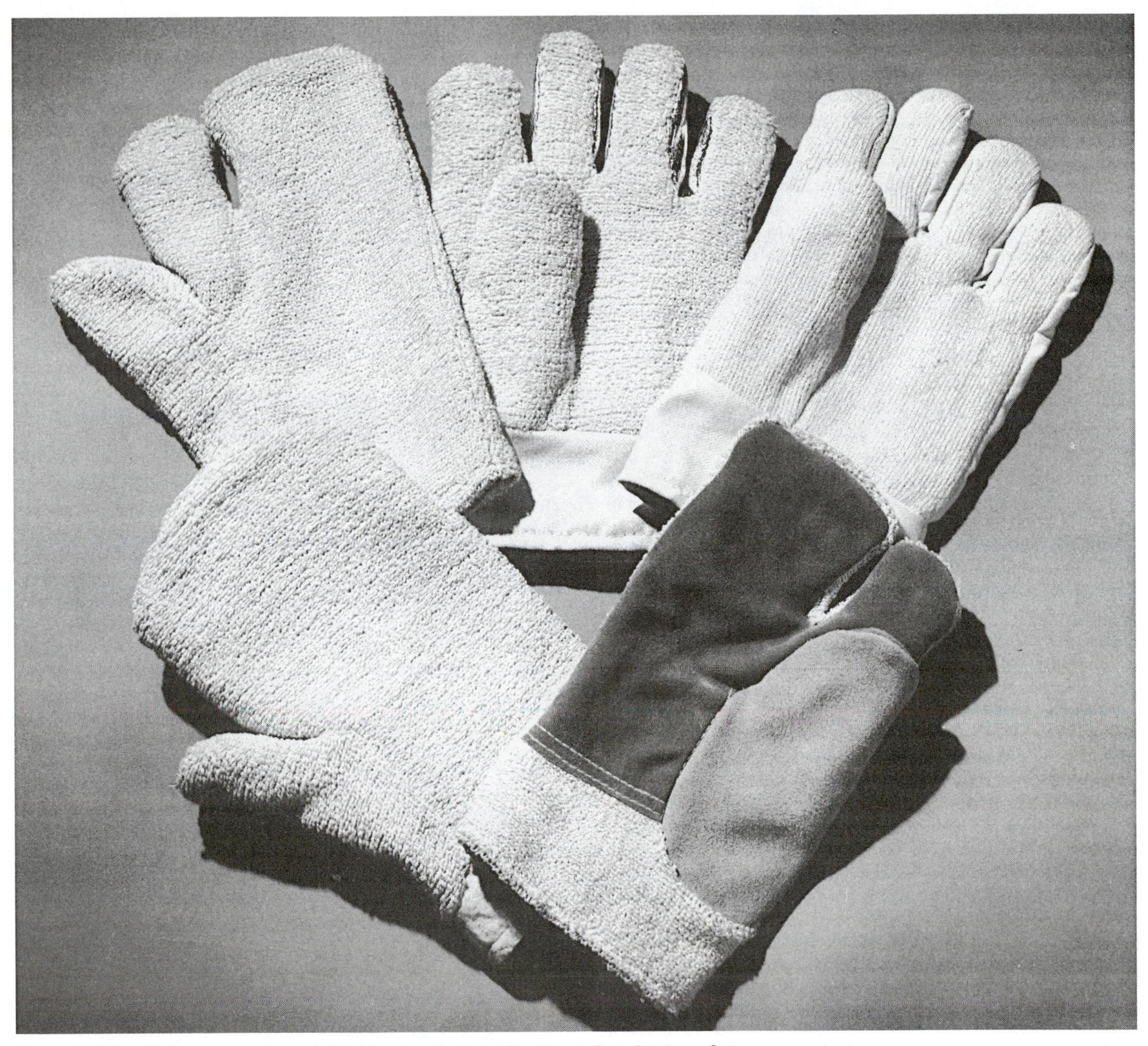

**Figure 17-26b.** Various types of specialized hand protectors. (Courtesy A-Best Products Co.)

clothes, are not impervious and should not be used where impervious materials are indicated.

**Gloves** coated with rubber, synthetic elastomers, polyvinyl chloride, or other plastics offer protection against all types of petroleum products, caustic soda, tannic acid, muriatic and hydrochloric acid. They are also recommended in the handling of sulfuric acid. Less deterioration takes place than with natural rubber. These gloves are available in varying degrees of strength to meet individual conditions.

Gloves should be long enough to come well above the wrists, leaving no gaps between the glove and the coat or shirt sleeve. Long, flaring gauntlets should be avoided unless they are equipped with locking devices to assure a snug fit about the wrist. Such gauntlets are especially desirable when acids and other chemicals are being poured.

In this operation, the chemicals may splash, and unless precautions are taken, harmful results may occur. When caustic substances and harmful solvents are being poured from large to small containers, sleeves should be worn outside gauntlets.

In many operations, rubber gloves with extra long cuffs have been used to advantage. The cuffs of these gloves are made with a heavy ridge near the top edge, which, when turned back, forms a trough to catch liquids running down the wrist or forearm. Some are made with beads near the cuff to hold inserts which form a liquid tight seal when inserted into a sleeve.

Gloves or mittens having metal parts or reinforcements should never be used around electrical apparatus.

Work by such persons as linemen and electricians on energized or high-voltage electric equipment requires specially made and tested rubber gloves.

Over-gloves of leather must be worn to protect the rubber gloves against wire punctures and cuts and to protect the rubber in the event of flash. Frequent testing and inspection of linemen's rubber gloves are essential, and those gloves failing to meet original specifications should be discarded.

Where acid may splash, rubber boots or rubber shoes also should be worn. The tops of the boots should be high enough to come beneath the edge of the apron. If shoes are worn, the tops should come inside the legs of impervious trousers. These precautions keep the liquid from draining off apron or trousers into the footwear.

**Procedural setups.** When personal protective equipment is used in a corrosive atmosphere, a rigid procedure should be set up for taking care of it after use to prevent contact with contaminated parts. Before the equipment is removed, whether or not it has come in contact with the corrosive chemical, it should be thoroughly washed with a hose stream. Boots, coats, aprons, and hats should then be removed, followed by removal of the gloves. This is the logical order of removal if the coat has been properly put on with the sleeves outside the cuffs of the gloves.

Hands should be washed thoroughly before face shield and goggles are removed. The hands and face should then be thoroughly washed again, but complete shower and change of clothing are much more desirable.

Gloves can be laundered to remove contaminants, prolong glove life, sanitize, and permit reissue. Gloves used in toxic chemical service must be laundered with special care to be sure the chemicals are thoroughly washed out.

For protection against exposure to oil and the various other compounds which rapidly attack ordinary rubber, all the equipment discussed can also be obtained in plastic and synthetic rubbers.

### Cold weather clothing

In recent years thermal insulating underwear has become popular among outdoor workers because of its lightweight protection from the cold. Thermal knit cotton patterned after regular underwear, quilted materials, or synthetic polyester fabric quilted between nylon are common types of construction.

While the polyester and nylon quilted material does not catch fire any easier than cotton, once it starts burning, it melts, forming a hot plastic mass, not unlike hot pitch, which will adhere to skin and cause serious burns. Quilted insulating underwear is now available that has been made fire retardant to combat this danger.

Other special fabrics available include a nylon material that chars at a relatively high temperature and does not melt, a glass fiber material for special uses, and a breathable fabric used with a sandwich of cotton or similar material to offer excellent cold weather protection.

In order to get a good idea of how warmly to dress in cold weather, it is frequently not enough to look at only the thermometer. It may read 35 F (1.7 C), but if there is also a wind of 45 mph (72.4 km/h) it will feel like −35 F (−37 C). (See Figure 16-4 in Planning for Emergencies.)

### Special clothing

Safety experts have been very ingenious in developing many highly specialized types of clothing for protection against special hazards. A partial list includes such items as:

**High visibility and night hazard clothing** for construction, utility, and maintenance workers, police and firefighters whose work exposes them to traffic hazards.

**Disposable clothing** made of plastic or reinforced paper is available for exposure to low level nuclear radiation, or for use in the drug and electronic industries, where contamination may be a problem.

**Leaded clothing** of lead glass fiber cloth, leaded rubber, or leaded plastic for laboratory workers and other personnel exposed to X rays or gamma radiation.

**Electromagnetic radiation suit,** which provides protection from the harmful biological effects of electromagnetic radiation found in high level radar fields and similar hazardous areas.

**Conductive clothing,** made of a conductive cloth, is available for use by linemen doing barehand work on extra-high voltage conductors. Such clothing keeps the worker at the proper potential.

For special applications, manufacturers have a vast number of materials they can draw upon to meet specific hazards.

## REFERENCES


American Conference of Governmental Industrial Hygienists, Cincinnati, Ohio 45201. *A Guide for Control of Laser Hazards,* 1976.

American National Standards Institute, 1430 Broadway, New York, N.Y. 10018.

*Identification of Air-Purifying Respirator Canisters and Cartridges,* K13.1.

*Method for Measurement of Real-Ear Protection of Hearing Protectors and Physical Attenuation of Earmuffs,* S3.19-R1979.

*Practice for Occupational and Educational Eye and Face Protection,* Z87.1-1979.

*Practices for Respiratory Protection,* Z88.2-1980.

*Protective Headwear for Industrial Workers,* Z89.1-1986.

*Safe Use of Lasers,* Z136.1-1986.

*Safety in Welding and Cutting,* Z49.1-1983.

*Safety Requirements for Window Cleaning,* A39.1-1969.

*Safety-Toe Footwear,* Z41-1983.

American Society for Testing and Materials, 1916 Race St., Philadelphia, Pa. 19103.

*Standard Specification for Rubber Insulating Gloves,* D 120-70, ANSI J6.6.

*Standard Specification for Rubber Insulating Sleeves,* D 1051-70, ANSI J6.5.

Compressed Gas Association, Inc., 1235 Jefferson Davis Hwy., Arlington, Va. 22202.

*Commodity Specification for Air,* G-7.1

*Oxygen,* C-4.

*Oxygen-Deficient Atmospheres,* SB-2.

Gasaway, Donald C. *Hearing Conservation: A Practical Manual and Guide.* Englewood Cliffs, N.J.: Prentice-Hall, Inc., 1985.

Industrial Safety Equipment Association, 1901 N. Moore St., Alexandria, Va. 22209.

Mack Publishing Co., 208 Northampton St., Easton, Pa. 18042.
*U.S. Pharmacopoeia.*

National Safety Council, 444 N. Michigan Ave., Chicago, Ill. 60611.
Industrial Data Sheets
*Flexible Insulating Protective Equipment for Electrical Workers,* 598.
*Respiratory Protective Equipment,* 734.
*Industrial Noise and Hearing Conservation.*
"Safety Award Clubs."
*Safety With the Laser,* NSNews Reprint No. 17.

Plog, B. A., ed. *Fundamentals of Industrial Hygiene,* 3rd ed. Chicago, Ill., National Safety Council, 1988.

*Respiratory Protective Devices Manual.* Committee on Respirators, P.O. Box 453, Lansing, Mich. 48901.

U.S. Department of the Interior, Washington, D.C. 20240. 30 C.F.R. Chapter 1, Subchapter B, Respiratory Protective Devices, Tests for Permissibility, Fees; Part 11. Note. The *Code of Federal Regulations* is available through the U.S. Government Printing Office, Washington, D.C. 20402. (See pages 56 and 57 in Chapter 2.)

U.S. Department of Health and Human Services, Public Health Service, Center for Disease Control, National Institute for Occupational Safety and Health. *NIOSH Certified Equipment List,* October, 1986, Pub. 87-102. Washington, D.C.: U.S. Superintendent of Documents, 1986.
*Respiratory Protection—An Employer's Manual.* National Institute for Occupational Safety and Health, Division of Technical Services. Cincinnati, Ohio, 1978.

# 18

# Industrial Sanitation and Personnel Facilities



THERE ARE FIVE INDUSTRIAL HEALTH AREAS that must be kept clean, sanitary, and well equipped for employee health and convenience:

1. Potable water supply for drinking, washing, and food preparation
2. Adequate disposal of sewage and garbage
3. Adequate personal service facilities
4. Sanitary food service
5. Satisfactory heating and ventilation.

These areas must be given the necessary attention if employees are to work efficiently, with the assurance that their health and welfare are well protected.

As with other industrial functions, maintaining a clean, sanitary work environment should rate a separately managed and comprehensive department if management and employees are to fully benefit. Sanitation, for example, must be properly managed and effectively integrated with production and maintenance if it is to be safe, efficient, orderly, and economical.

The general rules for sanitation include:

- An approved piping and storage system
- Good housekeeping—as clean as the nature of the work allows
- Personal cleanliness
- A good inspection system

Where wet processes are used, drainage must be maintained.

The director or supervisor responsible for maintaining the work environment must be at a level high enough in the organization to permit him to sustain his function against the pressures exerted by other departments, and to provide surveillance of the entire company or plant environment, in order to keep it at an appropriate and balanced level of cleanliness and order.

Some firms are appointing a manager of environment and safety, who has additional product safety responsibility.

## DRINKING WATER

Most plants receive water for drinking, washing, and food preparation from a municipal supply. As delivered to the plant meter, this water must meet U.S. Environmental Protection Agency (U.S.EPA) standards for contaminants that may be detrimental to health. The U.S.EPA requirements are monitored and enforced by a designated state agency.

### In-plant contamination

The fact that water is potable when delivered to the plant meter does not necessarily mean that it will be so when used, for there are many opportunities within a plant for water to become contaminated.

There are at least twelve infectious waterborne diseases commonly caused by drinking contaminated water (see Table 18–A). Sickness also can arise from chemically contaminated water.

One of the most common causes of contamination of the water supply is direct or indirect cross-connection with a source of nonpotable water. *Cross-connection* is the term for accidental entry of sewage or septic water into a drinking water supply. Often, sewer pipes and drinking water supplies run parallel—beneath sidewalks or next to a roadside ditch. Should both pipes develop leaks, allowing the two sources to cross-connect, contamination of the water main is possible. Contamination risk from cross-connection is greatest when utilities are badly installed, poorly maintained, or old.

Table 18-A. Infectious Waterborne Diseases Caused by Drinking Contaminated Water

| *Diseases* | *Incubation Period* | *Symptoms* | *Frequency* | *Mode* |
|---|---|---|---|---|
| Gastroenteritis | Variable | Lethargy, nausea, diarrhea, cramps, and other stomach ailments | Causes over ½ of waterborne disease | Sewage or chemicals in water |
| Bacterial shigellosis | 1–7 days | Fever, vomiting, stomach cramps, diarrhea | Serious in some cases. Common | Sewage in water |
| Salmonellosis | 6–72 hours | Abdominal pains, fever, vomiting, and nausea | Less common | Sewage in water |
| Typhoid fever | 1–3 days | Abdominal pains, fever, chills, diarrhea or constipation, and tearing of the intestines | Rare occurrences | Sewage in surface water |
| Giardiasis | 1–7 days | Chronic diarrhea, weight loss, intestinal and stomach gas, bloating, and anorexia | Outdoor enthusiasts commonly affected | Surface water and food |
| Hepatitis A | 14–45 days | Jaundice, nausea, anorexia, fever, and general physical discomfort | Rare in the U.S. | Drinking and swimming |

Other causes of contamination are improper maintenance of drinking and cooking facilities and improper installation of plumbing facilities permitting backsiphonage of used water. An example of a backsiphonage incident would be pumping well water into an application tank to dilute stock chemicals, such as pesticides, and having the pumping interrupted. A vacuum created by the pumping interruption (caused by a power outage) could backsiphon the solution—now a toxic chemical—into surrounding wells. This rarely occurs, however, because backflow prevention devices are available. Backsiphonage incidents also can affect users of publicly maintained water supplies.

The impact of a backsiphonage incident could be much worse if it happened at a food processing plant or other such establishment. Just a few disease organisms are capable of infecting an entire piping network if they are backsiphoned to the community well or pump station.

Because of the serious consequences of contamination, the integrity of the drinking water system must be maintained throughout the plant. If there are piping systems containing water used for other purposes, such as sprinklers and fire hydrants or manufacturing processes, each should be clearly identified, particularly at outlets. There should be no direct connection between drinking water and other water systems except with the installation of an approved backflow prevention device. Long dead-end runs of pipe that cannot be flushed or drained and which might serve as a reservoir for contaminated water should also be eliminated. The location of piping approved for potable water use should be easily identified.

Nonpotable water may be used for cleaning work premises (other than food preparation and personal service rooms), provided it does not contain concentrations of chemicals or fecal coliform bacteria.

Where there is a possibility of misuse or cross-connection of pipelines, all nonpotable water lines should be marked as being unsafe for drinking, washing the person or utensils, or food areas, personal service rooms, or clothes washing.

### Plumbing

Fixtures and faucets should be installed to prevent backsiphonage of contaminated water if the pressure drops in the supply line. Faucets and similar outlets should be at least 1 in. (2.5 cm) above the floodrim of the receptacle below. To prevent backflow into the drinking water supply, surge tanks and air gaps also may be required in the fill lines to process equipment.

Another common source of contamination is open joints in underground supply lines subject to seepage from ground water or water from leaky sewers. This condition can arise where pipes are subject to vibration or corrosion and the joints between pipes open mechanically or the pipe sections crack. Codes usually prohibit installation of sewer and drinking water lines in the same trench, unless the sewer line is placed at a much greater depth and a certain horizontal offset is provided.

Frequently, contamination of the water supply results when a system is opened for repair or alteration and the new pipe is not disinfected and properly flushed with clean water before being put back into service.

If the supply for sprinklers and fire hydrants is the same as that for drinking water, hydrant drains or "weeps" connected directly to sewer lines may be a source of contamination. An open standpipe or reservoir may also permit contamination.

Plastic pipe can be considered, but be sure to check local code requirements. Unplasticized PVC (polyvinyl chloride) is good for cold water lines. Hot water up to 165 F (75 C) and 100 psi (690 kPa) can usually be handled in pipe made of chlorinated polyester or unplasticized PVC. Only plastic pipe stamped with National Sanitation Foundation (NSF) approval is to be used for potable water.

The Safe Drinking Water Act restricts use of lead pipe, solder, or flux in the installation or repair of any public water system or in residential plumbing connected to a public water system. By U.S.EPA definition, an individual water system that furnishes 15 or more service connections or that supplies an average of 25 or more persons for over 60 days out of the year is considered a noncommunity public water system and is under the jurisdiction of the state public water systems control agency.

### Private water supplies

Industrial establishments in outlying districts commonly supply and treat their own water from private sources. Such installations should be made and operated under the supervision of a thoroughly trained and experienced sanitary engineer. The information in this and the next few sections is not meant to substitute for such supervision.

All underground and surface waters to be used for drinking purposes should be considered contaminated until proved otherwise. The water supplied from private sources for the personal use of plant personnel should meet the requirements of the U.S.EPA primary drinking water standards. As a rule, ground water collected from deep-drilled wells will be free of biologi-

**Table 18-B.** Recommended Limiting Concentrations of Contaminants in Drinking Water

| *Undesirable Substance* | *Concentrations above which water should not be used if other sources are available (mg/L)* | *Dangerous Substance* | *Concentrations above which water supply should be rejected (mg/L)* |
|---|---|---|---|
| Chloride | 250.0 | Arsenic | 0.05 |
| Copper | 1.0 | Barium | 1.0 |
| Fluoride | 2.0 | Cadmium | 0.01 |
| Iron | 0.3 | Chromium (hexavalent) | 0.05 |
| Manganese | 0.05 | Cyanide | 0.2 |
| Nitrate | 45.0 | Fluoride | 4.0 |
| Sulfate | 250.0 | Lead | 0.05 |
| Total dissolved solids | 500.0 | Nitrogen | 10.0 |
| Zinc | 5.0 | Selenium | 0.01 |
| | | Silver | 0.05 |

Data from U.S.EPA.

cal contamination but may be contaminated by various minerals. In contrast, shallow wells are more likely to have biological and man-made chemical contamination.

Surface water sources, such as lakes and streams, should always be treated with disinfectant and filtered for potable use. Many sources of contamination are possible from a surface water source, so professional advice is recommended in designing and operating a surface water treatment system.

Where there are several sources of water, the final choice will be influenced by (a) the daily water requirements, (b) the amount of treatment which water from each source will need to meet the purity standard, and (c) the potential each source has for additional contamination.

The daily per-person water requirements of an industrial plant can be estimated as follows: 15 to 20 gallons (55 to 75 liters) for drinking, lavatory, and toilet usage; 20 to 25 gallons (75 to 95 liters) per shower; and 5 to 10 gallons (20 to 40 liters) per meal if food is prepared on the premises.

## Water quality

The water supply source must be evaluated on the basis of the chemical and biological contaminants it may contain. Table 18-B lists the limiting concentrations of two classes of contaminants. The U.S.EPA has established two sets of standards relating to the quality of drinking water. *Primary standards* are mandatory and cover all contaminants which are considered health related. These fall into the general categories of: inorganic chemicals, organic chemicals, turbidity, microbial agents, and radioactivity. *Secondary standards* cover the aesthetic qualities of water, such as taste, odor, and color. Compliance with secondary standards is not mandatory, but it is strongly recommended that all potable water meet these criteria in order to be palatable and pleasing to use.

State agencies that have been delegated enforcement authority for the drinking water program by the U.S. EPA must establish state regulations at least as stringent as the current federal requirements. These regulations are under constant revision, so it is recommended that operators of all potable water supplies obtain a copy of the current state regulations and guidance materials.

The equipment necessary to treat water and make it potable depends on the degree of contamination and the likelihood that the source will become more heavily contaminated later. These factors can be evaluated only on the basis of a thorough sanitary survey of the water source. Such a survey will determine not only the type of treatment necessary, but also the nature and frequency of periodic laboratory tests of the source water and the treated water.

## Wells

The safest source of water is often a drilled well with intake below the water table. Such wells show a reliable yield and are free from bacterial and chemical contamination. If both well and city water are used, there should be no cross-connection between the two systems. Be sure to check the local code.

The wellhead should be located as far away as possible from sewage lines, septic tanks, and sewage drainage fields or process waste disposal systems. A 200-ft (61 m) separation is usually considered a minimum. Toxic chemicals leaking from pipelines, tanks, or lagoons, or that have been spilled or disposed of on the land surface can contaminate a well at much greater distance, and can cause a serious health threat. Pesticides and herbicides spilled on or applied to land in the area of a well can also contaminate a well at considerable distance, depending upon geology.

To prevent contamination of the underground water by seepage of surface waters, the space between the casing and the surrounding area should be sealed with a cement grout to a minimum depth of 10 ft (3 m) below the finished ground level or floor. As a further precaution, the casing should be grout-sealed to the lowest impervious stratum it passes through.

Well installations must meet the current state well code or standards for such design considerations as siting location criteria or type of materials that may be used. Most state codes require that the top of the well casing project at least 12 in. above the ground surface. The wellhead should not be covered over by paving or other material which would make access difficult.

Submersible, turbine, or jet pumps may be considered. Submersibles are located in the well and do not require a pumphouse. A safety factor is provided by two wells and two pumps. An automatic alternator can take effect if either unit fails.

## Disinfecting the water system

The pipes, reservoirs, standpipes, pump, and well casing of a new system should be thoroughly disinfected before being put into service. An old system carrying treated water for the first time following an extended outage also should be disinfected on the discharge side of the treatment plant, and a system that has been opened for repairs should be disinfected before being put back into service.

A drinking water system can more easily be disinfected by filling with water containing not less than 100 mg/liter of available chlorine. The solution should be allowed to remain for 24 hours in a new system or one which has not previously held treated waters. If the system previously has held treated water and is being put back into service following minor repairs, 12 hours probably will be sufficient. However, if *giardi lambia,* the most serious contemporary infectant, is suspected to be present other measures should be taken, such as filtering. Point-of-use treatment methods such as boiling or use of portable filters may be appropriate for temporary problems.

To determine the success of the disinfecting job, the residual chlorine in the solution is measured at the end of the required time. Test kits for this purpose are commercially available and are easy to use. If tests show residual chlorine, the biological chlorine demand of the system has been met and the system can be connected to the drinking water supply, flushed out, and put into service. If no residual chlorine is present, the system should be drained and recharged with new disinfectant solution and the procedure repeated.

If the system contains a standpipe or reservoir, the disinfectant solution can be added through it. Otherwise, the solution can be supplied in a temporary reservoir on the supply side of the system pump and injected through it. A solution containing 500 mg/liter available chlorine, applied with a fog nozzle, will disinfect standpipes and covered reservoirs.

Underground water supplies can become contaminated while being developed. If so, they too will have to be disinfected as follows. After the 24-hour yield of the well has been determined, the test pump should be run to clear the well of turbidity. A chlorine solution then should be added to the well to make, with the 24-hour yield, a solution of 50 mg/liter. The permanent pumping equipment then is connected to the wellhead and operated until the discharge has a distinct odor of chlorine.

There are several methods for uniformly distributing the disinfecting solution. The well casing can be sealed and the solution injected under pressure, or the solution can be added from a hose or a small pipe at several levels beneath the surface of water in the well. The chlorine solution should remain in the well for 24 hours, as mentioned.

## Water purification

Of the several methods of water purification available, filtration and chemical disinfection are the most practical for industrial private water supplies.

**Filtration.** This method of water purification primarily is used to clarify turbid waters, but it can also serve to remove some bacterial contamination. Filtering plants are of two types: slow filters and rapid filters.

Slow sand filters will clarify turbid waters when operated at a rate of 25 to 50 gallons per day per square foot (100 to 200 liters per day per square meter) of filter area. Such filters should be made with 0.25 to 0.35 mm sand and should be at least 20 in. (50 cm), but preferably 36 to 40 in. (90 to 100 cm), deep. During the operation of the filters, the film that accumulates on the surface of the sand must be kept below water level because this film increases the effectiveness of the filter.

Rapid sand filters, made with a uniform 0.4 to 0.5 mm sand with a depth of 30 in. (75 cm), will handle about 3,000 gallons of turbid water a day per square foot (12,000 liters a day per square meter) of filter surface.

Both types of filters should be made under competent engineering supervision and operated under continuous inspection. Depending upon the water source, filters can require a presedimentation basin for preliminary water treatment. Filters should be provided in pairs so one can be removed for cleaning and maintenance without disrupting the supply of filtered water.

**Disinfection.** Chlorine is generally the best available disinfecting agent for drinking water. It can be added to the water directly as a gas or as a soluble salt (calcium hypochlorite or chlorinated lime—refer to the Council's *Fundamentals of Industrial Hygiene,* for hazards of these chemicals).

Small-capacity chlorinators, which inject gaseous chlorine into a water system, are available and easy to operate. Injection pumps that supply high concentration chlorine solutions to the system at a proper rate are preferable, however, because of the ease and safety of operation.

Standby equipment should be maintained at all chlorinating stations, together with an adequate supply of spare parts. Gas maks that are effective against chlorine and a small bottle of ammonia to test for leaks should be kept just outside of areas in which chlorine is stored or used. Masks should be inspected at regular intervals, and authorized employees should be trained in emergency procedures.

The chlorinator should be adjusted to leave a chlorine residue of about 0.2 mg/liter in the water after 20 minutes of contact between chlorine and the untreated water. Test kits are available that will measure residual chlorine rapidly.

Small quantities of water for emergency use can be disinfected in one of several ways. Commercial preparations should be used according to the manufacturer's instructions. Boiling water for five minutes, or adding four drops of household bleach (hypochlorite solution, 4 percent available chlorine) or two or three drops of common tincture of iodine to a quart of water and allowing it to stand for 30 minutes also will produce safe drinking water. Its flatness and medicinal taste can be partially removed if it is aerated after disinfection by being poured from one container to another.

## Water storage

Reservoirs or standpipes for treated water should be completely enclosed and located so that accidental contamination is impossible. A reservoir large enough to hold a 48-hour reserve supply of treated water will provide adequate supply in the event of unusually heavy water use or during supply failure. Vents should be fitted with screened downspouts well above floor level. Entrance manholes should be enclosed by watertight frames at least 6 in. (15 cm) higher than the surrounding surface and fitted with watertight covers extending at least 2 in. (5 cm) down the outside of the frames. When not in use, the cover should be closed and locked.

A reservoir permits full use of a smaller well and pump and still provides a buildup for peak demand.

## SEWAGE, WASTE, AND GARBAGE DISPOSAL

In some outlying districts and rural areas, industrial plants must provide their own sewage disposal systems. State and/or local ordinances governing disposal must be adhered to. The disposal of process waste often requires separate facilities. All states and many local agencies require prior examination of all plans concerning waste treatment and disposal.

Provisions also should be made for the storage and collection or disposal of garbage and refuse.

### Building drains and sewers

The in-plant sanitary sewage collection system should conform to state and local codes.

Every fixture should be properly trapped and vented by means of drain(s) and stack(s) serving it to prevent the discharge of sewer gases into the building and to assure proper draining. Traps and especially grease interceptors (such as those placed in waste pipes serving the plant cafeteria and kitchen) and interceptors designed to collect other particulate foreign materials should be of adequate size, located for easy access, and should be cleaned periodically. Be sure *not* to (a) install a type of trap that is prohibited by the local code, or (b) place a trap in a prohibited location.

The building drain and sewer may be constructed of extra-heavy cast iron, bell-and-spigot pipe with drainage fittings. This material is less susceptible to clogging and much easier to clean out than pipes made from other materials, and it provides good "insurance" when installed under floors that would be expensive to tear up. Lead or other suitable material should be used for joints. The sewer should be tight under a 10-ft (3 m) head of water. Plastic and copper plumbing materials are also approved for drainage use in many areas.

A cleanout should be provided where the building drain passes through the building wall, and at other selected places as the code requires. Check local codes for trap size, permitted locations, and strainer requirements. In some areas, codes call for installation of backwater valves to prevent backup in the sewer line. These should be located where they are accessible for inspection and cleaning.

### Septic tanks

The main function of a septic tank is to separate solid from liquid wastes. It serves a secondary purpose of permitting aerobic bacteria to convert some solid waste matter into liquid waste.

The effluent from the dosing chamber of the septic tank should pass through a tight sewer over as short a distance as possible to the disposal field. The distribution box or boxes, which are the first unit of the disposal field, equalize the flow of liquid waste through the disposal field lines and serve as an inspection point where the quality of the effluent can be checked. Sludge must not be carried into the disposal field because it will clog the absorption system.

The disposal field should have enough discharge lines to permit the daily liquid waste to be absorbed and disposed of by aerobic bacteria. The number of lines and the length of each are based on the permeability of the soil to the liquid.

A septic tank should be located at least 200 ft (61 m) from any drinking water source. Surface drainage from the area around the septic tank should not be permitted to reach the water sources. The tank itself should be located well below neighboring water sources.

On-site wastewater treatment systems and portable wastewater holding tanks are available commercially for short-term projects. Methods of reducing water consumption should be strongly considered as a means of achieving more cost effective wastewater management. If required, they can be specified with a grinder and chlorination apparatus.

For land disposal in a septic system, the disposal field should be at least 100 ft (30 m) from any water supply, 50 ft from any stream, and 10 ft from any building or property line. A minimum distance of 50 ft between a disposal field and the head of a deep well is acceptable if the well has a watertight casing extending downward at least 50 ft.

A detailed map should be made of the sewage disposal system, showing the location of septic tanks, distribution boxes, and tile fields. The area covering the disposal field should be kept free from vehicular traffic. Sodium and potassium hydroxides and similar "conditioning" agents, as well as brine discharges, should not be emptied into the sewage disposal system or building drains. If they are, a clogged and useless disposal field can result. The effluent of the absorption field should be monitored frequently to assure the proper function of the system.

Accumulated solids should be removed from septic tanks on an annual or more frequent basis. If the tanks are allowed to become full, the solids will overflow into the drainage field, and, if these pipes become clogged, the field may have to be cleaned or replaced.

### Garbage disposal

Companies and plants having food services for their employees must provide for proper disposal of food wastes and refuse. There are several methods of disposal—local ordinances should be followed in each case.

Many plants collect garbage and store it for later pickup and disposal by municipal or private collection services. Large outside metal storage receptacles and compactors should be fenced in or covered and locked to prevent children from entering and playing in them and animals from strewing refuse about.

Garbage and refuse containers should be metal or plastic with tight-fitting covers to prevent the entry of insects or rodents. Containers should be easy to clean and handle, and washed with detergent-deodorant solutions. The use of plastic or polyethelene bags or liners is desirable and often required by many refuse haulers. The liner and contents can be removed easily and will help keep containers clean. Garbage containers in the plant should be located so employees will not throw food waste in waste baskets or other unsuitable receptacles. Collections should be frequent enough to prevent undue accumulation of garbage.

Discharge of ground food waste into the municipal sewage system is an acceptable method of disposal in many localities. Local ordinances should be checked. Installation of food waste disposers in the plant kitchen can provide a convenient, efficient way of eliminating the need for garbage storage and collection. Disposers should be located, grounded, and installed according to approved plumbing practices and local code requirements.

Kitchen employees should be instructed to use nonmetallic tampers, keep silverware out of the disposer, and clean the metal trap daily. The disposer should be stopped and the power disconnected before cleaning or clearing.

### Insect and rodent control

The bacterial problem often is the limiting factor for securing good sanitary procedures. We must recognize other potential biological hazards, the flying, leaping, crawling vermin, such as insects and rodents, which create esthetic and microbial problems.

In plants where insect or rodent infestation is a problem, it is always best to employ a professional exterminator. The hazard of poisonous chemicals and preparations should not be risked by personnel with little or no previous experience in exterminating. In many states, toxic rodent control chemicals may be handled only by persons with appropriate training and licensing by the state.

Communication with all plant departments, food vendors, and others concerned should be set up well in advance of the exterminator's visit. Outside signs can provide extra warning to neighboring plants or residents to protect children and pets from exposure.

To protect the workplace against entrance of vermin, wire mesh and metal screens can be used near the base of foundations.

### Refuse collection

Hazards in refuse collection vary with the type of equipment used and the various conditions surrounding the operation. A frequent cause of accidents involves packing blades that can cause partial loss of fingers, hands, arms, and feet. Other hazards arise from "booby traps" unwittingly laid by the companies refuse collectors serve—loose broken glass in a refuse container, lightweight trash cans filled with heavy objects (like chunks of concrete), heavy objects concealed by paper or other trash, hose or other obstacles strewn along the pathway to a rubbish can. Containers that are rusted through or have unserviceable handles increase the risk of job injury.

The incident rate for refuse collection for those units reporting injuries for 1985 to the National Safety Council was 32.6. This compares to about 6.3 for the 1985 all-industry average.

Cuts, lacerations, and punctures accounted for about 14 percent of the lost-time injuries; this compares favorably to industry in general where 17 percent of the workers suffered such injuries. Wearing heavy work gloves minimizes these types of injuries.

Refer to the booklet, *Operation Responsible—Safe Refuse Collection,* published by the U.S. Environmental Protection Agency (see References).

## PERSONAL SERVICE FACILITIES

Drinking fountains, washrooms, locker rooms, showers, and toilets—all personal service facilities contributing to employee comfort—should be conveniently located. These facilities make up an essential part of the occupational health program in most industries.

### Drinking fountains

Sanitary drinking fountains, one to about every 50 people, should be installed at convenient places throughout an industrial plant, in accordance with ANSI standard A112, *Specifications for Drinking Fountains,* and local code requirements. The fountain should have an angle jet and a lip guard (Figure 18-1). A waste container may be located at each fountain. It is important that the stream projector cannot be flooded or submerged in the event of stoppage, and that the stream be directed and projected so that it cannot be contaminated by the user. In dusty areas, fountains should be covered.

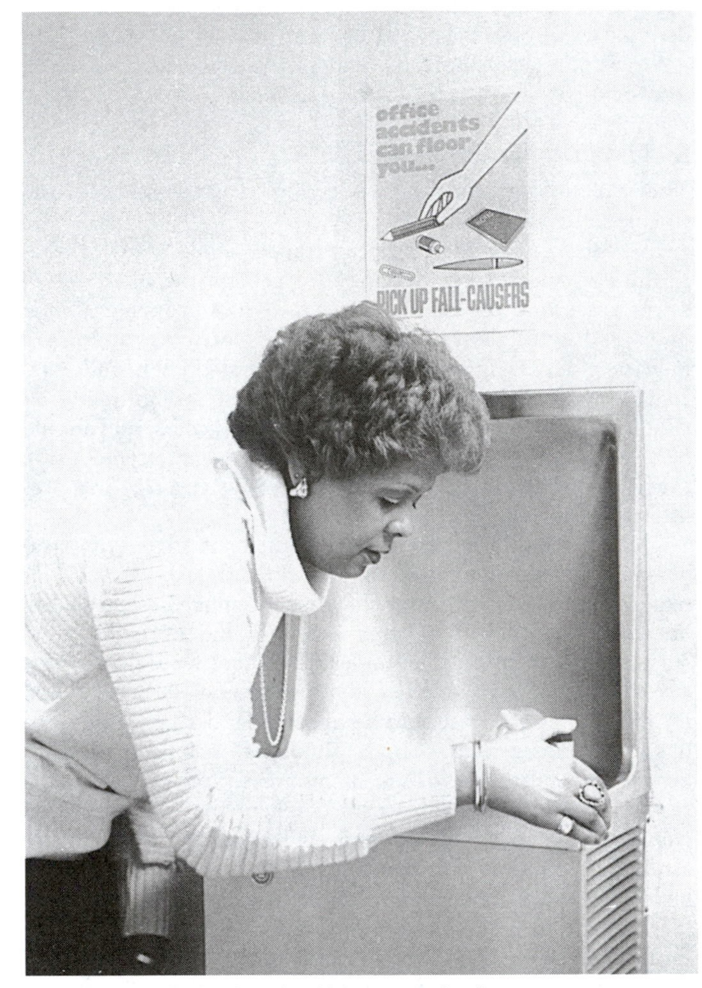


**Figure 18-1.** Fountains should be conveniently located so that employees can maintain their daily water intake. A safety poster can be placed near the fountain.

The water temperature should be 50 to 55 F (10 to 13 C) for heavy manual labor, or 45 F (7 C) for less-active office work. If ice is used, it should be in a separate compartment, without direct contact between the water and the ice.

Where city water is available on construction work, a water line can be extended to upper floors as the building is erected. A standard drinking fountain can be installed on each floor.

On some types of work, such as highway, pipeline, power line construction, and timber clearing, the drinking water source is so remote that it is impractical to pipe water to the job. Some companies have successfully solved this problem by using portable drinking fountains. These fountains have an insulated tank equipped with an angle jet drinking nozzle. The tank has an air pump and pressure release valve so that it can be pumped up to the necessary operating pressure. Containers should be kept scrupulously clean and should be sterilized daily with steam, boiling water, or chlorine solution.

Under no circumstances should use of a common drinking cup or ladle be permitted. If drinking cups are required, they should be single-service paper cups kept in a sanitary container at the drinking faucet, with a receptacle provided for disposal.

ANSI standard A117.1, *Buildings and Facilities – Providing Accessibility and Usability for Physically Handicapped People,* section 5.7, mentions modifications necessary for use by the handicapped.

Installation of drinking fountains in any toilet room usually is not permitted. Installation of bubblers as an integral part of– or connected to–another fixture, such as a lavatory or sink, usually is not permitted by codes.

Carafes (vacuum bottles) that are frequenty used in private offices are a potential source of bacterial contamination. They should be rinsed and refilled daily and cleaned periodically, using a sanitizer such as a cationic quaternary ammonium germicide.

## Salt tablets

The normal diet contains more than enough salt to replace that lost in perspiration. In extreme cases, where workers have not had time to become acclimatized (about 10 to 14 days), lightly salted (0.1%) water can be provided. Salt tablets are not recommended under any circumstances.

## Washrooms and locker rooms

ANSI Z4.1, *Minimum Requirements for Sanitation in Places of Employment,* serves as a guide to the types and sizes of washrooms, locker rooms, and accessories.

A large, single washroom and locker room for each sex may be sufficient for a compact plant or establishment employing fewer than 500 people. Washrooms in a large one-story plant generally are scattered throughout the building. If the plant consists of a series of separate buildings with only a few people working in each, all the facilities can be placed in a centralized building. This arrangement has been successful in such establishments as chemical plants, oil refineries, and railroad yards.

If the plant is relatively small, it is advisable to have the dressing rooms, lockers, and washrooms near the entrance. In a larger plant it is better to have these facilities in a single building centrally located or in several buildings near the work areas. In some industries, washing facilities are also used to prevent product contamination (Figure 18-2).

**All washing facilities** should be maintained in a sanitary condition. Each lavatory should have hot and cold water, or at least must have tepid running water, and hand soap or similar cleansing agent.

Waterless skin cleansers are not substitutes for soap and water, but are convenient for special use or where water is scarce.

One or more of the following means of drying hands and/or face must be convenient to the lavatories–individual paper or cloth handtowels or sections, clean individual sections of continuous cloth toweling, or warm air blowers.

Common-use towels should not be permitted. Paper towels should be soft enough not to cause irritation. They should be kept in a covered container with a disposal receptacle nearby.

Hot air hand driers should be well secured either to the floor or the wall to prevent loosening of the fixtures or the electric element. The equipment must be grounded and permanently installed without extension cords or plugs. Blowers must provide air at not less than 90 F (32 C), nor more than 140 F (60 C).

**Figure 18-2.** These employees must "prep" before they can enter critical, super-clean areas. Washing hands and rubber gloves, then drying them under air dryers prevents lint and dust contamination. (Printed with pemission from Western Electric Company, Allentown, PA.)

For industrial occupancies of up to 100 employees, one lavatory for each 10 employees is recommended; for more than 100 employees, one lavatory for each additional 15 employees is considered adequate. In industries where workers need additional washing time, one lavatory for every five employees is recommended.

Circular wash basins (Figure 18-3) of stainless steel, stoneware, enameled iron, or other materials impervious to moisture permit a number of persons to wash at the same time by means of center water sprays that are continuous or are controlled by a treadle. These basins are easily kept clean and sterile. Their construction prevents splashing and spilling of water.

To eliminate standing water, which can transmit disease from one employee to another, lavatories should have no stoppers. A mixing faucet or a spray will permit employees to wash in a flowing stream with controlled temperature. Knee-actuated water controls are available.

Wherever practicable, a thermostatic control should be installed in the hot water supply system in order to keep temperature below 140 F. Injecting live steam into tanks or lines of a cold water system (to make warm water) is dangerous, since failure of pressure in such a system could release steam through the taps.

A regular maintenance program for equipment should be in effect, and employees should be requested to report defective equipment. Broken faucets and valve handles may cause serious cuts or lacerations. Handles should be made of metal, not a breakable material, such as porcelain. If leaky faucets are repaired at once, employees will not develop the bad habit of turning valves off too tightly.

The proper type of soap is important, not only for ordinary

**Figure 18-3.** Lavatories should be supplied with running water at a controlled temperature. A sufficient number of wash-up facilities should be available. In a multiple-use lavatory, 24 in. (60 cm) of wash sink or 20 in. (50 cm) of circular basin, with water outlets for each space, is considered to be equivalent to one lavatory.

hygiene, but as a protection against dermatitis caused by the cleaning agent. The soap used should have no free alkali and should have a pH less than 10.5. It should be free of mineral abrasives. Individually dispensed paste, liquid, or powder (not bar soap) for common use should be provided. Liquid or powdered soaps are preferable, because they lend themselves to easy dispensing and are also an aid to housekeeping.

The practice of removing paint, dye, and other stains with solvents or other chemicals, and especially the practice of removing grease from the hands with naphthas, should be strongly discouraged. Solvents may cause a severe skin irritation.

Protective or barrier creams, if properly used and reapplied frequently, provide limited protection against hand and arm irritants. There are four common types of creams; no one cream is effective against all irritants. Repeated washing to remove the barrier cream is one principle benefit from its use.

**Lockers** should be perforated for ventilation and be large enough to permit clothing to be hung up to dry. If the clothing may be heavy or wet, it is desirable to provide forced circulation of hot air through the base of the lockers and out through the top, or to provide hangers on elevating chains so the work clothing can be dried between shifts.

Lockers should have sloped tops to prevent storage of material on the tops (see Figure 18-3). The multiple legs of lockers are serious impediments to floor cleaning; lockers should be placed on metal frames with a minimum of floor supports. They should be anchored together to prevent being overturned.

Persons working with highly toxic materials that are dusty or can otherwise contaminate clothing should have separate lockers for work clothing and street clothing. These lockers preferably should be in rooms on opposite sides of the shower room so employees will have to pass through the shower room when changing from work clothing to street clothing, and vice versa.

Benches in front of the lockers should be permanently fastened to the locker base, preferably on a hinged support so they can be turned up against the faces of the lockers while the aisles are swept. The benches should be checked at regular intervals and kept in repair, free from splinters, breaks, and other imperfections.

**Floors.** Washrooms and locker rooms should be well ventilated, kept warm and comfortable, and at 50 percent relative humidity. The heating equipment should be so installed to protect against burns and should comply with state codes.

The floors of washrooms and locker rooms should be of nonabsorbent material such as glazed brick, tile, or concrete. The floor material should be continued up into the walls as a cove for at least 6 in. (15 cm) before there is a joint. The walls should be connected to the floor cove with a tight joint and should be impervious to water to a height of at least 5 ft (1.5 m).

Flooring material should be selected for durability and sanitation and to minimize the hazard of slipping and falling. Terrazzo, tile, marble, and polished concrete floor surfaces are particularly hazardous when wet. For safety, a rigid cleaning and mopping schedule must be maintained to keep the flooring dry when in use.

Concrete floors can be made much less hazardous by covering the surface with a finishing layer of abrasive grain concrete. Abrasive strips can be helpful on old concrete floors, which have been worn smooth. Ceramic tiles with a nonskid, nonabsorbent, and watertight surface also are available. Mats also can be used.

A floor should be inspected frequently for watertightness. Leaky floors cause damage to joists and other structural members of the building, and, if organic materials collect in them, can attract vermin. Worn wood or concrete floors can be covered with a plastic material to obtain a watertight surface.

## Showers

Showers should be installed in establishments where workers become dirty, wet with perspiration, or are exposed to dust or vapors. The showers should be as close to the job as possible, preferably in a separate room adjacent to the dressing and locker rooms. Workers exposed to high temperature who come off the job wet with perspiration should not be exposed to cold weather when going to the shower and dressing rooms.

One shower should be provided for each ten employees of each sex, who are required to shower during the same shift. Each shower should be supplied with hot and cold water through a mixing fixture that the user can regulate. The maximum hot water temperature should be automatically maintained at 140 F (60 C).

Deluge showers, eyewash fountains, and similar installations for emergency use are discussed in Chapter 19, Occupational Health Services.

Body soap or other appropriate cleaning agents should be conveniently placed near the showers. Hot and cold water should feed a common discharge line. Employees who use showers should be provided with individual clean towels.

When employees are required by a particular standard to wear protective clothing because of possible contamination with toxic material, change rooms should be equipped with storage facilities for street clothes. As discussed earlier under Lockers, separate storage should be provided for protective clothing.

Where clothes are provided by the employer, and become wet, or are washed between shifts, provision should be made to insure the clothing is dry before reuse.

The floor of the shower room and of the individual compartments should be made of nonskid material to provide good footing when it is wet. Either abrasive grain concrete or concrete with a wood-float finish is a satisfactory surface.

Existing floors that were made smooth or have become smooth through long wear can be given a nonskid surface. Concrete floors should be thoroughly scrubbed with an abrasive pad using a synthetic detergent. Strips of abrasive material can be applied to other types of floors to provide a nonslip surface. The floor throughout the shower room area should slope toward drains, preferably at the back of the shower stalls. Curbs around the individual shower stalls are not necessary if the floor is properly sloped. They are a tripping hazard and, if used, should be dyed or painted a contrasting color.

Wood mats should not be used on shower room floors because of the tripping hazard and the probable exposure to splinters and loose joining members.

The pans of antiseptic solution commonly seen at the entrance of shower stalls or shower rooms are useless for killing organisms and a nuisance to keep clean.

As an item of general sanitation, shower rooms and stalls should be well ventilated and adequately lighted to prevent the formation of mold. The floor of the shower should be daily mopped with detergent, hot water and disinfectant to combat athlete's foot (fungus and ringworm infection). A foot-actuated spray can aid in controlling athlete's foot.

## Toilets

Wall-hung, elongated-bowl flush toilets with open-front seats should be provided according to the number of employees (Table 18-C). If persons other than employees are allowed to use toilet

**Table 18-C.** Minimum Toilet Facilities

| *Number of Employees* | *Minimum Number of Water Closets** |
|---|---|
| 1 to 15 | 1 |
| 16 to 35 | 2 |
| 36 to 55 | 3 |
| 56 to 80 | 4 |
| 81 to 110 | 5 |
| 111 to 150 | 6 |
| More than 150 | One additional fixture for each additional 40 employees. |

*Where toilet facilities will not be used by women, urinals may be provided instead of water closets, except that the number of water closets in such case shall not be reduced to less than ⅔ of the minimum specified.

Printed from OSHA Regulations, §1910.141(c)(1).

facilities, the number of toilets should be increased accordingly. Paper holders must be provided for every water closet. For each three toilet facilities, there should be at least one lavatory in the toilet room or adjacent to it.

Wall-hung units are easier to keep sanitary and to clean under. Codes prohibit any type that is not thoroughly washed at each discharge or that might permit siphonage of bowl contents back into the tank. Water supplied to tanks must have vacuum breakers or a positive air gap between the top of water in the tank and the water supply inlet.

Toilets should be placed not more than 200 ft (60 m) from any workplace. In multistory buildings, toilets should be not more than one floor above or below the work area. With toilets and lavatories at various points throughout the plant, the main locker room and shower room can be closed for cleaning during the work period, an advantage for the janitorial crew.

Toilet rooms and washrooms for women sometimes have an attendant on duty during use. Washroom attendants should not attempt to give first aid to women who become ill at work. Such aid is best given by the plant nurse or physician. Some states also require that women work no more than a certain distance from a woman's rest room. This should be checked during design stages. Some states also require installation of cots.

Ventilation is required for toilet rooms. If natural ventilation is relied upon, there should be windows or skylights having a ventilation area of 6 sq ft (0.5 $m^2$) for a room with one toilet, with an additional square foot (0.1 $m^2$) of window ventilation space for each additional toilet. If this amount of window space cannot be provided, forced ventilation should be supplied at a rate of three to four air changes per hour in the room.

Because windows and skylights generally do not afford sufficient light, light fixtures should be installed in all toilet rooms and washrooms. Switches, for the lights or for electric driers or other equipment, should be located where they cannot be operated by persons who are at the same time in contact with piping or other grounded conductors.

Individual wall-hung urinals should be provided in the men's room. These may be substituted to the extent of one-third or less of the number of stools specified. Trough urinals are poor substitutes for individual fixtures and are prohibited in many states. Approved urinals must have all surfaces that are subject to soiling accessible for cleaning. Integral screens over the discharge openings are the major cause of chronic toilet room odors because the decomposing soil under the screen cannot be removed by any practicable method. Blow-down washout urinals are the only acceptable type. Floor-type urinals, in which the drain pipe becomes chronically offensive, and wall-hung urinals with integral screens should be replaced by the approved sanitary type (blow-down washout), thus making room deodorants unnecessary.

Employees should be prohibited from lunching in toilet rooms, or in process areas where toxic or noxious materials are present. The habit of some workers to heat foods in molten lead reservoirs or other process heating equipment can be dangerous to their health and should be prohibited. This prohibition naturally implies the provision of proper lunchrooms or other eating facilities outside the toilet rooms or process areas.

Covered receptacles should be provided in plant lunchrooms for disposal of waste food and papers, and employees should be prohibited from disposing of such refuse in the toilet rooms. If cups of coffee or other drinks are carried from the lunchroom, they should be in covered containers or on trays to prevent spillage that could create unsanitary or slippery conditions.

Privies are unsatisfactory, but where no other method is feasible, privies and chemical closets should be approved by health authorities having jurisdiction. Portable toilets also are available. These often are necessary on construction jobs. The supplier can provide waste removal and maintenance.

### Janitorial service

As a part of an overall, managed plant sanitation function, a minimum, daily janitorial service should be provided for all personal service facilities. When properly designed, washrooms, shower rooms, and toilets can be thoroughly cleaned with little personal involvement in the process. Floors and fixtures should be mopped and cleaned with detergent and hot water, at least once daily. A sanitizing cleaner should be used as often as necessary. The occasional use of an acid-type cleaner may be necessary on toilet bowls and urinals.

Rubber gloves and goggles should be worn and the fixtures thoroughly flushed following use.

When floors are being mopped, the area should be blocked off by signs reading CAUTION—WET FLOOR, as a precaution against possible slipping accidents. This subject is also covered in Chapter 2, Construction and Maintenance of Plant Facilities, in the *Engineering and Technology* volume.

## FOOD SERVICE

### Nutrition

Nutrition, another factor in industrial health and safety, concerns the medical and safety departments of any company or plant. If a survey of the food service establishments in the neighborhood shows they cannot supply the nutritional needs of employees, then the plant is justified in establishing its own food service. With care and thought, adequate and balanced in-plant meals can be provided. Food must also be properly prepared and attractively served, with strict adherence to sanitary practices.

The company nurse, workng with the company or visiting physician, can provide employees with leaflets on better nutrition. The Council on Foods of the American Medical Association, and the American Dietetic Association (see References) have many excellent articles and materials available. A good breakfast, high in protein for timed energy release during the day, contributes to less fatigue and, consequently, less chance of accidents.

Workers usually need additional nutrition during the first half of a shift. Low blood sugar tends to be a health and accident hazard. A survey of vending machines will show that about two out of three food snack items are overbalanced in sweets. Overeating of sweets can contribute to low blood sugar. Snacks that contain higher protein, such as peanuts, meat sticks, and peanut butter foods, are desirable. Brown sugar rolls and buns, raisins, and sandwiches made of protein-rich breads are helpful.

Some organizations have dietitians review their food service menus and even talk to employees.

### Types of service

There are five main types of industrial food service:

1. Cafeterias preparing and serving hot meals
2. Canteens or lunchrooms serving sandwiches, other packaged foods, hot and cold beverages, and a few hot foods
3. Mobile canteens that move through the work areas, dispensing hot and cold foods and beverages from insulated containers
4. Box lunch service
5. Vending machines

Even using a mobile canteen to provide a midshift snack adds considerably to the nutrition of the average worker. If lunches are also served, the mobile canteen should carry both hot and cold foods and beverages.

The central cafeteria with a kitchen where full meals can be prepared and served is often the most satisfactory form of food service. In large plants, it may be economical to supply several cafeterias from a central kitchen.

**Vending machines and microwave ovens.** Self-service vending machines offer a wide variety of packaged, ready-to-eat foods. Some machines have ovens that let the user quick-cook meals. Two important safety and health precautions should be followed in the use of microwave ovens.

1. All repairs should be made by manufacturer's trained repair personnel.
2. Persons with pacemaker heart units should be warned against coming too close to microwave ovens.

Details are given in the Council's *Fundamentals of Industrial Hygiene.*

Proper installation and maintenance of food heating and refrigeration systems is important. Normal sanitary precautions of course apply to vending machines. Can openers in safe working order, sufficient utensils, and adequate waste disposal facilities, in both kitchens and eating areas, are other necessary provisions of a self-service operation.

## Eating areas

The cafeteria or lunchroom should be clean and attractive to encourage employees to eat away from their work area. Refrigerators for lunch storage also will help convince employees to eat in the proper area.

Minimum floor spaces for the number of people using the eating area at one time are given in Table 18-D. Where space is limited, lunch periods should be staggered so that employees do not have to eat on the job.

**Table 18-D.** Minimum Floor Space in Eating Areas

| *Number of People* | *Sq Ft per Person* |
|---|---|
| 25 or less | 13 |
| 26 through 74 | 12 |
| 75 through 149 | 11 |
| 150 and over | 10 |

Printed from *American National Standard Z4.1.*

## Kitchens

When a cafeteria kitchen is set up, the same attention should be paid to proper equipment and working conditions as would be in any other part of the plant. Food equipment should be of types approved by the National Sanitation Foundation. The layout should conform with public health food service codes, and be such that the various operations are segregated, with adequate walkways from point to point about the kitchen.

Floors should be made of impervious, water-resistant, nonskid material to minimize the hazard of slips and falls if water grease should be spilled on them, as often happens in a busy kitchen.

Ranges and other heat-producing equipment should be hooded and ventilated to carry away heat, combustion products, and vapors. Since these ventilating sytems get very greasy, the duct work should be accessible for cleaning. They should be made of heavy gage steel so any fire can be self-contained.

Sprinkler systems and portable extinguishers should be installed for fire protection. (See Chapter 17, Fire Protection, in the *Engineering and Technology* volume.) A fire blanket should be located near the ranges or in areas where clothing can be ignited.

Easily changeable racks for handtools, such as knives, cleavers, and saws, and storage racks or cabinets for utensils should be provided in convenient places.

## Controlling food contamination

Incorporated communities have detailed sanitary regulations for the installation and operation of industrial food service facilities. Regulations of the local authority having jurisdiction should be followed in detail. In unincorporated areas with no local authority, the state code or the recommendations of the USPHS *Food Service Sanitation Manual* should be followed.

Perishable food and drinks, particularly custard-filled and cream-filled pastry, milk and milk products, egg products, fish, meat, shellfish, gravies, poultry, stuffing, sauces, dressings, and salads containing meat, should be kept at or below 40 F (4 C) except while being prepared or served. If foods of this kind have been permitted to stand for some time at room temperature after preparation, reheating them is not a sufficient protection against bacterial poisoning.

The types of bacteria (staphylococcus) which may infect these foods produce a toxin that is not destroyed by normal cooking temperatures. If the bacteria have grown in the food during room temperature storage, they will be killed by reheating, but the toxin will remain and food poisoning will result.

To prevent bacterial food poisoning, the following suggestions are made.

1. Keep perishable foods under refrigeration until they are to be used.
2. Keep hot foods hot (160 F, 70 C, or above) and cold foods chilled (40 F or below).
3. Remove leftover foods from food-warming devices immediately after the last feeding period. Never hold hot foods in warmers from one meal to the next, or for several hours before a dinner is served.
4. Place leftover food under refrigeration as quickly as possible.
5. Instruct employees who handle food and utensils that they must wash their hands thoroughly with soap and water after using the rest rooms and before handling foods.
6. Eliminate flies, roaches, rodents, and other pests that can transmit disease. Consult a professional exterminator if necessary.
7. Never use galvanized or cadmium-plated containers for storage of moist or acid foods.
8. Consult your local health authorities if you have questions on sanitation.
9. Employees with open infections or communicable diseases should not handle food.

As a further precaution against the transfer of infections, no first aid material should be permitted in the kitchen. All conditions requiring first aid immediately should be seen by the company or plant nurse or the physician, and the individual should continue on the job only at the doctor's discretion.

To prevent cross infections in large dining rooms, the proper cleaning, sanitizing, and storing of containers, utensils, glassware, dishes, and silverware are highly important.

Use of single-service containers and utensils, however, can eliminate washing and handling.

It is generally easier to sanitize utensils by machine washing than it is by hand washing if the machine is kept clean and in top operating condition. In either case, one of the main requirements in maintaining adequate sanitation is proper training of employees.

Utensils should be carefully scraped and preferably prerinsed before being put into the detergent solution. They should be thoroughly washed with soap or a detergent, and well rinsed by a method which will destroy bacteria.

For thorough machine washing, the utensils must be stacked in the trays loosely enough so the cleansing agent gets to every part. The concentration of detergent in the wash water and the wash water temperature must be maintained. The wash water must be changed before it becomes excessively dirty. Spray nozzles in the dish washing machine must be cleaned daily to maintain proper flow and distribution.

Requirements for washing by hand are the same, except that each utensil must be individually scrubbed in all parts rather than simply stacked in a tray.

For rinsing, the cleaned utensils may be immersed in clean hot water at 170 F (77 C) for one-half minute. One problem is that of maintaining the temperature of the rinse water over long sessions of dishwashing, since a large volume of fresh hot water is required for this method. Water heaters should be of adequate size. Less water, however, will be needed if a chemical sterilizing agent is used. Hypochlorite solutions at a concentration of at least 50 ppm of available chlorine for an immersion time of one minute at 75 F (24 C) will provide adequate sterilization. Cationic quaternary ammonium germicides are also suitable.

When the utensils have been properly cleaned, they should be stored and handled so as to prevent contamination by the handler's fingers and from ordinary dust and dirt or from leakage from overhead pipes.

## REFERENCES


American Dietetic Association, 430 N. Michigan Ave., Chicago, Ill. 60611.

American Medical Association, Council on Foods, 535 N. Dearborn St., Chicago, Ill. 60610.

American Public Health Assn., 1015 15th St., Washington, D.C. 20005. *Standard Methods for the Examination of Water and Wastewater.*

American National Standards Institute, 1430 Broadway, New York, N.Y. 10018.
- *Air Gaps in Plumbing Systems,* A112.1.2.
- "Gas-Burning Appliances," Z21 Series.
- *Minimum Requirements for Non-Sewered Disposal Systems,* Z4.3.
- *Minimum Requirements for Sanitation in Places of Employment,* Z4.1.
- *Minimum Requirements for Sanitation in Temporary Labor Camps,* Z4.4.
- *National Electrical Code,* ANSI/NFPA 70.
- "Pipe Flanges and Fittings," B16 Series.
- *Specification for Drinking Fountains,* A112.
- *Buildings and Facilities—Providing Accessibility and Usability for Physically Handicapped People,* A117.1.

American Water Works Association, 6666 W. Quincy Ave., Denver, Colo. 80235.

Environmental Management Association, 1019 Highland Ave., Largo, Fla. 33540.

National Institute for the Foodservice Industry, 20 N. Wacker Drive, Chicago, Ill. 60606. *Applied Foodservice Sanitation* 3rd ed. 1985.

National Restaurant Association, 311 First St., N.W., Washington, D.C., 20001. "A Safety Self-Inspection Program for Food-Service Operators."

National Safety Council, 444 N. Michigan Ave., Chicago, Ill. 60611.
- *Fundamentals of Industrial Hygiene,* 3rd ed., 1987.
- Industrial Data Sheets
  - *Dusts, Fumes, and Mists in Industry,* No. 531.
  - *Industrial Skin Diseases,* No. 510.
  - *Refuse Collection in Municipalities,* No. 618.
  - *Your Guide to Safe Drinking Water,* 1985.

National Sanitation Foundation, P.O. Box 1468, 3475 Plymouth Rd., Ann Arbor, Mich. 48106.

Public Health Service, U.S. Department of Health and Human Services, 200 Independence Ave., S.W., Washington, D.C. 20201.
- *Food Service Sanitation Manual,* Publication No. FDA 78-2081.

U.S. Environmental Protection Agency, Water Supply Program Div., Washington, D.C. 20460.
- *Manual for Evaluating Public Drinking Water Supply.*
- "Operation Responsible—Safe Refuse Collection." Available through National Audio-visual Center, General Services Administration, Washington, D.C. 20409.

*Water Quality Criteria,* Report No. 3A. Sacramento, Calif., Dept. of General Services, Office of Procurement—Stores, Documents Section, 1015 North Highland, Sacramento, Calif. 95662.

# 19

# Occupational Health Services



OCCUPATIONAL HEALTH SERVICES range from the truly elaborate to the bare minimum required by OSHA. One establishment may have a fulltime staff of physicians, nurses, and technicians, housed in a model dispensary; another may have only the required first aid kit with an adequately trained person to provide first aid.

Ideally, modern occupational health programs, regardless of size, are composed of elements and services designed to maintain the health of the work force, to prevent or control occupational and nonoccupational diseases and accidents, and to prevent and reduce disability and the resulting lost time. A good program should provide for the following:

1. Maintenance of a healthful environment,
2. Health examinations,
3. Diagnosis and treatment,
4. Immunization programs,
5. Medical records,
6. Health education and counseling, and
7. Open communication between the plant or company physician and an employee's personal physician.

Any treatment of ill or injured persons has a bearing on the practice of medicine. All states, therefore, regulate medical and nursing practice and provide guidelines for the scope of practice. Ideally, all services related to health, injuries, first aid, and medication of any kind should be under the *supervision* of a licensed physician.

## OCCUPATIONAL HEALTH

Occupational health programs are concerned with all aspects of the employee's health and his or her relationship with the environment.

The American Medical Association's Council on Occupational Health's official guides to occupational health programs, *Scope, Objectives and Functions of Occupational Health Programs* and *Guide to Small Plant Occupational Health Programs,* state that the basic objectives of a good occupational health program should be:

1. "To protect employees against health hazards in their work environment,
2. "To facilitate placement and ensure the suitability of individuals according to their physical capacities, mental abilities, and emotional makeup in work that they can perform with an acceptable degree of efficiency and without endangering their own health and safety or that of their fellow employees,
3. "To assure adequate medical care and rehabilitation of the occupationally injured, and
4. "To encourage personal health maintenance."

"The achievement of these objectives benefits both employees and employers by improving health, morale, and productivity."

By applying occupational health principles, all workers, including the severely handicapped (see Chapter 20), are placed in jobs according to their physical capacities, mental abilities, and emotional makeup. It also assures continuing medical care and rehabilitation of occupationally ill and injured workers.

There is a relationship between accident prevention and occupational health. For example, some industrial chemicals, when improperly handled, present a variety of serious hazards to health, property, and/or the environment. Depending on conditions, the vapor from a chemical can ignite, explode or can cause dizziness or death when inhaled. Dermatitis can result from contact. (For details on the effects of specific chemicals, see the National Safety Council's *Fundamentals of Industrial Hygiene.*)

Safety professionals have ably demonstrated their ability to reduce accidental injury frequency by controlling many phases of the industrial environment, through worker education, and by improved supervisory techniques. A large part of the remainder of the problem rests within the physical and emotional characteristics of individual workers. Here lies the key to variations in job attitude, productivity, safety, and absence for personal health reasons.

The services and skills of several additional professions are also frequently needed if maximum results are to be attained.

- Medicine plays an important role, with physicians being specially trained in industrial and preventive medicine. They are assisted by specialists in orthopedic surgery, ophthalmology, radiology, surgery, dermatology, and psychiatry, and other areas.
- Occupational health nursing requires specialized knowledge, not only of good basic nursing procedures and health maintenance, but also of the legal, economic, social mores, and labor laws that form the parameters within which the nurse practices. Often, the nurse is the only fulltime medically trained employee in a company or plant.
- The industrial hygienist, a professional development of this technological age, serves as the analytical preventive engineering arm of occupational medicine by applying specialized knowledge to the recognition, evaluation, and control of health hazards in the work environment. For additional details, see the Council's *Fundamentals of Industrial Hygiene.* (See References.)
- Other phases of occupational health may use the expertise of the community and other professional disciplines.

Working both individually and collectively, these specialists have helped to improve the occupational health and safety record of many industries. In some companies, these professionals are so well organized and effective in the anticipation and correction of hazards that the employees are unquestionably safer and healthier at work than they are at home. In fact, the work environment is so well controlled, in many cases, that the problem now is how to get the workers to avoid the hazards of home life, recreation, and travel so that they will be able to return to work each day safe and sound.

A company's occupational health services should be involved in the off-the-job safety program without being obviously intrusive. Medical department personnel can be influential in extending this program beyond the facility to include off-the-job activities of employees and their families. The health and well-being of the employee's family have a direct bearing on efficiency and safety on the job. If employees are injured off-the-job, they are as much a loss to the operation as if they were injured on the job.

Some insurance carriers offer a consulting service which helps organizations set up an occupational health program suitable for their needs. The consultants usually know which doctors and clinics are available for this kind of service. The basic work, however, has to be done by the organization setting up a program.

The justification for these services lies in their accomplishments. Prevention is not only better than cure—it is easier and less expensive. Off-the-job safety programs and activities can benefit a company, plant, or shop of any size.

### Employee health services

Although good industrial medicine can be practiced in any logical location in the plant that is clean and private, experience indicates that the medical unit commands respect only if careful attention is paid to suitable and efficient housing, appearance, and equipment. The entire unit should be painted in light colors and kept spotlessly clean. The dispensary should have hot and cold running water and be adequately heated, ventilated, and illuminated. Proximity to toilet facilities is necessary. Suitable provision should be made for men and women, if both are employed.

**Health service office (dispensary).** A minimum of three rooms, consisting of a waiting room, a treatment room, and a room for physical examinations or consultation is recommended (Figure 19–1). Rooms for special purposes can be added according to the needs and size of the company. New employees who are waiting for physical examinations should not mingle with injured workers.

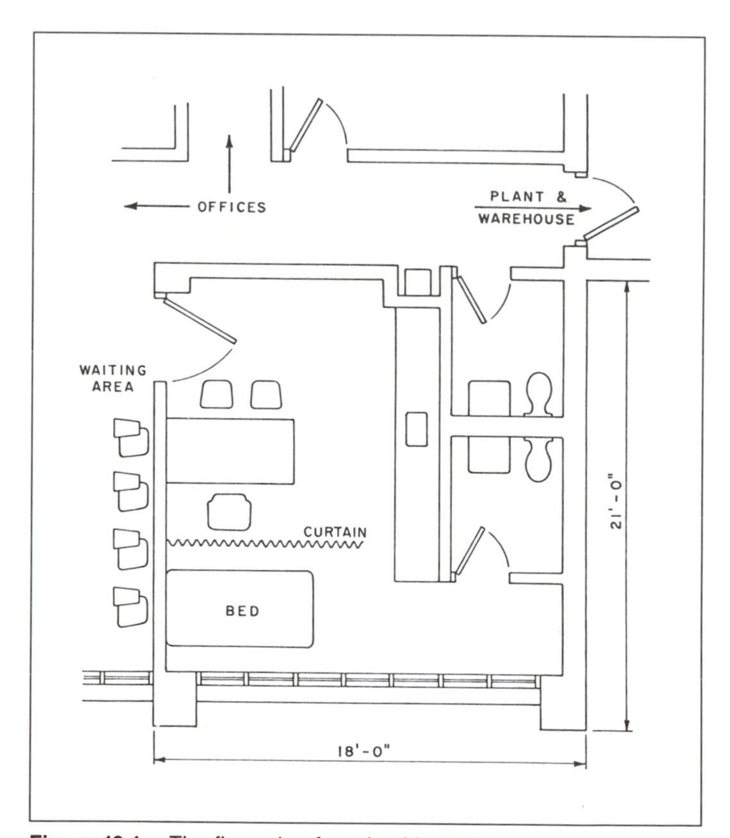


**Figure 19-1.** The floor plan for a health service office. The office has a waiting room, a treatment room, and a consultation room.

The surgical treatment room should be large enough to treat more than one person at a time—small dressing booths can be arranged to give some degree of privacy.

### First aid

Good administration of first aid is an important part of every safety program. It is recommended that a first-aid facility of some sort be set up in all establishments regardless of size. This may range from a deluxe first-aid kit to a well-staffed first-aid and medical facility.

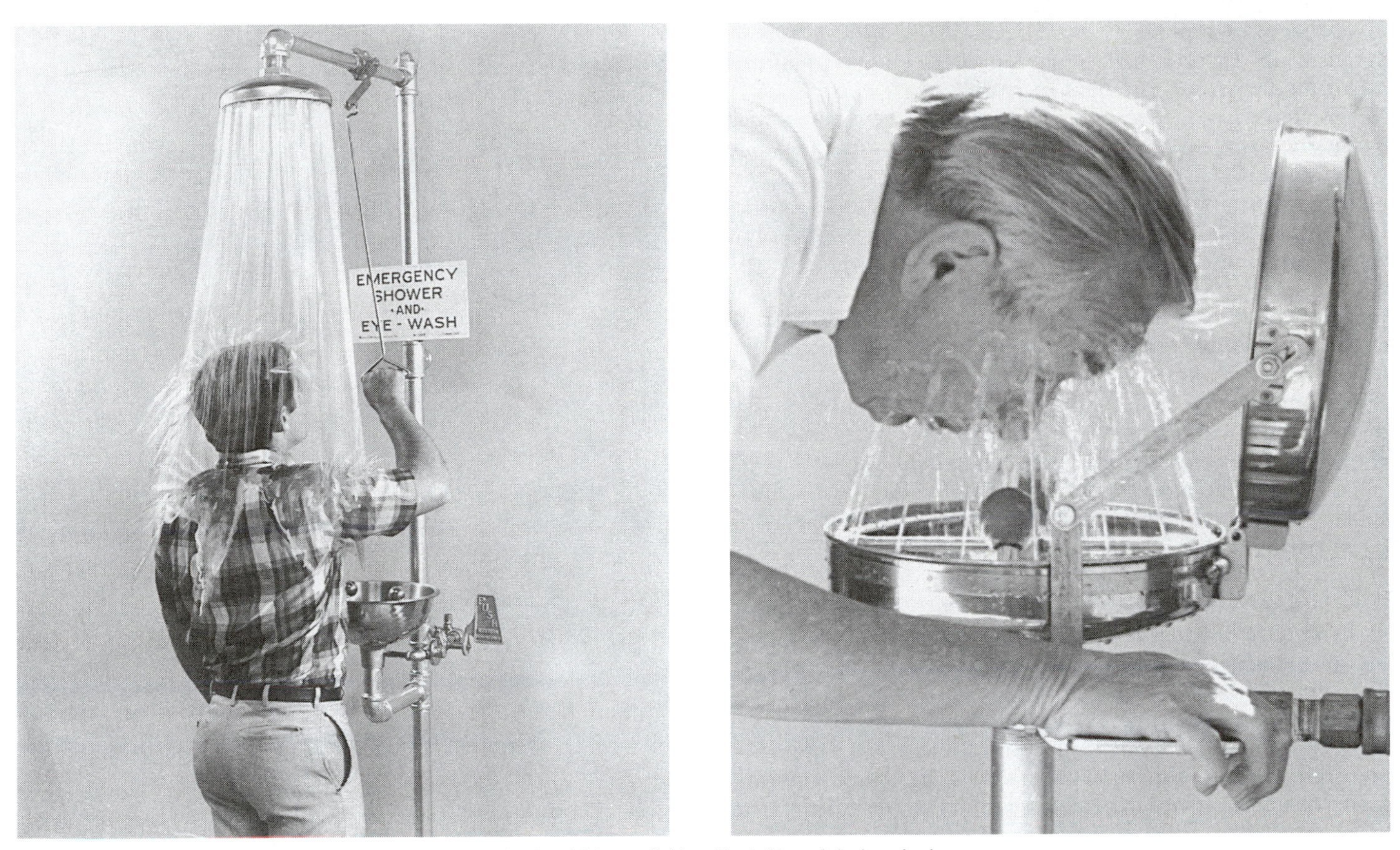


**Figure 19-2.** The emergency shower and eye wash should be well identified. Harmful chemicals are quickly washed away and clothing fires doused immediately by the drenching action of the shower (left). First-aid treatment to the eye must be prompt and consists of prolonged irrigation of the exposed eye with low-pressure water. (Left photo reprinted with permission from Western Drinking Fountains. Right photo reprinted with permission from Haws Drinking Faucet Co.)

First-aid kits and supplies should be kept in a central location so that they are readily accessible. Under no circumstances should medical supplies be spread about the plant for self-administration by employees. When no nurse is in charge, a supervisory employee for each shift should be delegated the responsibility for all medical supplies.

A careful record should be kept of each administration of first aid and an injury investigation report sent to the injured person's supervisor at the time first aid is administered.

Establishments that do not employ a full-time medical doctor should maintain good liaison with a physician, or physicians, designated to handle plant injuries. The physician should be invited to the plant occasionally to evaluate the quality of first-aid procedure, to make recommendations for improvement, and occasionally to tour the establishment. The company-designated physician (or physicians) should be well aware of the type of work done so they evaluate information from patients. The physician, nurse, or supervisor designated should routinely inspect first-aid supplies, stretchers, and stretcher locations.

**Definition and limitations.** In many small organizations and in field operations, it is neither practical nor justifiable to have qualified professional medical personnel available. In such cases, the best arrangement is to use suitable first-aid attendants who follow procedures and treatments outlined by a doctor. However, a doctor should be available on an on-call or referral basis to take care of serious injuries.

In some jurisdictions injured employees have their choice of a physician. In such cases, the employer should comply with this request, if possible.

There are two kinds of first-aid treatment:

- One is emergency treatment. According to the American Red Cross first-aid textbook, "First aid is the immediate, temporary treatment given in the case of accident or sudden illness before the services of a physician can be secured." Proper first-aid measures reduce suffering and place the injured person in a physician's hands in a better condition to receive subsequent treatment.
- The other kind of first aid is the prompt attention given to injuries, such as cuts, scratches, bruises, and burns, which are usually so minor that the injured person would not ordinarily seek medical attention.

Under OSHA recordkeeping procedures, first aid is defined as a one-time treatment plus any follow-up visit for observation of minor scratches, cuts, burns, splinters, and the like that do not ordinarily require medical care. By requiring that all employees immediately report for treatment when injured, regardless of the extent of the injury, much headway has been made in reducing infection and disability, and also in avoiding false claims of injury and disability.

A first-aid program should include:

1. Properly trained and designated first-aid personnel on every shift,
2. A first-aid unit and supplies, or first-aid kit,

3. A first-aid manual,
4. Posted instructions for calling a physician and notifying the hospital that the patient is en route,
5. Posted method for transporting ill or injured employees and instructions for calling an ambulance or rescue squad, and
6. An adequate first aid record system.

Because there is often a difference of opinion regarding proper treatment, first-aid procedures, approved by the consulting physician, should specify the type of medication, if any, to be used on minor injuries, such as cuts and burns. In areas where chemicals are stored, handled, or used, emergency flood showers and eye-wash fountains (Figure 19-2) should be available and clearly identified.

Equipment and supplies should be chosen in accordance with the recommendations of the physician, and service should be rendered only as covered by written standard procedures, signed and dated by the physician. If temporary relief for minor nonoccupational ailments, such as colds and headache, is to be given, the physician should specify the procedures to be followed. The limitations of first aid must be thoroughly understood.

The majority of states have medical practice acts under which a person is limited to a certain definite procedure when attending anyone who is sick or injured—except, of course, under the direct supervision of a physician. It is important, therefore, that anyone who is responsible for first aid have a full understanding of the limits which restrict the work. Because improper treatment might involve the company in serious legal problems, the first-aid attendant should be duly qualified and certified by the Bureau of Mines or the American Red Cross. These certificates must be renewed at specific intervals.

**First-aid training.** The American Red Cross first-aid textbook and the United States Bureau of Mines manual of first-aid instruction are recommended for teaching first aid. It is often found that accidents occur less frequently and, as a rule, are less severe among persons trained in first-aid work. It is therefore advisable that as many industrial workers as possible be given this training (Figure 19–3).

The National Safety Council publishes posters and pamphlets that can be used for training employees. Another valuable reference is *Emergency Care of the Sick and Injured,* by the Committee on Trauma, American College of Surgeons. (See References.)

**First-aid room.** It is always advisable to set aside a room at a convenient location for the sole purpose of administering first aid. It upsets morale to administer treatments in public and injured persons prefer privacy. Furthermore, the person administering first aid should have a proper place to work.

A good first-aid room should be similar to the dispensary. It should be equipped with the following items:

1. Examining table,
2. Cot for emergency cases, enclosed by movable curtain,
3. Dustproof cabinet for supplies,
4. Waste receptacle,
5. Small table,
6. Chair with arms, and one without arms,
7. Magnifying light on a stand,
8. Dispensers for soap, towels, cleansing tissues and paper cups,
9. Wheel chair,
10. Stretcher,

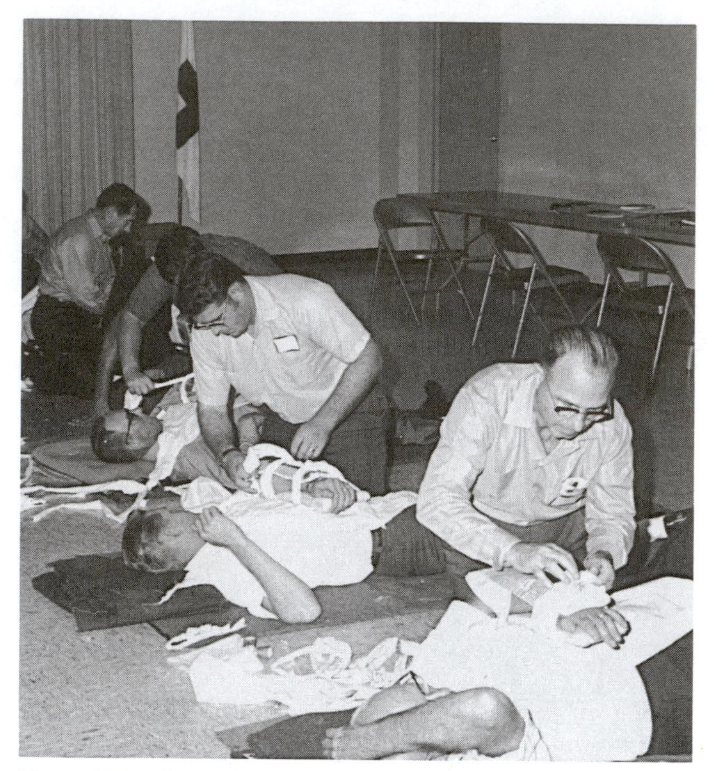

**Figure 19-3.** Demonstration and practice play an important role in first-aid training.

11. Blankets, and
12. Bulletin board on which are posted all important telephone numbers for emergencies.

Oxygen is beneficial in the treatment of many first-aid cases. Because of the danger of fire or explosion, smoking should be prohibited when oxygen is administered. Any type of resuscitating device should be used only by trained persons.

**First-aid kits.** There are many types of first-aid kits designed to fill every need, depending on the type of accident that might occur (Figure 19–4). Commercial or cabinet first-aid kits, as well as unit kits, must meet OSHA requirements.

Kits vary in size from the pocket model to what amounts to a portable first-aid room. The size and contents depend on the intended use and the types of injuries that are anticipated. For example, personal kits contain only essential articles for the immediate treatment of injuries. Departmental kits are larger to meet the needs of a group of workers so the quantity of material depends on the size of the working force. Trunk kits are the most complete (Figure 19–5)—they can be carried easily to an accident site, or can be stored near working areas that are distant from well-equipped emergency first-aid rooms. Although a trunk kit can include such bulky items as a wash basin, blankets, splints, and stretchers, it can still be carried by two persons.

Keeping first-aid kits strategically located throughout the plant seems to work best when such kits are supervised properly (Figure 19–6). A single, trained individual becomes responsible for the maintenance of each kit as well as for providing temporary treatment for serious injuries and first-aid measures for minor cuts and scratches. This employee never gives more than immediate, temporary care. By combining well maintained first-aid kits with a trained first-aid provider, many minor injuries that might otherwise have been ignored, receive prompt, proper treatment.

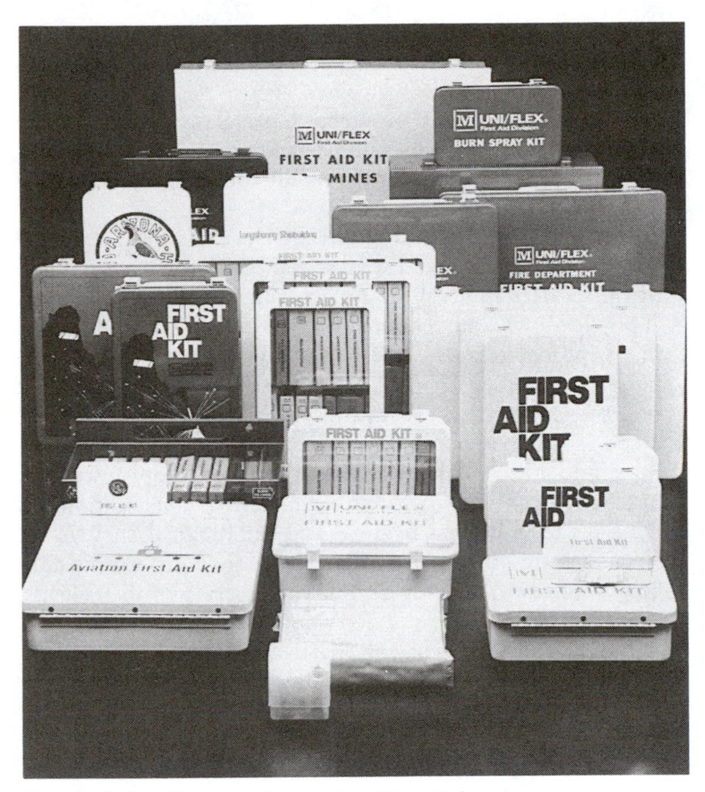


Figure 19-4. First-aid kits can be obtained for special purposes.

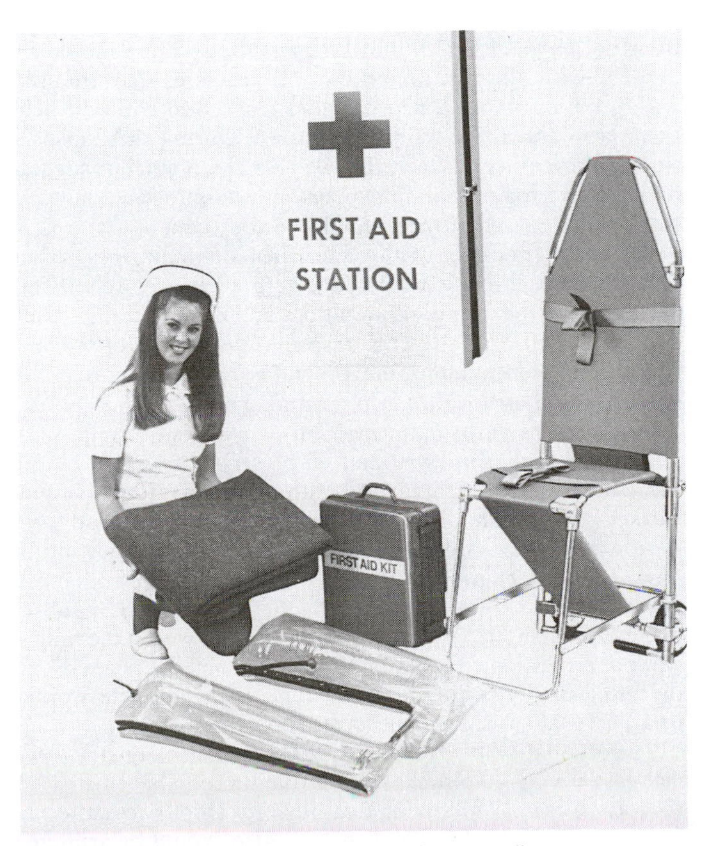


Figure 19-5. For areas that are distant from a well-equipped first-aid room, a trunk kit, as shown here, can be useful. (Reprinted with permission from Ferno-Washington Inc.)

Figure 19-6. First-aid kits should be well distributed throughout a plant. They should be supervised by a trained person so employees do not try to "doctor" themselves.

In industrial organizations such as mining and public utilities, whose activities are widely scattered, the use of first-aid kits and some self-medication by employees may be necessary. Under these circumstances, however, it is better if first-aid service can be controlled by having the attendant in charge properly instructed in first aid and by seeing that the service as a whole has medical supervision.

The maintenance and use of all first-aid kits should be supervised by medical personnel, and the materials provided should be approved by the consultant physician. A member of the medical services group should be assigned to inspect all first-aid materials regularly, and to submit a report of their content level and serviceability.

Maintaining quantities of materials in the first-aid kit is easier if each kit contains a list showing the original contents and the minimum quantities needed. All bottles or other containers should be clearly labeled and dated.

Recommended materials for first-aid kits are listed in the American Red Cross and the U.S. Bureau of Mines first-aid textbooks. Suggestions are also available from the American Medical Association, the American Petroleum Institute, and from manufacturers of first-aid materials.

**Stretchers.** Being able to quickly transport a seriously injured person from the scene of the accident to a first-aid room or a hospital can determine the gravity of the injury and perhaps even the victim's chances for survival.

The stretcher provides the most acceptable method of hand transportation for accident victims. It can be used as a temporary cot at the scene of the accident, during transit, and in the first-aid room or dispensary.

There are several types of stretchers. The commonly used army stretcher is satisfactory. However, when it is necessary to hoist or lower the injured person out of an awkward place, a specially shaped stretcher to which the patient may be strapped and kept immovable is better. A lightweight stretcher is shown in Figure 19–5.

Stretchers should be kept conveniently near places where employees are exposed to serious hazards. It is customary to keep stretchers, blankets, and splints in cabinets that are clearly marked and prominently located. Stretchers should be kept clean and ready to use at all times. They should be protected against destructive vapors or fumes, dust, or other substances, and mechanical damage. If the stretcher is made of materials that will deteriorate, it should be tested periodically for durability and strength.

## MEDICAL SERVICES

The medical department should be easily accessible. Locating it near the greatest number of employees facilitates the immediate reporting and treatment of all minor injuries. If possible, it should also be connected with the employment and safety departments, facilitating prompt physical examinations of applicants, the mutual use of clerical service, and the interchange of ideas and plans relative to employment, accident, and health problems.

Another possible location is near the plant entrance so that an ambulance can be brought to the door, if necessary; or so injured workers, who are off duty but under treatment, may come and go through a separate entrance. Also, try to put this department in a protected area so that a major company or plant disaster does not destroy the first-aid or dispensary facilities.

### The occupational physician

Physicians in industry may be employed at a number of different stations, (part time at each), dependent upon such considerations as the number of employees, the hazards in the operations, and the type and extent of the occupational health program. Arrangements may be made for full-time, part-time, on-call, or consulting service.

- If the service of a full-time physician is warranted, one should be employed. Some large organizations employ a full-time medical director on their headquarters staff, with part-time physicians serving their decentralized operations.
- Some physicians devote a scheduled number of hours, either daily or weekly, to the medical service needs of a company, while they are available at other times for emergencies. Others arrange their service by telephone on the basis of current needs of the company—when job applicants require examination, injured employees need medical care, or other medical problems arise.
- The on-call physician arrangement is most often used by companies (or establishments) having fewer than 500 employees, a low incidence of accidental injuries, or a minimum program of health services for employees. The on-call physician is usually located nearby and is available in emergencies. This physician cares for most injuries not requiring hospitalization and for those cases requiring hospitalization that fall within his or her competence. On-call physicians often use specialists as consultants.

  Frequently, the on-call physician has a similar arrangement with a number of companies in the vicinity—a particularly convenient system for small-plant or small-company clusters, such as in industrial parks.

  It is not unusual to find physicians specializing in the care of industrial injuries and diseases, and providing occupational medical services in some manufacturing centers.
- A physician who has a consulting service is not usually called in except for diagnosis and treatment of serious injuries, illnesses, toxicological problems, and special kinds of injuries or disorders (such as eye injuries) that require the services of a specialist.

State and local health departments often supply, without charge, medical, nursing, and engineering consultation, as well as make industrial hygiene and radiological surveys for industries within their areas. Many insurance and private consulting companies also provide this type of service to their clients.

**Duties of the occupational physician.** An effective medical service program should be planned by the medical department head (usually a physician). The full support of top management is critical to the success of the entire occupational health program. Only then is it possible to establish and maintain an adequate professional staff and facilities for examinations, emergency cases, and record storage.

Industrial physicians must be given enough authority so that workers will respect their judgment and follow their instructions on personal health and safety matters. They should be familiar with all jobs, materials, and processes that are used within the company. An occasional inspection trip will help them keep abreast of what is going on. During these inspections an inventory of hazards will help the physician suggest to the safety professional those potentially harmful environments from which the employees should be protected.

Physicians should be involved in other company services that relate to the health of workers, such as food service, welfare service, safety programming, sanitation, and mental health. They can also initiate and be responsible for sponsorship of company-wide immunization programs (against tetanus, polio, or flu) as well as for blood donor and chest X-ray programs.

Maintenance of the true physician-patient relationship (with fairness to both employee and employer) is essential to the success of any occupational health program—for example, worker patients should receive the same courtesy and professional honesty as do private patients. The first meeting of physician and employee usually occurs at the introductory physical examination. This meeting may be followed by subsequent examinations. The examining physician within the framework of professional discretion should acquaint the worker with the results of all examinations and, if necessary, refer the worker to a personal physician for correction of defects.

The occupational physician should provide emergency medical care for all employees who are injured or become ill on the job. Necessary follow-up treatment for employees suffering from occupational disease or injuries also should be arranged (Figure 19–7). The treatment of employee injuries or diseases not industrially induced is the function of private medical practice. Therefore, the occupational physician should abstain from

**Figure 19-7.** The industrial physician can provide emergency medical care for employees who are injured or who become ill on the job.

rendering such services except when independent facilities are not readily available, or the ailment or discomfort is so minor that the employee would ordinarily not seek medical attention, or the rendering of such service would enable an employee to complete a shift.

In most settings the occupational physician should not devote time or facilities to diagnose or treat dependents of employees. Instead, health education programs should be promoted for employees and their families so that they will be encouraged to seek proper private consultation.

Medical and surgical management in every case of industrial injury or disease should aim to restore disabled workers to their former earning power and occupation as completely as possible and without unnecessary delay. To help achieve this, concise, dependable medical reports should be promptly submitted to those agencies entitled to them. Furthermore, equitable administration of workers' compensation rests on medical testimony which adheres closely to reasonable scientific deductions regarding the injury or its possible consequences.

Maintenance of necessary records and reports is the responsibility of the industrial physician. These records act as a guide to management and keep both management and employees informed as to the success of the program. Records and reports are necessary to direct and evaluate preventive medical and safety engineering techniques, to chart progress in the reduction of accidents, and to meet the recordkeeping requirements of the Williams-Steiger Occupational Safety and Health Act of 1970 (see Medical records later in this section, and also refer back to Chapter 2, Government Regulation and Compliance, and Chapter 6, Accident Records and Incidence Rates).

The physician is also responsible for properly instructing nurses and paramedical personnel and directing their activities. Their duties, therefore, should be described in clear, concisely written directives, a copy of which should be posted in the medical department.

### The occupational health nurse

Occupational health nursing is a specialized branch of the nursing profession. The position requires a registered professional nurse who is licensed in the state where employed. In addition, it is desirable that the occupational health nurse have some knowledge of workers' compensation laws, insurance, health and safety laws, occupational diseases, sanitation, first aid, and recordkeeping.

**Figure 19-8.** The occupational health nurse can provide initial health care in case of injury or illness. (Reprinted with permission from Kwik Kold.)

The occupational health nurse contributes the most when working with the company physician who provides written directives that have been discussed and mutually understood and agreed upon. Working with the physician, the occupational health nurse can provide a variety of nursing services, such as initial care for injuries (Figure 19–8) or illnesses, counseling, health education, consultation about sanitary standards, and referral to community health agencies. The occupational health nurse also can participate in programs that evaluate employee health such as health examinations, or prevent disease, such as immunizations. Working with the physician, the plant nurse can perform excellent employee health education services by distributing literature on heart care, weight control, cancer and tuberculosis prevention, and the prevention and treatment of venereal disease.

When the physician is only employed part time, the nurse works with the safety director in planning and conducting accident prevention programs.

The occupational health nurse must maintain a confidential professional relationship with the employees in conformity with legal and ethical codes. The nurse may not divulge information contained in individual employee health records unless the employee gives signed permission. The medical files should be accessible only to medical personnel.

It must be clear that establishment of medical diagnosis and definition of treatment are the functions of the physician. The nurse is not a substitute for the physician. Each has a legally defined area of practice and responsibility. To have an effective medical program, a company should have the services of a physician and a nurse.

### Physical examination program

Surveillance of the health status of workers by qualified medical personnel is essential if an occupational health program is to obtain maximum benefits for both employer and employee. Therefore, a physical examination program should be established. The examining physician should discuss all significant findings with the worker. However, good judgment should be used to prevent raising unnecessary fears while emphasizing the importance of obtaining adequate personal medical care. With the employee's consent a transcript of the data may be supplied to the worker's personal physician or an insurance company. Courts, workers' compensation commissions, or health authorities may request this information by legal means, but employee consent is a more agreeable method. Certainly the confidential character of health examination records should be observed.

Because OSHA requires employers to keep accurate records of employee exposure to potentially toxic materials or harmful physical agents, the employer should be informed of a potentially harmful work environment detected through examination of persons subjected to it.

**Scope of the examination.** It is impossible to set forth what constitutes a complete examination, or even a suitable examination, because physicians have different opinions regarding the relative values of various test procedures, based on their own training and experience. Therefore, the scope of a physical examination should be determined by the company physician, who is familiar with the operations involved—the nature of the industry, its inherent hazards, and the variations in jobs, physical demands, and health exposures. The values of different test procedures and their cost in time and dollars must be assayed. Perhaps examinations should be different in scope for different jobs. For example, the required physical condition of an ironworker who is to be engaged in construction of a skyscraper is different from that of an office worker who will sit at a desk. However, there are basic physical examination considerations applicable to each.

The various kinds of examinations may be classified as follows—preplacement, periodic, transfer, promotion, special, and termination.

**Preplacement examinations.** The preplacement examination is done to determine and record the physical condition of the prospective worker to facilitate assignment to a suitable job in accordance with the worker's mental ability and physical capacity. The applicant (or the personal physician with the applicant's approval) should be advised of conditions that need attention. Medical department follow-up may be necessary. It must be a paramount principle that the purpose of the preplacement examination program is selection *and* placement—not merely selection of the physically perfect and rejection of all others.

From the public and occupational health standpoint, the only bar to immediate employment in nonhazardous occupations should be communicable disease, progressively incapacitating injury or disease, or incapacitating mental illness. It is obvious that communicable diseases must be controlled, and this may involve the assistance of public health officials. One of the values of an examination program is the detection of disease in its early stages, when it is most amenable to treatment. Applicants with incipient but still nondisabling disorders can often be employed while being treated by their personal physicians.

A significant percentage of persons have mental illness or emotional disturbance that impairs judgment or prevents them from performing normal work. These aberrations vary in degree and can be serious enough to bar employment. The trained physician can frequently detect them during the preplacement examination.

It is not the function, however, of the physician to inform applicants whether they are to be employed. This is the prerogative and duty of management, as there are other factors in addition to physical qualifications that bear upon suitability for employment.

**Periodic examinations** of all employees may be on a required or voluntary basis. A required program should be applied to workers who are exposed to health-hazardous processes or materials, or whose work involves responsibility for the safety of others, such as vehicle operators. Substances like lead or carbon tetrachloride that are capable of causing occupational disease are usually subjected to process controls that will keep the workers safe from poisoning. However, caution dictates the advisability of periodic examination of such workers to be certain that the engineering and hygiene controls are effective and continue to be so. Periodic examination also enables early detection of the hypersusceptible individuals and individual practices or procedures that are unsafe or circumvent safety devices.

Examination frequency will vary in accordance with the quality of the engineering control, the nature of the exposure (this is influenced by the rapidity of the action of the hazardous substance on the human body), and the findings on each examination. Thus, exposures to substances might require examinations or laboratory tests on a weekly, monthly, quarterly, annual, or biennial basis.

In many cases, laboratory tests of blood or urine will suffice as the major portion of a periodic examination program, with complete examinations being made less frequently. The type of special examination (laboratory, X-ray, etc.) necessary for any exposure and the interpretation of the results are decisions requiring expert medical personnel.

**Special examinations.** Employees having on-the-job difficulties that may be health related are often benefited by special examinations. Job transfer also may require medical evaluation.

Many organizations find it worthwhile to make "return to work" examinations of employees who have been absent more than a specified number of days as a result of a nonoccupational illness or injury. This is done to control communicable disease, as well as to determine suitability for return to work. There is a wide difference in the actual effects of the same disease on different persons, and an even greater divergence of personality reaction to the disorder. One person will go to bed at the first sign of discomfort, whereas another must be truly overwhelmed before failing to report for work.

When an employee returns to work following serious injury, either occupational or nonoccupational, a new evaluation of physical capacities may be necessary. Also the application of rehabilitation procedures may have reduced the disability and improved the range of employability.

**Exit examinations.** Upon termination of employment, some organizations examine employees and record the findings, particularly where operations involve definite exposure to health hazardous substances, such as lead, benzene (benzol), silica, and asbestos dust, or to harmful noise.

**Laboratory tests.** Urine and blood tests are a good investment in detecting liver and kidney disease, diabetes, and anemia, to name the more common diseases. Where there is to be an exposure to toxic substances, appropriate laboratory tests may be indispensable.

**X-ray tests.** A record of the condition of the lungs as shown by X ray is desirable for every applicant. Those who will be engaged in a dusty trade should have periodic chest X rays. Every X ray should be carefully identified with date and name and each should be compared with previous X-ray photographs.

X-ray screening tests, as part of the preplacement medical examination are of no significant value in predicting the onset of future physical conditions, such as back pain or injury. They are an unnecessary expense as well as an unnecessary exposure of the person to X-ray radiation.

**Vision tests.** In recent years, special devices have been developed for routine testing of several aspects of vision (Figure 19–9). Near, as well as far, vision should be recorded. The old method of testing only distance vision at 20 ft (6 m) is inadequate. Failure to compare the visual requirements of a job with the visual abilities of employees may result in employees becoming easily fatigued, inefficient, and involved in an accident. If the job demands it, color vision should also be tested.

If facilities are not available at the establishment, arrangements can be made with outside doctors for annual eye examinations and the fitting of prescription safety glasses if needed.

**Hearing tests.** Workers who will be exposed to hazardous noise levels should be examined for hearing acuity before placement to determine prior hearing loss, if any, and periodically thereafter to detect early loss due to noise. Personal protective devices can reduce noise reaching the auditory nerve, but wisdom dictates the value of tests and records. The audiometer is the accepted method of testing (Figure 19–10). A special booth will usually be required to administer these tests.

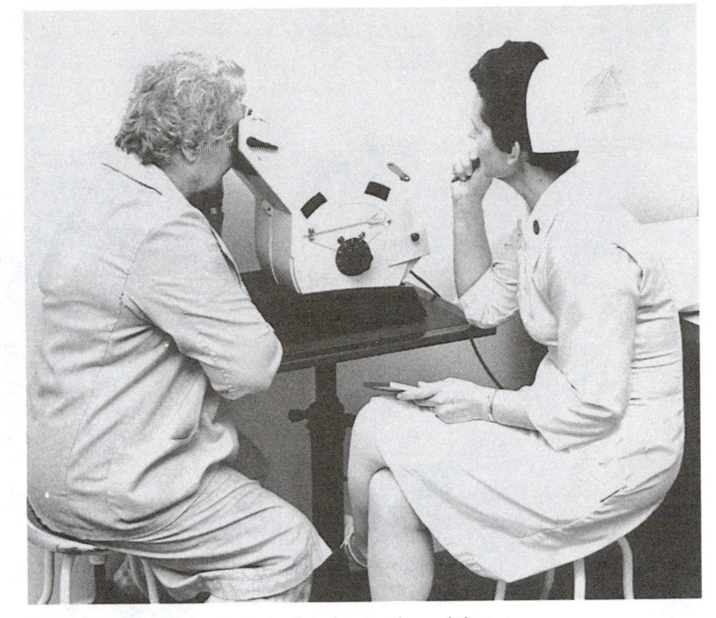

**Figure 19-9.** A typical device for testing vision. (Reprinted with permission from Titmus Optical Co.)

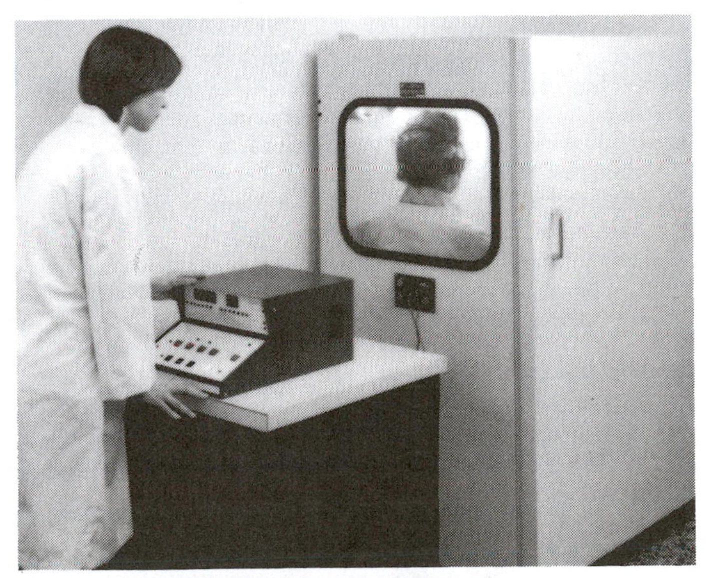

**Figure 19-10.** A hearing test using an automatic audiometer. The hearing test is performed in a booth to ensure a quiet environment without distractions. (Reprinted with permission from Environmental Technology Corp.)

**Health history.** A carefully taken personal health and occupational history may give as much information about the worker's health status as the physical examination. The history also will indicate the need for special tests and perhaps placement restriction. A nurse or other specially trained person can secure basic data from the examinee, such as height, weight, age, and the history. Nurses may also assist in other portions of the examination when specially trained and authorized to do so.

## Emergency medical planning

The industrial physician, nurse, and safety professional should confer with management to plan for emergency handling of large numbers of seriously injured employees in the event of

**Figure 19-11.** When speed is essential, a helicopter can be used to transfer a patient.

disaster, such as explosion, fire, or other catastrophe. These plans should be coordinated with community plans for such events. See Planning for Emergencies, Chapter 16.

Procedures should include the following:

1. Selection, training, and supervision of auxiliary nursing and other personnel;
2. Transportation and caring for the injured (Figure 19–11);
3. Transfer of seriously injured to hospitals;
4. Coordination of these plans with the safety department, guards, police, road patrols, fire departments, and other interested community groups.

### Medical records

Employers are required under OSHA to maintain accurate records and make periodic reports of work-related deaths, injuries, and illnesses. See Chapter 6, Accident Records and Incidence Rates.

A standard, effective August 1980, permits both the worker and the Occupational Safety and Health Administration access to employer-maintained medical and toxic exposure records. This standard, which also specifies conditions under which access is allowed, applies to all employers in general industry, maritime, and construction whose employees are exposed to toxic substances or harmful physical agents. According to the regulation, exposure records include records of the employee's past or present exposure to toxic substances or harmful physical agents, exposure records of other employees with past or present job duties or working conditions related to those of the employee, records containing exposure information concerning the employee's working conditions, and material safety data sheets.

Medical records contain an employee's medical history, examination and test results, medical opinions and diagnoses, descriptions or treatments and prescriptions, and employee medical complaints. Exposure records must be maintained for 30 years and medical records for the duration of employment plus 30 years under the new standard.

Industrial medical records also provide data for use in job placement, in establishing health standards, in health maintenance, in treatment and rehabilitation, in workers' compensation cases, in epidemiologic studies, and in helping management with program evaluation and improvement. Such data are collected in the history interview, from the preplacement examination and any subsequent examinations, and from all visits the worker makes to the dispensary or first-aid room—they establish a medical profile of each worker.

The key to accurate diagnosis and treatment often lies in the adequacy and completeness of this medical profile; therefore, its maintenance is a professional responsibility. Further, to compile a complete history, medical records should include absences caused by illness or off-the-job injury. Thus record keeping, nonoccupational as well as occupational, often uncovers chronic or recurrent conditions where treatment (referral to family doctors) and preventive measures pay dividends, assist in absenteeism control, and reduce accident rates.

Maintenance of health records, however, should not be so laborious a task that the occupational health nurse (or first-aid attendant) becomes a file clerk. It is important that the recording forms and filing systems be simple so that they are usable by and interpretable by a physician, nurse, or first-aid attendant.

Descriptions and illustrations of medical record forms are shown in Parts 1, 2, and 3 of the AMA's *Guide to Development of an Industrial Medical Records System* (see References). These forms can be modified by the medical director to suit the specific industry. Basically, the information should be important and useful, easily and accurately obtained, and the yield should justify the cost.

Although the employer should know the individual's limitations from a placement standpoint, only persons who have a need to know should have access to the records. These records are confidential. If a company does not have a resident medical director or nurse, its medical records are usually filed in the personnel department.

Reports to the insurance company and to the state authorities, as well as compensation payments, should, of course, be attended to promptly.

### Neck or wrist tags for medic alert

A universal symbol for emergency medical identification has been developed by the American Medical Association (Figure 19-12). The object of the symbol is to identify its wearer immediately as a person with a physical condition requiring special attention. If the wearer is unconscious or otherwise unable to communicate, the symbol will indicate that there are vital medical facts to be found on a health information card in the bearer's purse, wallet, or elsewhere. A telephone number is given on the identification tag for obtaining more detailed information. These details should be known by anyone before attempting to help an individual struck down by an accident or sudden illness. (See Medic Alert Foundation and Emergency Medical Identification in References.)

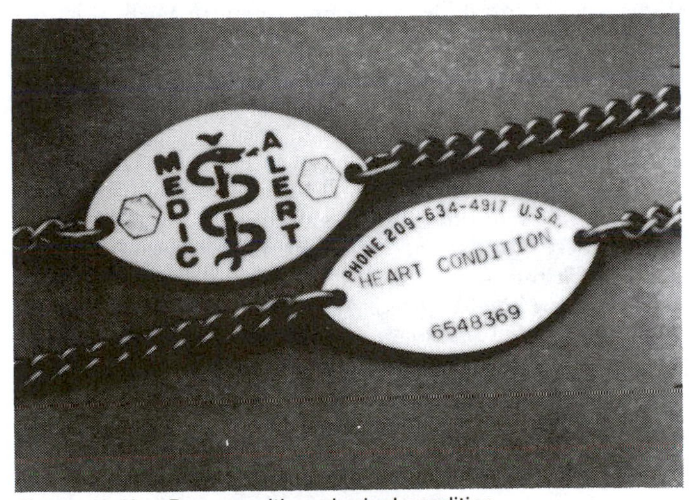


**Figure 19-12.** Persons with a physical condition for which emergency care may be needed should wear an identification tag. The tag provides a general indication of the problem and a phone number to be called for more details if the tag wearer is unable to supply them. (Reprinted with permission from Medic Alert Foundation, Turlock, CA.)

Another type of identification device is a card or medallion that contains a microfilm record of the person's problem. (See Lens-Card System and Medical History Inc. in References. The National Safety Council also offers this service.)

## SPECIAL HEALTH CONCERNS

### Health care of women

There is no reliable evidence to support the view that women are more susceptible than men to occupational health hazards, except during pregnancy.

Some companies used to dismiss women on learning of their pregnancy. As a result, many women concealed their condition until it became obvious, frequently until the last half of the pregnancy. This policy can work to the disadvantage of both employee and company inasmuch as the early months of the pregnancy present the greatest danger. It is desirable that women who work during pregnancy be under medical department supervision. The woman should consult her own doctor who should, in turn, report her condition to the plant or company physician so the advisability of her continuing to work can be determined.

The physician should examine both potential exposures and physical demands of the job. Chemical substances, such as carbon monoxide, chloroform, phosphorus, and mercury, may produce harmful effects on the fetus and lead to abortion. Ionizing radiation and biological agents are also of potential harm. Pregnancy may limit a woman's ability to do some physical work, since a pregnant woman may tire more readily, have poor balance, and respond more slowly to the physiological demands of strenuous physical work.

### Placement of women

State or provincial and federal laws and regulations governing the employment of women in industry should be studied carefully before female employees are placed. These laws and regulations should be rechecked before women are transferred to other types of work or are placed on different shifts.

The examining physician, at the time of employment, should furnish enough information to help place the applicant most advantageously from the standpoint of her health and safety. Periodic examinations and medical histories have indicated that properly trained women are capable of safely performing most types of work.

Methods for the prevention of industrial accidents among women are the same as those for men. However, when women are placed on machine jobs, it is important that adjustments be made, if necessary, to accommodate for size. For example, machine guards may have to be set so that smaller hands cannot enter the openings. Height of benches, distance away from parts, and foot or hand controls should be adjusted to conform to the worker's stature and reach. For some jobs, low platforms may be provided. Smaller hand tools may also be advisable. Some jobs may need to be broken down into simpler operations or mechanical aids may be needed.

More details are provided in Chapter 10, Ergonomics: Human Factors Engineering, and in the Council's book, *Fundamentals of Industrial Hygiene* (see References).

### Alcohol and drug control

Alcoholism and drug abuse continue to be among the nation's leading illnesses. Therefore, it is essential that the physician as well as the employer have a good understanding of the potential for helping persons with these illnesses in the work setting.

It is estimated that one out of every ten U.S. workers may have a drinking problem, resulting in an estimated $125 billion-a-year drain on the economy; but fewer than 10 percent of those who have drinking problems actually receive treatment. Managers and supervisors should be alert to employees whose work and performance are deteriorating because of an alcohol problem. These employees should be referred for medical care as speedily as possible.

When alcohol and drugs are combined, a variety of effects may result that severely impair a worker. Concentrations of alcohol and drugs remain in the blood stream much longer than most users realize, and the effects of this combination may arise unexpectedly.

Dependence on alcohol or other drugs is a major contributor to deterioration of family life, impaired job performance, morale and disciplinary problems, increased insurance rates, occupational accidents, increased absenteeism, and the rising crime rate. These illnesses know no boundaries. There is no "generation gap" among abusers—all ages, races, and socioeconomic groups are susceptible.

Every time an applicant or employee visits a physician or nurse, some health education and/or counseling should be given. Medical departments in industry provide an ideal opportunity for these services, which constitute an asset for the employees and for the employer as well. Employers are paying an increasingly larger part of the costs for health care for employees and their dependents. It is economically important for the employer that employees have an entry into the health care system when needed and, further, that the proper care is obtained.

## REFERENCES


American College of Surgeons, Committee on Trauma. *Emergency Care of the Sick and Injured.* Philadelphia, Pa., W.B. Saunders Company, 1982.

American Conference of Governmental Industrial Hygienists. *Guide to Health Records for Health Services in Small Industries.* Cincinnati, Ohio, ACGIH.

American Medical Association, Dept. of Occupational Medicine, 535 N. Dearborn St., Chicago, Ill. 60610

*Guide to the Development of an Industrial Medical Records System,* rev. 1984.

*Guide to Small Plant Occupational Health Programs,* rev. 1983.

*Guiding Principles of Medical Examinations in Industry,* rev. 1984.

Bond, M. B. "Occupational Medical Services for Small Employee Units," *Rocky Mt. Med. J.* 68:31-36.

Brown, M. L. "Occupational Health Nurse: A New Perspective." *Occupational Medicine—Principles and Practical Applications.* Chicago, Ill., Year Book Medical Publishers, Inc. 1975.

________. *Occupational Health Nursing.* New York, N.Y., Springer Publishing Company, Inc., 1981.

*Code of Federal Regulations,* 29. Washington, D.C., Government Printing Office, 1985.

*A Comprehensive Guide for Establishing an Occupational Health Service.* Atlanta, Ga., American Association of Occupational Health Nurses, Inc., 1987.

*Emergency Care and Transportation of the Sick and Injured,* Committee on Injuries, American Academy of Orthopedic Surgeons. Menasha, Wis., George Banta Co., Inc., 1971.

Emergency Medical Identification, American Medical Association, 535 N. Dearborn St., Chicago, Ill. 60610.

Hannigan, Larry, "Nurse, Health Assessment—The New 'Physical' in American Industry," *Occupational Health Nursing,* August 1982.

LaDou, Joseph, ed. *Introduction to Occupational Health and Safety.* Chicago, Ill., National Safety Council, 1986.

Lens-Card Systems, Inc., 2817 Regal Road, Plano, Texas 75075.

Medic Alert Foundation, Turlock, Calif. 95380.

Medical History Inc., 521 Fifth Ave., New York, N.Y. 10017.

*Occupational Health.* 9(No. 4): 1985.

Plog, B. A., ed. *Fundamentals of Industrial Hygiene,* 3rd ed. Chicago, Ill., National Safety Council, 1988.

# 20

# Workers with Disabilities



Today, almost every worker with a physical or mental impairment can qualify for some type of job. Industry and government surveys made during the past several decades prove it is good business to hire workers with permanent disabilities. They excel in promptness, can produce as well as the nonhandicapped, and their turnover rate is lower.

In addition, the law requires affirmative action for the hiring, upgrading, promotion, award of tenure, demotion, transfer, layoff, termination, right of return from layoff, and rehiring of qualified disabled individuals. The law also requires that an employer provide "reasonable accommodation" for these workers when necessary.

If the employer denies a disabled individual a specific job, the burden of proof is upon management to show the person is unqualified because of one or more of the following reasons:

- The job would put the individual in an unsafe situation
- Other employees would be placed in an unsafe situation if the person were put in the job
- The job requirements cannot be met by an individual with certain physical or mental limitations
- And (for all of the above) accommodation cannot reasonably be accomplished.

Affirmative action programs, including those for hiring workers with disabilities, are usually the responsibility of personnel other than the safety professional. The safety professional, however, should be a key resource person to such personnel and must play a critical role in job placement of the disabled workers. Safety evaluations for the worker must include access to and exit from the workplace as well as safety on the job.

This chapter is intended to assist in the safe and productive placement of disabled individuals in the work force. The professional and legal responsibilities are stated, and practical, how to do it information is given.

## HISTORY AND THE LAW

In the 1940s, special attention was given to employing workers with disabilities by a number of large companies that realized hiring these people was smart business practice. Although some companies employed workers with disabilities prior to the 1940s, three events occurred in that decade to stimulate these programs and cause other companies to initiate hire-the-handicapped programs.

- The first event was World War II. Many individuals with disabilities were hired to help fill the jobs vacated by employees who left for military service.
- The second contributing event was the restoration of World War II's wounded. For example, in the early 1940s, International Harvester Company undertook an affirmative action program to help each Harvester disabled veteran to become an employable person. Other companies established similar programs (see Figure 20-1).
- The third event was a study published by the U.S. Department of Labor that debunked myths about workers with disabilities being less productive, suffering more accidents, and losing more time from work than other workers. On the contrary, the Department of Labor study showed disabled workers were as productive as other workers, had lower frequency and severity lost-time injury rates, and were absent from work only one day more per year than other workers. In researching over 11,000 workers with disabilities for almost two years,

Figure 20-1. After losing his leg in combat, Felix "Phil" Sitkowski returned, as a clerk, to U.S. Steel South Works in South Chicago, Illinois. Later he applied for and was promoted to field lubricating analyst. Sitkowski's prosthesis allowed him to use ladders for making inspections. He is shown at the left with two of the eight millwright helpers he supervised. (Printed with permission from U.S. Steel South Works.)

the study's authors did not find a single disabling injury caused by a handicapped worker, to himself or to a fellow worker, that was the result of the handicap (see Figure 20-2).

Du Pont conducted a study in 1958 and updated it in 1981. The later Du Pont study of more than 2,700 of its workers with disabilities revealed that 96 percent of them rated average or better on safety performance; 92 percent rated average or better on turnover; and 85 percent rated average or better on attendance. After a decade or more of direct experience in hiring the disabled, the personnel files of many companies held indisputable proof of their programs' worth and the value of these employees to their companies.

### Occupational Safety and Health Act of 1970 (OSHAct)

The Occupational Safety and Health Act of 1970 has no special standards pertaining to employees with disabilities, nor does OSHA maintain any separate statistics on these workers. In fact, the requirements of some OSHA standards may be detrimental to certain types of handicapped employees and must be considered when placing such employees in jobs where these standards apply. (Refer to OSHA obstacles section of this chapter.)

Figure 20-2. Studies show that workers with disabilities, such as these employed at a candy factory near Peoria, Illinois, are productive, dependable, and have low work-injury rates. (Printed with permission from the Illinois Department of Rehabilitation Services.)

### Rehabilitation Act of 1973

The federal Vocational Rehabilitation Act of 1973 (Public Law 93-112), commonly referred to as the Rehabilitation Act, is the first major civil rights law protecting the rights of persons with disabilities. This law applies to federal contractors (Section 503) and recipients of federal assistance programs (Section 504). Therefore, all employers who do work for the federal government, or receive funds from the government, are subject to this law.

- Section 503 of the Act requires employers to take "affirmative action" to recruit, hire, and advance qualified individuals with disabilities. The law applies only to employers who have federal government contracts or subcontracts of $2,500 or more. Those holding contracts or subcontracts of $50,000 or more, with at least 50 employees, must prepare and maintain (review and update annually), at each establishment, an affirmative action program. The program, which sets forth the employer's policies, practices, and procedures regarding handicapped workers, must be available for inspection by job applicants and employees. This Section is enforced by the Office of Federal Contract Compliance Programs, which has issued extensive implementing regulations.
- Section 504 of the Act forbids acts of discrimination in employment against qualified disabled persons by employers who are recipients of federal funds. This Section is enforced by each department or agency that provides federal funds.
- In the Rehabilitation Act Amendments of 1974, Congress

amended, and thus broadened, the definition of "handicapped individual" for purposes of Section 504. With this amended definition, it became clear that Section 504 was intended to forbid discrimination against all handicapped individuals, regardless of their need for or ability to benefit from vocational rehabilitation services. Thus, Section 504 reflects a national commitment to end discrimination on the basis of handicap and establishes a mandate to bring persons with disabilities into the mainstream of American life.

- U.S. Department of Labor regulations (41 *CFR* 60-741), as last amended, became effective October 20, 1978, and were established to assure compliance with Section 503 of the Act. Department of Education regulations (34 *CFR* 104), effective May 4, 1980, were implemented to enforce the requirements of Section 504. Other federal departments and agencies also have issued regulations similar to those of the Department of Education, indicated above. For example, the U.S. Department of Labor has issued regulations (29 *CFR* 32), effective November 6, 1980, which implement Section 504 of the Act for the department. All these regulations require federal contractors and recipients of federal funds to make reasonable accommodations when necessary in employing the handicapped.

All records pertaining to compliance with the Act, including employment records and any complaints and actions taken as a result, must be maintained by the employer for at least one year. Failure to maintain complete and accurate records or failure to annually update the affirmative action program can result in the imposition of "appropriate sanctions" against the employer. The employer must permit access, during normal business hours, to its place of business, its books, records, and accounts pertinent to compliance with the Act, and all rules and regulations promulgated (pertaining to the Act) for the purposes of complaint investigations, and investigations of performance concerning affirmative action.

### Vietnam Era Veterans' Readjustment Assistance Act of 1974

This Act (Public Law 93-508), effective December 3, 1974, is the amended version of the 1972 Act (Public Law 92-540). A part (38 *USC* 2012) of the Act provides a legal incentive for America's business and industry to employ and advance disabled veterans of all wars and all veterans of the Vietnam Era. The law requires that all employers with federal government contracts or subcontracts of $10,000 or more must take affirmative action (similar to Section 503 of the Rehabilitation Act of 1973) to accomplish this endeavor.

U.S. Department of Labor regulations (41 *CFR* 60-250), last amended on February 12, 1980, were established to assure compliance with 38 *USC* 2012 of the Act.

### State and local laws

All 50 states and many local governments have now adopted building codes or legislation requiring barrier-free design or removal of barriers to the disabled person's mobility. Many of these codes require any public building or facility to be barrier free if the public is invited to use it for any normal purpose. This is defined as shopping, employment, recreation, lodging, or services.

If accessible facilities are to be identified, then the international symbol of accessibility shall be used (see Figure 20-3).

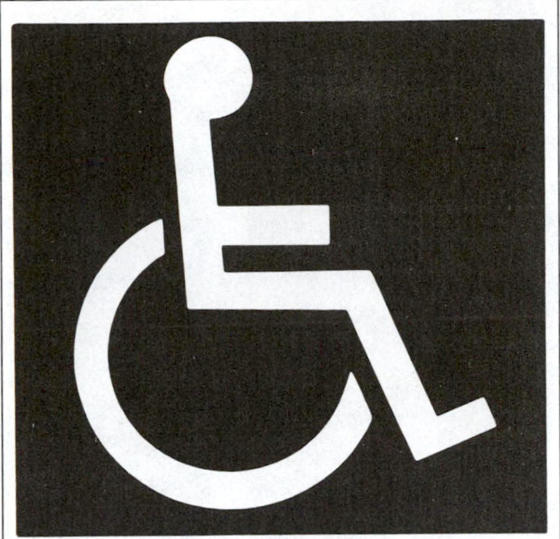

**Figure 2-3.** International symbol designates access for the disabled. The symbol is in white on a blue background. (Printed with permission from American National Standards Institute.)

## WHO ARE DISABLED JOB-SEEKERS?

The law defines three basic types of disabled persons seeking employment—the disabled individual, the disabled veteran, and the qualified disabled individual.

### 'Disabled individual'

The Rehabilitation Act of 1973, as amended, defines a "handicapped [disabled] individual" as any person who:

1. Has a physical or mental impairment that *substantially limits* one or more of the person's major "life activities," such as:
   Ambulation
   Communication
   Education
   Employment
   Housing
   Self-care
   Socialization
   Transportation
   Vocational training
2. Has a record of such impairment, or
3. Is regarded as having such an impairment.

The term "substantially limits," as used above, has to do with the degree to which the disability affects employability. A qualified person with a disability who, because of the disability, has difficulty obtaining an appropriate job or getting ahead on a job would be considered substantially limited.

The definition of a disabled individual adopted by the federal government is broad. Well-known conditions such as hearing and sight loss, musculo-skeletal impairments and mobility impairments are covered (see Figure 20-4). Also included are such conditions as mental illness, mental retardation, learning disabilities, neurological disorders, medical conditions such as cancer and diabetes, and substance abuse.

This broad definition would include these people:

- A blind person or a paraplegic person would have "an impairment substantially limiting one or more of life's major activities." So would a mentally retarded person.

**Figure 20-4.** As a result of being struck by a high-voltage highline wire, Charles Dannheim lost his legs and arms. Although he would be considered "substantially limited" and a disabled individual by definition, Dannheim works full-time as a cattle rancher with the aid of hooks and artificial legs. (Printed with permission from *Beef Magazine.)*

- A person who had been in a mental hospital and had been rehabilitated would have "a record of such impairment," even though now mentally sound. So would a person with a history of heart condition or cancer.
- A person who people think is disabled—for example a person who might seem mentally retarded but isn't—would be "regarded as having such an impairment."

### 'Disabled veteran'

A disabled veteran is a "special handicapped individual" who:

1. Is entitled to disability compensation under laws administered by the Veterans Administration for disability rated at 30 percent or more.
2. Was discharged or released from active duty due to a disability incurred or aggravated in the line of duty.

A veteran with nonservice-connected disabilities is not considered a special disabled veteran but may still qualify as a disabled individual under Sections 503 and 504 of the Rehabilitation Act of 1973.

The Vietnam War had the highest proportion of disabled service personnel of any war in history. A disabled veteran of the Vietnam Era is a person who was discharged or released from active duty for a service-connected disability if any part of such duty was performed between August 5, 1964, and May 7, 1975.

### 'Qualified disabled individual'

Is every handicapped person covered by the Rehabilitation Act of 1973? No. There is another crucial word: qualified. A person must be capable of performing a particular job—with reasonable accommodation to the disability (see Figure 20-5).

Is every disabled veteran and every Vietnam Era disabled veteran covered by either the Rehabilitation Act of 1973 or the Vietnam Era Veterans' Readjustment Assistance Act of 1974? No. The veteran also must be qualified for the job. That is, the veteran must be capable of performing a particular job, with reasonable accommodation to the disability.

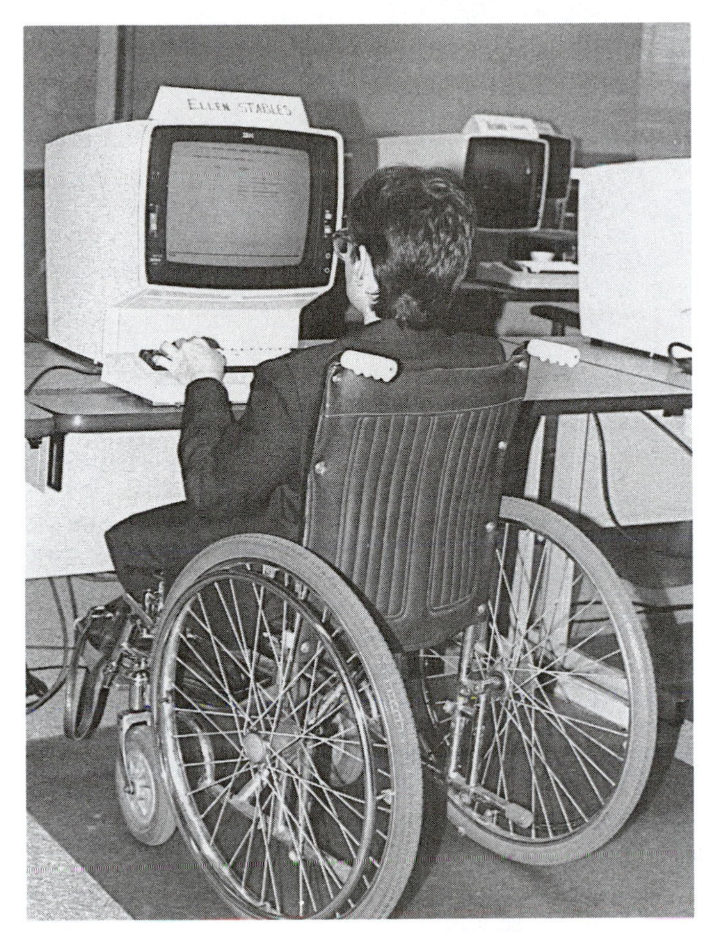

**Figure 20-5.** This woman will soon be a "qualified disabled individual," and very employable, because of the IBM computer programming training she is receiving at the El Valor Rehabilitation Facility in Chicago. (Printed with permission from the Illinois Department of Rehabilitation Services.)

## REASONABLE ACCOMMODATION

An employer shall make "reasonable accommodation" to the known physical or mental limitations of an otherwise qualified handicapped applicant or employee unless the employer can demonstrate that the accommodation would impose an undue hardship on him. The employer may not deny any employment opportunity to a qualified disabled employee or applicant if the basis for the denial is only the need to make a reasonable accommodation.

"Undue hardship" is determined by considering the following factors:

1. The overall size of the employer's operation with respect to number of employees, number and type of facilities, and size of budget
2. The type of operation, including the composition and structure of the work force
3. The nature and cost of the accommodation needed.

Reasonable accommodation may include, but is not limited to:

1. Making facilities used by employees readily accessible to and usable by disabled persons
2. Job restructuring, part-time or modified work schedules, acquisition or modification of equipment or devices, the provision of readers or interpreters, and other similar actions.

### Examples

Reasonable accommodation is demonstrated in these three examples:

- A construction equipment salesman, whose job description required him to climb onto the equipment and demonstrate its operation during sales presentations, was given a desk job after he suffered the amputation of his arm during an off-the-job motor vehicle accident. Although his prosthetic device enabled him to operate the equipment controls, the employer had considerable concern about the man's ability to climb on and off the equipment using the prosthetic device. This resulted in the job change. Upon enactment of the Rehabilitation Act of 1973, and its amendments, his employer reinstated the man as a salesman making a reasonable accommodation for his handicap. The accommodation consisted of providing him with a portable climbing device that enabled him to safely get on and off the equipment.
- A fork lift (powered industrial truck) mechanic became blind in one eye due to a nonindustrial health problem. Since his job description required the mechanic to test drive each fork lift truck after completing maintenance or repair on it and since the employer's standard safety policy required that all drivers of powered industrial trucks must have binocular vision (use of both eyes), the employer at first was going to switch the man to another job, which unfortunately would have lowered his earnings. Upon reviewing the requirements of the Rehabilitation Act of 1973, however, the employer provided the mechanic with a reasonable accommodation. The accommodation consisted of altering this mechanic's specific job description eliminating the requirement for him to test drive the vehicles and broadening the job description of the other mechanics to include the test driving of any vehicle repaired by the disabled mechanic.
- A disabled individual working for an electrical appliance firm was provided with a reasonable accommodation to assemble small parts. The accommodation consisted of minor readjustments of the work bench to accommodate the individual in a wheel chair.

The following example was considered to be unreasonable accommodation:

- An example of an unreasonable requirement would be the need to completely redesign or alter the circuitry or operating levers of a machine in order to accommodate a physically disabled individual.

### Accommodation is not new

Accommodations can include modifications of equipment or facilities and process or job description alterations.

Accommodation in employment is not a novel concept. The first applications of machine guards and ventilating fans were job accommodations. Also, the first hod carrier who lacquered and reinforced his bowler as a hard hat made a job accommodation. Job placement of employees based on medical examinations, when newly hired or returning to work after an illness of injury, is again an application of accommodation. This experience is common to every employer.

The safety professional routinely evaluates accommodations designed to reduce or eliminate hazards to employees. This professional, therefore, is preeminently qualified to evaluate reasonable accommodations of the workplace, its procedures and access for the physical or mental limitations of a disabled worker.

Job safety analysis and safe work procedures are a means of eliminating or reducing work hazards to minimize worker risk. They directly transfer to the process of accommodation. Training in safe work procedures will be important in accommodating the job for the person with a disability.

## ROLE OF THE SAFETY AND HEALTH PROFESSIONAL

Since affirmative action programs required by the government usually come under the responsibility of the EEO (Equal Employment Opportunity) manager or coordinator (or labor affairs personnel), the placement of qualified disabled individuals is normally under their overall jurisdiction.

The safety and health professional, nonetheless, should be a key resource person available to those responsible for job placement of qualified disabled individuals. This professional should be consulted before such placement is made and asked to evaluate any proposed reasonable accommodation. The following is an example of some of the responsibilities of the safety and health professional in relation to disabled employees.

### General responsibilities

- Maintain close liaison with the EEO manager-coordinator and the medical and personnel departments when they are placing disabled employees.
- Make job hazard analyses of existing jobs when employing, promoting, transferring, and selecting workers with disabilities requires that training programs be designed or updated.
- Make recommendations for safe modifications of machine tools, established processes and procedures, and existing facilities and workplace environment when reasonable provisions are being made to accommodate disabled employees.
- As required, cooperate with the plant or building engineer or mechanical engineer and the planning, production, and maintenance departments when disabled employee accommodations are being evaluated.

### Specific responsibilities

- Review the company's affirmative action program to acquire knowledge.
- Establish specific communication channels, pertaining to handicapped employees, with:
  1. *EEO manager-coordinator.* Let this person know you are part of the team and are available when a job needs evaluating.
  2. *Medical department.* Let them know you will be requesting their help when evaluating a job.
  3. *Employment department.* Let them know you are ready when necessary to assist them evaluate a job's safety, and remind them of basic safety considerations such as:
     a. Don't place a worker with a coronary condition in a job that would aggravate that condition.
     b. Don't place a person with a back problem in a job requiring heavy lifting.
     c. Make certain to place a handicapped employee in a job that would be safe for that person and that will not cause a hazard to others.

     NOTE: Individual judgments are based on a physician's evaluation with input from the safety professional.
  4. *Plant and mechanical engineers.* Reasonable accommodation does not necessarily mean reinstalling machines, but rather could mean minor relocating of a machine's controls so a handicapped employee could properly operate them. Therefore, advise the engineers you will evaluate all safety aspects of such an accommodation. Also advise them you are available for safety evaluations when they design reasonable accommodations into future facilities such as:
     a. Ramps for wheelchairs
     b. Wider door passages for wheelchairs
     c. Grab bars in accessible washroom facilities
     d. Braille numbers on elevators (Figure 20-6)
     e. Easy access to company facilities such as lunchrooms
     f. Elimination of curbed cross walks.
  5. *Production and maintenance departments.*
     a. Since reasonable accommodations also refer to job restructuring and modifications, tell the production and maintenance departments you will help by evaluating the safety aspects of such changes.
     b. When they delete any job specification that would arbitrarily and without justification screen out disabled individuals, you will be available if a safety evaluation is needed.

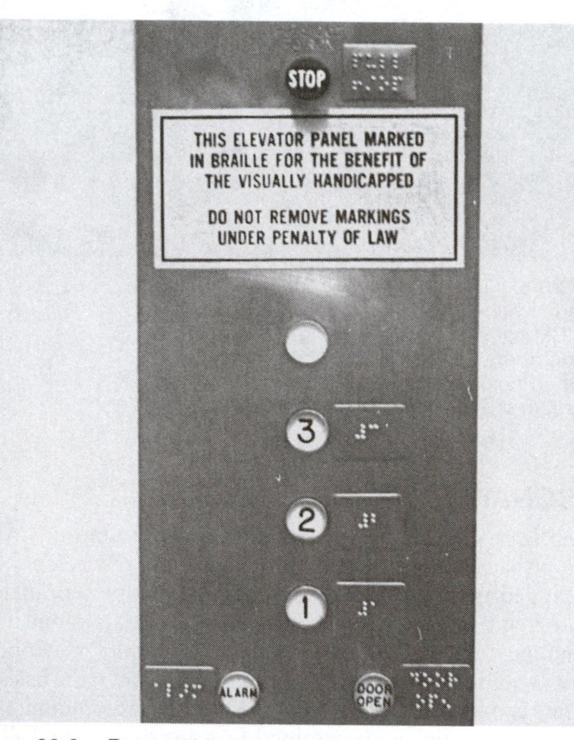


**Figure 20-6.** Reasonable accommodation for the disabled includes marking elevator buttons in braille. (Printed with permission from Governors State University, Park Forest South, Ill.)

- Refer all safety complaints (or hazards noted), involving a handicapped employee, to the EEO manager-coordinator.
- Conduct a safety evaluation of a disabled employee (in relation to the specific job or prospective job), and perform an entire job hazard analysis if needed.
- Conduct a safety evaluation whenever a reasonable accommodation is being *planned* for a handicapped employee.
- Coordinate with both the EEO manager-coordinator and the employment department to make certain any disabled employee being considered for a new position is qualified to

**HANDICAPPED EMPLOYEE SAFETY EVALUATION**

☐ Applicant:
☐ Employee: ______________________________________________ ________
(Last Name First Name M.I.) (Clock No.)

Handicap: ______________________________________________

Evaluation of ☐ Current job ☐ Prospective job

Job Title: ______________________________________________

Job Description (primary duties): ______________________________________________

______________________________________________

______________________________________________

Hazards to This Employee:
(State "none" if none) ______________________________________________

______________________________________________

Hazards to Other Employees:
(State "none" if none) ______________________________________________

______________________________________________

Proposed "reasonable accommodation" (if any): ______________________________________________

______________________________________________

CONCLUSION (Based on all known factors at this time):

☐ Job is safe for this employee:
☐ as is ☐ with proposed reasonable accommodation

☐ Job is unsafe for this employee:
☐ as is ☐ with proposed reasonable accommodation

☐ No hazard to other employees:
☐ as is ☐ only with proposed reasonable accommodation

☐ Hazard to other employees:
☐ as is ☐ with proposed reasonable accommodation

| Location: | Safety Supervisor (Print Name) |
| --- | --- |
| Date: | Safety Supervisor (Signature) |

NOTE: Complete two copies of this form and give one copy to local EEO manager-coordinator. Second copy is for the safety file.

**Figure 20-7.** This evaluation form is an excellent administrative tool, especially when a written report and other supporting documentation are attached.

safely and capably perform the job. This usually requires the safety and health professional to make a safety evaluation and observe the employee during training.

- Evaluate if reported harassment of a disabled employee is safety related. For example, name calling would not normally involve the safety of the employee but pranks by other employees could jeopardize the safety of the employee.
- Refer all questions about interpretation of government requirements to the personnel responsible for implementing the affirmative action program (usually the EEO manager-coordinator) or to the appropriate attorney.

The evaluation form (Figure 20-7) can help the safety and health professional perform safety evaluations for employees with disabilities, especially when reasonable accommodation is involved. A written report and supporting material (such as memos, blueprints, and photos) can be attached to the form to provide detailed information on why certain decisions were made.

It is recommended that such evaluations be kept for at least one year after the employee leaves the company. Records should be destroyed only after approval from the EEO manager-coordinator or other personnel responsible for the government-required affirmative action program.

## INSURANCE CONSIDERATIONS

There has been a general misconception of the effect of employ-

**PHYSICAL CAPACITIES FORM**

Leg Amputation 5″ below knee

Artificial leg—good fitting

Name ________ Sex M Age 31 Height 5′9½″ Weight 155

| PHYSICAL ACTIVITIES | |
|---|---|
| x 1 Walking | 16 Throwing |
| 0 2 Jumping | x 17 Pushing |
| 0 3 Running | x 18 Pulling |
| x 4 Balancing | 19 Handling |
| x 5 Climbing | 20 Fingering |
| 0 6 Crawling | 21 Feeling |
| x 7 Standing | 22 Talking |
| 8 Turning | 23 Hearing |
| 9 Stooping | 24 Seeing |
| 0 10 Crouching | 25 Color Vision |
| x 11 Kneeling | 26 Depth Perception |
| 12 Sitting | 27 Working Speed |
| 13 Reaching | 28 |
| x 14 Lifting | 29 |
| x 15 Carrying | 30 |

| WORKING CONDITIONS | |
|---|---|
| 51 Inside | x 66 Mechanical Hazards |
| 52 Outside | x 67 Moving Objects |
| 53 Hot | 0 68 Cramped Quarters |
| 54 Cold | 69 High Places |
| 55 Sudden Temp. Changes | 70 Exposure to Burns |
| x 56 Humid | 71 Electrical Hazards |
| 57 Dry | 72 Explosives |
| x 58 Wet | 73 Radiant Energy |
| 59 Dusty | 74 Toxic Conditions |
| 60 Dirty | 75 Working With Others |
| 61 Odors | 76 Working Around Others |
| 62 Noisy | 77 Working Alone |
| 63 Adequate Lighting | 78 |
| 64 Adequate Ventilation | 79 |
| 65 Vibration | 80 |

Blank Space=Full Capacity **x**=Partial Capacity 0=No Capacity

May work__hours per day__days per week. (IF TB, cardiac or disability requiring limited working hours.)

May lift or carry up to ** pounds.

DETAILS OF LIMITATIONS FOR SPECIFIC PHYSICAL ACTIVITIES:
Should not be required to walk, balance, climb, stand, kneel for prolonged periods of time.
**Should not lift heavy weights continuously.
Should not carry long distances.

**PHYSICAL DEMANDS FORM**

Job Title Data Processing Entry Occupational Code 4-44.110

Dictionary Title Data Processing Entry

Firm Name & Address ________

Industry ________ Industrial Code ________

Branch ________ Department ________

Company Officer ________ Analyst Wetzel Date ____

| PHYSICAL ACTIVITIES | |
|---|---|
| x 1 Walking | 16 Throwing |
| 2 Jumping | 17 Pushing |
| 3 Running | 18 Pulling |
| 4 Balancing | x 19 Handling |
| 5 Climbing | x 20 Fingering |
| 6 Crawling | 21 Feeling |
| 7 Standing | 22 Talking |
| 8 Turning | 23 Hearing |
| 9 Stooping | x 24 Seeing |
| 10 Crouching | 25 Color Vision |
| 11 Kneeling | 26 Depth Perception |
| 12 Sitting | 27 Working Speed |
| x 13 Reaching | 28 |
| x 14 Lifting | 29 |
| x 15 Carrying | 30 |

| WORKING CONDITIONS | |
|---|---|
| x 51 Inside | 66 Mechanical Hazards |
| 52 Outside | 67 Moving Objects |
| 53 Hot | 68 Cramped Quarters |
| 54 Cold | 69 High Places |
| 55 Sudden Temp. Changes | 70 Exposure to Burns |
| 56 Humid | 71 Electrical Hazards |
| 57 Dry | 72 Explosives |
| 58 Wet | 73 Radiant Energy |
| 59 Dusty | 74 Toxic Conditions |
| 60 Dirty | 75 Working With Others |
| 61 Odors | x 76 Working Around Others |
| 62 Noisy | 77 Working Alone |
| 63 Adequate Lighting | 78 |
| 64 Adequate Ventilation | 79 |
| 65 Vibration | 80 |

**DETAILS OF PHYSICAL ACTIVITIES:**
Sits at computer most of the day, reads copy, and fingers keyboard to enter data. Periodically walks short distances, reaches for, lifts, and carries small stacks of billing materials. Reaches for, handles, pushes, and pulls when organizing and filing paperwork at desk level.

**Figure 20-8.** Example of an employment service form used when matching workers to jobs.

ment of handicapped workers upon insurance rates. It has been thought by many that companies issuing workers' compensation insurance coverage increase the premiums where the physically handicapped are employed. This is not true. Rates are based on experience by the class of industry and modified in most cases by the individual plant experience. There is no indication that losses are increased when persons with disabilities are properly placed.

## JOB PLACEMENT

When properly placed and trained, and competing on an equal basis, workers with disabilities usually equal or are slightly better than other workers in production and safety; and their attendance and labor turnover records are usually superior to those without disabilities.

### General concepts

To properly place a disabled worker, the following requirements should be observed, where applicable, after receiving a physician's evaluation of the individual.

- The worker should meet physical qualifications of the job. When necessary, the worker should receive the support of reasonable accommodation.
- The worker should not be a personal hazard. For example, a person subject to dizzy spells should not work on a ladder or scaffold or around moving machinery, where injury or death could occur.
- The worker should not be a hazard to others. For example, a person with severe vision impairment should not drive a bus or operate an overhead crane, because the individual might cause injury to himself and to others.
- The task should not aggravate the degree of disability. A person with skin disease should not be exposed to skin irritants.
- To obtain valuable input, a conference with the individual should be held before job placement is made.

What proper placement does, then, is match the worker to the job on the basis of his ability to meet the qualifications of the job. (See Figure 20-8.) When this is done, the impairment disappears as a job factor. Moreover, it should be realized that most disabled persons have more ability than disability, because few jobs actually require all of a worker's ability. The job-employee match forms shown in Figure 20-8 are not only used to place disabled persons, some companies use them to place all new and transferred workers.

It is important to remember that each impairment can impose limitations on the activities in which the individual can engage and the working conditions and accident and health hazards to which this person can be exposed.

### OSHA obstacles

The safety and health professional should be aware that there are certain OSHA safety standards which, although promulgated for the protection of the average employee, can be detrimental to employees with disabilities. Some examples are:

- Standards referring to storage of flammable and combustible liquids, § 1910.106 (d)(6)(iii), include a requirement for a curb to capture spilled liquids. This would be a barrier for some disabled individuals.
- The "Means of Egress" standards, § 1910.37, are based on the ability of an individual to move 100 feet (30 m) in 30 seconds. Perhaps some employees with disabilities cannot move that fast.
- Respirators are required by the standards, for example, §1910.134, for certain jobs. Some individuals may have a physical impairment that can be affected by restricted breathing. If there is some indication of this problem, such employees must not wear a respirator until it is determined by a physician that it can be worn safely. This may preclude the individual from performing a specific job where a respirator is necessary.
- The permissible exposure levels (PELs) listed in the OSHA "Air Contaminants" standards, § 1910.1000, are based on the susceptibility of persons with normal breathing capacities to such contaminants. Some disabled individuals do not have normal breathing capacities.

  The safety professional must consider whether a disabled individual would be adversely affected by safety standards applying to the job being evaluated.

## ANALYSIS OF THE JOB

Each job must be evaluated to make sure the individual being considered can do it safely. The following areas should be taken into account when making the analysis; they have been adapted from a previous edition of this Manual.

### Physical classification

The labor market simply does not supply only "physically perfect" workers. In fact, the percentage of workers in perfect health is relatively low—the working population now includes more persons with disabilities than ever before. Advances in medical science prolong the lives of many who would have died of war injuries, or illnesses, such as smallpox, tuberculosis, diabetes, and heart disease. Accidents in industry, in traffic, and in the home continue to increase the number of persons with physical disabilities.

Because of proximity to the problem and ability to make regular plant inspections, the company's physician should have a better understanding of the job requirements than other physicians. Therefore, it is the physician's responsibility to provide management with clear evaluations of the employability of applicants for jobs. The physician's determinations must be made on the basis of job-worker compatibility.

Many systems of classification are now in use:

- Generally, however, these systems use broad statements, such as "physically fit for any work"; "defect that limits applicant to certain jobs" (the defect may or may not be correctable, but may require medical supervision); and "defect that requires medical attention and is presently handicapping." This last statement disqualifies a person for any type of employment.
- Another method provides a greater range in expression of limiting factors, allowing more alternatives for the individual case.
- Yet another method, which approaches the ultimate in functional evaluation of the individual, appraises capacities on a form with the identical terminology used in evaluating the physical or functional factors and working conditions of jobs. (See Figure 20-9.) This effective method of presenting information from a physical examination clearly indicates the specific work capability and limitations of the individual. Thus the medical report is more meaningful to the placement manager because the examining physician is responsible for determining the occupational significance of physical disorders.

  This makes proper analysis of each job important.

### Job appraisal

Employers must be aware of the physical requirements of jobs and the accident and health hazards involved. Each job-appraisal factor has a direct relationship that makes it either definitely unsuitable or potentially undesirable for one or more types of disability. The factors to be considered in job appraisal are physical requirements, working conditions, health hazards, and accident hazards.

**Physical requirements** include agility, strength, exertion, vision, hearing, talking, sitting, standing, walking, running, climbing, crawling, kneeling, squatting, stooping, twisting, lifting, and handling. They should be evaluated according to quality of ability and duration of activity. For example, a job involving a considerable amount of stair climbing is unsuitable for workers with heart disease, respiratory diseases, obesity, or lower limb orthopedic disorders. However, some of these people can tolerate a small amount of stair climbing.

**Working conditions** include indoors, outdoors, excessive heat or cold, excessive humidity or dryness, wetness, sudden temperature changes, ventilation, lighting, noise, and whether the work is performed alone, near others, with others, or as shift work or piece work. Some of these conditions could have a harmful influence upon certain disabilities. For example, work in excessive heat is generally unsuitable for persons who have had malaria, or for those with high blood pressure, heart disease, skin disease, and for the aged and the obese.

**Health hazards** include air pressure extremes; radiant energy (ultraviolet, infrared, radium emanations, and X rays); silica, asbestos, dusts, and skin irritants; respiratory irritants; systemic poisons; and asphyxiants. These hazards have serious effects and can aggravate a preexisting bodily defect. For example, a job might involve exposure to respiratory irritants of insignificant quantities to a normal person; yet this condition might aggravate the disability of a person who has chronic bronchitis. (These were discussed in the previous section, Job Placement.)

**Accident hazards** include danger of falls from elevations, working while on moving surfaces, slipping, and tripping hazards; exposure to vehicles or moving objects; objects falling from overhead; exposure to sources of foot injuries, eye injuries, cuts,

# JOB ANALYSIS
# FOR PHYSICAL FITNESS REQUIREMENTS

| TITLE OF POSITION | GRADE |
|---|---|
| | |

| NAME AND LOCATION OF ESTABLISHMENT | AGENCY |
|---|---|
| | |

Does establishment have medical supervision ☐ Yes ☐ No
Is there an industrial safety branch ☐ Yes ☐ No

*Refer to the manual for job analyses before using this form. Check all functional and working condition factors as well as acceptable disabilities whenever appropriate.*

## I. FUNCTIONAL FACTORS

L - *Little* M - *Moderate* G - *Great* O - *None*

| Hands - Fingers | L | M | G | O | Arms | L | M | G | O | Legs - Feet | L | M | G | O | Body - Trunk | L | M | G | O |
|---|---|---|---|---|---|---|---|---|---|---|---|---|---|---|---|---|---|---|---|
| 1. Reaching | | | | | 8. Reaching | | | | | 14. Walking or running | | | | | 22. Sitting | | | | |
| 2. Pushing or pulling | | | | | 9. Lifting | | | | | 15. Standing | | | | | 23. Bending | | | | |
| 3. Handling | | | | | 10. Pushing or pulling | | | | | 16. Sitting | | | | | 24. Reaching | | | | |
| 4. Fingering | | | | | 11. Carrying | | | | | 17. Carrying | | | | | 25. Lifting | | | | |
| 5. Climbing | | | | | 12. Climbing | | | | | 18. Climbing | | | | | 26. Carrying | | | | |
| 6. Throwing | | | | | 13. Throwing | | | | | 19. Jumping | | | | | 27. Jumping | | | | |
| 7. Touching | | | | | **Eyes** | | | | | 20. Turning | | | | | 28. Turning | | | | |
| | | | | | 30. Near vision | | | | | 21. Lifting | | | | | | | | | |
| **Voice** | | | | | 31. Far vision | | | | | **Ears** | | | | | | | | | |
| 29. Talking | | | | | 32. Color vision | | | | | 33. Hearing | | | | | | | | | |

## II. WORKING CONDITION FACTORS

| | | | | | | | | | | | | | | | | | | | |
|---|---|---|---|---|---|---|---|---|---|---|---|---|---|---|---|---|---|---|---|
| 34. Inside | | | | | 41. High humidity | | | | | 48. Odors | | | | | 55. Toxic conditions | | | | |
| 35. Outside | | | | | 42. Low humidity | | | | | 49. Body injuries | | | | | 56. Infections | | | | |
| 36. High elevations | | | | | 43. Wetness | | | | | 50. Burns | | | | | 57. Dust | | | | |
| 37. Cramped body positions | | | | | 44. Air pressure | | | | | 51. Electrical hazards | | | | | 58. Silica dust | | | | |
| 38. High temperature | | | | | 45. Noise | | | | | 52. Explosives | | | | | 59. Moving objects | | | | |
| 39. Low temperature | | | | | 46. Vibration | | | | | 53. Slippery surfaces | | | | | 60. Working with others | | | | |
| 40. Sudden temperature changes | | | | | 47. Oily | | | | | 54. Radiant energy | | | | | | | | | |

## III. ACCEPTABLE DISABILITIES

*Check appropriate square if acceptable*

A - *Amputation* D - *Disability* Y - *Yes* N - *No*

| Hands - Fingers | A | D | Arms | A | D | Legs - Feet | A | D | Body - Trunk | D |
|---|---|---|---|---|---|---|---|---|---|---|
| 1 or 2 on primary hand | | | 1 Arm | | | 1 Leg (high) | | | 1 Hip | |
| 1 or 2 on secondary hand | | | 2 Arms | | | 2 Legs (high) | | | 2 Hips | |
| More than 2 on primary hand | | | None ☐ | | | 1 Leg (low) | | | 1 Shoulder | |
| More than 2 on secondary hand | | | | | | 2 Legs (low) | | | 2 Shoulders | |
| 1 Hand | | | | | | 1 Foot | | | Back | |
| 2 Hands | | | | | | 2 Feet | | | None ☐ | |
| None ☐ | | | | | | None ☐ | | | | |

| Eyes | Y | N | Ears | Y | N | Cardio - Vascular | Y | N | Tuberculosis | Y | N |
|---|---|---|---|---|---|---|---|---|---|---|---|
| Blind | | | Deaf | | | Moderate tension | | | Minimal (healed, stable or arrested) | | |
| Industrially blind | | | Hard of hearing, 1 ear | | | High tension | | | Moderate (healed, stable or arrested) | | |
| Blind one eye | | | Hard of hearing, 2 ears | | | Organic heart disease compensated | | | Far advanced (healed, stable or arrested) | | |
| Color blind | | | Hearing aid acceptable | | | | | | Collapse therapy | | |
| Color blind for shades | | | | | | | | | | | |

**Figure 20-9.** This form is used when analyzing jobs for fitness requirments.

abrasions, and burns; mechanical and electrical hazards; and fire and explosion hazards. These hazards could have an unfavorable relationship to the disability of the person. For example, a job that may involve foot injury hazards is unsuitable for the diabetic because of his susceptibility to gangrene of the feet and slow healing of wounds and fractures.

## ACCESS TO FACILITIES

Safety considerations for hiring a disabled person begin at the very first step in the employment process, and at the very beginning of the work day for that employee. Is there reasonable access to the reception area, applicant-processing area, or work situation for the new employee? Curbs and stairs present barriers to persons who use canes, crutches, walkers, and wheelchairs, and also increase their chances of falling. A wheelchair user cannot safely negotiate even one step without assistance, and this can cause a slight risk to both the wheelchair user and the person assisting him. Wheelchair ramps are discussed under General access later in this section.

### Access to and from work station

Can disabled job applicants safely proceed to where they must go to complete an application? If already employed, can such individuals safely proceed to their work stations? Access and safety are interdependent factors that need to be reconciled when employing these people. Traditionally, the accessibility of premises has been the principal factor restricting work for disabled persons.

The crux of the safety problem is the means of escape in an emergency; considerations of this problem frequently restrict, or deny, the freedom of persons with disabilities to use premises as they would wish. However, in many establishments where persons with mobility disabilities are employed, safe evacuation plans have been satisfactorily worked out by the employer. Thus, effective safety management of the handicapped in many firms is already well established and provides a freedom of movement that is entirely compatible with principles of general safety. These measures include supervised use of the elevator for means of escape (this is discussed later), designated staff to assist in an emergency, strategic ramping of entrances-exits, and alarm systems suitable for blind and deaf workers.

In practice, for most firms employing disabled people, the need for adjusting safety procedures from the already established pattern is not evident. There is an important exception that concerns the development and use of the elevator as a means of escape.

Historically, and for compelling reasons, the elevator is not part of a fire escape route. These are some fairly simple refinements that can be made, such as protected circuitry, installation of a fireman's switch, and supervised use, which together may allow elevators to be used for the purpose of evacuating people who are not able to use the stairs.

Another means of safely evacuating wheelchair users and permanently or temporarily disabled persons is through use of an evacuation chair (see Figure 20-10). This chair is designed to ride on the ends of stair treads so one person can easily guide the disabled worker down fire stairs without putting either person under additional risk during an emergency evacuation. The evacuation chair is lightweight, folds flat, and can be unobtrusively stored on a wall bracket.

**Figure 20-10.** Using an evacuation chair, this woman safely guides a disabled, fellow worker down fire stairs without putting either of them under additional risk during an emergency. (Printed with permission from Evac+Chair Corporation, New York.)

In the absence of these possibilities, many disabled persons would be denied access to their places of employment. It is suggested that employers discuss this with fire prevention specialists and check state or provincial and local codes and regulations.

### General access

Do not overlook cafeteria, washroom, and restroom facilities, width of doors, and height of plumbing fixtures. These and related questions require careful analysis by the safety professional. When facilities are designed to be truly accessible to disabled employees, their feelings of dignity and independence will certainly enhance the morale of the entire work force.

With a minimum of expenditures, improvements and special considerations made for workers with disabilities also can benefit other employees. For instance:

- Wheelchair ramps are safer than steps—for everyone. However, the slope should be correct, not too steep and without sharp turns, to make negotiating safe and easy for wheelchair users. Ramp surfaces should be slip-resistant. Wooden ramps can become slippery and hazardous when wet. All ramps should be kept free of mud, snow, and ice.
- Clean and unobstructed aisles are necessary for safe wheelchair, cane, and crutch use, and they make the workplace safer for all employees and visitors. Good housekeeping improves traffic patterns and eliminates hazards.

## HUMAN FUNCTIONING DIMENSIONS

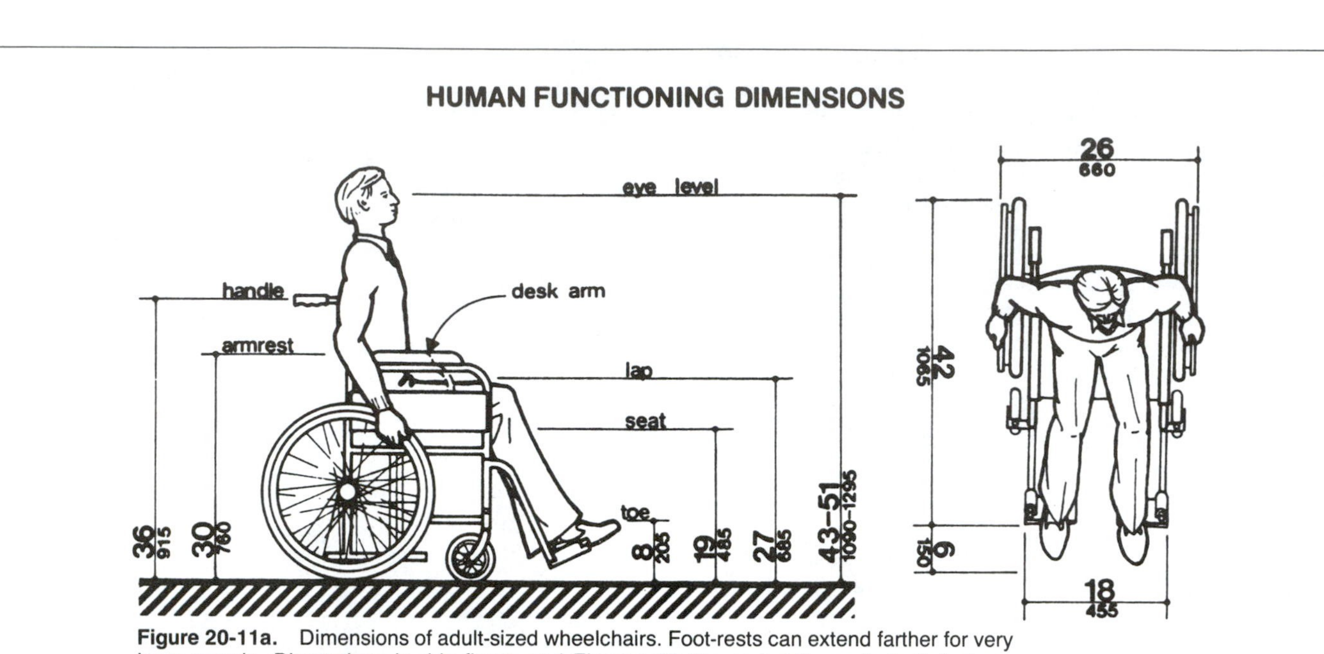


**Figure 20-11a.** Dimensions of adult-sized wheelchairs. Foot-rests can extend farther for very large people. Dimensions in this figure and Figures 20-11c and -11b are in both inches and millimeters. (Printed with permission from American National Standards Institute.)

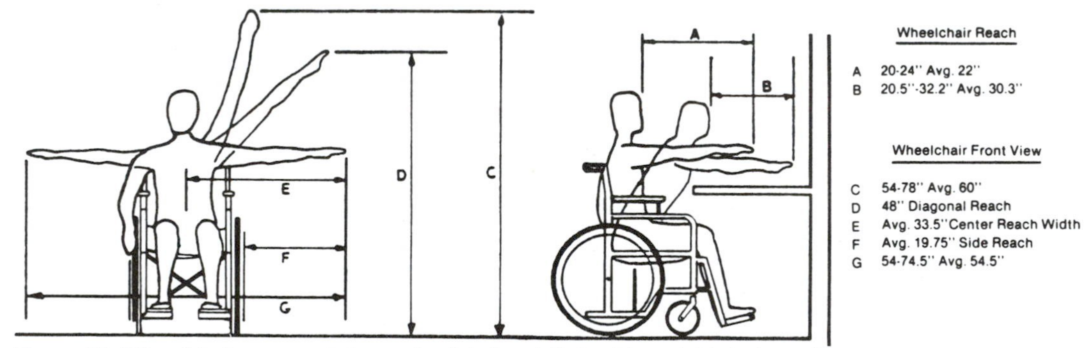


**Figure 20-11b.** Maximum reach from wheelchair— *left:* to sides; *right:* to front. (To convert to millimeters: 1 in. = 25.4 mm.) (Printed with permission from State Board of Barrier-Free Design, Columbia, S.C.)

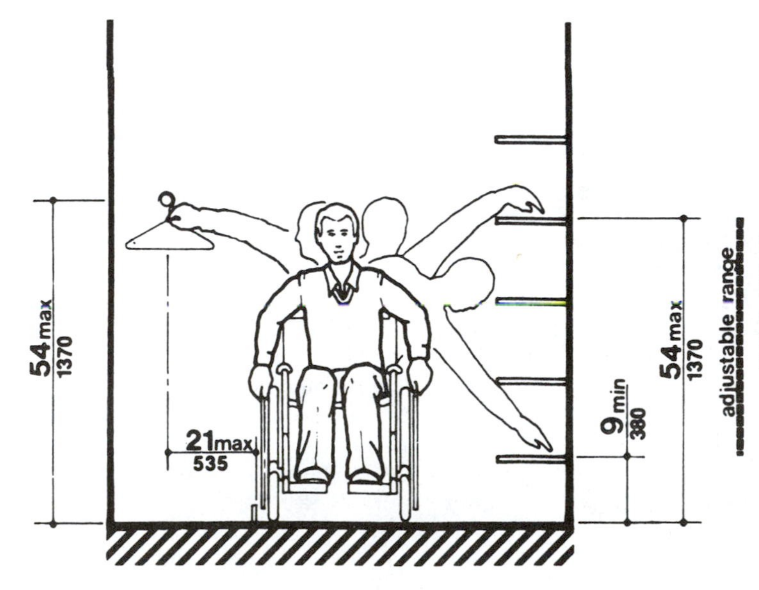


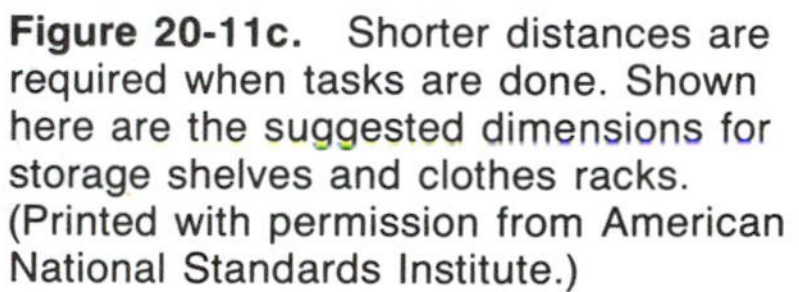

**Figure 20-11c.** Shorter distances are required when tasks are done. Shown here are the suggested dimensions for storage shelves and clothes racks. (Printed with permission from American National Standards Institute.)

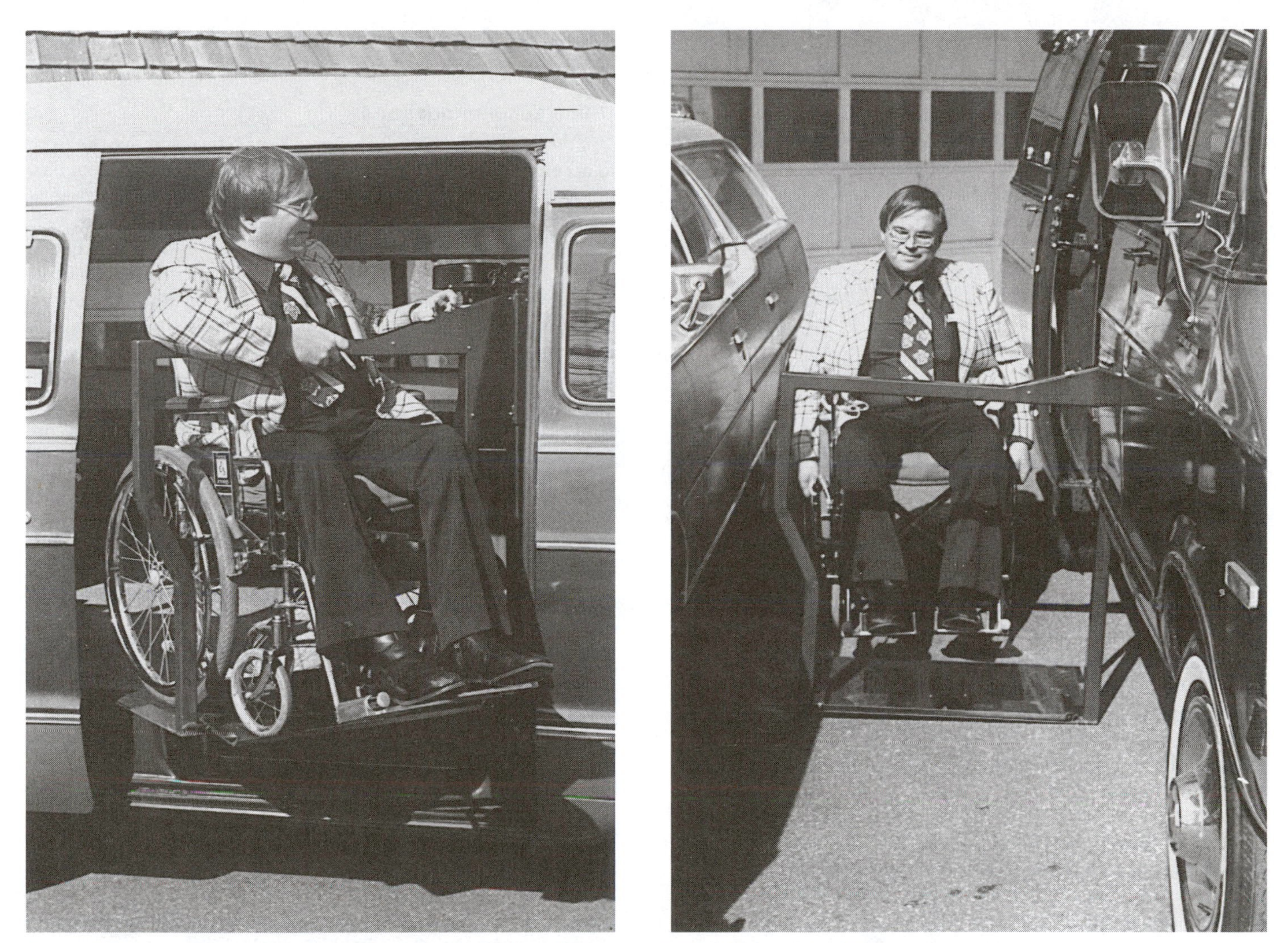

**Figure 20-12a.** Access aisle adjacent to a normal-width parking space is required for wheelchair clearance. Shown here is a chair lift that swings along side of a van. For return, a locking, outside control box opens sliding doors and allows the user to control the lift. (Printed with permission from ABC Enterprises, Inc.)

## WHEELCHAIR CHECKLIST

The space requirements of the average wheelchair are as follows: most wheelchairs are 36 in. high, 26 in. wide, and 42 in. long. (See Figure 20-11a.) They require at least 60 × 60 in. to make a 180- or 360-degree turn. However, 60 × 78 in. is preferred to make a smoother U-turn. All access aisles should be wide enough to allow a wheelchair user to make a smooth turn. "Accessibility Standards," published by the State of Illinois, contains illustrated information on maneuvering space requirements for wheelchairs (see References).

The average arm reach of people who use wheelchairs is usually 48 in. on the diagonal and 64½ in. to the side. The average reach directly upward is 60 in. (See Figure 20-11b.) The usual maximum downward reach from the chair is 10 in. (See Figure 20-11b.) Shorter distances may be needed, however, to accomplish certain tasks. (See Figure 20-11c.)

### Access to buildings

Eight-foot-wide parking spaces, adjacent to a five-foot-wide access aisle, should be reserved for automobiles driven by handi-

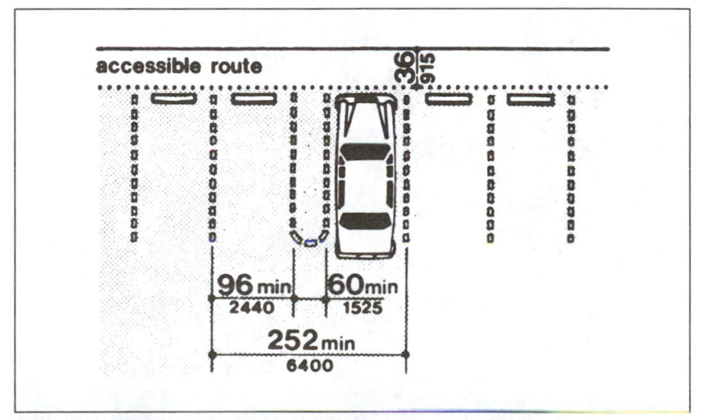


**Figure 20-12b.** Parking spaces for disabled persons can share a common access aisle. Aisle should be part of the accessible route to the building entrance. Overhangs from parked vehicles must not reduce the clear width of the accessible route, which must be the shortest possible distance to the entrance. (Printed with permission from American National Standards Institute.)

capped personnel and visitors. (See Figure 20-12a.) Two accessible parking spaces, however, can share a common access aisle. (See Figure 20-12b.) Wheelchairs require room to be removed and replaced in an auto and also for riding between aisles.

Parking spaces for disabled persons should be marked with an upright marker; symbols painted on the ground are often difficult to see or can become covered by snow or debris.

At least one accessible route and entrance to the building must be provided. The entrance width should be 32 in. (80 cm) and if a ramp leads to the entrance, it should be at least 36 in. (90 cm) wide. The ramp should be a maximum of 30 ft (9 m) long with an open, level area of at least 5 ft (1.5 m) at the bottom. Employees (and visitors) using wheelchairs, crutches, or canes can then move in and out of the building completely on their own.

Revolving doors are taboo for anyone in a wheelchair as well as for most people using crutches, wearing a leg cast, or even carrying bulky packages. These people need entry doors that are preferably of the time-delay type. When a knob is necessary, it should be 36 in. (90 cm) from the floor. It is better, however, to have a vertical grab handle on the door.

### Interior access

Both entry and interior doors should be a minimum of 32 in. (81 cm) wide. Interior doors should open by a single effort and have thresholds as level with the floor as possible.

To make a 180- or 360-degree turn, persons in wheelchairs need an open area of 60 in. (1.5 m) in a typical building corridor.

Restrooms should have at least one stall wide enough for wheelchair entry. The stall should be equipped with grab bars and other fixtures no higher than 36 in. (90 cm) above floor level. (See Figure 20-13.) The grab bars should be at 33 in. (83 cm). Stall doors should be at least 32 in. (81 cm) wide.

Controls, switches, fire alarms, and other devices that might be used by a disabled individual must be within convenient reach of a wheelchair user.

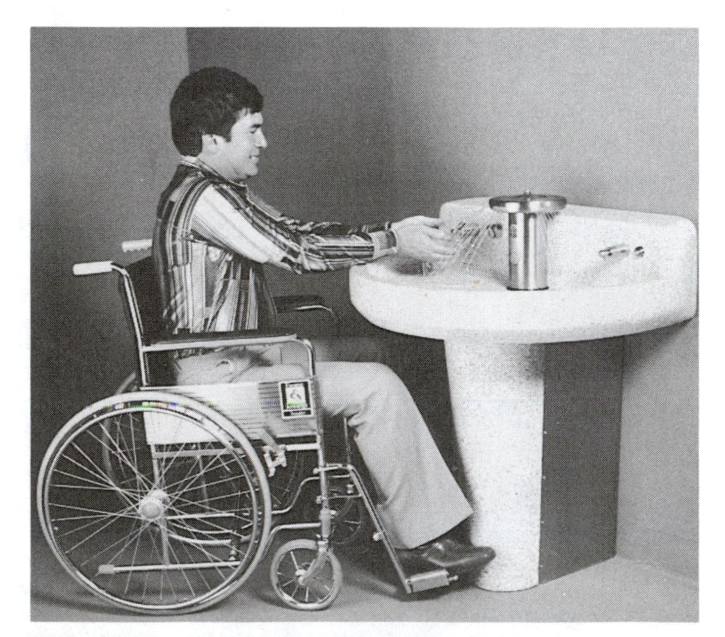

**Figure 20-13.** A semicircular washstand no higher than 3 ft (0.9 m) allows this wheelchair user easy access. (Printed with permission from Bradley Corporation.)

Restrooms should be located on each of the floors where disabled persons work.

### Office accommodations

Desk tops should be no less than 28 in. (70 cm) above the floor to accommodate wheelchairs. Metal desks usually have adjustable feet that can be raised to the maximum. If more room is needed, the desk can be raised with additional blocks.

Chairs can be regular height, but they should be sturdy and have arms to enable disabled people to lift themselves up more easily. Although some individuals need a chair that will not easily move so they can stand without the chair sliding out from under them, casters placed on the bottom of chairs may be desirable for other handicapped workers. Casters make for easier mobility and, if necessary, allow the chair occupant to pull himself from one piece of furniture to another, thereby avoiding constant movement in and out of the chair.

Business machines should be placed, if possible, in such a way so they will not become a barrier or obstruct traffic.

The following accommodations may sometimes apply, depending upon the individual worker and the job requirements:

- Files should be placed so drawers can be reached from both the front or side. This eliminates awkward reaches from those who must use crutches, a cane, or a wheelchair.
- If books, reports, or other bulky objects must be carried from place to place, a shopping-type cart on wheels should be kept handy so the materials can be loaded on to it and pushed.
- Venetian blind, window shade, and drapery cords should be long enough to be reached easily.

## SOURCES OF HELP

The overall goal of all employers should be to hire qualified disabled individuals and place them in available and safe occupations. One of the goals of the safety and health professional is to assist the employer in this worthwhile endeavor.

The following list includes government and private agencies that can help achieve this goal.

Alabama Institute for the Deaf and Blind, Talladega, Ala. 35160.

George Washington University, Job Development Laboratory, Rehabilitation Research and Training Center, 2300 I St., NW, Washington, D.C. 20052.

Mainstream, Inc., 1200 15th St., NW, Washington, D.C. 20005.

The National Institute for Rehabilitation Engineering, 97 Decker Rd., Butler, N.J. 07405.

Paralyzed Veterans of America, 801 18th St., NW, Washington, D.C. 20006.

President's Committee on Employment of the Handicapped, 1111 20th St., NW, Washington, D.C. 20036.

The Rehabilitation Institute of Chicago, 345 E. Superior St., Chicago, Ill. 60611.

## REFERENCES

American National Standards Institute, 1430 Broadway, New York, N.Y. 10018. *Specifications for Making Buildings and Facilities Accessible to and Usable by Physically Handicapped People,* A117.1.

Brooks, Wayne T. "Supervising Handicapped Workers for Safety." *Transactions of the National Safety Congress—Industrial Subject Sessions,* 1978.

Capital Development Board, *Accessibility Standards Illustrated.* Reprint of 1978 ed. with all revisions to March 1, 1985 and Environmental Barriers Act, Public Act 84-948. Springfield, Ill.: State of Illinois, 1985.

Hill, Nina, et al. "The Merits of Hiring Disabled Persons," *Business & Health,* February 1987.

——."Hiring the Handicapped: Overcoming Physical & Psychological Barriers in the Job Market," *Journal of American Insurance,* Third quarter, 1986.

President's Committee on Employment of the Handicapped, Washington, D.C. 20036.

"Affirmative Action to Employ Handicapped People—A Pocket Guide."

"Supervising Handicapped Employees."

Superintendent of Documents, U.S. Government Printing Office, Washington, D.C. 20402.

Occupational Safety and Health Act of 1970 (P.L. 91-596).

*Occupational Safety and Health Act Regulations,* Title 29, *CFR,* Chapter XVII, Part 1910.

Rehabilitation Act of 1973 (P.L. 93-112).

Rehabilitation Act Amendments of 1974 (P.L. 93-516).

Vietnam Era Veterans' Readjustment Assistance Act of 1974 (P.L. 93-508), Section 38 *USC* 2012.

U.S. Department of Human Services, Washington, D.C. 20202. "Nondiscrimination on the Basis of Handicap in Programs and Activities Receiving or Benefiting from Federal Financial Assistance" (45 *CFR* 84).

U.S. Department of Labor, Employment Standards Administration, Office of Federal Contract Compliance Programs, Washington, D.C. 20210. "Affirmative Action Obligations of Contractors and Subcontractors for Handicapped Workers" (41 *CFR* 60-741).

Woodward, Robert E. "Industry Unlocks Its Doors to the Handicapped." *Plant Facilities,* Vol. II, No. 2 (February 1979).

——. *Comprehensive Barrier-Free Design Standard Manual.* Columbia, S.C., State Board for Barrier-Free Design, P.O. Box 11954. 1979.

# Nonemployee Accident Prevention



Good management is seen in the conduct of the routine operation of a successful business, and in areas that have a direct bearing on the public relations between management, personnel, and customers.

One of these areas, nonemployee accident exposure, affects not only nonemployees, but also employees and the quality of the product or service they provide or sell. Operations stop (or at least slow down) when an accident occurs, no matter who has it, or who is involved. With the increasing consumer product safety legislation (see the following chapter) and greater cost of claims, more attention must be given to this area of business loss.

Today, more businesses and people are involved in serving and selling to the public than are engaged in making the products that are sold. Just look at the types of businesses, specialty shops, department stores, shopping center complexes, restaurants, fast-food service operations, hotels and motels, automotive service and dealerships, hardware and building supply stores, amusement parks, banks and office buildings, to name a few, that serve the public. "Franchise" businesses provide services and market products to their patrons. These patrons, guests, or visitors are the source of nonemployee injuries.

These service and selling operations must constantly survey patron activities even though they cannot control personal activities or habits of their patrons.

Inebriated hotel, motel, or store guests or patrons can be a threat when they improperly dispose of smoking materials or their instability results in slips, falls, or other problems. Children can also be involved in accidents in the business environment.

Loss control plans must prepare for nonemployee accidents. These unplanned events can be minor or catastrophic. A firm can be liable for damage or for injuries from the minute someone enters the property (including the parking lot). A firm can be liable for the actions of its employees that result in damage or personal injury when that employee is sent off the company property on business.

## THE LEGAL SIDE OF NONOCCUPATIONAL INJURIES

Customer or product claim cases often wind up in a courtroom. The legal terminology and aspects of the law that deal with accident claims, in situations in which one is likely to become involved, include the following definitions and explanations. These are not intended to be all-inclusive, but to provide a quick review of some principal considerations.

**Tort.** A private or civil wrong or injury—a violation of a right not arising out of a contract. It may be either (1) a direct invasion of some legal right of the individual, (2) the infraction of some public duty by which special damage accrues to the individual, or (3) the violation of some private obligation by which like damage accrues to the individual. Torts deal with negligence, accidents, trespass, assault, battery, seduction, deceit, conspiracy, malicious persecution, and many other wrongs or injuries.

**Negligence.** Failure to exercise that degree of care which an ordinarily careful and prudent person ("reasonable man") would exercise under similar circumstances. To establish a proper claim of negligence, however, there must be (1) a legal duty to use care, (2) a breach of that duty, and (3) injury or damage.

**Degree of care.** The degree of attention, caution, concern, diligence, discretion, prudence, or watchfulness depends upon the circumstances. For example, a high degree of care is demanded from people who invite others onto their premises by formal, verbal, or implied invitation. All sales and service enterprises must exercise a high degree of care for the safety of their patrons. As long as a business is open, it assumes a responsibility to its clientele.

**Invitee.** One whose presence on the premises is upon the invitation of another, such as a patron at a sports stadium or a person who visits an exhibition hall even though no admission is charged.

**Licensees.** Licensees are neither "invitees" nor "trespassers." They have not been specifically invited to enter upon the property but they have a reasonable excuse (by permission or by operation of the law) for being there. These could be vendors, delivery personnel, people visiting executives and purchasing agents for business purposes, and the like. Policemen and firefighters who enter property in the course of their duties have sometimes been held by the courts to be invitees (patrons) and sometimes licensees (nonpatrons).

**Contributory negligence.** Not every injury gives rise to a claim for damages. If a defendant can prove that the plaintiff (claimant) contributed to the injury by not exercising ordinary care, the resultant damages may be considerably reduced or negated. If it can be proved that the plaintiff was even only slightly negligent in a manner that contributed to the harm or damage, damages might not be allowed.

This points up the importance of a detailed and thorough accident investigation procedure. It can strengthen a company's ability to reduce the overall cost of doing business and help prevent future lawsuits.

**Assumption of risk.** Claimants cannot collect damages when the law presumes they were aware of peril or danger, yet were willing to proceed with their original intention and undertook the action. "That to which a person assents is not regarded by law as an injury." For example, a skier who falls while descending a slope is said to assume the normal risks that can happen when participating in this sport, unless there was special negligence in the design or maintenance of the slope and its environment, since there is no assumption of risk when the owner or operator is negligent. However, an injury involving a chair lift or tow rope could be of a mechanical nature and could be costly.

**Hold-harmless.** A clause in a contract agreeing for one party to assume all liabilities or losses or expenses involved, thus reimbursing the other party held liable, is a hold-harmless agreement. For example, a department store may have a hold-harmless agreement with a manufacturer who supplies a particular type of merchandise. Should a consumer claim injury arising out of the use of that product, the manufacturer would reimburse the store, should it be held liable, and will pay for legal and other expenses incurred by the store in defending itself. Even though the consumer purchased the item from the department store, a hold-harmless agreement may relieve the store of the financial effect of direct liability.

**Attractive nuisance.** Liability growing out of a dangerous condition, generally to children. It excuses trespassing and penalizes for failure to keep children away or for failure to protect or eliminate a hazard that may reasonably be expected to attract them to the premises.

**Burden of proof.** The injured party must prove injury or damage and its causal relation to the event or item that resulted in the accident. The defense is not liable if it is without fault. Proof must be established by facts, *not* opinion, suspicion, rumor, hearsay, gossip, or emotional reaction. Proof is the conclusion drawn from the evidence.

Honest and sincere witnesses convey different impressions from the same evidence attested to by dishonest witnesses. Thus we see how important promptness is when assembling and preserving the evidence. Signed statements taken shortly after the accident or an all-important photograph can often make the big difference. In liability claims, the burden of proof rests upon the plaintiff (claimant).

This chapter cannot cover all possible nonemployee involvements. Instead it will describe prominent ways that nonemployees can get hurt. Accident prevention techniques will be broad and varied. It will be up to each health and safety professional to examine the company's operations in light of these guidelines.

## PROBLEM AREAS OF ACCIDENTS

When starting a nonemployee accident prevention program, look over the records of past nonemployee injuries to identify accident trouble spots, so these hazards can be mitigated. Next, choose general accident trouble spots, such as those due to lighting, interior design, and traffic flow (to name a few), and do the same thing. Change facilities and/or operations so that both employees and nonemployees are given a safe business environment.

The following sections review some major problem areas and discuss typical accidents that occur in modern business facilities.

### Glazing

Modern office building, store, and plant design often uses glass extensively in doors, show windows, panels, and enclosures. Such areas can result in a confusing pattern, especially to the first-time visitor or patron.

Unmarked glass panels and doors can lead to severe injuries and cuts. Be sure to follow recommended practices and standards for glazing and strength of glass. Some states legally require safety glass and marked identification of doors and panels. The U.S. Consumer Product Safety Commission has regulations regarding glazing (Title 16—*Code of Federal Regulations,* Chapter II; see References).

Some plastic glazing offers similar advantages to glass but without the hazard of damage or injury. In fact, some plastics have higher impact resistance; but, conversely, plastic glazing tends to scratch or be more easily abraded than glass. Plastic, however, is harder to break than glass if escape through a window is necessary.

A few precautions will allow the continued enjoyment of the beauty of glass and lessen the chances for injury:

1. Make glass doors more visible to adults and children by placing decals or pressure-sensitive tape at their respective eye

levels. Sand-blasted or etched designs serve the same purpose.
2. Decals or pressure tapes will also prevent glass panels from appearing to be doorways. An attractive and tall plant, placed in the center of the panel, will also identify the true purpose of the panel.
3. Installing safety bars reduces the size of the open glass areas and lessens the chance of glass breakage. The bars should be at the door-handle level on sliding doors and should be on both sides of a swinging door.
4. Keep doorways and areas that are close to glass panels clear of tripping hazards, such as scatter rugs and toys. Indeed, this is a good rule for all areas of occupancy.
5. Enforce no play, no running rules especially where there is a glass hazard.

## Parking lots

Most shopping or business centers have parking lots for patron use; usually these are self-service. Bicycles, mopeds, and motorcycles as well as automobiles, vans, and trucks also require parking facilities. If these self-service parking lots are well designed and attractive, they nearly eliminate parking damage to cars.

Parking lot problems depend to some extent on how the lots are used. Self-service parking lots for industrial plant employees, for instance, usually differ from public lots at shopping centers, theaters, stadiums, stores, schools, restaurants, or motels. Usually, public lots have a more steady flow of traffic, have children present, and/or are obstructed by shopping carts. More importantly, they are used by persons whose driving skills are varied and who may be less experienced with the lot's layout.

Some public parking lots at places such as theaters, sports arenas, and schools, frequently have the combined hazards of extreme traffic fluctuation and great variations in driver skill and alertness.

An easy-to-use layout, adequate signs, and conspicuous markings help make a lot safe and attractive. These are the first steps in reducing hazards caused by disadvantageous location or property configuration.

More than any other single safety measure, enclosing the parking lot with a curb or fence so that cars cannot enter or leave traffic unexpectedly will reduce accidents in the area. Parking lots should have separate, well-marked entrances and exits, placed so that they favor right-hand turns. Single lane entrances should be at least 15 ft (4.5 m) wide; exits should be at least 10 ft (3 m) wide. Where entrance and exit must be combined, the double-lane drive should have at least 26 ft (8 m) of usable width. When the double-lane combination is necessary, it should have median curbs or strips to positively control the flow of traffic.

In general, entrances and exits should be:

- At least 50 ft (15 m) from intersections
- Away from heavily traveled highways or streets
- Well marked and well illuminated
- As few in number as possible.

Pedestrian traffic also must be considered in parking lots. Stairways and ramps should be constructed in accordance with ANSI Standard A64.1. They should be well marked, well illuminated, and have handrails. The needs of the handicapped must also be considered in the design of the lot and in any walkway to the building. See Chapter 20, Workers with Disabilities.

Parking aisles should be perpendicular to the buildings, so that pedestrians will walk down the aisles rather than between parked cars. Where possible, marked lanes, islands, or raised sidewalks should be provided between rows, particularly where there are heavy concentrations of pedestrian traffic. These walks should be wide enough so that car bumpers overhanging them do not restrict pedestrians who are walking to or from their cars.

Angle parking has both advantages and disadvantages. Fewer cars may fit in the lot, but, on the other hand, angle parking is easier for customers, and it does not require a lot of room for sharp turns.

The area allowed per car in parking lots varies from 200 to more than 300 sq ft (18.5–28 m$^2$), if aisles are included. (See Chapter 1 in the *Engineering and Technology* volume.)

**Parking stalls.** The design of parking stalls depends on (1) the size and shape of the lot, (2) the traffic pattern in the lot, and (3) the type of driver and pedestrian who use the lot. (See Figure 20-12A and B in Chapter 20 for stall dimensions for the handicapped.)

Bumpers on stalls have these advantages:

1. They prevent drivers from driving forward through facing stalls and proceeding in the wrong direction in one-way aisles.
2. They encourage drivers to pull forward against the bumper, thereby preventing rear overhang which might reduce aisle width.
3. They break up huge expanses of an open lot which may tempt drivers to cut across aisles and endanger pedestrians and other drivers.
4. They block cars from accidentally rolling down inclines or running through walkway areas.

Stall bumpers also have some disadvantages:

1. They may require maintenance.
2. They may interfere with drainage or snow removal.
3. They may cause pedestrians to trip and fall. (Painting them a bright yellow or other distinctive contrasting color may assist in reducing such incidents.)
4. They may restrict some desirable flexibility of traffic flow.

**Signs and lighting.** Traffic signs in parking lots should conform to recommended standards, and should be similar to other street and highway signs (see *Manual on Uniform Traffic Control Devices for Streets and Highways,* ANSI Standard D6.1; see References).

Stop or yield signs should be installed at main crosswalks for pedestrians, where exits cross public sidewalks, and where exits enter main thoroughfares.

A well-lit parking lot reduces accidents and discourages crime. The amount of light recommended for parking at night usually ranges from 0.5 to 1.0 foot candles (decalux) per square foot at a height of 36 in. (0.9 m). Lights are usually mounted 30 to 35 ft (9-10 m) high on poles that have bases protected against impact by cars.

**Supervision.** Effective supervision helps prevent parking lot accidents. However, supervision of traffic in public lots is not always easy. The shopping center caters to the public, and it must weigh the possibility of claims against the effect of control enforcement on public relations. In many areas, police have no jurisdiction in privately owned parking lots. Perhaps the most effective controls, then, are those that are resulting from good parking lot design.

**Design.** Built-in bumps on lot surfaces have been used to deter speeding, but because they may cause damage to cars their use seems questionable. Another device is to keep straight lanes short and to provide sharp curves which require reduced speed.

Parking lot operators definitely should try to control unauthorized use of lots. Weekend parking, scooter races, games, and other unauthorized activity should be forbidden. Local police or company security personnel should check parking areas after hours. If lots cannot be fenced or entrances cannot be protected against unauthorized entry, prominent warning signs will help minimize the risks of unauthorized use and vandalism.

Close cooperation between local traffic authorities and those responsible for the supervision of lots is extremely desirable. Cooperation must be developed with police and sheriff's departments responsible for traffic control and crime prevention, and also with fire departments that need access to fire hydrants in a lot and to buildings served by the lot. Zoning boards are also interested in problems connected with parking lots.

**Shopping carts.** In order to deep shopping carts out of parking lots, customers could drive their cars to a loading point where an attendant would put their goods in the car, or an attendant might wheel the cart to the car, unload it, and bring it back.

If, however, shopping carts are permitted in the parking lot, they should be collected frequently and temporarily stored in a space allocated for them. Do not leave carts to accumulate in stalls, or worse still, in heavily trafficked aisles. Cart-collecting areas should be well marked, well illuminated and, preferably, separated from traffic by barriers or bumpers in the pavement. Encourage customers to leave carts in such areas by installing signs and access lanes. To keep carts from rolling into traffic lanes install low curbs or a shallow depression.

People who supervise the lot should check that children do not play with the empty carts or use them as scooters and racers.

**Other vehicles.** Using bicycles, mopeds, and motor cycles as commuting vehicles is increasing. More people would use these vehicles if (1) the routes were safer and (2) secure parking were available. Current parking facilities are not generally adequate for these vehicles. Like the motorist, the bicycle, moped, or motorcycle operator prefers to park near his or her destination. So in planning for such parking, two major problems are faced: (1) convenience and (2) security.

Bicycle storage is fairly easy, but the problem of security is more difficult. Twelve to 15 can be stored in the space it takes to park one car. A lesser number of mopeds or motorcycles will also fit in the same area.

Three basic types of bicycle parking facilities are in current use. The common bicycle rack comes in many types and shapes and requires chaining or clamping the bike to the rack. Unless the bicycle frame and both wheels are locked, the operator can lose part of the bike. Second, a hitching post has a chain secured to lock the bike. Finally, a key-operated locker is much like baggage lockers. These are totally closed to both view and weather—an advantage where weather protection is a concern, especially for all-day parking.

Moped owners and motorcyclists face most of the same convenience, security, and weather problems as bicyclists.

## Entrances to the building

This discussion is from *Loss Control—A Safety Guidebook for Trades and Services,* by George J. Matwes and Helen Matwes. Copyright © 1973 by Litton Educational Publishing, Inc. Reprinted with permission of Van Nostrand Reinhold Company.

Entrances to department stores and office buildings must be of number and size to meet all building codes. Revolving doors should have governors that limit their speed to 12 rpm. All worn weatherstripping should be replaced. Sidewalks and driveways should be kept in good repair in order to avoid tripping and falling hazards. Entrance lighting should be a minimum of ten footcandles.

In order to provide passage from the sidewalk to the customer's car, it is customary in shopping centers to construct a ramp. The best type of ramp is an indented ramp; one that is actually cut into the sidewalk itself with an easy slope. Rails on each side prevent someone from falling into one of the recessed sides if they are sufficiently deep enough to be a hazard. Do not paint the ramp because paint seals the concrete surface to the point where it becomes excessively slippery.

Entrances to buildings are of considerable importance from a safety viewpoint. Often entrance and exit doors are automatic and are activated when anyone steps onto a carpet. Actually each door has two carpets: one on the sidewalk and one on the floor inside the door. For the "in" door, the door swings inward when the outside carpet is stepped on. The carpet on the inside of the door acts as a safety mat, i.e., should anyone, even a child, step onto the safety mat, it will automatically deactivate the outside carpet thereby eliminating the possibility of the door swinging inward and striking the person on the other side. The same principle, in reverse, applies to the "out" door.

The sidewalk carpet installation in front of the entrance door is usually flanked by two handrails to avoid activation when someone is merely walking by.

If the vestibules are glass-enclosed, decals should be applied to alert people against walking into glass panels, as discussed earlier.

## Walking surfaces

Slips and falls are the most likely source of nonemployee injuries. These injuries can occur almost anywhere at any time. There are few surfaces that can be ignored and the dangerous ones include everything from asphalt roads, concrete walks, wooden, tiled or rug-covered floors, and special surfaces on stairs and conveyances (moving sidewalks, escalators, elevators), to bridges and catwalks.

The natural properties of any surface can change substantially when people track in mud, snow, dirt, and moisture. Moisture-absorbant mats, runners, or rugs are designed to reduce such hazards. Floor maintenance requires special attention to eliminate the hazard of torn or curled-up floor coverings. (See Figure 21-1.) Floors, stairs, and other walking surfaces should be kept nonslippery, clear, and in good repair.

Slip meters developed by testing agencies and insurance companies measure the slipperiness of floors. One type of instrument (see Figure 21-2), mounted on three leather "feet," is pulled across the floor by a motorized winch. The dial on top measures the intensity of the pull required to start moving it. This is converted into a "slip index," or the coefficient of friction (0.5–0.6 is ideal). While very important, this "slip index" is not foolproof. Floor slipperiness may increase because of moisture, oil, grease, foreign or waste materials, and incorrect cleaning or waxing.

Falls on floors occur in various ways and from various causes. A person may slip and thus lose traction, or trip over an open

**Figure 21-1.** Cotton and other fabric mats used in entrances absorb moisture and dust, but require constant watching to minimize tripping hazards. Someone may catch a toe or heel in the raised fold and trip or fall. *Top:* The first person's toe or heel will catch in the mat. *Bottom:* The next person will either trip on the raised portion or dislodge it further, building up the hazard. The remedy is to lay the mat flat and fasten it securely.

drawer, box in the aisle, or other object. The primary mechanical causes of falls on floors are unobserved, misplaced, or poorly designed movable equipment, fixtures, or displays; poor housekeeping; and defective equipment. The condition of shoes or type of footwear soles and heels is likely to be a major contributing factor.

Inadequate illumination also can cause falls. Light values at floor level should be uniform with no glare or shadows. Also, there should be no violent contrasts in light levels between floor areas, such as from bright sunlight outside the entrance to a dimly lit lounge or restaurant.

Some other causal factors are patron-related: age, illness, emotional disturbances, fatigue, lack of familiarity with the environment, and poor vision. Because these cannot be readily controlled, it becomes doubly important to make the walking surface as safe as possible. For example, mirrors and other distracting decorations should not be placed in areas visible from steps or from approaches to steps or escalators.

**Types of floor surfaces.** A wide variety of floor surfaces are available. In office buildings, hotels, mercantile, and similar establishments, masonry (terrazzo, cement, or quarry tile) floors are common at entrances, in lobbies, on stairways, and sometimes throughout the ground floor and in upper floor corridors. Decorative materials such as terrazzo, marble, and ceramic tile most often are used for interiors while concrete and granite are generally considered more practical for exterior use. Details are in Tables 21–A and B.

In other public areas of these buildings, the base floor, usually of concrete or wood, is generally surfaced with one or more of the popular resilient floor covering materials. Carpeting is commonly used in limited areas of various stores and hotels. Elsewhere, asphalt, linoleum, rubber, or plastic in either sheet or tile form, will usually be found. Obviously, safety, initial cost, durability and maintenance costs are some factors that govern the choice of floor covering.

Most flooring materials, whether wood, masonry, or the resilient types, are reasonably slip resistant in their original untreated condition. Exceptions are found among some of the masonry materials. A highly polished marble, terrazzo, or ceramic tile, used to achieve an ornamental effect, can be slippery when dry. If moisture is present, its slipperiness will be greatly increased. Improper surface treating preparations and improper cleaning materials and methods also increase slipperiness. Unless a nonslip material is added to the aggregate during construction, the only preparation that should be used on such floors is a penetrating slip-resistant sealer.

**Floor coverings and mats.** Reduce the possibility of slips and falls by using good carpeting, bound edging, and flush floor-level mats and runners (see Figure 21-3). Whenever possible, provide a contrasting color on carpeted areas that meet and continue on treads and risers (or treads only) of stairways. If material such as an extruded metal runner is used to provide self-cleaning removal of snow, ice, or mud at entryways, it should be flush and not present a tripping hazard. Care also must be taken with rubber mats, rug runners, and the like to prevent them from becoming tripping hazards. Oftentimes the edges become rumpled, corners and ends are torn or do not lie flat, and excess wear causes tears. Mats and runners should be replaced at the first sign of such unsafe conditions.

Floor mats, runners, and carpeting are used wherever water,

**Figure 21-2.** Slip meters, ranging from the motorized type (shown here) to a simple spring scale and heavy block pulled by hand, can be used to gauge the slipperiness of floors. (Reprinted with permission from Liberty Mutual Insurance Company.)

Table 21-A. Physical Properties of Floor Finishes

| Types of Finish | Resistance to Abrasion | Resistance to Impact | Resistance to Indentation | Quality of Slipperiness | Quality of Warmth | Quality of Quietness | Quality of Ease of Cleaning |
|---|---|---|---|---|---|---|---|
| Portland cement concrete *in situ* | VG-P | G-P | VG | G-F | P | P | F |
| Portland cement concrete precast | VG-G | G-F | VG | G-F | P | P | F |
| High-alumina cement concrete *in situ* | VG-P | G-P | VG | G-F | P | P | F |
| Magnesite | G-F | G-F | G | F | F | F | G |
| Latex-cement | G-F | G-F | F | G | F | F | G-F |
| Resin emulsion cement | G-F | G-F | F | G | F | F | G-F |
| Bitumen emulsion cement | G-F | G-F | F-P | G | F | F | F |
| Pitch mastic | G-F | G-F | F-P | G-F | F | F | G |
| Wood block (hardwood) | VG-F | VG-F | F-P | G-F | F | F | G |
| Mastic asphalt | VG-F | VG-F | VG-F | VG | G | G | G-F |
| Wood block (softwood) | F-P | F-P | F | VG | G | G | G-P |
| Metal tiles | VG | VG | VG | F | P | P | G-F |
| Clay tiles and bricks | VG-G | VG-F | VG | G-F | P | P | VG |
| Epoxy resin compositions | VG | VG | VG | VG | F | F | VG |

Code: VG—Very Good; G—Good; F—Fair; P—Poor; VP—Very Poor.

**Figure 21-3.** A heavyweight rubber or plastic mat with a nubby finish or raised design and beveled edges tends to lie flat and stay in place. Rotating the mat distributes the wear and minimizes "bald" spots in high-traffic locations.

oil, food, waste, and other material on the floor might make it slippery. For example, they are commonly used at entrances, around swimming pools, in shower stalls, around drinking fountains and vending machines, on boat decks, in garages, and in factories.

Definite procedures should be set up for placing, cleaning, removing, and storing mats. Those who put mats in place during inclement weather should have clear-cut instructions as to when and where mats should be put down and removed. Failure to install the mats promptly and close enough to the door may result in slippery entrance-ways and in water and dirt being carried beyond the entrance-way creating a hazard and maintenance problem in another location.

Definite procedures for inspecting and checking the condition of mats and for maintaining them in safe condition should be followed.

Stair rails, treads, and surfaces should be in good repair and checked frequently for defects. Nothing should be stored on the

**Table 21-B.** Guide to Floor Materials and Surfacings

| *Floor types** | *Characteristics* | *Use of Abrasives* | *Dressing Materials* |
|---|---|---|---|
| Asphalt tile | Composed of blended asphaltic and/or resinous thermoplastic binders, asbestos fibers, and/or other inert filler materials and pigments. | Abrasive materials of various types may be used to reduce slipperiness of floors. Colloidal silica** can be incorporated in wax and synthetic resin floor coatings. | **Wax or wax-base products**—For most purposes, wax has several advantages. This is especially true of Carnauba wax, an ingredient generally used in so-called wax products. This wax, a Brazilian palm tree product, dries in place with a very hard and glossy finish, but with a characteristically slippery surface. Because of its many good qualities, it is widely used as a base for floor surface preparations, both in paste and emulsion forms. Other waxes, notably petroleum wax and beeswax, have their place in floor dressing formulas; they are softer and less slippery than Carnauba, but are still slippery to a degree depending on the formulation. |
| Linoleum | Cork dust, wood flour, or both, held together by binders consisting of linseed oil or resins and gum. Pigments are added for color. | | |
| Rubber | Vulcanized, natural, synthetic, or combination rubber compound cured to a sufficient density to prevent creeping under heavy foot traffic. | Slip-resistant except when wet. | |
| Vinyls | Composed of inert, nonflammable, nontoxic resins compounded with other filler and stabilizing ingredients. | Adhesive fabric with ingrained abrasives can be used. They are patterned in strips, tiles, and cleats. | |
| Terrazzo | Consists of marble or granite chips mixed with a cement matrix. | Silicon carbide or aluminum oxide can be included in mix when floor is laid. Also an abrasive-reinforced plastic coating can be painted on. | **Slip-resistant sealant** will typically improve slip-resistant quality if renewed periodically. |
| Concrete | Made of portland cement mixed with sand, gravel, and water and then poured. | | |
| Mastic | Like asphalt tile in composition but is heated on the job and troweled onto the floor to form a seamless flooring. Such floors are often used over concrete to give a new durable, resilient surface. | (Same as asphalt tile) | **Synthetic resins**—These preparations, known as "synthetics," "resins," or "polishes," are intended to supply the desirable characteristics of wax without producing the same degree of surface slipperiness. They include soaps, oils, resins, gums, and other ingredients, compounded in various ways to produce the desired result. |
| Wood | May be either soft or hard, in a variety of thicknesses and designs. | Metallic particles and artificial abrasives in varnish or paint give good nonslip qualities to various floors. | |
| Cork tile | Made of molded and compressed ground cork bark with natural resins of the cork to bind the mass together when heat cured under pressure. | (Same as asphalt tile) | **Other materials**—Paint products (paint, enamel, shellac, varnish, plastic) are semipermanent finishes used principally on wood and concrete floors. They do not materially increase the slipperiness of the base. |
| Steel | Iron containing carbon in any amount up to about 1.7 percent as an alloying constituent, and malleable when used under suitable conditions. | Surface can be touched up with an arc welding electrode so the shape of raised places on the surface resembles angle worms. Also an abrasive reinforced plastic coating can be painted on to any desired thickness, dries hard as cement, and has a sandpaper like finish. If a temporary nonskid surface is needed, two uses of mats can be employed: (1) flexible rubber mats made from old automobile tires; (2) rubber or vinyl runners. | |
| Clay and quarry tile | Kiln-dried clay products are similar to bricks and are extensively used in areas requiring wet cleaning. | Typically resistant to abrasives. | May be treated by etching. May be formulated as nonslip by adding carborundum or aluminum oxide when mixing the clay before kilning. |

*Floors and stairways should be designed to have slip-resistant surfaces insofar as possible; adhesive carborundum strips may be used on stair treads or ramps and at critical concrete areas. Etching with mild hydrochloric (muriatic) acid solution will lessen slip problems.

**Colloidal silica is an opalescent, aqueous solution containing 30 percent amorphous silicon dioxide and a small amount of alkali as a stabilizer.

stairways and landings that could contribute to falls. See also Chapter 15, Office Safety.

### Merchandise displays

This section was adapted from *Loss Control—A Safety Guidebook for Trades and Services,* by George J. Matwes and Helen Matwes. Copyright © 1973 by Litton Educational Publishing, Inc. Reprinted with permission of Van Nostrand Reinhold Company.

Displays must be constructed to prevent breakables and other articles from being dislodged so they could fall and strike, injure, or trip the customer. Sharp or broken edges of displays and counters should be repaired and smoothed so they do not cut or catch the passersby.

Marketing people in some stores (like supermarkets) try to

**Figure 21-4.** Point-of-purchase display hangers must be located in such a way that the human eye cannot contact them (such as shown *at left* where both the upper and lower shelves project beyond the hangers). Shelf hangers must be safeguarded if they project into the aisle (*center*). Locating projections above eye level is another safe method (*right*).

stack as much merchandise as possible on a display table for its obvious eye appeal. However, if these displays are allowed to be stacked too high they become impractical for the shopper. When reaching for an item a cascade of boxes, cans, bottles, or whatever else being displayed on the shelf or table can come tumbling down.

The average woman shopper is five foot four inches (1.6 m). If heavy juice cans or glass bottles or jars are stacked on the top tiers of shelves or displays, it is quite obvious why accidents from falling items occur. Furthermore, when a product is stacked too high, the customer may need to step onto a lower shelf to reach it. Shelves should be stacked evenly by layers placing heavy items on the lower shelves and lighter items on the top shelves (Figure 21-4, center).

Nonfood accessories such as hardware items, notions, and kitchen equipment are often carded and hung on pegboard panels. These panels or sections should be adequately recessed to accommodate the extended J-hooks (with minimum J-radius of 1 m). Remember shoppers will be bending over to reach lower items and the hooks above should not extend out so far as to cause contact with a person's eyes or face. Specially safeguarded extenders are available (Figure 21-4, center).

Merchandise with sharp or cutting edges should not be in open displays unless the edges are covered or otherwise protected. Protective sleeves or plastic coatings or covers serve to protect the edges from customer handling.

All electrical and mechanical display elements should have adequate protective features.

**Display platforms** should be of color(s) or lighting that contrasts with the floor or carpet and should not obstruct aisles (Figure 21-5). Corners of platforms should be rounded or clipped. Displays and manikins should be at least 6 in. (15 cm) off the floor so that they will not be tipped over. The display or the manikin should be fastened to this base.

Floor displays should be at least 3 ft (0.9 m) high to be seen while not becoming a tripping hazard. They should not be at the ends of the aisles where shopping carts can dislodge them.

**Hanging displays** should be at a safe height. If hooks are installed at 90-degree points around a column, approximately 18 in. (50 cm) down from the ceiling, this then becomes part of the fixturing and allows for easy change of seasonal displays. It can also eliminate the need for taping, stapling, and other makeshift methods that are not always safe.

If displays are hung from the ceiling, make sure that ceilings are structurally sound and that all code requirements have been met.

A duplex electrical receptacle should be located 18 in. from the ceiling on certain columns with another outlet 12 to 18 in. from the floor.

### Housekeeping

In mercantile establishments, it is estimated that fixtures, displays, and other portable equipment are involved in over 40 percent of customer falls. It is, therefore, essential that management provide safe equipment and that the accident control program particularly emphasize safe placement and use of that equipment.

Dress racks and stock trucks should be removed from the sales area and returned to the stock room as soon as they have been emptied.

All electrical wiring and extension cords hooked to store machines, displays, special decorations, and the like should be run so they do not lie on the floor. Where necessary, wires or cords may be installed in low-profile channels.

Poor housekeeping accounts for one-third or more of all customer falls. Each employee needs to realize that it is part of his or her responsibility to maintain good housekeeping in the work area and to promptly report unsafe floor conditions, such as tears in carpets and holes in the floors. Spills must be wiped up immediately, or barricaded until the hazard can be removed. A special warning sign can be used.

**Figure 21-5.** The display platform covering should contrast with the floor or carpet; manikins should be securely fastened so they cannot tip.

## ESCALATORS, ELEVATORS, AND STAIRWAYS

Many business establishments move people between floors by escalators, elevators, stairways, and ramps.

Escalators can be operated from a low of 70 feet per minute (fpm) to a speed of 125 fpm (0.36-0.64 m/s). The average recommended speed is 90 fpm (0.46 m/s). A 4 ft (1.2 m) wide escalator can move 4,000 to 8,000 passengers per hour. Operating it faster or slower than anticipated often is a source of injury to the young or the elderly.

See also the discussion in Chapter 7, Plant Railways and Elevators, in the *Engineering and Technology* volume.

### Escalators

Accidents may occur when passengers are entering or leaving escalators or while they are riding them. Common problems that can result in such accidents are:

1. Unsafe floor conditions and poor housekeeping at landings;
2. Sales counters, manikins, display bases, and similar units hampering the movement of passengers;
3. Lights or spotlights facing passengers as they step on or off at landings;
4. Mirrors near escalator landings causing passengers to misjudge their step and stumble;
5. Merchandise signs or displays distracting riders, causing them to bump into one another or to fall;
6. Failure of passengers to step on the center of a step tread causing them to fall, possibly against others;
7. Overshoes, particularly the thin plastic type, sneakers, or other objects catching in the comb-plate when pressed against a riser, or catching at the side of the moving steps;
8. A passenger "riding" his hand on the handrail beyond the combplate and back into the handrail return may receive an injury if his fingers run into the handrail guard;
9. If it is not protected against accidental contact, a passenger (or object) accidentally pressing the emergency STOP button can cause the escalator to halt unexpectedly;

10. A package, stroller, or other conveyance placed on the escalator and jamming between the balustrades or slipping from the grasp of the person trying to hold it;
11. Parts of the body becoming involved with moving parts of the escalator resulting in falls, lacerations, and amputations of the toes, fingers, or other parts of the body. (This might occur if patrons are barefooted, as they may be in tropical areas;)
12. A passenger walking or running up or down a moving escalator;
13. Children sitting on escalator steps;
14. A passenger who fails to hold onto a moving handrail and stumbles or falls.

**Escalator standards and regulations.** When installing or modifying escalators, check local and state ordinances. Also refer to National Safety Council Data Sheet No. 516, *Escalators.* The American National Standard A17.1, *Elevators, Escalators, and Moving Walks,* is known as the Elevator Code. Some of its important points are:

- Hand and finger guards are to be protected at the point where the handrail enters the balustrade.
- Prominent caution signs should be displayed (see Figure 21-6).
- Comb plates with broken teeth should be replaced immediately.
- Strollers, carts, and the like must be prohibited on escalators.

The Elevator Code also states that balustrades must be provided with handrails moving in the same direction and substantially at the same speed as the steps.

City, state or provincial, and local code requirements also should be followed.

**Inspections.** Examine all escalators from landing to landing every day. This includes riding them before store opening to find out if any visual- or sound-indicated defects exist.

- Once a week, inspect for the following:
  Step treads and comb plates should not have any broken treads or fingers;
  Examine handrails for damage;
  Check balustrades for loose or missing screws and for damaged or misaligned trim.
- Semiannually, inspect:
  Step chain switches
  Governor
  Top and bottom oil pans
  Skirt clearances
  Step treads and risers
  Steps
  Machine brake
  Skirt switch
  Handrail brushes.

**Start and stop controls.** Escalators should have an emergency STOP button or switch accessibly located at the top and the bottom landing. These must stop, but not start the escalator. Placing a STOP button at the base of a newel, with either recessed design or a cover, will help avoid having the button activated unnecessarily. All employees should be trained in the use of emergency switches and where they are located.

## Elevators

Business establishments with elevators face other accident problems. One of the most common is a customer being struck by the door. This is particularly common in self-operating elevators.

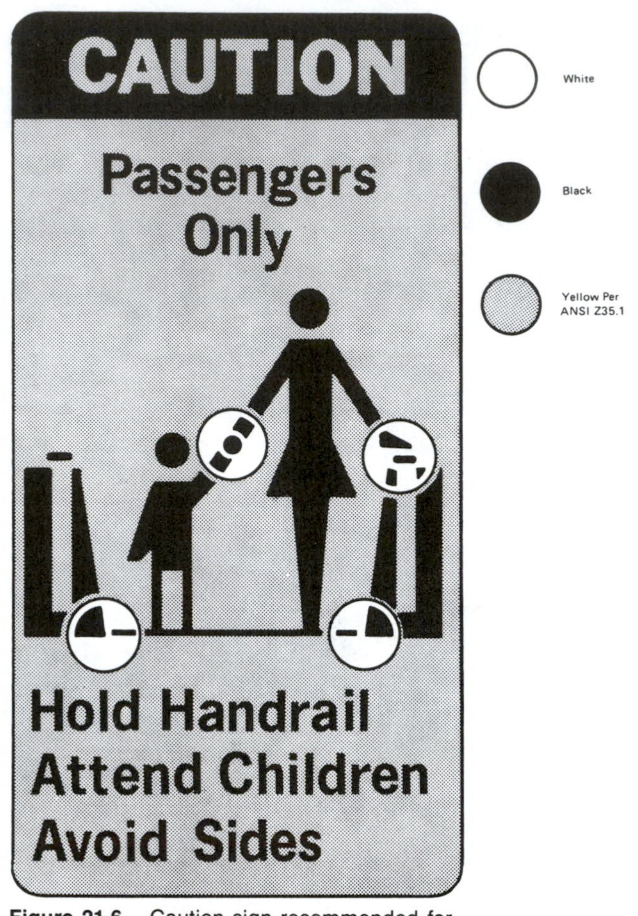


**Figure 21-6.** Caution sign recommended for escalators by ANSI/ASME standard A17.1d-1984. Sign should be 4 in. (102 mm) wide by 7¾ in. (197 mm) high.

**Inspections.** A logbook containing the following information should be kept:

Day, month, year, and time of inspection;
Observations by mechanics or inspectors;
All breakdowns, including causes and corrective action(s);
Entries should be initialed and dated.

City, state or provincial, and local codes should be followed. American National Standard A17.1, the Elevator Code referred to under Escalators, should be used.

At least every three years, all elevators should have a balance test and a contract load test. Make any required adjustments.

Once a year, hand test the following controls:

Governors
Governor cable grip jaws
Gripping jaws of car and counterweight
Releasing carrier
Cutoff switches
Tail rope and trip rod drums
Safety rails.

Spot check automatic elevators at the start of each day for level floor stops, brakes, and other mechanical operations. Test the alarm bell to be sure it rings and that its signal registers in the maintenance department.

See Chapter 7, Plant Railways and Elevators, in the *Engineering and Technology* volume.

### Stairways

Because stairways are so important in emergency egress from buildings, they are covered in detail in the National Fire Protection Association's publication *Life Safety Code,* NFPA 101, Chapter 5, Means of Egress. Also covered in that chapter are ramps, exit passageways, smokeproof towers, outside stairs, fire escape stairs and ladders, illumination, exit marking, and escalators and moving walks. Only inside stairways are included in the following discussion.

The typical stair accident occurs when someone slips while descending a stairway. Most fall victims tend to look at the treads a great deal less than other people do. See the U.S. Department of Commerce study "Guidelines to Stair Safety" (listed in References at the end of this chapter).

The major areas of concern are listed next, along with corrective actions.

- Because stairways are not level walking surfaces they require different walking habits. Their use should be minimized by posting signs directing people to the escalators and elevators. Exits to and from stairs should never be chained or otherwise locked in such a way that they cannot be used in an emergency.
- Treads and handrails should be highlighted so they are immediately and easily distinguished from the riser and wall surfaces. Adequate lighting is essential; NFPA specifies a minimum of one foot-candle (decalux), measured at the floor. Arrange lights so they do not create glare surfaces or temporarily blind stair users. Arrange them so that the failure of one unit will not leave any area in darkness. If one side of the stairwell is open to adjoining space, it is a good idea to close off that view to prevent distractions that result in falls.
- The edge of the step or tread (nosing) should be easy to see. If possible, use contrasting carpeting to distinguish the approach to the steps from the stairs themselves. If not, special nosing may be installed to provide definition between the steps. Nosings must be securely fastened. Uncarpeted stairs should be edgemarked.
- A continous handrail must be provided; see details in Chapter 15 under "Aisles and stairs." The railing preferably should be of a lighter color, because people seem to be more inclined to use a lighter colored railing than a dark one that looks dirty and greasy. The railing should be kept clean. The handrail should extend to the top and bottom of the staircase so that it may be grasped before stepping on the first step or leaving the last step of the flight. Be sure that the railing does not extend into a passageway in such a way as to create a hazard.
- Stair treads need to be stable and provide good traction. Outdoor stairs need extra slip resistance and should have adequate water runoffs that lead away from other walking surfaces.
- Worn or defective treads or other parts should be replaced immediately.
- Stairs should be clear of obstructions. Sharp handrail and guardrail ends should be removed or covered to prevent injury. Glass areas adjacent to stair landings or at either end of the stairway should be clearly marked and protected to prevent people from walking through them. Fixtures that project into the stairway should be moved.

## PROTECTION AGAINST FIRE, EXPLOSION, AND SMOKE

If fire breaks out or an explosion occurs, the quick and orderly evacuation of the premises will protect all persons including visitors. Base all evacuation plans on the premise that visitors will be on the property for the first time. Such persons need to be protected by ample and special direction signs, even though location of exits might be a well known fact to regular employees.

Businesses that generally attract large numbers of customers, guests, or patrons should provide well marked exits, emergency lighting, and ample direction signs inscribed to direct people to safe exits.

In a fire, dense, penetrating smoke can be as deadly as the heat or flame. Lungs can be seared quickly.

Enclosed stair wells provide the best fire escape routes. Doors to such stair wells must never be obstructed, locked, or propped open.

A public address system, staffed by a qualified and trained person, can be used to direct the building's evacuation and issue lifesaving instructions.

Usually the early detection of a fire and the use of a good evacuation plan can prevent panic and personal injuries. More details are given later in this chapter and in Chapter 16, Planning for Emergencies.

Installation of a fire-detection and suppression system for life safety should be considered by every business. Architects should be held accountable for proper design for new construction, especially in the high-rise buildings.

### Fire detection

Properly engineered fire detection systems are sound investments. But the best installation is useless if there is no response to the alarm. Systems should have a direct connection to the local fire department or to some alarm center.

**The four stages of fire.** Fire is a chemical combustion process created by the rapid combination of fuel, oxygen, and heat. A full discussion is found in Chapter 17, Fire Protection, in the companion volume *Engineering and Technology.*

Most fires develop in four distinct stages and detectors are available for each.

- INCIPIENT STAGE. No visible smoke, flame, or significant heat develops but a significant amount of combustion particles is generated over a period of time. These particles, created by chemical decomposition, have weight and mass, but are too small to be visible to the human eye. They behave according to gas laws and quickly rise to the ceiling. Ionization detectors respond to these particles.
- SMOLDERING STAGE. As the incipient condition continues, the quality of combustion particles increases to the point where they become visible—this is called "smoke." No flame or significant heat has developed. Photoelectric detectors "see" visible smoke.
- FLAME STAGE. As the fire condition develops further, ignition occurs and flames start. The level of visible smoke decreases and the heat level increases. Infrared energy is given off; this can be picked up by infrared detectors.
- HEAT STAGE. At this point, large amounts of heat, flame, smoke and toxic gases are produced. This stage develops very quickly, usually in seconds. Thermal detectors respond to heat energy.

**Burning plastics.** Some fuels such as plastic waste receptacles can produce a great deal of toxic smoke when they burn. Therefore nontoxic and noncombustible materials (such as metal cans) are preferrable.

For example, polyvinyl chloride (PVC) in a single foot of one-inch size PVC rigid nonmetallic conduit involved in a fire:

1. Can produce a sufficiently heavy, dense smoke to obscure 3,500 cu ft (100 $m^3$) of space, and
2. Can generate enough hydrogen chloride to provide a lethal concentration of HCl in approximately 1,650 cu ft (45 $m^3$) of space.

**Engineering and control procedures.** The best fire detection system is only as good as its weakest component. The services of a fire protection consultant should be obtained in engineering the system and establishing the control procedures. Here are four steps to consider.

1. Select the proper type detector(s) for the hazard areas. For example, a computer area may involve ionization or combination detectors. A warehouse may have infrared and ionization detectors. In low-risk areas, thermal detectors or combinations of detectors may be used.
2. Determine the spacing and locations of detectors to provide the earliest possible warning.
3. Select the best control system arrangement to provide fast identification of the exact source of alarm initiation.
4. Assure notification of responsible authorities who can immediately respond to the alarm and can take appropriate action. Every detection system must have an alarm signal transmitted to a constantly supervised point. If this cannot be assured on the premises, the signal must be transmitted to a central station, fire department, or other reporting source.

## Response methods

Early warning systems are as important during hours of occupancy as they are when the premises are vacant.

When detection pinpoints a trouble spot, immediate response by a responsible trained company representative is all important. Shaving seconds and minutes can mean the difference between lives lost or saved, and fire confined or allowed to spread out of control.

Here are some systems that are used.

**Twenty-four hour supervisory service.** If an installation has 24-hours-a-day, seven-days-a-week supervision at some point in the building, or building complex, then alarm, trouble, and zone signals should be hooked into this location.

**Less than 24-hour supervisory service.** For periods when an installation does not have responsible personnel to respond to the alarm, a backup annunciation system should be provided. The National Fire Protection Association advises connecting to a central station supervisor's service or other service.

Such systems should include a way to initiate fire and trouble signals to the central station transmitter equipment. In the case of a trouble signal, a representative can be dispatched immediately to investigate the trouble and notify proper representatives of the property under surveillance.

**If central station tie-in is not available.** In areas where no central station supervision is available, the local fire or police department may accept installation of a remote fire alarm panel at their headquarters or firehouse.

**If central station or telephone-leased line tie-in is not available.** If none of the foregoing possibilities is available, then consideration should be given to using qualified and licensed telephone answering services. Automatic dialing units connected to responsible officials of the property is another alternative.

## High-rise building fire and evacuation controls

Just what is a high-rise building? The General Services Administration, at a conference in Warrenton, Virginia, in April of 1971, established four basic criteria to designate a high-rise. First, the size of the building made personnel evacuation impossible or not practical. Second, part or most of the building was beyond the reach of fire department aerial equipment. Third, any fire within the building must be attacked from within because of building height. Fourth, the building had the potential for "stack effect."

Today's huge high-rise megastructures all have one thing in common—they are intended to house people. As buildings go higher, the population density per square foot of ground area increases, posing a whole new set of problems concerning the health, safety, and welfare of their occupants. Actually, each building is a sealed life-support system. Present engineering approaches facilitate heating and cooling but are extremely wasteful of energy. Because they are more airtight than ever, these buildings house an increasing danger from smoke and toxic combustion by-products.

The ladders on most fire department aerial equipment are limited to a height of approximately 85 ft (26 m). This means that a building higher than about 8 to 10 stories cannot be served by this equipment.

Due to the height problem and increased floor areas, fires must be fought from within the building. Automatic sprinklers, hose standpipes, and portable extinguishers, as well as hose lines from the building exterior, are used for this purpose.

An evacuation checklist is given in Figure 21-7.

Occupational Safety and Health Standards Subpart E—Means of Egress (Code of Federal Regulations, Title 29, Chapter XVII, Part 1910, Subpart E; Revised as of July 1, 1979; Amended by 45 FR 60203, September 12, 1980) mandates in Section 1910.38 (a)...The emergency action plan shall be in writing. (Exception for location with 10 or fewer employees.)

(a) (5) Training (ii) The employer shall review the plan with each employee covered by the plan at the following times:

(A) Initially when the plan is developed,
(B) Whenever the employee's responsibilities or designated actions under the plan change, and
(C) Whenever the plan is changed.

Although OSHA is oriented to the employees, obviously such a plan includes a system of responsible individuals and actions to accomplish an evacuation of both employees and nonemployees in an emergency.

**Stack effect.** Every building has its own peculiarities for creation of a "stack effect," or spread from one area to another of fire, smoke, and toxic fumes. Among the factors are structure configuration, height, number and size of openings, wind velocities, temperature extremes, number and location of mail chute openings, and elevator shafts, all of which create varying air flows that tend to accelerate and intensify an interior fire. Unprotected air conditioning systems are an open invitation to catastrophe. If there is no automatic smoke and heat detection,

**EVACUATION PREPAREDNESS CHECKLIST**

All questions in this checklist should be answered with "yes," "no," "NA" (not applicable), or "U" (undetermined.) For all answers that are not "yes," or "NA," the persons responsible for the specific areas in question that need correction should be noted.

**Floor Diagrams**
Are floor plans prominently posted on each floor?
Is each plan legible?
Does the plan indicate every emergency exit on the floor?
Does a person looking at the plan see an "X" indicating "you are here now"?
Are room number identifications for the floor as well as compass directions given?
Are directions to stairwells clearly indicated?
Are local and familiar terms used on the diagram to define directions to emergency exit stairwells? For example, are particular areas identified, such as mail room, cafeteria, personnel department, wash rooms?

**Exit paths to stairwells**
If color coding of pillars and doors, or stripes and markings on floors are used, are they properly explained?
Is additional clarification needed?
Are paths to exits relatively straight and clear of all obstructions?
Are proper instructions posted at changes of direction en route to an emergency exit?
Are overpressure systems and venting systems operative?

**Elevators**
Are signs prominently posted at and on elevators warning of the possible dangers in using elevators during fire and emergency evacuation situations?
Do these signs indicate the direction of emergency exit stairwells which are available for use?

**Elderly and physically handicapped**
Are there elderly or physically handicapped persons who will need assistance during a fire and emergency evacuation of premises?
What provision is made for their removal during an emergency?
Who will assist? How will the handicapped be moved?

**Emergency exit doors**
Are all emergency exits properly identified?
Are exit door location signs adequately and reliably illuminated?
Do exit doors open easily and swing in the proper direction (open out)?
Are any exit doors blocked, chained, locked, partially blocked, obstructed by cabinets, coat racks, umbrella stands, packages, etc.?
Note: Blockage must be removed immediately and subsequently prohibited.
Are all exit doors self-closing?
Are there complete closures of each door?
Are all exit doors kept closed, or are they occasionally propped open for convenience or to allow for ventilation?
Note: This practice must be prohibited.

**Emergency stairwells**
Are stair treads and risers in good condition?
Are stairwells free of mops, pails, brooms, rags, packages, barrels, or any other obstructing material?
Are all stairwells equipped with proper handrails?
Does each emergency stairwell go directly to the grade floor exit level without interruption?
Does the stairwell terminate at some interim point in the building?
If so, are there clear directions at that point which show the way to completion of exit?
Is there provision for directing occupants to refuge areas out of and away from the building when they reach the ground floor?
Are directions provided where evacuees can congregate for a "head count" during and after the evacuation has been completed?
Is there adequate lighting in the stairwell?
Are any bulbs and/or fixtures broken or missing?
Where? Describe locations.
Are exits properly identified?
Are they illuminated for day, night, and power-loss situations?
Are any confusing non-exits clearly marked for what they are?
Are floor numbers displayed prominently on both sides of exit doors?

**Emergency lighting**
In the event of an electrical power failure or interruption of service in the building, is automatic or manually operated emergency lighting available?
If not, what will be used?
Where are stand-by lights kept?
Who controls them?
How would they be made available during an emergency?
Is there an emergency generator in the building?
Is it operable?
Is it secured against sabotage?
Is a "fail-safe" type of emergency lighting system available for the exit stairwells that will function automatically in event of total power failure?
How long can it provide light?
Is the emergency lighting tested on a regular monthly basis with results recorded? Who maintains such records?

**Communications**
How should occupants of the building be notified that an emergency evacuation is necessary?
Is one or more communication systems available to each floor? (P.A. system, Muzak, stand-pipe phones, battery-operated "pagers," etc.)
If messengers must be used, have they been properly instructed?
Is the communication system(s) in good working condition?
Under what emergency conditions is it used and who operates it?
Can announcements be prerecorded by someone with a calm but authoritative voice?
Is the communications system protected from sabotage?
Do all occupants know how to contact building control to report a dangerous situation?
Is the building's emergency communications system tested monthly? By whom and to what extent?

**Figure 21-7.** Reprinted from *National Safety Council Industrial Data Sheet 656,* Evacuation System for High-Rise Buildings.

no automatic fan shutdown, and no automatic fire dampers, smoke and toxic fumes can be quickly drawn into the exhaust or return air duct system and promptly distributed to all other floors and areas of the building served by the air conditioning system.

The National Fire Protection Association recognizes this potential and in its Standard No. 90A, *Installation of Air Conditioning and Ventilating Systems,* states that in systems of over 15,000 cfm (7 $m^3/s$) capacity, smoke detectors *shall* be installed in the main supply duct downstream of the filters. These detectors shall automatically shut down fans and close smoke dampers to stop the recirculation of the smoke—or they may incorporate automatic exhaust.

In planning evacuation, assume that children and the elderly and handicapped will be involved. In addition, some people panic in a fire situation. The quantity and size of staircases will undoubtedly prohibit complete evacuation. Tests indicate that with an occupant load of 240 persons per floor, total evacuation of an 11-story building can take up to 6½ minutes, while an 18-story building can take up to 7½ minutes. Exits are just not designed to handle all occupants simultaneously.

Most codes do not consider elevators to be an exit component and prohibit their use during fire emergencies. But codes generally also require that one or more elevators be designated and equipped for firefighters. Key operations transfer automatic elevator operation to manual and bring the elevator to the street floor for use by the fire service. The elevator shall be situated so as to be readily accessible by the fire department.

Many elevators use capacitance-type call buttons, which may bring them to a stop on the fire-involved floor. Then they cannot move because smoke interrupts the light beam and keeps the doors open. Other possibilities include the inadvertent arrival of the elevator at the fire-involved floor by a passenger not knowing the fire exists and wishing to get off, or a person pushing the call button and then, in panic, using the staircase for exit. With problems of this magnitude, it can be assumed that complete evacuation is impossible.

### Crowd and panic control

Any commercial establishment may be faced with an unruly crowd because of an emergency, panic, or even a planned demonstration. Self interest dictates that preplanning is needed to protect the facility and its employees, as well as patrons and bystanders. Different measures are required depending on whether the panic occurs in the building or outside of it and who is involved. In shopping areas it is good to have directional signs displayed at many areas in the building. Exit signs are especially important. But these alone do not reduce the higher risk potential inherent when a great number of people are gathered, such as at sports events, entertainments, schools, and other places.

While it is best to prevent it from starting, in almost every emergency, there could be panic. Employees need periodic drills and practice in handling emergency situations with customers, some of whom may be confused and others who may be handicapped. The threat of an unusual occurrence (such as a riot, bombing, and the like) cannot be overlooked and plans must be made to handle such a possibility.

High-rise buildings pose new and special problems but other public places such as theaters and amusement and recreation facilities must also provide well-planned emergency procedures. Panic and the press of frantic, hysterical people have caused wholesale destruction of life in many emergencies. Oftentimes, such losses could have been prevented through the strict observance of building and fire codes that eliminate physical hazards and "death traps" because of improper design and lack of firefighting and disaster-control measures.

Civil strife and sabotage are covered in Chapter 16, Planning for Emergencies.

**Demonstrations.** The following procedure concerns demonstration control at a store, but can easily be adapted to the needs of any establishment. These suggestions are taken from *Loss Control—A Safety Guidebook for Trades and Services,* by George J. Matwes and Helen Matwes. Copyright © 1973 by Litton Educational Publishing, Inc. Reprinted with permission of Van Nostrand Reinhold Company.

- *Demonstration outside of store building.* Advise employees to call Security and/or management. Security should telephone police, advise them of the situation, and follow their instructions.

  Arrange for two key personnel to assume previously assigned positions at all store entrances and other key points; they should know Security's telephone number in order to relay information and receive instructions. They should never leave their assigned posts unless relieved or advised accordingly. Caution them to remain calm and not to interfere with the entrance or egress of customers or employees.

  Those employed in portable, high-valued merchandise departments (diamonds, furs, etc.) should arrange to have such merchandise placed in an assigned secure area. Those employees working in departments selling firearms, knives, axes, straight razors, bows and arrows, and even meat cutlery, should have them removed from the selling floor to a secure area. Proceed with "business as usual" in all other departments.

  Because rumors can create panic situations or problems among employees and customers present, all employees should be advised of two or three emergency interior telephone numbers, to verify information and squelch rumors.
- *Demonstration moving into store.* (People carrying signs, groups linking arms across aisles and taunting employees, fights between individuals.)
  1. Advise all employees to avoid any comments, antagonisms, or physical contact with marchers, to answer all queries courteously, and above all to keep calm.
  2. In areas where demonstrations are taking place, have employees stop selling, lock their registers, remain in their areas keeping as calm as possible, and await further instructions from their supervisors.
  3. In areas where "business as usual" is being maintained, arrange for frequent cash pickups.
  4. Key personnel and employees should take their assigned places, as discussed above.
  5. In the event such a demonstration turns into group looting or group "hit and run" stealing, employees should not attempt to make any apprehensions. Security personnel will follow previous orders for such conditions as advised by management.

Remember, personal safety is more important than property protection.

**Panic,** one of the most serious emotional reactions that can grip people, can result in otherwise calm people taking irrational means for self-preservation. Often panic leads to the injury of many persons. An evacuation plan, well rehearsed with supervisors and employees, is needed for every business. (A good discussion of panic and techniques for handling it is given in the National Safety Council's publication *Supervisors Guide to Human Relations.*)

### Self-service operations

The best-known self-service operation is found in gasoline stations. Here, customers put the gasoline in their vehicles without employee assistance. The employee merely sees that customers observe the safety rules and follow the prescribed procedure, usually posted on the pump housing or on a nearby sign.

All states require that the vehicle engine be turned off and that there be no smoking or open lights.

### Evacuation of the handicapped

The evacuation of handicapped persons from hotels, stores, and other facilities is an added problem for both management and the safety professional. Communication, especially, is a problem. Special written instructions can be given to people with an auditory impairment; verbal instructions can be given the blind.

More details on how to help the handicapped are presented in Chapter 20, Workers with Disabilities.

## TRANSPORTATION

Some businesses by their very nature have special, and often prominent accident control problems. Much of their loss control efforts must be directed at protecting the nonemployee from harm.

All types of transportation—from commercial airlines, railroads, marine, and buses to local transit, taxi, and school bus operations—must be vitally concerned with injury prevention. Not only is maintenance and vehicle or unit operation involved, but the problem also extends into areas around it, such as terminals, stations, school bus loading areas, and the like.

### Courtesy cars

When a company provides transportation for customer courtesy and convenience (such as a hotel or motel courtesy car or bus), the same concern and precautions used in commercial operations should be followed. These include providing the safest vehicle, maintaining it in proper working condition (meeting all local, state or provincial, and federal requirements), and operating it with a professional driver who is trained and skilled in all facets of the vehicle's operation.

### Company-owned vehicles

Another source of damage and injury claims arises out of the operation of company motor vehicles by employees. Because the odds are that one out of every four drivers will have a collision in any given year, liability from such accidents is a constant threat and often a very real dollar drain for insurance protection and claim settlement. The question of liability resulting from an employee's use of his or her own car on company business must also be considered.

## PROTECT ATTRACTIVE NUISANCES

There are many opportunities for every business to be involved in losses caused by the public's curiosity. Some examples follow.

Any unattended vehicle or machine that is left in an operable condition is attractive to the young (and the so-called "young in heart.") "Tamper-proof" locks discourage the unauthorized use of vehicles or machines and, if necessary, watchmen and security guards should be employed.

Frequently, partially finished road repairs, other construction or storms and adverse weather can "booby trap" a vehicle. It is best to barricade such hazards and warn motorists away from them. Refer to ANSI Standard D6.1, *Manual on Uniform Traffic Control Devices for Streets and Highways.*

Many contractors or builders provide special, but safe, observation facilities for public sidewalk superintendents. Local authorities and insurance engineers should be consulted for regulations and control measures.

### Swimming pools

Swimming pools are found in many hotels, motels and public areas (Figure 21-8). Pool accidents usually result from inadequacy or lack of protective barriers around pools, absence of lifeguards or qualified adult supervision, disregard for the rules of good pool conduct, and the failure to teach youngsters drowning-prevention technics.

Pool precautions should include:

1. Pools must be screened, fenced, or otherwise enclosed in order to control admittance. A tamper-proof lock should be provided. A pool alarm may provide additional protection.
2. Accurately mark pool depths on both pool deck and poolside. Indicate in feet or meters.
3. Unless the pool has been constructed and staffed for diving, a diving board is not recommended.
4. Basic lifesaving equipment should include a lightweight but strong pole with blunt ends at least 12 ft (3.7 m) long, or a ring buoy to which a long throwing rope has been attached.
5. The person in charge of pool operation should be familiar with swimming pool management. A lifeguard should be on duty whenever the pool is in use.
6. A telephone should be nearby, such as in the bathhouse or changing room. Emergency telephone numbers should be on hand—the nearest available physician, ambulance service, hospital, police, and the fire and/or rescue unit.
7. Decks around the pool should be kept clear of debris and breakable bottles should never be allowed in the area. Make sure all cups and dishes used at poolside are unbreakable. Provide litter baskets and replace defective matting.
8. Electrical equipment used for the pool must conform to local regulations and the latest *National Electrical Code* requirements. Any electrical appliance used near the pool must be protected by a ground fault circuit interrupter.
9. No one should be allowed in the pool during a thunderstorm.
10. All pool appliances and equipment should be maintained properly. Periodic safety checks should be made.
11. Sensible pool rules should be established, posted, and enforced.

## REFERENCES

American National Standards Institute, 1430 Broadway, New York, N.Y. 10018.

*Buildings and Facilities—Providing Accessibility and Usability for Physically Handicapped People,* A117.1.

*Elevators, Escalators, and Moving Walks,* ANSI/ASME A17.1

*Manual on Uniform Traffic Control Devices for Streets and Highways,* D6.1

*Requirements for Fixed Industrial Stairs,* A64.1

*Guidelines to Stair Safety,* Department of Commerce, U.S. Government Printing Office, Washington, D.C. 20402.

Hannaford, Earle S. *Supervisors Guide to Human Relations,* 2nd ed. Chicago, Ill., National Safety Council, 1976.

Matwes, George J., and Matwes, Helen. *Loss Control: A Safety Guidebook for Trades and Services.* New York, N.Y., Van Nostrand Reinhold Co., 1973.

National Fire Protection Association, Batterymarch Park, Quincy, Mass. 02269.

**Figure 21-8.** Basic lifesaving equipment must be available at every pool. Shown here at each lifesaving platform are ring buoys with the throwing rope attached. (Reprinted with permission from National Spa and Pool Institute.)


*Installation, Maintenance, and Use of Auxiliary Protective Signaling Systems,* NFPA 72B.
*Installation, Maintenance, and Use of Central Station Signaling Systems,* NFPA 71.
*Installation, Maintenance, and Use of Local Protective Signaling Systems,* NFPA 72A.
*Installation of Air Conditioning and Ventilating Systems,* NFPA 90A.
*Life Safety Code,* NFPA 101.
*Installation, Maintenance, and Use of Proprietary Protective Signaling Systems,* NFPA 72D.
*Installation, Maintenance, and Use of Remote Station Protective Signaling Systems,* NFPA 72C.
*National Electrical Code,* NFPA 70.

National Safety Council, 444 North Michigan Ave., Chicago, Ill. 60611.
Industrial Data Sheets
*Carbon Monoxide,* 415.
*Escalators,* 516.
*Evacuation System for High-Rise Buildings,* 656.
*Falls on Floors,* 495.
*Fire Prevention in Stores,* 549.
*Floor Mats and Runners,* 595.
*Sidewalk Sheds,* 368.
National Safety News Reprints

National Spa and Pool Institute, 2111 Eisenhower Ave., Alexandria, Vir. 22314.
"Tips and Information."
"Minimum Standards for Public Swimming Pools."

"Property Conservation Engineering and Management." *Record,* 50:3 (May-June 1973). Factory Mutual System, Norwood, Mass. 02062.

Superintendent of Documents, U.S. Government Printing Office, Washington, D.C. 20402. Commercial Practices, Title 16, *Code of Federal Regulations,* Chapter II—Consumer Product Safety Commission.

# Product Safety Management Programs



INJURIES RESULTING FROM THE USE (or often the misuse) of products are the basis for an increasing number of product liability lawsuits. These suits cost industry millions of dollars each year. The best way the manufacturer can prevent or defend such claims is by manufacturing a reasonably safe and reliable product, and, where necessary, by providing instructions for its proper use. The key to achieving a reasonably safe and reliable product and, at the same time, reducing the product liability exposure is to build *in* product safety. This is done by establishing and auditing of an effective product safety management program.

This chapter gives an overview of a series of complex, interrelated product safety and liability prevention management program issues. A more comprehensive treatment of the same subject can be found in the National Safety Council's *Product Safety Management Guidelines Manual.*

From the start, a product safety management program must be designed to encompass all product management personnel and product manufacturing processes in order to determine what actions are necessary to produce a safe product and reduce product liability potential.

To carry out the establishment and auditing of a product safety management (PSM) program, two functions are of prime importance:

The selection of a program coordinator

The selection of a program auditor.

The same person sometimes serves as both the program coordinator and auditor. This is determined by the size, needs, and structure of the individual company.

## ESTABLISHING AND COORDINATING THE PROGRAM

Regardless of a company's size, establishing and coordinating a satisfactory PSM program requires a comprehensive systems analysis of all facets of operation and production, from the design stage through manufacturing, quality control, and shipping. To keep all these functions organized and pulling together, someone must be selected to coordinate the program, either alone or assisted by a committee.

### Program coordinator

Because the success of the PSM program requires the cooperation and coordination of all departments, the program head or committee chairman should be carefully chosen. The individual must be appropriately qualified to exert stringent and continuous control over all phases of product development from initial product design through eventual product sale and distribution.

It is not critically important whether the position is full- or part-time, and, if part-time, what the coordinator's primary function is within the organization. What is important is the PSM coordinator's level of authority. Can the coordinator take needed action without having to go through several supervisory levels? That is, does the coordinator have access to top-level management? Within reason, is the coordinator permitted to implement program plans or suggestions?

If the business organization has a PSM committee to coordinate program activities, the head of the committee must have a similar level of authority, and committee members should have the authority to speak for their respective departments.

**Responsibilities.** A program coordinator must have enough authority to take any required action, and carry out the responsibilities below:

Function in a staff capacity to corporate management
Assist in setting general PSM program policy
Recommend special action regarding:
- Product recall
- Field modification
- Product redesign
- Special analyses

Conduct and review complaint, incident, or accident analyses
Coordinate appropriate PSM program documentation
Assure the adequate flow of verbal and written communications
Develop sources of product safety and liability prevention data for use by operating personnel
Maintain liaison with business, professional, and governmental organizations on all matters pertinent to product safety and liability prevention
Conduct PSM program audits, where appropriate.

**Ground rules.** The following ground rules must also be clearly defined by management if the PSM program coordinator is to be effective:

- The purpose of the PSM program coordinator must be clearly defined.
- The authority and responsibility of the PSM program coordinator must be clearly specified by top management and understood by the PSM coordinator.

As previously noted, a PSM program will involve most of the departments in a company and will require the application of many diverse disciplines. Consequently, *coordination* must be provided by the PSM program coordinator in order to assure a thorough and systematic approach to the implementation of a PSM program.

**Committee coordination.** Whenever a PSM program committee is used, the size of the group must be maintained within manageable limits. This is generally no more than five or six members. In a large corporation, it may be more desirable to appoint a small corporate committee or individual corporate coordinator and also appoint a separate program coordinator in each corporate division or department.

A committee's makeup will vary according to the organizational structure of the particular company. However, key members of the committee will almost always represent the following departments:

Design or engineering
Manufacturing
Quality control
Service, marketing, or installation
Legal.

In addition, sales, advertising, insurance, personnel, public relations, plant safety, and purchasing department representatives should be designated and be prepared to serve as consultants to the PSM program committee when the expertise of these various departments is required.

### Program auditor

Program audit is perhaps the most important factor in a successful product safety management program. Because the role is so important, discussion of the program auditor's responsibilities is the main topic of this chapter.

Although the program auditor can be a private consultant or an insurance company product safety or loss control specialist, a company's PSM program probably will be most effective if the program auditor is a member of company management. Again, the program auditor may or may not be the same individual as the program coordinator; this depends on the needs of the company.

The program auditor's main duty is to evaluate the adequacy of the organization's PSM program activities in relation to actual and potential exposures. This evaluation determines what the organization should do to prevent product-related losses by comparing what *is* being done against what *should be* done.

**Improve program effectiveness.** When auditing the organization's product management control system, the program auditor should strive to maximize PSM program effectiveness. This can be accomplished by doing the following:

- Determine the organization's potential product liability exposures.
- Determine the organization's PSM program deficiencies (for example, evaluation of the organization's ability to control any product liability exposures found).
- Develop concise, realistic, corrective procedures to minimize product liability exposures.
- Motivate management to implement the proposed corrective measures (to reduce losses, improve good records, improve the level of product control, comply with government regulations, for example).
- Transmit all PSM prevention program audit information to management for its review.

Product safety management prevention program audits can be successfully conducted in situations where the organization does not have a formal program, as long as the auditor has satisfactory knowledge of product management control practices.

Coordination adequacy should be measured by the auditor as each departmental activity is evaluated through observation and questioning during the PSM program audit.

**System analysis.** An organization's PSM program philosophy should be one of continuous overview or systems analysis of the entire product management control system. In performing the systems analysis, the PSM program auditor must:

- Evaluate management's commitment to product safety and liability prevention.
- Determine the effectiveness of the organization's PSM program coordinator as being the most tangible indicator of management's interest in manufacturing a safe, reliable product. Depending on the organization, there may be no need for a PSM committee and there may be no full-time PSM program coordinator. What must be determined is whether someone, regardless of title, has really been assigned the *responsibility* for coordinating all activities associated with the manufacture of safe, reliable products.
- Evaluate the PSM program coordinator's effectiveness in coordinating all activities associated with the manufacture of safe, reliable products. When one person serves as both the auditor and coordinator, a personal but, hopefully, objective evaluation must be made of the program coordination functions.

**Duties.** The PSM program auditor must determine the business organization's ability to manufacture safe, reliable products and must evaluate how adequate its capabilities are for controlling existing product exposures, eliminating potential exposures, and determining uncontrollable exposures.

In addition to seeking purely quantitative responses, the program auditor must also be receptive to broad generalized impressions that may be gained during the course of evaluating the organization's product safety and liability prevention program activities.

For example, the program auditor should:

- Be alert for obvious deficiencies that reflect management's neglect.
- Watch to see, when a deficiency is noted and immediate corrective action is needed, whether the organization is honestly interested in manufacturing a safe, reliable product. Remember, actions speak louder than words. At the very least, when a deficiency is noted, responsible management should determine the cause of the problem and indicate what corrective action will be taken.
- Substantiate the organization's actual performance with regard to each deficiency or problem noted by preparing and submitting an appropriate report. Recommendations should be prepared and submitted as required.
- Discuss obvious deficiencies with responsible management personnel to be certain that all the facts have been given. Often as a result of making deficiencies known to responsible management personnel, additional information is forthcoming that can cause the program auditor to modify a previous conclusion.
- Be aware of oversolicitous members of management who may be attempting to hide deficiencies.
- Ask for substantiating documentation of actions taken, particularly if the responses received were not direct.
- Consider a failure to receive the undivided attention of responsible management during the course of the audit as an indication of a lack of concern for product safety and liability prevention on their part.
- Consider the possibility, when unable to elicit information from responsible management in response to specific questions, that the organization is deficient in the areas questioned and note this observation in the PSM program audit report.

The function of the PSM program auditor is to interview all key management personnel, to observe the actual manufacturing operation, and to investigate, question, and verify performance.

## CONDUCTING A PSM PROGRAM AUDIT

This section explains what the PSM program auditor should look for and evaluate during an audit. The major goal of the PSM program audit is to substantially reduce the causes of product liability exposure. Some of these causes are:

Product designs not being reasonably safe (failure to review product design safety)
Inadequate manufacturing and quality control procedures
Inadequate preparation and review of warnings and instructions
Misleading representation of product or services.

The purpose of a PSM program is to develop a means to perform (1) an evaluation of the product during manufacture, distribution, and sale and consumer use; and (2) to control any accident and hazard potential through good product safety management techniques.

### Preliminary procedures

The following preliminary activities are performed by the auditor before beginning the formal audit. They are intended to facilitate and enhance the subsequent performance of the audit.

Clearly explain the purpose of the audit to management
Arrange mutually agreeable audit dates with involved members of management
Clearly define how the audit's operational plan (schedule) will involve members of management
Review appropriate PSM program-related documentation
Review appropriate product-related procedures
Review appropriate product-related technical and standards information
Acquire all necessary product-related printed instructional and precautionary materials.

If these preliminary PSM program audit procedures are carried out, the program auditor usually will receive complete cooperation and necessary information from involved management personnel.

Carrying out the PSM program requires a procedure of checks and audits be followed in each of the organization's major departments. To get the most out of the procedure, it is necessary to have the cooperation of management in general and the safety professional in particular.

### Management commitment

Wholehearted commitment and support of the PSM program by all levels of management provides the spark and impetus to get the program moving and keep it rolling.

Just as with an occupational safety and health program, management must have a written policy recognizing its responsibilities to provide support, to set basic objectives, and to establish priorities for a product safety and liability prevention program within the organization. Some evidence of management's commitment to the program might include:

A record of shop conversations on the subject
A posted letter or bulletin
Evidence of specific meetings
The distribution of brochures
A special mailing to employees
A formal statement of management policy on the subject
Use of a PSM program coordinator and committee
Use of a PSM program auditor.

Wholehearted management support has to be given to the PSM program if it is to succeed. All key people within the organization must be told by management that they have an important role to play in the program and they must commit the time and effort required to make sure the program is successful. In short, management must clearly communicate to all employees by word and deed that the control of product losses is a key company objective.

Ideally, the chief executive officer will issue a written policy clearly stating the company's commitment to product safety and liability prevention. It should be distributed to management, all company departments, and to each employee. There always should be some tangible indication that management is truly committed to product safety and liability prevention and that employees have been advised of this fact.

In addition, the auditor must work with and cultivate the respect and cooperation of supervisory personnel. Supervisors and middle-management personnel are very important to the PSM program because of their direct contact with the employees who are making the products, and, hopefully, following the PSM program guidelines for product safety.

### Role of the safety professional

The role played by the safety professional in a PSM program will vary in proportion to the size of the company. However, in all cases the safety professional plays a vital role in implementing a successful program. In a small to medium-sized company it is entirely possible that the safety professional, because of broad experience in areas of safety technology, will be selected by management as either PSM program coordinator or auditor. However, as the company product line increases, management may select the head of the engineering or design department for this assignment with the safety professional as assistant. Some companies do, however, hire a full-time product safety director.

Regardless of the specific role of the safety professional within the company's PSM program, the auditor should evaluate the safety professional to determine if he or she is, in fact, adequately contributing to the company's overall PSM program.

Following are some of the contributions the safety professional can make:

- The safety professional, because of knowledge of plant operations and general safety expertise, can evaluate and offer comments on the company's PSM program.
- The safety professional, because of experience in safety training, can evaluate and comment on the product safety-related training programs developed under the PSM program.
- The safety professional, because of knowledge of accident investigation techniques, can assist those who will be performing product accident investigations.
- Members of the safety department can provide product safety surveillance in production areas. This helps to prevent errors and product-related accidents. These individuals should be advised that all product safety-related complaints or problems uncovered as a result of their surveillance must be discussed with the plant engineer and the production department and formally documented, with copies sent to the PSM program coordinator.
- Because of its past experience in developing and implementing employee safety programs, the safety department often is aware not only of potential product hazards but also of ways in which customers misuse those products. Therefore, a knowledgeable safety department representative should be used as a consultant to the team performing design reviews, hazard analyses, and product safety audits.

## DEPARTMENTAL AUDITS

Although virtually all departments are involved in the PSM program, certain departments play more significant roles. For purposes here, only those departments will be discussed.

### Engineering or design department functions

The primary function of the engineering or design department should be to design saleable, reliable products that can be used with reasonable safety. It is more practical and usually much less costly for the manufacturer to build reliability and safety into the product than to suffer the consequences of catastrophic product liability losses.

Products should be reasonably safe during (1) normal use, (2) normal service, maintenance, and adjustment, (3) foreseeable uses for which the product is not intended, and (4) reasonably foreseeable misuses. The courts are saying today that a manufacturing company has the responsibility of making sure that its products are safe for any reasonable foreseeable use or misuse to which the customer might put them.

The engineering or design department is a critical area within PSM program activity; consequently, the program auditor should check the following:

- Does the organization evaluate product hazards prior to production?
- Is anyone within the organization formally assigned this responsibility?
- Is there a formal written or prepared design review procedure, even if it is not referred to as such?

At certain predetermined points in the manufacture of all complex products, tests or inspections of the products are made to determine compliance with the manufacturing requirements. Why? Because it has been found that problems detected at predetermined points in the process can be corrected more quickly and economically than problems that are detected after manufacture. This is true of all complex processes, including the product engineering design process.

At predetermined points in the design process, the design should be checked for compliance with its requirements, the objective being to make sure that the optimum product design is achieved.

The appropriate check for the product engineering design process is called the *formal design review.*

**Formal design review.** The formal design review (FDR) is a scheduled systematic review and evaluation of the product design by personnel not directly associated with the product's development but who, as a group, are knowledgeable in and have a responsibility for all elements of the product throughout its life cycle, including design, manufacture, packaging, transportation, installation, use and maintenance, and final disposal.

**Codes and standards.** The PSM program auditor must determine whether products conform to all applicable safety standards (including state, or provincial, federal codes and regulations, testing or inspection laboratory requirements, industry standards, technical society standards, and machine safeguarding standards). These standards, in most cases, should be considered as being the minimum for safety and reliability. In some cases, where no applicable formal standards apply, it should be determined whether or not in-house design standards are being used and, if so, what criteria are used before a decision is made as to the adequacy of the standard. Prior to each survey, the auditor should become familiar with the applicable standards.

**Human factors.** The PSM program auditor must determine if human factors have been considered in product designs. (See description in Chapter 10, Human Factors Engineering.) Some of the human factors that should be considered are the physical, educational, and mental limitations of people who will use these products. What the program auditor also tries to find out is whether consideration has been given to the possibility that customers might use the product in ways other that it was designed to be used—but might be considered reasonable by law.

**Critical parts evaluation.** *Critical parts or components* are defined as those "whose failure could cause serious bodily injury, property damage, business interruption, or serious degradation of product performance." Individual organizations, however, may have different criteria for defining a critical part, such as "one that is unusually expensive, difficult to acquire, or requires lengthy order lead time." When analyzing critical parts, the program auditor must be certain that everyone is informed of and uses the appropriate definition.

The auditor determines whether products are being analyzed for critical parts or components. If so, are the critical items receiving any special attention? Have they been field tested and designed to outlast the product itself? If this is impractical, is a special effort made to warn customers and users of possible hazards of critical-part failure? Has any effort been made to instruct customers and users in inspection techniques to detect impending failures? Have maintenance procedures been outlined for critical parts or components on the product itself and in operating and maintenance instructions? Has careful consideration been given to the expected life of the product?

All parts or components must have a life expectancy compatible with the life expectancy of other parts and the total product. A part or component whose life expectancy is shorter or not compatible with that of the remainder of the product should be considered "critical" because of the possibility of its being the basis for a product liability lawsuit, an expensive product recall, or a field-modification program.

**Packaging, handling, and shipping.** The PSM program auditor determines if the products have packaging adequate to prevent deterioration, corrosion, or damage. Requirements for packaging must cover conditions affecting the product while at the manufacturing site, during transit, and under normal conditions of storage by the customer. Packaging methods should be established for each individual product. Product packages should be marked to indicate special requirements—DO NOT STORE IN HOT AREAS, THIS END UP, USE NO HOOKS.

Special containers and transportation vehicles should be used as necessary to prevent damage.

The program auditor also must carefully evaluate shipping procedures because a product shipped without proper instructions or with hidden damage may cause an accident when used. The reason for selecting the packaging materials, cartons, and carrier should be based not only on economics, but also on the need for delivering a product in perfect condition to the distributor or customer.

The product, as shipped, must agree with the purchase order. All appropriate labels, manuals, warnings, and descriptive materials must be included and should be checked at the time of shipping.

**Warning labels.** It is not enough to design and make a satisfactory product; an organization also has the responsibility to label its products correctly and to warn potential consumers and users of any dangers involved. The duty of a manufacturer to label products and warn of potential danger is increasingly the basis for product liability lawsuits. This is especially true in chemical, drug, and food cases.

The program auditor must carefully examine the product instructions and labeling to be sure they conform to pertinent regulations and recent court decisions affecting a company's field of operations.

The basic rule, as stated in *Restatement of Torts, Second Series* (see References), No. 388, is that "a manufacturer or supplier must exercise reasonable care to inform its consumers of a product's dangerous condition or the facts which make it likely to be dangerous if he knows or has reason to know that the product is likely to be dangerous for the use for which it is supplied and has no reason to believe that those for whose use the product is supplied will realize its dangerous condition."

Generally, a manufacturer has no duty to warn its consumers of danger if the danger is well known. However, duty to warn and the adequacy of warning is a jury question and their interpretation of "duty" and "adequacy" will vary as a function of the circumstances of product use.

In general, the courts have tended to find that insufficient warning is the same as no warning. Therefore, a company must not only give clear and easily interpreted instructions on how to use the product, but must also give very clear and specific warnings of any inherent dangers or possible misuses that could result in injury.

The courts have consistently held that a manufacturer has a duty to warn with respect to any *reasonably foreseeable* use by a user or consumer and that for a warning to be considered adequate, it must advise the user of the following:

The hazards involved in the product's use
How to avoid these hazards
The possible consequences resulting from a failure to heed warnings.

Instruction manuals accompanying the product should repeat hazard warnings as well as how the hazards can be reduced or avoided. In addition, instructions should be given on how to inspect the product upon receipt and how to assemble, install, and inspect it periodically. If the manuals include troubleshooting hints, the dangers involved should be explained; that is, use of unauthorized parts, do not remove back panel, high-voltage hazards, to name a few, as well as comments pertaining to the preventive maintenance program required.

Finally, the program auditor must determine whether sales brochures and product advertising should be reviewed by the engineering or design department to be sure the product's capabilities are accurately depicted and show only safe operating and maintenance procedures.

Conversely, the program auditor also must determine whether all warning labels, hazards, and instructions developed by the engineering or design department have been reviewed by the company's legal counsel to make sure that product users are receiving adequate instructions for use, warnings about potential product hazards, and instructions for proper maintenance.

## Manufacturing department

After a reasonably safe and reliable product has been designed, the manufacturing department must turn the design specifications into a finished product. If this is not satisfactorily done, manufacturing errors could result in an unsafe and possibly unreliable product.

The manufacturing department can contribute to a company's overall PSM program in many ways. The most important are listed here. By evaluating each of the following (as it applies to a particular situation), the program auditor can effectively measure the manufacturing department's contribution.

- Motivate manufacturing employees by taking steps to make sure each employee understands that he or she is making a vital contribution to a quality product.

- Instill each employee with pride in his or her work and in the company's product by:
  Using up-to-date manufacturing equipment
  Keeping the plant's manufacturing capacity within bounds
  Providing adequate room in which to work
  Avoiding cramped working conditions
  Keeping the workplace clean and well lit
  Implementing good equipment maintenance practices.
- Provide standardized foreman-supervised on-the-job training procedures.
- Implement zero defects or error-free performance or other error-elimination programs.
- Design a program or procedure to identify and eliminate all production trouble spots. This should be accomplished in conjunction with the inspection and testing personnel of the quality control department.
- Prevent unauthorized deviations from design specifications and work procedures.
- Participate in safety audits on new product designs. This is often accomplished in the course of serving on the company's PSM committee, if one exists.

Sometimes product specifications are not realistic for the actual machines and equipment in the shop. When this occurs, the production department must advise the engineering department that there is a problem. No attempt should be made to maintain cost or schedules at the expense of deviating from the specifications without consulting the engineering department. If deviations are necessary or unavoidable, they should be made only with the approval of the engineering or design department.

**Recordkeeping.** Manufacturing records should be kept for the life of all the products and particularly for their critical parts or components. These records are absolutely necessary in the event of product recall or field-modification programs and for successful defense in product liability suits. Records on critical components should be complete enough to identify the batch or lot or supplier of the raw material and the finished products in which they were used.

**Discontinued and new products.** Another important area of manufacturing department exposure is the existence of discontinued products and the development of new products or product lines.

The PSM program auditor must evaluate and determine if the existence of discontinued products (products on the market but no longer being manufactured) or the manufacture of new products could result in unnecessary loss exposure situations.

### Service department

Most companies fulfill service contracts for their customers, either through the dealer or distributor or through service subcontractors. Consequently, a company's product liability exposure may be significantly increased because the company has extended its exposure beyond the confines of its controlled environment—the manufacturing facility.

The PSM program auditor must carefully evaluate the company's service department controls to accurately determine not only the adequacy of these controls and their contribution to the company's PSM program activities, but also the adequacy of the company's use of service department information feedback.

In many companies (depending upon the type of product), service department personnel are required to maintain close contact with customers. Consequently, they, more than any other employees, are familiar with the customers. They are most likely to hear any customer complaints, reactions, and compliments regarding the company's products. They see product misuses and usually have some knowledge of any incidents and accidents that have occurred. In fact, service department personnel are often the first to hear of product accidents.

### Legal department

The legal department in any organization should play a significant role in both the prevention and defense of product liability cases. Too often, companies only use their legal department in those situations when product safety problems or product liability litigation is imminent. They very often fail to recognize the importance of having legal personnel become involved in the day-to-day aspects of product safety and liability prevention.

The legal department should be assigned the following responsibilities in order to maximize its contribution to the company's PSM program activities.

**Legal personnel** should act as legal advisors to the PSM coordinator.

**Review for potential liability.** Legal personnel should provide legal review for potential liability of all product-related literature.

**Coordination.** Legal personnel should provide coordination of all product investigations and product claim defenses.

### Marketing department

Unfortunately, after a "reasonably" safe product has been designed and manufactured, product claims can be unnecessarily incurred because of the way in which the product was represented to customers and users. Customers often must rely on the company's sales personnel, advertising and sales brochures, and operating, service, and maintenance instructions for their knowledge of the product's capabilities and hazards.

Consequently, if the customer is led to believe that the product has capabilities it does not possess or if the instructions do not adequately warn of the product's hazards, an injured customer may have legal cause for action where none existed before. Therefore, the PSM program auditor must review the company's advertising and sales materials and evaluate whether:

They are clear and accurate
They overstate the product's capabilities
They encourage the customer to believe the product has uses for which it was not designed or intended
The product can safely do all that the company's advertising and sales material says it can
Only safe operating procedures are depicted
The company's product is illustrated only with safety devices in place
Advertising and sales materials have been reviewed and approved by the engineering or design and legal departments with regard to accuracy and potential legal liability exposures.

Warranties and disclaimers developed by the company must also be reviewed by the program auditor to determine whether they are:

Reasonable and practical for the uses intended
Included with each of the company's products

Clear and concise
Prominently displayed and easily recognizable by the customer or product user
Thoroughly reviewed by the company's legal department or legal counsel.

Another important duty of the marketing department that must be verified by the program auditor is the retention (for the life of the product) of sales and distribution records that can be used to identify purchasers. These records are essential if product recall or field-modification programs are to have any chance of success. In addition, these records should indicate, whenever possible, the use to which the company's products will be put, particularly when the company is selling to subcontractors or assemblers.

The company also must be able to verify that its sales personnel and dealers have been instructed to accurately describe the capabilities of the products they are selling or distributing and thus avoid incurring undesired implied or expressed warranties. For example, sales personnel must be instructed not to exaggerate the capabilities of the company's products.

### Purchasing department

Although some companies are not large enough to have a formal purchasing department, every company has someone responsible for acquiring the raw materials and components needed for the manufacture of the product.

This is a significant activity with respect to the company's product liability potential. Consequently, the PSM program auditor must give careful consideration to the company's purchasing activities and determine whether or not the company is performing acceptably in this area.

Some of the primary responsibilities of the purchasing department are to:

- Become familiar with all material specifications set by the engineering, design, and manufacturing departments.
- Always purchase quality raw materials, parts, and components meeting the specifications set by the cognizant departments.
- Evaluate, in conjunction with the company's quality control department, the capabilities and reliability of suppliers, using vendor rating systems. Compile a list of approved suppliers.

### Personnel department

Employee job qualifications and work attitudes have a considerable influence on the quality of a company's products. Employees who lack proper job skills or who are not interested in their work increase the possibility of defective products.

It is essential that the company's personnel department be diligent in (1) selecting, training, and placement of new and transferred employees, and (2) continually seeking to upgrade the morale and performance goals of present employees.

### Insurance department

The insurance specialist or department is often considered only as the department that buys insurance and reports claims to the insurance company. In reality, because of its experience in handling product claims, the insurance department usually has a background of valuable knowledge of and experience with the factors that can generate product liability lawsuits. Consequently, the insurance department should be used as an information clearing house in conjunction with the PSM program coordinator.

The insurance department usually reports liability claims to the insurance carrier and coordinates any accident investigations with the carrier. There are occasions when the insurance department will coordinate the legal defense (if one is necessary), between the legal department and insurance carrier. Depending upon the company, the legal department may perform these functions instead.

### Public relations department

Although a public relations department, like the safety department, generally plays a smaller role in a company's PSM program activities than many of the other departments, the program auditor should not overlook it in the overall evaluation of the PSM program.

Product accidents, product recalls, or product field-modification programs may be noticed by the news media and require that the company prepares press releases. If these statements are skillfully prepared, they can help prevent unfavorable publicity. In fact, correct handling of publicity on product accidents, product recall, or product field-modification programs can be skillfully used by the public relations department to display to the public the company's diligent efforts to design, manufacture, and sell a safe, reliable product.

The public relations department should also be capable of generating favorable product-related publicity for media consumption when the company has legitimately incorporated safety advances into its products.

## QUALITY CONTROL AND TESTING DEPARTMENT

The term *quality control* refers broadly to that function of management where calculated actions are taken to assure that manufactured, assembled, and fabricated products conform to design or engineering requirements. At one time, the terms quality control and statistical quality control were considered synonymous because statistical techniques were considered to be the major tools of quality control. Over the years quality control has taken on a larger meaning that includes whatever actions are necessary to assure products conform to design and other requirements and achieve customer acceptance and satisfaction.

Before discussing the evaluation of a company's quality control activities, it must be pointed out that quality control policy, department organization, function, and responsibility will vary according to a company's size, management policy and organization, type of product, plant location, number of plants, economic resources, and other variables.

When evaluating a company's quality control department, the important points to determine are:

**Is the quality control program adequate** to carry out the company's quality objectives? The program auditor must determine if the company's quality control program is satisfactory as a function of the company's products and their attendant exposures. The lack of one or more functions or an incomplete function does not necessarily mean that the system is inadequate. A necessary function can be included as part of a subsequent function or not required at all because of the type of product, manufacturing process, or company size.

**Does the program function as planned?** To answer this question, the program auditor must go beyond the quality control manual and review the actual implementation of the system.

When performing an audit, the PSM program auditor must ask: How? Why? When? Where? What? Who? Then, follow through with "Explain how it works." "Let's see examples." "Show me the records."

Asking these questions may indicate the company has instituted a quality control program, but it does not prove the system has actually been installed throughout the plant, nor that it is functioning as planned.

The PSM program auditor must see evidence of the implementation and functioning of the quality control program. To get a true picture of the program, paper work, procedures, instructions, and records must be reviewed.

The program auditor should spend some time with the quality control shop personnel, if possible. These individuals can expound on the cons as well as the pros of the program, thus providing information that is relevant to the program's strengths and weaknesses not otherwise obtainable.

The basic quality control-assurance program functions that must be evaluated by the PSM program auditor are as follows:

- Organization and manuals
- Engineering/product design coordination
- Evaluation and control of suppliers/vendors
- Evaluation of manufacturing (in-process and final assembly) quality
- Evaluation of special process control
- Evaluation of measuring equipment calibration system
- Sample inspection evaluation
- Evaluation of nonconforming material procedures
- Evaluation of material status/storage system
- Evaluation of error analysis and corrective action system
- Evaluation of record retention system.

### Manuals

One of the most important requirements for effective operation of a quality control program is positive interest and concern. This begins at the top level of management. In most instances, quality control responsibilities must be delegated, and the assignment of responsibility and authority should be clearly defined throughout all levels of management, supervision, and operation.

There should be a quality control manual, the form and content of which will vary according to a company's requirements. The manual is usually divided into two sections: policy and procedures.

- Policy:
  States company quality control policy and objectives
  Establishes organizational responsibilities
  Establishes systems for implementing quality control policy
- Procedures:
  States operational responsibilities
  Gives detailed operating instructions.

The policy section of the manual is general in nature, while the procedures section contains the daily, detailed operations of the quality control department.

### Engineering and product design coordination

The quality control department can, by looking at things from a different viewpoint, assist the engineering or design department in its research, development, design, and specifications functions.

Quality control normally is not involved in engineering or design or research unless it has accumulated beneficial data. Quality control assists by performing functions such as inspection testing, recording, maintenance, equipment calibration, and data accumulation.

### Evaluation and control of suppliers

It is just as important to control the quality of purchased materials and services as it is to establish and enforce such controls internally. The degree and extent of the quality controls established for a supplier will depend on the complexity and quantity of the supplier's product and its quality history. The most effective requirement is to choose suppliers who can maintain adequate quality. Another vital factor is open, active, and adequate flow of information between the company and its suppliers.

### Manufacturing (in-process and final assembly) quality

Quality control of items produced in-plant is accomplished by planning, inspection, process control, and equipment calibration.

**Manufacturing planning.** Planning in advance of production assures uniformity of product and manufacturing instructions. It normally is a joint effort of the manufacturing department and quality control department. The manufacturing personnel initiate the internal paperwork, including all data necessary to produce the product. Quality control personnel review the paperwork to confirm that the data conform to established parameters, and include quality requirements.

**Manufacturing work instructions.** All work affecting quality should be supported by documented, how-to-work instructions that are appropriate to the nature of the work, the situation under which the work is to be done, and the skill level of the personnel doing the work. Each instruction also must include quantitative and qualitative means for determining that each operation has been satisfactorily done. The depth of detail will depend upon the skill level of the worker and upon the complexity of the task.

The instructions to production personnel should detail the steps necessary to produce the products.

**Inspection instructions.** Instructions must be given to inspection personnel to assist them in making sure products conform to design specifications.

**Manufacturing inspection.** Product inspection determines whether the product conforms to established requirements. It usually includes one or more of the following general methods:

Visual inspection
Dimensional inspection
Hardness testing
Functional testing
Nondestructive testing (NDT)
Chemical-metallurgical testing.

When product inspection is conducted on an in-process basis and not upon completion of all operations, the inspection must not be conducted until it has been determined that all previous inspections have been made as required and the product up to this point is acceptable.

**Records.** Records of the manufacturing operations, including incoming materials, assemblies or processes, and final inspection results, should be maintained.

### Special process control

Processes used by manufacturing to change the physical, mechanical, chemical, or dimensional characteristics to make a product include, but are not limited to, the following:

Heat treating
Plating
Fusion welding
Stamping and forming
Batch mixing
Chemical mixing
Adhesive bonding.

Because processes greatly influence the quality of the completed product, it is imperative that they be positively, yet economically controlled. A 100% inspection is not feasible because it might require destructive testing to determine conformance. The processes, therefore, require self-monitoring controls to assure that similarly processed items will be of the desired quality.

### Calibration of measuring equipment

Measuring and process control equipment and instruments used to assure that manufactured products and processes conform to specified requirements must be calibrated periodically. In other words, any equipment used, either directly or indirectly, to measure, control, or record should be calibrated against certified standards so the equipment can be adjusted, replaced, or repaired before it becomes inaccurate.

### Sample inspection evaluation

A sample inspection is often used by quality control to ascertain the quality of a lot without inspecting all items in the lot. It can be very beneficial if used correctly, but the converse is true if it is not used correctly.

Even when a 100% inspection is performed under the most favorable conditions, it is only 85-90% effective because of human and other error. A sampling inspection is not necessarily 100% effective, but it can approach that level. The sampling disadvantage, which is far outweighed by its advantages, is that occasionally the sample gathering procedure for a lot does not give a true picture of the lot quality—good lots can be rejected or bad lots can be accepted. The goal, however, is to make the acceptance of good lots much more likely than the acceptance of bad lots.

### Nonconforming material procedures

The term *nonconforming material* is usually applied to products that are rejected because they do not meet established requirements. Raw material, parts, components, subassemblies, and assemblies become nonconforming material at whatever manufacturing stage they become discrepant.

The company should have a system to control nonconforming material that includes the following:

Identification
Segregation
Disposition
Reinspection
Customer notification
Supplier reporting
Records.

### Material status and storage

The company must have a tangible means of knowing whether a product (or lot) has not been inspected, has been inspected and approved, or has been inspected and rejected. All raw materials, components, subassemblies, assemblies, and end products should bear identification of process, inspection, and test status.

The material storage function involves the temporary holding of raw materials and recently purchased or in-process parts and assemblies. Stores personnel must have a control system that tells at all times the status and the condition of stored raw materials, components, and products.

The program auditor must evaluate the effectiveness of the controls used by the company to make sure the stock on hand is the stock ordered, and that the system used to identify the materials in the storeroom is adequate.

### Error analysis and corrective action system

The error analysis and corrective action system is a followup of the nonconforming material system. The objective of this system is to use the discrepancies reported in the nonconforming material system to perform the following:

Analyze manufacturing errors
Initiate corrective action requests to the party responsible for the errors, evaluate responses, and determine the effectiveness of the action taken
Analyze manufacturing rejection rate and report to management the dollar losses due to scrappage, rework or repair, and reinspection costs.

### Recordkeeping system

Records are the backbone of a quality control program because they reflect the history of the product. Good records indicate that the company's quality control system is functioning as planned, substantiate that a product was inspected, and often provide the actual inspection findings. To repeat what has been said before—records are vital in the defense of a product liability lawsuit.

Good quality control records also contribute to product traceability and should include:

Lists of raw materials from which products are produced
Inspection results for purchased and manufactured products
Inspection results from each inspection station
Special processes control data
Calibration data
Sample inspection data
Nonconforming material data
Error analysis and corrective action data
Shipping data.

These records should be stored in metal cabinets in a low hazard area and kept for the life of the product.

## FUNCTIONAL ACTIVITY AUDITS

Most of the remainder of this chapter is devoted to a discussion of two interrelated activities: *recordkeeping* and *field-information systems.* These important activities are referred to as functional to differentiate them from the departmental categorizations previously used (design, quality control, insurance, and others).

The term functional implies that these are activities that, if properly accomplished (from a PSM program standpoint), must be performed by several or possibly all of the departments within a company, rather than just one department.

### Recordkeeping

In today's products liability environment, a company must not only manufacture and market safe, reliable products, it also must be able to prove in court, if the need arises, that the products it manufactures are safe.

Complete, accurate records can be extremely convincing in a court of law. Consequently, records should be retained that are pertinent to all phases of a company's manufacturing, distributing, and importing activities, from the procurement of raw materials and components, through production and testing, to the marketing and distribution of the finished products.

In addition to their usefulness in demonstrating to courts and juries that a company manufactures safe, reliable products, comprehensive PSM program records also permit a company to readily identify and locate products that its data collection and analysis system indicates may have reached the customer in a defective condition. Should a product recall or field-modification program become necessary, comprehensive PSM program records can result in successful implementation.

Both federal regulations and internal PSM program requirments significantly affect the form and content of a company's recordkeeping program, specifically, the types and quantity of records retained and period of retention.

**Federal regulations.** A company should be aware of federal regulations that will have an effect on recordkeeping requirements.

Some of the more important federal regulations that must be considered when developing a PSM recordkeeping system are:

Consumer Product Safety Act (PL 92-573)
Federal Hazardous Substances Act (15 USC 1261)
Federal Food, Drug and Cosmetic Act (21 USC 321)
Poison Prevention Packaging Act (PL 91-601)
Occupational Safety and Health Act (PL 91-596)
Child Protection and Toy Act (PL 91-113)
Magnuson-Moss Warranty-Federal Trade Commission Improvement Act (PL 93-637).

**Internal PSM program requirements.** A company must maintain all pertinent information related to the design, manufacture, marketing, testing, and sales of its products. Of particular value are records documenting design decisions, since faulty design is an almost universal allegation in product liability lawsuits. It is an especially effective defense to be able to introduce evidence proving that a certain design or material was chosen with the safety of the consumer in mind. Also, records of why certain production techniques were chosen, such as having a part forged rather than cast, should be retained. Records also can establish the testing to which a product was subjected and the reasons for modifications and improvements. When a hazard cannot be eliminated, a company's records should show conclusively why it cannot.

An important means of product liability protection, in the absence of a fault-proof product, is a complete, up-to-date set of records that can establish the care taken by a company to market safe, reliable products. The PSM program auditor must bear in mind that records that should be kept by a company are determined by the company's size, product, and organizational structure.

**Period of retention.** It is a difficult problem for a company to determine what records should be retained and for how long. Product defects may be alleged many years after manufacture or sale.

Generally speaking, unless a company can accurately define the life of its product, it should be prepared to retain all PSM records in perpetuity. If this creates record storage or maintenance difficulties, a company should use a microfilm or computerized recordkeeping system. The PSM program coordinator must make sure that key personnel fully understand why records are being maintained, what they contain, and for how long they should be retained.

**Reasons for keeping records.** There are several guidelines that should be applied to the implementation of a PSM recordkeeping program. Some of the reasons for keeping records are:

- To comply with federal regulations covering the design, manufacture, and sale of the company's products
- To demonstrate commitment on the part of company management to market a quality product
- To establish how much care is needed to produce and sell a safe, reliable product
- To enable tracing the product or customer
- To establish a sound data base for items such as insurance costs, sources of supply, and product recall or field-modification expense requirements.

### Field-information system

Because a company must receive information from the field about product performance, the PSM program auditor must be able to determine the effectiveness of its field-information system in relation to the type of product and its distributing system. What the program auditor must thoroughly evaluate is (1) the capability to identify and trace a product from the raw material through final sales and distribution stages, (2) the ability to acquire and use field data (complaints, incidents, and accidents), and (3) the desire to use these two factors to implement product field-modification or recall action, where appropriate.

**Data collection and analysis of complaints, incidents, and accidents.** Every company should have a reporting system that will permit it to acquire and subsequently evaluate product information from the ultimate testing laboratory—the customer. This information comes from a variety of sources both within and without the organization (service personnel, salespeople, repair, distributors, and retailers).

A reasonably detailed report used with a data collection and analysis system can accurately evaluate each complaint, incident, and accident received and realistically assess any problem as it develops.

For maximum effectiveness, data should be directed to one individual in the company, the PSM program coordinator.

The program auditor also should be able to verify, through company records, the results of any field data analysis and what corrective action, if any, was taken.

Diversification among manufacturers makes it impossible to create one data collection and analysis system applicable to all companies. However, every data reporting system must provide the answers to the following questions:

Who is the customer?

What type of product is involved?
What is the problem?
How is the product being used?

**Analysis of causes.** The complaint report should provide answers to questions such as:

Is the product being used in the manner for which it was intended?
How is it mounted or installed?
What environmental conditions has the product been subjected to?
Has the product been altered or modified by the customer?

When it is determined that a potential hazard exists, a company should be prepared to promptly instigate field investigations in order to determine its source (such as design, manufacturing, or quality control).

The causes, as determined by the field investigation, should then be analyzed to determine if a trend is developing or if the product problems are unrelated. This information should be provided to the PSM program coordinator, who will decide whether to recommend further action be taken.

The type of action taken might include changes in design, manufacturing, quality control, or advertising procedures, or the implementation of a product field-modification or recall program. If *no* action is taken by a company, it is saying the overall data collection and analysis system has no value in the manufacture of safe, reliable products.

The surest means available to evaluate a company's data collection and analysis system is for the auditor to request a review of the company's complaint, loss, field data, and similar files. If the company does not have any, or if they are poorly coordinated, then it can safely be assumed that the company does not significantly use this source of information to help it manufacture safe products.

### Product recall or field modification

The need for product recall or field modification can become a reality for the manufacturer of almost any product and can also involve those who supply component parts, materials, and services for products. Therefore, the PSM program auditor must be prepared to evaluate the company to determine whether it is capable of implementing a successful product recall or field-modification program should one become necessary because of information developed by the company's data collection and analysis system.

In the event substantial performance or safety defects are determined to exist in some or all of a company's shipped products, it is incumbent upon the company to recall or field-modify these products.

The identification of the specific lots, batches, quantities, or units to be recalled or modified in the field can be a difficult procedure unless an adequate tracing system has been established during the product planning and manufacturing stages. Every company's recordkeeping system should be extensive enough to provide adequate product traceability from design through production and delivery of the product to the customer. In addition, recall or modification should be considered in those cases where the company is aware of or has itself developed an improvement or changed its product (for example, added a guard or a fail-safe). If the improvement has been developed subsequent to the sale of earlier product models, a company must be prepared to advise its customers of the available product improvement, the hazard it eliminates, and what they must do to incorporate it into their product models.

If a product recall or field-modification program does become necessary, the quantity of product that must be recalled or modified and the cost to do so can be minimized if the exact number of defective products, parts, or components can be pinpointed.

**Field-modification program plan.** A product recall or field-modification program plan includes specific general procedures. Each company's needs will, of course, be different and must be defined by the PSM program auditor after evaluating the recordkeeping program and product traceability systems capability.

Based on information acquired from complaints, incidents, and accident reports from customers, distributors or dealers, and state (or provincial) or federal agencies, a company must be capable of determining immediately if a substantial hazard exists. Techniques for making this determination include: on-site investigation of the complaint, incident, or accident; hazard and failure analyses of the unit involved in the complaint, incident, or accident by the company or an independent laboratory; analyses of other units of the same product batch; evaluation of in-house test, or other, records.

If a company finds that a substantial product hazard does exist, the appropriate product recall or field-modification plan should be implemented.

### Independent testing

As an integral part of the PSM program, a company should either have adequate internal product testing facilities or have an affiliation with an independent testing laboratory that can provide whatever testing is required.

In certain instances, companies maintaining bona fide laboratory or adequate testing facilities should be encouraged to certify on their own authority and reputation that their products comply with existing standards, or in those instances where no standards exist, that their products comply with an appropriate standard that the company has developed.

## PSM PROGRAM AUDIT REPORT

A great deal of general information on PSM program evaluation has been provided in the preceding pages of this chapter. This information covers all the program activities that an auditor must consider. In addition, the program auditor will generate a sizable body of information pertaining to the program. This information must be assembled, organized, and presented to management in a clear and concise format.

For management's benefit during review, the audit report should have a summary of findings as its initial section. This summary section must clearly express the program auditor's opinion of the company's overall PSM program and whether it is necessary to prepare recommendations to correct deficiencies detected during the audit. If recommendations are made, they must be adequately supported in the body of the report and should be practical and simple to comply with, if possible. In other words, recommendations should attempt to use the company's already existing systems and procedures. Recommendations must be directed toward correcting specific problems. Management must be clearly advised as to the degree of importance of each recommendation and the possible consequences if the company fails to comply.

The program auditor should estimate a realistic time frame for recommendation compliance. This estimate is extremely useful in conjunction with the program auditor's evaluation of the degree of importance of each recommendation.

It must be clear to management how the program auditor has formulated an opinion about the company's PSM program. Otherwise, the value and impact of the program auditor's efforts to qualitatively evaluate and materially improve an existing PSM program or control system will be seriously mitigated.

## REFERENCES

American National Standards Institute, 1430 Broadway, New York, N.Y. 10018, "Catalog of American National Standards" (issued annually).

Baldwin, S. et al. *The Preparation of a Product Liability Case.* Boston, Mass.: Little, Brown and Company, 1981.

Canavan, M. *Product Liability for Supervisors and Managers.* Reston, Va.: Reston Publishing Company, 1981.

Eads, G., and Reuter, P. *Designing Safer Products—Corporate Responses to Product Liability Law and Regulation.* Santa Monica, Calif.: Rand Institute for Civil Justice, 1983.

Hammer, W. *Product Safety Management and Engineering.* Englewood Cliffs, N.J.: Prentice-Hall, Inc., 1980.

Kolb, J., and Ross, Steven. *Product Safety and Liability.* New York, N.Y.: McGraw-Hill Book Company, 1980.

National Bureau of Standards, Washington, D.C. 20234, *An Index of U.S. Voluntary Engineering Standards,* Spec. Pub. No. 329, Standards Information Service.

National Safety Council, 444 N. Michigan Ave., Chicago, Ill. 60611.

Industrial Safety Data Sheets (listing available).

*Product Safety Management Guidelines Manual,* 1987.

Product Safety Management Training Course, *Establishing and Auditing an Effective Product Safety Management Program.*

*Product Safety Reference Notes,* 1987.

"Product Safety Up-To-Date" (published bimonthly).

Product Liability Prevention Technical Committee. *Product Recall Planning Guide.* Milwaukee, Wisc.: American Society for Quality Control, 1981.

*Restatement of Torts, Second Series.* Philadelphia, Pa.: American Law Institute, 1965.

Schaden, R., and Heldman, V. *Product Design Liability.* New York, N.Y.: Practising Law Institute, 1982.

Seiden, R.M. *Product Safety Engineering for Managers: A Practical Handbook and Guide.* Englewood Cliffs, N.J.: Prentice-Hall, Inc., 1984.

Thorpe, J.F. *What Every Engineer Should Know About Product Liability.* New York, N.Y.: Marcel Dekker, Inc., 1979.

Underwriters Laboratories Inc., 333 Pfingsten Rd., Northbrook, Ill. 60062, *Standards for Safety Catalog.*

Weinstein, A.S. *Products Liability and the Reasonably Safe Product.* New York, N.Y.: John Wiley and Sons, 1978.

# 23 Motorized Equipment



Motorized equipment discussed in this chapter includes trucks, passenger cars, buses, motorcycles, and off-the-road equipment such as bulldozers, cranes, and road-graders. Powered industrial handtrucks and trucks are covered in the *Engineering and Technology* volume in chapters on Manual Handling and Material Storage, and Powered Industrial Trucks, respectively.

Safe vehicle operation is the result of planning and action, not chance. Often insufficient attention is paid to the problem of ensuring that vehicles are safely operated. The reason may be lack of awareness of the problem or the difficulties of organizing an adequate safety program and providing good supervision.

In about 85 or 90 percent of all motor vehicle accidents, unsafe acts of drivers can be identified as the cause; only 10 to 15 percent are due to mechanical failure of vehicles or improper maintenance of equipment. Contemporary vehicle accident prevention efforts focus attention on these two principal accident factors—driver failure and vehicle failure—because both can be controlled.

Experience has shown driver failure can be controlled by a carefully planned program of driver selection, training, and supervision, and vehicle failure can be reduced by systematic preventive maintenance. The unsupervised fleet also will have higher accident costs than the supervised one.

## COST OF VEHICLE ACCIDENTS

The total cost of a vehicle accident far exceeds the amount recovered from the insurance company. Accident control in a large motor transportation fleet is critical because increased insurance premiums mean reduced profit. Insurance premiums can fall or rise with the accident frequency and dollar losses sustained by a fleet.

As discussed in Chapter 7, Accident Investigation, Analysis, and Costs, the cost of insurance is only one of the costs resulting from an accident. There also are indirect (uninsured) costs. As with most work-related accidents, these may be several times the direct costs. Typical indirect costs include the following:

1. Salary paid and loss of service of the employee injured in an accident. The cost of losing a key employee at a critical time is incalculable.
2. Added workers' compensation costs resulting from a disabling injury.
3. Loss of the vehicle's commercial value while it is being repaired or replaced.
4. Cost of supervisory time spent in investigating, reporting, and in cleaning up after the accident.
5. Cost of repairing the company vehicle.
6. Cost of repairing or replacing other company property.
7. Poor customer and public relations resulting from a company vehicle having been involved in an accident.
8. Cost of replacing or retraining an injured employee.
9. Time lost by co-workers while discussing the nature of the accident and the extent of the injury.

In addition to appealing to the profit motive, the most telling argument for controlling accidents is the company's moral obligation to the employee and to the public. The employer through his authority to hire, supervise, discipline, and discharge employees exercises a high degree of control over their driving performance. Motor vehicle collisions are, in fact, wasteful errors traceable to poor management of the fleet operation.

Management can exercise that control by employing the simple and proven techniques of driver safety education and supervision.

## VEHICLE SAFETY PROGRAM

A vehicle safety program should provide for the following:

1. A definite safety policy, originated, supported, and emphatically enforced by management, with delegated authority
2. A safety director, full- or part-time, to advise management
3. A driver safety program, including driver selection procedure, driver training, and interest-sustaining activities. Proper supervision and instruction are mandatory for success.
4. An efficient system for accident investigation, reporting, and analysis; determination of appropriate corrective action; and followup procedures to help prevent recurrences.
5. Preventive maintenance procedures.

An example of a company safety policy statement is shown in Figure 23-1.

Policy Statement

The efficiency of any operation can be measured by its ability to control loss. Accidents resulting in personal injury and damage to property and equipment represent needless suffering and waste. Management's responsibility is to provide the safest conditions and equipment for all employees. The company policy on safety is:

1. The safety of the employee, the public, and the operation is paramount and every attempt will be made to reduce the possibility of accidental occurrence.
2. Safety will take precedence over expediency and shortcuts.
3. The company intends to comply with all safety laws and ordinances.

Every employee will be expected to demonstrate an attitude that reflects this policy as outlined in the company safety program.

**Figure 23-1.** This American Trucking Associations' recommended safety policy can be adapted to fit individual company needs. (Printed with permission from *Fleet Safety Programs, Policy and Procedures for the Safety Director.* American Trucking Associations, Department of Safety, Washington, D.C.)

### Responsibility

The first requirement of a good, driver safety program is that management, from the president to the immediate supervisor, must accept responsibility for safe operation of company vehicles. As in the case of other desirable qualities of job performance, management first must demand a safe operation, define standards of acceptable performance, and then organize means for the evaluation and correction of job performance to meet these standards.

Management must evolve and enforce the policy that practical accident prevention is a requirement of employment.

A safety professional is usually responsible for supervising the company's program for the safe operation of all automotive equipment. In a small industrial concern, this may be a part-time assignment. Some of the duties of the safety professional are to:

1. Advise management on accident prevention and safety.
2. Develop and promote safety activities and work injury prevention measures throughout the fleet.
3. Study and recommend fleet safety policy in relation to equipment and facilities, personnel selection and training, and other phases of fleet operation.
4. Evaluate driver performance.
5. Conduct and arrange for effective safety training.
6. Review accidents for determination of cause. Compile and distribute statistics on accident-cause analysis and experience. Identify problem persons, operations, and locations.
7. Maintain individual driver-safety records, and administer the Safe-Driver Award incentive program.
8. Procure or prepare and disseminate safety education material.

### Driver safety program

A driver-safety program for a fleet should include the following five basic accident prevention procedures.

1. Set up an in-service driver training program.
2. Discuss preventability of each accident with persons concerned.
3. Require immediate reporting of every accident.
4. Compute and publicize the fleet accident record.
5. Maintain an accident-record card for each driver.

The planning and administration of a safety program for motor transportation fleets are described in greater detail in the National Safety Council's *Motor Fleet Safety Manual,* see References.

An accident can be defined as:

> any incident in which the company vehicle comes in contact with another vehicle, person, object, or animal, which results in death, personal injury, or property damage, regardless of who was injured, what property was damaged or to what extent, where it occurred, or who was responsible. (Another definition appears in the Preface to this volume.)

This definition includes even minor accidents involving little more than a fender scratch. All vehicle accidents, major and minor, are of importance to the safety supervisor since he is primarily concerned with the eradication of faulty driving habits or attitudes. Minor accidents, just as must as the more spectacular ones, provide clues to such faults.

### Accident reporting procedure

Each driver should be required to make out a complete accident report on a standard accident report form for every accident in which his or her vehicle is involved. If possible, this report should be turned in to the supervisor on the day the accident occurs. National Safety Council Form Vehicle 1 is a good example of an accident report form. (See the Council's *Motor Fleet Safety Manual* for illustrations of forms.)

Drivers should be told how to accurately and intelligently fill out the accident report form. Failure to report an accident, no matter how slight, or falsification of data on an accident report should be made cause for disciplinary action against the driver.

Drivers should be required to complete the report at the scene of the accident, if possible, and then promptly send it to their supervisor.

The failure to report an accident can be controlled by requiring a vehicle inspection before and after each trip, and noting any changes in the condition of the vehicle.

An accident report packet that can be carried in a glove compartment is useful. This packet (Figure 23-2) should provide for a memorandum report of the accident such as National Safety Council Form Vehicle 2. In addition to serving as a checklist, it should contain a pencil, plain paper, courtesy cards, and telephone numbers of company officials and insurance representatives. The necessary note-taking material thus will be available at the scene of an accident.

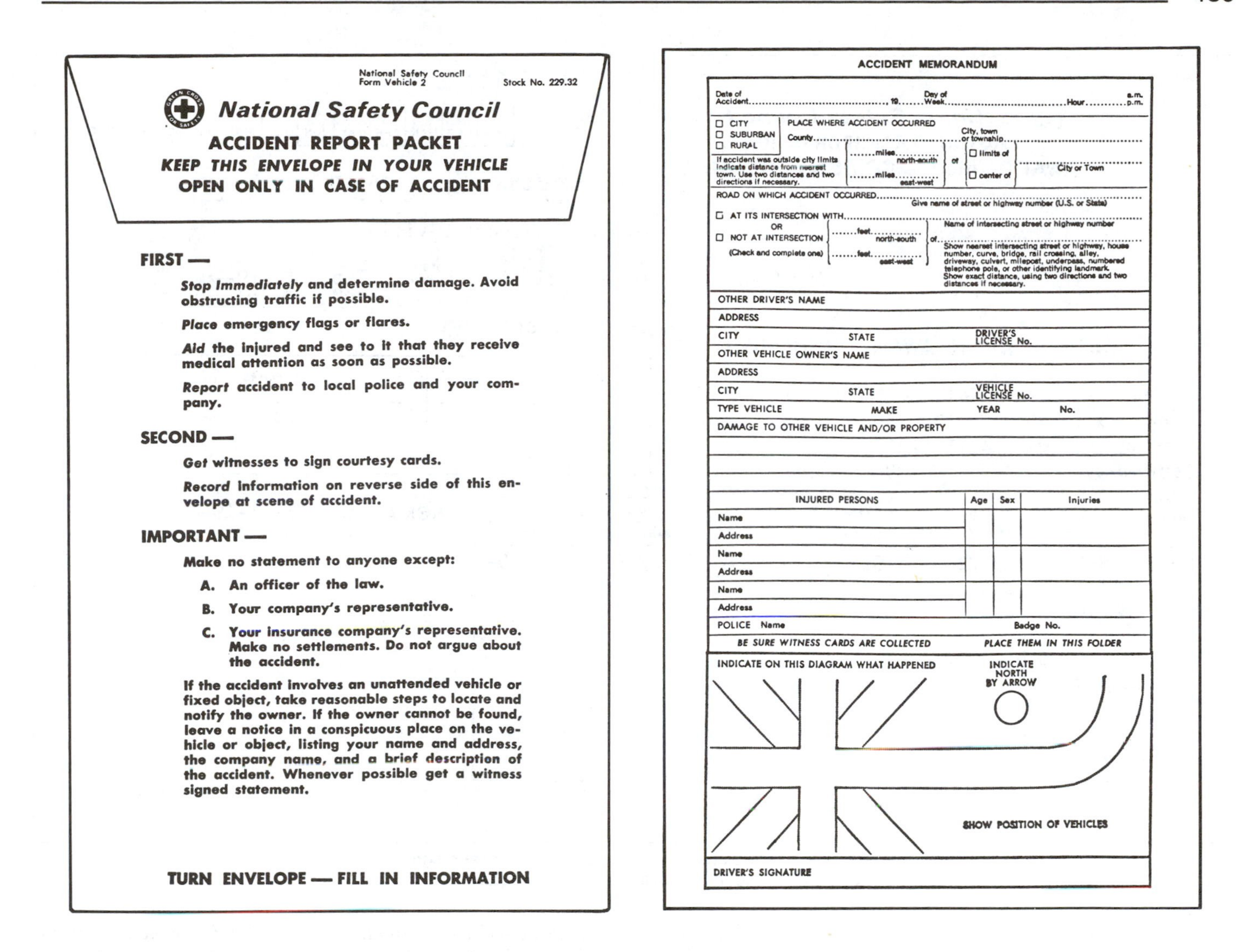

National Safety Council
Form Vehicle 2
Stock No. 229.32

**National Safety Council**

**ACCIDENT REPORT PACKET**

**KEEP THIS ENVELOPE IN YOUR VEHICLE**

**OPEN ONLY IN CASE OF ACCIDENT**

**FIRST —**

*Stop Immediately* and determine damage. Avoid obstructing traffic if possible.

*Place* emergency flags or flares.

*Aid* the injured and see to it that they receive medical attention as soon as possible.

*Report* accident to local police and your company.

**SECOND —**

*Get* witnesses to sign courtesy cards.

*Record* information on reverse side of this envelope at scene of accident.

**IMPORTANT —**

Make no statement to anyone except:

A. An officer of the law.

B. Your company's representative.

C. Your insurance company's representative. Make no settlements. Do not argue about the accident.

If the accident involves an unattended vehicle or fixed object, take reasonable steps to locate and notify the owner. If the owner cannot be found, leave a notice in a conspicuous place on the vehicle or object, listing your name and address, the company name, and a brief description of the accident. Whenever possible get a witness signed statement.

**TURN ENVELOPE — FILL IN INFORMATION**

**ACCIDENT MEMORANDUM**

Date of Accident ........................ 19...... Day of Week ........................ Hour .......... a.m. p.m.

☐ CITY ☐ SUBURBAN ☐ RURAL — PLACE WHERE ACCIDENT OCCURRED — County ........................ City, town or township ........................

If accident was outside city limits indicate distance from nearest town. Use two distances and two directions if necessary. .......miles....... north-south .......miles....... east-west of ☐ limits of ☐ center of ........ City or Town

ROAD ON WHICH ACCIDENT OCCURRED ........................ Give name of street or highway number (U.S. or State)

☐ AT ITS INTERSECTION WITH ........................ Name of intersecting street or highway number

OR

☐ NOT AT INTERSECTION (Check and complete one) .......feet....... north-south .......feet....... east-west of ........................ Show nearest intersecting street or highway, house number, curve, bridge, rail crossing, alley, driveway, culvert, milepost, underpass, numbered telephone pole, or other identifying landmark. Show exact distance, using two directions and two distances if necessary.

OTHER DRIVER'S NAME

ADDRESS

CITY STATE DRIVER'S LICENSE No.

OTHER VEHICLE OWNER'S NAME

ADDRESS

CITY STATE VEHICLE LICENSE No.

TYPE VEHICLE MAKE YEAR No.

DAMAGE TO OTHER VEHICLE AND/OR PROPERTY

| INJURED PERSONS | Age | Sex | Injuries |
|---|---|---|---|
| Name | | | |
| Address | | | |
| Name | | | |
| Address | | | |
| Name | | | |
| Address | | | |

POLICE Name Badge No.

*BE SURE WITNESS CARDS ARE COLLECTED* *PLACE THEM IN THIS FOLDER*

INDICATE ON THIS DIAGRAM WHAT HAPPENED

INDICATE NORTH BY ARROW

SHOW POSITION OF VEHICLES

DRIVER'S SIGNATURE

**Figure 23-2.** Accident Report Packet is available as NSC Form Vehicle 2. At left is the front side of an envelope that has almost everything a person needs to complete an accurate report and obtain the identification of witnesses.

Specially printed courtesy cards will save time and will help the driver get names and addresses of key witnesses (Figure 23-3). Drivers should be impressed with the importance of identifying as many witnesses as possible.

In the case of serious accidents, especially those resulting in a fatality or personal injury, a representative of the company—the manager, safety director, or claim agent—should make a personal investigation at the scene of the accident as quickly as possible.

The purpose of such an investigation is to verify the accuracy of information submitted by the driver on the accident report form and to obtain other data which might prove valuable for accident prevention work or for defense against unjust claims.

WILL YOU KINDLY FAVOR THE DRIVER BY FILLING IN THIS CARD?

ACCIDENT AT ________________________ (show street no. or intersection)

DATE ____________ Time ____________ ☐ A.M. ☐ P.M.

DID YOU SEE THE ACCIDENT HAPPEN? ☐ YES ☐ NO

DID YOU SEE ANYONE HURT? ☐ YES ☐ NO

WERE YOU RIDING IN A VEHICLE INVOLVED? ☐ YES ☐ NO

IN YOUR OPINION WHO WAS RESPONSIBLE?

☐ OUR DRIVER ☐ OTHER DRIVER ☐ PASSENGER ☐ PEDESTRIAN

NAME ________________________

ADDRESS ________________________

CITY & STATE ____________ TELEPHONE ____________

**National Safety Council** THANK YOU

Rep. 20M 86003 Stock No. 229.38
Form Vehicle 12

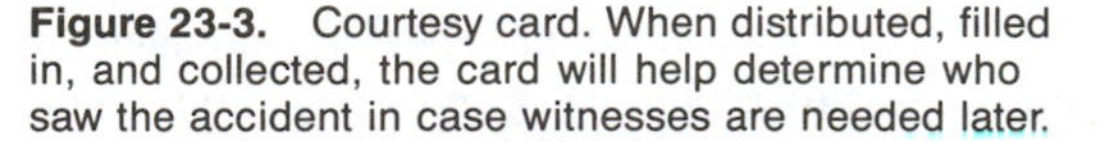

**Figure 23-3.** Courtesy card. When distributed, filled in, and collected, the card will help determine who saw the accident in case witnesses are needed later.

## Corrective Interviewing

As soon as possible after an accident, the company driver involved should be interviewed by the safety professional to determine whether or not the driver might have prevented it. If, after reading the report and discussing the matter with the driver, the safety professional feels that the driver could have helped prevent the accident, it should be classified preventable. In this case, the safety professional must explain to the driver what actions contributed to the accident and make sure that the driver understands what to do to prevent similar accidents in the future.

**National Safety Council**
MOTOR TRANSPORTATION
AWARD AND ACCIDENT RECORD

| JAN | FEB | MAR | APR | MAY | JUN | JUL | AUG | SEP | OCT | NOV | DEC |
|---|---|---|---|---|---|---|---|---|---|---|---|
| | | | X | | | | | | | | |

Company: Highway Express

Address: 123 (Number) Home St., (Street) Any Town

Name: Smith, (Last) Bill (First) E. (Middle)

Location: Midwest Division

Badge Number: 123 Date Employed: Feb. (Month) 1 (Day) 19-- (Year) Age 29 (At Emp.)

SAFE DRIVER AWARD RECORD

Earned 2 Year N. S. C. Award during period from 3-2-19-- to 3-2-19-- with Company Overnight Freight, Inc. Certified by F. J. O'Connell, Supt.

DRIVING TESTS

| Date | Score | Remarks |
|---|---|---|
| 2-4-19-- | 90 | Slow reaction time |
| | | |

| Award earned | Certificate Number | Date Award earned | Preventable Accidents | | Non-Driving Time | |
|---|---|---|---|---|---|---|
| | | | File No | Date | From | To |
| 3 | 468950 | 3/2/19-- | 1016 | 4/10.19-- | 12/1/19-- | 1/1/19-- |
| 4 | 501850 | 3/2/19-- | | | | |
| 5 | 600100 | 4/2/19-- | | | | |

**REMARKS:** 12-1-19-- took on month leave of absence.

Form Vehicle 6 PRINTED IN U.S.A. STOCK No. 229.36

**Figure 23-4.** Award and accident record should be kept as part of each driver's personnel record. Form is useful in administering a company award plan and in counseling accident repeaters.

If, in the safety professional's opionion, the driver did everything possible to prevent the accident, it should be classified nonpreventable.

Responsibility for the prevention of accidents includes more than careful observance of traffic rules and regulations. *Drivers must drive to prevent accidents,* regardless of faulty driving or nonobservance of traffic laws on the part of other drivers.

### Driver record cards

A record card should be maintained for each employee who drives a company vehicle. This card furnishes not only a record of accidents, but also the information needed for safe driver award plans or other forms of recognition (Figure 23-4). The date of each accident and the accident category, preventable or nonpreventable, should be entered on this card. The safety professional should review the cards at least every six months and make note of those drivers having had accidents.

When a driver becomes an accident repeater, management should make every effort to rehabilitate him through counseling, retraining, closer supervision, or reassignment.

When all efforts fail to curb preventable accidents, discharge or assignment to nondriving duties will be in the best interests of both the firm and the employee.

### Fleet accident frequency

A useful accident control tool is monthly or quarterly computing the fleet's accident frequency rate per 1,000,000 vehicle miles. Vehicle miles should be computed from odometer readings of all vehicles and not left to rough guesses based on route mileages unless the operations of the fleet are stable from day to day. The standard formula for figuring a fleet accident rate is:

$$\text{Fleet accident frequency rate} = \frac{\text{No. of accidents} \times 1{,}000{,}000}{\text{miles driven}}$$

By keeping a monthly record of frequency rates, the safety professional can:

1. Analyze changes in group safety performance
2. Compare records from several years to find seasonal, or other, trends
3. Compare the fleet's performance with similar fleets.

The fleet safety program can be planned based on an analysis of the accident frequency rate and trends. Accident frequency rates also are of interest to management since they can prove the safety program is effective.

The National Fleet Safety Contest, conducted annually by the National Safety Council, issues a quarterly bulletin giving the accident frequency rates of all participating fleets, thus providing a means of comparison between similar companies.

### Selection of drivers

For some jobs requiring vehicle driving, such as sales representative or technician, other qualifications unfortunately can outweigh a safe driving record. Many times the prospective employee's competence as a driver is investigated in an incomplete fashion, because the interviewer does not regard driving as an important element of the job.

However, when a person is to be hired for a job in which driving is a regular or even occasional function, every effort must be made to select an individual who can be expected to drive safely. In making that selection, the following factors should be considered.

**Experience.** Individuals who have a record of frequent involvement in vehicle accidents should not be assigned to drive company vehicles. An individual's safety record should be investigated (1) in a personal interview, (2) by consulting with former employers, and (3) by checking the state motor vehicle department for accident reports, and (through them) the National Driver Register service for out-of-state or two license revocations.

There are a number of private services that can provide quick access to motor vehicle records for many states.

**Attitude.** Dissatisfied, timid, cocky, troublesome, or otherwise temperamental or unstable individuals often do not make good drivers.

**Personal traits.** A close relationship has been noted between the ability to drive safely and such personal traits as dependability, judgment, courtesy, pleasant personality, and the ability to get along harmoniously with other people.

Conversely, persons who tend to be antisocial, argumentative, and impulsive are suspect as drivers.

The Federal Aviation Agency reported the following undesirable personality traits in *Pilot Judgment Training and Evaluation:*

Anti-authority—"Don't tell me."
Impulsivity—"Do *something*—quickly."
Invulnerability—"It won't happen to me."
Macho—"*I* can do it."
Resignation—"What's the use."

Individuals differ in their ability to act safely. The fleet safety program therefore must begin at the employment office.

In determining standards of selection, a careful analysis of the job and of driver qualifications should be made. Results of the job analysis are embodied in a job description in which the separate tasks involved in the job are completely and accurately described.

After the job has been analyzed, the next step is to decide what qualifications the applicant must have to satisfactorily perform the job. There should be a sound reason for each qualification imposed. It may be helpful to study the employees who are performing the job in an average or better than average manner. Their qualifications should indicate the requirements expected in new employees. Interstate carriers must comply with driver qualification regulations of the U.S. Department of Transportation.

Safe driving always should be paramount. Otherwise, accident losses might completely offset any advantages gained through a driver's other special abilities.

### Information-gathering techniques

After job essentials and qualifications have been determined, the next step is to develop methods for gathering and sifting employment data about each applicant. Standard employment procedure includes:

Application form
Personal references
Interview
Psychological tests
Driving tests
Physical examination

Illustrations of the forms and more descriptive detail will be found in the Council's *Motor Fleet Safety Manual;* see References.

**The application form** is a printed or duplicated form on which the applicant submits details of past employment and other personal data. What a person has done in the past is a good indication of what he can be expected to do in the future. The completed form also saves time during the interview, since essential data are made available to the interviewer at a glance.

Because of antidiscrimination laws and possible lawsuits, personal information should not be requested on the written application or during the interview, unless it can be proven the answers are necessary to establish occupational qualification.

Questions on the form should cover the basic qualifications for the job, and should be arranged in logical sequence. Specifically, questions about driving experience should include mileage and years spent as a driver and types of vehicles operated, seasons of the year and geographical areas in which vehicles were operated, preventable and nonpreventable accidents experienced, number of convictions for traffic and other violations, number and type of driver's licenses held, and safe driver awards received.

**Personal references.** The applicant should furnish the names and addresses of previous employers in the space provided on the application form, and these references should be checked. Additional reference sources are the police department and state motor vehicle department, as mentioned before.

To guard against any legal liability resulting from personnel investigation activities, all applicants should first sign a release that contains the statement:

> I hereby release the __________ company or any agency it may designate or any persons the __________ company or agency may contact in the course of its investigation, from any liability which may result from the conduct of such investigation or from the result of the investigation.

The results of these reference checks are so important that it is very risky to start an applicant on the job before they are received. To be practical, the interviewer will usually make telephone checks, and rely on a letter of personal reference only as a last resort. Applicants cannot be kept waiting until a written check comes back; but putting the driver to work unchecked is foolhardy and can lead to expensive accidents.

**A properly conducted interview** should reveal additional facts about the applicant's employment experience, knowledge of traffic regulations, attitude, personality, appearance, family life, and general background.

The interview should be conducted in private, and the applicant should be seated and put completely at ease. The interviewer should keep in mind at all times the inventory of basic qualifications. A checklist of these may be made up to serve as a guide. After the interview, the applicant can be rated on each of the qualifications listed.

**Psychological tests** are devices for obtaining samples of behavior under controlled conditions. They can be useful in the selection process if the behavior traits sampled are known to be related to job success and if interpretation of the results is made by a qualified person. However, since tests cannot predict the future, they should not be the only basis for selection. The personnel officer never should use tests as a substitute for other ways of securing employment information.

Many of the large trade associations have had personnel specialists develop standardized personnel selection procedures for their member organizations. These procedures usually include psychological tests that have been found applicable to driver selection.

Small fleets that are not members of an association providing this service should be able to retain, for a reasonable fee, a qualified personnel psychologist for advice on this matter. The psychology departments of some state universities give valuable advice on this problem through their extension services.

**Driving tests.** Each applicant should undergo an actual driving test or a perceptual-motor test as part of selection procedure. Some firms rely on a perceptual-motor test for indication of driving ability and on an extensive driver training course following employment for developing that ability.

Driving tests are of two kinds, the driving range type and the in-traffic type. The requirements of a good driving test in traffic can be listed as follows:

1. The road test should be long enough to fairly sample a number of typical driving situations. Certainly 20 minutes should be considered minimum.
2. The test should include typical maneuvers in heavy traffic as well as on freeways to really test a driver's ability. Almost anyone can successfully drive around the block.
3. A standard scoring procedure and predetermined test route should be used so the test will be the same for all drivers examined.
4. The examiner should check definite items concerning the driver's performance (in order to reduce subjective judgment) and point out driving faults that may be corrected by proper training.

**Driver performance measurement test.** Michigan State University has developed the Driver Performance Measurement Test. This test can help assess safe driving ability during selection by pointing out the precise unsafe habits likely to lead to later accidents. Information can be obtained from Michigan State University, Lifelong Education Highway Traffic Safety Programs, 70 Kellogg Center, East Lansing, Mich. 48824. The American Trucking Associations, 2200 Mill Road, Alexandria, Va. 22314, can supply information on driving range test requirements.

The system requires the preparation of a thoroughly calibrated test course on public roads. As many as 50 traffic situations may be charted and correct-incorrect driving procedures determined for each. After the course is validated, it can be used by all nearby companies to test drivers, once individual company observers have been trained and certified.

While the driving range type test course obviously requires a lot of organization and test time, it does give a standardized measure of driving performance. Any company hiring large numbers of drivers should investigate this test.

Some companies maintain a truck "roadeo" type of driving course, where pure maneuvering skill can be measured. This may be important to measure in a city driver who, for instance, may be backing into tight alleys and loading docks from his first day on the job.

Skill course diagrams, tests, and obstacle specifications can be found in the *Driver Trainer Manual,* available from the Private Truck Council of America, Inc., 2022 P Street, NW., Washington, D.C. 20036.

In some trucking operations, job applicants may be tested for skill in nondriving tasks. Experienced tank truck driver applicants may be asked to hook up hoses to pumps. A driver claiming experience with doubles trucks may be required to hook up a set of trailers, thus showing his knowledge of procedures as well as his driving skill.

The applicant's driving tests performance will indicate if driver training is needed, and will indicate weaknesses that can be corrected during the training period. Tests for interstate drivers must meet the requirements of the U.S. Dept. of Transportation.

The U.S. Department of Transportation (DOT), National Highway Traffic Safety Administration, has developed a series of tests for tractor-trailer drivers based on the DOT Model Curriculum for Training Tractor-Trailer Drivers. This system of tests is known as TORQUE; the specific tests making up TORQUE are as follows:

*Tractor-Trailer Operator Road Test (TORT).* A road test designed to provide an objective and valid means of assessing the skills needed to operate tractor-trailers in the highway traffic environment.

*Tractor-Trailer Operator Skill Test (TOST).* A test of skill in operation of tractor-trailers. TOST is designed for administration in an off-street environment or practice driving range.

*Tractor-Trailer Operator Knowledge Examination (TOKE).* This multiple-choice, written test samples critical tractor-trailer knowledge including those making up the content of TOM.

*Tractor-Trailer Operator Manual (TOM).* This manual communicates the information and develops the knowledge to operate tractor-trailers on public streets and highways.

**Physical examination.** Applicants should be examined by a qualified physician before being placed. For firms engaged in interstate commerce, physical examination of all new drivers is mandatory and drivers must be reexamined every two years. The regulations prescribed by the U.S. Department of Transportation for bus and truck operators coming under its jurisdiction (see References) require physical examinations for all drivers and provide that motor carriers must have on file a certificate showing each driver to be physically qualified.

Most drivers are, in addition, given a series of psycho-physical tests of vision, depth perception, and hearing. Substandard findings are submitted to competent medical authority for evaluation. Drivers are told of their weaknesses and how best to compensate for them.

**Acceptance interview.** All applicants who pass the various requirements should be called in and either told that they are hired or being placed on a waiting list. This acceptance interview should reinforce applicants' enthusiasm for the job. It is important applicants be made to feel that, although company employment standards are rigorous, they have attributes the company sincerely wants. They should be welcomed to the company as valued new employees. They should be given further orientation about the company and told what training they will receive, and when and where to report. Wages and working conditions should be finalized at this time.

### Legal and social restrictions

The motor transportation fleet is a constructive social and economic force in the community in which it operates not only because of the transportation service it renders, but also because of the jobs it provides. Placing a person on the payroll, therefore, has social and economic significance not only to the one who is hired, but also to the fleet and to the community.

In this context, the employer today must abide by certain hiring restrictions. Laws throughout our nation forbid discriminating against job applicants. Society often takes the view that the employer who wishes to use the local labor market must do so with due regard for every applicant's right to fair consideration for the job.

These are not unreasonable restrictions. They certainly should not—in any way—force the employer to hire the unqualified. It is no favor to an applicant to be hired for a position for which he or she is not qualified and risks accidents, possible injuries, and an unfavorable work record. The hiring officer has a moral obligation to the applicant, to the company, and to the community to choose the best-qualified and safest drivers. The hiring officer knows the demands of the job and the qualifications required of the applicant. The hiring decision should always represent a choice that is beneficial to the applicant and to the fleet.

### Driver training

Individual training by a skilled instructor for all employees assigned to drive company motor vehicles is a highly desirable objective. Various types of courses can be given.

BASIC, for new employees
REMEDIAL, for drivers who get into trouble
REFRESHER, for periodic updating of all drivers
SPECIAL, for operators of specialized equipment

A training course must be planned to fit each job, training materials assembled, and classroom facilities provided. The objective, however, is well worth the investment. The driver training course should cover the following points.

- **State and municipal driving rules.** Most state (or provincial) motor vehicle departments publish the rules and regulations of the road for those seeking drivers' licenses. The training course should cover the salient points found in such booklets.
- **Company driving rules.** Each company has rules governing the use of company vehicles: how the vehicles may be obtained, where they may be operated, where parked and under what conditions, speed to be observed, and so on. These rules should thoroughly be covered in the training course.
- **What to do in case of an accident.** This topic should include instructions on how to make out company accident report forms, how individual driver records affect the employee, what to do in case of a vehicle accident away from the plant, and similar points.
- **Defensive driving.** This concept embraces all the common-sense rules of safe and courteous driving. Defensive driving instruction seeks to build in the prospective driver a high sense of responsibility not only for the safety of his own vehicle, but for the safety of other street and highway users who are less skilled and who have had less training and practice (Figure 23-5). At this time, the company's safe driver award or incentive plan should be explained. The National Safety Council conducts defensive driving courses for both truck and automobile drivers.

Figure 23-5. Training session in National Safety Council "Defensive Driving Course" uses many visuals.

### Safe driving incentives

One of the basic assumptions of a safety program is that most drivers believe that they know how to drive much better than they really do. Proceeding on this assumption, then, most safety professionals regard it as part of their job to provide effective motivation in various ways so drivers will use more of their driving skill more of the time. To supply this motivation directly or indirectly, the safety professional should:

1. Require a detailed report of every accident
2. Interview the driver after each accident to determine whether it could have been prevented
3. Keep a record of each driver's safety performance
4. Provide continuous safety instruction and reminders. Use all media: company newsletters and bulletins, booklets, posters and bulletin board displays, and meetings and direct personal conversation.
5. Recognize safe driving performance. Safe driver awards and cash or merchandise prizes for driving for stated periods without a preventable accident are strong motivations. However, the safety director should recognize that a recognition program *supports* a safety program, but is not its substitute.

### Safety devices

Many accidents involving motor vehicles are due to lack of safety devices and particularly to inadequate maintenance. Therefore, fundamental requirements for safe operation are that all company vehicles be equipped with the necessary safety devices and that vital parts, such as tires, brakes and steering mechanisms, and headlights, taillights, and horn be maintained in first-class condition (see Figure 23-6).

Safety devices include the following:

**DRIVER'S VEHICLE INSPECTION REPORT**

AS REQUIRED BY THE D.O.T. FEDERAL MOTOR CARRIER SAFETY REGULATIONS

CARRIER: ______________________

DATE ____________ TIME: ____________ A.M. ____________ P.M.

Check any defective item and give details under "Remarks"

TRACTOR NO. ______________

| | | |
|---|---|---|
| ☐ Air Compressor | ☐ Front Axle | ☐ Springs |
| ☐ Battery | ☐ Fuel Tanks | ☐ Steering |
| ☐ Body | ☐ Heater | ☐ Tachograph |
| ☐ Brake Accessories | ☐ Horn | ☐ Tires |
| ☐ Brakes | ☐ Lights | ☐ Transmission |
| ☐ Clutch | ☐ Loses Water | ☐ Wheels |
| ☐ Defroster | ☐ Mirrors | ☐ Windows |
| ☐ Door Handles | ☐ Oil Pressure | ☐ Windshield Wiper |
| ☐ Drive Line | ☐ Radiator | |
| ☐ Engine | ☐ Rear End | ☐ OTHER |
| ☐ Fifth Wheel | ☐ Safety Equipment | |

TRAILER(s) NO.(s) ______________

| | |
|---|---|
| ☐ Brake Connections | ☐ Roof |
| ☐ Brakes | ☐ Springs |
| ☐ Coupling Chains | ☐ Tarpaulin |
| ☐ Coupling (King) Pin | ☐ Tires |
| ☐ Doors | ☐ Wheels |
| ☐ Hitch | |
| ☐ Landing Gear | ☐ OTHER |
| ☐ Lights | |

☐ CONDITION OF THE ABOVE VEHICLES IS SATISFACTORY

REMARKS: ______________________

Driver's Signature ______________

☐ Above defects corrected ☐ Above defects need not be corrected for safe operation of vehicle

Mechanic's Signature ______________

Driver reviewing repairs ______________ (Signature)

**Figure 23-6.** Although maintenance personnel are responsible for giving the driver a vehicle that is in top mechanical condition, it is the driver who must assure himself at the start of each day that the vehicle is in good condition. Use of a checklist makes sure that no point is forgotten. (Courtesy of J. J. Keller & Associates, Inc., Neenah, Wis. Used with permission.)

Directional signals
Windshield wipers
Windshield defroster
Fire extinguisher
Power steering
Low air-pressure warning system
Rock guards over the drive tires
Adequate outside mirrors
Backup light
Audible backup signal for heavy-duty trucks
Nonslip surfacing on fenders, floors, and steps
Safety belts
High-quality tires
Automatic sander
Anti-jackknife device
Reflective markings

In addition, the following devices are recommended for dump trucks:
A light or indicator to show when the body is in a raised position
A Caution sign on the rear of packer-loader trucks
Cab protector or canopy
A built-in body prop

**Loading and unloading of trucks.** To reduce the danger to the driver from falling material while the truck is being loaded, the truck should be spotted so the load does not swing over the cab or seat. (See Figure 23-7.) If a truck cannot be so located and does not have a protective canopy over the cab, the driver should dismount and stand clear of the truck.

Certain hazardous materials and cargo require placarding and special precautions especially when loading and unloading. The DOT and EPA regulations must be followed where applicable.

Accidental injuries incurred in the loading and unloading of materials, such as lumber, pipe, equipment, and supplies, are especially numerous, but can be avoided if these precautions are followed:

1. The bulk and weight capacity of the truck should be observed.
2. Loads that may shift should be blocked or lashed. Tiedowns (ropes, chain, boomers) should be tightened on the right side or top of the load.
3. If material extends beyond the end of the tailgate, a red flag (or, at night, a red lamp) should be fastened to the end of the material. No material should extend over the sides.
4. Before loading or unloading a truck, the brakes must be securely set or the wheels blocked to protect the workers both on the truck and on the ground.
5. A truck should not be moved until all workers are either off the truck or properly seated on seats provided and are protected from injury if the load should shift during transit.
6. To avoid falling when unloading a flatbed truck, employees should keep away from the sides of the truck, especially when shoes, floors, and loads are wet or muddy.
7. Be alert for pinch points when loads are being pulled, hauled, or lifted.
8. All safe practices for material handling, such as using mechanical handling equipment, getting sufficient help, and so on, should be observed.
9. Specialized training depending on the classification of the cargo (for example, flammable, corrosive, radioactive) should be included.

**Detached trailers.** When loading and unloading detached trailers with a lift truck, be sure the wheels are adequately blocked.

It is important to place the chock properly when trailers are at the dock being loaded and unloaded. Preferred location of chocks is under the rear set of wheels. If a trailer is not properly blocked, the vehicle may move because of an incline or be set in motion by the loading or unloading operation.

The trailer nose can be supported by screw or hydraulic jacks—one on each side of the nose—in order to strengthen the support of the landing gear assembly. Under a heavy forklift load, landing gears have collapsed from the weight because of dolly metal rust or fatigue, defective struts, or some other cause.

### Preventive maintenance

Well-managed motorized equipment, both highway and off-the-

**Figure 23-7.** The truck and loader are properly positioned for safe loading of materials. Loader is approaching the driver's side of the truck, not the blind side, and the driver is in the cab of the truck, not on the ground next to it. (Note the protective canopy over the cab.) The bucket is well positioned to drop the load in the center of the truck bed, reducing spillage. The loader operator has a full view of the truck, the bucket, and the material. Note the excellent housekeeping—loose material is leveled around the entire loading operation and the haul road is well maintained. (Courtesy National Safety Council *Construction Newsletter.*)

road, is covered by an extensive and more-or-less complicated preventive maintenance program, the primary considerations of which are safety, economy, and efficiency. Such a program, based on either the mileage or the operating hours of the equipment, as recommended by the manufacturer, determines when oil will be changed, tires rotated or replaced, and minor and major overhaul jobs undertaken.

The objectives of such a program are:

1. To prevent accidents and delays
2. To minimize the number of vehicles down for repair
3. To stabilize the work load of the maintenance department
4. To save money by preventing excessive wear and breakdown of equipment.

Such a program should, as a matter of course, cover all mechanical factors relating to safe operation of all motorized equipment, such as brakes, headlights, rear and stop lights, turn signals, tires, windshield wipers, muffler and exhaust system, steering mechanism, glass, horn, and rearview mirrors. (These are discussed more fully a little later in this chapter.) The manufacturer of the equipment can help with the maintenance specifications.

If at all possible, each driver or operator should be assigned a specific vehicle and be given responsiblity for reporting defects. This encourages drivers and operators to take better care of their vehicles.

Drivers and operators can play an important role in a preventive maintenance program if they are properly instructed and motivated. Because they are most familiar with the vehicle and how it normally operates, they are usually the first to notice when minor, as well as major, mechanical defects develop.

Drivers in interstate commerce are required by "DOT Motor Carrier Safety Regulations" to perform a pre-trip vehicle inspection. It is recommended that drivers of all types of vehicles thoroughly inspect their vehicles before beginning the workday. Any defects must be reported and corrected *before* that vehicle is used (see Figures 23-8 and -9).

Vehicle Condition Report (VCR) forms should be furnished all drivers and operators for the inspection. A copy of the current VCR should be kept in the vehicle cab. The VCR acts as a reminder as well as a checklist, and should normally cover the following items:

BRAKES. Brakes should apply evenly to all wheels so a vehicle does not swerve when they are applied. This even application also gives maximum braking effectiveness.

HEADLIGHTS should function and be properly aimed to avoid blinding other motorists and to give maximum road lighting efficiency. The dimming switch and upper and lower beams should work properly.

CONNECTING CABLES on a combination vehicle should have connections strong enough not to be affected by vehicle vibration. All other cables, such as brake and electrical, should be free of defects.

STOP LIGHTS, turn lights, rear lights, and side-marker lights should work.

TIRES should be inflated to manufacturer's recommended pressure, and regularly checked for adequacy of tread and for cuts or breaks. Dual tires should be well matched.

WINDSHIELD WIPERS must wipe clean and not streak.

STEERING WHEEL should be free from excessive play. Front wheels should be properly aligned.

GLASS should be free from cracks, discoloration, dirt, or unauthorized stickers that might obscure vision.

HORN should respond to a light touch.

REARVIEW MIRRORS should give the driver a clear view. So outside rearview mirrors can provide maximum sight advantage, portions can be conventional and convex.

STALLING PROBLEMS should be investigated and corrected immediately.

INSTRUMENTS should be in good working order; they are essential to safe and economical operation.

EXHAUST SYSTEM should be checked to protect against carbon monoxide gas leaks. The exhaust manifold, pipe connections, and muffler should periodically be inspected, and leaky gaskets replaced.

EMERGENCY EQUIPMENT in every vehicle should include a fire extinguisher, essential tools for road repairs, spare bulbs, flares, reflectors, flags, and such other equipment deemed necessary in case of fire, accident, or road breakdown. These items should periodically be checked to make sure of their availability and usability. Interstate vehicles must be equipped with emergency items required by the U.S. Department of Transportation.

Many states and cities require periodic safety tests and inspections for all vehicles. The maintenance superintendent should know the applicable inspection standards. The preventive maintenance policy of the company should require that all vehicles meet these requirements.

## REPAIR SHOP SAFE PRACTICES

The vehicle safety supervisor also should take an active interest in the work habits of automotive repair shop employees to determine their safety attitudes, and should cooperate with the plant safety supervisor. An effective program must make sure that all employees engage in safe work practices. Also, that all federal and state (provincial) regulations are followed.

### Servicing and maintaining equipment

Serious injuries occur in servicing and maintaining trucks. Heavy equipment requires mechanical aids for handling heavy parts. Hoists for lifting parts in and out of trucks and for moving parts about the shop not only prevent accidents, but also make work easier and save time. See the discussion in Chapter 3, Manual Handling and Material Storage, in the *Engineering and Technology* volume.

Serious injuries are likely to occur from unexpected movement of equipment undergoing repair. Brakes should be set and wheels should be blocked. If work must be done under a raised body, the body must be secured or blocked against coming down in case the hoist or jack control levers or pedals are inadvertently struck and the load released.

A jack often is used to raise equipment, and then it becomes a support in an unstable position. Because serious injuries occur when a truck falls off a jack, it is important the jack be set on a firm foundation and be exactly perpendicular to the load. To help prevent the jack from slipping, a thin wood block between the top of the jack and the load is recommended. When the truck has been raised to the desired height, it should be supported by stanchions, blocking, or other secure support.

No work should be done near the engine fan or other exposed moving parts until the engine has been stopped. If the engine must be run to inspect or check on moving parts, keep a safe distance away and do not attempt an adjustment. Jewelry, especially rings, should not be worn.

Close-fitting unfrayed clothing, safety shoes, and goggles are essential for repair people.

Burn injuries happen frequently when servicing trucks. To check or fill a radiator, the employee should use a heavy work glove, bleed-off any steam, and then remove the cap. Gasoline or alcohol used near hot engines and spilled on them can cause a serious fire. Suitable funnels and safety containers should be used.

### Tire operations

A particularly serious hazard in inflating truck tires is the possibility that the locking ring may blow off at a high pressure. Use of a tire safety rack will greatly reduce the hazard. All truck tires should be checked to make sure that the valve has been removed and the tire fully deflated prior to disassembly. When removing a tire from a dual wheel, fully deflate both tires of the pair before removing the tire to be repaired. Tires must be inflated in steel cages that will restrain flying objects should a blowout occur. A locking ring must be seated properly and must not be yanked free by being twisted. A defective locking ring or rim should be replaced with a sound one. Ring and rim seats should be clean. Parts, rims, and rings for various types of wheels should be carefully segregated to eliminate the possibility of a mismatch.

It is advisable to use inflators that can be preset and have locking attachments so the worker does not have his hand or arm in the danger area, even if the tire is in a cage.

Blowouts can occur because of overinflation of the tire, improper placement of the tire on the rim or wheel (causing

**Figure 23-8.** Here's what a scraper looked like after it rolled down a canal berm (*left*). Operator was belted in and was only slightly hurt; if he had tried to jump, he would have been thrown into the path of the rolling scraper. The close-up photo (*above*) shows that ROPS (rollover protective structures) also protected the windshield and operator's area from damage as well as probably saving the life of the operator. (Photographs courtesy of U.S. Dept. of Interior, Bureau of Reclamation.)

pinching or chafing of the tire or tube), or improper mounting of lock-rings or rims. Records show that most accidents involving truck tires occur while tires are being inflated.

Only employees well trained in tire repair and thoroughly familiar with the hazards and safe methods involved in handling tire equipment should inspect, install, repair, and replace tires and rims.

Other hazards are strain or hernia injuries resulting from lift-

**Figure 23-9.** Beware of the "crunch zone." All equipment operators must signal when backing up. All other persons must give plenty of room to vehicles, especially those with limited operator visibility—for example, this large frontend loader where the driver cannot see anything closer than 45 ft to the rear. This photograph also illustrates the safe practice of lowering the bucket or blade on all equipment not being used. (Courtesy of the Inland Steel Corporation.)

ing heavy tire assemblies. Mechanical lifting and moving devices should be provided so workers are not required to lift heavy tires.

Rubber cement and flammable solvents used for patching inner tubes, and casing compounds used for filling tire cuts, should be kept in safety cans.

Electric heating elements used for vulcanizing or branding tires should be regularly inspected. Defective wiring should be replaced.

Where power-driven rasps or scrapers are used for casings or inner tubes, the operators must wear eye protection and a dust mask. A local exhaust system should be applied to these machines to keep the fine rubber dust out of the workroom air.

### Fire protection

Because of fire hazards heavy-duty trucks should be equipped with Type B-C fire extinguishers listed by the Underwriters Laboratories for use on burning oil, gasoline, grease, and electrical equipment. The extinguisher should be placed in a convenient location in the cab or on the running board, and the driver should be taught how to operate it. Monthly inspection of firefighting equipment is advisable.

The repair shop should also have an adequate number of fire extinguishers of the ABC type and employees who are trained in their use.

If cutting, burning, or welding must be done near fuel oil tanks, an extinguisher should be at hand. A tarpaulin should be used to cover fuel, oil tanks, or combustible materials to protect them against sparks and excessive heat. Such work should not be performed on a fuel tank or other container until it has been drained and thoroughly purged of vapors. Use of the "hot work" permit system is recommended.

Fueling requires certain precautions to avoid fires. The engine should be stopped. Smoking should be prohibited. Safety containers and a grounded fuel hose should be used for fueling. When the tank is being filled, the metal spout of the hose should firmly contact the tank to ensure grounding and to neutralize static charges sufficient to ignite fuel vapors and cause an explosion or fire.

Fires occur in shops each year because gasoline and similar flammable solvents are used for cleaning parts. Safe, nontoxic cleaning liquids that are nonflammable and do not injure the skin are available and should be used.

The likelihood of a fire in a shop also can be reduced by good housekeeping, especially the disposal of oily waste and similar materials in covered metal containers.

Fire prevention procedures apply to all motorized equipment, such as power cranes, shovels, bulldozers, as well as trucks.

Details are covered in the *Engineering and Technology* volume.

### Grease rack operations

In greasing operations, a person may slip and fall because of accumulated grease and oil on the floor, injure his hands on sharp or rough edges on the vehicle, incur strains in trying to rock the vehicle to make grease penetrate into stiff bearings or springs, inhale sprayed or atomized oils used for spring lubrication, or suffer hand and head injuries from high-pressure guns.

Floors should be kept free of grease and oil to prevent slips and falls. Spills which occur during the working day should immediately be cleaned up or covered with an oil-absorbent compound.

Remind workers to keep their hands away from sharp or rough edges and to obtain immediate first aid treatment for all cuts and scratches.

Warn workers against putting their hands in front of the grease gun nozzle when the handle is pulled. Instances have been reported in which quantities of grease have been forced under the skin of workers by high-pressure grease guns.

Tops of grease cylinders should be securely fastened into place; otherwise, covers may blow off and seriously injure anyone who is nearby.

All equipment should be inspected weekly and repairs should be made when needed.

Workers should be warned of the danger of inhaling sprayed or atomized oils. They should stand clear of the lubricant spray, which settles quickly, and must not direct the spray at other employees.

### Wash rack operations

When washing vehicles, a person can slip and fall on wet floors, incur cuts or abrasions from the sharp or rough edges of the vehicle, or suffer burns from careless use of hot water or steam.

The concrete floor of the wash rack should be rough troweled to produce a nonslip surface. While washing vehicles, employees should wear safety toe rubber boots, preferably with nonslip soles and heels, and a rubber coat or apron.

Workers should never point the high velocity steams of hot or cold water at another person because serious injuries can result.

Workers should direct the hose, particularly when washing under the vehicle, in such a way as to avoid being struck by a backlashing stream of water and dirt.

Where a hot water hose is used, cover the metal parts to avoid skin contact and consequently prevent burns. Heavy-duty gloves and face shields should be used when necessary. A portable fan may be needed to blow steam away, so that the operator can see the work. Washers should be alert for sharp and rough edges on the vehicle which might cause cuts and abrasions. A periodic scheduled cleanup of the entire wash rack and associated equipment is recommended.

### Battery charging

Although most battery-operated vehicles now are recharged by means of on-board chargers and plug-in cords, there still are some vehicles requiring out-of-vehicle recharging. Other circumstances occasionally call for battery removal and charging such as when servicing or replacing. Proper safety procedures must be observed whenever a battery is serviced or charged. Do not charge a frozen battery.

The principal hazards of battery charging operations are acid burns during filling, back strains from lifting, electric shocks, slips, falls, and explosions.

Employees should wear safety apparel suitable for battery shops; this includes splash-proof eye and face protection and acid-proof gloves, aprons, and boots with nonslip soles. (Rubber boots and aprons must be worn when batteries are being filled. Goggles and face shield should be worn when working around batteries to prevent acid burns to the eyes and face.)

A wood-slat floor should be used and kept in good condition, to prevent slips and falls and to protect against electric shocks from batteries being charged.

Fire doors should be installed between charging rooms and other areas where flammable liquids are handled and stored.

The manufacturer's recommendations about charging rates for various size batteries should be closely followed in order to prevent rapid generation of hydrogen. Potentially explosive quantities of oxygen and hydrogen are developed in cells of batteries. This is particularly true if the battery is defective or if a heavy charge has been or is being applied. The lower the water level in the battery, the greater the cavity for the accumulation of gas.

Care should be taken to prevent arcing while batteries are being charged, tested, or handled. Tools and loose metal (and even lifting hoist chains) should not be in such a position that they may fall on batteries and cause a short circuit, which in turn can result in serious burns or an explosion.

When manual lifting is necessary, sufficient help should be provided to prevent strains, sprains, or hernias. Hand carts for transporting batteries are commercially available or can be made in a company shop.

Acid carboys should be handled with special care to prevent breakage and possible injury due to splashing of acid. Acid carboys should never be moved without their protecting boxes. They should not be stored in excessively warm locations or in the direct rays of the sun. Carboy tilters can be used.

A summary of recommendations for changing and charging batteries is given in the *Engineering and Technology* volume.

- **First aid for chemical burns.** Many batteries contain an acid electrolyte; some, such as the nickel-iron battery, contain an alkali solution. Whether it be acidic or alkaline, if electrolyte gets on a person's skin, it must be immediately washed off with large quantities of running water. Neutralizing agents are so often mishandled, they often can do more harm than good—only use them if first aid directions are available on labels or through company or plant directions. Get medical aid at once.
- **First aid for burns of the eye.** Irrigate the eye thoroughly with large amounts of clean water. Place a sterile dressing over the eye to immobilize the lid and get medical aid at once. Well-marked supplies of appropriate neutralizing agents can be kept close at hand for immediate use. Check with company physician. A safety shower and eyewash fountain are required by OSHA wherever acids or caustics are used. See Chapter 19, Occupational Health Services.

### Gasoline handling

The handling and storing of gasoline should comply with the provisions of the National Fire Protection Association *Flammable and Combustible Liquids Code,* NFPA 30, which is discussed in the *Engineering and Technology* volume.

Use of gasoline must be prohibited for all cleaning. Solvents with higher flash points are available and are equally effective and much safer. Even when higher flash point solvents are used, if carburetor or gas line parts are cleaned, the solution should be regularly changed since the admixture of small quantities of gasoline will tend to lower the flash point, and increase the danger of fire and explosion.

Grease, oil, and dirt may be removed from metal parts by nonflammable solutions, or by high-flash point solvents in special degreasing tanks with adequate ventilating facilities.

Use of gasoline to remove oil and grease from garage floors must be prohibited. Nonflammable cleaning compounds are commercially available and should be used.

Gasoline should not be used for removing oil and grease from hands. Soaps are available that will effectively remove greasy dirt from the skin without danger of injury. There are also protective creams and ointments which, if applied before starting work, will protect the skin from dirt and grease.

In some shops, employees use gasoline to clean work clothes. This unsafe practice must be prohibited.

If gasoline is spilled, it should be taken up immediately; there are substances available for absorbing oil spills. If gasoline in quantity gets into the sewage system, the fire department should be notified so the sewers can be flushed. Because gasoline vapor is heavier than air, it collects in low spots, such as basements, elevator pits, and sumps. These places should be kept ventilated whenever gasoline vapors are present.

### Other safe practices

**Using jacks and chain hoists.** Vehicles jacked up or hung on chain hoists should always be blocked with stanchions, pyramid jacks, or wood blocks (which have first carefully been inspected). The best jack for general garage use is the hydraulic-over-air type—if one system fails, the other prevails. Ordinary pedestal jacks are not to be used, especially the type supplied for passenger cars, as the vehicle may be tipped or jarred off the jacks and cause injury.

When a person is working under a blocked-up vehicle, other employees should not work on the car in such a manner that the car may be knocked from its blocks.

Employees who work under vehicles should be safeguarded from danger when their legs protrude into passageways. Barricades should be used for protection, or else the worker's entire body should be under the vehicle.

A frequent cause of injury to employees who work under vehicles is dirt and metal chips falling into the eyes. To protect the eyes, employees working under these conditions, should wear suitable eye protection—goggles or plastic eye shields. When necessary, the fog-resistant type should be used.

**Removing exhaust gases.** Repair shop employees should use local exhaust and ventilating facilities to prevent accumulation of vehicle exhaust gases within the shop.

**Repairing radiators.** Where radiators are boiled out or tested for leaks, the operator should be provided with both chemical goggles and a face shield of clear plastic. The entire face needs protection.

**Cleaning spark plugs.** All mechanics using sandblast spark plug cleaners should wear goggles or face shields.

**Replacing brakes.** The use of air pressure when cleaning around brake drums and backing plates while replacing brakes can send asbestos filings from the brake area into the breathable air and cause respiratory problems. Rather than using compressed air to clean, vacuums, chemical wash solutions, or steam cleaners should be used. Approved respiratory protective equipment should be worn to prevent inhaling airborne asbestos fibers when working on brake components.

**Controlling traffic.** Movement of vehicles inside shops and garages should be regulated by rigidly enforced traffic rules. Traffic lanes and parking spaces should be painted on the floors and the direction of traffic flow indicated. Vehicles with air brakes should not be moved until sufficient air pressure has been built up.

Every driver should stop his vehicle, then sound the horn before passing through the entrance or exit door. Signs requiring this procedure should be posted in conspicuous places. Mirrors should be installed at blind corners.

Vehicles should be moved in low gear and at low speed inside shop areas, especially up and down ramps.

**Other sources of injury** are jumping across open inspection pits, falling off ladders, hurting backs while trying to move supplies and equipment, and using hand tools improperly. Prevention of all work injuries requires proper selection and training of employees, careful supervision of their work habits, review of all injury causes, and the creation of safety-mindedness in all employees.

### Training repair shop personnel

Apprentices and new employees should be trained to do each job in the most efficient manner. Job instruction should include the safety rules and regulations pertaining to each job and the reason for such rules. The new mechanic should be thoroughly indoctrinated concerning the company's policy toward safety. He should understand the organization of the safety program and the part he is expected to play in it.

Having been indoctrinated and trained to work safely, the new employee must be kept actively interested in observing accepted safe practices in the conduct of his job. There are many devices available to the safety director to accomplish this end, including safety supervision, safety contests, safety meetings, posters, safety bulletins, and pamphlets. See Chapter 9, Safety Training.

More details on mechanical and chemical safety can be found in the *Engineering and Technology* volume and in *Fundamentals of Industrial Hygiene,* respectively, the other two books in this series.

## OFF-THE-ROAD MOTORIZED EQUIPMENT

Heavy-duty trucks are mentioned again here because they are extensively used for special off-the-road operations in industries such as quarrying, mining, and construction. When on the road, they are, of course, governed by the same safe-driving practices as other types of automotive equipment.

The use of heavy-duty trucks, mobile cranes, tractors, bulldozers, and other motorized equipment in quarrying, mining, and construction presents the possibility of accidents. Workers near equipment can be struck, run over, and killed; equipment sometimes slips over embankments, injuring the operator and other people (Figure 23-8). Even personnel who are involved in servicing and maintaining equipment can find it hazardous.

Many accidents, even those that do not injure anyone, result in costly damage to equipment, loss of efficiency and production, and high maintenance costs.

In general, prevention of accidents to heavy equipment requires:

1. Safety features on equipment
2. Systematic maintenance and repair
3. Trained operators
4. Trained repair personnel.

Many operators of heavy equipment are injured while boarding and unboarding heavy-duty vehicles. Slippery steps generally are the cause of such accidents. The driver training program needs to address this issue.

Safe and proper equipment operation instructions will be found in manufacturers' manuals. Many driving practices are the same as those necessary for the safe operation of highway vehicles. Off-the-road driving, however, involves special hazards and requires special training and safety measures.

### Haul-roads

Roadway improvements pay for themselves because they reduce accidents and lower maintenance costs.

Both temporary and permanent roads often are too narrow for heavy equipment and other, oncoming traffic, especially at curves and fills. Enough space must be provided at curves so large trucks need not cross the centerline of the road. Curves should be banked toward the outside.

Both temporary and permanent roadways require regular patrolling and maintenance. Too often, serious accidents, breakdowns, delays, and unnecessary maintenance expense can be traced to neglected roadways. Members of road patrols should be provided with protective equipment, such as barricades, warning signs, red flags, flagmen, and flares.

Seasonal conditions create road hazards that require prompt attention. Some companies provide sprinkler trucks to protect their employees against harmful dusts, discomfort, and the possibility of accidents during dry and windy periods. Others keep dust down by spraying oil with an asphalt base, on the road surface.

Skidding on snow and ice is a serious hazard during the winter. Snow and ice should be removed by means of snowplows or blade graders as promptly and as completely as possible.

When roadways are built close to high banks, the slopes of the banks should be inspected for loose rocks, especially after rain and freezing or thawing weather. Loose rock should be barred down, that is, pried out with a steel bar.

Where trucks enter public highways, signs should be installed warning both the highway traffic and off-road vehicles. Design, color, and placement of the signs should be in accordance with U.S. Department of Transportation, Federal Highway Administration, Washington, D.C., *Manual on Uniform Traffic Control Devices for Streets and Highways,* also published as American National Standard D6.1 (see References). If operations are conducted at night, these signs should be made of a light-reflecting material or directly lighted. In situations where temporary roads cross railroad tracks, especially when high speed trains are involved, contact should be made with the railroad representatives and a flagman should be used at the crossing.

### Driver qualifications and training

The modern heavy-duty vehicle or other off-the-road equipment is a carefully engineered and expensive piece of equipment and warrants operation only by drivers who are qualified physically, mentally, and by training and experience. The physical and mental qualifications for an efficient and safe operator of heavy over-the-road equipment (discussed earlier in this chapter) apply also to drivers of off-the-road vehicles.

No driver should be allowed to work until his knowledge, experience, and abilities have been determined. The amount of time varies for a prospective driver or operator to become thoroughly acquainted with the mechanical features of the truck or piece of equipment, safety rules, driver reports, and emergency conditions. Even an experienced operator should not be permitted to operate equipment until the instructor or supervisor is satisfied with his abilities.

Because accidents caused by unsafe practices outnumber those resulting from unsafe condition of equipment and roadways, the time required for thorough checking and training is well warranted. After an employee has been trained, continuing supervision is required to make sure that he continues to operate in the way in which he was instructed.

### Operating vehicles near workers

Workers are exposed to the danger of being struck or run over by vehicles, particularly around power shovels, concrete mixers, and other equipment, in garages, shops, dumps, and construction areas.

**Backing—the "crunch zone."** The most dangerous movement is backing (see Figure 23-9). Some companies require drivers to blow three blasts of the horn for a back-up signal.

An automatic audible signaling device to warn workers when a vehicle is backing is required by OSHA regulations.

Where a number of employees are working, the driver should ask another employee to signal whether the path is clear before the vehicle is backed or moved. The person giving the signals should always take a position within sight of the driver. Also, a standard set of signals should be devised to ensure proper communication.

**Moving forward.** Serious accidents also occur during forward movements. The hazard to workers increases with the greater height and capacity of trucks. A driver often fails to see workers crossing from the right immediately ahead of the truck. Thus, drivers often are required to blow two blasts on the horn before starting forward. Construction vehicles with attachments such as front-end loaders and dozers should position their attachments to provide for maximum visibility whenever changing positions.

### Procedures for dumping

Vehicle operations on dumps and banks involve the danger of the vehicle going over the crest while dumping a load. A person trained in proper dumping procedures is probably the best insurance against loss of life and damage to equipment. Drivers are required to follow his instructions and signals, especially in backing to dump. Prearranged signals must be used at all times.

The person responsible for dumping must know how close to the edge a vehicle can safely approach under various weather conditions. The helper should be positioned on the driver's side of the vehicle (1) so the signals can be easily seen, (2) so the driver will have the signaller in sight, and (3) so the signaller will be clear of the backing truck and falling material. To protect the helper further, the driver should turn and look over the left shoulder when backing to have a maximum view of the area into which the rear of the truck is moving.

Left-hand driving also reduces the danger of going over the crest, especially in the operation of side dump trucks, since the driver is on the crest side.

To avoid hitting overhead lines or other low clearances, the dump box should be lowered as soon as the load is dumped.

To help prevent the crest from caving in, stockpiles and dumps frequently are graded toward the crest so that vehicles back up the slope. Loads also may be dumped a safe distance from the crest and then leveled by a grader or bulldozer.

Strongly built cabs and cab protectors on canopies and especially safety belts are effective in preventing injuries if vehicles overturn. The inclination to jump clear of a vehicle that is beginning to roll or slide over an embankment is ill advised. It is far safer to remain in the cab.

Holes, ruts, and similar rough places on dumps, and roadways may cause the front wheels of a truck to cramp so the steering wheel spins, injuring fingers, arms, and ribs, particularly if the truck is not provided with power steering. Gripping the wheel on the outside and not by the spokes, driving at reduced speed, and observing the ground ahead for rough places will help the driver avoid such injuries.

Floodlighting during night operations helps to prevent accidents.

### Protective frames for heavy equipment

All bulldozers, tractors, and similar equipment used in clearing operations must be equipped with substantial guards, shields, canopies, and grilles to protect the operator from falling and flying objects.

Crawlers and rubber-tired vehicles, self-propelled pneumatic-tired earth movers, water tank trucks, and similar equipment must be equipped with steel canopies and safety belts in order to protect operators from the hazards of rollover (see Figures 23-8 and -11a). Drivers should be trained and required to wear the safety belt.

A canopy and its support should be designed and made to support not less than two times the weight of the prime mover. This calculation is to be based on the ultimate strength of the metal and integrated loading of support members, with the resultant load applied at the point of impact. In addition, there should be a vertical clearance of 52 in. (132 cm) from the deck to the canopy where the operator enters or leaves the seat.

For more details, see National Safety Council Industrial Data Sheet 622, *Tractor Operation and Roll-over Protective Structures.*

### Transportation of workers

Some jobs—for various reasons—require workers be transported to and from the work site. However, transporting employees to and from work can involve special risks and consequently special precautions must be taken. Where such transportation is a regular occurrence, a bus or other vehicle designed to transport passengers can avoid many of the particular risks.

Hazards are more likely to exist when transportation of workers is performed on an irregular basis. Often, under these conditions, an open cargo truck of some kind is used as the transporting vehicle. If so, workers should be advised of the following safety procedures:

- Getting on. Look before and where you step. Use every handhold available, even a helping hand from someone already on the truck. Step squarely; never at an angle. No one should ever attempt to board a moving vehicle—regardless of how slow it may be traveling.
- If possible, benches should be provided. In no case, however, should the passengers remain standing while the vehicle is in motion. If necessary, they should sit on the truck bed.
- Avoiding horseplay. Of all the many negative actions that a group of people may take, this is one of the most stupid and dangerous. Some companies will go as far as firing anyone caught indulging in this type of activity. Horseplay is inherently dangerous, and in a moving vehicle, it may be fatally so.
- Getting off. Workers should again look before they step, then get off slowly and easily, using every possible handhold. Under no circumstance should anyone attempt to jump off a moving or stopped vehicle. Many injuries occur as a result of jumping off a vehicle, whether it is moving or not.

### Towing

Towing is a hazardous operation, especially when coupling or uncoupling the equipment. Workers can be crushed when a truck or other piece of equipment moves unexpectedly while they are between the two pieces of equipment.

The following safe practices are essential to prevent accidents in the coupling or uncoupling of motorized equipment:

1. No one should go between the vehicles while either one is in motion.
2. Vehicles must be secured against movements by having the brakes set, the wheels blocked, or both.
3. A driver should not move his vehicle while someone is between it and another vehicle, a wall, or anything else that is reasonably solid and immovable. In fact, before moving, the driver should receive an all-clear signal.
4. Tow bars are usually safer than towing ropes. If ropes, which may be more convenient to use under certain circumstances, are employed, they must be in good condition and of sufficient size and length for the towing job.
5. Equipment towed on trainers should be secured to the trailer.

### Power shovels, cranes, and similar equipment

Safe operation of power shovels, draglines, and similar equipment begins with machine purchase. A good policy is to spell out in the equipment specifications that guards must cover gears, and that safe oiling devices, handholds, slip-resistant steps, and other safeguards be provided. In any case, before equipment is put into operation, it should be thoroughly inspected and necessary safety devices should be installed.

To keep workers from being injured when caught between truck frames, crawler tracks, cabs, and counterweights of cranes and shovels, a barricade can be used to warn employees who are working near operating equipment that they are close to a hazardous area. Signs, flashing lights, and other warning devices can also be used to alert people to the hazards. Barricades are easily moved when equipment is moved.

Both operators and maintenance personnel, whether experienced or not, should be instructed in manufacturer-recommended procedures pertaining to lubrication, adjustments, repairs, and operating practices, and should be required to observe them. A preventive maintenance program for shovels and other equipment is essential for safety and efficiency. Frequent and regular inspections and prompt repairs are the bases for effective preventive maintenance.

Generally, the operator is responsible for inspecting the mechanical condition of such items as holddown bolts, brakes, clutches, clamps, hooks, and similar vital parts.

Wire ropes should be kept lubricated in accordance with the manufacturers' instructions, and should be inspected daily since rope failures can cause serious accidents. Ropes are particularly likely to develop weakness at the fastenings, at crossover points on drums, and in the sections that are in frequent contact with the sheaves.

**Grounding systems.** In order to prevent electrical shock, electrically powered equipment requires a good earth grounding system to protect workers from electrical faults in trailing cables or at the machine. Though the cable may make close physical contact with the surface of the ground, the resistance to the flow of current from the frame of the equipment and cable to the earth usually is high because of wire insulation. A machine-to-ground fault resistance of 100 ohms and a current of 10 amperes means an electric shock hazard of about 1,000 volts. A leakage fault current of 1/50 ampere is very painful and can result in the loss of muscular control. As little electric current as 1/10 ampere through the body can result in death. (See chapter on Electrical Hazards in the *Engineering and Technology* volume.) Since these low leakage currents can be forced through wet skin by commercial 120 volts AC, no employee should be exposed even momentarily to this electrical hazard.

A good earth-ground system may be made by driving copper-clad steel rods, or electrodes, into suitable soil for a distance of at least 8 ft (2.4 m). Rods, which are available commercially for this purpose, lower the earth resistance and provide a better ground. Since the number of rods and their spacing depend considerably on soil conditions, the conductivity of the soil may have to be increased by application of common salt, sodium nitrate, copper sulfate, or similar chemicals which then are carried into the soil by rain. A grounding system having a total resistance of 1 ohm, including the cable, can be obtained in many areas by proper design and construction.

The pole line ground wire should have at least the same wire gage size as the power wires. Wherever power is tapped from the power line, a connection is made from the pole ground wire to a ground wire in the cable. A good cable has metal shielding outside the insulating material around each conductor, and the ground conductor is in full electrical contact with the shielding. The shielding should have electrical continuity, and, if broken, if should be bridged.

The ground wire in the cable is connected to the equipment frame where there should be good metal-to-metal contact. If necessary, paint or other covering should be scraped off to achieve good contact. A resistor between the pole line ground wire and the transformer neutral limits the amount of current to not more than 50 amperes, eliminating dangerous voltages at the shovel and permitting sufficient current to open the circuit breakers.

The neutral ground system permits the operation of all equipment except the machine where the fault occurs. The machine is segregated from the rest of the system by the immediate operation of a circuit breaker actuated by the fault. Suitable switching equipment in the neutral grounded system eliminates the danger from several faults existing at the same time in different phases at different locations, except for the interval required for the circuit breakers to open.

The equipment ground should not be connected to the substation ground in any way, to avoid energizing the equipment by a fault in the power supply system.

Circuit breakers and other devices in the grounding system should be inspected monthly. The resistance of the ground rods or system also should be regularly checked. A megohmmeter test of the cable insulation is a recommended part of cable inspection. Defective cable insulation should not be taped—the cable should be replaced.

Workers should be provided with rubber gloves and insulated tongs or hooks for handling trailing power cables.

Minimum wear and damage to trailing cables is important for safety. They should be protected from blasting operations as much as possible but kept as close to operations as practical so that a minimum length of cable is required. Tripods and wooden construction horses can be used to keep cable off the ground. Tripods are preferable to trenches where cables cross roads.

The electrical parts of shovels and similar equipment, including trailing cables, should be inspected regularly and maintained by an electrician.

**Maintenance practices.** If a shovel is operating in a deep excavation, repairs or adjustments should be made with the shovel in a safe position where it will not be endangered by falling or sliding rocks or earth.

When repairs are to be made, the operator is responsible for setting the brakes, securing the boom, lowering the dipper or bucket to the ground, taking the machine out of gear, and before leaving the machine, exercising similar precautions to prevent accidental movement.

Before repairing any vehicle, maintenance personnel should notify the operator about its nature and location. If the work is to be done on or near moving parts, the controls should be locked out and tagged, and the lock and tag should be removed only by authorized persons. This precaution is essential to prevent the operator from starting the equipment inadvertently.

Parts that must be in motion while workers are working on them should be turned slowly, by hand if possible, in response to guidance or on signal if two or more persons must be involved. This precaution applies particularly to those who work around gears, sheaves, and drums. Workers who grasp ropes just ahead of the sheaves risk having their hands jerked into the sheaves. To prevent hand injuries, a rope being wound on a drum should be guided with a bar.

If guards must be removed for convenience in making repairs, the job cannot be considered complete until the guards, plates, and other safety devices have been replaced.

Repair personnel should wear snug-fitting clothing, eye protection, and safety shoes. Gloves should not be worn when working on or near any moving parts of the machines.

**Operating practices.** Slides of rock and material from high faces and banks result in some of the worst accidents involving power shovels. In some instances, the shovel and operator have been buried, and in other cases, workers have been struck and injured while working around the equipment.

Some quarries that have high faces limit the height of banks to 25 ft (7.6 m) by benching and by blasting procedure that forces the rock out from the face sufficiently to reduce the height of the pile. The shovel operator is thus able to maintain the back at a safe slope. When loading under a high face, the operator should swing the shovel to the sight side and away from the face, thereby providing a better view and reducing the injury potential.

**Undercutting banks** of earth, sand, gravel, and similar materials is dangerous, especially during winter and spring months. Freezing and thawing can result in a collapse of the overhanging material. To maintain a safe slope, the overhanging material may be blasted.

The shovel operator has responsibility for the safety of other employees whose duties take them into the vicinity of the shovel. These workers may be struck by falling rock, squeezed between the shovel and the bank or similar pinch points, or struck by the dipper. No worker should enter a dangerous location without first notifying the operator who, in turn, should not move the equipment.

Dippers should be filled to capacity but not overflowing, to prevent falling material from endangering workers and to eliminate excessive spillage. Insofar as possible, loading should be done from the blind side. The operator should not swing a load over a vehicle nor load a truck until its driver has dismounted and is in the clear, unless the truck is provided with a canopy designed for the protection of the driver. Rail cars and motor trucks should be loaded evenly so earth or rocks do not overhang the sides.

Housekeeping on and around the shovel should be stressed. The operator should keep tools in a definite place and keep the cab floor free of grease and oil. Ice and snow should be removed promptly, and a bulldozer should keep the area around the shovel free of rocks and ruts.

One should get on or off a shovel or dragline only after having notified the operator, who, in turn, should swing the platform so the handhold can be grasped and the steps or tread used. No one should get on or off by jumping onto the tread, either while the operator is making a swing or while the equipment is stationary.

No unauthorized person should be permitted on a shovel or dragline.

**Mobile cranes.** The outstanding characteristic of accidents involving crawler and similar types of cranes is the severity of the injuries. Although these accidents occur with relative infrequency, the injuries are about twice as serious as those resulting from accidents involving other types of heavy equipment. For this reason, a crane operator particularly should be selected for his intelligence, stability, and willingness to follow instructions.

The operator is largely responsible for the safe condition of the crane and should make regular inspections of brakes, ropes and their fastenings, and other vital parts, and promptly report worn, broken, and defective parts. Like other equipment, mobile cranes should be maintained on a regular schedule.

The operator is responsible for the safety of the oiler and also has a large measure of responsibility for preventing injuries to hookers or riggers and others working around the equipment. However, anyone working in the vicinity of a crane has the responsibility to stay clear of the boom. In no case should anyone work or cross under the boom.

Some of the worst accidents result from overloading cranes. In no case should the manufacturer-specified load limits for various positions of the boom be exceeded. These load limits should be conspicuously posted in the crane cab. If there is doubt about the weight of a load, the safe crane capacity should be tested by first lifting the load slightly off the ground. Operating a crane on soft or sloping ground is dangerous. The crane should always be level before it is put into operation. Outriggers give reliable stability only when used on solid ground. The use of makeshift methods to increase the capacity of a crane, such as timbers with blocking, is too dangerous to be permitted.

Boom stops limit the travel of the boom beyond the angle of 80 degrees above the horizontal plane and prevent the boom from being pulled backwards over the top of the machine by the boom-hoisting mechanism or the sudden release of a heavy load suspended at a short radius. Either of these occurrences usually result in serious damage to the equipment and injuries to the operator or other workers.

Accidents usually occur when the operator is performing more than one operation and becomes confused or distracted and excited. Also, clutch linings may swell during wet weather, and the master clutch or the boom clutch, or both, may drag and cause the boom to be pulled over backwards. Clutches should be tested before starting work on rainy days and the clearances adjusted if necessary. Another accident cause is the sudden release of a load when the boom angle is high, for example, from the parting of a sling.

Boom stops are best suited to medium-size cranes (the 5- to 60-ton range). Boom stops should disengage the master clutch or kill the engine and stop the boom before it reaches the maximum permissible angle. One type of stop meeting these requirements has a piston and cylinder; it is spring or pneumatically actuated, and is mounted on the A-frame to intercept the boom as high above the boom hinges as possible. By positive displacement of an actuator mounted on the A-frame, the boom action disengages the master clutch (or ignition breaker or compression release) by means of light rope reeved over a few small sheaves.

When a mobile crane must be operated near electric power lines, the power company should be consulted to determine whether the line can be deenergized. Most fatalities have resulted from contact with power lines, and often the power company's service is seriously disrupted. Various states and OSHA have enacted legislation specifying the distances booms and wire ropes must be kept from power lines. A minimum of 10 ft (3 m) is often specified; however, the recommendations of the power company and legal requirements should be observed. (See Figure 23-10.)

An experienced operator working with an untrained or relatively inexperienced hooker or rigger should direct the details of lifts, such as the type of sling and hitch to be used. Although the operator usually can rely on the knowledge of an experienced rigger, the operator has the right to question the safety of a lift and have his supervisor make a decision.

**Figure 23-10.** When mobile cranes must be operated in the vicinity of electric power lines, first consult with the utility company to determine whether the lines should be deenergized. OSHA and many state regulations require that booms and wire ropes be kept a minimum of 10 ft (3 m) away from the power lines. (Courtesy of Washington State Department of Transportation.)

The following safe practices are essential when handling loads.

1. The hook must be centered over the load to keep it from swinging when lifted.
2. Employees should keep their hands out of the pinch point when holding the hook or slings in place while the slack is taken up. A hook, or even a small piece of board, may be used for the purpose. If a worker must use his hand, the sling should be held in place with the flat of the hand.
3. The hooker, rigger, and all other people must be in the clear before a load is lifted.
4. Tag lines should be used for guiding loads.
5. Hookers, riggers, and others working around cranes also must keep clear of the swing of the boom and cab.
6. No load should be lifted or moved without a signal. Where the entire movement of a load cannot be seen by the operator, as in lowering a load into a pit, someone should be posted to guide him. To avoid confusion in signals, only standard hand signals should be used.

### Graders, bulldozers, and scrapers

Many of the basic safety measures recommended for trucks also apply to graders and other types of earth-moving equipment. All machines should be regularly inspected by the operator, who also should promptly report any defects and malfunctioning systems or parts. The safety and efficiency of the equipment are increased by scheduled maintenance.

Only physically and mentally qualified individuals should be selected as operators and trained in correct operating practices, as specified in the manufacturer's manual and by company requirements. Prevention of injury in the servicing and repairing of machines requires special precautions in addition to the observance of general safe procedures applying to other types of motorized equipment.

**Maintenance.** Brakes, controls, engine, motors, chassis, blades, blade holders, tracks, drives, hydraulic mechanisms, transmission, and other vital parts require regular inspection. Wheel and engine-mount bolts likewise require frequent checking for tightness.

Making adjustments and repairs with the engine running is a dangerous practice, particularly when work is being done near the fan of the engine or when a clutch of a tractor is being adjusted. Refueling should be done only with the engine stopped.

The danger from a locking ring blowing off when the tire of a truck is being inflated applies equally to a tractor tire. (See details under Tire operations, earlier in this chapter.)

When cutting edges are to be replaced, the scraper bowl or dozer blade should always be blocked up. After the scraper has been lifted to the desired height, blocks are placed under the bottom near the ground plates. Apron arms are raised to the extreme height and a block is placed under each arm, so the apron can drop enough to wedge each block firmly in place.

Before receiving wire rope on a drum or through sheaves, the operator should disengage the master clutch, idle the engine, and lock the brakes. The engine should be at a complete stop before working with the rope on a front-mounted drum.

If an operator is assisting a repair worker and working behind the scraper with the tail gate in the forward position, a block should be placed behind the tail gate so it cannot fall. This precaution is necessary in case someone should release the power control until brake permitting the tail gate to come back.

When ropes are to be replaced on scrapers, the tail gate should be back at the end of its travel.

**General operating practices.** The operator must look to the front, sides, and rear before moving his machine and be constantly alert for employees on foot when operating near other equipment, offices, tool and supply buildings, and similar places.

Speeds are largely governed by conditions. Slow speeds are essential in driving (1) off the road and beyond the shoulder, on steep grades, and at rough places to avoid violent tilting that may throw the driver off the machine or against levers and cause serious injury, (2) in congested areas, and (3) under icy and other slippery conditions. No one other than the operator should be permitted on the vehicle at the same time.

Jumping from a standing machine can result in sprained ankles and back and other injuries. The safe practice is to step down after looking to make sure footing is secure and there is no danger from other vehicles. Ice, mud, round stones, holes, and similar conditions cause many falls. For the same reason, deck plates and steps on equipment should be free of grease and other slipping hazards.

An operator should not drive the equipment onto a haul road without first stopping and looking both ways, regardless of whether the place of entry is marked with a stop sign. Generally, loaded equipment is given the right of way on job or haul roads.

Before an operator leaves his equipment, even for a short time, the bucket or blade should be lowered to the ground and the engine stopped (see Figure 23-9). A safe parking location is on level ground, off a roadway, and out of the way of other equipment.

The operator should never leave his equipment on the inclined surface or on loose material with the engine running—the vibration can put the equipment in motion.

**Procedures on roadways.** When graders, scrapers, and other earth-moving equipment are in operation along a section of a road, the precautions discussed next will help prevent accidents to the public, employees, and equipment.

Traffic must be warned, by barrier signs at both ends of the road section undergoing construction, that there is danger ahead. Primary warning signs, such as Road Under Construction or Barricade Ahead, should be placed 1,500 ft (460 m) from the end point of operation.

Orange flags or markers at the ends of blades, which may project beyond the tread of a machine, serve to warn persons and other equipment operators.

An orange flag on a staff projecting at least 6 ft (1.8 m) above the rear wheel of a blade grader or mowing vehicle is recommended for operation in hilly country. (See Figure 23-11a.)

**Figure 23-11a.** Mowers should have rollover protective structures (ROPS), slow-moving vehicle emblem, and an orange flag projecting at least 6 ft (1.8 m) high. (Courtesy Navistar International Corp.)

Slow-moving vehicle emblems must be affixed at the rear of these vehicles if they are to drive even short distances on public roads (see Figure 23-11b).

**Figure 23-11b.** Close-up of a slow-moving vehicle (SMV) emblem.

Operators of motor graders should keep to the right side of a roadway. When blading against traffic is necessary, flags and barricades should be used to warn traffic. Warning signs must be placed at a considerable distance from the work area. This distance increases as the highway speed increases. Suggestions are given in the DOT *Manual on Uniform Traffic Control Devices for Streets and Highways,* ANSI D6.1. Most states use the latter standard (DOT) as minimum requirements, with additional traffic control where conditions dictate.

Where operations are extensive, flagmen should be placed at each end of the working area so they are visible to oncoming traffic for at least 500 ft (152.5 m).

Where earth-moving equipment is stopping, turning, or backing at curves, crests of hills, and similar dangerous locations, flagmen must be stationed. For safety, such movements generally require a clear view of approaching traffic for a distance of about 1,000 ft (305 m).

Flagmen must also be used where the working area is congested by other equipment, workers, building, excavation, and similar hazards. See Figure 23-12 for hand-signaling procedures.

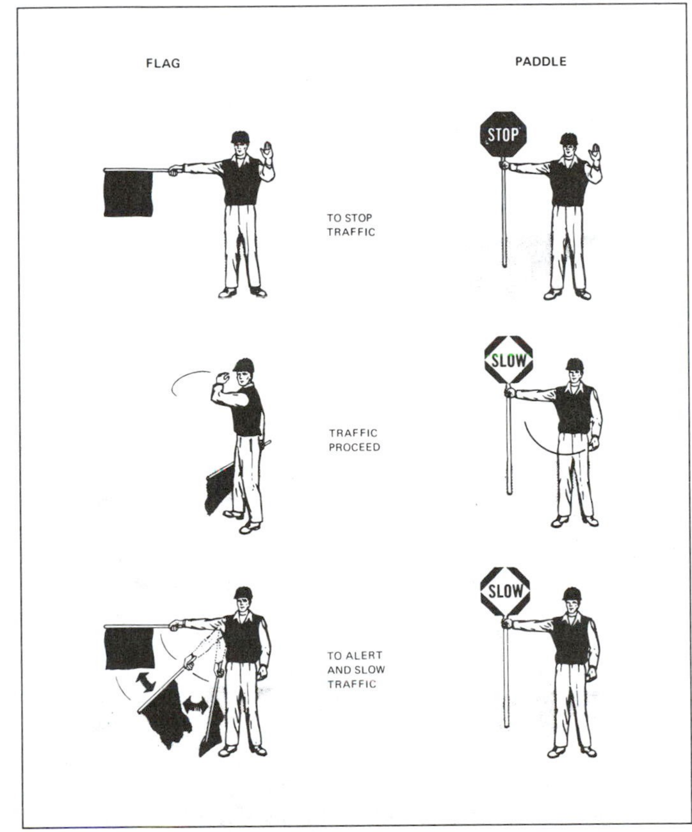


**Figure 23-12.** Hand-signaling devices used by flagmen. Flag and background of STOP sign are bright red; diamond of SLOW sign, orange. (From American National Standard D6.1-1978.)

**Coupling and towing equipment.** An operator should not back up to couple a tractor to a scraper, sheepsfoot roller, or other equipment without having first checked to make sure that everyone is in the clear. If the operator is assisted by a person on the ground, he should not move the equipment until signaled.

Before an employee is allowed to couple the trailing equipment, the tractor should be stopped, the shift lever put in neutral, and the brakes set. The wheel of the equipment to be coupled should be blocked.

All equipment being towed should be secured by a safety chain attached to the pulling unit, in addition to the regular hitch or drawbar, since a drawbar failure can result in a serious accident.

When a scraper is towed from one job to another, the operator should use a scraper bowl safety latch, or place a safety bolt in the beam to given maximum clearance for road projections such as at crossings. This precaution prevents the bowl from striking the ground or pavement and injuring persons or damaging equipment.

**Clearing work.** Work requiring exposure to low limbs of trees or to high brush involves serious hazards which can be readily overcome by suitable protective measures and safe practices.

When using a bulldozer, it is best to equip it with a heavy, well supported, arched steel mesh canopy to protect the operator. See section on Protective frames for heavy equipment, earlier. Goggles should be worn to protect the eyes from whipping branches.

Head protection provides protection from falling branches. When a bulldozer shoves hard against the butt of a large dead tree, the tree may crack in the middle or limbs may fall onto the machine. Dead branches or tops also can drop from live trees. A safe procedure to eliminate the danger is to cut the roots on three sides and then apply the power to the fourth side. A long rope can be used to pull over large trees, but it must be determined in advance that the tractor and operator will be in the clear when the tree falls.

Operators have the responsibility of seeing that all workers in the area are in the clear before pushing over any trees, bulldozing rock, and rolling logs.

**Special hazards.** Fatalities can easily occur while equipment is operated on dumps and fills, near excavations, and on steep slopes.

The bulldozer blade should be kept close to the ground for balance when the machine is traveling up a steep slope.

When a tractor-dozer is to be driven down a slope, three or four loads of dirt should be dozed to the edge of the slope and kept in front of the blade.

If the dirt is lost on the way down, the operator should not lower the blade to regain the load because of the danger of overturning. Using the blade as a brake on a steep slope should only be done in cases of extreme emergency.

How close to an excavation or the crest of a dump a machine can be safely operated depends on ground conditions. Wet weather requires equipment to operate a greater distance from the edge or crest. Someone to signal the driver is especially essential when the ground is treacherous.

Sometimes employees, the public, livestock, and property are endangered when material is pushed over the edge in side hill work. In such cases, sufficient clearance below must be provided before the work begins.

## REFERENCES

Alliance of American Insurers, 1501 Woodfield Rd., Schaumburg, Ill. 60195. *Code of the Road.*

American Automobile Association, 8111 Gatehouse Road, Falls Church, Va. 22047. "Driver Training Equipment" (catalog).

American National Red Cross. *Standard First Aid and Personal Safety Manual,* 2nd ed. Washington, D.C.: American National Red Cross, 1979.

American National Standards Institute, 1430 Broadway, New York, N.Y. 10018.
*Method of Recording Basic Facts Relating to Nature and Occurrence of Work Injuries,* ANSI Z16.2-1962(R1969).
*Method of Recording and Measuring the Off-the-Job Disabling Accidental Injury Experience of Employees,* ANSI Z16.3-1973.
*Method of Recording and Measuring Work Injury Experience,* ANSI Z16.1-1967(R1973).
*Motor Vehicle Traffic Accidents, Manual on Classification of,* ANSI D16.4-1977.
*Recordkeeping for Occupational Injuries and Illnesses, Uniform,* ANSI Z16.4-1977.
*Powered Industrial Trucks,* ANSI/NFPA 505-1981.
*Powered Industrial Trucks—Low Lift and High Lift Trucks, Safety Standard for,* ANSI/ASME B56.1-1983.
*Recording and Measuring Motor Vehicle Fleet Accident Experience and Passenger Accident Experience, Method,* ANSI D15.1-1976.

American Society of Safety Engineers, 1800 E. Oakton, Des Plaines, Ill. 60018.
*Dictionary of Terms Used in the Safety Profession.*
*Photographic Techniques for Accident Investigation.*
*Profitable Risk Control.*
*Safety Law—A Legal Reference for the Safety Professional.*

American Trucking Associations, Inc., 2200 Mill Road, Alexandria, Va. 22314.
*ATA Hazardous Materials Tariff,* 1983.
*Bulletin Advisory Service* (3 vols.), 1983.
*Effective Truck Terminal Planning and Operations,* 1980.
*Fundamentals of Transporting Hazardous Materials,* 1982.
*Fundamentals of Transporting Hazardous Waste,* 1980.
*National Truck Driving Championship,* 1986.

Associated General Contractors of America, Inc., 1957 E St. NW., Washington, D.C. 20006. *Manual of Accident Prevention in Construction.*

Association of American Railroads. *Loading, Blocking and Bracing of Freight in Closed Trailers for Trailer or Flat Bed Service, Suggested Methods for.* Washington, D.C.: Association of American Railroads.

Association of Casualty and Surety Companies, 110 William St., New York, N.Y. 10038.
*Guide Book, Commercial Vehicle Drivers*
*Truck and Bus Drivers Rule Book*

Baker, J. S. *Traffic Accident Investigation Manual.* Chicago: Traffic Institute, 1975.

Heavy Construction Contractors Association, P.O. Box 505, Merrifield, Va. 22116. (General.)

*Kirk-Othmer Encyclopedia of Chemical Technology,* 3rd ed. New York: Wiley Interscience, 1984.

National Committee for Motor Fleet Supervisor Training, *Motor Fleet Safety Supervision, Principles and Practices.* East Lansing, Mich.: National Committee for Motor Fleet Supervisor Training, 1983.

National Fire Protection Association, Batterymarch Park, Quincy, Mass. 02269.
*Flammable and Combustible Liquids Code,* NFPA 30.
*Fire Prevention Code,* NFPA 1.
*Fire Protection Handbook,* 15th ed., 1981.
*Hazardous Chemical Data,* NFPA 49.
*Hazardous Materials Transportation Accidents,* 1978.
*Life Safety Code Handbook,* 1981.
*National Electrical Code Handbook,* 1984.

National Safety Council, 444 N. Michigan Ave., Chicago, Ill. 60611.
*Accident Facts.* (Published annually.)
*Alcohol and the Impaired Driver.*
*Aviation Ground Operators Safety Handbook.*
*Chemical Hazard Fact Finder.*
Defensive Driving Program Materials.
*Driver and Home Safety Manual.*
*Fleet Accident Rates.* (Published annually.)
Industrial Data Sheets
*Barricades and Warning Devices for Highway Construction Work,* 239.
*Berms in Pits and Quarries,* 680.
*Falling or Sliding Rock in Quarries,* 332.
*General Excavation,* 482.
*Grounding Electric Shovels, Cranes, and Other Mobile Equipment,* 287.
*Lead-Acid Storage Batteries,* 635.
*Liquefied Petroleum Gases for Industrial Trucks,* 479.
*Motor Graders, Bulldozers, and Scrapers,* 256.
*Motor Trucks for Mines, Quarries, and Construction,* 330.
*Mounting Heavy Duty Tires and Rims,* 411.
*Operation of Power Shovels, Draglines, and Similar Equipment,* 271.
*Ready-Mixed Concrete Trucks,* 617.
*Snow Removal and Ice Control on Highways,* 638.
*Tractor Operation and Roll-over Protective Structures,* 622.
National Fleet Safety Contest.
*Motor Fleet Safety Manual.*
*Public Employee Safety Guides*
"Street and Highway Maintenance"
"Vehicular Equipment Maintenance"
Safe Driver Award Program.
*School Buses and Operations, Standards for.*
*Supervisors Safety Manual.*

New York University Center for Safety Education, Washington Square, New York, N.Y. 10003. Publications list.

North American Association of Alcoholism Programs, 1611 Devonshire Drive, Columbia, S.C. 29204.

*Pilot Judgment Training and Evaluation,* Volumes I-III (DOT/FAA/CT-82-56), Bunnell, Fla.: Embry-Riddle Aeronautical University, June, 1982.

Private Truck Council of America. *Driver Training Manual.* Washington, D.C.: Private Truck Council of America, 1981.

U.S. Department of Defense, Department of the Army, Washington, D.C. 20310.
*Driver Selection and Training,* TM 21-300.
*Drivers' Manual.* TM 21-305.
*General Safety Requirements.* EM 385-1-1, U.S. Army Corps of Engineers.
"Methods of Teaching."
*Motor Transportation, Operation,* FM 25-10.

U.S. Department of the Interior, Bureau of Mines, Washington, D.C. 20240.
*Minerals Yearbook.*
Also various handbooks, miners' circulars, and other publications.

U.S. Department of Labor, 200 Constitution Avenue NW., Washington, D.C. 20210.

*Occupational Safety and Health Standards.*
*OSHA Compliance Guide,* Volume 3.
*OSHA Compliance Operations Manual.*
*OSHA Recordkeeping Requirements.*
*OSHA Job Hazard Analysis.*

U.S. Department of Transportation, 400 Seventh Street SW., Washington, D.C. 20590.
*Bureau of Motor Carrier Safety Regulations.*
*Hazardous Materials Emergency Response Guidebook.*
*Manual on Uniform Traffic Control Devices for Streets and Highways.* (Also identified as American National Standard D6.1).
*Model Curriculum for Training Tractor-Trailer Drivers.*
Title 49. *Code of Federal Regulations, Parts 390-397, Motor Carrier Safety Regulations.*

U.S. Government Printing Office, North Capital and H Streets NW., Washington, D.C. 20402.
*Code of Federal Regulations:*
Title 29–"Labor"
Title 40–"Protection of the Environment"
Title 49–"Transportation"

# Sources of Help



THE SAFETY PROFESSIONAL FREQUENTLY NEEDS highly specialized or up-to-the-minute, unpublished information. The sources for obtaining this information are numerous. Professional societies and trade associations are excellent sources of help; however, their charters of responsibility are varied. As an aid in the safety professional's search for information, this chapter selects some sources and defines their functions.

On a particularly difficult problem, it may be necessary to contact a number of sources before an effective solution can be obtained. Governmental authorities and specific organizations can furnish the minimum requirements under the law or applicable standards. Insurance companies or their associations may offer assistance through their knowledge of a similar problem. The trade association in the industry may have developed materials and aids in solving the problems.

The National Safety Council, through its resources and membership, can usually provide added input to the development of an effective countermeasure.

## SERVICE ORGANIZATIONS

### National Safety Council
444 North Michigan Avenue
Chicago, Ill. 60611

The National Safety Council is the largest organization in the world devoting its entire efforts to the prevention of accidents. It is nonprofit and nonpolitical. Its staff members work as a team with more than 2,000 volunteer officers, directors, and members of various divisions and committees to develop and maintain accident prevention material and programs in specific areas of safety. These areas include industrial, traffic, home, recreational, and public. Volunteer-staff teams are also organized in such areas as public information, publications, membership extension, and field organization. Council headquarters' facilities include one of the largest safety libraries in the world.

At the Chicago headquarters, a staff of more than 350, about half of whom are engineers, editors, statisticians, writers, educators, data processors, librarians, and other specialists, carry out the major activities. In addition to its main office in Chicago, the Council has a regional office at 1050 17th Street, NW., Washington, D.C. 20036.

**Industrial division.** Recognizing that industry's safety problems often require specialized treatment, the Council has divided its industrial effort into sections guided by the Industrial Division. Each section is administered by its own executive committee, nominated and elected from the membership within that industry. Each executive committee consists of a general chairman, a vice-chairman, a secretary, a newsletter editor, subcommittee chairmen, and others elected at the annual meeting held at the National Safety Congress.

The industrial membership of the Council is organized according to the following industries. These sections are designed to provide special help for all facets of the industrial section, as shown below.

AEROSPACE
Missile and aircraft manufacture, related components
AIR TRANSPORT
Ground safety; personnel and equipment

Automotive, Tooling, Metalworking, and Associated Industries
Machining, fabrication, assembly, general manufacturing
Cement, Quarry and Mineral Aggregates
Quarrying, processing, manufacturing, production
Chemical
Manufacturing compounds and substances
Coal Mining
Underground and open pit
Construction
Highway, buildings, heavy, home, specialty
Electronic and Electrical Equipment
Manufacturing and assembly
Fertilizer and Agricultural Chemical
Manufacturing, storage, transportation, retailing
Food and Beverage
Process foods, dairies, brewers, confectioners, distillers, canners and freezers, grain handling and processing, meat packing and processing, restaurants and fast food
Forest Industries
Logging, manufacturing, converting, pulp and paper, plywood, furniture, and related products
Glass and Ceramics
Manufacturing; flat, containers, miscellaneous products, fiber, refractories, molds
Health Care
Hospital, patient, employee, visitor safety, security, emergency operations
Marine
Deep water and inland waterway; crew, passenger, vessel safety, stevedoring, shipbuilding, repair, cargo
Metals
Foundries, manufacturing ferrous and nonferrous, fabricating, steel service centers
Mining
Metals and minerals
Petroleum
Exploration, drilling, production, pipeline, marketing, retail
Power Press and Forging
Metal stamping and forming, forging
Printing and Publishing
Letterpress, offset, newspaper, bindery
Public Employee
City, county, state, federal (employees and governments)
Public Utilities
Communications, electric, gas, water, construction
Railroad
Employee, passenger, public safety, freight
Research and Development
Safety; laboratory, physical, fire, health
Rubber and Plastics
Tires, molded products, belts, footwear, synthetics
Textile
Manufacturing and fabrication, natural and synthetic fibers, ginning
Trades and Services
Food service and retailers, other retailers, hotels, motels, mercantile and automotive, leather, financial institutions, warehouses, offices, recreational facilities

The Council assigns a staff professional to each section to assist with programs, membership, organization, and the preparation of informational materials. In this way, each major industry group is assured of representation in the affairs of the Council and has the means to develop material and services to meet its needs.

Section executive committees meet several times a year. Improvement of Council services through new technical materials and visual aids takes a major portion of each committee's meeting time. Since increased membership can improve these services, membership solicitation is an ongoing program. Planning of National Safety Congress programs is given careful consideration. Special committees are often assigned to work on problems unique to an industry and on which there are no ready program materials.

To coordinate the entire Council industrial program, a committee representing all industries, the Industrial Division, meets three times a year. The function of this Division is to review current occupational safety and health problems and to determine on a national scale the best procedures to follow in providing increasingly beneficial programs to industry.

The Industrial Division is made up primarily of industrial members of the Council. It comprises 26 section general chairmen, 26 vice-general chairmen, and other members at large, drawn from business and industry member organizations, governmental agencies, insurance organizations, professional and trade associations, and other groups.

The Industrial Division is divided into 16 standing subcommittees that cover various segments of the occupational safety and health area to assist the 26 Industrial Sections and the Industrial Department in carrying out its responsibilities for occupational safety and health business and industry. The Council's Industrial Department manager is assigned as the staff representative to assist the Industrial Division members in their activities.

**Program materials.** The following Council publications have proven to be particularly useful for industrial and off-the-job safety programs. (Unless otherwise stated, they are published monthly.)

*Family Safety & Health* (quarterly)
*Today's Supervisor*
*Safety & Health*
*Regulatory Affairs Update*
"Product Safety Up-to-Date"
Section "Newsletters" (one for each of the 26 sections—six issues a year)
*Safe Driver* (issued in three editions—Truck, Passenger Car, and Bus)
*Safe Worker*
*Traffic Safety* (six issues a year)

In addition, the following statistical materials are also available from the Council:

*Accident Facts* (annually)
Section Contest Bulletins (monthly)
"Work Injury and Illness Rates" (annually)

**Technical materials** (see current Council "General Materials Catalog" for a complete listing):

*Accident Prevention Manual for Industrial Operations* (this book)
*Administration and Programs* volume
*Engineering and Technology* volume
*Aviation Ground Operations Handbook*
*Fundamentals of Industrial Hygiene*

"Industrial Data Sheets" (a series; listing available)
*Industrial Noise and Hearing Conservation*
*Motor Fleet Safety Manual*
"National Directory of Safety Films"
"Power Press Safety Manual"
*Safeguarding Concepts Illustrated*
*Safety Guide for Health Care Institutions*
"Safety Handbook for Office Supervisors"
"Small Fleet Guide"
*Supervisors' Guide to Human Relations*
*Supervisors Safety Manual*

**Training and motivational materials:**
Banners
Booklets
Calendars
Films
Posters
Safety and health slides
Supervisory training pamphlets
Video tapes
Teleconference materials

**Meetings.** The National Safety Council sponsors the annual Congress and Exposition, one of the largest conventions held anywhere. Nearly 300 general or specialized sessions, workshops, and clinics are held covering the full range of safety and health topics.

The annual meetings or special business meetings of a dozen of allied organizations and associations are also held concurrent with the Congress, greatly enhancing the exchange of views and information in safety and health.

**Special services.** Special services available through the Council to support industrial safety programs include:

*Library.* With a collection of more than 100,000 documents, of which 70,000 are indexed on an in-house computerized data base, the National Safety Council's Library is one of the largest and most comprehensive safety and occupational health libraries in the world.

*Statistical.* The Council's statisticians provide a highly refined statistical capability. They are a recognized source of reliable, accurate, and authoritative data within the safety community. Equipped with data processing and research tools, they study various types of accident data in the continuing search for clues on the causes of accidents.

*Consulting.* In 1980, the Council introduced a loss control management consulting service providing a full range of services in safety and occupational health management.

**Safety training.** The National Safety Council's Safety Training Institute conducts the following training courses:

- Principles of Occupational Safety and Health
- Safety Training Methods
- Safety Management Techniques
- Practical Aspects of Industrial Hygiene
- Fundamentals of Hospital Safety
- Safety in Chemical Operations
- Laboratory Safety
- OSHA Hazard Communication Standard
- Motor Fleet Accident Investigation Workshop

See the annual "Safety Training Institute Course Schedule" for dates of presentation, course content, and other courses available.

The Institute also offers three Home Study Courses: "Supervising for Safety," "Supervisors Guide to Human Relations," and "Protecting Workers' Lives." These courses are recognized by the National Home Study Council.

Three supervisor training courses are also available for purchase from the Safety Training Institute and have been designed for presentation by in-house training personnel. These programs include the "Management Development Program," which is oriented to the new supervisor; "Supervisor's Development Program" for individuals with six or more months' experience; and the "Human Relations Course for Supervisors," directed toward the more experienced supervisor. (Complete information on these and other courses can be found in the National Safety Council's "General Materials Catalog.")

The "Forklift Truck Operators Training Course" is an eight-hour program designed to help a company comply with training regulations. The course can be presented in the standard form or can be tailored to fit an individual company's facilities and needs.

Several of the courses offered by the Safety Training Institute are also presented by local safety councils.

The Safety Training Institute can tailor courses and seminars to customer needs and present them at customer locations. Information on costs, scheduling, and specifics is available from the Safety Training Institute.

Safety courses are also sponsored by other educational institutions. See the current "College and University Safety Courses" pamphlet, which is published by the Council. See also the discussion under Educational Institutions at the end of this chapter.

**The Labor Division.** Members of the Council's Labor Division represent unions and governmental Labor agencies. This division and its members represent labor and its safety and health viewpoints in many of the Council's areas of work, such as construction, public utilities, coal mining, legislation, defensive driving, and vocational education.

Specific problems requiring labor's review or consensus input may be referred to one of the following committees of the division, subject to approval of the Executive Committee:

Awards
Construction safety
Ergonomics (ad hoc)
Federal and postal employees
Industrial
Membership
Nominating
Occupational Driver Safety (ad hoc)
Occupational and Environmental Health
Program Planning
Promotion of Safety Training and Education
Public Safety
Research, Government, and Standards
Substance Abuse (ad hoc)
Transportation

These committees may refer to the resources of the labor organizations affiliated with the Council for such areas as support, clarification, and resources.

**Environmental health and safety.** To serve as a communications medium between the public, who demand a larger role in the management of community environmental risks, and those public and private risk managers who make the decisions, the National Safety Council established the Environmental Health and Safety Institute in 1986. This special-purpose organization will be guided by the Council's Board of Governors and operate mostly through philanthropic fundings from concerned corporations, labor unions, foundations, and individuals. Its goal is to develop accurate and objective information on environmental and public health risks, improve public knowledge about these risks, and disseminate this information to the public.

**Local safety councils.** More than 100 local safety councils throughout the United States and Canada have been chartered by the National Safety Council. These councils work under the leadership of public-spirited citizens, commercial and industrial interests, responsible official agencies, and other important groups. They are non-profit, self-supporting organizations whose purpose is to reduce accidents through accident prevention education.

Accredited councils operate under the guidance of a full-time executive and staff and receive support services from the National Safety Council. (A list of accredited councils can be obtained from the Safety Council Relations Department.)

These organizations give assistance to local safety engineers and others concerned with occupational safety and health. Safety professionals, in turn, render substantial service through participation in the local council's work.

Many of the local councils offer the following services:

- Act as a clearinghouse of information on safety and health problems, and maintain a library of audio and visual and other communication aids;
- Provide a forum for exchange of experience through regularly scheduled meetings of supervisory personnel;
- Sponsors a variety of safety and health courses including many presently being offered by the National Safety Council's Safety Training Institute;
- Conduct annual, area-wide safety and health conferences;
- On request, provide assistance with safety problems and programs;
- Stimulate and assist in the development of accident prevention programs for all employers;
- Conduct safety contests with awards for outstanding safety records.

The scope and extent of activities of each council depend, of course, upon local conditions and available resources.

### American National Red Cross
17th and "D" Streets, NW.
Washington, D.C. 20006

The American National Red Cross, through its more than 2900 chapters, offers free courses in first aid, cardiopulmonary resuscitation (CPR), swimming, lifesaving, and small craft handling.

Experience in industry shows that first aid and safety training contributes to the reduction of accidents—both on and off the job—by creating an understanding of accident causes and effects and by improving attitudes toward safety. In addition, this training prepares individuals to give proper emergency care to accident victims. In some situations, immediate action may mean the difference between life and death.

Arrangements for first aid training can be made through local Red Cross chapters. Most industries prefer to select key personnel to receive training as volunteer instructors who, in turn, can conduct classes for fellow employees. Others may wish to arrange for employee training by instructors provided through a local chapter.

Texts, instructor's manuals, charts, and other teaching materials and visual aids, such as films and posters, are available through local chapters.

### Industrial Health Foundation, Inc.
34 Penn Circle W.
Pittsburgh, PA 15206

The foundation, a nonprofit research association of industries, advocates industrial health programs, improved working conditions, and better human relations.

The foundation maintains a staff of physicians, chemists, engineers, biochemists, and medical technicians.

Activities fall into three major categories:

- To give direct professional assistance to member companies in the study of industrial health hazards and their control.
- To assist companies in the development of health programs as an essential part of industrial organization.
- To contribute to the technical advancement of industrial medicine and hygiene by educational programs and publications.

Activities are classified as follows:

*Medical:*
Organization and administrative practices
Opinions on doubtful X-ray pictures
Surveys of health problems
Specific industrial medical problems
Epidemiology studies.

*Chemistry, toxicology, industrial hygiene:*
Field studies—plant or industry basis
Toxicity of chemicals, physical agents, or processes
Sampling and analytical procedures.

*Engineering:*
Ventilating systems
Exhaust hoods.

*Education:*
Training courses in occupational health and safety for:
Physicians
Nurses
Industrial hygienists and engineers.
Symposia on special subjects of current interest in these fields.

The foundation holds an annual meeting of members, conferences of member company specialists, and special conferences on problems common in a particular industry.

The following publications are issued:
*Industrial Hygiene Digest,* monthly (abstracts).
Bibliographies on current interest subjects.
Technical bulletins.
Proceedings of symposia.

### National Society to Prevent Blindness
500 East Remington Road
Schaumburg, Ill. 60173

The NSPB is the oldest voluntary health agency nationally engaged in the prevention of blindness through a comprehensive program of community services, public and professional education, and research. The Society's industrial service program is guided by its Advisory Committee on Industrial Eye Health and Safety, comprised of experts in the fields of industry, education, medicine, nursing, and accident prevention.

Activities and programs of the department are:

- Promotes and administers the Wise Owl Club of America, eye-safety incentive program, among industrial, military, municipal, and educational organizations. Membership in the Club is for those who protect their vision by wearing proper eye-protection both on and off-the-job.
- Promotes state-wide eye safety for all school and college laboratory and shop students, and their teachers and visitors. Both voluntary and legislative means are used. Provides counseling to school administrators and teachers in establishing and implementing eye safety programs. Encourages amendments to state school eye safety laws which presently exclude compliance by private schools.
- Participates as member of American National Standards Institute studies concerning topics related to illumination, vision, and eye protection.
- Promotes universal usage of nonflammable prescription and plano safety eyeglasses and sunglasses by the general public through voluntary and legislative means, and encourages cooperation by the eye care professions and the ophthalmic industry.
- Promotes periodic vision testing in industry and technical schools to determine visual defects and their relationship to visual requirements for various tasks and occupations. Provides counsel on improvement of visual working conditions through proper use of illumination and color.
- Provides exhibits on eye health and safety topics at industrial and educational meetings. Distributes literature and films on eye safety.
- Stimulates relationships with organizations, groups, and individuals with a related interest in sight conservation. Addresses industry, safety and educational meetings to project NSPB recommendations and program aims.

## STANDARDS AND SPECIFICATIONS GROUPS

### American National Standards Institute

1430 Broadway
New York, N.Y. 10018

The American National Standards Institute (ANSI) coordinates and administers the federated voluntary standardization system in the United States, which provides all segments of the economy with national consensus standards required for their operations and for protection of the consumer and industrial worker. It also represents the nation in international standardization efforts through the International Organization for Standardization (ISO), the International Electrotechnical Commission (IEC), and the Pacific Area Standards Congress (PASC).

ANSI is a federation of some 1,200 national trade, technical, professional, labor, and consumer organizations, government agencies, and individual companies. It coordinates the standards development efforts of these groups and approves the standards they produce as American National Standards when its Board of Standards Review determines that a national consensus exists in their favor. A catalog of safety standards is issued annually and is distributed without charge.

Many American National Standards, as well as other national consensus standards, have taken on additional importance since the passage of the Williams-Steiger Occupational Safety and Health Act of 1970. In promulgating standards under the act, the Occupational Safety and Health Administration has stated a definite preference for basing its regulations on consensus standards that have proved their value and practicality by use.

Under ANSI procedures, the responsibility for the management of specific standards projects is divided according to subject matter and assigned to an ANSI Safety and Health Standards Board. Standards dealing with safety fall under the jurisdiction of the Standards Management Board. Many American National Standards on safety and health are developed by ANSI-accredited Standards Committees.

### American Society for Testing and Materials

1916 Race Street
Philadelphia, Pa. 19103

ASTM is the world's largest source of voluntary consensus standards for materials, products, systems, and services. There are currently more than 8,000 ASTM standards.

ASTM membership is drawn from a broad spectrum of individuals, agencies, and industries concerned with materials, products, and systems. The 30,000 members include engineers, scientists, researchers, educators, testing experts, companies, associations and research institutes, governmental agencies, and departments (federal, state, and municipal), educational institutions, consumers, and libraries.

ASTM standards are published for such categories as:

Nonferrous metals
Ferrous metals
Cementitious, ceramic, and masonry materials
Medical devices
Security systems
Energy
Construction
Chemicals and products
Environmental effects
Occupational safety and health
Protective equipment for sports
Electronics
Transportation systems
Business supplies

ASTM sponsors committees on geothermal resources and energy, quality control, food service equipment, and protective coatings for power generation facilities. The committee on consumer product safety has helped develop standards that will assist in protecting the public by reducing the risk of injury associated with the use of consumer products such as cigarette lighters, bathtubs, and shower structures, children's furniture, trampolines, and nonpowered guns.

ASTM standards are of interest to the safety professional since they identify areas of hazards and establish guidelines for safe performance. The society also publishes standards for atmospheric sampling and analysis, fire tests of materials and construction, methods of testing building construction, nondestructive testing, fatigue testing, radiation effects, pavement skid resistance, protective equipment for electrical workers, and others.

These constitute basic reference materials for safety professionals who will frequently be confronted with ASTM standards in processes, as with power plant installations in which vessels, piping, valves, and other component parts are designed and fabricated according to these standards. ASTM standards may also be factors in the raw materials used in protective equipment or other devices.

**Canadian Standards Association**
178 Rexdale Boulevard
Rexdale, Ontario M9W 1R3, Canada

The CSA was chartered in 1919. Until 1944, it was known as the Canadian Engineering Standards Committee. It is a private, not-for-profit organization serving as a standards developer and certifier. It publishes some 1,200 standards, including the *Canadian Electrical Code.*

There is a Standards Steering Committee in each of the 38 broad areas of standardization.

**Other standards groups.** In addition to the American National Standards Institute, many governmental and other agencies have established specifications used by safety professionals. Many industries through their trade associations have also established either (1) codes covering operations in their own plants or (2) safe practices to be followed in the use of their products. Some of these groups are:

American Society of Mechanical Engineers
American Welding Society
Compressed Gas Association
Department of Transportation
General Services Administration
Industrial Safety Equipment Association
Interstate Commerce Commission
National Association of Plumbing and Mechanical Officials
National Board of Boiler and Pressure Vessel Inspectors
National Bureau of Standards
National Fire Protection Association
Occupational Safety and Health Administration
U.S. Bureau of Mines

## FIRE PROTECTION ORGANIZATIONS

Many safety professionals are also responsible for fire prevention and extinguishment in addition to other aspects of safety and health.

The following organizations offer help in this field; also see Insurance Associations, later in this chapter.

**Factory Mutual System**
1151 Boston-Providence Turnpike
Norwood, Mass. 02062

The Factory Mutual System is the world's largest mutual industrial insurance group and a world leader in loss control engineering and research. The System is an outgrowth of the philosophy of positive protection instead of just sharing the risk. In other words, recognition of the good risk through rate (premium) reduction is considered preferable to merely allowing the good risks to help pay for the bad.

The System consists of the following companies: Allendale Mutual Insurance (Johnston, R.I.), Arkwright Mutual Insurance Company (Waltham, Mass.), Protection Mutual Insurance (Park Ridge, Ill.), Factory Mutual International (London, England), and Factory Mutual Engineering and Research (Norwood, Mass.).

Factory Mutual Engineering and Research provides loss prevention services for policyholders. Its aim is to make properties and production facilities safe from fire, explosion, wind, water, and many other perils for which coverage is provided, including damage to boilers, pressure vessels, and machinery. Engineering services include evaluation of hazards and protection through property inspections by Factory Mutual consultants located in major industrial centers. Research services involve the evaluation of fire protection devices and equipment for approval and the development of recommendations based on tests and loss experience for the prevention of loss. The Test Center in Rhode Island provides full-scale fire testing, simulating industrial conditions with regards to height, weight, and protection of major industrial storage occupancies.

Loss control training services are available to policyholders, and assistance is given to special technical problems relating to loss prevention as well as the human element aspect. The source of these services is the Factory Mutual Training Resource Center.

Training courses include Practicing Property Conversion, Boiler and Machinery Preventive Maintenance (both developed specifically for insureds), and Designing for Firesafety and Hazard Control (open to all architects and design professionals). These courses are taught by Factory Mutual engineers, research scientists, and training experts.

Factory Mutual's publications include the *Record, Approval Guide, Handbook of Property Conservation, Loss Prevention Data Books,* and *Factory Mutual Resources—A User's Catalog.* The *Record* is an internationally recognized bimonthly management magazine dealing with property conservation. The *Approval Guide* is a manual which lists industrial fire protection equipment that has been tested and approved by Factory Mutual laboratories. The *Handbook of Property Conservation* and the *Loss Prevention Data Books* cover recommended practices for protection against fire and related hazards. The *Factory Mutual Resources—A User's Catalog* lists all available FM publications, films, training aids, and workshop kits.

**Industrial Risk Insurers**
85 Woodland Street
Hartford, Conn. 06102

The IRI is an association of more than 40 insurance companies that provides underwriting and advisory loss-control services to industry. It maintains a staff of loss prevention personnel with representation in key industrial centers.

The fire prevention laboratory in Hartford contains many types of fire protection equipment for examination and demonstration under working conditions. This laboratory is used for the basic and advanced training of IRI loss prevention personnel and for training plant protection personnel of IRI policyholders and of other insurance organizations who are given short courses in the proper use of fire protection devices.

**National Fire Protection Association**
Batterymarch Park
Quincy, Mass. 02269

The National Fire Protection Association is the clearinghouse for information on the subject of fire protection, fire prevention and firefighting. It is a nonprofit technical and educational organization with a membership of some 32,000 companies and individuals.

The technical standards issued as a result of NFPA committee work are widely accepted by federal, state, and municipal governments as the basis of legislation, and widely used as the basis of good practice. More than 50 are used as OSHA regulations. Constantly revised and updated, 240 standards are currently issued by NFPA, which are available in separate booklets.

Many of them supply authoritative guidance to safety engineers. Representative subjects include:

*Industrial Fire Loss Prevention*
*Portable Fire Extinguishers*
*Sprinkler Systems Organization and Training of Private Fire Brigades*
*Flammable and Combustible Liquids Code*
*Hazardous Chemicals Data*
*Cutting and Welding Processes*
*Storage and Handling of Liquefied Petroleum Gases*
*Prevention of Dust Explosions in Industrial Plants*
*National Electrical Code and Handbook*
*Lightning Protection Code*
*Air Conditioning and Ventilating Systems*
*Life Safety Code and Handbook*
*Safeguarding Building Construction Operations*
*Protection of Records*
*Truck Fire Protection*
*Powered Industrial Trucks*

The standards are also published as the "National Fire Codes" in 16 volumes totaling 12,800 pages.

Other publications of interest, such as the employee training course *Introduction to Fire Protection,* many items on fire safety in health care facilities, and the *NFPA Inspection Manual,* are available to safety professionals from the association.

The *Fire Protection Handbook* is also published by NFPA. An authoritative encyclopedia on fire and its control, the 1,300-page handbook is divided into 18 sections and 131 chapters.

The current NFPA "Price List of Publications and Visual Aids" is available from the NFPA Publications Sales Division.

### Underwriters Laboratories Inc.
333 Pfingsten Road
Northbrook, Ill. 60062

Underwriters Laboratories, a not-for-profit organization, maintains laboratories for the examination and testing of devices, systems and materials to determine their compliance with safety standards.

UL publishes annual directories of manufacturers whose products have met the criteria outlined in appropriate standards and whose products are covered under UL's Follow-Up Services program.

These directories are:

"Building Materials"
"Fire Protection Equipment"
"Fire Resistance"
"Recognized Component"
"Electrical Appliance and Utilization Equipment"
"General Information from Electrical Construction Materials and Hazardous Location Equipment"
"Hazardous Location Equipment"
"Marine"
"Automotive, Burglary Protection and Mechanical Equipment"
"Gas and Oil Equipment"
"Appliances, Equipment, Construction Materials and Components Evaluated in Accordance with International Publications"

UL Follow-Up inspectors make unannounced visits to production sites where UL-Listed products are made at least four times a year to determine the effectiveness of the manufacturer's quality control program for continued compliance with UL safety standards.

Safety professionals often specify the UL Mark when they purchase fire, electrical, and other equipment that falls in categories covered under the services of Underwriters Laboratories.

Engineers should be aware, however, that UL Listings apply only within the scope of its standards for safety, and may have no bearing on performance or other factors not involved in the UL investigation. The markings on the products should tell a safety professional what he needs to know about the function of the device or material. UL tests are conducted under conditions of installation and use that conform to the appropriate standards of the NFPA or other applicable codes. Any departure from these standards by the user may affect the performance qualifications found by Underwriters Laboratories.

## INSURANCE ASSOCIATIONS

In addition to the insurance associations listed under Fire Protection Organizations, there are a number of insurance federations with accident prevention departments that produce technical information available to safety people.

### Alliance of American Insurers
1501 Woodfield Road
Schaumburg, Ill. 60195

The Alliance of American Insurers is a national organization of leading property-casualty insurance companies. Its membership includes more than 150 companies that safeguard the value of lives and property by providing protection against mishaps in workplaces and losses from fires, traffic accidents, and other perils. Alliance member companies have a tradition of loss prevention which dates from the organization of the first American mutual insurance company in 1752.

Through its Loss Control Department, the Alliance makes a concerted effort to reduce accidents, fires, and other loss-producing incidents. Under the guidance of its loss control advisory committee, sound safety engineering, industrial hygiene, fire protection engineering, and other loss prevention principles are promoted. Major activities include the dissemination of information on safety subjects, conduct of specialized training courses for member company personnel, sponsorship of research, cooperation in the development of safety standards, development of visual aids, and the publication of technical and promotional safety literature. A catalog of safety materials is available without charge.

Notable among the publications issued by the Alliance are:

*Safe Openings for Some Point-of-Operation Guards*
*Wood Working Circular Saws, Protection for Variety and Universal Types*
*Spreaders for Variety and Universal Saws*
*Nip Hazards on Paper Machines*
*Material Handling Manual*
*Handbook of Organic Industrial Solvents*

*Judging the Fire Risk*
*Tested Activities for Fire Prevention Committees*
*Exit Drills in the Home*
*Handbook of Hazardous Material*
*Safety Memos for Fleet Supervisors*

Other Alliance departments also regularly issue a variety of bulletins, reports, research reports, and similar materials that often may relate to and support work safety and accident prevention. The communications department, for example, publishes the general-interest magazine, *Journal of American Insurance,* as well as leaflets, brochures, and other informational and educational materials. Inquiry is invited.

The Alliance cooperates extensively with trade associations, professional societies, and other organizations with similar interests, such as the National Safety Council, the National Fire Protection Association, the American National Standards Institute, Inc., the American Society of Safety Engineers, and the American Industrial Hygiene Association.

**American Insurance Services Group, Inc.**
Engineering and Safety Service
85 John Street
New York, N.Y. 10038

The Engineering and Safety Service of American Insurance Services Group, Inc. is dedicated to providing information, education and consultation to the technical and loss control personnel of its participating property-casualty insurance companies. This is accomplished through training programs, conferences, seminars, and a series of reports and bulletins.

Engineering and Safety Service's stock in trade is the latest and most beneficial information, gathered and prepared in a variety of ways for loss control professionals. The focus is always on safety and risk reduction in industrial, commercial, and residential settings within the following technical fields: occupational safety and health, pollution control, product safety, special and chemical hazards, fire protection, crime prevention, environmental science, building technology, industrial hygiene, construction hazards, commercial fleets, and boilers and machinery. In addition, Engineering and Safety Service represents the insurance industry on numerous national standards-making committees of American National Standards Institute, National Fire Protection Association and others.

The Engineering and Safety Service develops and publishes a wide variety of safety-related materials which it makes available to the public, in addition to those it produces for its subscriber companies.

## PROFESSIONAL SOCIETIES

**American Society of Safety Engineers**
1800 East Oakton
Des Plaines, Ill. 60018

The American Society of Safety Engineers is the only organization of individual safety professionals dedicated to the advancement of the safety profession and to foster the well-being and professional development of its members.

In fulfilling its purpose, the Society has the following objectives:

Promote the growth and development of the profession
Establish and maintain standards for the profession
Develop and disseminate material which will carry out the purpose of the Society
Promote and develop educational programs for obtaining the knowledge required to perform the functions of a safety professional
Promote and conduct research in areas which further the purpose and objectives of the Society
Provide forums for the interchange of professional knowledge among its members
Provide for liaison with related disciplines.

An annual professional development conference is conducted for members.

The Society is actively pursuing its Professional Development Programs including the development of curricula for safety professionals, accreditation of degree programs, member education courses, additional publications, and defining research needs and communications to keep safety practitioners current. In addition, the Society has increased its participation and activity in national government affairs.

The Society has established a separate research corporation, the American Society for Safety Research, because of its concern for the need for greater research efforts in the prevention of accidents and injuries. The Society believes that increased research will improve accident and injury control techniques, thus will serve the interests of its members as well as contributing to the economy of our nation and to the health and welfare of all persons.

The Society was also instrumental in establishing a separate corporation to develop a certification program for safety professionals. This corporation is known as Board of Certified Safety Professionals of America. (See BCSP later in this chapter.)

The members of the Society receive the monthly *Professional Safety* as part of their membership. (Nonmembers may also subscribe.) Articles on new developments in the technology of accident prevention are included, as well as information on the activities of the Society, its chapters and members. The Society also publishes other technical or specialized information, such as "A Selected Bibliography of Reference Materials in Safety Engineering and Related Fields," a glossary of terms used in the safety profession, and a series of monographs.

Founded in 1911 as the United Association of Casualty Inspectors, it grew from the original enrollment of 35 to more than 20,000 members internationally in 1986.

Chapters engage in a number of activities designed to enhance the professional competence of their members. Most hold monthly meetings featuring speakers, demonstrations, workshops, and discussions designed to help members keep abreast of developments in their professional field. A list of chapters grouped by states and listed by name and by location can be obtained from the national office.

**American Association of Occupational Health Nurses, Inc.**
50 Lenox Pointe
Atlanta, Ga. 30324

As the national professional organization for the registered nurse working in business and industry, the association strives to raise the qualifications for occupational health nurses, improve nursing services and standards, and provide educational programs for nurses in this special field. *AAOHN Journal* and *AAOHN NEWS* are published monthly.

The annual meeting is held in conjunction with the American

Occupational Health Conference, which AAOHN cosponsors with the American Occupational Medical Association.

**American Board of Industrial Hygiene**
4600 West Saginaw, Suite 101
Lansing, Mich. 48917

This specialty board is authorized to certify properly qualified industrial hygienists. The overall objectives are to encourage the study, improve the practice, elevate the standards, and issue certificates to qualified applicants.

**American Chemical Society**
1155 16th Street, NW.
Washington, D.C. 20036

This society is devoted to the science of chemistry in all its branches, the promotion of research, the improvement of the qualifications and usefulness of chemists, and the distribution of chemical knowledge.

Articles on safety appear in the monthly publication *Industrial and Engineering Chemistry,* the weekly publication *Chemical and Engineering News,* and in the *Journal of Chemical Education.* Digests of papers dealing with aspects of industrial hygiene appear monthly in *Chemical Abstracts.* The environment is discussed in *Environmental Science and Technology.*

The society has a Committee on Chemical Safety and a Division of Chemical Health and Safety. It also has a Chemical Health and Safety Referral Service which is accessible by telephone at 202/872-4515.

**American College of Surgeons**
55 East Erie Street
Chicago, Ill. 60611

The American College of Surgeons, in addition to its role in the Joint Action Program with the National Safety Council and the American Association for the Surgery of Trauma, is launching a series of Advanced Trauma Life Support courses for physicians given through the auspices of the College's Committee on Trauma. The object, as always, is improved care for the injured patient.

**American Conference of Governmental Industrial Hygienists**
6500 Glenway Avenue, Bldg. D-7
Cincinnati, Ohio 45211

A professional association composed of industrial hygiene personnel in government (federal, state, county, or municipal government) or working under a government grant. ACGIH was organized in 1938 by a group of governmental industrial hygienists as a medium for the exchange of ideas, experiences, and the promotion of standards and techniques in occupational health.

ACGIH's wide scope of activities are accomplished through the work of its standing committees.

Particularly valuable to safety professionals is *Industrial Ventilation—A Manual or Recommended Practice,* by the Committee on Industrial Ventilation. ACGIH annually publishes a table of recommended threshold limits for chemical substances and physical agents.

A list of publications will be sent on request to ACGIH.

**American Industrial Hygiene Association**
475 Wolf Ledges Parkway
Akron, Ohio 44311

AIHA is a professional association for persons practicing industrial hygiene in industry, government, labor, academic institutions, and independent organizations. Established in 1939 by leading industrial hygienists, the purpose of AIHA is to promote the recognition, evaluation, and control of environmental stresses arriving in or from the workplace or its products, and to encourage increased knowledge of industrial and environmental health by bringing together specialists in this professional field. Membership currently is more than 7,000.

AIHA publishes a full range of occupational health guides, manuals, monographs, and texts that are available for purchase by members and other interested professionals. Included in AIHA publications are: *AIHA Recommendations for Asbestos Abatement Projects; Noise and Hearing Conservation Manual,* 4th edition; *Respiratory Protection Monograph; Occupational Exposure Limits—Worldwide;* and *A Guide to Product Health and Safety and the Right to Know.* AIHA also publishes a "Hygienic Guide" Series that includes more than 180 information sheets on individual materials encountered in industrial environments. The "Workplace Environmental Exposure Level (WEEL) Guide Series" presents the available toxicological and safe handling procedures for which no "standards" have been set. A complete brochure, "Information Resources," can be obtained by contacting AIHA.

The American Industrial Hygiene Conference, co-sponsored by AIHA and the American Conference of Governmental Industrial Hygienists, is the largest international conference for industrial hygienists. This meeting consists of technical sessions, professional development courses, and exhibits of industrial hygiene products and services.

**American Institute of Chemical Engineers**
345 East 47th Street
New York, N.Y. 10017

Founded in 1908, the AIChE is an individual member society serving 60,000 chemical engineers. The AIChE offers corporations and government agencies the opportunity to join in the cooperative sponsorship of research and development projects that they could not afford to do individually.

The Center for Chemical Process Safety was founded by AIChE early in 1985; its first major project to reach completion was publication of *Guidelines for Hazard Evaluation Procedures.* The Center is working to provide chemical engineers and other interested parties with the tools that can help them identify potentially hazardous situations in existing and new facilities and either eliminate these hazards or substantially reduce them. Engineering practices to prevent or mitigate episodic events (such as the accident at Bhopal, India) tend to be of a "high-tech" nature—vapor cloud modeling, fault tree analysis, and quantitative risk assessment, to name a few. The Center focuses on scientific and engineering practices relative to preventing the accidental releasing of chemicals.

**American Institute of Mining, Metallurgical, and Petroleum Engineers**
345 East 47th Street
New York, N.Y. 10017

The AIME promotes the advancement of knowledge of the arts and sciences involved in the production and use of useful minerals, metals, energy sources, and materials and to record and dissseminate developments in these areas of technology for the benefit of mankind.

Publications are *Mining Engineering, Journal of Metals, Journal of Petroleum Technology, Iron and Steelmaker,* all issued monthly and *Transactions of the Society of Mining Engineers* (quarterly), *Transactions of the Metallurgical Society* (bimonthly), and *Transactions of the Society of Petroleum Engineers of AIME* and *Society of Petroleum Engineers Journal* (quarterly).

**The American Medical Association**
535 North Dearborn Street
Chicago, Ill. 60610

The American Medical Association, with a current membership of about 200,000 physicians, has a long record of involvement in public health.

The Department of Environmental, Public, and Occupational Health (DEPOH) was organized in 1970 as a combination of AMA's Departments of Occupational Health and Environmental Health. In general terms, it is concerned with the well-being of all population groups, whether they be in the work place or in the community. Thus the expertise of the staff covers the health effects of air, water, chemical, and physical stresses; communicable diseases; population growth; injuries; epidemiology; preventive medicine; sports medicine; and problems of the aged and the handicapped.

The department is principally an authoritative source of information for inquiries about environmental, public, and occupational health. In addition, it plans appropriate conferences and courses, prepares authoritative publications for physicians and the public, carries out special assignments and studies, and acts as the AMA liaison in federal health programs and legislation.

Since 1939, AMA has sponsored and organized annual congresses on occupational health. Intended especially to benefit the part-time occupational health practitioner, these congresses have addressed such topical subjects as: the reproductive disorders of workers, mental health in industry, group practice in occupational health, occupational pulmonary diseases, diseases and injuries of the back, and toxic chemicals in the community.

National conferences on the medical aspects of sports, which focus on the prevention and treatment of athletic injuries, are intended for team physicians, coaches, trainers, and administrators. Proceedings of the conferences are published.

The AMA continues to have an active voice in the regulatory process: for example, identifying to the National Institute of Occupational Safety and Health (NIOSH) those AMA physicians who have the expertise to review NIOSH Criteria Documents. And when the Occupational Safety and Health Administration (OSHA) proposed a rule (in July 1978) to give OSHA and NIOSH complete access to employee's medical records, department staff protested in the public hearings.

The *Journal of the American Medical Association* (JAMA) is published weekly and frequently contains articles on some aspect of occupational health.

**American Nurses' Association, Inc.**
2420 Pershing Road
Kansas City, Mo. 64108

The American Nurses' Association is the voluntary membership organization for all registered nurses. The Division on Community Health Nursing Practice is the component of the ANA which provides authoritative information about the practices of occupational health nursing. The objectives of the division are to improve community health nursing practice including occupational health nursing practice for better employee health care.

Publications of interest are: "Standards of Community Health Nursing Practice," "Concepts of Community Health Nursing," and "A Statement on Certification of Occupational Health Nurses." Titles of other brochures, pamphlets, guides, and articles are included in the association's publications list which will be sent on request. (It should be noted that these other publications do not deal specifically with Community Health Nursing Practice.)

**American Occupational Medical Association**
2340 South Arlington Heights Road
Arlington Heights, Ill. 60005

The American Occupational Medical Association fosters the study and discussion of problems peculiar to the practice of industrial medicine and surgery, encourages the development of methods adapted to the conservation and improvement of health among workers, and promotes a more general understanding of the purpose and results of employee medical care.

Some of the association committees and sections are:
Academic Occupational Medicine Section
Alcoholism and Drug Abuse
Annual Scientific Meeting
Arts Medicine Section
Dermatology Section
Education Council
Energy Technology
Ethical Practice in Occupational Medicine
Health Achievement in Industry Award
Health Education
Maritime Medical Affairs Section
Medical Center Occupational Health Services
Medical Information Systems
Medical Practice in Small Industries
Microcomputers in Occupational Medicine Users Section
Noise and Hearing Conservation
Nuclear Industry Physicians' Association Section
Occupational and Clinical Toxicology
Occupational Medical Practice
Psychiatry and Occupational Mental Health
Railroad Medicine Section
Trauma Education.

The official publication of the association is the monthly *Journal of Occupational Medicine.*

An annual American Occupational Health Conference is held, usually in April, in collaboration with the American Association of Occupational Health Nurses.

**American Psychiatric Association**
1400 K Street, NW.
Washington, D.C. 20005

This association of 33,000 members is concerned with research in all phases of mental disorders, standards of psychiatric education and treatment, medico-legal aspects of psychiatric practice, and the promotion of mental health for all citizens. Publications: *American Journal of Psychiatry* monthly, *H&CP (Hospital and Community Psychiatry)* monthly, *Psychiatric News* every two weeks.

**American Public Health Association**
1015 15th Street, NW.
Washington, D.C. 20005

The American Public Health Association is a multidisciplinary, professional association of 30,000 health workers. Through its two monthly periodicals, *The American Journal of Public Health* and *The Nation's Health* and other publications, APHA disseminates health and safety information to those responsible for state and community health-service programs. Program area Sections are devoted to injury control and emergency health services, and to occupational health and safety.

**American Society for Industrial Security**
1655 North Fort Myer Drive
Arlington, Va. 22209

The American Society for Industrial Security, an international professional society of more than 25,000 industrial security executives in both the private and public sector, has 185 chapters worldwide. Committee activities that would be of interest to the safety professional are safeguarding proprietary information, physical security, terrorist activities, disaster management, fire prevention and safety, and investigations.

The Society publishes the magazine *Security Management* monthly to which safety professionals may subscribe and issues a newsletter to its members bimonthly.

The Society certifies security professionals through its Certified Protection Professional (CPP) program. Other member services include the use of the security library and job placement. The Society also produces numerous workshops and educational programs and an annual seminar open to both members and nonmembers.

**American Society of Mechanical Engineers**
345 East 47th Street
New York, N.Y. 10017

This society, the professional mechanical engineers' organization, encourages research, prepares papers and publications, sponsors meetings for the dissemination of information, and develops standards and codes under the supervision of its Policy Board.

The society developed the following safety codes under the procedures meeting the criteria of American National Standards Institute:

*Safety Code for Elevators*
*Safety Code for Mechanical Power-Transmission Apparatus*
*Safety Code for Conveyors, Cableways, and Related Equipment*
*Safety Code for Cranes, Derricks, and Hoists*
*Safety Code for Manlifts*
*Safety Code for Powered Industrial Trucks*
*Safety Code for Aerial Passenger Tramways*
*Safety Code for Mechanical Packing*
*Safety Code for Garage Equipment*
*Safety Code for Pressure Piping*
*Safety Standards for Compressor Systems*

The ASME Boiler and Pressure Vessel Committee is responsible for the formation and revision of the ASME *Boiler and Pressure Vessel Code.*

The society publishes the *Transactions of the American Society of Mechanical Engineers* and the monthly publications *Mechanical Engineering* and *Applied Mechanical Review.*

**American Society for Training and Development**
Box 1433, 1630 Duke Street
Alexandria, Va. 22313

Professional society of persons engaged in the training and development of business, industrial, and government personnel.

**Board of Certified Hazard Control Management**
8009 Carita Court
Bethesda, Md. 20817

Founded in 1976, the Board evaluates and certifies the capabilities of practitioners engaged primarily in the administration of safety and health programs. Levels of certification are senior and master, with master being the highest attainable status indicating that the individual possesses the skill and knowledge necessary to effectively manage comprehensive safety and health programs. The Board offers advice and assistance to those who wish to improve their status in the profession by acquiring skills in administration and combining them with technical safety abilities. Establishes curricula in conjunction with colleges, universities, and other training institutions to better prepare hazard control managers for their duties. Publications are the *Hazard Control Manager* and the *Directory.*

**Board of Certified Safety Professionals**
208 Burwash Avenue
Savoy, Ill. 61874

The Board of Certified Safety Professionals (BCSP) was organized as a peer certification board in 1969 with the purpose of certifying practitioners in the safety profession. The specific functions of the Board, as outlined in its charter, are to evaluate the academic and professional experience qualifications of safety professionals, to administer examinations, and to issue certificates of qualification to those professionals who meet the Board's criteria and successfully pass its examinations. Contact the BCSP for details on the requirements for certification.

**Certified Healthcare Safety Professionals**
8009 Carita Court
Bethesda, Md. 20817

A program to certify health-care safety professionals has been established by the International Healthcare Safety Professional (HSP) Certification Board. The HSP program is designed to raise the competence, status, and recognition of health care safety professionals and assist in the transfer of technology and the exchange of ideas for improving safety in health-care activities.

The Healthcare Safety Professional (HSP) certification program has the following objectives:

- Evaluating the qualifications of persons engaged in hazard control activities in hospital/health-care facilities.
- Certifying as proficient individuals who meet the level of competency for this recognition.
- Increasing the competence and stimulating professional development of practitioners.
- Providing recognition and status for those individuals who by education, experience, and achievement are considered qualified.

- Facilitating the exchange of ideas and technology that will improve performance.

There are three levels of certification: the Master, requiring four years' experience and a baccalaureate degree; the Senior, requiring four years' experience and an associate of arts degree; and the Affiliate, which requires assignment as a safety officer, completion of high school or equivalent, plus specified safety courses over a period of two years.

**The Chlorine Institute**
70 West 40th Street
New York, N.Y. 10018

Founded in 1924, the institute provides "a means for chlorine producers and firms with related interests to deal constructively with common industry problems—especially in safety, transportation, regulations and legislation, and community relations."

Results of committee deliberations are distributed worldwide to chlorine producers, consumers, and other interested groups and persons. In addition to the *Chlorine Manual* and audiovisual programs, some 70 engineering and design recommendations, specifications, and drawings are available.

**Flight Safety Foundation, Inc.**
5510 Columbia Pike
Arlington, Va. 22204-3194

The Flight Safety Foundation is an international membership organization dedicated solely to improving the safety of flight. Nonprofit and independent, it serves the public interest by actively supporting and participating in the development and implementation of programs, policies, and procedures affecting safety by stimulating research into ways and means of eliminating accident-inducing factors, by in-depth appraisals of actual and potential problem areas in flight and ground safety and by developing possible solutions to those problems. As a vehicle of information interchange, it cooperates with all other organizations and individuals in the field of aviation safety in educating all segments of the aviation community in the principles of accident prevention.

To further these objectives, the foundation acts as a clearinghouse for the collection, analysis, and dissemination of safety information. Its personnel participate in safety discussions and safety studies and conduct safety seminars for aviation personnel. It presents awards and otherwise encourages the growth of safety programs. It acts as a catalytic agent in drawing attention to needed improvements and changes in safety techniques.

Perhaps most well-known of the FSF's activities are its annual safety seminars, held in successive years in various parts of the world. These seminars constitute a major gathering of FSF's membership and guests to review and discuss papers presented by prominent world aviation safety experts and to share ideas for safety improvement in an informal, neutral environment.

The Corporate Aviation Safety seminar is held annually in North America. It provides a forum for discussion and review of air safety matters. The programs feature leaders of industry, operator and user experts, as well as government and university researchers.

**National Association of Suggestion Systems**
230 North Michigan Avenue
Chicago, Ill. 60601

The National Association of Suggestion Systems, incorporated as a nonprofit organization, encourages suggestion system activity in industry, commerce, finance, and government. Specific objectives are:

- To increase appreciation of the usefulness of employee suggestion systems.
- To encourage study of the elements necessary to successful use of employee thinking.
- To provide an opportunity for the personal development of those who represent member institutions.
- To gather and disseminate useful information through meetings, publications, factual surveys, and the like.
- To promote personal contacts between suggestion system administrators and the leaders of various industries.

**National Safety Management Society**
3871 Piedmont Avenue
Oakland, Calif. 94611

The National Safety Management Society, founded in 1966, is a nonprofit corporation that seeks to expand and promote the role of safety management as an integral component of total management by developing and perfecting effective methods of improving control of accident losses, be they personnel, property, or financial. Membership is open to those having management responsibilities related to loss control.

**SAFE Association**
15723 Vanowen Street
Van Nuys, Calif. 91406

The SAFE Association, a nonprofit professional association, is dedicated to the preservation of human life. SAFE members represent the fields of engineering, psychology, medicine, physiology, management, education, industrial safety, survival training, fire and rescue, law, human factors, equipment design, and many subfields associated with the design and operation of aircraft, automobiles, buses, trucks, trains, spacecraft, and watercraft.

Regional chapters sponsor meetings and workshops to provide an exchange of ideas, information on members' activities, and presentations of new equipment and procedures encompassing governmental, private and commercial safety and survival applications. SAFE publishes a quarterly journal, periodic newsletters, and an annual Proceedings of the SAFE Symposium.

**System Safety Society**
14252 Culver Drive
Irvine, Calif. 92714

The System Safety Society is a nonprofit organization of professionals dedicated to safety of products and activities by the effective implementation of the system safety concept. This concept is, basically, the application of appropriate technical and managerial skills to assure that a systematic forward-looking hazard identification and control function is made an integral part of a project, program, or activity at the conceptual planning phase, continuing through design, production, testing, use, and disposal phases. The objectives include:

To advance the state-of-the-art of system safety.
To contribute to a meaningful management and technological understanding of system safety.
To disseminate newly developed knowledge to all interested groups and individuals.
To further the development of the professionals engaged in system safety.

Through its local chapters, committees, executive council, publications, and meetings, the society provides many opportunities for interested members to participate in a variety of activities compatible with society objectives. In addition to its operating committees, society activities include publication of *Hazard Prevention,* the official society journal. Published five times a year, it keeps members informed of the latest developments in the field of system safety.

International System Safety Conferences are sponsored biennially.

**Veterans of Safety**
203 North Wabash Avenue
Chicago, Ill. 60601

Membership numbers 1,700 safety engineers with 15 or more years of professional safety experience. Founded in 1941, the objective of Veterans of Safety is to promote safety in all fields. Activities include: (*a*) Safety Town USA, to educate preschool and elementary school children in pedestrian safety; (*b*) Most Precious Cargo Program, to improve school bus safety; and (*c*) Unified Emergency Telephone Numbers Program, to establish nationwide unformity in emergency telephone numbers to contact fire, police, and medical aid. Gives annual awards for best technical safety papers published. Maintains placement service.

## TRADE ASSOCIATIONS

**American Foundrymen's Society**
Golf and Wolf Roads
Des Plaines, Ill. 60016

The American Foundrymen's Society is the only technical society that serves the interest of the foundry industry. It disseminates information on all phases of foundry operations, including safety, hygiene, and air pollution control.

The following publications are available through the society and should be of interest to the safety professional:

American National Standard Series Z241 for sand preparation, molding and coremaking, melting and pouring, and cleaning and finishing of castings.
*Control of the Internal Foundry Environment*
*Control of the External Foundry Environment*
"Foundry Health and Safety" Series
*Foundry Landfill*
*State-of-the-Art Noise Control for Foundries*
*Solid Waste Disposal*

**American Gas Association**
1515 Wilson Boulevard
Arlington, Va. 22209

The association, through its Accident Prevention Committee, serves as a clearinghouse and in an advisory capacity to persons responsible for employee safety and to safety departments of its member companies. Its purposes are to study accident causes, recommend corrective measures, prepare manuals, and disseminate information to the gas industry that will help reduce employee injuries, motor vehicle accidents, and accidents involving the public. The committee meets several times a year and conducts a Safety and Health Workshop for Supervisors each year. It also provides speakers for regional gas associations and other gas industry meetings.

The committee has task committees on each of the following aspects of employee safety: awards and statistics, distribution and utilization, education, motor vehicles, publications, gas transmission, and promotional and advisory.

Published material includes suggested safe practices manuals and quarterly and annual reports on the industry's accident experience and programs.

The association also maintains laboratories in Cleveland and Los Angeles, where gas appliances are tested and design certified.

**American Iron and Steel Institute**
1000 16th Street, NW.
Washington, D.C. 20036

AISI represents companies accounting for more than 80 percent of raw steel production in the U.S.

The Occupational Health and Safety Committee investigates matters relating to the working environment of employees in the iron and steel industry to enhance the health, hygiene, and safety of steel industry personnel, both in and out of the workplace. This includes recognition and evaluation of occupational health and safety hazards with respect to existing and new technology. The Committee also develops industry information and data that may be used for a variety of purposes, and participates in the development of pertinent national and international consensus standards.

**American Mining Congress**
1920 N Street, NW.
Washington, D.C. 20036

Congress membership is from coal, metal, and nonmetal mining companies. This association has a Coal Mine Safety Committee, a Noncoal Mine Safety Committee, and an Occupational Health Committee.

The monthly *Mining Congress Journal* regularly carries articles and news about safety and health.

**American Paper Institute, Inc.**
260 Madison Avenue
New York, N.Y. 10016

The American Paper Institute, through the Safety and Health Subcommittee of its Employee Relations Committee, conducts a broad safety and health education and information service for the paper industry. The Subcommittee sponsors workshops and seminars on various safety and health subjects. Ad hoc task forces are also established to deal with specific problems requiring rapid response, specialized knowledge, or concentrated effort.

In addition, the institute's Employee Relations Department issues a quarterly "Safety and Health Report," which covers the latest developments in OSHA, NIOSH, paper industry safety issues, and other items of current interest.

The department compiles an annual "Summary of Occupational Injuries and Illnesses" and also has an industry-wide Safety Award program, which recognizes individual mills that have outstanding safety records.

**American Petroleum Institute**
1220 L Street, NW.
Washington, D.C. 20005

The objective of the Committee on Safety and Fire Protection of the American Petroleum Institute is to reduce the incidence of accidental occurrences, such as injuries to employees and the public, damage to property, motor vehicle accidents, and fires. To attain this objective, the Committee on Safety and Fire Protection:

- Provides statistical reports, pamphlets, data sheets, and other publications to assist the industry in the prevention of accidents and the prevention, control, and extinguishment of fires.
- Provides a means for the development and exchange of information on accident prevention and fire protection to be used for education and training in the industry.
- Provides a forum for discussion and exchange of information concerning safe practices and the science and technology of fire protection and safety engineering.
- Promotes research and development in the fields of accident prevention and fire protection for the benefit of the petroleum industry as a whole.
- Maintains contact and cooperates with association and code writing bodies such as NFPA and ANSI.

Safety and fire protection manuals have been published on such subjects as:

*Protection Against Ignitions Arising Out of Static, Lightning and Stray Currents*
*Safe Operation of Inland Bulk Plants*
*Safe Practices in Gas and Electric Cutting and Welding in Refineries, Gasoline Plants, Cycling Plants, and Petrochemical Plants*
*Cleaning Mobile Tanks in Flammable or Combustible Liquid Service*
*Cleaning Petroleum Storage Tanks*
*Guidelines for the Application of Water Spray Systems in the Petroleum Industry*
*A Guide for Controlling the Lead Hazard Associated with Tank Entry and Cleaning*
*First Aid Training Guide*
*Guides for Fighting Fires in and Around Petroleum Storage Tanks*
*Fire Hazards of Oil Spills on Waterways*
*Guidelines for Confined Space Work in the Petroleum Industry*
*Evaluation of Firefighting Foams as Fire Protection for Alcohol-Containing Fuels*
*Fire Protection in Refineries*
*Inspection for Accident Prevention*
*Inspection for Fire Protection*
*Service Station Safety*
*Overfill Protection for Petroleum Storage Tanks*

**American Pulpwood Association**
1025 Vermont Avenue, NW., Suite 1020
Washington, D.C. 20005

This association fosters study, discussion, and action programs to guide and help the pulpwood industry in growing and harvesting pulpwood raw material for the pulp and paper industry. The safety and training programs of the Association are served through six regional Technical Divisions.

Available literature includes training guides, notebooks, safety alerts, and technical releases that describe items of personal protective equipment, safe working procedures, and other pertinent accident control items.

**American Road and Transportation Builders**
525 School Street, SW.
Washington, D.C. 20024

ARTBA is a national federation of public and private interests concerned with transportation construction issues. Its members are involved in highway design, construction, management, signing, lighting, and other areas related to road safety.

In 1986, ARTBA organized a national forum on work zone (road construction site) safety. A brochure summarizing the recommendations of the forum is available upon request. The organization conducts ongoing public awareness activities on this topic.

ARTBA publishes a membership newsletter 36 times per year, a quarterly membership magazine, and periodic special reports.

**American Trucking Associations, Inc.**
2200 Mill Road
Alexandria, Va. 22314

American Trucking Associations is the national federation representing the trucking industry. Through its 3,000-member Council of Safety Supervisors, standards for the selection, training, and supervision of truck fleet personnel have been developed—these form the foundation for the ATA Safety Service, a basic safety program for truck fleets.

Guidebooks, forms, and a driver safety program are available through the ATA Department of Safety as are monthly mailings of safety bulletins, driver letters, and safety posters. In addition to providing these services and materials, and acting as secretariat for the Council of Safety Supervisors, the Department of Safety Security is the trucking industry's liaison with federal agencies and national organizations concerned with safety of highway truck operations.

Membership in the Council of Safety Supervisors is available to any person concerned with truck safety. In addition to regional and national meetings, committees work on such problems as employee selection, training and supervision, accident investigation and reporting, transportation of hazardous materials, physical qualifications, and injury control.

There are 47 state councils of safety supervisors and eight councils concerned with safety of tank truck operations and of automobile transporters. Such councils conduct monthly and quarterly meetings, and engage in safety engineering activities that relate to their particular interests according to geographic location or type of operation.

**The American Waterways Operators**
1600 Wilson Boulevard
Arlington, Va. 22209

The American Waterways Operators is the trade association representing the national interests of operators of towboats, tugboats, and barges engaged in domestic trade on the inland and coastal waters of the United States, as well as those of the shipyards that build and repair these vessels.

AWO has a safety committee that pursues three primary objectives:

- Promotion of individual member company's safety programs,
- More participation by members in the Barge and Towing Vessel Industry Safety Contest, which the association co-sponsors with the National Safety Council, and
- Preparation and distribution of AWO safety posters, which the association issues to its members each year, and for which the association conducts a safety poster contest.

The association's manual, *Basic Safety Program for the Barge and Towing Vessel Industry,* is designed either (a) to be adopted as a complete company program, or (b) to be used as a guide to develop or supplement an individual company's program. The manual covers methods and techniques for accident prevention, and contains valuable guidelines on all aspects of personnel safety in the barge and towing vessel industry.

**American Water Works Association**
6666 West Quincy Avenue
Denver, Colo. 80235

The Association, through its Loss Control Committee, develops loss control programs for the water utility industry. Programs presently available include (1) collecting, analyzing, and compiling annual statistics in both employee and motor vehicle accidents; (2) a safety award program; (3) an audiovisual library; (4) one- and two-day loss control seminars for supervisors; (5) loss control audit program; and (6) a safety poster program. In addition to these programs, the Loss Control Committee has three subcommittees on Accident Prevention, Health Maintenance, and Risk Management that study and disseminate information that will assist the water utility industry in increasing its concepts of total loss control.

Publications available include a manual on "Safety Practices for Water Utilities" and "Safety Talks," a compilation of 52 weekly safety talks. Slide/script programs available include: Work Area Protection; Cave-In Protection; Motor Vehicle Accident Prevention; Safe Handling of Water Treatment Chemicals; Entering and Working in Confined Spaces; and Safe Operation of Heavy Equipment.

Regional safety meetings are held with AWWA section safety chairmen to promote safety throughout the water utility industry. These meetings are designed to provide an exchange of ideas and experiences to enhance water utility safety.

**American Welding Society**
P.O. Box 351040
Miami, Fla. 33135

The society is devoted to the proper and safe use of welding by industry. Through its Safety and Health Committee, the society coordinates safe practices in welding by promoting new and revising existing standards.

"Safe Practices for Welding and Cutting Containers that Have Held Combustibles" is one booklet produced by this Committee and is available to industry. Some recent reports deal with noise, fumes and gases, and health effects.

The society also sponsors American National Standards Institute Committee Z49, whose publication, *Safety in Welding and Cutting,* is the authoritative standard in this field. It deals with the protection of workers from accidents, occupational diseases, and fires arising out of the installation, operation, and maintenance of electric and gas welding and cutting equipment.

Frequent articles on safety in welding appear in the official publication of the society, the *Welding Journal.*

**Associated General Contractors of America, Inc.**
1957 E Street, NW.
Washington, D.C. 20006

This association of contractors specializes in the building, highway, municipal utility, and heavy construction fields. All areas of the United States are served by the association's 113 chapters, which carry on their own programs and render assistance to their members.

The association has had better than 50 years of continuing interest in the activities of its Safety and Health Committee, to which member contractors have freely contributed their time and effort. The committee's *Manual of Accident Prevention in Construction* is revised periodically. Each year, the national organization presents awards to members and chapters for significant achievement in accident prevention.

**Association of American Railroads**
American Railroads Building
50 F Street, NW.
Washington, D.C. 20001

All Class I railroads (those with an annual revenue in excess of $50 million) are members of the AAR. The following divisions and committees of the association are concerned with safety:

Communication and Signal Section
Engineering Division
Mechanical Division
Medical Section
Operating Rules Committee
Police and Security Section
Safety Section
State Rail Programs Division.

The Safety Section holds annual meetings; it issues a monthly newsletter, produces posters, and publishes pamphlets on railroad safety.

The Bureau for the Safe Transportation of Explosives and Other Dangerous Articles is also located at the above address.

**Associations Council of the National Association of Manufacturers**
1776 F Street, NW.
Washington, D.C. 20006

The NAM safety and health activities are carried on under the aegis of its Occupational Safety and Health Committee which has a dual function: (a) promoting sound health and safety policies and programs in industry; and (b) working with the federal government to assure that present regulation of health and safety practices in industry and proposals for new regulations and legislation are realistic from industry's viewpoint.

**Bituminous Coal Operators' Association**
303 World Center Building
918 16th Street, NW.
Washington, D.C. 20006

The Bituminous Coal Operators' Association has a Safety Department to furnish health and accident services to its members.

**Chemical Manufacturers Association, Inc.**
2501 M Street, NW.
Washington, D.C. 20037

Formerly known as the Manufacturing Chemists Association, Inc., one of CMA's most important services is dissemination of information on the handling, transportation, and use of chemicals.

The association supports a Health and Safety Committee composed of medical directors, safety directors, and other health-related managers selected from its member companies. This committee meets monthly, supervising the activities of numerous specialized task groups and developing chemical safety and health programs and information for use by member companies, state and federal agencies, other organizations, and the public. The committee holds two open meetings each year on a broad range of topics, with frequent specialized seminars being sponsored by task groups. The committee is also charged with the development and management of the Chemical Awareness and Emergency Response (CAER) program.

Other committees of the association that include safety in their program are the Environmental Management Committee, Engineering Advisory Committee, and the Distribution Committee.

**Compressed Gas Association, Inc.**
1235 Jefferson Davis Highway
Arlington, Va. 22202

The major purpose of the Compressed Gas Association is to provide, develop, and coordinate technical activities in the compressed gas industries, in the interest of safety and efficiency in the U.S., Canada, and Mexico.

Most of the work of the association is done by more than 40 technical committees, made up of representatives of member companies who are highly qualified technically in their respective areas:

Acetylene
Ammonia
Atmospheric gases: nitrogen, oxygen, argon and the rare gases
Carbon dioxide
Chlorine
Cryogenic and low-temperature gases
Ethylene
Gas specification
Halogenated hydrocarbons: propane, butane, etc.
Hydrogen
Hydrogen sulfide
Medical gases: nitrous oxide, cyclopropane, ethylene
Methyl chloride
Natural gas
Petroleum hydrocarbon gases
- Refrigerants: ammonia, fluorocarbons
- Aerosol propellants
- Poisonous gases: hydrogen cyanide, phosgene, etc.
- Other gases: anhydrous ammonia, chlorine, methylamines, sulfur dioxide, carbon monoxide, fluorine

Safety device
Specialty gases
Sulfur dioxide

The association publishes the *Handbook of Compressed Gases,* which contains complete descriptions of 49 widely used gases, and gives the safest recognized methods for handling and storing them. Nearly 80 standards and bulletins are published with audiovisual training aids. A list is available on request.

**Edison Electric Institute**
1111 19th Street, NW.
Washington, D.C. 20036

The Edison Electric Institute is the association of the nation's investor-owned electric utilities. Its members serve 99.6 percent of the customers serviced by the investor-owned segment of the industry.

The Safety and Industrial Health Committee is dedicated to improving working conditions through development of safe work practices. The Committee meets twice a year and compiles reports and publications on subjects of interest to safety professionals in the utility industry. Also prepared by the committee are videotape cassettes and sound-slide films.

**Graphic Arts Technical Foundation**
4615 Forbes Avenue
Pittsburgh, Pa. 15213

GATF has a membership of over 6,000 in the graphic communications industries. Its Education Council is a coordinating organization, the membership of which is made up of large national and various local printing and allied trade associations. Individual companies are also members of the council.

The Environmental Conservation Board of the Graphic Arts Industry (ECB) is a member-supported corporation based at GATF headquarters. The ECB is responsible for the publication by GATF of the *Environmental Control Report/Health and Safety News,* and the "ECB Newsletter."

The *Environmental Control Report,* issued quarterly, discusses the regulations and standards published by various governmental agencies to implement the laws. It assists members of the industry in complying with the Occupational Safety and Health Act and environmental legislation (Clean Air Act, Clean Water Act, Resource Conservation and Recovery Act, and Toxic Substances Control Act). The "ECB Newsletter" is a monthly flyer (one page, two sides) of brief notes and items pertaining to environmental legislation and the Occupational Safety and Health Act.

**Health Physics Society**
1340 Old Chain Bridge Road
McLean, Va. 22101

Organized in 1955 and incorporated in 1961, the HPS has as its objectives: (1) to aid and advance health physics research and applied activities, (2) to encourage dissemination of information between individuals in this and related fields, (3) to improve public understanding of the problems and needs in radiation protection, (4) to initiate and develop programs for training of health physicists, and (5) to promote the health physics profession.

*Health Physics* is the official journal of the society.

**Human Factors Society**
P.O. Box 1369
Santa Monica, Calif. 90406

HFS is a society of psychologists, engineers, physiologists, and other related scientists who are concerned with the use of human factors in the development of systems and devices of all kinds. *Human Factors* is the official journal.

**Illuminating Engineering Society of North America**
345 East 47th Street
New York, N.Y. 10017

The society is the scientific and engineering stimulus in the field of lighting. The work of the Industrial Lighting Committee and its numerous subcommittees for various specific industries should be of particular interest to industrial safety professionals. Many other projects, such as street and highway, aviation, and office lighting, may also be of interest.

Through the society, safety personnel can obtain reference material on all phases of lighting, including authoritative treatise on nomenclature, testing, and measurement procedures.

The society publishes a monthly magazine, *Lighting Design and Application;* a quarterly, *Journal of the Illuminating Engineering Society;* and the *IES Lighting Handbook,* a reference guide. In addition, the society publishes approximately 50 other publications on specific lighting areas such as mining, roadway, office, and emergency lighting.

**Industrial Safety Equipment Association, Inc.**
1901 North Moore Street
Arlington, Va. 22209

This association has represented manufacturers of industrial safety equipment since 1934. It is devoted to the promotion of public interest in safety and encourages development of efficient and practical devices and personal protective equipment for industry.

It is umbrella-like in nature, providing technical improvement through the constant activities of its 12 separate product groups. Of outstanding importance is the broad representation of its members on numerous American National Standards Institute standards committees engaged in promulgation of industry-wide standards of performance for specific types of personal protective equipment. Safety professionals can be guided by these codes and can be assured that certified products will conform to published standards.

**Institute of Makers of Explosives**
1120 19th Street, NW.
Washington, D.C. 20036

The Institute is the safety association of the commercial explosives industry in the United States and Canada. Founded in 1913, IME is a nonprofit, incorporated association whose primary concern is safety and its application to the manufacture, transportation, storage, handling, and use of commercial explosive materials used in blasting and other essential operations. The member companies of IME produce over 75 percent of the commercial explosive materials consumed annually in the United States or some four billion pounds.

Safety library publications available from IME are:

"Construction Guide for Storage Magazines"
"American Table of Distances"
"Suggested Code of Regulations"
"Warnings and Instructions for Consumers in Transporting, Storage, Handling, and Using Explosive Materials"
"Glossary of Commercial Explosive Industry Terms"
"Handbook for the Transportation and Distribution of Explosive Materials"
"Safety in the Transportation and Distribution of Explosive Materials"
"Safety Guide for the Prevention of Radio Frequency Radiation Hazards in the Use of Electric Blasting Caps"
"Destruction of Commercial Explosive Materials" (A statement of policy—not a "how to" publication)
"Recommendations for the Safe Transportation of Detonators in the Same Vehicle with Certain Other Explosive Materials"
"Trade Name Loading Guide for IME 22 Container"

**International Association of Drilling Contractors**
P.O. Box 4287
Houston, Texas 77210

This association works to improve oil well drilling contracting operations as a whole and to increase the value of oil well drilling as an integral part of the petroleum industry.

The association holds an annual safety clinic and has standing and special safety committees of contractor representatives to study current problems.

Safety meetings for tool pushers, drillers, crew members, and safety directors are sponsored by the association; they are conducted throughout the country in locations where these people normally reside.

A Supervisory Accident Prevention Training Program has been developed to instruct drillers and tool pushers in how to establish and maintain effective accident prevention programs. Six to eight professional safety instructors personally conduct these programs anywhere in the world where 18 to 25 people wish to enroll. Many other schools of either two-day or five-day duration are available through the association.

Safety award certificates, cards, safety hat decals, and plaques are given to member personnel and rigs that have completed one or more years without a disabling injury.

The group has produced safety manuals for the industry, inspection reports, color codes, safety signs, studies on protective clothing, and other publications. They have produced color films, film strips, and slides on specific drilling rig safety practices. The association also produces safety posters keyed to the hazards of the drilling industry.

**International Association of Refrigerated Warehouses**
7315 Wisconsin Avenue
Bethesda, Md. 20814

The association's Safety Committee conducts a program specifically aimed at reducing accidents and injuries in refrigerated warehouses. The program includes periodic industry surveys to determine types of injuries being experienced, their causes, frequency and severity, safety bulletins, awards, and information on how to establish and operate a safety program. Members are encouraged to submit problems to the Safety Committee for study and suggested solutions.

**Metal Casting Society**
455 State Street
Des Plaines, Ill. 60016

The Metal Casting Society is a trade association that represents gray, ferrous and nonferrous foundries in the United States and Canada. Founded in 1975 through a merger of the Gray and Ductile Iron Founders' Society and the Malleable Founders' Society, it is a nonprofit, voluntary membership organization with administrative headquarters in Des Plaines, Ill. It is governed by an elected board of directors which is assisted by eleven standing committees in the formulation of programs and policy to promote the progress of its members and the industry.

A booklet developed for the Cast Metals Federation defines safe working conditions in ferrous foundries. Booklets entitled "How You Can Work Safely" are available for foundry employees in either English or Spanish language editions.

**National Constructors Association**
1101 15th Street, NW.
Washington, D.C. 20005

NCA is made up of more than 50 national and regional design-construction firms that build large industrial facilities for oil, steel, power, and chemicals. Its Safety and Health Committee carries out many programs to enhance the physical welfare of employees and the public. Other activities include labor and employee relations, government and international affairs, and taxes, insurance, and legal matters.

**National Health Council, Inc.**
622 Third Avenue
New York, N.Y. 10017

For nearly six decades, the National Health Council has provided a national focus for sharing common concerns and evaluating needs and pooling ideas, resources, and leadership services for national organizations in the health field.

Today 86 major organizations, including voluntary health agencies, professional and other membership associations, insurance companies, business corporations, and federal government agencies, are members.

Like its constituents, the Council exists to improve the health of Americans. Its principal functions are (a) to help member agencies work together more effectively in the public interest to identify and promote the solution of national health problems of concern to the public; and (b) to improve further governmental and private health services for the public at the state and local levels.

The tradition of the National Health Council is one of thoughtful exploration of the nation's critical health problems and of taking action on solutions where possible.

**National LP-Gas Association**
1301 West 22nd Street
Oak Brook, Ill. 60521

Founded in 1931, the association is a nonprofit, cooperative group of producers and distributors of liquefied petroleum gas (LP-gas), manufacturers of LP-gas equipment, and manufacturers and marketers of LP-gas appliances. NLPGA promotes technical information and industry standards in its special field.

Its Safety Committee develops and maintains educational programs to train the public and industry in the safe handling and use of LP-gas and in safe practices for the installation and maintenance of equipment and appliances.

An Educational Committee working closely with the Safety Committee arranges training schools and conferences for dealers and distributors.

The association distributes informational, technical and legislative bulletins and publishes a weekly newsletter for members. It holds an annual meeting and sectional meetings with a definite portion of each program devoted to safety.

The Association publishes a *Safety Handbook,* training guides, audiovisual training programs, and distributes consumer education leaflets for members to help educate its customers.

**National Petroleum Refiners Association**
1899 L Street, NW.
Washington, D.C. 20036

Primarily a service organization for the petroleum refining and petrochemical manufacturing industries, the NPRA also serves as a clearinghouse for new ideas and developments for its membership.

Among the several meetings that are sponsored by NPRA are those of the trade group's ten Fire and Accident Prevention Groups. These are one-day regional meetings that are conducted at different refinery locations and are primarily for first-line supervisors and plant safety and fire protection personnel. These meetings, in ten geographical areas, were established to promote the exchange of information and experiences pertaining to fire and accident prevention in refining and petrochemical operations.

The association also prepares and distributes various safety and fire protection bulletins, information on OSHA and NIOSH activities, and an annual summary of industry data dealing with injury and illness experiences. In an effort to promote safety in plant operations, the NPRA has also established a comprehensive safety awards program.

**National Restaurant Association**
311 First Street, NW.
Washington, D.C. 20001

The National Restaurant Association, through its Technical Services, Public Health, and Safety Department, carries on a program to reduce accidents and hazards that affect the safety of food service employees and patrons.

A major association activity is the preparation and distribution of educational materials to the membership. These materials include self-inspection guidelines on general safety concerns, OSHA requirements, and fire protection, as well as posters and audiovisual programs. These materials are also available for purchase through the association's Information Services Department.

The association conducts research to substantiate industry positions on DOE regulations and on OSHA.

Safety-related information appears in the association's *Washington Weekly* report and monthly magazine, *Restaurants USA*.

The Educational Foundation of the National Restaurant Association (formerly the National Institute for the Foodservice Industry) develops and markets educational and certification programs to industry members. The Educational Foundation is located at 20 North Wacker Drive, Chicago, Ill. 60606.

**National Rural Electric Cooperative Association**
1800 Massachusetts Avenue, NW.
Washington, D.C. 20036

The association, through its Retirement Safety and Insurance Department, promotes a vigorous and diversified program of accident prevention among rural electric cooperatives. The following is a brief resume of safety activities:

**Job Training and Safety Fund.** A fund is distributed to state safety committees based on the proportionate amount of premium developed within each state in the casualty dividend pool.

**Publications and film.** Safety articles are published each month in the association's magazine, *Rural Electrification.* Safety releases are issued several times a year to job training and safety instructors and state safety committees. Safety films have been produced and made available to member systems.

**Meetings.** NRECA publicizes the National Job Training and Safety Conference; staff members participate each year.

**National Sanitation Foundation**
3475 Plymouth Road
Ann Arbor, Mich. 48106

The National Sanitation Foundation (NSF) is an independent,

not-for-profit organization of scientists, engineers, technicians, educators, and analysts. It is a trusted neutral agency, serving government, industry, and consumers in achieving solutions to problems relating to public health and the environment. Standards and criteria are developed in selected public health and environmental areas; and research, testing, and education are provided in the fields of public health and the environment. Services are organized into three major areas: Listing, Certification, and Assessment.

- Listings
  Food Service Equipment
  Plastics Piping Components
  Swimming Pool Equipment
  Special Categories of Equipment and Products
- Registries
  Bottled Water
  Drinking Water Laboratory Accreditation
  Vessel Sanitation Inspection Program
- Standards and Criteria
- Technical Information, Educational Materials, and Reports
- Facts About:
  NSF
  Listing Services
  Certification Services
  Assessment Services

**National Soft Drink Association**
1101 Sixteenth Street, NW.
Washington, D.C. 20036

NSDA's Safety Committee periodically prepares information bulletins concerning OSHA regulations and citations, educational safety procedures, and training films.

**New York Shipping Association, Inc.**
80 Broad Street
New York, N.Y. 10004

New York Shipping Association is composed of American and foreign flag ocean carriers and contracting stevedores, marine terminal operators, and other employers of waterfront labor within the bi-state Port of New York and New Jersey.

Safety by NYSA is supervised by a director who is appointed by the association president. The safety director maintains contact with federal, state, and other agencies involved with industrial safety and health regulations. In this regard, the safety director disseminates relevant data to the member companies of NYSA to assist them in reducing accidents on piers and at marine facilities.

Further, he coordinates industry activity and maintains liaison with the various stevedoring and marine terminal companies who operate their own company safety and health programs.

**Portland Cement Association**
5420 Old Orchard Road
Skokie, Ill. 60077

PCA, devoted to research, educational, and promotional activities to extend and improve the use of portland cement, is supported by more than 45 U.S. and Canadian member companies that operate more than 130 cement manufacturing plants. Occupational safety and health have been considered important by the association since its formation in 1916. The Occupational Safety and Health Services Department works closely with a committee of member company representatives to provide activities, services and materials that are responsive to the needs of those companies.

Consultation on technical safety and health matters is provided and association staff members visit plants on request to perform safety audits, safety program evaluations, and health exposure surveys. Laboratory analysis of substance samples, the use of sampling equipment, and training for plant personnel in noise and dust monitoring procedures also are services available to member companies.

PCA expresses the views and opinions of its member companies to governmental organizations regarding proposed legislation and regulations and other issues affecting the cement industry.

To supplement materials used in plant accident prevention programs, a variety of promotional, informational, and educational items are prepared. Included among these are letters, memos, pamphlets, manuals, summaries, periodicals, and visual aids. Knowledge and experience are shared through meetings and conferences.

Outstanding individual and group performance are recognized through an award program. Worker awards include the Distinguished Safety Service award and Silver Honor Roll certificate. Plant recognition is provided by the PCA Safety Trophy, Thousand-Day Club, and Certificate of Merit.

An injury/illness reporting program enables the association to accumulate data and identify significant causal and circumstantial factors. This information is used to define the nature and extent of cement industry injury/illness experience and in the preparation of educational and informational materials.

**Printing Industries of America, Inc.**
1730 North Lynn Street
Arlington, Va. 22209

Printing Industries of America, an association of printers' organizations, actively sponsors the development of safety in the graphic arts through its affiliated local organizations and through its participation in the Graphic Arts Technical Foundation (described earlier).

**Scaffolding, Shoring, and Forming Institute, Inc.**
c/o Thomas Associates, Inc.
1230 Keith Building
Cleveland, Ohio 44115

The institute has a deep interest in safety; members try to do everything possible to improve this situation in the construction industry. Publications involving safety rules are: "Scaffolding Safety Rules," "Steel Frame Shoring Safety Rules," "Horizontal Shoring Beam Safety Rules," "Single Post Shore Safety Rules," "Suspended Powered Scaffolding Safety Rules," "Flying Deck Form Safety Rules," "Rolling Shore Bracket Safety Rules."

The Safety Procedures publications are:

"Guide to Scaffolding Erection and Dismantling Procedures,"
"Guide to Safety Requirements for Suspended Powered Scaffolds,"
"Guide to Steel Frame Shoring Erection Procedures,"
"Guide to Horizontal Shoring Beam Erection Procedures for Stationary Systems," and
"Guide to Safety Procedures for Vertical Concrete Formwork."

Slide series are also available.

**Steel Plate Fabricators Association, Inc.**
1250 Executive Place
Geneva, Ill. 60134

The association has an active safety committee which prepares publications on safety for member companies and their employees. Some of these are:

"Supervisor's Accident Prevention Manual for Field Erection and Construction,"
"Basic Safety Rules for Fabricating Shops," and
"Basic Safety Rules for Field Erection and Construction."

The association conducts a monthly steel plate fabricators safety contest.

## EMERGENCY INFORMATION BY PHONE

Although it is best to have a prepared plan of action that anticipates the emergency, sometimes this is not possible. The following source has its phones manned 24 hours a day.

### CHEMTREC

Emergency information about hazardous chemicals involved in transportation accidents can be obtained 24 hours a day. It is the Chemical Transportation Emergency Center (CHEMTREC), and it can be reached by a nationwide telephone number—(800) 424-9300. The Area Code 800 WATS line permits the caller to dial the station-to-station number without charge. CHEMTREC will provide the caller with response/action information for the product or products and tell what to do in case of spills, leaks, fires, and exposures. This informs the caller of the hazards, if any, and provides sufficient information to take immediate first steps in controlling the emergency. CHEMTREC is strictly an emergency operation provided for fire, police, and other emergency services. It is not a source of general chemical information of a nonemergency nature.

Signs are available, see Figure 24-1, that can be mounted on vehicles carrying hazardous materials.

FOR CHEMICAL EMERGENCY
Spill, Leak, Fire, Exposure or Accident
CALL CHEMTREC - DAY OR NIGHT
800-424-9300

**Figure 24-1.** Black numbers on bright yellow panels, made of heavy-duty film are designed to be mounted on vehicles carrying hazardous materials. The CHEMTREC number is clearly displayed for police officers, firefighters, or others at the scene of the accident to call for information about the hazards of the chemical or product that is involved. (Courtesy of the Signmark Division, W. H. Brady Co.)

## ON-LINE DATA BASES FOR COMPUTER ACCESS

There are more than 3,000 on-line data bases available for access by users from remote computer terminals and microcomputers. These data bases contain information on a wide range of subject areas that can be used to meet both general and specific needs. To help users identify data bases of particular interest, the latest edition of the directories listed below can prove useful:

*Data Base Directory*
Knowledge Industry Publications, Inc.
701 Westchester Avenue
White Plains, N.Y. 10604

*Directory of On-line Data Bases*
Cuadra/Elsevier
52 Vanderbilt Avenue
New York, N.Y. 10017

*Federal Data Base Finder*
Information USA, Inc.
4701 Willard Avenue
Chevy Chase, Md. 20815

For convenience, let's divide the data bases that are available into two categories: governmental and privately held.

### Governmental information sources

The U.S. government collects volumes of information on almost every subject imaginable. Fortunately under our Constitution, our government is, technically, subject to the control of the citizens of our country. As a result, more and more agencies are computerizing their files in order to permit public access. Frequently, one can obtain searches and cited documents free of charge; also, many of the staff are experts who can guide you as well as answer specific questions, refer you to other sources, and provide printed materials.

Some of the major governmental sources that are of interest to safety and health people are described next. Without a question, the governmental agency that provides the most data bases that are of interest to safety and health professionals is the National Library of Medicine, Bethesda, Md., which has the following data bases. If you want to access one or more of them, call the information desk at (800) 638-8480.

### MEDLARS on-line network

The most useful data bases for the safety professional who needs to know more about industrial hygiene and health subjects is MEDLARS On-line Network, which is an on-line network of approximately 20 bibliographical data bases covering worldwide literature in the health sciences. Those of special value are described next. For more information, contact:

National Library of Medicine
MEDLARS Management Section
8600 Rockville Pike
Bethesda, Md. 20209

**AVLINE.** AVLINE (Audio Visuals on-Line) contains references to 11,000 audiovisual instructional packages in the health sciences. All of these materials are professionally reviewed for technical quality, currency, accuracy of subject content, and educational design. AVLINE enables teachers, students, librarians, researchers, practitioners, and other health science professionals to retrieve citations which aid in evaluating audiovisual materials with maximum specificity.

**CANCERLIT.** CANCERLIT (Cancer Literature) is the National Cancer Institute's on-line data base dealing with all aspects of cancer. The data base contains information on more than 300,000 references dealing with cancer; the sources of this data base include more than 3,000 U.S. and foreign journals, as well as books and other reference sources.

**CHEMLINE.** CHEMLINE (Chemical Dictionary on-Line) is the National Library of Medicine's on-line, interactive chemical dictionary file created by the Specialized Information Services in collaboration with Chemical Abstracts Service (CAS). It provides a mechanism whereby more than 635,000 chemical substance names, representing nearly 100,000 unique substances, can be searched and retrieved on-line. This file contains CAS Registry Numbers; molecular formulas; preferred chemical index nomenclature; generic and trivial names derived from the CAS Registry Nomenclature File; and a locator designation which points to other files in the NLM system containing information on that particular chemical substance. For a limited number of records in the file, there are Medical Subject Headings (MeSH) terms and Wiswesser Line Notations (WLN). In addition, where applicable, each Registry Number record in CHEMLINE contains ring information including—number of rings within a ring system, ring sizes, ring elemental composition, and component line formulas.

**MEDLINE.** MEDLINE (Medical Literature Analysis and Retrieval System on-Line) is a data base maintained by the National Library of Medicine; it contains references to approximately a million citations from 3,000 biomedical journals. It is designed to help health professionals find out easily and quickly what has been published recently on any specific biomedical subject. Medline is accessed from a variety of typewriter-like terminals connected to computers in Bethesda, Md., and Albany, N.Y., via ordinary telephone lines and nationwide communications networks.

**TOXLINE.** TOXLINE (Toxicology Information on-Line) is the National Library of Medicine's extensive collection of computerized toxicology information containing references to published human and animal toxicity studies, effects of environmental chemicals and pollutants, adverse drug reactions, and analytical methodology.

### NIOSH's on-line network

NIOSH's on-line computerized research and reference data base, called NIOSHTIC, contains bibliographic abstracts of approximately 130,000 documents in many subject areas, such as toxicology, occupational medicine, industrial hygiene, and personal protective equipment. Additional details can be obtained from:

National Institute for Occupational Safety and Health
Technical Information Branch
4646 Columbia Parkway
Cincinnati, Ohio 45226

### RTECS

As an on-line data base, RTECS is available as a real time, interactive computer data base that permits the user to search the RTECS (Registry of Toxic Effects of Chemical Substances) for special data or subsets of data and to compile RTECS subfiles tailored to their particular needs. Updates to these systems are provided on a quarterly basis.

RTECS may be accessed through either of two vendors: Fein-Marquardt Company and Information Consultants Incorporated (ICI).

The RTECS is also available as a computer tape from National Technical Information Service (NTIS), 5285 Port Royal Road, Springfield, Va. 22161.

## PRIVATELY OWNED INFORMATION SOURCES

Two nongovernmental organizations make their data bases available to their members.

### HAZARDLINE

The HAZARDLINE data base provides regulatory, handling, identification and emergency care information for over 3,000 hazardous substances. The information is gathered from regulations issued from state and federal agencies, court decisions, books and journal articles, in order to assemble a comprehensive record for each substance.

For more information contact:
Occupational Health Services, Inc.
P.O. Box 1505
400 Plaza Drive
Secaucus, N.J. 07094

### SRIS

The National Safety Council's Safety Research Information Service (SRIS) includes more than 11,000 basic research documents and abstracts. SRIS is part of the Council's library data base which includes more than 70,000 books and documents regarding all aspects of safety and health.

For more information, contact:
Library
National Safety Council
444 North Michigan Avenue
Chicago, Ill. 60611

## U.S. GOVERNMENT AGENCIES

An overwhelming amount of safety information is available from the federal government concerning all aspects of safety and health, environmental problems, pollution, statistical data, and other industry problems.

Because of the constant change in government agency activities and frequent reorganizations, it is recommended that the reader consult the *United States Government Organization Manual,* published by the Government Printing Office, Washington, D.C. 20402. It can be found in most libraries.

Information on the Occupational Safety and Health Administration, the Mine Safety and Health Administration, the National Institute for Occupational Safety and Health, the Environmental Protection Agency, the Public Health Service, and the Environmental Protection Agency can be found elsewhere in this Manual.

## DEPARTMENTS AND BUREAUS IN THE STATES AND POSSESSIONS

It is important that safety professionals have a good working knowledge of the state agencies responsible for the enforcement of safety and health laws. They should, therefore, contact the proper groups in their state's labor departments or other pertinent agencies and find out how the various boards, divisions, and services function.

Some difficulty can be avoided if safety professionals are familiar with the labor legislation and safety and health codes under which they work. Codes and laws vary widely in the different states and provinces, and those persons who have safety jurisdiction in plants in a number of places must understand these differences.

In many cases, the standards set up by the code may serve only as a minimum, and safety professionals will want to compare them with American National Standards or other regulations to establish more rigid rules for their own plants. They should also know the jurisdiction rights of the factory inspectors, so that they can better understand the job inspectors have to do and how they can help them in the performance of their duties.

### Labor offices

Labor offices in the several states perform many functions, generally depending on the number and kind of labor problems.

A listing of state and provincial labor offices and the title of the chief executive of each agency or subdivision to whom inquiries should be addressed is given in the U.S. Department of Labor Bulletin "OSHA Onsite Consultation Project Directory." The bulletin, revised periodically, is available from the Occupational Safety and Health Administration, Washington, D.C. 20210.

For the convenience of the reader, the addresses that were published on the June 1986 list are in this chapter; also see list of regional offices in Chapter 2, Governmental Regulation and Compliance.

### Health and hygiene services

Departments or boards of health and industrial hygiene services are integral parts of the organization of each of the states, the District of Columbia, and the autonomous territories of the United States.

It is to these organizations that safety professionals must look for their state's specific standards and recommendations on such points as occupational health, food and health engineering, disease control, water pollution, and other facets of the overall field of industrial hygiene.

Industrial hygiene units usually function full time or, in several states, on a limited basis. In addition to the units that operate under state health departments, a number of other industrial hygiene units are run by municipalities or other local authorities.

In addition to direct industrial hygiene services, these state units are able to bring to industry a more or less complete health program by integrating their work with that of other divisions in the state government, such as sanitation and infectious disease control.

### OSHA On-site Consultation Program

State and local programs coordinate their efforts with:

Directorate of Federal-State Operations
Office of Consultation Programs
Room N3476
200 Constitution Avenue, NW.
Washington, D.C. 20210
Phone: (202) 523-8902

They also cooperate with medical societies and nurses' associations. (See descriptions earlier in this chapter.)

The names and addresses of such state, commonwealth, or territorial agencies with which the safety professional may need to communicate are as follows (as of June 1986):

**State (Region); Office/Address**

*Alabama (IV)*
7(c)(1) Onsite Consultation Program
P.O. Box 6005
University, Alabama 35486
(205) 348-7136

*Alaska (X)*
Division of Consultation & Training LS&S/OSH
Alaska Department of Labor
3301 Eagle Street, Suite 303
Pouch 7-022
Anchorage, Alaska 99510
(907) 264-2599

*Arizona (IX)*
Consultation and Training
Division of Occupational Safety & Health
Industrial Commission of Arizona
P.O. Box 19070
800 West Washington
Phoenix, Arizona 85005
(602) 255-5795

*Arkansas (VI)*
OSHA Consultation
Arkansas Department of Labor
1022 High Street
Little Rock, Arkansas 72202
(501) 375-8442

*California (IX)*
CAL/OSHA Consultation Service
525 Golden Gate Avenue, 2nd Floor
San Francisco, California 94102
(415) 557-2870

*Colorado (VIII)*
Occupational Safety & Health Section
Institute of Rural Environmental Health
Colorado State University
110 Veterinary Science Building
Fort Collins, Colorado 80523
(303) 491-6151

*Connecticut (I)*
Division of Occupational Safety & Health
Connecticut Department of Labor
200 Folly Brook Boulevard
Wethersfield, Connecticut 06109
(203) 566-4550

*Delaware (III)*
Occupational Safety and Health
Division of Industrial Affairs
Delaware Department of Labor
820 North French Street, 6th Floor
Wilmington, Delaware 19801
(302) 571-3908

*District of Columbia (III)*
Office of Occupational Safety & Health
District of Columbia Department of Employment Services
950 Upshur Street, NW.
Washington, D.C. 20011
(202) 576-6339

*Florida (IV)*
7(c)(1) Onsite Consultation Program
Bureau of Industrial Safety & Health
Florida Department of Labor & Employment Security
LaFayette Building, Room 204
2551 Executive Center Circle, West
Tallahassee, Florida 32301
(904) 488-3044

*Georgia (IV)*
7(c)(1) Onsite Consultation Program
Georgia Institute of Technology
O'Keefe Bldg.—Rm. 23
Atlanta, Georgia 30332
(404) 894-3806

*Guam (IX)*
OSHA Onsite Consultation
Government of Guam
P.O. Box 23548, Guam Main Facility
Agana, Guam 96921-0318
9-011 (671) 646-9446

*Hawaii (IX)*
ATTN: Consultation Program Manager
Division of Occupational Safety & Health
677 Ala Moana Blvd., Suite 910
Honolulu, Hawaii 96813
(808) 548-2511

*Idaho (X)*
Safety & Health Consultation Program
Boise State University
Dept. of Comm. & Env. Health
Boise, Idaho 83725
(208) 385-3283

*Illinois (V)*
Division of Industrial Services
Illinois Department of Commerce & Community Affairs
100 West Randolph St.—Suite 3-400
Chicago, Illinois 60601
(312) 917-2337

*Indiana (V)*
Bureau of Safety, Education & Training
Indiana Division of Labor
1013 State Office Building
Indianapolis, Indiana 46204
(317) 232-2688

*Iowa (VII)*
7(c)(1) Consultation Program
Iowa Bureau of Labor
307 East Seventh Street
Des Moines, Iowa 50319
(515) 281-5352

*Kansas (VII)*
7(c)(1) Consultation Program
Kansas Department of Human Resources
512 West 6th Street
Topeka, Kansas 66603
(913) 296-4386

*Kentucky (IV)*
Education and Training
Occupational Safety & Health Program
Kentucky Department of Labor
U.S. Highway 127 South
Frankfort, Kentucky 40601
(502) 564-6895

*Louisiana (VI)*
(No services available.)

*Maine (I)*
Division of Industrial Safety
Maine Department of Labor
Labor Station 82
283 State Street
Augusta, Maine 04333
(207) 289-2591

*Maryland (III)*
7(c)(1) Consultation Services
Division of Labor & Industry
501 Saint Paul Place
Baltimore, Maryland 21202
(301) 659-4218

*Massachusetts (I)*
7(c)(1) Consultation Program
Division of Industrial Safety
Massachusetts Department of Labor and Industries
100 Cambridge Street
Boston, Massachusetts 02202
(617) 727-3463

*Michigan (Health) (V)*
Special Programs Section
Division of Occupational Health
Michigan Department of Public Health
3500 North Logan
P.O. Box 30035
Lansing, Michigan 48909
(517) 335-8250

*Michigan (Safety) (V)*
Safety Education & Training Division
Bureau of Safety and Regulation
Michigan Department of Labor
7150 Harris Drive
P.O. Box 30015
Lansing, Michigan 48909
(517) 322-1809

*Minnesota (Safety) (V)*
Consultation Division
Minnesota Department of Labor and Industry
444 Lafayette Road, 4th Floor
St. Paul, Minnesota 55101
(612) 297-2393

*Minnesota (Health) (V)*
Consultation Unit
Department of Public Health
717 Delaware, S.E.
Minneapolis, Minnesota 55440
(612) 623-5100

*Mississippi (IV)*
7(c) (1) Onsite Consultation Program
Division of Occupational Safety & Health
Mississippi State Board of Health
P.O. Box 1700
Jackson, Mississippi 39205
(601) 982-6315

*Missouri (VII)*
Onsite Consultation Program
Division of Labor Standards
Missouri Department of Labor and Industrial Relations
1001-D Southwest Blvd.
P.O. Box 449
Jefferson City, Missouri 65102
(314) 751-3403

*Montana (VIII)*
Montana Bureau of Safety & Health
Division of Workers' Compensation
5 South Last Chance Gulch
Helena, Montana 59601
(406) 444-6401

*Nebraska (VII)*
Division of Safety, Labor and Safety Standards
Nebraska Department of Labor
State Office Building
301 Centennial Mall, South
Lincoln, Nebraska 68509
(402) 471-2239

*Nevada (IX)*
Training and Consultation
Division of Occupational Safety & Health
4600 Kietzke Lane, Bldg. D-139
Reno, Nevada 89502
(702) 789-0546

*New Hampshire (I)*
Onsite Consultation Program
New Hampshire Department of Labor
19 Pillsbury Street
Concord, New Hampshire 03301
(603) 271-3170

*New Jersey (II)*
Division of Workplace Standards
New Jersey Department of Labor
110 East Front Street
Trenton, New Jersey 08625
(609) 292-2313

*New Mexico (VI)*
OSHA Consultation
Occupational Health & Safety Bureau
1190 St. Francis Drive—Rm. 2200 No.
P.O. Box 968
Santa Fe, New Mexico 87504
(505) 827-8949

*New York (II)*
Division of Safety and Health
New York State Department of Labor
One Main Street
Brooklyn, New York 11201
(718) 797-7645

*North Carolina (IV)*
North Carolina Consultative Services
North Carolina Department of Labor
Shore Memorial Building
214 West Jones Street
Raleigh, North Carolina 27603
(919) 733-4880

*North Dakota (VIII)*
North Dakota State Department of Health
Division of Environmental Engineering
1200 Missouri Avenue, Room 304
Bismarck, North Dakota 58502
(701) 224-2348

*Ohio (V)*
Division of Onsite Consultation
Ohio Department of Industrial Relations
P.O. Box 825
2323 West 5th Avenue
Columbus, Ohio 43216
(800) 282-1425
(Toll-free in State)
(614) 466-7485

*Oklahoma (VI)*
OSHA Division
Oklahoma Department of Labor
1315 Broadway Place—Room 301
Oklahoma City, Oklahoma 73103
(405) 235-0530 X240

*Oregon (X)*
Voluntary Compliance, Consultative Sec.
Accident Prevention Division
Oregon Department of Workers' Compensation
Labor and Industries Bldg., Room 115
Salem, Oregon 97310
(503) 378-2890

*Pennsylvania (III)*
Indiana University of Pennsylvania
Safety Sciences Department
Uhler Hall
Indiana, Pennsylvania 15705
(800) 382-1241
(Toll-free in State)
(412) 357-2561/2396

*Puerto Rico (II)*
Occupational Safety & Health Office
Puerto Rico Department of Labor and Human Resources
505 Munoz Rivera Avenue, 21st Floor
Hato Rey, Puerto Rico 00918
(809) 754-2134/2171

*Rhode Island (I)*
Division of Occupational Health
Rhode Island Department of Health
206 Cannon Building
75 Davis Street
Providence, Rhode Island 02908
(401) 277-2438

*South Carolina (IV)*
7(c) (1) Onsite Consultation Program
Consultation and Monitoring
South Carolina Department of Labor
3600 Forest Drive
P.O. Box 11329
Columbia, South Carolina 29211
(803) 734-9599

*South Dakota (VIII)*
S.T.A.T.E. Engineering Extension
Onsite Technical Division
South Dakota State University
P.O. Box 2218
Brookings, South Dakota 57007
(605) 688-4101

*Tennessee (IV)*
OSHA Consultative Services
Tennessee Department of Labor
501 Union Building, 6th Floor
Nashville, Tennessee 37219
(615) 741-2793

*Texas (VI)*
Division of Occupational Safety and State Safety Engineer
Texas Department of Health
1100 West 49th Street
Austin, Texas 78756
(512) 458-7287

*Utah (VIII)*
Utah Safety & Health Consultation Service
P.O. Box 45580
Salt Lake City, Utah 84145
(801) 530-6868

*Vermont (I)*
Division of Occupational Safety & Health
Vermont Department of Labor and Industry
120 State Street
Montpelier, Vermont 05602
(802) 828-2765

*Virginia (III)*
Virginia Department of Labor and Industry
P.O. Box 12064
205 N. 4th Street
Richmond, Virginia 23241
(804) 786-5875

*Virgin Islands (II)*
Division of Occupational Safety & Health
Virgin Islands Department of Labor
Lagoon Street
Frederiksted
Virgin Islands 00840
(809) 772-1315

*Washington (X)*
Washington Department of Labor and Industries
P.O. Box 207
814 East 4th
Olympia, Washington 98504
(206) 753-6500

*West Virginia (III)*
West Virginia Department of Labor
State Capitol, Bldg. 3, Room 319
1800 E. Washington Street
Charleston, West Virginia 25305
(304) 348-7890

*Wisconsin (Health) (V)*
Section of Occupational Health
Wisconsin Department of Health and Social Services
1414 E. Washington Avenue—Room 112
P.O. Box 309
Madison, Wisconsin 53701
(608) 266-0417

*Wisconsin (Safety) (V)*
Division of Safety and Buildings
Wisconsin Department of Industry, Labor and Human Relations
1570 East Moreland Boulevard
Waukesha, Wisconsin 53186
(414) 521-5063

*Wyoming (VIII)*
Occupational Health and Safety
State of Wyoming
604 East 25th Street
Cheyenne, Wyoming 82002
(307) 777-7786

**Consultation Training Coordination**
Occupational Safety and Health Administration
Training Institute
1555 Times Drive
Des Plaines, Illinois 60018
(312) 297-4810

## CANADIAN DEPARTMENTS, ASSOCIATIONS, AND BOARDS

In all provinces of Canada there is a Workmen's Compensation Board or Commission. Some of these handle accident prevention directly. In other provinces there are provisions similar to Section 110 of the Quebec Workmen's Compensation Act, which stipulates "that industries included in any of the classes under Schedule I may form themselves into an Association for accident prevention and formulate rules for that purpose. Further, the Workmen's Compensation commission, if satisfied that an Association so formed sufficiently represents the employers in the industries included in the class, may make a special grant toward the expense of any such Association."

It is under these provisions that the various safety associaions were organized and are functioning. In some provinces, accident prevention is directly assumed by the board itself by establishing a safety department.

Furthermore, all provinces have legal safety requirements which are administered by the Department of Highways, Department of Labor, and the Department of Mines. These sources can be contacted by writing to the deputy minister of the department located in the capital of each province.

### Governmental agencies

**Federal:** Director, Occupational Safety and Health Branch, Labour Canada, Ottawa, Ontario, Canada KIA OJ2.

This Branch is responsible for the implementation and administration of the Canada Labour Code Part IV and pursuant regulations which became effective January 1, 1968, and deals primarily with Occupational Safety and Health. It is also responsible for the development of Occupational Safety and Heath regulations, procedures, and standards for regulating all work places under federal jurisdiction.

### Accident prevention associations

Canada Safety Council, 1765 Blvd. St. Laurent, Ottawa, Ontario K1G 3V4

**Provincial associations.**

Alberta Safety Council, 201-10526 Jasper Ave., Edmonton T5J 1Z7
British Columbia Safety Council, 4500 Dawson St., Burnaby, V5C 4C1
Nova Scotia Safety Council, 2745 Dutch Village Rd., Halifax B3L 4G7
Quebec Safety League Inc., 6785 St. Jacques Ouest, Montreal H4B 1V3
Saskachewan Safety Council, 348 Victoria Ave., Regina S4N 3RI
Industrial Accident Prevention Association of Ontario, 2 Bloor St. W., Toronto M4W 3N8

Included in this association are ten class associations:

Woodworkers Accident Prevention Assn.
Ceramics & Stone Accident Prevention Assn.
Metal Trades Accident Prevention Assn.
Chemical Industries Accident Prevention Assn.
Grain, Feed & Fertilizer Accident Prevention Assn.
Food Products Accident Prevention Assn.
Leather, Rubber & Tanners Accident Prevention Assn.
Textile & Allied Industries Accident Prevention Assn.
Printing Trades Accident Prevention Assn.
Ontario Retail Accident Prevention Assn.
Quebec Forest Industrials Safety Assn., Inc., 3350 Wilfrid Hamel Blvd., Quebec, Quebec G1P 2J9
Forest Products Accident Prevention Assn., P.O. Box 270, North Bay, Ontario P1B 8H2
Ontario Pulp & Paper Makers Safety Assn., 91 Kelfield St., Rexdale M9W 5A4
Ontario Safety League, 82 Peter St., Toronto, Ontario M5V 2G5
Mines Accident Prevention Assn., Box 1468, North Bay, Ontario P1B 8K6
Transportation Safety Assn. of Ontario (Inc.), 80 Bloor St. W., Toronto, Ontario M5S 2V1
Construction Safety Assn. of Ontario, 74 Victoria St., Toronto M5C 2A5
Quebec Pulp and Paper Safety Assn. Inc., 425 St. Amable, Quebec, Quebec G1R 2K2
Quebec Logging Safety Assn. Inc., 425 St. Amable, Quebec, Quebec G1R 5E4
Workmen's Compensation Board—Accident Prevention Department, P.O. box 1150, Halifax, N.S. B3J 2Y2

## INTERNATIONAL SAFETY ORGANIZATIONS

### Inter-American Safety Council (Consejo Interamericano de Seguridad)

33 Park Place
Englewood, N.J. 07631

The Inter-American Safety Council was founded and incorporated in 1938, as a noncommercial, nonpolitical, and nonprofit educational association for the prevention of accidents. It is the Spanish and Portugese language counterpart of the National Safety Council.

The Council is the first and only association of its kind rendering services to all industries and agencies in the Latin American countries and Spain. The objectives are to prevent accidents—to reduce the number and severity of accidents in every activity, both on the job and off the job. The services which the Council provides for its members are paid by membership dues and sales of the Council's monthly publications and other educational materials. All of its work is done from the headquarters in New Jersey.

Membership is open to all industries, organizations, institutions, or other groups with two or more employees, interested in accident prevention in Latin America and Spain. More than 1,800 plants or work locations in 22 countries are members or are using the materials and services of the Council. In addition, over 300 universities, technical schools, public libraries, and the like receive the monthly publications free of charge.

Among the services available to members are: monthly publications, annual contest, special awards, consultation, statistical service, reproduction and translation rights, and participation in the election of Council officers. In addition, the Council acts as a clearing house of accident prevention materials available in the United States. A catalog is available from the organization.

The Council's monthly publications include two magazines and safety posters:

Noticias de Seguridad (Safety News)
El Supervisor (The Supervisor)
Safety posters in sizes 8½ by 11 and 17 by 22 in.

In addition, the Council publishes translations of publications, films, safety slides, training programs and other materials of the National Safety Council and other accident prevention organizations.

### International Association of Industrial Accident Boards and Commissions

P.O. Box 79109
Jackson, Miss. 39236

This group, composed of American, Canadian, Australian, New Zealand, and Philippine members, is concerned with worker's compensation and safety.

### International Labor Organization

International Labor Office, CH1211
Geneva 22, Switzerland
1750 New York Avenue NW.
Washington, D.C. 20006

The International Labor Organization (ILO), a specialized agency associated with the United Nations, was created by the Treaty of Versailles in 1919 as part of the League of Nations. Its purpose is to improve labor conditions, raise living standards, and promote economic and social stability as the foundation for lasting peace throughout the world. To this purpose, one of ILO's functions is "the protection of the worker against sickness, disease, and injury arising out of his employment."

The organization consists of about 140 member countries, including the United States (which joined in 1934). ILO functions through an annual conference of member states, a governing body, advisory committees, and a permanent office, the International Labor Office. ILO is distinctive from all other international agencies in that it is tripartite in character—that is, the conference, the governing body, and some of the committees are composed of representatives of governments, employers, and workers.

In the field of safety and health, the International Labor Office maintains a permanent international staff of medical doctors, engineers, and industrial hygienists. Assistance in specific fields is given by panels of consultants, drawn from all parts of the world to act in an advisory capacity and to discuss problems, draft regulations, or render help in emergencies. The office also maintains an Occupational Safety and Health Information Center (CIS) which analyses and provides abstracts of relevant articles appearing in official publications and journals throughout the world.

The United States has several members on the panels and has been represented on all the temporary expert committees and special conferences.

The main tasks of the ILO in the field of occupational safety and health are:

International instruments. These include conventions and recommendations, and also model safety codes and codes of practice.

An *Encyclopedia of Occupational Health and Safety* has been prepared in English and French, to succeed *Occupation and Health,* which was published in 1930. This is designed to provide guidance to a wide range of people concerned with health, safety, and welfare at work. Although problems are reviewed from an international angle, special account is taken of the needs of developing countries.

The compilation of technical studies.

The publication of medical and technical studies.

Direct assistance to governments, by furnishing experts, drafting regulations, supplying information, etc.

Collaboration with other international organizations, the World Health Organization, and the International Organization for Standardization.

Assistance to national safety and health organizations, research centers, employers' associations, trade unions, etc., in different countries.

In general, keeping in touch with the safety and health movement throughout the world and assisting the movement by all the means in its power.

**Pan American Health Organization**
Pan American Sanitary Bureau
525 23rd Street, NW.
Washington, D.C. 20037

Originally established as the International Sanitary Bureau in 1902, the Pan American Health Organization serves as the regional office for the World Health Organization for the Americas. The purposes of the PAHO are to promote and coordinate the efforts of the countries of the Western Hemisphere to combat disease, lengthen life, and promote the physical and mental health of the people.

Programs encompass technical collaboration with governments in the field of public health, including such subjects as sanitary engineering and environmental sanitation, eradication or control of communicable diseases, and maternal and child health.

**World Health Organization**
Avenue Appia
1211 Geneva 27
Switzerland

The World Health Organization (WHO) is a specialized agency of the United Nations with primary responsibility for international health matters and public health. Through this organization, which was created in 1948, the health professions of its 166 member states exchange their knowledge and experience with the aim of making possible the attainment by all citizens of the world by the year 2000 of a level of health that will permit them to lead a socially and economically productive life.

Through technical cooperation with its member states, WHO promotes the development of comprehensive health services, the prevention and control of diseases, the improvement of environmental conditions, the development of health manpower, the coordination and development of biomedical and health services research, and the planning and implementation of health programs.

WHO also plays a major role in establishing international standards for biological substances, pesticides, and pharmaceuticals; formulating environmental health criteria; recommending international nonproprietary names for drugs; administering the International Health Regulations; revising the International Classification of Diseases, Injuries, and Causes of Death; and collecting and disseminating health statistical information.

Authoritative information on the various fields covered by WHO is to be found in its many scientific and technical publications. Of particular interest to health and safety professionals is the Environmental Health Criteria series, prepared by the International Programme on Chemical Safety, of which there are some 70 volumes so far. Each book in the series reviews all available information on a selected chemical, environmental pollutant, or method for testing toxicity and carcinogenicity, in order to provide guidance on prevention of health hazards and the setting of safe exposure limits. In 1987, WHO is launching a new series of Health and Safety Guides, based on the Criteria series and designed to enhance workers' awareness of precautions needed in the handling of individual potentially dangerous chemicals. Current titles in occupational health and safety include *Early Detection of Occupational Diseases* (1986), *Epidemiology of Occupational Health* (1986), *Recommended Health-Based Limits in Occupational Exposure to Selected Mineral Dusts (Silica, Coal)* (1986), and *Psychosocial Factors at Work and Their Relation to Health* (1987). Catalogs of new publications are available on request.

## EDUCATIONAL INSTITUTIONS

Many colleges and universities offer formal courses in industrial safety. In a publication "Educational Opportunities in

Occupational Safety and Health," compiled by the American Society of Safety Engineers, accredited four-year colleges and universities are grouped by those that offer a degree program with concentration on industrial safety and those that offer one or more credit courses as an elective within engineering or education curricula. About one-half of the 1,200 four-year colleges and universities in the U.S. provided catalogues for a recent study in occupational safety and health offered by post-secondary educational institutions.

Four-year and two-year degree programs and credit courses are listed in the ASSE report. Courses in water safety, first aid and safety, firefighting, and driver or traffic education are not listed. The report shows approximately 200 four-year institutions offer one or more courses in the five main occupational categories. Write to the ASSE for a copy of this publication. (Address is listed under Professional Societies earlier in this chapter.)

## BIBLIOGRAPHY OF SAFETY AND RELATED PERIODICALS

### Safety

National Safety Council
444 North Michigan Ave.
Chicago, Ill. 60611
See details of publications in descriptive listing earlier in this chapter.

*Accident Analysis and Prevention* (quarterly)
Pergamon Press Inc.
Fairview Park
Elmsford, N.J. 10523

*Canadian Occupational Safety* (bimonthly)
222 Argyle Ave.
Delhi, Ontario N4B 2Y2
Canada

*Hazard Prevention* (5 times a year)
System Safety Society
7345 South Pierce
Littleton, Colo. 80123

*Health and Safety at Work* (monthly)
McLaren House
P.O. Box 109
19 Searbrook Rd.
Croydon, Surrey CR9 1QH
England

*Human Factors* (bimonthly)
The Human Factors Society, Inc.
P.O. Box 1369
Santa Monica, Calif. 90406

*Journal of Occupational Accidents* (quarterly)
Elsevier Scientific Publishing Company
52 Vanderbilt Ave.
New York, N.Y. 10017

*Mine Safety and Health* (bimonthly)
Dept. of Labor, Mine Safety and Health Administration
Superintendent of Documents
U.S. Government Printing Office
Washington, D.C. 20402

*Nuclear Safety* (bimonthly)
Superintendent of Documents
U.S. Government Printing Office
Washington, D.C. 20402

*Occupational Hazards* (monthly)
Penton Publishing Inc.
1100 Superior Avenue
Cleveland, Ohio 44114

*Occupational Safety and Health* (monthly)
Royal Society for the Prevention of Accidents
Cannon House, The Priory
Queensway
Birmingham B4 6BS
England

*Professional Safety* (monthly)
American Society of Safety Engineers
1800 East Oakton Street
Des Plaines, Ill. 60018

*Protection* (monthly)
Alan Osborne & Associates
Seager Bldgs., Brookmill Rd.
London SE8
England

*The Record, The Magazine of Property Conservation* (bimonthly)
Factory Mutual System
1151 Boston-Providence Turnpike
Norwood, Mass. 02062

*Safe Journal* (quarterly)
Safe Association
15723 Vanowen Street
Van Nuys, Calif. 91406

### Industrial hygiene and medicine

*Archives of Environmental Health* (bimonthly)
Heldref Publications
4000 Albemarle Street, NW.
Washington, D.C. 20016

*Journal of the American Medical Association* (weekly)
American Medical Association
535 North Dearborn Street
Chicago, Ill. 60610

*American Industrial Hygiene Association Journal* (monthly)
American Industrial Hygiene Association
475 Wolf Ledges Parkway
Akron, Ohio 44313

*American Journal of Nursing* (monthly)
American Nurses Association
55 West 57th Street
New York, N.Y. 10019

*American Journal of Public Health* (monthly)
American Public Health Association
1015 15th Street, NW.
Washington, D.C. 20005

*British Journal of Industrial Medicine* (monthly)
British Medical Association
Tavistock Square
London WC1
England

*CIS Abstracts* (8 times a year)
International Occupational Safety and Health Information Center (CIS)
International Labor Office
1211 Geneva 22
Switzerland

*Industrial Hygiene News Report* (monthly)
Flournoy Publishers, Inc.
1845 West Morse Avenue
Chicago, Ill. 60626

*Industrial Hygiene Digest* (monthly)
Industrial Health Foundation
34 Penn Circle West
Pittsburgh, Pa. 15206

*Journal of Occupational Medicine* (monthly)
Flournoy Publishers, Inc.
1845 West Morse Avenue
Chicago, Ill. 60626

*AAOHN Journal*
American Association of Occupational Health Nurses
3500 Piedmont Road, NE.
Atlanta, Ga. 30305

### Fire

*Fire Command!* (monthly)
*Fire Journal* (bimonthly)
*Fire News* (monthly)
*Fire Technology* (quarterly)
National Fire Protection Association
Batterymarch Park
Quincy, Mass. 02269

*Fire Engineering* (quarterly)
Technical Publishing Co.
875 Third Avenue
New York, N.Y. 10022

*Fire Prevention* (quarterly)
Fire Protection Association
Aldermary House
Queen Street
London EC4N 1TJ
England

*Fire Surveyor*
Paramount Publishing Ltd.
17-21 Shenley Road
Borghamwood
Herts WD6 1RT
England

Many of the state departments of labor and federal agencies publish periodicals which are available upon request. With the exception of the state agencies, most of the others are mentioned under the subject headings in this section.

A number of trade journals have sections devoted to industrial safety. The safety professional should become acquainted with those servicing the industry in which he or she is primarily interested.

# Index

## A

## B



## C

## I



## J

## P

## W